Statistical Research Methods
in the Life Sciences

Statistical Research Methods
in the Life Sciences

P. V. Rao

University of Florida

An Alexander Kugushev Book

Duxbury Press
An Imprint of Brooks/Cole Publishing Company
I(T)P®   An International Thomson Publishing Company

Pacific Grove • Albany • Belmont • Bonn • Boston • Cincinnati • Detroit • Johannesburg • London
Madrid • Melbourne • Mexico City • New York • Paris • Singapore • Tokyo • Toronto • Washington

Sponsoring Editor: *Alex Kugushev*
Assistant Editor: *Cynthia Mazow*
Marketing Team: *Carolyn Crockett/Romy Taormina*
Editorial Assistants: *Martha O'Connor/Rita Jaramillo*
Production Editor: *Tessa A. McGlasson*
Manuscript Editor: *Bernard Gilbert*
Cover Design: *Harry Voigt*
Cover Image: *Spike Walker/Tony Stone Images*

Interior Design: *Cloyce J. Wall*
Interior Illustration: *Suffolk Technical Illustrators*
Art Editor: *Lisa Torri*
Indexer: *Ted Laux*
Typesetting: *SuperScript*
Cover Printing: *Phoenix Color Corp.*
Printing and Binding: *Quebecor/Fairfield*

For more information, contact Duxbury Press at Brooks/Cole Publishing Company:

BROOKS/COLE PUBLISHING COMPANY
511 Forest Lodge Road
Pacific Grove, CA 93950
USA

International Thomson Editores
Seneca 53
Col. Polanco
11560 México, D.F., México

International Thomson Publishing Europe
Berkshire House 168-173
High Holborn
London WC1V 7AA
England

International Thomson Publishing GmbH
Königswinterer Strasse 418
53227 Bonn
Germany

Thomas Nelson Australia
102 Dodds Street
South Melbourne, 3205
Victoria, Australia

International Thomson Publishing Asia
221 Henderson Road
#05-10 Henderson Building
Singapore 0315

Nelson Canada
1120 Birchmount Road
Scarborough, Ontario
Canada M1K 5G4

International Thomson Publishing Japan
Hirakawacho Kyowa Building, 3F
2-2-1 Hirakawacho
Chiyoda-ku, Tokyo 102
Japan

You can request permission to use material from this text through the following phone and fax numbers:
Phone: 1-800-730-2214; Fax: 1-800-730-2215.

Printed in the United States of America

10   9   8   7   6   5   4   3

Library of Congress Cataloging-in-Publication Data

Rao, P. V. (Pejaver Vishwamber), [date]-
    Statistical research methods in the life sciences / P.V. Rao.
        p.   cm.
    Includes bibliographical references (p.   ) and index.
        ISBN 0–534–93141–3
        1. Life sciences--Statistical methods.   I. Title.
QH323.5.R34   1997
570'.72--dc21   97–15435
                                                            CIP
                                                            AC

To the memory of my parents, *Janaki and Shridhar Rao*
To the love and friendship of my wife, *Premila*

PREFACE

This book is primarily intended as a text for a one- or two-semester course on statistical research methods for graduate students in biology, agriculture, and related life sciences. The book contains a variety of real examples and exercises drawn from these areas, including many from the author's consulting experience.

Features

The book introduces statistical models early and uses a model based approach in the development of the statistical methods. The main focus is on planning and analyzing designed experiments. Point and interval estimation take priority over statistical hypothesis testing. Particular attention is paid to methods of constructing and interpreting one-at-a-time and simultaneous confidence and prediction intervals (one- and two-sided). Methods and guidelines for determining sample sizes get more emphasis in this text than is normally found in books written for similar audiences. The important difference between hypothesis testing and confidence interval approaches to sample size calculation is emphasized, and guidelines are given on how to use SAS to determine sample sizes.

Even though the book emphasizes interpretation of results over computational details, computational details are given when it is felt that such details help in the interpretation of the computed quantities. Printouts from popular statistical software— SAS and StatXact—are frequently used to display the results in the worked examples in the text.

Also available from the publisher is the *SAS Companion for Statistical Research Methods in the Life Sciences*, authored by Mary Sue Younger (1997), which shows how to use the statistical computing software SAS to perform the calculations described in this book.

Level

The book, with the exception of Appendix B, is written at a mathematical level typical among first-year graduate students in life sciences. The ability to manipulate simple mathematical formulas with symbols and interpret graphs of simple functions is expected, but knowledge of calculus or a background in statistics is not necessary. Some familiarity with upper division linear algebra and calculus will be needed if covering Appendix B.

Organization

The book can be divided into seven parts. The first part, consisting of the first three chapters, is devoted to a discussion of some basic concepts and definitions central to the study of statistics. Throughout this initial part, the focus is on the theme that statistics deals with methods of collecting and using information obtained in samples to draw conclusions about populations. The notion of population, sample, sampling distribution, estimation, hypothesis testing, and prediction are introduced in this part. In Part 2 (Chapters 4, 5, and 6) the most commonly occurring situations—in which the researcher is interested in making inferences about one or two populations—are used to describe statistical methods of estimation, hypothesis testing, and prediction. Part 3 (Chapters 7, 8, and 9) discusses some general issues pertaining to designing research studies and presents methods based on one-way analysis of variance (ANOVA) for the design and analysis of comparative experiments; that is, experiments in which the objective is to compare several treatments with each other. Part 4 contains Chapters 10 and 11 and is devoted to a discussion of regression methods. A detailed discussion of simple and multiple linear regression is included in these chapters. Chapter 12 is the fifth part, in which ANOVA models and regression models are treated as special cases of the general linear models. Analysis of covariance is treated as an example of the use of general linear models. Chapters 13, 14, and 15 constitute Part 6, in which the one-way ANOVA is extended to cover experiments involving multiple factors with fixed and random effects. Finally, Part 7 (Chapter 16) provides an introduction to analysis of repeated measures designs, an important application of the methods based on general linear models involving random and fixed effects. In this text, split-plot experiments are treated as special cases of repeated measures studies.

Suggested Use

Most of the material in the book has been class tested in graduate-level applied statistics courses taught by the author to graduate students in the life sciences and in statistics. The book contains sufficient material to cover a two-semester sequence of courses. Depending on the emphasis of the course, some material may be omitted or used as extra reading assignments. For example, for students with some statistics background, Chapters 1, 2, and 3, may be covered very quickly; class time might be spent on only the main definitions with the remaining material used as reading assignments. In some courses, one or more of the chapters and sections dealing with ordinal data, categorical data, and sample sizes may be skipped. One possible division of the text material into two semesters will cover the ANOVA and regression topics in the first semester with the second semester devoted to ordinal data, categorical data, and factorial (fixed and mixed models) analyses. Such a division will correspond to the following coverage of the topics in the text:

Semester 1:   Chs. 1, 2, 3, 4, 7, 8 (skip 8.8 and 8.9), 9, 10 (skip 10.10), and 11
Semester 2:   Chs. 5, 6, Secs. 8.8, 8.9, 10.10, Chs. 12, 13, 14, 15, and 16

The book, supplemented with selected topics from Appendix B, can also be used as the text for a one- or two-semester applied statistics course for first-year graduate students majoring in statistics. Appendix B collects a number of theoretical results that a masters-level statistician should be familiar with.

Supplements

In addition to Mary Sue Younger's *SAS Companion*, which provides SAS software instruction for the examples in *Statistical Research Methods in the Life Sciences*, the following supplementary materials are available: the *Solutions Manual* contains complete solutions for all the problems in the text; the *Student Solutions Manual* contains complete solutions for all the odd-numbered problems in the text; and a data disk (attached to the inside back cover of the text) contains data sets for the problems in the text. The data sets are formatted for SAS, StataQuest, Minitab, and in ASCII.

Acknowledgments

I wish to thank many colleagues and students who contributed directly and indirectly to the development of this book. In particular, I would like to thank my colleague Dennis D. Wackerly, whose many insightful comments have helped me a great deal in the arduous task of completing this writing project. Thanks are also due to Randy L. Carter, for his careful review of an early draft of the first four chapters. I am indebted to Victor Chew, for providing me with data that were used as examples in this book, and to Geoff Vining, for class-testing parts of the book. Thanks are also due to Yoko Tanaka for helping me with the artwork in the text, and to several reviewers for their valuable feedback: Dale O. Everson, University of Idaho; Marvin Lentner, Virginia Tech; Frank G. Martin, University of Florida; Michael Martin, Stanford University; Deborah Rumsey-Johnson, Kansas State University; Mack C. Shelley, Iowa State University; and Mary Sue Younger, University of Tennessee. Finally, and most importantly, I wish to thank my wife, Premila, my daughter Anita and her husband, Ralph, and my son Anil, for their warmth, love, and understanding during the several years when this book was being written.

P. V. Rao

CONTENTS

1

Statistics:
Its Objectives and Scope

1.1
Introduction

The word *statistics* means different things to different people. In everyday usage, statistics are numbers used to summarize information about objects or phenomena. We have statistics pertaining to the performance of a football team, statistics to describe physical aspects of a human being, statistics to summarize characteristics of groups of human beings, statistics to describe weather conditions, and statistics to measure the effectiveness of a drug, for example. In this book, we will use the term in a broader sense: Statistics refers to a body of scientific principles and methodologies that are useful for obtaining information about a phenomenon or a large collection of items. Statistical methods are techniques for using limited amounts of information to arrive at conclusions—called statistical inferences—about the phenomenon or the collection of items of interest. The use of statistics to make inferences is best illustrated by means of some examples.

EXAMPLE **1.1**   For residents of West-Central Florida, many of whom live in homes built on reclaimed phosphate mine lands, the possibility of above-normal indoor radiation is a matter of great concern. A regulatory agency investigating the possible health implications of living in these homes—for example, the Environmental Protection Agency (EPA) or the Florida Department of Health and Rehabilitative Services (HRS)—seeks answers to the following questions.

1   What is the average indoor radiation level in homes built on reclaimed phosphate mine lands? How do indoor radiation levels vary between homes within a region?

2   What is the average level of indoor radiation exposure for residents of reclaimed phosphate mine lands? How do these exposure levels vary among residents?

3   What proportion of the residents of these homes are being exposed to radiation levels considered hazardous to human health?

Obviously, the answers to these questions will lead to some inferences about the radiation levels in all homes built on reclaimed lands. As is often the case in scientific investigations, it is virtually impossible to obtain all of the data required for a complete and precise answer to these questions. Measuring indoor radiation levels and determining the exposure levels of occupants in every home would be impractical in terms of time, equipment, and personnel. Consequently, the needed information is obtained on the basis of measurements from relatively few homes and residents; that is, the required inference is based on information contained in a selected subset of homes and residents in the region of interest.   ■

EXAMPLE **1.2**   The effect of harsh environmental conditions in tropical and subtropical areas is such that many commonly used tropical grasses (for example, Pensacola Bahia grass) are unable to meet the maintenance requirement of grazing livestock. Studies based on chemical analyses and digestibility (quantity digested as a percent of total intake) in laboratory experiments have shown that Mott dwarf elephant grass has the potential to support high levels of animal performance throughout the grazing season. In a recent study, Caceres (1990) compared, for sheep, the digestibility of Mott dwarf elephant grass harvested in June and September with the digestibility of Pensacola Bahia grass harvested in June and September. The questions addressed by Caceres include the following.

1   What is the average digestibility of Pensacola harvested in June? How do digestibility values for Pensacola harvested in June vary from sheep to sheep?

2   Can the average digestibility for Mott harvested in June be expected to be higher than that for Pensacola harvested in June? How much higher?

3   What can be said about the differential digestibility between June- and September-harvested Pensacola compared to the differential digestibility between June- and September-harvested Mott?

Answers to these questions require inferences about the future performance of two varieties of grass harvested at two different times. These inferences must be based on measured digestibility values for a selected number of animals fed each of the four types of grass—June-harvested Pensacola (JP), September-harvested Pensacola (SP), June-harvested Mott (JM), and September-harvested Mott (SM). On that basis, inferences are made about four sets of measurements: (1) the future digestibility values for all sheep fed JP; (2) values for all sheep fed SP; (3) values for all sheep fed JM; (4) values for all sheep fed SM.   ■

EXAMPLE 1.3   An investigator is interested in evaluating the relationship between age, blood sugar level, and blood cholesterol level of insulin-dependent diabetics who are on a special experimental diet. The investigator wants to answer the following questions, among others:

1   How does the blood cholesterol level change with age and blood sugar level?

2   Are higher cholesterol levels associated with higher sugar levels?

3   Do older diabetics tend to have higher sugar and cholesterol levels?

These questions relate to a collection of measurements for all insulin-dependent diabetics who are on the special diet. Each measurement in this collection consists of three values—the age, blood sugar level, and blood cholesterol level of a patient.   ■

As Examples 1.1–1.3 illustrate, scientific inferences involve two distinct types of collections of measurements: the collection about which information is desired; and the collection from which information is derived. These collections of measurements are called the population and sample, respectively, and are described in more detail in the next section.

1.2
Population and sample
Population

A *population* (sometimes referred to as a statistical population) is a collection (or aggregate) of measurements about which an inference is desired. Example 1.4 contains descriptions of several populations.

EXAMPLE 1.4   The regulatory agencies in Example 1.1 would like to obtain information about indoor radiation levels in homes built on reclaimed phosphate mine lands.

Suppose that there are 4000 homes built on reclaimed lands and that a total of 15,200 persons reside in them. Then, the first question concerning the indoor radiation levels pertains to a population of 4000 radiation-level measurements, one for each home in the group of homes under consideration. The second question in Example 1.1 refers to a population formed by the collection of 15,200 measurements of exposure levels of the residents in the 4000 homes. The third question asks for information about the proportion of residents of the 4000 homes who are being exposed to a hazardous level of radiation. What population does this question refer to? Suppose that, for every individual residing in one of these homes, we place the measurement 1 in the population if the resident is exposed to an unsafe level of radiation; otherwise, we place the measurement 0 in the population. The resulting population is a collection of 15,200 measurements consisting of 0s and 1s. Figure 1.1 shows a 0–1 population of 200 individuals. In this population 1 and 0 represent, respectively, an individual who was and was not exposed to an unsafe level of radiation. Notice that since the population contains 10 measurements equal to 1, 10% of the population have been exposed to unsafe radiation levels. The third question in Example 1.1, then, refers to the proportion of 1s in the 0–1 population.

Now consider Example 1.2. In this example, the scientist is interested in making inferences about the digestibility of four types of forage—JP, SP, JM, and SM—in sheep. The first question refers to the population of digestibility measurements for all sheep who are fed JP. Unlike the three populations in Example 1.1, this population exists only conceptually, because it consists of a set of measurements to be observed in the future. The second question in Example 1.2 refers to two conceptual populations: the potential digestibility measurements for sheep fed JP; and the digestibility measurements for sheep fed JM. The third question in Example 1.2 inquires about the averages of four conceptual populations representing the future digestibility measurements from sheep fed each of the four forage grasses.

Finally, in Example 1.3, the population of interest is a collection of measurements—each of which consists of three values (age, blood sugar level, and blood cholesterol level)—for an insulin-dependent diabetic who is on the experimental diet. ■

FIGURE 1.1

Statistical population representing exposure levels of 200 individuals

```
            0 0 0 0 0 0 0 0 0 0 0 0 0 0
        0 0 0 0 0 0 0 0 0 0 0 0 0 0 0 0 0 0 0 0
      0 0 0 0 0 0 0 0 0 0 0 0 0 0 0 0 0 0 0 0 0 0 0 0
    0 0 0 0 0 0 0 0 0 0 0 0 0 0 0 0 0 0 0 0 0 0 0 0 0 0 0 0
    0 0 0 0 0 0 0 0 0 0 0 0 0 0 0 0 0 0 0 0 0 0 0 0 0 0 0 0
    0 0 0 0 0 0 0 0 0 0 0 0 0 0 0 0 0 0 0 0 0 0 0 0 0 0
      0 0 0 0 0 0 0 0 0 0 0 0 0 0 0 0 0 0 0 0 0 0 0
        0 0 0 0 0 0 0 0 0 0 0 0 0 0 1 1 1 1 1 1 1 1
              1 1
```

Note that, in statistics, a measurement is one of the elements that form the population. In certain populations, each measurement may consist of several values. Populations in which each measurement is a single value are called *univariate* populations. The 0–1 population associated with Example 1.1 and the population of digestibility values in Example 1.2 are univariate populations. A population in which each measurement consists of more than one value is called a *multivariate* population. The population of measurements of insulin-dependent diabetics is a multivariate population.

Example 1.4 illustrated several statistical populations. Because it is not feasible or practical to examine every measurement in a population, statistical methods of inference utilize information contained in appropriately chosen subsets of the populations of interest. Such subsets are called samples.

Sample and sample size

A *sample* consists of a finite number of measurements chosen from a population. The number of measurements in a sample is called the *sample size*.

EXAMPLE **1.5**   To obtain information about the average radiation level in homes built on reclaimed phosphate mine lands in Example 1.1, an investigator might decide to measure the indoor levels in 15 selected homes. The results form a sample of size $n = 15$ from the population of 4000 radiation-level measurements. To make inferences about the proportion of residents exposed to dangerous levels of indoor radiation, the investigator might select a sample of, say, $n = 100$ measurements from the 0–1 population of 15,200 measurements described in Example 1.4. Such a sample can be selected by examining 100 residents of the 4000 homes and recording a 1 if an individual has been exposed to a hazardous level of radiation and a 0 otherwise.

Inference about the digestibility of the four types of forage grasses in Example 1.2 can be based on information obtained from four samples. These four samples should be selected from the four conceptual populations denoting the digestibility values of sheep fed on one of the four grasses. Caceres (1990) based his digestibility study on samples of size $n = 6$ from each of the four populations under consideration. To obtain his samples, Caceres chose 24 mature wethers and divided them into four groups of six wethers each. (A wether is a male sheep that was castrated before it reached maturity.) Each of the four groups was fed a different forage type, and digestibility measurements were made over a period of seven days.

Answers to the questions about associations between the age, blood sugar level, and blood cholesterol level of diabetics in Example 1.3 can be based on measurements made on a sample of, say, $n = 40$ treated insulin-dependent diabetics. Such a sample is a collection of 40 measurements, each of which consists of three values: the age, blood sugar level, and blood cholesterol level of a treated patient.   ■

Examples 1.1–1.5 provide just a few instances of the important role of statistics in information gathering, inference making, and scientific investigations. Many such examples will be seen in later chapters.

Exercises

1.1 To determine whether a new tumor-inhibiting drug is effective in retarding the growth of a particular type of tumor in rats, a researcher wants to compare the rate of tumor growth in rats treated with the drug and the rate for untreated rats. Describe the populations of interest in this study.

1.2 A forester wants to estimate the proportion of diseased trees in a stand of pines. Describe the population about which the forester is interested in making an inference.

1.3 An agronomist wishes to compare yields from four methods of fertilizing a crop. Describe the populations of interest to the agronomist.

1.4 The Environmental Protection Agency (EPA) is concerned about the possibility that land reclaimed after phosphate mining in Florida may have radiation levels exceeding the safe levels for human habitation. The EPA wishes to estimate the radiation levels inside and outside of the homes built on reclaimed land. Describe the populations of interest to the EPA.

1.5 An entomologist plans to compare the effects of two nematocides on the number of nematodes in soil around citrus trees three months after application. Describe the population(s) about which the entomologist wants to make inferences.

1.6 A herd of 85,000 beef cattle is for sale. To calculate a fair price for the herd, a potential buyer would like answers to the following questions.
a   What is the average price of the cattle in the herd?
b   What is the average weight of the cattle in the herd?
c   What proportion of animals in the herd are fecund females?
Describe the populations that are of interest to this individual.

1.7 A researcher is curious about the relationship between the age of a pregnant mother and the birth-weight of her child. Describe the population of interest to the researcher.

1.8 An investigator wants to know how the weight gain of individuals on a special diet is related to the individuals' gender, initial body weight, and age. Describe the population of interest to the investigator.

1.9 A behavioral scientist wants to compare the proportions of male and female fourth-grade children who exhibit the symptoms of attention deficit disorder (ADD). Describe the population(s) of interest to the behavioral scientist.

1.3
Statistical components of a research study

A typical research study consists of three stages. The statistical techniques useful in these three stages are commonly known as *statistical methods in research,* and can be divided into three groups:

1   Methods for designing the research study

2   Methods for organizing and summarizing data

3   Methods for making inferences

Let's consider briefly how these three groups of statistical methods are used in a research study.

Designing research studies

A design for a research study is a protocol for selecting samples from populations. The design must take into consideration not only the principal objectives of the investigation, but also the need to ensure that (1) the measurement process is simple; (2) the study produces reliable and useful data; (3) the study cost is reasonable; and (4) the study can be concluded in a timely manner. In other words, the study must be designed in such a way that the needed information can be obtained in the most efficient way.

Consider an agronomist who wants to compare the yield of a standard variety of barley with the yield of a new variety. There are three key steps in the design of such a study.

1   To specify, as clearly as possible, the research questions that the study is supposed to answer

2   To choose the sample sizes

3   To decide how the samples will be selected

The first step involves clearly identifying not only the populations that are to be studied but also the type of information about these populations that will be needed to answer the study questions. Obviously, the agronomist is interested in two conceptual populations: (1) all yields for the standard variety; (2) all yields for the new variety. However, it is not clear what the agronomist means by saying that he or she wants to compare the two varieties. Does the agronomist want to compare the average yields for the two varieties? In that case, the research question concerns the difference between the averages of the two populations of interest. On the other hand, if the agronomist wants to know whether the yields from the standard variety are more variable than the yields from the new variety, the research question requires a comparison of appropriate measures of variability in the two populations. Methods of assessing the variability in a population and other tools for the statistical formulation of research questions will be developed in subsequent chapters.

Failure to identify clearly the populations about which the study questions are being asked is one of the most common causes of inadequate study designs. In Chapter 7, we will discuss how statistical models can be used to describe precisely the populations of interest in a research study. The following example illustrates why correct identification of the populations of interest is so important.

EXAMPLE **1.6**   Let's consider a research study to assess the average level of indoor radiation in homes built on reclaimed lands. Obviously, in such a study, indoor radiation levels will be measured in a sample of homes built on reclaimed lands. The resulting set of measurements is a sample of indoor radiation-level measurements from the population of all such measurements. Suppose we decide to select 30 homes for radiation-level monitoring. How do we select those homes?

Assume that there are three types of homes—slab-on-grade homes, homes with crawl space, and apartments—and that the radiation level in a home may depend on its type. Should we select a representative sample of 30 homes from all homes under consideration or a representative sample of ten homes from each of the three types of homes? In the former case, the resulting data constitute a sample from a single population—the population of indoor radiation levels in all homes built on reclaimed lands. Inferences from such a sample will pertain to the average indoor radiation level in all homes built on reclaimed land.

In comparing the average level in one type of home with that in another type, we must be aware that information about different types of homes available from our data will depend on the number of homes of each type included in the sample. For instance, in a sample consisting of two apartments, 20 slab-on-grade homes, and eight homes with crawl space, there is much less information about radiation levels in apartments than about the levels in slab-on-grade homes.

If a primary objective of the study is to assess the difference between the average indoor radiation levels in different types of homes built on reclaimed land, the study should focus on three different populations corresponding to the radiation levels prevailing in the three types of homes. Selecting a sample of ten homes from each of these populations will ensure that the same amount of information about the average indoor radiation level is available for each of the three types of homes.　■

After identifying the two populations of interest, the agronomist must decide on the sample sizes—the number of measurements to be selected from the populations—that will provide information with desired reliability about the difference between the varieties. In practice, the decision on the sample sizes is usually a compromise between the cost and reliability of the information to be collected. Larger sample sizes, while providing more information about the population differences, will be more expensive and time-consuming.

Once the sample sizes have been chosen, the next step is to decide how the samples will be selected. How can the experimenter make sure that the observed yields in the samples are typical for the two varieties? Does the sampling technique ensure that the observed differences in the samples reflect the corresponding differences in the populations?

Methods for selecting the sample sizes and collecting appropriate samples will be described in detail in subsequent chapters. For the present, it is important to realize that these problems need careful attention. Specification of sample sizes and the procedures for sample selection needed to attain the stated research objectives are two major components of any well-planned research study.

Organizing and summarizing data

The design of a research study determines the protocol for collecting data. The next task in a statistical inference-making procedure is to describe and analyze the data. In one form or the other, this stage involves careful summarization of the information contained in the samples. The appropriate analysis of the data in a particular set of samples depends upon the type of inference desired. For example, to estimate the average indoor radiation level in the homes in Example 1.1, it is reasonable to expect the analysis to include, among other things, the calculation of the average of the measured radiation levels in the sample of homes.

Making inferences

The third and final stage of a statistical inference procedure is to use the information in the samples to make conclusions about populations. The key statistical issue in such inferences is their accuracy. For instance, suppose that the average indoor radiation level in a sample of 15 homes is 0.032 WL, where WL stands for working level, the common unit used for monitoring radon in uranium mines. Then 0.032 WL could be regarded as an estimate of the average indoor level in all homes built on reclaimed phosphate mine lands. How accurate is this estimate? Since the estimate is calculated for a sample of only 15 homes, we cannot be certain that the average for all homes is exactly 0.032 WL. However, if the sample could be regarded as representative of the population, it would be reasonable to expect that the difference between the estimated value of 0.032 and the true mean radiation level for all homes will be small, but we cannot be certain about the actual magnitude of this difference. The natural question, therefore, is whether it is possible to assess, with reasonable certainty, the magnitude of the error in our estimate. For example, can we say, with a reasonable degree of confidence, that the average level for the population of all homes will be within 0.001 WL of the average value calculated from sample homes? In other words, if the average of the sample is 0.032 WL, is it reasonable to conclude that the average for all homes will be between $0.032 - 0.001 = 0.031$ WL and $0.032 + 0.001 = 0.033$ WL?

In the following chapters, we will see that, when the samples are selected in a suitable manner, the probability of a given magnitude of error in an estimate may be controlled by changing the sample size. For the present, we want to emphasize that any statistical inference, by its very nature, is subject to error. The objective of statistical procedures is not only to make inferences about populations but also to assess the reliability of such inferences.

Exercises

1.10   Refer to Exercise 1.1.

   a Formulate a research question that addresses the primary objective of the study.

   b Suppose it is decided to conduct the study by measuring the tumor volumes at the beginning and at the end of a four-week period on ten treated and ten untreated experimental rats. Describe the sample(s) and explain how they might be used to answer the research question formulated in (a).

1.11   Refer to Exercise 1.2. Describe a typical set of sample measurements that can be used to make inferences about the proportion of diseased trees.

1.12   Refer to Exercise 1.3. Describe a typical set of samples that can used to compare the four methods of fertilizing the crop.

1.13   Refer to Example 1.6. Suggest a study design that can be used to compare the radiation levels inside and outside the homes built on reclaimed lands if the study objective is:

a   to compare the inside and outside levels for each of the three types of homes;
b   to make an overall comparison of the inside and outside levels.

1.14   Refer to Exercise 1.5. Suggest a study design for selecting the samples needed to make inferences related to the entomologist's research objective.

1.4
Using computers for data analysis

Calculators and computers are indispensable tools for statistical analysis of data. Modern-day calculators are efficient and inexpensive for many statistical calculations with small data sets. Many of these calculators have built-in programs that will perform standard statistical calculations at the touch of a single key.

However, a desktop or mainframe computer is essential when analyzing large data sets or when using complex statistical procedures. Numerous statistical software packages are available for use with personal and mainframe computers. With a little practice, these packages are quite easy to use, and a person who knows what needs to be done can perform the complex calculations needed for statistical analyses of large sets of data in a relatively routine manner.

Statistical software packages such as SAS, Splus, StatXact, BMDP, GLIM, MINITAB, and SPSS can be used not only to perform complicated computational tasks but also to draw the graphical displays necessary for interpreting and reporting the results of statistical analyses.

With some practice, using a statistical software package becomes straightforward, if not simple. The toughest tasks in data analyses are (1) selecting the statistical procedures that will be used in the analyses; (2) preparing the data sets in a format suitable for use in the statistical package; (3) interpreting the results. To perform the actual computations, the user simply invokes the software system and enters the appropriate computer codes. The computer codes needed to perform a task vary from one package to another, but are usually easy to master and execute. For instance, the simple code PROC MEANS; will instruct the SAS software package to calculate the averages and many other quantities for all the variables whose values are contained in the data set. Each of the seven software packages listed here has its own users' manual, which contains detailed directions about the codes necessary for implementing various procedures.

This is not a text on how to use the computer to do statistical calculations. Rather, statistical procedures commonly used in life sciences research are described so that the reader may (1) select from the variety of statistical methods available in statistical computing packages and (2) interpret the results of the selected statistical analyses.

In this book, we will use printouts from SAS and StatXact to illustrate applications of statistical methods. SAS is one of the most powerful and most frequently used statistical software packages. For readers who are not familiar with the use of SAS, we recommend the companion text by Younger (1997), in which the use of SAS to perform the statistical analyses described in this text is considered in detail. StatXact is a powerful statistical software package specially designed for handling ordinal and categorical data. It is menu-driven and has an excellent manual, which the reader will find very helpful when performing calculations.

2

Describing Statistical Populations

2.1 Introduction

In Chapter 1, we described how the notion of statistical populations enters into the design, analysis, and inference-making stages of scientific investigations. We emphasized that the objective when making statistical inferences is to use information in samples to draw conclusions about populations. In this chapter, we develop the necessary terminology and describe some convenient methods for summarizing information about statistical populations.

2.2
Types of populations

Statistical populations can be classified into categories depending upon the characteristics of the measurements contained in them.

Univariate and multivariate populations

In Chapter 1, we gave several examples of univariate and multivariate statistical populations. In a *univariate* population, each measurement consists of a single value; in a *multivariate* population, measurements consist of more than one value. A multivariate population in which each measurement is characterized by k values will be called a *k-variate population*. For instance, the population of 4000 radiation-level measurements in Example 1.4 is a univariate population, whereas the population of interest in Example 1.3, in which each measurement consisted of three measured values—age, blood sugar level, and blood cholesterol level—is a 3-variate population. Most of our discussions in this book will be about univariate populations, which we will refer to simply as populations.

Real and conceptual populations

As we see from the populations described in Example 1.4, a population may be real or conceptual. The population of 4000 indoor radon levels is a real population, whereas the population of digestibility values for sheep fed June-harvested Pensacola Bahia grass is a conceptual population.

Finite and infinite populations

A population may only contain a finite number of measurements, as in the case of the population of indoor radon levels of 4000 homes, or it may have infinitely many measurements, as in the case of a conceptual population of potential digestibility measurements, in which every value in the interval [0%, 100%] is a possible value of a measurement in the population.

Quantitative and qualitative populations

A measurement is said to be *quantitative* if its value can be interpreted on a natural and meaningful numerical scale. A measurement is *qualitative* if its value serves the sole purpose of identifying an object or a characteristic. The value of a qualitative measurement has no numerical implications.

EXAMPLE **2.1** Identifying the distribution of the pest population is a key step in a pest management program. Suppose that there are five types of pests in a given region. The population of pest types can be considered as a collection of measurements with five distinct values—1, 2, 3, 4, and 5. Each measurement represents one type of pest. The numerical value of a measurement in this population does not measure any quantity such as age, weight, or yield. Rather, it is an indicator of the quality (pest type) of the item being measured. Thus, in this population, the measurements are qualitative. By contrast, the population of indoor radon levels in Example 1.4 is a quantitative population. ■

Discrete and continuous populations

Most populations of practical interest can be classified as discrete or continuous. A population is said to be discrete if the distinct values of the measurements contained in it can be arranged in a sequence.

We have already seen examples of discrete populations. The 0–1 population of radiation exposure levels in Example 1.4 is a discrete population. The distinct values of the measurements in this population can be arranged as the finite sequence of two terms: 0, 1. The following example describes a discrete population in which the distinct values of the measurements form an infinite sequence of terms.

EXAMPLE **2.2** In a study of the number of bacterial colonies growing on an agar plate, the conceptual population of interest consists of several 0s, representing cases where no colony is found; several 1s, representing cases where exactly one colony is found; several 2s, representing cases where exactly two colonies are found; and so on. Accordingly, this is modeled, in practice, as a discrete population in which the distinct values of the measurements are the numbers in the infinite sequence 0, 1, 2, ... ■

A distinguishing feature of a discrete population is that the distinct values of the measurements in the population can be represented as a sequence. A continuous population, on the other hand, consists of measurements that take all the values in one or more intervals of a real line. For example, the conceptual population of digestibility values for forage grasses in Example 1.2 may be regarded as a continuous population with measurements taking values in the interval [0, 100], where 0 and 100 correspond, respectively, to 0% and 100% of the intake digested by an animal.

EXAMPLE **2.3** Time to relief is an important variable in drug evaluation studies. It is the time elapsed between the administration of the drug and the time at which the individual experiences complete relief. The conceptual population of all measurements of time to relief is often regarded as a continuous population with values in the interval $[0, +\infty)$. ■

EXAMPLE **2.4** Predicting the movement of chemicals through soil is of great interest because of the risk that agricultural application of pesticides and fertilizers will lead to groundwater contamination. When investigating the movement of a chemical in soil, one important question concerns the depth (vertical distance) traveled by a chemical particle in a given time. Consider a particle currently located at a point that is 1 m below the soil surface and 20 m above the groundwater level. Assume that the particle may move a maximum of 1 m upward or a maximum of 20 m downward. Thus, all possible values of the depth attained by the particle within a fixed time form a continuous population of measurements that have values in the interval $[-1, 20]$. Note that we are considering vertical distances above the current location as negative depths. ∎

Exercises

2.1 Classify the populations in Exercise 1.1 into categories on the basis of the following criteria. Give reasons for your choices.

    a Univariate versus multivariate   **b** Real versus conceptual
    c Finite versus infinite   **d** Quantitative versus qualitative
    e Discrete versus continuous

2.2 Repeat Exercise 2.1 for the population in Exercise 1.2.

2.3 Repeat Exercise 2.1 for the population in Exercise 1.3.

2.4 Repeat Exercise 2.1 for the population in Exercise 1.4.

2.5 Repeat Exercise 2.1 for the population in Exercise 1.5.

2.6 Repeat Exercise 2.1 for the population in Exercise 1.6.

2.7 Repeat Exercise 2.1 for the population in Exercise 1.7.

2.8 Repeat Exercise 2.1 for the population in Exercise 1.8.

2.9 Repeat Exercise 2.1 for the population in Exercise 1.9.

2.3
Describing populations using distributions
The probability function of a discrete population

For a given discrete population, let $f(y)$ denote the proportion of measurements that have the value y. Then $f(y)$ is called the *relative frequency* of y, and the set of all possible values of $(y, f(y))$ is called the *probability distribution* of the population.

EXAMPLE **2.5** The 0–1 population in Example 1.4 consists of the exposure status (1, unsafe level; 0, safe level) of 15,200 residents of homes built on reclaimed mine lands. Suppose that there are 5016 residents with unsafe exposure levels. Then the population contains 5016 measurements with the value $y = 1$ and $15,200 - 5016 = 10,184$ measurements with the value $y = 0$. The proportions (also called relative frequencies) of 0s and 1s in the population are

$$f(0) = \frac{10,184}{15,200} = 0.67 \text{ and } f(1) = \frac{5016}{15,200} = 0.33,$$

respectively. Thus, the probability distribution of the population is specified by the two pairs (0, 0.67) and (1, 0.33). ∎

A probability distribution can be represented as a *table*, a *graph*, or a *formula*. In Example 2.5, a typical measurement y in the population can have two values, $y = 0$ and $y = 1$, with the corresponding proportions $f(0) = 0.67$ and $f(1) = 0.33$. Table 2.1 shows a tabular representation of the probability distribution in Example 2.5. Note that all possible values of the pair $(y, f(y))$ are listed in this table.

In a graphical representation of the probability distribution, the values of y are plotted along the horizontal axis and a perpendicular line of length $f(y)$ is placed at each value of y. Figure 2.1 shows a graphical representation of the probability distribution in Table 2.1.

TABLE 2.1

Tabular representation of a probability distribution

y	$f(y)$
0	0.67
1	0.33

FIGURE 2.1

Graphical representation of a probability distribution

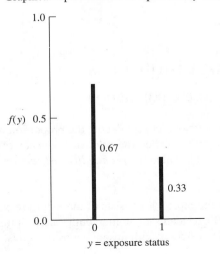

The third method of representing a probability distribution is to use a formula. The formula specifying the proportions in a discrete population is called a *probability function*. It is easily verified that the formula

$$f(y) = (0.33)^y(0.67)^{1-y}, \qquad y = 0, 1, \tag{2.1}$$

can be used to determine the proportions in the 0–1 population of Table 2.1. For example, substituting $y = 0$ and $y = 1$ in Equation (2.1), we get

$$f(0) = (0.33)^0(0.67)^{1-0} = 0.67 \quad \text{and} \quad f(1) = (0.33)^1(0.67)^{1-1} = 0.33.$$

Thus, Equation (2.1) is the probability function of the 0–1 population in Table 2.1. As will be seen in Section 2.5, Equation (2.1) is an example of a widely used probability function known as binomial probability function.

The choice of a particular method for representing a probability distribution depends upon the purpose for which the distribution is being determined. As a general rule, the graphical method will be preferred where a visual description of the population distribution is required. The locations of the perpendicular lines identify the values of the measurements, while the heights of the lines show the relative frequency with which they occur. Representation of a distribution using a probability function, while compact and useful from a theoretical point of view, has its greatest advantage in the case of a population containing infinitely many distinct values.

EXAMPLE **2.6**   The population of bacterial colony counts in Example 2.2 contains infinitely many distinct values. Consequently, neither the tabular nor the graphical representation is ideally suited for its description. If the biological process that generates the bacterial colonies satisfies the conditions described in Section 2.5, a *Poisson probability function* can be used to compute the relative frequencies. If λ is the average number of colonies per agar plate, then according to the Poisson probability function

$$f(y) = \frac{e^{-\lambda}\lambda^y}{y!}, \qquad y = 0, 1, \ldots, \tag{2.2}$$

is the relative frequency of y. As an example of the use of Equation (2.2), suppose that $\lambda = 2$; in other words, there are, on average, two colonies per agar plate. Then the Poisson probability function in Equation (2.2) can be expressed as

$$f(y) = \frac{e^{-2}2^y}{y!} \qquad y = 0, 1, \ldots.$$

Letting $y = 2$ and $y = 4$, we get

$$f(2) = \frac{e^{-2}2^2}{2!} = 0.2707 \quad \text{and} \quad f(4) = \frac{e^{-2}2^4}{4!} = 0.0902,$$

so that, on the basis of the given probability function, we can say that 27.07% and 9.02% of the agar plates will have two and four bacteria colonies, respectively. Proportions for other values of y, the number of colonies in an agar plate, can be computed in a similar manner. Of course, the appropriateness of Equation (2.2) to model the bacterial colony distribution in a particular application depends upon whether the conditions in Section 2.5 can be regarded as reasonable.   ∎

Properties of probability functions

Every probability function satisfies two conditions. First of all, since a nonzero proportion is always positive, a probability function is always positive; that is, $f(y) > 0$ for all y. Secondly, since the sum of the proportions of all distinct values in a population must be equal to 1, a probability function must satisfy the equation: $\Sigma f(y) = 1$. In particular, the sum of the heights of all vertical lines in a graphical representation of a probability distribution must equal 1.

EXAMPLE **2.7**   In the pest-type population in Example 2.1, each measurement represents one type of pest. Let $f(y)$ denote the relative frequency of pest type y, where $y = 1, 2, 3, 4, 5$. A possible probability distribution of such a population is given in Table 2.2. Note that the relative frequencies in Table 2.2 are all positive and add to 1.

TABLE   2.2
Probability distribution of pest types

Pest type y	Relative frequency $f(y)$
1	0.16
2	0.18
3	0.10
4	0.46
5	0.10

Figure 2.2 shows a graphical representation of the probability distribution of the pest population in Table 2.2. The horizontal axis in Figure 2.2 represents the values of the measurements in the population. At each of the five possible distinct values of the measurements in the population, a vertical line shows the proportion of

FIGURE   2.2
Probability distribution of pest types

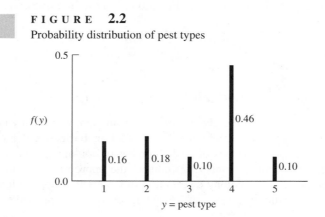

measurements having that value. Notice that the heights of the five vertical lines will add to 1.

In a discrete population, the proportion of measurements with values in a given interval equals the sum of the heights of the vertical lines within the interval. From Figure 2.2, the relative frequency of pests of types 2, 3, and 4 is the sum of the heights of the vertical lines at $y = 2$, $y = 3$, and $y = 4$; that is, the proportion of pest types 2, 3, and 4 in the population is equal to

$$f(2) + f(3) + f(4) = 0.18 + 0.10 + 0.46$$

$$= 0.74. \tag{2.3}$$

Therefore, 74% of the pests in the population are of type 2, 3, or 4. ∎

The discrete distributions described thus far are univariate distributions. The probability function of a discrete k-variate population is a function $f(y_1, \ldots, y_k)$ of k arguments $y_1, \ldots, y_k$. In Exercises 2.15–2.17, you will have an opportunity to work with probability functions of discrete bivariate populations.

The probability density function of a continuous population

Just as a probability function describes a discrete population, a *probability density function* describes a continuous population. The symbols $f(y)$ and $f(y_1, \ldots, y_k)$ will also be used to denote the probability density function of univariate and k-variate continuous populations. Either a formula or a graph is used to represent a probability density function. Figure 2.3 shows a possible probability density function to represent the time-to-relief population in Example 2.3.

FIGURE 2.3

Probability density function for the time-to-relief population

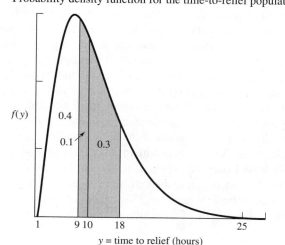

$f(y)$ 0.4

0.1 0.3

1 9 10 18 25

y = time to relief (hours)

In a continuous population, the proportion of measurements with values in a given interval equals the area contained below the probability density function within that interval. In Figure 2.3, the shaded area equals the proportion of drug users who obtain relief in the time interval from 9 to 18 hours. Since the proportion of all measurements in the population will equal 1, the total area under a probability density curve will also equal 1.

The probability density function of a continuous population can be used in essentially the same manner as the probability function of a discrete population. The only difference is that, while the appropriate relative frequencies are added to determine proportions in discrete populations, appropriate areas under the probability density curve are used to determine the proportions in continuous populations. The next example illustrates how a probability density function can be used to describe a continuous population.

EXAMPLE **2.8**   Let's see how the probability density function in Figure 2.3 can be used to summarize information about the time-to-relief population in Example 2.3. For the purpose of this example, assume that the shaded area equals 0.40 and that the area to the right of 10 hours is 0.50.

First of all, since the proportion of measurements with values less than 25 is almost equal to 1, we can say that the drug will provide relief within 25 hours in almost all cases. Also, the area under the curve to the left of 1 is virtually negligible, leading us to the conclusion that the drug rarely provides relief in less than 1 hour. Since 50% of the total area under the curve is to the left of 10 hours, 50% of those treated by the drug will obtain relief within 10 hours and 50% will require more than 10 hours. Finally, 40% of those treated by the drug will obtain relief in a period lasting 9–18 hours.   ∎

Random variables and probability distributions

As we will see in Chapter 3, some familiarity with the notions of a random variable and its probability distribution is needed to understand the principles underlying inferential methods of statistics. Those notions are briefly introduced here; a more detailed account will be found in Appendix B (Section B.2).

Let's begin with some terminology. A measurement is said to be chosen at random from a statistical population if it is selected in such a way that every measurement in the population has an equal chance of being selected. A *random variable* is a variable that takes the value of a measurement selected at random from a population. The random variable is multivariate if the corresponding measurement is multivariate; otherwise, it is univariate. When there is no possibility of confusion, a univariate random variable will simply be referred to as a random variable.

Very often, it is important to distinguish between the name of a random variable and its observed values. Whenever it is important to do so, we will use capital letters to denote the names of the random variables; specific observed values will be denoted by the corresponding lowercase letters. In Example 2.5, let Y be the random variable denoting the safety level of the exposure of a randomly selected resident. An observed value of Y will be denoted by y, where y can take one of the two values: $y = 0$ if the selected resident has safe exposure level; and $y = 1$ otherwise. Similarly, in Example 1.2, let Y be the random variable denoting the digestibility value for an experimental sheep. An observed value of Y is a value y in the interval [0, 100].

Let Y be a random variable. The probability that the random variable Y will have the value y, denoted by $\Pr\{Y = y\}$, is a number that indicates the level of confidence with which we can predict that, in a single observation, the observed value of Y will equal y. The degree of confidence can be defined in many ways; in this book, we use the relative frequency definition. According to this definition, $\Pr\{Y = y\}$ is an approximation to the proportion of times the value $Y = y$ will be observed in a long sequence of independent observations of Y.

EXAMPLE **2.9** Let Y be the random variable that equals 1 if a particular coin lands face up in a single toss. According to the relative frequency interpretation of probability, the statement $\Pr\{Y = 1\} = 0.50$, meaning that there is a probability of 0.50 that a particular coin will land face up in a single toss, should be interpreted as saying that, in a long sequence of repeated tosses, the coin will land face up approximately 50% of the times. Similarly, if Y is the random variable that equals 1 if a patient treated by a particular drug is cured, the statement $\Pr\{Y = 1\} = 0.90$ is the same as the statement that approximately 90% of a large number of patients treated by the drug will be cured.   ■

Every statistical population can be modeled by a random variable. There is a close connection between the distribution of a population and the probabilities of the associated random variable. This connection is most easily seen in the case of a discrete population, as illustrated in the next example.

EXAMPLE **2.10** Consider the 0–1 population of Example 2.5, in which the measurements indicate the safety level of radiation exposures of persons living in homes built on reclaimed lands. Recall that 5016 of the 15,200 residents of these homes were exposed to unsafe radiation levels, so that the probability distribution of the population has the relative frequencies: $f(0) = 0.67, f(1) = 0.33$.

Let Y denote the safety level of exposure of a randomly selected resident. Then the random variable Y can take two values: 0 and 1. Every resident in the population has the same chance of being selected, and so we can expect that, in a long series of repeated observations of Y, the proportion of times that $Y = 1$ will be observed is the same as the proportion of residents in the population who are exposed to an unsafe level of radiation. In other words, the probability that the random variable will take the value 1 is equal to $f(1) = 0.33$. Similarly, the probability that the random variable Y will take the value 0 is equal to $f(0) = 0.67$.   ■

The relationship between the probability function of a discrete population and the probability that a randomly selected measurement has a specific value can be expressed as follows. Consider a discrete population with probability function $f(y)$. Let Y be the random variable that denotes the observed value of a randomly selected measurement. Then the probability that Y has the value y is $f(y)$; in symbols

$$\Pr\{Y = y\} = f(y). \tag{2.4}$$

The probability interpretation implied by Equation (2.4) for a single specific value extends directly to the interpretation of a specific collection of values. For any given numbers a and b $(a \le b)$, the probability that the observed value of Y is between a and b is equal to the proportion of the values in the population that are between a and b.

EXAMPLE **2.11**   Let Y be the random variable that denotes the value of a measurement selected at random from the population of pest types in Example 2.7. Then, the probability that Y will take the value 2, 3, or 4 is equal to the proportion of measurements with values 2, 3, or 4 in the population

$$\Pr\{2 \le Y \le 4\} = \Pr\{Y = 2\} + \Pr\{Y = 3\} + \Pr\{Y = 4\}$$
$$= f(2) + f(3) + f(4)$$
$$= 0.18 + 0.10 + 0.46$$
$$= 0.74.$$

Thus, the probability that a randomly selected pest is of type 2, 3, or 4 is 0.74.   ∎

The probability interpretation of the probability function of a discrete population can be extended in a straightforward manner to the probability density function of a continuous population. If Y is a random variable denoting the value of a measurement randomly selected from a continuous population, then the probability that Y will take a value between a and b $(a \le b)$ is equal to the area between a and b under the probability density function. For example, inspecting the probability density function of the time-to-relief population in Figure 2.3, we can say that the probability that a randomly selected subject treated by the drug will have relief within 10 hours is 0.50. In symbols, if Y is the random variable that denotes the time to relief, then

$$\Pr\{Y \le 10\} = 0.50.$$

Further inspection of Figure 2.3 shows that

$$\Pr\{9 \le Y \le 18\} = 0.40.$$

As noted earlier, a random variable can be univariate or multivariate depending upon whether the associated population is univariate or multivariate. The random variables described thus far in this section are univariate. Some examples of multivariate random variables will be found in Appendix B (Section B.2) and in Exercises 2.15–2.19.

A note on terminology

A statistical population can be described using the distribution of its measurements or in terms of the probabilities of the values of a random variable. For instance, the 0–1 population of radon exposure levels in Example 1.4 can be visualized in two equivalent ways: as a collection of measurements containing 67% 0s and 33% 1s; or as the population associated with the random variable that takes the values 0 and 1 with probabilities 0.67 and 0.33, respectively.

Exercises

2.10   Suppose that the litter-size distribution for a species of hogs is known to have the following probability distribution.

y = litter size	1	2	3	4
$f(y)$ = relative frequency	0.2	0.3	0.3	0.2

  a   Draw a graph showing the probability function of the litter-size distribution.
  b   Explain why the litter-size probability function satisfies $f(1) + f(2) + f(3) + f(4) = 1$.
  c   Use the relative frequency interpretation to interpret the value of $f(y)$ when $y = 3$.
  d   Use the relative frequency interpretation to interpret the value of $f(1) + f(2)$.

2.11   Let Y be the litter size of a hog randomly selected from the population in Exercise 2.10. On the basis of the given probability function, what can you say about the probability that the size of a randomly selected litter will be:

  a   less than 4?   **b**   3 or more?

2.12   The following table shows the probability distribution of the number of leaves in a variety of tobacco plant.

y = number of leaves	15	16	17	18	19	20	21	22
$f(y)$ = relative frequency	0.02	0.03	0.20	0.25	0.25	0.20	0.03	0.02

  a   Draw a graph showing the probability function of the distribution of number of leaves per plant.
  b   Calculate the proportion of plants that have more than 19 leaves.
  c   Use the relative frequency interpretation to interpret the value of $f(15) + f(16) + f(17)$.

2.13   Let Y be the number of leaves in a randomly selected plant from the population of tobacco plants in Exercise 2.12.

    a   Calculate the probability that an observed value of Y is

       i   more than 18;   **ii**   between 17 and 20 (both inclusive);

       iii   less than 18.

    b   Use the relative frequency interpretation to interpret each of these three probabilities.

2.14   Suppose that there is a probability of 0.80 that a seed selected at random from the conceptual population of all seeds of a particular variety of plant will germinate. Let Y be the random variable that takes the value 1 if a randomly selected seed germinates and 0 otherwise.

    a   Describe the probability distribution of Y in tabular form.

    b   Describe the probability distribution of Y in graphical form.

    c   Describe the probability distribution of Y using a formula.

2.15   The following are the proportions of individuals in various categories of a large population classified according to age and blood pressure:

Age	Blood pressure		
	Low	Normal	High
Young	0.01	0.16	0.03
Middle age	0.06	0.32	0.08
Old	0.04	0.16	0.14

Suppose that an individual is selected at random from this population. Let $Y = (Y_1, Y_2)$ be the bivariate random variable such that Y_1 is the individual's age classification and Y_2 is the individual's blood pressure classification. Describe the probability distribution of Y in tabular form. [Hint: Let $f(y_1, y_2)$ denote the relative frequency of the value $y = (y_1, y_2)$ in the population. Construct a table showing $f(y_1, y_2)$ for all possible values of (y_1, y_2).]

2.16   Refer to Exercise 2.15. In each of the following cases, determine the probability that a randomly selected individual will be:

    a   a middle-aged person with low or normal blood pressure;

    b   a young or middle-aged person with high blood pressure;

    c   a person with normal blood pressure.

2.17   Consider the univariate population of all values of the random variable Y_1 defined in Exercise 2.15.

    a   Describe the probability function of this population.

    b   Use the probability distribution of Y_1 to calculate the probability that a randomly selected individual will be middle-aged or old.

2.18   Consider the univariate population of all values of the random variable Y_2 defined in Exercise 2.15.

    a   Determine the probability function of this population.

b Use the probability function of Y_2 to calculate the probability that a randomly selected individual will not suffer from high blood pressure.

2.19 Let $Y = (Y_1, Y_2)$, where Y_1 is the number of errors in test 1 (out of a maximum of 2) and Y_2 is the number of errors in test 2 (out of a maximum of 3), be a bivariate random variable denoting the results of a pair of tests administered to a randomly selected child from a large school district. The following table shows the probability function of Y:

y_1	0	0	0	0	1	1	1	1	2	2	2	2
y_2	0	1	2	3	0	1	2	3	0	1	2	3
$f(y_1, y_2)$	0.04	0.03	0.06	0.04	0.09	0.24	0.14	0.08	0.04	0.06	0.14	0.04

a Calculate the probability that a randomly selected child will make the same number of errors in both tests.

b Calculate the probability that a randomly selected child will make more errors in test 1 than in test 2.

c Calculate the probability that a randomly selected child will make no more than one error in each of the two tests.

2.20 The accompanying figure shows the probability density function of the percentage impurity of rainwater in a region on a randomly selected rainy day. Use the given areas under the probability density curve to answer the following questions.

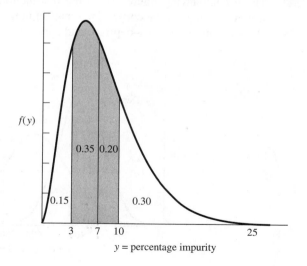

a Explain how you will interpret the shaded area under the probability density function.

b Suppose we determine the impurity of rainwater on a randomly selected rainy day. What is the probability that the impurity will be between 3% and 7%?

c Determine a value y such that the percent impurities will exceed y in 50% of the randomly selected rainy days.

2.21   The accompanying graph shows the probability density functions of the yields (bushels per acre) of two varieties of wheat.

   a   From the point of view of increased yield, which variety is better? Why?

   b   Which variety produces yields that are less variable?

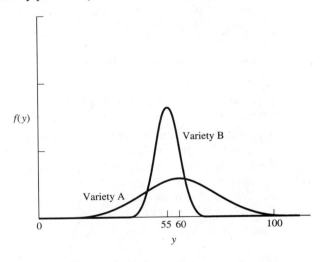

2.22   The accompanying figure shows the distributions of hypothetical populations of weights (in mg) of the cortex (grey matter) in the brains of rats reared in two types of social environments—deprived and enriched.

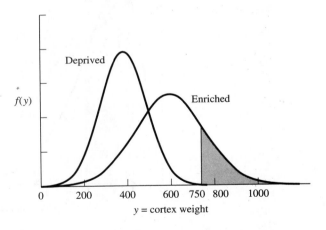

   a   On the basis of the given distributions, which environment seems to be more conducive to a heavier cortex? Explain.

   b   Interpret the shaded area as the probability of an event associated with the cortex weights of rats reared in an enriched environment.

   c   Which environment produces cortex weights with a higher variability? Explain.

2.4
Describing populations using parameters

A number calculated using all the measurements in a population is called a *parameter* of that population. Just as a probability distribution is used to summarize information about a population, a parameter is used to summarize information about specific characteristics of a population. In this section we describe two types of parameters: the *location parameters* used to locate the center of a univariate population; and the *dispersion parameters* used to assess the variability in a univariate population.

Location parameters

The population mean and the population median are two of the most frequently used location parameters.

The population mean

As the name implies, the population mean is the arithmetic mean of all measurements in a quantitative population. The symbol μ, with an appropriate subscript when necessary, is usually used to denote a population mean.

The population mean μ is a measure of location because it lies somewhere between the smallest and the largest values in the population. In other words, the measurements in a population are located both below and above the population mean. For example, if the average IQ in the population of school-age children in a particular region is known to be $\mu = 115$, then, except in the unlikely case where every child has IQ $= 115$, some children will have an IQ less than 115 and some will have an IQ greater than 115.

Because a population may contain infinitely many measurements, the population mean must be calculated using formulas specifically designed for that purpose. For instance, if the measurements in a discrete population have the distinct values $y_1, y_2,$ ..., with the corresponding relative frequencies $f(y_1), f(y_2), \ldots$, then the population mean can be computed as

$$\mu = y_1 f(y_1) + y_2 f(y_2) + \cdots = \Sigma y f(y). \tag{2.5}$$

The use of Equation (2.5) is illustrated in the next example.

EXAMPLE 2.12   The spatial patterns of organisms occupying discrete habitable sites are often described using the distribution of a discrete population. The following table shows the distribution of the number of caterpillars occupying the shoots of a tree. For each y, the table shows $f(y)$, the relative frequency of shoots containing y caterpillars.

y	$f(y)$
10	0.15
11	0.15
12	0.10
13	0.10
14	0.15
15	0.15
60	0.20

The mean number of caterpillars in the population can be computed using Equation (2.5):

$$\mu = 10 \times 0.15 + 11 \times 0.15 + \cdots + 60 \times 0.20$$

$$= 22.$$

Thus, on the average, 22 caterpillars occupy a shoot in this tree. Some shoots will be occupied by more than 22 caterpillars and some by fewer than 22 caterpillars.   ∎

The population median

A population median is another measure of location. It is a number η such that no more than half the measurements have values less than η and no more than half the measurements have values greater than η. A population median indicates the location of the middle value in a population.

In practice, there are two reasons for using the median instead of the mean as the location parameter for a population. First, it turns out that the population mean is not a theoretically well-defined concept for certain population distributions, known as *heavy-tailed distributions*. Heavy-tailed distributions contain a relatively large proportion of measurements with extreme (very large or very small) values relative to distributions called normal distributions. (The properties of normal distributions will be described in detail in Section 2.6.) Figure 2.4 shows how the shape of a heavy-tailed distribution compares with the shape of a normal distribution. There are heavy-tailed populations for which it is theoretically meaningless to consider the mean as a measure of location.

Second, the mean is not helpful when modeling a population that has a skewed distribution, with a few very large or very small values. Consider the following example.

EXAMPLE **2.13**   For the caterpillar population in Example 2.12, *any* value between 13 and 14 satisfies the definition of a median. To see this, select a specific value between 13 and 14— say $\eta = 13.3$. Then 50% of the measurements (corresponding to $y = 10, 11, 12,$ and 13) in the population have values less than 13.3, and 50% of the measurements (corresponding to $y = 14, 15$ and 60) have values greater than 13.3. Thus $\eta = 13.3$ satisfies the definition of the median, because no more than half of the measurements are less than 13.3 and no more than half of the measurements are greater than 13.3.

FIGURE 2.4

Heavy-tailed and normal distributions

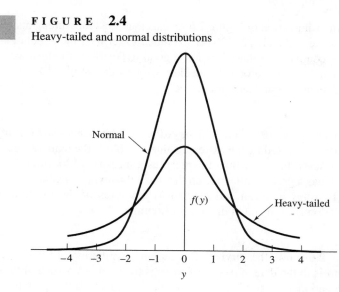

Similarly, any other value of η between 13 and 14—say, $\eta = 13.9$—will also satisfy the definition of a median. The interval (13, 14) is called the *median interval*. Any value in the median interval is a median of the caterpillar population. Thus, the caterpillar population has more than one median. If a unique value is desired, the midpoint 13.5 of the median interval can be taken as the median of this population.

Figure 2.5 shows the distribution of the caterpillar population, along with the values of the population mean and median.

FIGURE 2.5

Distribution of the number of caterpillars

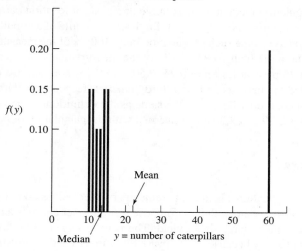

An examination of Figure 2.5 reveals that most measurements in this population have values (10, 11, 12, 13, 14, 15) much lower than the population mean $\mu = 22$. Consequently, the value 22 is not a good indicator of the location of the measurements in this population. The median, with a value of $\eta = 13.5$, is a better indicator of the center of the distribution.   ■

In Exercise 2.25, you will be asked to argue that skewing of the caterpillar population (caused by the extreme value $y = 60$) is the main reason why the median turned out to be better than the mean as a measure of location. The extreme value $y = 60$ has a greater influence on the mean than on the median.

The roles of mean and median as measures of the location of a continuous population are illustrated in the next example.

EXAMPLE **2.14**   Figure 2.6 shows the survival distribution (distribution of the time from the date of diagnosis to the date of death) of patients treated for a particular type of cancer. The mean and median survival times for the individuals in this population are $\mu = 10.5$ years and $\eta = 5$ years, respectively. Notice that, while most of the survival times are in the range 0–15 years, there is a group of patients who survive between 15 and 25 years. The mean survival time of 10.5 years, which is inflated because of the few individuals with large survival times, does not really represent the location of the bulk of the population. The median survival time of 5 years, on the other hand, is not affected by the magnitudes of the few large measurements in the population. On the basis of the median, we can say that 50% of the patients survive less than 5 years. On the other hand, the value $\mu = 10.5$ is hard to interpret in the absence of additional information about the survival distribution.   ■

The population percentiles

The population median is an example of a class of location parameters called *population percentiles*. For $0 \leq p \leq 1$, the $100p$ percentile of a population, denoted by η_p, is defined as a value such that no more than $100p\%$ of the measurements are less than η_p and no more than $100(1 - p)\%$ of the measurements are greater than η_p. The 25, 50, and 75 percentiles ($p = 0.25, 0.50,$ and 0.75) are the most commonly used and are called the first, second, and third *quartiles*, respectively. Note that the definition of the second quartile $\eta_{0.5}$ is the same as the definition of the population median η. Exercises 2.29 and 2.30 are concerned with the calculation of population quartiles.

Dispersion parameters

In most applications, location parameters by themselves do not provide adequate information about a population. Suppose that a farmer chooses a new variety of barley because it has a large mean yield. The new variety may be less desirable than the standard variety because of high variability in its yield, as illustrated in Figure 2.7.

FIGURE   2.6
Survival distribution of individuals with cancer

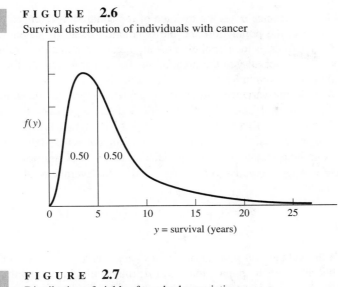

FIGURE   2.7
Distribution of yields of two barley varieties

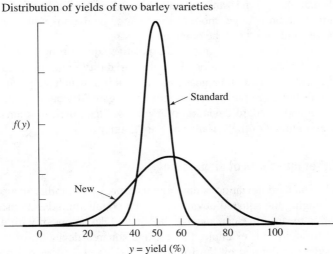

The yields for the standard variety show a greater concentration near its mean $\mu_S = 50$, while the new variety, though having a larger mean yield $\mu_N = 60$, shows more variation in its production capacity.

A useful measure of the variability of values in a population is the population standard deviation.

Population standard deviation

The *standard deviation* of a quantitative population, usually denoted by σ, is defined as the positive square root of a parameter called the *population variance*. The population variance σ^2 is the average of the squared differences between the measurements in the population and the population mean. If y denotes a value in the population and μ is the population mean, then the population variance is the average of $(y - \mu)^2$, over all y in the population. The general formula for the variance of a population is given in Section B.3 (Appendix B).

Why do the variance and standard deviations provide meaningful measures of variability in a population? Let's consider a discrete population of distinct values y_1, y_2, ... and respective relative frequencies $f(y_1)$, $f(y_2)$, ... If μ is the population mean, then $(y_1 - \mu)^2$, $(y_2 - \mu)^2$, and so on are the squared differences between the values in the population and the population mean. A proportion $f(y_i)$ of the squared differences will be equal to $(y_i - \mu)^2$, and so the average of all squared differences is equal to

$$\sigma^2 = (y_1 - \mu)^2 f(y_1) + (y_2 - \mu)^2 f(y_2) + \cdots$$
$$= \Sigma (y - \mu)^2 f(y). \tag{2.6}$$

If the values in a population do not vary much, they tend to cluster in a close neighborhood of the population mean μ; this results in small values of the squared deviations $(y - \mu)^2$ and a small value of σ^2. If the variation among the values in a population is large, more of the squared deviations will be relatively large, and a large value of σ^2 will be obtained.

The standard deviation is the positive square root of the variance and so contains the same information about the variability in the population. In contrast to the variance, however, it is measured in the unit in which the population values are measured. For example, if the variable Y = gain in weight is measured in grams, the mean and standard deviation of Y are both measured in grams, whereas the unit of measurement of the variance of Y is $(\text{grams})^2$.

Other measures of dispersion

The standard deviation is only one measure of variability in a population. Instead of averaging the squared deviations of the population measurements y from the population mean μ, we can average the squared deviations of y from some other measure of location. For example, the average squared deviation of y from the population median η, defined by the formula $\Sigma (y - \eta)^2 f(y)$, is also a population measure of variability.

Using μ and σ to describe populations

Among the wide variety of parameters suitable for describing a population, the population mean μ and the population standard deviation σ are the most popular. One reason for their widespread use to describe statistical populations is a property known as the *empirical rule*. As its name implies, the validity of the empirical rule cannot be established on a strict mathematical basis, but practical experience indicates that it is applicable to many (but not all) populations encountered in practice. The rule works particularly well for populations that are mound-shaped; it is stated in Box 2.1. A more conservative rule—called Tchebycheff's rule—can be used to describe any population in terms of μ and σ; this rule is described in Exercise 2.34.

BOX **2.1**

The empirical rule

Approximately 68% and 95% of the measurements in a population are, respectively, within a distance of one and two standard deviations from the population mean. Nearly all of the measurements are within a distance of three standard deviations of the population mean.

According to the empirical rule, the intervals $(\mu - \sigma, \mu + \sigma)$, $(\mu - 2\sigma, \mu + 2\sigma)$, and $(\mu - 3\sigma, \mu + 3\sigma)$ contain, approximately, 68%, 95%, and 100% of the measurements, respectively. The next example illustrates how the empirical rule may be used to obtain information about the distribution of a population.

EXAMPLE **2.15**   Suppose that the distribution of the percentage impurity of rainwater in Exercise 2.20 has mean $\mu = 4\%$ and standard deviation $\sigma = 0.2\%$. Then, in view of the empirical rule, we can conjecture that, for approximately 95% of the rainy days, the rainwater will contain between $4 - (2)(0.2) = 3.6\%$ and $4 + (2)(0.2) = 4.4\%$ impurities. Furthermore, on almost all days, the rainwater impurity level will be no less than $4 - (3)(0.2) = 3.4\%$ and no more than $4 + (3)(0.2) = 4.6\%$.   ∎

Coefficient of variation

In spite of its usefulness as a measure of variability, the standard deviation must be used with caution when comparing variabilities of different populations. A magnitude of variation considered small in one population may not be considered small in another. For instance, a standard deviation of $\sigma = 1$ m would be considered small when measuring heights of trees in a population with an average height $\mu = 40$ m. The same standard deviation is too large when measuring heights of one-week-old seedlings.

The *coefficient of variation* measures the variability in the values in a population relative to the magnitude of the population mean. For a population with mean μ and standard deviation σ, the coefficient of variation is defined as

$$\text{CV} = \frac{\sigma}{|\mu|} \tag{2.7}$$

provided $\mu \neq 0$; that is, the coefficient of variation is the standard deviation of the population expressed in units of μ.

Since the standard deviation and the mean are measured using the same unit, the CV is a unit-free number and can be used as an index of population variability.

Often, the CV is expressed as a percentage: $CV = 100(\sigma/|\mu|)$. Thus a CV of 10% implies that the standard deviation of the measurements in the population is 10% of the population mean.

Mean and variance of 0–1 populations

In subsequent chapters, populations in which each measurement is either a 0 or a 1 will play an important role. In Exercise 2.38, you are asked to verify that the mean and variance of such populations can be expressed in terms of the relative frequencies of 0 and 1. Box 2.2 contains these expressions.

BOX **2.2**   ***The Mean and Variance of a 0–1 Population***

Let $f(0)$ and $f(1)$ denote, respectively, the relative frequencies of 0 and 1 in a population containing 0s and 1s. Let μ and σ^2 denote the mean and variance of this population. Then

$$\mu = f(1)$$

$$\sigma^2 = f(1)f(0).$$

Often, it is convenient to express the mean and variance of a 0–1 population in terms of π, the proportion of 1s in the population. Then since $f(1) = \pi$ and $f(0) = 1 - \pi$, the results in Box 2.2 can be expressed as

$$\mu = \pi$$

$$\sigma^2 = \pi(1 - \pi). \tag{2.8}$$

Covariance

Let $Y = (Y_1, Y_2)$ be a quantitative bivariate random variable, and let σ_1^2 and σ_2^2 denote, respectively, the variances of the univariate populations associated with Y_1 and Y_2. The variability in the bivariate population associated with Y can be characterized in terms of three parameters: Two measure the individual variabilities of Y_1 and Y_2, while the third measures their joint variability. As might be expected, the variabilities of Y_1 and Y_2 are measured by σ_1^2 and σ_2^2, respectively. The third parameter is called the *covariance* of Y_1 and Y_2 and is denoted by σ_{12}. A general definition of the covariance is given in Appendix B. In the special case where Y is a bivariate discrete random variable with probability function $f(y_1, y_2)$, the covariance of Y_1 and Y_2 is defined as

$$\sigma_{12} = \Sigma(y_1 - \mu_1)(y_2 - \mu_2)f(y_1, y_2), \tag{2.9}$$

where μ_1 and μ_2 are the means of the univariate random variables Y_1 and Y_2, respectively, and the sum is taken over all possible combinations of the values of Y_1 and Y_2.

The uses and interpretations of covariances will be discussed in more detail in Chapter 10. Here we use Figure 2.8 to present an intuitive explanation of the definition in Equation (2.9). Figure 2.8 shows the set of all possible values of the bivariate random variable (Y_1, Y_2) divided into four quadrants of the y_1–y_2 plane. The quadrants are named A, B, C, and D and bounded by perpendicular straight lines drawn through the point (μ_1, μ_2). If we use the population mean as the central value and regard any value larger than the mean as a large value and any value less than the mean as a small value, then A, B, C, and D denote regions of (Y_1, Y_2) values such that y_1 is small but y_2 is large, both y_1 and y_2 are large, y_1 is large but y_2 is small, and both y_1 and y_2 are small, respectively. Thus, if there are more values of (y_1, y_2) in B and D than in A and C, we can conclude that, in the bivariate population, large values of Y_1 tend to be paired with large values of Y_2 (or, equivalently, small values of Y_1 tend to be paired with small values of Y_2). We can then say that Y_1 and Y_2 tend to vary in the same direction or, equivalently, tend to increase or decrease together. Similarly, if there are more values of (y_1, y_2) in A and C than in B and D, then we have a case where there is a tendency for Y_1 and Y_2 to vary in the opposite directions.

Now, if a measurement (y_1, y_2) is in D, then $y_1 < \mu_1$ and $y_2 < \mu_2$, so that $y_1 - \mu_1$ and $y_2 - \mu_2$ are both negative and hence the product $(y_1 - \mu_1)(y_2 - \mu_2)$ is positive. Similarly, if (y_1, y_2) is in B, both $y_1 - \mu_1$ and $y_2 - \mu_2$ are positive, so that the product $(y_1 - \mu_1)(y_2 - \mu_2)$ will again be positive. On the other hand, if (y_1, y_2) is in A or C, the product $(y_1 - \mu_1)(y_2 - \mu_2)$ will be negative. Consequently, when there is a tendency for Y_1 and Y_2 to vary in the same direction, most of the terms in Equation (2.9) will be positive, and a large positive value for σ_{12} is obtained. Thus, a positive covariance is an indication that, in the bivariate population, the values of Y_1 and Y_2 tend to increase or decrease together. Similarly, a negative covariance is an indication that the Y_1 and Y_2 values vary in the opposite direction.

FIGURE 2.8

Interpreting covariance

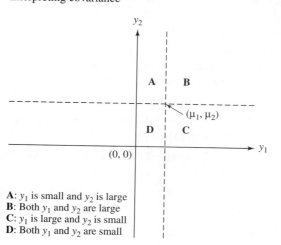

A: y_1 is small and y_2 is large
B: Both y_1 and y_2 are large
C: y_1 is large and y_2 is small
D: Both y_1 and y_2 are small

Exercises

2.23 **a** Calculate the mean of the litter-size population in Exercise 2.10 and interpret this number.

 b Argue that the mean of this population is also its median.

2.24 Show that the mean number of leaves in the population described in Exercise 2.12 is also a median of this population.

2.25 In Example 2.12, modify the caterpillar population by deleting all measurements with $y = 60$.

 a Show that the distribution of the modified population is

y	$f(y)$
10	0.1875
11	0.1875
12	0.1250
13	0.1250
14	0.1875
15	0.1875

 b Argue that either the mean or the median (defined as the center of the median interval) can serve as a reasonable measure of the location of the modified population.

 c Argue that the presence of the extreme value $y = 60$ is the main reason for the failure of the mean as a measure of location of the unmodified caterpillar population.

2.26 Make appropriate changes in the relative frequencies in Exercise 2.12 to obtain:

 a a population in which the mean is smaller than the median;

 b a population in which the mean is larger than the median.

2.27 Notice that the distribution of the population in Exercise 2.12 is symmetric about the population mean but the distributions of the populations in Exercise 2.26 are not symmetric about the mean. Explain how the asymmetry of a distribution determines whether or not the mean exceeds the median.

2.28 Consider a conceptual population of 0s and 1s, where 1 denotes a patient who responds to a drug and 0 denotes a patient who does not. Suppose you were told that the mean of this population is 0.63. How would you interpret this number?

2.29 Refer to the distribution of the number of leaves per shoot in Exercise 2.12.

 a Show that any number between 17 and 18 is a first quartile of this distribution.

 b Show that any number between 19 and 20 is a third quartile of this distribution.

2.30 Modify the distribution in Exercise 2.12 so that the first and third quartiles are within the intervals (16, 17) and (20, 21), respectively.

2.31 The accompanying figure shows a possible distribution of the incomes of farmers in a particular region.

 a Determine the three quartiles of the distribution and interpret them.

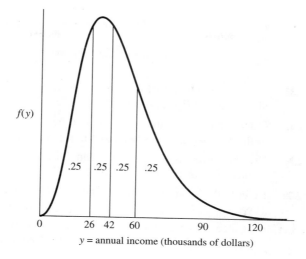

$f(y)$

.25 .25 .25 .25

0   26 42 60   90   120

y = annual income (thousands of dollars)

b   On the basis of the given distribution, is it possible to decide whether or not the mean income is higher than the median income? Explain.

2.32   Suppose it is known that the mean yield for a variety of barley is $\mu = 120$ bushels per acre, with a standard deviation of $\sigma = 10$ bushels per acre. Use the empirical rule to make statements about the yields that will be observed for the new variety.

2.33   Suppose that the distribution of scores in an aptitude test has a mean of 350 and a standard deviation of 40. Use the empirical rule to argue that:

a   approximately 95% of those who take the test will have scores between 270 and 430;

b   almost all of those who take the test will have scores between 230 and 470.

2.34   Recall that the empirical rule is a statement about the proportion of measurements contained within a distance of one, two, and three standard deviations from the mean of a population. The empirical rule is applicable to many (but not all) of the populations commonly encountered in practice. An alternative to the empirical rule is Tchebycheff's rule: For any number $k > 1$, at least $100(1 - 1/k^2)\%$ of the measurements in a population will be within a distance of k standard deviations from the population mean. In symbols, if Y is the random variable characterizing a population with mean μ and standard deviation σ, then

$$\Pr\{\mu - k\sigma \le Y \le \mu + k\sigma\} \ge 1 - \frac{1}{k^2}.$$

Unlike the empirical rule, Tchebycheff's rule is a precise mathematical statement that is applicable to all populations.

a   Use Tchebycheff's rule to obtain lower bounds to the proportion of measurements within one, two, and three standard deviations of a population mean.

b   Compare the lower bounds in (a) with the estimates based on the empirical rule. Which rule is more conservative? Why?

2.35   Use the information in Exercise 2.32 and Tchebycheff's rule to answer the following questions.

   a   What can you say about the proportion of yields between 100 and 140 bushels per acre?

   b   What can you say about the proportion of yields between 105 and 135 bushels per acre?

2.36   Calculate the standard deviation and the coefficient of variation for the litter-size distribution in Exercise 2.10.

2.37   Calculate the standard deviation and the coefficient of variation for the distribution of the number of leaves per shoot in Exercise 2.12.

2.38   Use the formulas for the mean and variance of a discrete population to verify the expressions for μ and σ^2 in Box 2.2.

2.39   Refer to the bivariate error distribution in Exercise 2.19.

   a   Calculate the mean and standard deviation of Y_1, the number of errors in test 1.

   b   Calculate the mean and standard deviation of Y_2, the number of errors in test 2.

   c   Calculate the covariance between the random variables Y_1 and Y_2.

2.5
Some discrete distributions

When selecting the appropriate statistical inferential procedures, researchers are guided by certain assumptions that can be made about the population from which the samples will be drawn. Usually, these assumptions are stated in terms of the characteristics of the population distributions. The assumptions are either based on the past experience of investigators in the field or derived from theoretical considerations about the mechanism generating the population of measurements. For example, Pielou (1969) describes conditions under which the distribution of the number of caterpillars occupying the shoots of a tree can be modeled using the probability function of a Poisson distribution.

In this section, we describe two discrete distributions—the binomial distribution and the Poisson distribution—that are often used for modeling discrete populations. Some useful continuous distributions will be described in the next section.

The binomial distribution

The binomial distribution is best introduced in the context of the binomial experiment described in Box 2.3.

BOX **2.3**   ***The binomial experiment***

A binomial experiment is an observational process that can be described as follows.

1   There are n identical trials.

2   Each trial can result in one of two outcomes—a success or a failure.

3   The probability of success in any single trial is π, which stays the same from trial to trial.

4   The trials are independent; that is, the outcome of one trial is not influenced by the outcome of any other trial.

5   The measurement of interest is the number of observed successes in n trials.

The population of measurements from a binomial experiment consists of $n + 1$ distinct values: from 0, which corresponds to 0 successes and n failures, and 1, which corresponds to 1 success and $n - 1$ failures, up to n, which corresponds to n successes and 0 failures. Let Y be the random variable that denotes the number of successes in a binomial experiment. If the experiment consists of n trials, each with probability of success π, then it can be shown that the probability of observing the value y for Y is

$$\Pr\{Y = y\} = f(y) = \frac{n!}{y!(n-y)!}\pi^{y}(1 - \pi)^{n-y}, \qquad y = 0, 1, \ldots, n, \qquad \textbf{(2.10)}$$

where $n!$ denotes the product of the first n integers; that is, $n! = 1 \times 2 \times \cdots \times n$. By convention, we let $0! = 1$.

The expression for $f(y)$ in Equation (2.10) is called a *binomial probability function*. The random variable Y is called a binomial random variable with parameters n and π. We will use the symbol $B(n, \pi)$ to refer to a population with a binomial distribution or its associated random variable.

The following is an example of a binomial experiment.

EXAMPLE **2.16**   In order to estimate the germination rate of a new seed variety, a plant breeder decided to plant six seeds in six identical pots and observe the number of germinating seeds. Let π denote the germination rate for the seed variety; that is, suppose that, on the average, we expect $100\pi\%$ of the seeds to germinate. Under the assumption that the germination of each seed is independent of the other seeds, this is a binomial experiment with $n = 6$ trials; π is the the probability that a seed will germinate. If Y is the number of germinating seeds, then Y is a binomial random variable with parameters $n = 6$ and π.

As an example, suppose that the germination rate is 70% ($\pi = 0.70$). Then the probability that y out of $n = 6$ seeds will germinate is

$$f(y) = \frac{6!}{y!(6-y)!}0.70^{y}(1 - 0.70)^{6-y}, \qquad y = 0, 1, 2, \ldots, 6.$$

For example, the probability of exactly two seeds germinating is

$$f(2) = \frac{6!}{2!4!}0.70^{2}0.30^{4} = 0.0595,$$

whereas the probability of at least five seeds germinating is

$$f(5) + f(6) = \frac{6!}{5!1!}0.70^5 0.30^1 + \frac{6!}{6!0!}0.70^6 0.30^0 = 0.4201.$$

Thus, when the germination rate is 70%, there is a 5.95% chance that exactly two out of six seeds will germinate and a 42% chance that at least five out of six seeds will germinate. ■

Box 2.4 shows the formulas for computing the mean and variance of a given binomial population.

BOX **2.4**   ***Mean and variance of a binomial population***

The mean and variance of a binomial population with parameters n and π are as follows:

$$\mu = n\pi$$
$$\sigma^2 = n\pi(1 - \pi).$$

EXAMPLE **2.17**   Refer to Example 2.16. Suppose that the germination rate is 60% and we are interested in Y, the number of germinating seeds, when $n = 15$. In this case, Y is a binomial random variable with parameters $n = 15$ and $\pi = 0.60$. The mean and variance of the associated population can be determined as in Box 2.4. We have

$$\mu = n\pi = 15(0.60) = 9,$$
$$\sigma^2 = n\pi(1 - \pi) = 15(0.60)(0.40) = 3.6.$$

Correspondingly, the standard deviation is

$$\sigma = \sqrt{3.6} = 1.9.$$

Thus, on average, nine out of 15 seeds will germinate. The standard deviation of 1.9 provides an assessment of the variability in the observed number of germinations in repeated plantings of 15 seeds. ■

The Poisson distribution

The Poisson distribution is often used to model a population in which each measurement is a count of the occurrences of a particular event over a specified region of space or during a specified period of time. Typical examples of counts that have Poisson distributions include the number of bacterial colonies in a specimen sample (for instance, an agar plate) and the number of radioactive particles emitted during a specific interval of time. Poisson distributions might also serve as the models for conceptual populations in which, for example, each measurement is the number of insects occupying a randomly selected plot of land, the number of black spots on a randomly selected orange, the number of caterpillars occupying a randomly selected shoot of a tree, the number of dark spots in a randomly selected patient's lung, and the number of deaths due to a particular disease during a randomly selected time interval of fixed length.

Roughly speaking, the binomial distribution provides a model for the outcomes of experiments that consist of n identical trials, each of which can result in a success or a failure. The Poisson distribution, on the other hand, describes the outcome of experiments consisting of infinitely many independent identical trials (conducted over time or space), each of which can result in a success or a failure. For example, when a Poisson distribution is used to model the number of radioactive particles emitted by a source, each emitted particle is regarded as a success in the trial that is being conducted at that particular instant.

Whether a Poisson distribution is suitable for modeling a particular population will depend on the chance mechanisms that govern the occurrence of the events of interest. The notion of a *Poisson process,* described in Box 2.5, provides a practical means for deciding whether a Poisson distribution is a reasonable model for a given population of counts.

BOX **2.5**

The Poisson process

A Poisson process is a probabilistic mechanism responsible for the occurrence of events in a specific region of space or a specific interval of time. A Poisson process over space can be characterized as follows.

1   The probability that the event will occur in a region of infinitesimally small area A (volume V) is equal to $\lambda A (\lambda V)$, where $\lambda > 0$ is a constant.

2   The probability that more than one event will occur in a given region of infinitesimal area (volume) is negligible.

3   The occurrence of an event in any one region will in no way influence the occurrence of an event in any other region; that is, events over disjoint regions are independent.

Poisson processes occurring over time are characterized similarly, but the properties are stated in terms of the length T of a short time interval, rather than the area A (or volume V) of a small region.

The constant λ is called the *rate parameter* of the Poisson process.

Consider an event that can occur over a region of space (for instance, a bacterial colony in an agar plate) or over an interval of time (for instance, the number of gene mutations during a given time interval). Assume that the event is generated by a Poisson process with rate parameter λ. Let Y be the random variable that denotes the number of occurrences of the event of interest. Then it can be shown that the probability that Y will take the value y is given by

$$\Pr\{Y = y\} = f(y) = e^{-\lambda K}\frac{(\lambda K)^y}{y!}, \qquad y = 0, 1, \ldots, \qquad \text{(2.11)}$$

where K equals the area of the region or the length of the interval. The probability that exactly y events will occur in a region of unit area (or unit length) can be obtained by setting $K = 1$ in Equation (2.11)

$$f(y) = e^{-\lambda}\frac{\lambda^y}{y!}, \qquad y = 0, 1, \ldots. \qquad \text{(2.12)}$$

The population whose distribution is specified by the probability function in Equation (2.12) is called a Poisson population with parameter λ. The distribution of the population is called a Poisson distribution with parameter λ. The following example illustrates how a Poisson distribution can be used to describe a population of counts of events occurring over a region.

EXAMPLE **2.18**   In Example 2.6, the Poisson distribution was suggested as a possible model for the number of bacterial colonies growing on an agar plate. The appropriateness of such a model depends upon whether it is reasonable to assume that the occurrence of bacterial colonies is governed by the law of a Poisson process—that is, whether the following assumptions are reasonable.

1   The probability of finding a colony in an infinitesimally small area on the agar plate is proportional to the area.

2   The probability of finding more than one colony in an infinitesimally small area of the plate is negligible.

3   Each occurrence of a colony in the different areas of the agar plate is an independent event.   ∎

Box 2.6 shows the mean and variance of a Poisson distribution.

BOX **2.6**   *Mean and variance of a Poisson distribution*

The mean and variance of a Poisson distribution with parameter λ are as follows

$$\mu = \lambda$$

$$\sigma^2 = \lambda.$$

Notice that the rate parameter λ in a Poisson distribution equals the mean number of occurrences (that is, the rate of occurrence) of the event of interest. The following example shows how this interpretation of λ can be used to model the probability function of a Poisson distribution.

EXAMPLE 2.19 Assume that the number of bacterial colonies found in an agar plate follows a Poisson distribution and that, on average, an agar plate with an area of 100 cm^2 will contain 12 colonies. Calculate the probability that five colonies will be found in a 10-cm^2 agar plate.

   The number of colonies in a 10-cm^2 agar plate can be regarded as a random variable associated with a Poisson population in which the measurements denote the number of colonies found in a 10-cm^2 agar plate. Since, on average, a 100-cm^2 agar plate will contain 12 colonies, the average number of colonies in a 10-cm^2 agar plate is $12/10 = 1.2$. Therefore, the number of colonies found in a 10-cm^2 agar plate will have a Poisson distribution with parameter $\lambda = 1.2$. The probability of finding five colonies can be calculated by setting $\lambda = 1.2$ and $y = 5$ in the formula for the Poisson probability function

$$f(5) = \frac{e^{-1.2}1.2^5}{5!} = 0.006.$$

Thus we can expect six out of 1000 ($6/1000 = 0.006$) 10-cm^2 agar plates to contain five colonies. ∎

The Poisson approximation to a binomial distribution

Computations involving binomial probability functions can be quite complicated at times, because they involve expressions such as $n!$, $y!$, and π^y. In certain situations, a Poisson probability function can be used as an approximation for the binomial probability function. Recall that a binomial probability function specifies the distribution of the population consisting of counts of the number of successes in independent binomial trials. It can be shown that, as n gets large and π approaches 0 in such a way that the product $n\pi$ approaches a value $\lambda > 0$, the binomial probability function approaches a Poisson probability function. More specifically, if the number of trials is large $(n \geq 100)$ and the probability of success in any single trial is relatively small $(n\pi < 5)$, then the probabilities determined by the binomial probability function in Equation (2.10) will be close to the corresponding probabilities determined by the Poisson probability function

$$f(y) \cong e^{-n\pi} \frac{(n\pi)^y}{y!}, \qquad y = 0, 1, 2, \ldots . \tag{2.13}$$

The next example illustrates a situation where the Poisson distribution can be used to calculate good approximate values of binomial probabilities.

EXAMPLE 2.20 Refer to Example 2.16. Suppose that $n = 200$ seeds with a small germination rate—say, 2%—were planted. If the seeds germinate independently of each other, the random variable Y, the number of germinating seeds, will have a binomial distribution with $n = 200$ and $\pi = 0.02$. The probability that $Y = y$ can be calculated using the binomial probability function with $n = 200$ and $\pi = 0.02$

$$f(y) = \frac{200!}{y!(200-y)!}\pi^y(1-\pi)^{200-y}, \qquad y = 0, 1, \ldots, 200. \qquad \textbf{(2.14)}$$

Computation of $f(y)$ using Equation (2.14) can be laborious. Since $n \geq 100$ is large and $n\pi = 4 < 5$, a Poisson approximation with $\lambda = n\pi = 200(0.02) = 4$ can be used to obtain a computationally simple expression for $f(y)$. According to the Poisson approximation for a binomial, the probability $f(y)$ in Equation (2.14) is approximately equal to

$$f(y) = e^{-4}\frac{4^y}{y!},$$

so that the probability that six out of 200 seeds will germinate can be approximated as

$$\Pr\{Y = 6\} = e^{-4}\frac{4^6}{6!} = 0.1042.$$

Figure 2.3 in the companion text by Younger (1997) shows how SAS software may be used to compute the Poisson probabilities. ∎

In Example 2.20, we saw how the Poisson probabilities can be used to approximate the binomial probabilities. The next example shows how the Poisson approximation can be used to motivate the Poisson model for a discrete population.

EXAMPLE **2.21**   An experiment to determine the effect of a nematocide on the nematodes found in the soil around citrus trees was performed as follows. Three months after application of the nematocide to ten potted trees, a soil sample was taken from the soil around each tree. The nematodes found in each of the ten soil samples were counted.

The investigator was interested in the conceptual population of nematode counts in soil samples taken three months after treatment with the nematocide. Such a population is a discrete population containing measurements with values $0, 1, 2, \ldots$.

A Poisson distribution may be adopted for the population of nematode counts if it is reasonable to assume that the nematodes occur according to a Poisson process. That will be the case if the following assumptions are reasonable.

A1 The probability of finding a single nematode in an infinitesimally small volume of soil sample is proportional to the volume of the sample.

A2 The probability of finding more than one nematode in an infinitesimally small soil sample is negligible.

A3 Each occurrence of a nematode in the different soil samples is an independent event.

The nematode count population can also be approximated by a Poisson distribution when it is reasonable to regard the number of nematodes found in a soil sample as the number of successes in a binomial experiment with a large number of trials. Consider the following set of assumptions.

B1 Each soil sample consists of a large number n of locations that can be occupied by individual nematodes.

B2 The probability π that a specific location will be occupied by a nematode is small.

B3 The nematodes are randomly distributed over locations, so that the occupation of each location by an individual is an independent event.

If we regard occupation of a site by a nematode as a success, then assumptions B1–B3 imply that the number of nematodes in a soil sample can be regarded as the number of successes in a binomial experiment with n trials and a probability of success π. Since n is large and π is small, the probability of observing y nematodes in a given soil sample can be approximated by the Poisson probability function with parameter $\lambda = n\pi$. ∎

The binomial and Poisson distributions are only two of the many distributions used to model discrete populations. Which particular distribution is reasonable in a given problem will depend upon the assumptions that can be made about the populations being studied. For more information on discrete distributions, see Johnson, Kotz, and Kemp (1992) and Pielou (1969).

Exercises

2.40 Suppose that five trees are treated with a fumigant known to be effective in keeping 80% of the plants disease-free for one year. Let Y denote the number of disease-free trees one year after treatment.

    a Describe conditions under which Y will have a binomial distribution.
    b Assuming the conditions in (a), write a formula for the probability function of Y.
    c Assuming the conditions in (a), create a table showing the probability function of Y.
    d Calculate the mean and standard deviation of the distribution in (b).

2.41 It is claimed that a new drug for hypertension (high blood pressure) is effective in 70% of the treated patients. Let Y denote the number who respond among seven treated patients.

    a Describe a set of conditions that will ensure that Y can be regarded as a binomial random variable.
    b Under the conditions described in (a), calculate the probability that at least five out of seven patients will respond to the treatment.

2.42 In a study to evaluate the effectiveness of a new agent for treating a type of cancer, each of the ten patients treated with the drug was classified as responding (indicating a positive response) or nonresponding (indicating no response or a negative response). Let π denote the response rate (the proportion of treated patients who respond) and Y be the observed number of responding patients. Consider the following procedure for making inferences about π.

Observed response	Conclusion
$Y < 3$	Response rate is less than 30% ($\pi < 0.30$)
$Y \geq 3$	Response rate is at least 30%

a Calculate the probability that the procedure will conclude that $\pi < 0.30$ when $\pi = 0.10, 0.20, 0.30, 0.40, 0.50, 0.60, 0.70, 0.80, 0.90$.

b On the basis of your calculations in (a), what can you say about the procedure regarding:

    i the probability of concluding that the response rate is at least 30% when the true response rate is less than 10%?

    ii the probability of concluding that the response rate is less than 30% when the true response rate is more than 40%?

2.43 In a microbial mutagenesis assay, a plate of bacteria is exposed to a test compound, and the number of revertants is counted after incubation. Suppose that, in the assay of a particular compound A, the number of revertants has a Poisson distribution with $\lambda = 9$. Calculate the probability that an assay of the compound will produce:

a no revertants;   **b** three revertants;

c more than five revertants.

2.44 Refer to Exercise 2.43. Let Y denote the ~~probability of finding~~ number of assays with no revertants in five independent assays of the compound.

a Argue that Y can be regarded as a binomial random variable.

b Determine the values of n and π for the binomial distribution in (a).

c Calculate the probability of finding at least one revertant in two out of five assays.

d Use the Poisson approximation to the binomial to determine the probability of finding five or more revertants in 100 assays.

2.45 Let Y denote the number of caterpillars occupying the shoots of a tree.

a Describe a set of assumptions that will justify treating Y as a Poisson random variable.

b Suppose that Y in (a) has a Poisson distribution with parameter λ. Explain what it means to say that $\lambda = 2$.

c Assuming that $\lambda = 2$, calculate the proportion of shoots that are not occupied by caterpillars.

2.6
Some continuous distributions

In this section, we describe the two continuous distributions that are most frequently used to model continuous populations: the normal distribution and the log-normal distribution.

The normal distribution

The normal distribution was originally derived as a model for a continuous population in which each measurement can be considered as a true value plus an error, where the error is the sum of infinitely many small errors that represent the effects of infinitely many independent causes. In practice, a normal distribution will serve as a good model to describe a population if the errors in its measurements are the sum of many factors. For instance, when measuring the dry weight of a specimen of grass, there will be several sources of error, such as fluctuations in temperature and humidity in the room, variation in the calibration of the weighing machine, and error in reading and recording the weight. If each of these errors contributes relatively little to the measured value and the actual error is the sum of many such errors, then the distribution of the population of possible measurements of a given specimen can be approximated by a normal distribution.

The probability density function of a normal distribution is determined by two parameters, the population mean μ and the population standard deviation σ

$$f(y) = \frac{1}{\sigma\sqrt{2\pi}} e^{-\frac{(y-\mu)^2}{2\sigma^2}}, \quad -\infty < y < +\infty.$$

A graph of a normal distribution is shown in Figure 2.9.

FIGURE 2.9

The normal distribution with mean μ and standard deviation σ

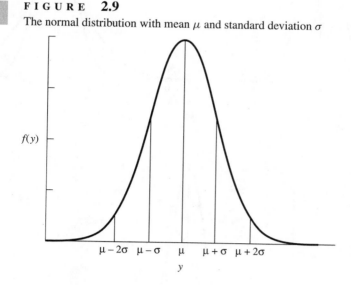

$f(y)$

$\mu-2\sigma$ $\mu-\sigma$ μ $\mu+\sigma$ $\mu+2\sigma$

y

As can be seen from Figure 2.9, the normal distribution is symmetric and bell-shaped, with its center of symmetry at μ. Because the distribution is symmetric, the mean and median of a normal distribution coincide.

The location and shape of a normal distribution are determined by the values of μ and σ. In Figure 2.10a, we show a pair of normal distributions, one with $\mu = 30$, $\sigma = 20$, and the other with $\mu = 40$, $\sigma = 20$; in Figures 2.10b and 2.10c, we show pairs of normal distributions with parameter values $(\mu, \sigma) = (35, 20)$, $(35, 10)$ and $(\mu, \sigma) = (30, 10)$, $(40, 15)$, respectively.

Notice that the only difference between the two distributions in Figure 2.10a is their locations. Population B ($\mu = 40$, $\sigma = 20$) can be obtained from population A ($\mu = 30$, $\sigma = 20$) by a simple shift of 10 units to the right. On the other hand, populations A ($\mu = 35$, $\sigma = 20$) and B ($\mu = 35$, $\sigma = 10$) in Figure 2.10b have a common center of location. Because the standard deviation is larger for population A than for population B, population A is more dispersed and has a flatter distribution. Finally, in Figure 2.10c, the two populations not only have different locations, but also have different dispersions.

FIGURE 2.10

Normal distributions with different (μ, σ)

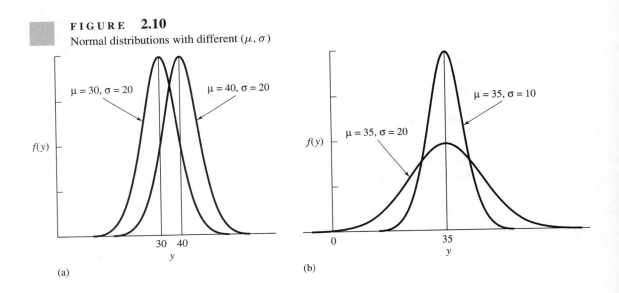

(a)

(b)

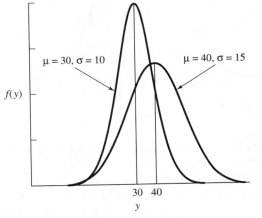

(c)

The notation $N(\mu, \sigma^2)$ will be used to denote a normal distribution with mean μ and standard deviation σ (variance σ^2). A $N(0, 1)$ distribution will be called a *standard normal distribution*. Thus, a standard normal distribution has mean $\mu = 0$ and variance $\sigma^2 = 1$.

In Box 2.7 we present a method of transforming the measurements in a $N(\mu, \sigma^2)$ population to a standard normal population and vice versa. Because areas under normal distributions play a key role in statistical theory and applications, the transformation in Box 2.7 is frequently used in statistics.

BOX 2.7

A property of normal distributions

If every measurement y in a $N(\mu, \sigma^2)$ population is transformed by the transformation

$$z = \frac{y - \mu}{\sigma},$$

the population of all transformed values will have a $N(0, 1)$ distribution. In other words, if Y is a random variable with a $N(\mu, \sigma^2)$ distribution, then

$$Z = \frac{Y - \mu}{\sigma}$$

is a random variable with a $N(0, 1)$ distribution.

The z-score

The transformation in Box 2.7 replaces each y in a normally distributed population with its distance from the population mean μ measured in units of population standard deviation σ. The z corresponding to a y is called the z-score of y. Notice that the relationship between z and y can be expressed as:

$$z = \frac{y - \mu}{\sigma} \quad \text{or} \quad y = \mu + z\sigma.$$

Let $z_1 = (a - \mu)/\sigma$ and $z_2 = (b - \mu)/\sigma$ be the z-scores of some numbers a and b ($a \leq b$). The property described in Box 2.7 implies that the proportion of measurements in a $N(\mu, \sigma^2)$ distribution within the interval (a, b) is the same as the proportion of measurements in a standard normal distribution within the interval (z_1, z_2). Furthermore, since $a = \mu + z_1\sigma$ and $b = \mu + z_2\sigma$, the property stated in Box 2.7 also implies that the proportion of measurements in a $N(\mu, \sigma^2)$ distribution within the interval $(\mu + z_1\sigma, \mu + z_2\sigma)$ is the same as the proportion of measurements in a standard normal distribution within the interval (z_1, z_2).

For several selected positive values of z, Table C.1 in Appendix C gives the area to the right of $\mu + z\sigma$ in a $N(\mu, \sigma^2)$ distribution, which is shown as the region A_1 in Figure 2.11.

If we set $\mu = 0$ and $\sigma = 1$ in Figure 2.11, we get Figure 2.12, which shows the areas under a $N(0, 1)$ distribution. Since the areas A_1, A_2, A_3, and A_4 in Figure 2.11

FIGURE   **2.11**

Areas under a $N(\mu, \sigma^2)$ distribution

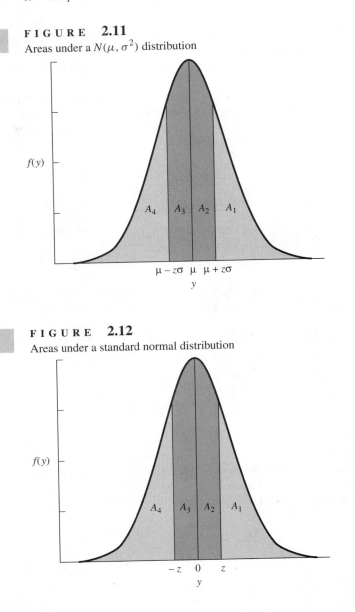

FIGURE   **2.12**

Areas under a standard normal distribution

are the same as the corresponding areas in Figure 2.12, Table C.1 can also be used to find the area to the right of $\mu + z\sigma = 0 + z(1) = z$ under a standard normal curve.

Since the total area under a probability density function is unity and a normal distribution is symmetric about its mean, the area A_1 can be used to find several other areas under the curve. For example, because the area under the curve to the right of the mean is 0.5, the area A_2 must equal $0.5 - A_1$. In other words, for a $N(\mu, \sigma^2)$ distribution, the proportion of values between μ and $\mu + z\sigma$ is found by subtracting from 0.5 the proportion of measurements larger than $\mu + z\sigma$. Symmetry considerations show that we must also have $A_1 = A_4$ and $A_3 = A_2$.

The use of Table C.1 to find the probabilities associated with a normal distribution is illustrated in the next two examples.

EXAMPLE **2.22**   Let Y be a random variable that denotes the observed value of a measurement selected at random from a $N(\mu, \sigma^2)$ population. For $z = 1.62$, the area A_1 in Figures 2.11 and 2.12 equals the probability that the observed value of Y will be 1.62 standard deviations above the mean

$$\Pr\{Y \geq \mu + 1.62\sigma\} = A_1.$$

The value of A_1 is the entry in Table C.1 that is located at the intersection of the row with heading 1.6 and the column with heading 0.02. Thus, $A_1 = 0.0526$; there is a 5.26% chance that a randomly selected measurement from a normal distribution will exceed the population mean by at least 1.62σ.

The area $A_2 + A_3$ in Figure 2.11 equals the probability that a measurement selected at random from a $N(\mu,\ \sigma^2)$ distribution will be within 1.62 standard deviations from the mean

$$\Pr\{\mu - 1.62\sigma \leq Y \leq \mu + 1.62\sigma\} = A_2 + A_3.$$

Because of the symmetry of a normal distribution, A_2 must equal A_3. Since $A_2 = 0.50 - A_1$, and $A_1 = 0.0526$, it follows that the required area is $A_2 + A_3 = 2A_2 = 2(0.50 - 0.0526) = 0.8948$. ∎

EXAMPLE **2.23**   Suppose that the yield of a new variety of corn is normally distributed with mean $\mu = 100$ bushels per acre and standard deviation $\sigma = 20$ bushels per acre. What is the probability that an observed yield for this variety will be less than 60 bushels per acre? What is the probability of an observed yield between 90 and 130 bushels per acre?

Let Y be the random variable denoting an observed yield of the new variety. Because the yields are normally distributed, the required probability is the area under a $N(100, 400)$ distribution within the interval $(-\infty, 60)$. In order to use Table C.1 to find this area, we first note that the endpoint 60 is $z_2 = (60 - \mu)/\sigma = (60 - 100)/20 = -2.0$ standard deviations away from μ. The negative sign of z_2 means that the value 60 is below μ. In other words, saying that 60 is 40 units below the population mean of $\mu = 100$ is the same as saying that the value 60 is two standard deviations ($\sigma = 20$) below the population mean, which, in turn, is the same as saying that $z_2 = -2.0$. Similarly, the endpoint $-\infty$ is $z_1 = (-\infty - 100)/20 = -\infty$ standard deviations below the population mean. Therefore, the endpoints of the interval of interest may be written as

$$-\infty = \mu + z_1\sigma = \mu + (-\infty)\sigma; \qquad 60 = \mu + z_2\sigma = \mu - 2\sigma.$$

The required probability is the area within the interval $(-\infty, -2)$ under a standard normal distribution. This area is shown as A_1 in Figure 2.13.

FIGURE   2.13

Area within the interval $(-\infty, 60)$ under a $N(100, 400)$ distribution

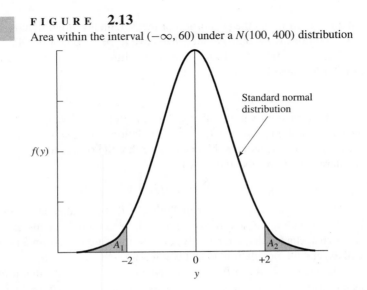

From Table C.1 with $z = 2.0$, we get $A_2 = 0.0228$. From symmetry considerations we know that $A_1 = A_2$, and so the required probability is 0.0228. Therefore, approximately 2.3% of the yields of the new corn variety will be less than 60 bushels per acre.

To determine the probability that the yield will be between 90 and 130 bushels, we first calculate the z-scores for 90 and 130: $z_1 = (90 - \mu)/\sigma = (90 - 100)/20 = -0.50$ and $z_2 = (130 - \mu)/\sigma = (130 - 100)/20 = 1.50$ and determine that the endpoints of the interval (90, 100) are, respectively, 0.5 standard deviations below and 1.5 standard deviations above the population mean. Therefore, the endpoints can be written as

$$90 = \mu - 0.50\sigma; \qquad 130 = \mu + 1.50\sigma.$$

Thus, the required probability is equal to the sum of the areas A_1 and A_2 in Figure 2.14.

From Table C.1 with $z = 1.50$, we get $A_1 = 0.50 - 0.0668 = 0.4332$. Note that, because of the symmetry of a normal distribution, A_2 is equal to the area under the curve within the interval $(\mu, \mu + 0.50\sigma)$. For $z = 0.50$ in Table C.1, we get $A_2 = 0.50 - 0.3085 = 0.1915$, so that the required probability is $A_1 + A_2 = 0.4332 + 0.1915 = 0.6247$. Thus 62.47% of the yields from the new corn variety can be expected to be between 90 and 130 bushels per acre. ∎

The z-scores and areas under normal distributions

Example 2.23 indicates that the area within an interval (a, b) under a $N(\mu, \sigma^2)$ distribution can be determined in two steps. In the first step, the distances between population mean μ and the endpoints a and b are expressed in units of the population

FIGURE   2.14

The area within the interval (90, 130) under a $N(100, 400)$ distribution

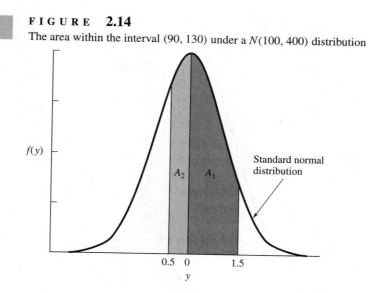

standard deviation σ. This is done by converting the endpoints of the interval to their z-scores using the transformation

$$z_1 = \frac{a - \mu}{\sigma}, \qquad\qquad z_2 = \frac{b - \mu}{\sigma}.$$

In the second step, the z-scores are used in Table C.1 to determine the desired area.

EXAMPLE 2.24   The serum iron content (mg/100 ml) in a population of subjects is known to be normally distributed with a standard deviation of $\sigma = 5$ mg/100 ml. Also, 33% of the population is known to have a blood serum level of at least 115 mg/100 ml. We are interested in determining the mean serum level for the population.

   We know that the area to the right of 115 in a normal distribution with $\sigma = 5$ is equal to 0.33. This is area A_1 in Figure 2.15. In Table C.1, the area $a = 0.33$ corresponds to $z = 0.44$. Therefore, the area to the right of $\mu + (0.44)\sigma$ is equal to 0.33; this implies that 115 is 0.44 standard deviations to the right of the population mean μ. Hence, $115 = \mu + 0.44\sigma$ or, equivalently, $\mu = 115 - 0.44\sigma = 115 - (0.44)(5) = 112.8$. Therefore, the mean serum level for the population is 112.8 mg/100 ml.   ■

EXAMPLE 2.25   An extensive survey of migrant workers' wages shows that the average daily wage is $\mu = \$15.00$, with a standard deviation $\sigma = \$12.00$. Is it reasonable to assume that the migrant workers' wages are normally distributed?

   Suppose, for the time being, that the migrant workers' wages have a normal distribution, with mean $\mu = 15$ and $\sigma = 12$. Then the proportion of wage-earning workers in the population is the same as the proportion of workers earning a positive wage. This proportion is equal to the area to the right of 0 in a $N(15, 144)$ distribution

FIGURE   2.15

Distribution of serum levels

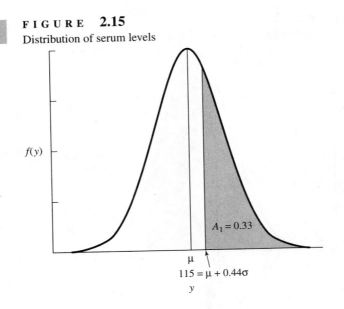

$f(y)$

$A_1 = 0.33$

μ

$115 = \mu + 0.44\sigma$

y

(Figure 2.16). The z-score for 0 is $z = (0 - 15)/12 = -1.25$. Using this result and Table C.1, we find that the area to the right of 0 is 0.8944. Thus, if the normal distribution is used as a model for the migrant workers' wages, $(1 - 0.8944)$ of the workers, or approximately 11% (the shaded area in Figure 2.16), receive less than $0 per day. Since negative wages for more than 10% of the population is not a

FIGURE   2.16

Distribution of migrant workers' wages

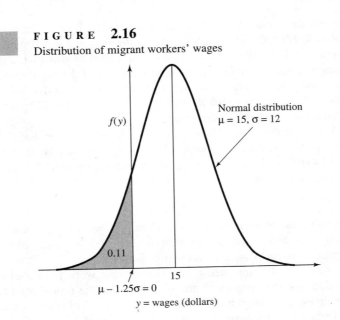

$f(y)$

Normal distribution
$\mu = 15$, $\sigma = 12$

0.11

15

$\mu - 1.25\sigma = 0$

$y = $ wages (dollars)

reasonable assumption, the normal distribution does not appear to be a good model for describing the wage distribution of migrant workers.   ∎

Normal approximation to a binomial

In addition to their use for modeling continuous populations, the normal distributions can also be used to approximate binomial distributions. The approximation works well provided that the minimum of $n\pi$ and $n(1-\pi)$ is at least 5. The actual method of approximation is described in Box 2.8.

BOX **2.8**

> ### *Normal approximation to a binomial*
>
> Let Y be a binomial random variable with parameters n and π such that $\min\{n\pi, n(1-\pi)\} \geq 5$. The probability that the observed value of Y is between a and b (both inclusive), where a and b are integers such that $0 \leq a \leq b \leq n$, can be approximated by the area within the interval $(a - 0.5, b + 0.5)$ under a normal distribution with mean $\mu = n\pi$ and standard deviation $\sigma = \sqrt{n\pi(1-\pi)}$.

Let $f(y)$ denote the probability function of a $B(n, \pi)$ random variable Y. Then the probability that Y will take a value between a and b, both inclusive, is

$$\Pr\{a \leq Y \leq b\} = f(a) + f(a+1) + \cdots + f(b). \tag{2.15}$$

Recall from Box 2.4 that the mean and standard deviation of a $B(n, \pi)$ distribution are $\mu = n\pi$ and $\sigma = \sqrt{n\pi(1-\pi)}$, respectively. Therefore, the method described in Box 2.8 approximates the sum of binomial probabilities in Equation (2.15) as the area within the interval $(a - 0.5, b + 0.5)$ under a normal distribution with the same mean and variance as the $B(n, \pi)$ distribution. Figure 2.17 illustrates the approximation when $n = 10$, $\pi = 0.5$, $a = 6$, and $b = 8$.

EXAMPLE **2.26**   To obtain a quick estimate of the damage to an orange crop after a hard freeze, 150 oranges were selected at random from a grove and examined for freeze damage. In all, 65 oranges were found to be damaged.

Let's assume that a proportion π of the oranges in the grove suffered freeze damage. Think of each selection of an orange as a trial; its outcome is a success if the selected orange suffered freeze damage and a failure otherwise. Then we have an experiment with $n = 150$ trials. If the number of oranges in the grove is large, so that the outcome of any trial is not affected by the outcome of any other trial, the experiment could be regarded as a binomial experiment with $n = 150$ trials and a probability of success π.

FIGURE   2.17
Approximating $B(10, 0.5)$ by a normal distribution

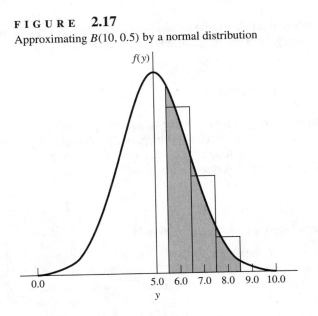

Under the assumption that a binomial model for the number of damaged oranges is reasonable, the probability of finding 65 or more damaged oranges in a random sample of 150 oranges is the same as the probability that a $B(150, \pi)$ random variable will take a value between $a = 65$ and $b = 150$, both inclusive. Thus, the required probability is

$$\Pr\{65 \leq Y \leq 150\} = f(65) + f(66) + \cdots + f(150),$$

where Y is a random variable with a $B(150, \pi)$ distribution and

$$f(y) = \frac{150!}{y!(150 - y)!}\pi^y(1 - \pi)^{150-y}$$

is the probability function of Y.

Computation of this probability is tedious, on account of the need to calculate quantities such as $150!$, $(150 - y)!$, and π^y. Of course, a number of statistical software packages, including SAS and MINITAB, can be used to compute these probabilities. Under suitable conditions, the normal approximation in Box 2.8 gives good results.

Suppose $\pi = 0.40$. Then the mean and standard deviation of the approximating normal distribution are, respectively, $\mu = n\pi = (150)(0.40) = 60$ and $\sigma = \sqrt{n\pi(1 - \pi)} = \sqrt{(150)(0.40)(0.60)} = 6$. Thus, the required probability can be approximated by the area within the interval $(a - 0.5, b + 0.5) = (64.5, 150.5)$ under a $N(60, 36)$ distribution. The z-scores for the endpoints 64.5 and 150.5 are, respectively

$$z_1 = \frac{64.5 - 60}{6} = 0.75 \quad \text{and} \quad z_2 = \frac{150.5 - 60}{6} = 15.08.$$

Using Table C.1 (Appendix C), the required probability is found to be 0.2266. Therefore, if 40% of the oranges in the grove are damaged, the chance of finding 65 or more damaged oranges in a sample of 150 oranges is approximately 22.26%. ∎

The log-normal distribution

A log-normal distribution describes a continuous population of measurements of positive values. Just as a normal distribution is useful for describing a set of measurements subject to additive errors, a log-normal distribution is useful when the errors act multiplicatively on the measurements. In particular, a log-normal distribution is indicated when each measurement can be expressed as the product of two positive quantities, a true value and an error, where the error is the product of a large number of small positive errors due to independent factors. A population is said to have a *log-normal distribution* if the logarithms of the values in the population have a normal distribution. Thus, if Y is the random variable associated with a log-normal distribution, the population of observed values of log Y has a normal distribution. Figure 2.18 illustrates the relationship between log-normal and normal distributions.

In view of the mathematical relationship between the values in normal and log-normal populations, inferences about log-normal populations can be carried out by transforming the sample values using the log-transformation and regarding the transformed sample as a sample from a population with a normal distribution. Example 2.27 illustrates the idea.

FIGURE   2.18
Log-normal (a) and normal (b) distributions

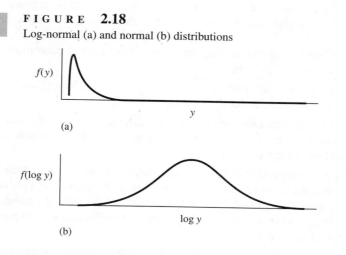

EXAMPLE **2.27**   Soil properties such as bulk density, organic matter content, clay content, and soil water content are generally characterized by a normal distribution (Rao, Rao, Davidson, & Hammond, 1979). However, flow-related properties such as air permeability, saturated hydraulic conductivity, and soil-water flux have been reported to be log-normally distributed.

The following data are the soil-water fluxes at 200-cm depth measured on 20 randomly selected plots in a large experimental region.

> 0.5080, 0.6090, 0.6230, 0.6860, 0.7350, 0.7500, 0.7520,
> 0.8690, 0.8890, 0.8890, 0.8990, 0.9370, 0.9820, 1.0220,
> 1.0370, 1.0880, 1.1130, 1.2260, 1.3310, 1.5230

In Exercise 3.6, you will be asked to verify, by means of suitable graphical procedures, that it is reasonable to regard these flux measurements as a sample from a log-normal population. Consequently, the data obtained by log-transforming the soil-water flux values could be regarded as a sample from a normal population. Inferential techniques for samples from normal populations, described in later chapters, can be used with the transformed data to make inferences about the population of the soil-water flux values in the experimental region. ■

Exercises

2.46 Suppose that the diastolic BP (the blood pressure existing during the relaxation phase between heart beats) in adult males in a particular age group has a normal distribution, with mean 83 mm Hg and standard deviation 14 mm Hg.

  a A diastolic BP above 100 is considered to be above normal. Determine the proportion of individuals in the population who have abnormal BP.

  b Determine an interval of BP values such that the probability that the BP value of a subject randomly selected from this population will be within this interval is 0.95.

  c Calculate the probability that a person selected at random from this population will have a BP value:

  i higher than 150;   **ii** lower than 70;
  iii between 75 and 85.

2.47 Suppose that the distribution of the logarithm of the survival times (years) of patients treated for a particular type of cancer is a normal distribution with mean 1.1 and standard deviation 0.2.

  a What proportion of the treated patients will survive more than two years?

  b What proportion of the treated patients will survive less than one year?

  c Determine a number t_0 such that 90% of the treated patients will survive less than t_0 years—that is, such that the probability of a treated patient surviving less than t_0 years is 0.90.

2.48 Assume that the scores in a particular examination have a normal distribution, with mean 80 and standard deviation 5. The instructor wants to assign letter grades in such a way that those in the top 10% will get an A, the next 20% will get a B, the next 30% will get a C, and the next 10% will get a D. Determine the cutoff scores that the instructor should use when assigning the letter grades in this examination.

2.49 Recall that, for a given z, Table C.1 shows the proportion of measurements in a $N(\mu, \sigma^2)$ population that exceed $\mu + z\sigma$. Let Y be a random variable with a $N(\mu, \sigma^2)$ distribution. Show that Y will exceed $\mu + z\sigma$ if and only if the z-score $Z = \frac{Y - \mu}{\sigma}$ exceeds z—that is, that

$$Y \geq \mu + z\sigma \text{ if and only if } \frac{Y - \mu}{\sigma} \geq z.$$

2.7
The critical values of a probability distribution

Recall that a $100p$-percentile of a distribution is a value η_p such that no more than $100p\%$ of the measurements in the population will be less than or equal to η_p and no more than $100(1 - p)\%$ of the measurements in the population will be greater than or equal to η_p. As will be seen in the next chapter, the population percentiles play a key role when implementing many of the standard statistical inferential procedures. For example, an important step in statistical procedure known as hypothesis testing involves the comparison of appropriate population percentiles with quantities calculated from samples. In the statistical procedure called interval estimation, population percentiles are combined with the quantities calculated from samples to arrive at intervals that can be regarded as the ranges of plausible values for population parameters. When used in statistical procedures, population percentiles are often referred to as *critical values*. In the remainder of this book, we shall use the definition of the α-level critical value in Box 2.9.

BOX **2.9**
The α-level critical value of a distribution

Let α be a value such that $0 \leq \alpha \leq 1$. The α-level critical value of the distribution of a population is a value $y(\alpha)$ such that exactly $100\alpha\%$ of the measurements in the population will be at least as large as $y(\alpha)$.

If Y is the random variable associated with a given population, the α-level critical value of that population will satisfy the relationship

$$\Pr\{Y \geq y(\alpha)\} = \alpha. \tag{2.16}$$

For a continuous distribution, the α-level critical value is the same as the area under the probability density curve within the interval $[y(\alpha), +\infty)$. Figure 2.19 shows that, for a continuous population, the α-level critical value is the same as the $100(1 - \alpha)$ percentile; that is, $y(\alpha) = \eta_{1-\alpha}$.

FIGURE   **2.19**

Alpha-level critical value and the area under the distribution

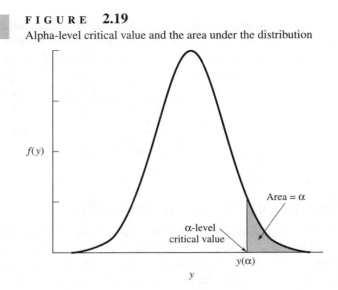

For a continuous population, there is an α-level critical value corresponding to each value of α such that $0 < \alpha < 1$. If the population is discrete, however, there will be values of α that do not correspond to a critical value. For instance, consider the binomial distribution with $n = 6$ and $\pi = 0.70$. Calculations similar to those in Example 2.16 can be used to verify that the frequency distribution of this population is as in the first two columns of Table 2.3. The third column of Table 2.3 shows, for each distinct value y, the proportion of measurements in a $B(6, 0.70)$ population that are at least as large as y. For example, the value 0.7442 corresponding to $y = 4$ is obtained as

$$\Pr\{Y \geq 4\} = f(4) + f(5) + f(6)$$

$$= 0.3241 + 0.3025 + 0.1176 = 0.7442, \tag{2.17}$$

where Y is a random variable with a $B(6, 0.70)$ distribution.

TABLE   **2.3**

Critical values of a binomial distribution with $n = 6$ and $\pi = 0.70$

y	$f(y)$	Proportion of values that are at least y
0	0.0008	1.0000
1	0.0103	0.9992
2	0.0595	0.9889
3	0.1852	0.9294
4	0.3241	0.7442
5	0.3025	0.4201
6	0.1176	0.1176

Comparing Equations (2.16) and (2.17), we see that the 0.7442-level critical value of a $B(6, 0.70)$ distribution is $y(0.7442) = 4$. Therefore, the entries in the first column of Table 2.3 are the critical values corresponding to the levels in the third column. Clearly, the levels shown in the third column are the only values of α for which there exist corresponding critical values.

In the remainder of this book, the α-level critical value of the standard normal distribution will be denoted by $z(\alpha)$. Table C.1 in Appendix C can be used to determine these critical values.

EXAMPLE **2.28** The 0.05-level critical value of a $N(0, 1)$ distribution is a value $z(0.05)$ such that the area under the standard normal distribution and above the interval $(z(0.05), +\infty)$ is 0.05. From Table C.1, we see that $z(0.0505) = 1.64$ and $z(0.0495) = 1.65$. Therefore $z(0.05)$ is a value between 1.64 and 1.65. We may take $z(0.05) = 1.645$. ∎

Exercises

2.50 Determine the α-level critical values for a standard normal distribution when

  a $\alpha = 0.10$;   **b** $\alpha = 0.05$;   **c** $\alpha = 0.025$;   **d** $\alpha = 0.001$.

2.51 The following table shows the probability distribution of a discrete population.

y	0	1	2	3	4	5	6	7
$f(y)$	0.0012	0.0124	0.1240	0.3421	0.2422	0.1231	0.1403	0.0147

  a In the following list, circle the values of α such that there is an α-level critical value for the distribution

$$\alpha = 0.01, \ \alpha = 0.0347, \ \alpha = 0.05, \alpha = 0.15, \ \alpha = 0.155.$$

  b Determine the α-level critical value corresponding to each of the circled values in (a).

2.8
Using computers to describe populations

Most of the statistical software packages permit the easy computation of critical values of the binomial, Poisson, normal, and other common statistical distributions. These computing packages can also be used to determine the probability that a measurement selected at random from one of these distributions will take a value in a specified interval. For instance, the probability that at least five seeds will germinate in Example 2.16 can be computed using a function in SAS called PROBBNML. Also, recall that in Example 2.22, we used Table C.1 to determine the probability that an observed value of a measurement selected at random from a $N(\mu, \sigma^2)$ population will be 1.62 standard deviations above μ. This probability can be calculated using the SAS function PROBNORM. Finally, the value of z such that the area to the right of z under a standard normal distribution equals a given value (see Example 2.24) can be determined using the SAS function PROBIT. Details of how to conduct these calculations using SAS are discussed in Chapter 2 of the companion text by Younger (1997).

Exercises

Work the following exercises using appropriate (e.g., SAS) statistical computing software.

2.52   Fifteen patients are treated with a drug known to have a cure rate of 80%. Compute the probabilities of the following events.

a   Exactly 12 patients are cured.   **b**   At least 9 patients are cured.
c   At most 7 patients are cured.

2.53   Assume that the number of deaths due to a rare disease (e.g., TB) over a period of one year is known to have a Poisson distribution with mean 1.5. Compute the following probabilities.

a   Exactly 10 deaths in 15 years.   **b**   No more than 3 deaths in 2 years.

2.54   Compute the following quantities for a standard normal distribution.

a   Area under the curve and within the interval $(-1.2, 1.6)$.
b   The 0.036-level critical value.

2.9
Overview

In Chapter 1, we defined statistical populations as collections of measurements about which statistical inferences are made. This chapter is devoted to methods for describing statistical populations.

The populations encountered in practice may be univariate or multivariate, real or conceptual, finite or infinite, quantitative or qualitative, and discrete or continuous.

A population can be described using a distribution or a set of parameters. While the distribution of a population provides a complete characterization of the values of its measurements, parameters—such as the mean μ, the median η, and the standard deviation σ—provide information on selected characteristics of its measurements. The empirical rule and Tchebycheff's rule provide two examples of the use of μ and σ to characterize statistical populations.

We have considered the binomial, the Poisson, the normal, and the log-normal distributions. Other examples of population distributions will be encountered in later chapters.

The notion of a random variable is a convenient tool for describing what happens when we observe a randomly selected measurement from a statistical population. Given a statistical population, we define a corresponding random variable as the variable that takes the value of a randomly selected measurement from that population. There is a close connection between the distribution of a population and the probability distribution of the corresponding random variable. Equation (2.4) summarizes this relationship for discrete populations.

As we will see in subsequent chapters, answers to questions arising in a wide variety of scientific investigations can be cast in terms of the parameters of suitably defined statistical populations. The primary objective in the remainder of this book is to describe statistical procedures that are useful for making inferences about parameters of interest to researchers in life sciences.

3 Statistical Inference: Basic Concepts

3.1
Introduction

As we noted in Chapter 1, a wide variety of problems encountered in scientific research can be stated as problems of inference about statistical populations. The primary objective in this chapter is to describe some basic statistical concepts and terminology.

3.2
Simple random samples

Since statistical inferential procedures utilize samples to draw conclusions about populations, the reliability of a statistical inference will depend upon the extent to which the samples are representative of the populations. Consider the following example.

EXAMPLE **3.1**   Muck is a rich, highly organic type of soil that serves as the growth medium for most vegetation in the Florida Everglades. Because muck can be destroyed by a variety of natural causes, the estimation of muck depth is important in the management of the Everglades. The population of interest in this example is the infinite set of measurements of muck depths at all locations in the Everglades. Information about this population must be based on a sample of muck depths at sites selected in such a way that the depths at the sample sites could be regarded as typical for the population of all sites. How is such a sample selected?   ■

There are a number of methods for selecting samples from populations. The most important is simple random sampling. A simple random sample of size n is a collection of n measurements selected in such a way that every subset of n measurements in the population has an equal chance of being included in the sample.

Note that selecting a simple random sample of size $n = 1$ is the same as selecting a measurement at random, as described in Section 2.3.

Selecting simple random samples from populations containing only a finite number N of identifiable measurements is straightforward. First, we assign the numbers $1, 2, \ldots, N$ to the N measurements in the population. Then we write the N numbers on N pieces of paper, thoroughly mix, and select n pieces. The numbers on the n selected pieces are the measurements in a simple random sample of size n.

This method of selecting a simple random sample assumes that the population consists of a finite number of measurements and that each of these measurements can be identified and enumerated. Unfortunately, most of the populations encountered in practice, like the population of muck depths in Example 3.1, are not only infinite but also continuous (nonenumerable). Any method of selecting simple random samples from such populations is at best a good approximation. For instance, an investigator who wants to select a random sample of $n = 10$ blood pressure measurements for patients treated with a particular drug may obtain the sample by treating ten typical patients and measuring their blood pressures. Similarly, a researcher interested in selecting a random sample of $n = 15$ yields of a particular corn variety grown under a specific set of conditions may select the sample by observing the yields of the variety grown under the desired conditions in 15 field plots. In practice, the method of selecting random samples varies from investigator to investigator and from population to population. The practical objective is always to ensure that the sample meets the criterion for a random sample as closely as possible.

Exercises

3.1   Suggest how random samples can be selected in the growth-rate comparison study described in Exercise 1.1.

3.2   Suggest how a random sample can be selected in order to estimate the proportion of diseased trees in a stand of pine trees.

3.3   Suggest a method of selecting random samples in order to estimate the radiation levels inside homes built in a particular region.

3.4   Suggest a method of selecting random samples in order to study the relationship between the age of pregnant mothers and the birth-weights of their babies.

3.5   A sociologist is interested in knowing the average age of citizens in a particular county who received social security checks during January 1995. Suggest a method for selecting a random sample of social security recipients that can be used to obtain the information needed by the sociologist.

3.3
Describing samples

Like a population, a sample is a collection of measurements. Consequently, the tabular, graphical, and numerical methods discussed in Chapter 2 can be used to describe samples.

Tabular and graphical descriptions of samples

Just as a probability distribution describes a discrete population, a *frequency distribution* describes a sample. A frequency distribution can be in tabular or graphical form. The tabular form is a table that gives the proportions of measurements that fall into each of a set of *r* mutually exclusive and exhaustive categories. The graphical form is a bar-graph in which the lengths of the bars represent the proportions in the various categories. The proportions of measurements contained in the categories are called the *relative frequencies*. The bar-graph of a sample frequency distribution is sometimes called a *histogram*.

EXAMPLE **3.2**   The disease-free survival (DFS) time of a treated cancer patient is defined as the length of elapsed time between the time at which the patient goes into remission (becomes free of cancer) and the time at which the patient relapses (cancer recurs).

The following data show the DFS times, in months, of a random sample of $n = 20$ breast cancer patients.

$$2, 2, 3, 6, 6, 7, 9, 11, 12, 12, 13, 15, 19, 19, 21, 23, 24, 30, 44, 45$$

A frequency distribution of the DFS times can be constructed as follows. First, we divide the sample values into a number of mutually exclusive categories. The definition of categories can be arbitrary, except that the number of groups should be small enough to make the distribution easily comprehensible but large enough to avoid obscuring important features of the sample.

The sample DFS times range from a minimum of 2 to a maximum of 45. Let's divide these values into $r = 10$ mutually exclusive categories by dividing the interval $(0, 50)$ into ten subintervals, each of length 5. Then the first category will contain all values y such that $0 \leq y < 5$, the second category will contain all values y such that $5 \leq y < 10$, and so on.

Since there are three values—$y = 2$, $y = 2$, and $y = 3$—in the first category, the relative frequency in this category is $f_1 = 3/20 = 0.15$. Similarly, the four values in the third category $(10 \leq y < 15)$ yield a relative frequency $f_3 = 4/20 = 0.20$. Table 3.1 represents the resulting frequency distribution of the DFS times.

The histogram of the data in Table 3.1 is shown in Figure 3.1. ∎

Interpretation of the relative frequencies in a frequency distribution is similar to that in a probability distribution. The only difference is that a relative frequency in a probability distribution refers to the proportion among all population values, whereas in a frequency distribution it refers to the observed sample values. For instance, the relative frequency $f_4 = 0.15$ in the fourth category of Table 3.1 implies that 15% of the DFS times in the sample are within the interval $15 \leq y < 20$.

The manner in which the intervals are selected can have a marked effect on the way information is summarized by a histogram. Particularly in small data sets, this

TABLE   3.1

Frequency distribution of DFS times

Interval numbers i	Class interval	Frequency f_i	Relative frequency p_i
1	$0 \leq y < 5$	3	0.15
2	$5 \leq y < 10$	4	0.20
3	$10 \leq y < 15$	4	0.20
4	$15 \leq y < 20$	3	0.15
5	$20 \leq y < 25$	3	0.15
6	$25 \leq y < 30$	0	0.00
7	$30 \leq y < 35$	1	0.05
8	$35 \leq y < 40$	0	0.00
9	$40 \leq y < 45$	1	0.05
10	$45 \leq y < 50$	1	0.05

F I G U R E   3.1

Histogram of the frequency distribution in Table 3.1

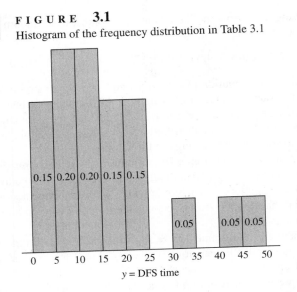

effect can be severe. Too much information about individual values is likely to be lost. For example, Figure 3.2 shows a histogram of the DFS times based on $r = 5$ categories. Compare Figure 3.2 with Figure 3.1 and notice how drastically the shape of the histogram changes with changing class intervals.

An alternative to a histogram is a *stem-and-leaf plot* (or simply stem-leaf plot), which typically consists of two columns of numbers: the stem portion and the leaf portion. Let's consider how this plot may be applied to the DFS data in Example 3.2.

F I G U R E   3.2

Histogram of survival times based on five categories

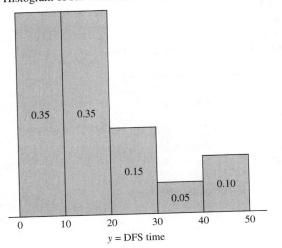

Each DFS value may be regarded as a two-digit number. (We write 2 as 02, and so on.) Then, as we see in Table 3.2, the stem portion of the plot lists the left-hand digit in each sample value (for example, 0 from 02 and 4 from 45) arranged in increasing order from top to bottom. In any given row, the leaf portion then lists the right-hand digits of all the sample values (for example, 2 from 02 and 5 from 45) that have the left-hand digit given in the stem portion. Thus, the leaves 1, 3, and 4 at stem level 2 correspond to the DFS times 21, 23, and 24, respectively.

T A B L E   3.2

Stem-and-leaf plot of DFS times

Stem	Leaf
0	2, 2, 3, 6, 6, 7, 9
1	1, 2, 2, 3, 5, 9, 9
2	1, 3, 4
3	0
4	4, 5

Of course, there are many ways in which a set of sample values can be divided into stem and leaf portions. For example, the stem and leaf portions may be divided at the decimal point. In such a plot, the number 2.23 will have 2 in the stem portion and .23 in the leaf portion, and the number 30.14 will have 30 in the stem portion and .14 in the leaf portion. (See Table 3.3.) In practice, it is important to consider different possibilities when constructing the stem-and-leaf plot of a given sample. For more information about stem-and-leaf plots and other methods of describing and exploring data, see Tukey (1977) and Hoaglin, Mosteller, and Tukey (1983). Chapter 3 of Younger (1997) shows how the SAS procedures CHART and UNIVARIATE may be used to construct histograms and stem-and-leaf plots.

T A B L E   3.3

Stem-and-leaf plot of a sample of $n = 15$ measurements

Stem	Leaf
0	.23, .35, .66
1	.36, .48
3	.06, .36, .71, .91
4	.62, .80
6	.45, .64
7	.90
9	.31

Numerical description of samples

Recall that parameters can be used to provide numerical descriptions of statistical populations. The sample analog of parameters are called statistics. A *statistic* is a function of the measurements in a sample, in the sense that the sample measurements determine its value. The difference between a parameter and a statistic is that, while a parameter is calculated using all the measurements in a population, the statistic uses only the sample values. Let's look at some examples of statistics.

Sample mean

Just as the population mean μ is the average of all measurements in a population, the sample mean is the average of all measurements in a sample. In this book, the sample mean will be denoted by placing a bar over a letter of the English alphabet. Thus, the mean of a sample consisting of n measurements, $y_1, y_2, \ldots, y_n$, is

$$\bar{y} = \frac{y_1 + y_2 + \cdots + y_n}{n}. \tag{3.1}$$

Sample standard deviation

The sample analog of σ^2, the population variance, is the sample variance

$$s^2 = \frac{(y_1 - \bar{y})^2 + (y_2 - \bar{y})^2 + \cdots + (y_n - \bar{y})^2}{n - 1}. \tag{3.2}$$

The positive square root of the sample variance is the sample standard deviation

$$s = \sqrt{\frac{(y_1 - \bar{y})^2 + (y_2 - \bar{y})^2 + \cdots + (y_n - \bar{y})^2}{n - 1}}.$$

The reader may be puzzled by the use of the divisor $n - 1$ instead of n in Equation (3.2). As we will see in Section 3.8, the divisor $n - 1$ is needed to correct the tendency of s^2 to underestimate σ^2; the degree of underestimation is not severe if n is large, but could be serious for small sample sizes.

The sample variance is often computed using the *short-cut formula*

$$s^2 = \frac{1}{n - 1} \left\{ y_1^2 + y_2^2 + \cdots + y_n^2 - \frac{(y_1 + y_2 + \cdots + y_n)^2}{n} \right\}$$

$$= \frac{1}{n - 1} \left\{ \Sigma y^2 - \frac{(\Sigma y)^2}{n} \right\} \tag{3.3}$$

The calculation of the sample mean and standard deviation is illustrated in the next example.

EXAMPLE **3.3**   The weights (in pounds) of a random sample of $n = 10$ rainbow trouts in a lake are as follows

$$1.19, 0.93, 2.40, 1.71, 0.89, 1.74, 1.06, 1.16, 1.47, 1.15$$

The mean of the sample is

$$\bar{y} = \frac{1}{10} (1.19 + 0.93 + \cdots + 1.15) = \frac{13.7}{10} = 1.37.$$

The variance can be calculated using either Equation (3.2) or Equation (3.3). If we use Equation (3.2), we get

$$s^2 = \frac{1}{10-1} \left\{ (1.19 - 1.37)^2 + (0.93 - 1.37)^2 + \cdots + (1.15 - 1.37)^2 \right\} = 0.2187.$$

Alternatively, using Equation (3.3), we get

$$s^2 = \frac{1}{10-1} \left\{ 1.19^2 + 0.93^2 + \cdots + 1.15^2 - \frac{(1.19 + 0.93 + \cdots + 1.15)^2}{10} \right\}$$

$$= 0.2187. \quad \blacksquare$$

A property of variance useful for simplifying its calculation is that its value will not change if a constant is subtracted (added) from each measurement. To get an intuitive sense of this property, note that the variability in a set of measurements will not be altered if every measurement is increased or decreased by the same amount. In Exercise 3.8, you will be asked to verify that the variance of the sample of rainbow trout weights will remain the same if an arbitrary constant is added to each of the values in the sample.

Sample median

Any value $\tilde{y}$ such that no more than half the measurements are less than or equal to $\tilde{y}$ and no more than half the measurements are greater than or equal to $\tilde{y}$ is a *sample median*. For a given sample, more than one value may satisfy the definition of sample median. A formula for computing the sample median can be written using the notion of the *order statistics* of a sample.

The i-th order statistic, denoted by $y_{(i)}$, is defined as a measurement occupying the i-th position when the sample measurements are arranged in increasing order. Thus, $y_{(1)}$, the smallest value in the sample, is the first order statistic; $y_{(2)}$, the next largest value, is the second order statistic; and so on.

EXAMPLE 3.4 Consider the following sample of size $n = 5$

$$y_1 = 4, \ y_2 = 16, \ y_3 = 8, \ y_4 = 12, \ y_5 = 8.$$

Arranging the values in increasing order of magnitude, we get

$$y_1 < y_3 \leq y_5 < y_4 < y_2$$

so that the order statistics of the sample are

$$y_{(1)} = 4, \ y_{(2)} = 8, \ y_{(3)} = 8, \ y_{(4)} = 12, \ y_{(5)} = 16. \quad \blacksquare$$

A median of a sample can be calculated as the middle order statistic if the sample size is an odd number (that is, if $n = 2k + 1$ for some integer k) and the average of two middle order statistics if the sample size is an even number (that is, if $n = 2k$ for some integer k). In symbols

$$\tilde{y} = \begin{cases} y_{(k+1)} & \text{if } n = 2k + 1 \\ \frac{1}{2}(y_{(k)} + y_{(k+1)}) & \text{if } n = 2k. \end{cases}$$

(3.4)

EXAMPLE 3.5   For the rainbow trout weights in Example 3.3, the sample size is $n = 10$. Thus, $n = 2k$, where $k = 5$. The sample median is the average of the two middle order statistics

$$\tilde{y} = \frac{1}{2}\left\{y_{(k)} + y_{(k+1)}\right\} = \frac{1}{2}\left\{y_{(5)} + y_{(6)}\right\} = \frac{1.16 + 1.19}{2} = 1.175.$$

In Example 3.4, the sample size is $n = 5$. Thus, $n = 2k + 1$, where $k = 2$. Therefore, the median of the sample is the middle order statistic $y_{(3)}$; that is, $\tilde{y} = 8$ for this sample. ■

To each parameter of a population, there corresponds a statistic that has a similar interpretation. The sample mean, the sample variance, and the sample median are some examples; others are the *sample quartiles* and the *sample coefficient of variation*. Sample quartiles are defined analogously to the sample median. Just as the sample median divides the sample into two halves, the sample quartiles divide the sample into four parts. The second quartile $y^{(2q)}$ is the same as the sample median, so that $y^{(2q)} = \tilde{y}$. The first sample quartile $y^{(1q)}$ is defined as that value in the sample such that no more than one-fourth of the measurements have values less than $y^{(1q)}$ and no more than three-fourths of the measurements have values greater than $y^{(1q)}$. The third sample quartile is defined similarly. As with sample medians, more than one set of values may satisfy the definition of the sample quartiles. The following formulas can be used to compute first and third quartiles of a given sample

$$y^{(1q)} = \begin{cases} y_{(k+1)} & \text{if } n = 4k + 2 \text{ or } 4k + 3 \\ \frac{1}{2}(y_{(k)} + y_{(k+1)}) & \text{if } n = 4k \text{ or } 4k + 1 \end{cases}$$

$$y^{(3q)} = \begin{cases} y_{(n-k)} & \text{if } n = 4k + 2 \text{ or } 4k + 3 \\ \frac{1}{2}(y_{(n-k)} + y_{(n-k+1)}) & \text{if } n = 4k \text{ or } 4k + 1. \end{cases}$$

(3.5)

Calculation of sample quartiles is illustrated in Example 3.6.

Box-plot

We have already seen how a histogram or a stem-and-leaf plot can be used to summarize the measurements in a sample. The box-plot was developed by Tukey (1977) for graphical display of the measurements in a sample. While a stem-and-leaf plot emphasizes the individual values within a set of measurements, the box-plot attempts to highlight the sample's location and dispersion characteristics. There are many variations of a box-plot; for details, see Tukey (1977) and Hoaglin, Mosteller, and Tukey (1983). Here we will present the essential features of a typical box-plot.

The purpose of a box-plot is to display the main distributional characteristics of a data set. To interpret a box-plot properly, the sample values should be visualized as points on an imaginary vertical line located at the center of the figure. Larger values in the data correspond to higher points on the vertical line.

There are three key components to a box-plot:

1 The *box*, which contains 50% of the sample values, starting at the first sample quartile and ending at the third sample quartile.

2 The two *whiskers*, which extend above and below the box up to the locations of the largest and the smallest sample values that are within a distance of 1.5 times the interquartile range.

3 The *outliers*, the sample values located outside the whiskers.

Figure 3.3 displays the essential components of a typical box-plot.

The box, which is represented by a rectangle in Figure 3.3, shows the relative location of the middle 50% of the values. Any value outside the whiskers is tagged as an outlier because it can be shown that, in random samples from normally distributed populations, such a value will occur with a very small probability. The relative location of the median and the relative lengths of the whiskers are important indicators of the symmetry of the sample values. Ideally, symmetric data will have a median located at the center of the box, and the two whiskers will be of approximately equal lengths. The difference between the lengths of the upper and lower whiskers provides information about the difference between the lengths of the left and right tails of the sample frequency distribution. The presence of outliers indicates either that there are some values that are not consistent with the rest of the data or that the sample has been selected from a population containing measurements with extreme values (relatively large or small values).

Notice that, in the box-plot in Figure 3.3, the median is fairly close to the center of the box, but the lower whisker is longer than the upper whisker; this indicates a higher concentration of data at the lower end. There are two outliers below the median, and none above the median.

F I G U R E   3.3

Components of a typical box-plot

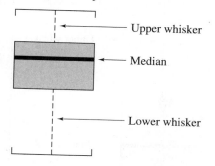

EXAMPLE **3.6**   Let's construct a box-plot for the survival data of Example 3.2. The first step is to determine the sample quartiles. To calculate the sample quartiles, we express the sample size n as a multiple of 4 plus a remainder ($n = 4k$, $n = 4k + 1$, $n = 4k + 2$, or $n = 4k + 3$) and apply Equation (3.5). There are $n = 20$ measurements in the sample, and so $n = 4k$, where $k = 5$. Therefore, from Equation (3.5)

$$y^{(1q)} = \frac{1}{2}(y_{(5)} + y_{(6)}) = \frac{1}{2}(6 + 7) = 6.5;$$

$$y^{(2q)} = \tilde{y} = \frac{1}{2}(y_{(10)} + y_{(11)}) = \frac{1}{2}(12 + 13) = 12.5;$$

$$y^{(3q)} = \frac{1}{2}(y_{(20-5)} + y_{(20-5+1)}) = \frac{1}{2}(y_{(15)} + y_{(16)})$$

$$= \frac{1}{2}(21 + 23) = 22.$$

The lower and upper edges of the box in the box-plot will be located at heights of 6.5 and 22, respectively. The sample interquartile range is equal to the value $22 - 6.5 = 15.5$, so that the two whiskers will extend up to the farthest data points that are within a distance of $(1.5)(15.5) = 23.25$ from the edges of the box. Thus, the upper whisker will extend from the point 22 to the largest value that is less than $22 + 23.25 = 45.25$, and the lower whisker will extend from the point 6.5 to the smallest data value that is above $6.5 - 23.25 = -16.75$. The largest and the smallest values in the sample are, respectively, 2 and 45, and so all the values in the sample are contained within the range of the whiskers; there are no outliers in this sample. The lengths of the upper and lower whiskers are, respectively, $45 - 22 = 23$ and $6.5 - 2 = 4.5$. Thus, the upper whisker is much longer than the lower whisker. Figure 3.4 shows the box-plot of the disease-free survival times.   ■

F I G U R E   3.4
Box-plot of disease-free survival (DFS) times

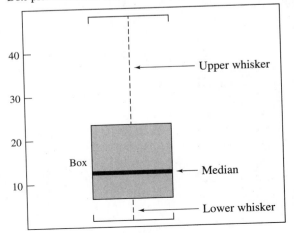

A box-plot can be drawn by hand, but it is best to use a computer. A variety of statistical software, including SAS, MINITAB, and Splus, can be used to draw a box-plot; the box-plot in Figure 3.4 was drawn using Splus. We conclude this section by reproducing a portion of the output obtained by applying the SAS procedure called UNIVARIATE to the DFS times in Example 3.2.

PORTION OF OUTPUT FROM UNIVARIATE PROCEDURE

Numerical and graphical description of the sample of dfs times
Univariate Procedure

Variable=TIME

```
                    Moments
          N           20
          Mean        16.15
          Std Dev     12.42779

          Variance    154.45
```

```
                  Quantiles(Def=5)

          100% Max     45
          75% Q3       22
          50% Med      12.5
          25% Q1       6.5
          0% Min       2
```

Variable=TIME

```
      Stem Leaf                #        Boxplot
        4 5                    1           |
        4 4                    1           |
        3                                  |
        3 0                    1           |
        2                                  |
        2 134                  3        +-----+
        1 599                  3        |  +  |
        1 1223                 4        *-----*
        0 6679                 4        +-----+
        0 223                  3           |
         ----+----+----+----+
      Multiply Stem.Leaf by 10**+1
```

This output has three parts. The first part shows the sample size ($N = 20$), the sample mean ($\bar{y} = 16.15$), the sample standard deviation ($s = 12.43$), and the sample variance ($s^2 = 154.45$). The second part shows the largest ($y_{(20)} = 45$) and the smallest ($y_{(1)} = 2$) order statistics along with the three quartiles $y^{(q1)} = 6.5$, $y^{(q2)} = 12.5$, and $y^{(q3)} = 22$. The third part shows stem-and-leaf and box-plots. Notice that in the stem-and-leaf plot, the sample values are divided by 10 and rounded off to one decimal place. Thus, the sample value 45 is replaced by 4.5. Similarly, the sample value 9 is replaced by 0.9. There are two stem-level entries (3 and 2) with no corresponding leaf column entries. SAS introduces these entries to ensure that every stem level occurs equally often (two in this case) in a stem-and-leaf plot. For details about how to use the SAS UNIVARIATE procedure to describe a sample, the reader should refer to Chapter 3 in the companion text by Younger (1997).

Exercises

3.6   Refer to the sample of $n = 20$ given in Example 2.27.

   a   Use statistical software such as SAS to construct a histogram, a stem-and-leaf plot, and a box-plot for the given sample.

   b   On the basis of (a), do you think that the soil-water flux measurements constitute a random sample from a normal distribution?

   c   Use statistical software such as SAS to construct a histogram, a stem-and-leaf plot, and a box-plot for the logarithms (to the base e) of the measurements in the sample.

   d   Compare the histograms, stem-and-leaf plots, and box-plots in (a) and (c). Is it reasonable to assume that the soil-water flux distribution is closer to a log-normal distribution than to a normal distribution?

   e   Calculate the means, the quartiles, and the standard deviations of (i) the soil-water flux measurements, and (ii) the logarithms of the soil-water flux measurements.

   f   Summarize the information about the shape of the soil-water flux distribution that can be gleaned from (e).

3.7   The following data summarize the total milk production (in pounds) during the first lactation of $n_1 = 14$ cows on a control diet and $n_2 = 14$ cows on a supplemented diet.

Control diet	10,887	10,579	11,606	10,961	9746	11,886	9518
	10,900	13,818	10,992	12,397	9769	10,778	12,753
Supplemented diet	12,799	13,721	11,262	12,229	11,069	10,279	12,524
	11,426	12,607	12,648	14,871	12,241	11,037	11,129

   a   Construct the frequency distributions in tabular form for the two samples and draw the corresponding histograms.

   b   Visually compare the two histograms to answer the following questions about the cows' milk production under the two diet supplements.

    i   Do you think the populations of milk yields have normal distributions?

    ii Are the population means different?

    iii Are the population variances different?

c Calculate the means and standard deviations of the two samples. Compare these values to draw conclusions about the milk production under the two diets. (Note that calculation of the standard deviation will be simplified if the data are coded by subtracting a constant from each value.)

d Use statistical computing software such as SAS

    i   to draw histograms, box-plots, and stem-and-leaf plots of the data for the two diets;

    ii to calculate the means, medians, standard deviations, and the interquartile ranges of the two samples.

e On the basis of the plots and statistics computed in (d), what conclusions can you draw about the shape and variability of the distributions of the two populations?

3.8 By means of the data on rainbow trout weight in Example 3.3, verify the properties of the standard deviation—specifically, that:

a the standard deviation of a sample is unaltered by adding (subtracting) the same constant to each measurement;

b if each measurement in a sample is multiplied by a constant c, then the standard deviation of the sample is multiplied by $|c|$, where $|c|$ is the absolute value of c (for example, $|5| = |-5| = 5$).

To verify these properties, first add an amount—say, 15—to each measurement and then calculate the standard deviation. Next, multiply each measurement by a constant—say, $c = -5$—and then calculate the standard deviation.

3.9 The accompanying figure shows the box-plots of the standardized scores in a mathematics test given to random samples of children in four school districts. Use the box-plots to compare the characteristics of the four distributions of the scores of children taking tests in the four school districts.

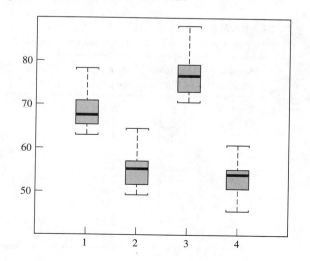

3.4
Sampling distributions

Sampling distributions are frequently used to make statistical inferences. In this section, we illustrate the use of sampling distributions with a hypothetical population consisting of six measurements. Of course, we realize that such a small population is unrealistic, but it will simplify our discussion here.

EXAMPLE **3.7**   Consider a hypothetical population containing only six measurements: {1, 2, 3, 5, 6, 8}. This is a discrete population in which the measurements have six distinct values 1, 2, 3, 5, 6 and 8, each with relative frequency 1/6. The probability distribution of this population may be expressed in tabular form as follows:

y	1	2	3	5	6	8
$f(y)$	$\frac{1}{6}$	$\frac{1}{6}$	$\frac{1}{6}$	$\frac{1}{6}$	$\frac{1}{6}$	$\frac{1}{6}$

Using Equations (2.5) and (2.6), the mean and variance of this population can be shown to be $\mu = 4.17$ and $\sigma^2 = 5.81$, respectively.

Suppose we select a random sample of size $n = 3$ from this population and calculate the sample mean $\overline{Y}$. What can we say about the observed value of $\overline{Y}$? The sampling distribution of $\overline{Y}$ can be used to answer this question.

Since the sample will be selected at random, the statistic $\overline{Y}$ is a random variable (see Section 2.3) whose observed value is the same as the mean of the randomly selected sample. Thus, the set of possible values of $\overline{Y}$ is the same as the set of all sample means that can be obtained by selecting samples of size $n = 3$ from the given population. As can be seen from Table 3.4, there are 20 different possibilities for a sample of size $n = 3$, so that the means of these 20 samples, listed as $\overline{y}$ in Table 3.4,

TABLE   **3.4**
All possible samples of size $n = 3$ and the population of $\overline{Y}$

Sample number	Sample values	Sample mean ($\overline{y}$)	Sample number	Sample values	Sample mean ($\overline{y}$)
1	1, 2, 3	2.00	11	1, 5, 6	4.00
2	1, 2, 5	2.67	12	1, 5, 8	4.67
3	1, 2, 6	3.00	13	2, 5, 6	4.33
4	1, 2, 8	3.67	14	2, 5, 8	5.00
5	1, 3, 5	3.00	15	3, 5, 6	4.67
6	1, 3, 6	3.33	16	3, 5, 8	5.33
7	1, 3, 8	4.00	17	1, 6, 8	5.00
8	2, 3, 5	3.33	18	2, 6, 8	5.33
9	2, 3, 6	3.67	19	3, 6, 8	5.67
10	2, 3, 8	4.33	20	5, 6, 8	6.33

constitute the population of values that can be observed for the random variable $\overline{Y}$ (the population of $\overline{Y}$). The distribution of the population of $\overline{Y}$ is called the *sampling distribution* of $\overline{Y}$.

As we see from Table 3.4, the population of $\overline{Y}$ is a discrete population containing measurements with 12 distinct values. Table 3.5 shows the sampling distribution of $\overline{Y}$ in tabular form.

Selecting a random sample of size $n = 3$ from the population $\{1, 2, 3, 5, 6, 8\}$ and determining its mean $\overline{y}$ is the same as selecting a random sample of size $n = 1$ from the population of $\overline{Y}$. The probability that the mean of the selected sample will equal a given value $\overline{y}$ is $f(\overline{y})$

$$\Pr\{\overline{Y} = \overline{y}\} = f(\overline{y}).$$

The sampling distribution of $\overline{Y}$ provides a probabilistic description of the possible values of the mean of a random sample of size $n = 3$. Such a description can be used to assess the performance of a procedure in which the mean of a random sample of size $n = 3$ is used to make inferences about the population. As an example, suppose that we decide to use the observed sample mean as an estimate of the population mean μ. Then we might be interested in knowing the probability that our estimate will differ from μ by an amount less than 0.25, say. Since $\mu = 4.17$, the desired probability is the same as the probability that the observed value of $\overline{Y}$ will be within the range $(4.17 - 0.25, 4.17 + 0.25) = (3.92, 4.42)$. The sampling distribution of $\overline{Y}$ in Table 3.5 shows that the only possible values of $\overline{Y}$ inside the range are 4.00 and 4.33, so that the required probability is

$$\Pr\{3.92 < \overline{Y} < 4.42\} = \Pr\{\overline{Y} = 4.00\} + \Pr\{\overline{Y} = 4.33\}$$

$$= f(4.00) + f(4.33) = 0.10 + 0.10 = 0.20.$$

Thus, if we use the mean of a random sample of size $n = 3$ to estimate the population mean, the probability that the difference between our estimate and the actual value of μ will be less than 0.25 is 0.20. In other words, when estimating μ by $\overline{Y}$ based on a random sample of size $n = 3$, the probability that the margin of error will be less than 0.25 is only 20%. There is an 80% chance that the margin of error will be larger than 0.25. ■

Example 3.7 illustrates one use of the sampling distribution of the sample mean. We will encounter many other applications in the subsequent chapters.

TABLE 3.5
Sampling distribution of $\overline{Y}$

$\overline{y}$	2.00	2.67	3.00	3.33	3.67	4.00
$f(\overline{y})$	0.05	0.05	0.10	0.10	0.10	0.10

$\overline{y}$	4.33	4.67	5.00	5.33	5.67	6.33
$f(\overline{y})$	0.10	0.10	0.10	0.10	0.05	0.05

Obviously, the sample mean is not the only statistic that can be used to make inferences about populations. There are many others, including the sample variance and the sample median. A sampling distribution can be readily constructed for any given statistic. A formal definition is presented in Box 3.1.

BOX **3.1**

Sampling distribution of a statistic

Let U be a statistic based on a random sample of size n from a given population. The distribution of the conceptual population of all values of U that can be generated by selecting random samples of size n from the population is called the sampling distribution of U.

EXAMPLE **3.8**

Returning to the population of six measurements {1, 2, 3, 5, 6, 8} in Example 3.7, we recall that its variance was $\sigma^2 = 5.81$. How well will the variance S^2 of a random sample of size $n = 3$ estimate the population variance σ^2?

The sampling distribution of the statistic S^2 can be used to answer this question. As stated in Box 3.1, the sampling distribution of $U = S^2$ is the distribution of the population consisting of the variances of the 20 possible samples of size $n = 3$ in Table 3.4. It can be verified that the population of $U = S^2$—that is, the set of variances of the 20 samples—is as follows:

1.00, 4.33, 7.00, 14.33, 4.00, 6.33, 13.00, 2.33, 4.33, 10.33,
7.00, 12.33, 4.33, 9.00, 2.33, 6.33, 13.00, 9.33, 6.33, 2.33

The sampling distribution of S^2 can be used to determine the probability that the variance of a random sample of size $n = 3$ will have a specified value. For example, it can be verified (see Exercise 3.12) that the probability that the difference between σ^2 and an observed value of S^2 will be less than 1 is 0.15.   ■

The statistical populations from which the samples are selected will be called *target populations*. The distributions of the target populations determine the sampling distributions of statistics based on random samples selected from the populations. In the following sections, we will encounter three ways in which sampling distributions may be used to make inferences about statistical populations.

In Section 3.8, sampling distributions are employed in a statistical procedure called interval estimation. In this procedure, a random sample is used to determine a range (interval) of plausible values that can be expected to contain the unknown value of a specific parameter, with a given degree of certainty. In Section 3.9, we describe the use of sampling distributions to develop statistical procedures called hypothesis tests. In hypothesis-testing procedures, the observed value of a statistic is examined to determine the probability that the sample has arisen from a population with a hypothesized distribution. If this probability is small, the hypothesis is considered unlikely to be true. Finally, the use of sampling distributions in statistical prediction is described in Section 3.10.

Exercises

3.10 Refer to Example 3.7.

  a Let $\mu_{\bar{Y}}$ and $\sigma_{\bar{Y}}$ denote the mean and standard deviation of the sampling distribution of $\bar{Y}$. Determine $\mu_{\bar{Y}}$ and $\sigma_{\bar{Y}}$.

  b Explain how the mean and standard deviation calculated in (a) can be used in the empirical and Tchebycheff's rules to evaluate the desirability of using $\bar{Y}$ to estimate μ.

3.11 Refer to Example 3.7.

  a Determine the sampling distributions of $\bar{Y}$ with $n = 3$ when the target populations are {0, 1, 2, 6, 7, 9} and {4, 8, 12, 20, 24, 32}.

  b Graphically display the sampling distributions of $\bar{Y}$ for three target populations: the one in Example 3.7 and the two described in (a). Comment on how the shape of the sampling distribution changes with the shape of the target population.

  c Determine the means and standard deviations of the sampling distributions in (b). Comment on the relationships of the means and standard deviations of the sampling distributions with the corresponding population parameters.

3.12 Refer to Example 3.8.

  a Determine the sampling distribution of S^2.

  b Let μ_{S^2} and σ_{S^2} denote the mean and standard deviation of the sampling distribution of S^2. Determine μ_{S^2} and σ_{S^2}.

  c Suppose that you are thinking of estimating the variance of this target population by the variance of a random sample of size $n = 3$. How can the mean and standard deviation in (b) be used to evaluate the desirability of doing so?

  d Calculate the probability that the error in estimating σ^2 by S^2 based on a sample of size $n = 3$ is less than 1.

3.13 Refer to the target population in Example 3.7.

  a Determine the sampling distribution of the median of a random sample of size $n = 3$.

  b Calculate the mean and standard deviation of the sampling distribution of the sample median.

  c Use the sampling distributions to compare the performances of the sample mean and the sample median in estimating μ, the population mean.

3.5
Sampling distribution of the sample mean

Many commonly used statistical procedures are based on statistics that can be expressed as the means (at least approximately) of random samples. Some key properties of the sampling distribution of the mean of a random sample are presented in Box 3.2.

BOX **3.2**

The sampling distribution of $\overline{Y}$

Let $\overline{Y}$ be the mean of a random sample of size n from a population containing infinitely many measurements. Let μ and σ^2 denote, respectively, the mean and variance of the population.

I   The mean and variance of the sampling distribution of $\overline{Y}$ are, respectively

$$\mu_{\overline{Y}} = \mu$$

$$\sigma^2_{\overline{Y}} = \frac{\sigma^2}{n}.$$

II If the sample size is large (as a rule of thumb, $n \geq 30$), the sampling distribution of $\overline{Y}$ can be approximated by a normal distribution with mean $\mu_{\overline{Y}}$ and variance $\sigma^2_{\overline{Y}}$, regardless of the shape of the target population. In other words, the sampling distribution of $\overline{Y}$ can be approximated by a $N(\mu_{\overline{Y}}, \sigma^2_{\overline{Y}})$ distribution. This normal approximation improves as n increases.

III If the target population has a normal distribution, then the sampling distribution of $\overline{Y}$ is a $N(\mu_{\overline{Y}}, \sigma^2_{\overline{Y}})$ distribution, regardless of the sample size.

Relationship between mean, variance, and sample size

The results in Box 3.2 refer to two populations: the target population from which the sample is selected and the population of $\overline{Y}$ consisting of the means of all possible samples of size n. The mean and standard deviation of the target population are μ and σ, while $\mu_{\overline{Y}}$ and $\sigma_{\overline{Y}}$ are the mean and standard deviation of the population of $\overline{Y}$.

The mean of the sampling distribution of $\overline{Y}$ is the same as the mean of the target population, but the standard deviation of the sampling distribution is lower (if $n > 1$) than the standard deviation of the target population. Indeed, the standard deviation of the sample mean is inversely proportional to the square root of the sample size. For example, if the mean and standard deviation of the target population are $\mu = 40$ and $\sigma = 10$, the mean and standard deviation of the sampling distribution of $\overline{Y}$ when the sample size is $n = 4$ are $\mu_{\overline{Y}} = 40$ and $\sigma_{\overline{Y}} = \frac{10}{\sqrt{4}} = 5$. Thus, the standard deviation of the sampling distribution of $\overline{Y}$ is proportional to $1/\sqrt{4} = 1/\sqrt{n}$, the inverse of the square root of the sample size. If the sample size is increased from $n = 4$ to $n = 16$, the mean of the sampling distribution will remain unaltered at 40 but the standard deviation will be reduced from 5 to 2.5.

For $n = 5$, 15, and 30, Figure 3.5 shows histograms of the sampling distributions constructed from the means of 1000 random samples each of size n selected from two target populations: A symmetric population with a standard normal distribution; and a population with a distribution that is skewed to the right because it contains some very large values. The first row of Figure 3.5 shows the shapes of the two target populations. In Figure 3.5, the tendency of sample means to cluster closer to the means of the target populations as the sample size is increased from $n = 5$ to $n = 30$ illustrates the fact that the variance of the sampling distribution of $\overline{Y}$ decreases with increasing sample.

FIGURE 3.5

Sampling distributions of sample means

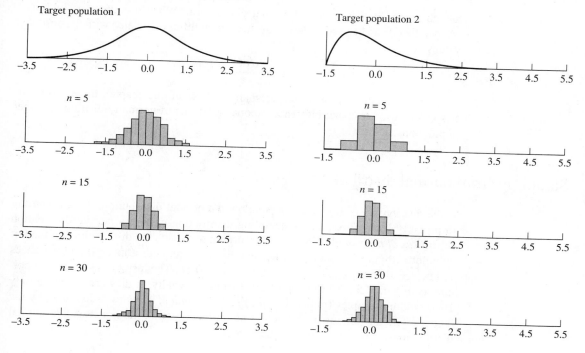

Central limit theorem

Notice in Figure 3.5 that, with increasing sample size, the shape of the sampling distribution tends to resemble the shape of a normal distribution. This tendency is described by a well known result in probability theory known as the central limit theorem. The essential implication of the central limit theorem can be summarized as in Box 3.3.

BOX **3.3**

> ### The Central limit theorem
>
> The sampling distribution of the mean $\overline{Y}$ of a large random sample of size n selected from a population with mean μ and standard deviation σ can be approximated by a normal distribution with mean μ and standard deviation $\sigma_{\overline{Y}} = \sigma/\sqrt{n}$.

Box 2.7 implies that if Y has a $N(\mu, \sigma^2)$ distribution, then the z-score of Y, $Z = (Y - \mu)/\sigma$, has a $N(0, 1)$ distribution. Therefore, the central limit theorem implies that, for large n, the z-score of $\overline{Y}$

$$Z = \frac{\overline{Y} - \mu_{\overline{Y}}}{\sigma_{\overline{Y}}} = \frac{\sqrt{n}(\overline{Y} - \mu)}{\sigma}$$

has an approximately standard normal distribution. When σ is known, Table C.1 (Appendix C) can be used to determine the probabilities associated with the observed values of

$$Z = \frac{\sqrt{n}(\overline{Y} - \mu_0)}{\sigma},$$

for any given value μ_0 of μ. As we shall see later in this chapter, these probabilities can be used to make inferences about the means of normally distributed target populations.

Sampling from normal distribution

If the distribution of the target population is a normal distribution, then, according to property III in Box 3.2, the sampling distribution of $\overline{Y}$ is a normal distribution *regardless of the sample size*. Thus, according to theory, the distributions in the left column of Figure 3.5 should be normal distributions with a mean 0 and variances 1, $1/\sqrt{5} = 0.45$, $1/\sqrt{15} = 0.26$ and $1/\sqrt{30} = 0.18$. The histograms in the left column of Figure 3.5 tend to support this theory especially as they are based on only 1000 random samples (rather than *all possible random samples* as assumed by the theory). The histograms in the right column of Figure 3.5 do not resemble normal distributions except when the sample size is large.

Calculating probabilities associated with $\overline{Y}$

The results in Box 3.2 can be used to calculate the probabilities associated with the observed values of $\overline{Y}$. These probabilities can be calculated at three levels of precision depending on the sample size and the type of information available about the target population.

The lowest level corresponds to the case where we know only the mean and variance of the target population; we do not know whether the population distribution is normal. In that case, we can use property I in Box 3.2, along with either Tchebycheff's rule (see Exercise 2.34) or the empirical rule in Box 2.1 to compute bounds for the required probabilities.

The next level corresponds to the case where the sample size is large and the mean and variance of the target populations are known. In that case, we can use property II in Box 3.2 to determine an approximation for the sampling distribution of $\overline{Y}$. This approximation can then be used to calculate approximate probabilities of events associated with the observed values of $\overline{Y}$. Finally, if the target population is known to have a normal distribution with known mean and variance, we can use property III in Box 3.2 to determine the exact form of the sampling distribution of $\overline{Y}$. Then exact values of the probabilities associated with $\overline{Y}$ are computationally feasible.

Example 3.9 illustrates how the results in Box 3.2 can be used to make statistical inferences.

EXAMPLE **3.9**   A horticulturist is interested in estimating the average height μ of a new variety of ornamental plant. On the basis of previous experience with similar plants, the horticulturist assumes that the standard deviation of the population of plant heights is $\sigma = 5.5$ cm. If the plant is grown in 36 experimental plots and the average height $\overline{Y}$ of the $n = 36$ plants is used as an estimate of μ, what is the probability that the observed value of $\overline{Y}$ will be within 2 cm of μ?

If we regard the observed heights of 36 experimental plants as a random sample of size $n = 36$ from the population of plant heights, the relevant question concerns the observed value of the mean of a random sample of size $n = 36$. The horticulturist is interested in the probability that the value of $\overline{Y}$ is within the interval $(\mu - 2, \mu + 2)$.

First, according to property I in Box 3.2, the sampling distribution of $\overline{Y}$ has mean μ and standard deviation $\sigma_{\overline{Y}} = \sigma/\sqrt{n} = 5.5/\sqrt{36} = 0.92$ cm. Tchebycheff's rule (see Exercise 2.34) can be used to compute a bound for the required probability.

Applying the rule to the population of $\overline{Y}$, we find that for any $k > 1$

$$\Pr\{-k\sigma_{\overline{Y}} \le \overline{Y} - \mu \le +k\sigma_{\overline{Y}}\} \ge 1 - (1/k^2).$$

Set $k\sigma_{\overline{Y}} = 2$, so that $k = 2/\sigma_{\overline{Y}} = 2/0.92 = 2.17$. Then

$$\Pr\{-2 \le \overline{Y} - \mu \le +2\} \ge 1 - 1/(2.17)^2 = 0.79.$$

Thus, there is at least a 79% chance that the mean height calculated from 36 experimental plots will be within 2 cm of the true mean height. In other words, in more than 79% of instances, the procedure of estimating μ by means of $\overline{Y}$ based on a random sample of $n = 36$ plants will yield results within 2 cm of the correct value of μ.

The lower bound of 79% obtained using Tchebycheff's rule can be improved on the basis that the sample size $n = 36$ can be regarded as large. A good approximation to the required probability can be calculated using the central limit theorem stated in Box 3.3. According to the central limit theorem, the sampling distribution of $\overline{Y}$ is approximately a normal distribution with mean μ and standard deviation $\sigma_{\overline{Y}} = 0.92$. Using this theorem and the areas under the normal distribution given in Table C.1 (Appendix C), we find that the probability of observing a value of $\overline{Y}$ within 2 cm of μ is approximately 0.97.

Finally, if the target population of plant heights is known to have a normal distribution, then property III in Box 3.2 implies that the sampling distribution of $\overline{Y}$ is an exact normal distribution with mean μ and standard deviation 0.92. In that case, the probability that an observed value of $\overline{Y}$ will be within 2 cm of μ is exactly 0.97.   ■

Exercises

3.14   Refer to Example 3.2, where a random sample of $n = 20$ disease-free survival (DFS) times of breast cancer patients was given. Suppose, for the purpose of this exercise, that the mean and standard deviation of the population of DFS times are known to be $\mu = 15.5$ months and $\sigma = 3.5$ months, respectively.

   a   Let $\overline{Y}$ denote the mean of a random sample of $n = 20$ DFS times. Determine the mean and standard deviations of the sampling distribution of $\overline{Y}$.

   b   From practical experience, it is known that the distribution of the population of survival times is usually not normal. On the basis of this information, is it justified to assume that the sampling distribution of $\overline{Y}$ is approximately a normal distribution? Explain.

   c   Use the empirical rule to obtain the approximate probability that the mean of $n = 20$ DFS times will estimate the true mean survival time μ with a margin of error of less than 6 months.

   d   Answer (c) for a sample size $n = 50$.

   e   Calculate the probability that the mean of a random sample of $n = 50$ DFS times will estimate the true mean survival time with a margin of error of less than 2 months.

3.15   Refer to Example 3.3, where a random sample of $n = 10$ weights of rainbow trout in a lake is given. Assume that the rainbow trout weights have a normal distribution, with mean $\mu = 1.5$ pounds and $\sigma = 0.4$ pounds.

   a   Describe the sampling distribution of the mean of the weights of a random sample of $n = 15$ rainbow trout.

   b   Calculate the probability that the mean of a random sample of $n = 15$ rainbow trout weights will be within 0.25 pounds of the population mean.

3.6
Some useful sampling distributions

In the preceding section, we described conditions under which the sampling distribution of a sample mean will be a normal distribution and discussed how normal distributions can be used to make inferences about the mean of a target population.

Other uses of the normal distribution in statistical inferences will be encountered in subsequent chapters.

In addition to the normal distribution, three other distributions, known as t-, χ^2-, and F-distributions, are frequently used to make statistical inferences; we briefly describe these distributions in the present section.

The t-distribution

A t-distribution with ν degrees of freedom, also known as a $t(\nu)$-distribution, is a continuous distribution that is bell-shaped and symmetric about zero. In subsequent chapters, the t-distributions will be encountered as sampling distributions useful for making inferences about populations under a variety of conditions. For example, if $\overline{Y}$ and S are the mean and standard deviation of a random sample from a normally distributed population with a known mean μ, then it can be shown that the sampling distribution of the statistic

$$t = \frac{\sqrt{n}(\overline{Y} - \mu)}{S} \tag{3.6}$$

is a $t(n-1)$ distribution. Notice that the statistic t defined in Equation (3.6) is the same as the statistic Z defined when discussing the Central Limit Theorem in the previous section, except that the divisor in Z is the population standard deviation σ, whereas the divisor in t is the sample standard deviation S. Thus, unlike Z, the statistic t can be calculated without knowing the value of the population variance σ^2. In Chapter 4, we will see how the statistic in Equation (3.6) can be used to make inferences about the mean of a normally distributed population with an unknown standard deviation.

The main difference between a t-distribution and a standard normal distribution is that, in general, the t-distribution is flatter (that is, has heavier tails) than the $N(0, 1)$ distribution. As the degrees of freedom ν increases, the shape of the t-distribution approaches that of a standard normal distribution. Figure 3.6 shows the shapes of $t(\nu)$-distributions for selected values of ν.

Recall that the α-level critical value of a distribution is the same as its $1 - \alpha$ percentile. The α-level critical value of a $t(\nu)$-distribution will be denoted by $t(\nu, \alpha)$. Table C.2 in Appendix C gives $t(\nu, \alpha)$ for selected values of ν and α. Figure 3.7 shows the relationship between $t(\nu, \alpha)$ and α. The critical values of t-distributions can be found using the SAS procedure TINV. Example 3.10 illustrates how Table C.2 can be used to find the critical values of t-distributions.

EXAMPLE **3.10**   Let's find the 0.025-level and 0.975-level critical values of a t-distribution with $\nu = 12$ degrees of freedom. In Table C.2 we refer to the intersection of the row for 12 degrees of freedom (denoted by df in Table C.2) and the column for $\alpha = 0.025$, and determine that $t(12, 0.025) = 2.1788$. Therefore, 2.5% of the measurements in a $t(12)$-distribution will exceed 2.1788. Also, because of symmetry, 2.5% of the measurements in a $t(12)$-distribution will have values less than -2.1788, so that 97.5%

F I G U R E   3.6

Selected $t(v)$-distributions

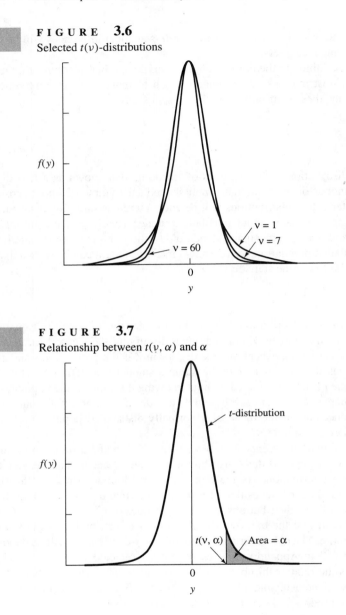

F I G U R E   3.7

Relationship between $t(v, \alpha)$ and α

of the measurements will be larger than -2.1788. Thus, the 0.975-level critical value of a t-distribution with $v = 12$ degrees of freedom is $t(12, 0.975) = -2.1788$.   ∎

It can be shown that a $t(\infty)$-distribution is the same as a $N(0, 1)$ distribution, so that Table C.2 can also be used to find the critical values of a $N(0, 1)$ distribution. For example, for degrees of freedom ∞ in Table C.2, we find that $t(\infty, 0.05) = 1.645$

and $t(\infty, 0.025) = 1.960$. The reader can verify from Table C.1 that these are the 0.05- and 0.025-level critical values of a $N(0, 1)$-distribution.

The χ^2-distribution

A χ^2-distribution with ν degrees of freedom—or simply a $\chi^2(\nu)$-distribution—is an asymmetric distribution that can be used to model a population containing no negative measurements. Like the normal distribution and the t-distribution, the χ^2-distribution appears as the sampling distribution of a wide variety of statistics used to make inferences. For instance, if S^2 is the variance of a random sample of size n from a normal population with a variance σ^2, then the sampling distribution of the statistic

$$\chi^2 = \frac{(n-1)S^2}{\sigma^2} \tag{3.7}$$

is known to be a $\chi^2(n-1)$-distribution. In Chapter 4, we will see how the statistic χ^2 can be used to make inferences about the variance of a normally distributed population. Figure 3.8 shows $\chi^2(\nu)$-distributions for several values of ν.

The α-level critical values of $\chi^2(\nu)$-distributions, denoted by $\chi^2(\nu, \alpha)$ can be found using the SAS procedure CINV. Table C.3 in Appendix C lists $\chi^2(\nu, \alpha)$ for selected values of ν and α.

FIGURE   3.8
Selected $\chi^2(\nu)$-distributions

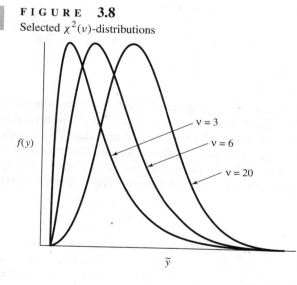

EXAMPLE 3.11   Let us find the 0.05- and 0.95-level critical values of a χ^2-distribution with $\nu = 12$ degrees of freedom. Entering Table C.3 at the row corresponding to $\nu = 12$, we find the critical values corresponding to $\alpha = 0.05$ and $\alpha = 0.95$ in the columns with headings 0.05 and 0.95, respectively. Therefore, $\chi^2(12, 0.95) = 5.226$ and

$\chi^2(12, 0.05) = 21.026$. Notice that, as can be seen from Figure 3.9, 5% and 95% of the measurements in a population with a $\chi^2(12)$-distribution will have values larger than 21.026 and 5.226, respectively. Therefore, 90% of the measurements will have values between 5.226 and 21.026.   ■

F I G U R E   3.9
Areas under a $\chi^2(12)$-distribution

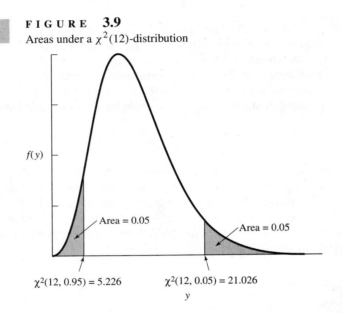

$\chi^2(12, 0.95) = 5.226$ $\chi^2(12, 0.05) = 21.026$

y

The *F*-distribution

An *F*-distribution with ν_1 degrees of freedom for the numerator and ν_2 degrees of freedom for the denominator—or simply an $F(\nu_1, \nu_2)$-distribution—is a continuous asymmetric distribution of a population that contains no negative values. Suppose that S_1^2 and S_2^2 are the variances of independent random samples of sizes n_1 and n_2 from two normally distributed populations. It can be shown that, if the variances of the two populations are equal, the sampling distribution of the statistic

$$F = \frac{S_1^2}{S_2^2}$$

is an $F(n_1 - 1, n_2 - 1)$-distribution. In Chapter 4, we will see how the critical values of an $F(n_1 - 1, n_2 - 1)$-distribution can be used to compare the variances of two normally distributed populations. Figure 3.10 shows some $F(\nu_1, \nu_2)$-distributions for various values of ν_1 and ν_2.

The α-level critical value of an $F(\nu_1, \nu_2)$-distribution will be denoted by $F(\nu_1, \nu_2, \alpha)$. The SAS function FINV can be used to determine these values. Table C.4 in Appendix C lists $F(\nu_1, \nu_2, \alpha)$ for selected values of ν_1, ν_2, and α. For

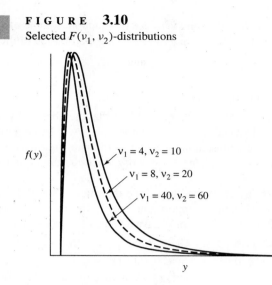

FIGURE   3.10
Selected $F(v_1, v_2)$-distributions

reasons of space, Table C.4 contains critical values for $\alpha \leq 0.10$. The relationship

$$F(v_1, v_2, \alpha) = \frac{1}{F(v_2, v_1, 1 - \alpha)} \tag{3.8}$$

can be used to find critical values for $\alpha \geq 0.90$.

EXAMPLE 3.12   Let's determine the 0.05- and 0.95-level critical values of an $F(5, 12)$-distribution. From Table C.4, we see that $F(5, 12, 0.05) = 3.11$. To find $F(5, 12, 0.95)$, we use Equation (3.8):

$$F(5, 12, 0.95) = \frac{1}{F(12, 5, 1 - 0.95)} = \frac{1}{F(12, 5, 0.05)} = \frac{1}{4.68} = 0.21. \quad \blacksquare$$

Exercises

3.16   Refer to Table C.2 in Appendix C.

a   Determine $t(12, 0.05)$ and $t(20, 0.001)$.

b   Determine two values t_1 and t_2 such that the area between them under a $t(10)$-distribution is 0.95.

c   Determine a value t_0 such that the area below t_0 under a $t(13)$-distribution is 0.025.

d   What can you say about the probability that a randomly selected value from a $t(8)$-distribution will exceed 2? What about the probability that it will be less than -1.5?

3.17 Let $\overline{Y}$ and S^2 denote the mean and variance of a random sample of size $n = 15$ from a $N(\mu, \sigma^2)$-distribution. Recall that the sampling distribution of the statistic

$$t = \frac{\sqrt{n}(\overline{Y} - \mu)}{S}$$

defined in Equation (3.6) has a $t(14)$-distribution.

a Use Table C.2 to determine the probability that the observed value of t—that is, the value of t calculated from a random sample—will be larger than 3.79.

b Suppose that a random sample of size $n = 15$ from a normal distribution yielded $\overline{Y} = 30$ and $S = 4$. Calculate the value of t, on the assumption that the population mean is 25.

c Use the probability calculated in (a) and the value of t calculated in (b) to argue that the observed sample supports the conclusion that the actual value of μ is larger than 25. [Hint: The sample gave a value of t larger than 3.79.]

3.18 Refer to Table C.3 in Appendix C.

a Determine $\chi^2(12, 0.05)$ and $\chi^2(20, 0.001)$.

b Determine two values χ_1^2 and χ_2^2 such that the area between them under a $\chi^2(10)$-distribution is 0.95.

c Determine a value χ_0^2 such that the area below χ_0^2 under a $\chi^2(13)$-distribution is 0.025.

d What can you say about the probability that a randomly selected value from a $\chi^2(8)$-distribution will exceed 3? What about the probability that it will be less than 1?

3.19 An investigator selected a random sample of size $n = 20$ from a normally distributed population and observed that the sample variance $s^2 = 10$. The investigator wondered whether this observed value for S^2 justifies the conclusion that $\sigma^2 > 4$.

a Use Table C.3 to determine the 0.001-level critical value of a $\chi^2(19)$-distribution.

b Calculate the value of

$$\chi^2 = \frac{(n-1)S^2}{\sigma^2},$$

on the assumption that $\sigma^2 = 4$.

c Use the critical value determined in (a) and the value calculated in (b) to argue that the observed value of 5 for S^2 makes it reasonable to conclude that $\sigma^2 > 4$. [Hint: The observed value of χ^2 is larger than the 0.001-level critical value.]

3.20 Refer to Table C.4 in Appendix C.

a Determine the critical values $F(8, 15, 0.05)$ and $F(6, 10, 0.95)$.

b Determine two values F_1 and F_2 such that the area between them under an $F(5, 7)$-distribution is 0.90.

c Determine the probability that a randomly selected value from an $F(6, 9)$ distribution will be larger than 10.

3.21 Using the SAS procedures described in Section 2.3 of the Companion text by Younger (1997), or otherwise, compute the following.

a The 0.034-level critical value of a t-distribution with 13 degrees of freedom.

b   The 0.84- and 0.06-level critical values of a χ^2-distribution with 9 degrees of freedom.

c   The 0.75- and 0.12-level critical values of an F-distribution with 6 degrees of freedom for the numerator and 11 degrees of freedom for the denominator.

3.7
Types of statistical inferences

The statistical inferential problems discussed in this book can be classified into one of three types:

- estimating parameters of target populations
- testing hypotheses about target populations
- predicting responses that will be observed in the future

The next example illustrates the essential differences between these three types of inferential problems.

EXAMPLE 3.13   In a study to evaluate a new treatment for protecting pine trees against a particular disease, an investigator is interested in planning a study to obtain information about θ, the proportion of treated trees that remain disease-free for at least one year.

The investigator can formulate three different types of problems, depending on the desired inference about θ. In an *estimation problem,* the objective is to determine a set of numbers that can be considered as plausible values (in some sense) for θ. For instance, on the basis of the results of the study, the investigator might decide that the value of θ is likely to be between 0.60 and 0.80.

In a *hypothesis-testing problem,* the objective is to verify a claim—a hypothesis—about the true value of θ. For example, the investigator might be interested in seeing whether the claim that more than 70% of treated trees will be disease-free for at least one year is justified. In that case, the investigator's objective is to verify the hypothesis (claim) that $\theta > 0.70$.

In a *prediction problem,* the investigator is interested in predicting the response to the new treatment that will be actually observed on a future occasion. For example, instead of making an inference about the population proportion θ, the investigator may want to use the data to predict the proportion of trees that will be disease-free for at least one year in a particular stand of trees that is to be treated in the future.   ■

Details of estimation, hypothesis testing, and prediction in various practical settings are covered in subsequent chapters. In the remainder of this chapter, we briefly introduce the key principles that form the basis of these procedures.

We begin with a description of the notation used in the remainder of the book. The reader may find the notation unnecessarily complicated at first, but with some practice it will become easy to use. A patient reader will soon find that using the notation helps unify a wide variety of statistical techniques under a single logical scheme.

Notation

A parameter is a quantity that describes a population characteristic. Thus, a population mean μ, a population standard deviation σ, a population proportion π, and the standard deviation $\sigma_{\overline{Y}} = \sigma/\sqrt{n}$ of the population of $\overline{Y}$ are all examples of parameters. As a general rule, parameters will be denoted by Greek letters.

A statistic is a sample characteristic. We will use either the uppercase letters of the English alphabet (such as $\overline{Y}$ for a sample mean, S for a sample standard deviation, and P for a sample proportion) or Greek letters with a hat (such as $\hat{\mu}$, $\hat{\sigma}$, $\hat{\pi}$) to denote statistics.

Usually, the observed values of statistics serve as estimates of the corresponding population parameters. A statistic whose observed value is used as an estimate of a parameter is called the *estimator* of that parameter. Whenever it is convenient to do so, we will use the corresponding lowercase letter to denote the observed value of an estimator. For instance, an observed value $\overline{y}$ of the estimator $\overline{Y}$ is an estimate for the population mean μ.

The hat notation is a convenient method of emphasizing that the statistic in question is an estimator of the parameter represented by the Greek letter. For example, the symbol $\hat{\mu}$ denotes a statistic used to estimate the population mean μ. Thus, if the sample mean is used as an estimator of the population mean, as is often the case, then $\hat{\mu} = \overline{Y}$.

Occasionally, it will be necessary to make exceptions to the convention of using Greek letters for parameters and English letters for statistics. For example, the symbol SE (for standard error) will sometimes be used to denote the standard deviation of the sampling distribution of a statistic. Thus, $SE(\overline{Y})$ is the same as $\sigma_{\overline{Y}}$.

It will sometimes be necessary to use multiple hats to indicate that multiple stages are involved in the computation (estimation) of a given quantity. Thus, $\widehat{SE}(\hat{\mu})$ denotes the standard error of the estimator of μ.

Note that, very often, there will be many symbols for a particular quantity. For example, if we estimate the population mean μ by the sample mean $\overline{Y}$ (that is, $\hat{\mu} = \overline{Y}$) and the population variance σ^2 by the sample variance S^2 (that is, $\hat{\sigma}^2 = S^2$), there are five different ways of denoting the estimated standard error of the sample mean

$$\hat{\sigma}_{\overline{Y}} = \widehat{SE}(\overline{Y}) = \widehat{SE}(\hat{\mu}) = \frac{\hat{\sigma}}{\sqrt{n}} = \frac{S}{\sqrt{n}}.$$

Exercises

3.22   An animal scientist wants to collect data to answer the following questions concerning a new diet supplement for dairy cows.

    a   Is the average milk production under the new diet higher than that under the standard diet?

b What is the average milk production under the new diet?

c What can be said about the milk production of a particular cow that will be fed the new diet?

Categorize, giving reasons, each of the three questions as a question involving estimation, hypothesis testing, or prediction. In each case, describe the relevant populations and parameters, if any.

3.23 A behavioral scientist wants to collect data to answer the following questions concerning a behavior disorder known as ADHD (attention deficit hyperactivity disorder) among children of ages 4–7 years.

a What is the proportion of children in this age group with ADHD?

b Does the proportion of children with ADHD depend on gender? In other words, do the proportions differ between males and females?

c In a random sample of 25 children of ages 4–7, how many will have ADHD?

Categorize, giving reasons, each of the three questions as a question involving estimation, hypothesis testing, or prediction. In each case, describe the relevant populations and parameters, if any.

3.24 An environmental scientist wants to collect data to answer the following questions concerning environmental effects on the population of birds in a particular region.

a Has the number of species of birds decreased over the past ten years?

b How many birds of a particular species will be there on a particular day in the future?

c Ten years from now, how many distinct species will be found in the region?

Categorize, giving reasons, each of the three questions as a question involving estimation, hypothesis testing, or prediction. In each case, describe the relevant populations and parameters, if any.

3.8
Estimating parameters

Let θ denote a parameter such as the population mean, the population standard deviation, and the difference between two population means.

Point estimation

A *point estimator* (or simply an estimator) of θ is a statistic $\hat{\theta}$ whose observed value (that is, the value calculated from samples) serves as an estimate of the value of θ. For instance, the sample mean $\overline{Y}$ $(\hat{\theta})$ is an intuitively reasonable point estimator of the population mean μ (θ).

EXAMPLE 3.14 Nitrogen is the most common fertilizer element applied to mineral soils. In tropical areas with warm temperatures and heavy rainfall, only a part of the applied nitrogen is used by the crops. Information about μ, the mean percentage nitrogen loss (N-loss), is important for research on optimal growth conditions for plants.

The following data represent the amount of nitrogen lost (expressed as a percentage of the total amount of nitrogen applied) over a 16-week period when Urea + N-Serve (UN) was used as the fertilizer for sugarcane:

$$10.8, \ 10.5, \ 14.0, \ 13.5, \ 8.0, \ 9.5, \ 11.8, \ 10.0, \ 8.7, \ 9.0, \ 9.8, \ 13.8, \ 14.7, \ 10.3, \ 12.8$$

Let $\overline{Y}$ denote the mean of a random sample of $n = 15$ measurements of N-loss when UN is used as the nitrogen fertilizer for sugarcane. Then $\overline{Y}$ is a natural point estimator of μ, the mean N-loss in the population of all sugarcane plants treated with UN. This particular sample yielded the observed value $\overline{y} = 11.15$ for $\overline{Y}$. Thus, the point estimate of μ based on the observed sample is 11.15%.   ∎

For a given point estimator, different samples could give different point estimates. For instance, in Example 3.14, another random sample of percentage N-loss measurements will most likely give a value of $\overline{Y}$ that differs from 11.15%. The sampling distribution of a point estimator $\hat{\theta}$ can be used to evaluate the performance characteristics of $\hat{\theta}$ as the estimator of θ. Two of the most frequently used criteria for evaluating a point estimator are bias and mean square error.

Bias in point estimators

A point estimator $\hat{\theta}$ is said to be an *unbiased estimator* of θ if the mean of the sampling distribution of $\hat{\theta}$ equals θ; that is, $\hat{\theta}$ is an unbiased estimator of θ if $\mu_{\hat{\theta}} = \theta$. An appealing property of an unbiased estimator $\hat{\theta}$ is that, if samples are selected repeatedly from the target population and the value of the unbiased estimator $\hat{\theta}$ is calculated from each selected sample, then the *average* of all the calculated estimates will equal the true value of the parameter. Thus, in repeated use of an unbiased estimator $\hat{\theta}$, some overestimates ($\hat{\theta} > \theta$) and underestimates ($\hat{\theta} < \theta$) may be obtained, but the average of all the estimates will be equal to θ.

Expected value

Since the mean of the sampling distribution of a random variable Y is the average of all values of Y that can be observed from repeated samples, the mean of Y is often referred to as its *expected value* and denoted by the symbol $\mathcal{E}(Y)$. Thus $\hat{\theta}$ is an unbiased estimator of θ if the expected value of $\hat{\theta}$ equals θ. That is, if $\mathcal{E}(\hat{\theta}) = \theta$.

We have already seen an example of an unbiased estimator. In Box 3.2, it was noted that $\mu_{\overline{Y}} = \mu$, so that the sample mean $\overline{Y}$ is an unbiased estimate of the population mean μ. Thus, in repeated samples, some samples will have $\overline{Y}$ values larger than μ, and some will have $\overline{Y}$ values smaller than μ, but the average of all $\overline{Y}$ values will be exactly equal to μ.

An estimator $\hat{\theta}$ is said to be a *biased estimator* of θ if the expected value of $\hat{\theta}$ is not equal to θ; that is, $\mathcal{E}(\hat{\theta}) \neq \theta$ for a biased estimator. The difference $\mathcal{E}(\hat{\theta}) - \theta$ is called the *bias* of $\hat{\theta}$.

EXAMPLE 3.15   In Section 3.3, we noted that, if the denominator of the sample variance S^2 is changed from $n - 1$ to n, the resulting estimator

$$S'^2 = \frac{(Y_1 - \overline{Y})^2 + \cdots + (Y_n - \overline{Y})^2}{n}$$

tends to underestimate the population variance σ^2. A precise description of this tendency to underestimate can be given in terms of the concept of bias. If μ_{S^2} and $\mu_{S'^2}$ are the means of the sampling distributions of S^2 and S'^2, respectively, then as shown in Example B.1 in Appendix B

$$\mathcal{E}(S^2) = \sigma^2 \text{ and } \mathcal{E}(S'^2) = \sigma^2 - \frac{\sigma^2}{n},$$

so that the bias of S^2 is $\sigma^2 - \sigma^2 = 0$ and the bias of $S'^2 = (\sigma^2 - \frac{\sigma^2}{n}) - \sigma^2 = -\frac{\sigma^2}{n}$. Thus, S^2 is an unbiased estimator of σ^2 and S'^2 has a negative bias. The bias of S'^2 increases as σ^2 increases and decreases as n increases.   ■

Even though it is desirable for an estimator to be unbiased, an unbiased estimator is not always preferable to a biased estimator. As an example, consider two statistics U_1 and U_2 such that U_1 is an unbiased estimator of $\theta (\mathcal{E}(U_1) = \theta)$ and U_2 is an unbiased estimator of θ^* ($\mathcal{E}(U_2) = \theta^*$), a value larger than θ. Then, as an estimator of θ, the statistic U_2 is biased upwards, because the mean of its sampling distribution is larger than θ. Is U_1 better than U_2 for estimating θ?

To answer this question, compare the sampling distributions of the two estimators. Suppose, for example, that the sampling distributions of U_1 and U_2 are as in Figure 3.11. Notice that the area under the frequency density function between $\theta - 1$ and $\theta + 1$ is higher for U_2 than for U_1. Thus, the value of U_2 is more likely to be within 1 unit of θ than is the value of U_1.

FIGURE 3.11
Sampling distributions of two estimators of θ

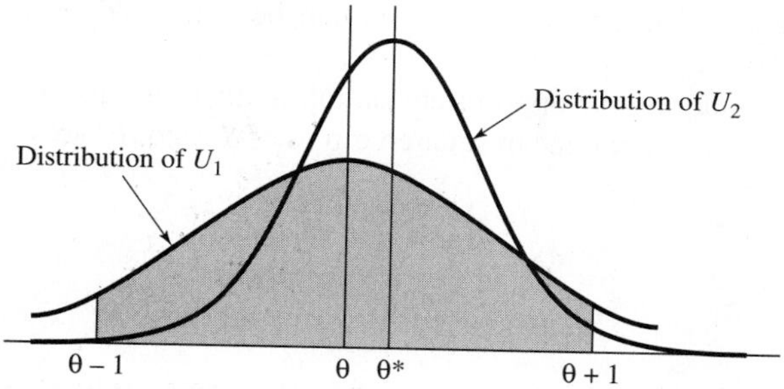

Efficiency of point estimators

Let $\hat{\theta}$ be a point estimator of θ and let $\mu_{\hat{\theta}}$ denote the mean of the sampling distribution of $\hat{\theta}$. The *mean squared error* of $\hat{\theta}$—often written as $MSE(\theta)$—is defined as the mean of the population of all values of the squared difference $(\hat{\theta} - \theta)^2$ between $\hat{\theta}$ and θ. The mean squared error of an estimator provides a measure of its accuracy. It can be shown that the smaller the mean squared error, the more likely it is that the magnitude of the difference between the observed value of the estimator and the true value of the parameter will be less than a specified bound.

Often, estimators are compared by looking at their mean squared errors. If $\hat{\theta}_1$ and $\hat{\theta}_2$ are two possible estimators of θ, the *efficiency* of $\hat{\theta}_1$ relative to $\hat{\theta}_2$ is defined as the ratio

$$E = \frac{MSE(\hat{\theta}_2)}{MSE(\hat{\theta}_1)}. \tag{3.9}$$

Since smaller mean squared errors are associated with greater accuracy, large values of E favor $\hat{\theta}_1$. In particular, $E > 1$ implies that $\hat{\theta}_1$ is more efficient than $\hat{\theta}_2$.

In Exercise 3.25, you will see examples of the important property that, if $\hat{\theta}$ is an unbiased estimate of θ, then

$$MSE(\hat{\theta}) = \sigma_{\hat{\theta}}^2, \tag{3.10}$$

where $\sigma_{\hat{\theta}}^2$ is the variance of $\hat{\theta}$. Thus, comparing the mean squared errors of two unbiased estimators of θ is the same as comparing their variances. It follows that, if $\hat{\theta}_1$ and $\hat{\theta}_2$ are unbiased estimators of θ, the efficiency of $\hat{\theta}_1$ relative to $\hat{\theta}_2$ is

$$E = \frac{\sigma_{\hat{\theta}_2}^2}{\sigma_{\hat{\theta}_1}^2}. \tag{3.11}$$

The next example illustrates the use of mean squared errors to compare estimators.

EXAMPLE **3.16**   Suppose we are interested in estimating the variance σ^2 of a normally distributed population. Two possible estimators of σ^2 are $\hat{\theta}_1 = S^2$, the sample variance with denominator $n - 1$, and $\hat{\theta}_2 = S'^2$, the sample variance with denominator n. The performance of these two estimators can be compared by comparing their mean squared errors.

By considering the appropriate sampling distributions, it can be shown that, for a sample of size n, the mean squared errors of S^2 and S'^2 are

$$MSE(S^2) = \frac{2}{n-1}\sigma^4,$$

$$MSE(S'^2) = \frac{2n-1}{n^2}\sigma^4.$$

Thus, from Equation (3.9), the efficiency of S'^2 relative to S^2 is

$$E = \frac{MSE(S^2)}{MSE(S'^2)} = \frac{\frac{2}{n-1}\sigma^4}{\frac{2n-1}{n^2}\sigma^4} = \frac{2n^2}{(n-1)(2n-1)}.$$

In Exercise 3.28, the reader will be asked to verify that $E > 1$ for all values of n and that its value approaches 1 as n increases. Thus, for small sample sizes, MSE favors S'^2. However, due to its lack of bias and other desirable properties, S^2 is usually used for estimating σ^2. ■

Interval estimation

An *interval estimator* of a parameter θ is an interval $(\hat{\theta}_L, \hat{\theta}_U)$, with endpoints $\hat{\theta}_L$ and $\hat{\theta}_U$ that are statistics calculated from random samples. The endpoints are selected in such a way that the interval will contain the true value θ with a predetermined probability. The probability that the interval estimator constructed from an observed sample will contain the true value of the parameter is called the *confidence coefficient*. The confidence coefficient is usually expressed as $1 - \alpha$, where α is a number between 0 and 1. Thus, a confidence coefficient of 0.9 corresponds to $\alpha = 0.1$. An interval estimator with confidence coefficient $1 - \alpha$ is called a $100(1 - \alpha)\%$ confidence interval estimator or simply a $100(1 - \alpha)\%$ confidence interval.

A $100(1 - \alpha)\%$ confidence interval implies a rule (a procedure) for using information in random samples to calculate an interval estimate for the parameter of interest. If a large number of $100(1-\alpha)\%$ confidence intervals are constructed using the same procedure from a large number of random samples, then approximately $100(1-\alpha)\%$ of the intervals so constructed will contain the true value of the parameter θ. For example, approximately 95% of a large number of 95% confidence intervals can be expected to contain the true value of the parameter.

Figure 3.12 illustrates the meaning of the confidence coefficient associated with an 80% $(1 - \alpha = 0.80 \Leftrightarrow \alpha = 0.20)$ confidence interval estimator. The vertical axis represents the possible values for the parameter θ. The horizontal line at the center locates the true value of θ. Each vertical line represents one 80% confidence interval for θ. A vertical line that intersects the horizontal line corresponds to an interval that contains the true value of θ. Thus, we should expect approximately 80% of the vertical lines to intersect the horizontal line. In Figure 3.12, four out

FIGURE 3.12

Twenty 80% confidence intervals

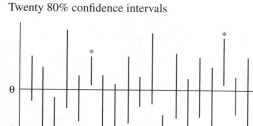

of the 20 intervals do not contain θ. These intervals are indicated by asterisks. The remaining 16 (16 out of 20 is 80%) intervals do contain θ.

Of course, a typical user of confidence intervals constructs only one interval. The confidence level selected by the user determines the probability that the interval constructed using one random sample will actually contain the true value of the parameter. The investigator wants to choose an interval with a high confidence coefficient, just as a consumer wants to buy a specific brand of refrigerator if a high percentage of that brand is known to be good. There is no guarantee that every confidence interval will contain the true value of the parameter. All we can say is that the higher the confidence level, the higher will be the probability that the interval covers the true value. For this reason, the set of values covered by a calculated confidence interval may be regarded as the set of "acceptable" or "plausible" values of the parameters of interest.

In order to see how sampling distributions can be used to construct confidence intervals, consider the case where the population mean ($\theta = \mu$) is to be estimated by the mean $\overline{Y}$ of a random sample of size n. To keep the discussion simple, assume that σ^2, the variance of the target population, is known and that the sample size is large.

A procedure for constructing a 95% confidence interval for μ can be developed as follows.

1   The Central limit theorem implies that the sampling distribution of $\overline{Y}$ is a normal distribution with mean $\mu_{\overline{Y}} = \mu$ and standard deviation $\sigma_{\overline{Y}} = \sigma/\sqrt{n}$.

2   From Table C.1 (Appendix C), it can be seen that 95% of the values in a $N(\mu, \sigma^2)$ distribution are within the interval ($\mu - 1.96\sigma$, $\mu + 1.96\sigma$). Consequently, 95% of the observed values of $\overline{Y}$ will be within the interval ($\mu - 1.96\sigma_{\overline{Y}}$, $\mu + 1.96\sigma_{\overline{Y}}$). Let's call this interval I.

3   Let $\overline{y}$ denote an observed value of $\overline{Y}$ within the interval I. Figure 3.13 shows the interval I along with two possible locations of $\overline{y}$. In Figure 3.13a, $\overline{y}$ is located below μ; in Figure 3.13b, $\overline{y}$ is above μ. Notice in Figure 3.13 that $\overline{y}$ will be

F I G U R E   3.13
Interval covering 95% of the area under the sampling distribution of $\overline{Y}$

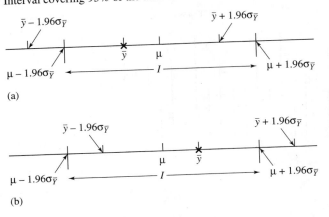

(a)

(b)

within I if and only if the distance between $\bar{y}$ and μ is less than $1.96\sigma_{\bar{Y}}$. In other words, the value $\bar{y}$ of $\bar{Y}$ will be within I if and only if μ is within the interval whose endpoints are obtained by adding and subtracting $1.96\sigma_{\bar{Y}}$ to $\bar{y}$.

4 Thus, an observed value $\bar{y}$ of $\bar{Y}$ will be within I if and only if the interval $(\bar{y} - 1.96\sigma_{\bar{Y}}, \ \bar{y} + 1.96\sigma_{\bar{Y}})$ contains μ.

5 Since 95% of the observed values of $\bar{Y}$ will be within I, then 95% of the intervals with the endpoints $\hat{\mu}_L = \bar{Y} - 1.96\sigma/\sqrt{n}$, $\hat{\mu}_U = \bar{Y} + 1.96\sigma/\sqrt{n}$ will contain the value of μ.

6 The interval with the endpoints

$$\hat{\mu}_L = \bar{Y} - z(0.025)\sigma_{\bar{Y}}, \qquad \hat{\mu}_U = \bar{Y} + z(0.025)\sigma_{\bar{Y}}$$

is a 95% confidence interval for μ.

The above arguments can easily be extended to derive expressions for the endpoints of a $100(1 - \alpha)\%$ confidence interval for μ. All that is necessary is to replace 95% and 1.96 by $100(1 - \alpha)\%$ and $z(\alpha/2)$ respectively. Therefore, just as in the case of a 95% confidence interval, we can conclude that

$$\hat{\mu}_L = \bar{Y} - z(\alpha/2)\sigma_{\bar{Y}}, \qquad \hat{\mu}_U = \bar{Y} + z(\alpha/2)\sigma_{\bar{Y}} \tag{3.12}$$

are the endpoints of a $100(1 - \alpha)\%$ confidence interval for μ.

A $100(1 - \alpha)\%$ confidence interval with two calculated endpoints, $\hat{\theta}_L$ and $\hat{\theta}_U$, is called a *two-sided interval*. A two-sided interval provides simultaneous upper and lower bounds for the parameter of interest. In some situations, we may be interested in only one bound. For example, when estimating the nitrogen loss associated with a fertilizer regimen, the primary focus usually is to see that the average N-loss is not excessive. Thus, an upper bound on the true average N-loss would be of interest to the investigator. On the other hand, in a study to evaluate the effectiveness of an insecticide, estimation of the lower bound on the average number of insects killed will be of primary interest.

A confidence interval containing only one calculated endpoint is called a *one-sided interval*. A $100(1 - \alpha)\%$ upper confidence interval is an interval $(-\infty, \hat{\theta}_U)$ with a calculated upper endpoint such that the probability that the interval contains the value of θ is $1 - \alpha$. The upper endpoint $\hat{\theta}_U$ is called a $100(1 - \alpha)\%$ upper bound for θ.

A $100(1 - \alpha)\%$ lower confidence interval is an interval $(\hat{\theta}_L, +\infty)$ with a calculated lower endpoint constructed in such a way that the interval will contain the value of θ with probability $1 - \alpha$. The lower endpoint $\hat{\theta}_L$ is called a $100(1 - \alpha)\%$ lower bound for θ.

Under the assumption that the sampling distribution of the sample mean is normal, formulas analogous to Equation (3.12) can be derived for the endpoints of $100(1 - \alpha)\%$ one-sided confidence intervals for the population mean μ. Thus, the $100(1 - \alpha)\%$ lower confidence bound (LCB) for μ is

$$\hat{\mu}_L = \bar{Y} - z(\alpha)\sigma_{\bar{Y}}; \tag{3.13}$$

the $100(1 - \alpha)\%$ upper confidence bound (UCB) for μ is

$$\hat{\mu}_U = \bar{Y} + z(\alpha)\sigma_{\bar{Y}}; \tag{3.14}$$

and the $100(1 - \alpha)\%$ confidence bounds (CB) for μ are

$$\hat{\mu}_L = \overline{Y} - z(\alpha/2)\sigma_{\overline{Y}},$$

$$\hat{\mu}_U = \overline{Y} + z(\alpha/2)\sigma_{\overline{Y}}. \tag{3.15}$$

The next example illustrates the construction and interpretation of the three types of confidence intervals.

EXAMPLE 3.17 Let's compute a 95% confidence interval for μ, the true mean percentage N-loss, in Example 3.14. For present purposes, we'll assume that the N-loss data are a random sample from a normally distributed population with variance $\sigma^2 = 4$. Of course, the assumption that the variance of the population is known is unrealistic in most practical applications. The question of what to do when σ^2 is unknown will be addressed in the next chapter.

The required confidence interval can be constructed using Equation (3.15) for the endpoints $\hat{\mu}_L$ and $\hat{\mu}_U$. We have $n = 15$, $\overline{y} = 11.15$, and $\sigma_{\overline{Y}} = 2/\sqrt{15} = 0.5164$. Also, for a 95% confidence interval, $\alpha = 0.05$, and from Table C.1 (Appendix C) we get $z(\alpha/2) = z(0.025) = 1.96$. Therefore, the endpoints of a 95% confidence interval for the true mean percentage N-loss are

$$\hat{\mu}_L = \overline{y} - z(0.025)\sigma_{\overline{Y}}$$

$$= 11.15 - 1.96(0.5164) = 10.14,$$

$$\hat{\mu}_U = \overline{y} + z(0.025)\sigma_{\overline{Y}}$$

$$= 11.15 + 1.96(0.5164) = 12.16.$$

We can conclude with 95% confidence that the true mean percentage N-loss is between 10.14% and 12.16%. We have 95% confidence in this because the procedure used for constructing the interval is expected to produce intervals containing the true mean percentage N-loss in 95% of samples of size $n = 15$ drawn from the target population. Of course, there is no guarantee that the particular interval we have constructed actually contains the true value of μ.

At the 95% confidence level, any value between 10.14% and 12.16% is an acceptable (based on the observed sample) value for the true mean N-loss rate. Any value outside the interval (10.14, 12.16) is not acceptable for μ at the 95% confidence level. Thus, 10.50% and 11% are both acceptable at the 95% confidence level as the values of μ, but 18% is not an acceptable value for the true mean percentage N-loss.

One-sided confidence intervals for μ can be constructed using Equations (3.13) and (3.14). From Table C.1 (Appendix C), we get $z(0.05) = 1.645$, so that the 95% lower and upper bounds for the true mean percentage N-loss are

$$\hat{\mu}_L = 11.15 - (1.645)(0.5164) = 10.30,$$

$$\hat{\mu}_U = 11.15 + (1.645)(0.5164) = 12.00.$$

Thus, 12.00% is a 95% upper confidence bound and 10.30% is a 95% lower confidence bound for the true percentage N-loss. In Exercise 3.32, you are asked to interpret these bounds and argue that a two-sided interval whose lower and upper endpoints are the 95% lower and upper confidence bounds for μ (for example, $10.30 \leq \mu \leq 12.00$), respectively, has confidence level 0.90. ∎

Exercises

3.25 Refer to Exercise 3.11, where you were asked to determine the sampling distributions of $\overline{Y}$ when random samples of size $n = 3$ are selected from three populations each containing six measurements.

  a Verify that the sample mean is an unbiased estimator of the population mean in all three cases.

  b Verify that the mean squared error of the sample mean is the same as the variance of the sample mean in all three cases.

3.26 Answer the following questions for the three populations referred to in Exercise 3.25.

  a For each population, check to see if the sample median is an unbiased estimator of the population mean.

  b For each population, calculate the mean squared error when the sample median is used to estimate the population mean.

  c For each population, use the appropriate mean squared errors to compare the sample mean and the sample median as estimators of the population mean.

3.27 Let $\hat{\theta}$ be an unbiased estimator of θ. Use the definitions to argue that the variance of $\hat{\theta}$ is the same as its mean squared error.

3.28 Refer to Example 3.16, where an expression was obtained for the efficiency of S'^2 relative to S^2 for estimating the variance σ^2 of a normal distribution. Calculate the efficiency of S'^2 relative to S^2 when the sample size is $n = 2, 3, 4, 5, 10, 100$, and 1000. Use the results of your calculations to argue that S'^2 is a more efficient estimator of σ^2 than is S^2.

3.29 The mean height of a random sample of $n = 15$ Douglas fir trees was $\overline{y} = 143$ cm. For the purpose of this problem, assume that the tree height population has a normal distribution with standard deviation $\sigma = 18$ cm.

  a Construct a 95% lower bound for μ, the average height of the population of all Douglas fir trees.

  b State the conclusion that can be drawn from the lower bound computed in (a).

  c On the basis of the lower bound you calculated, is it reasonable to say that the average tree height of the population is at least 130 cm? Explain.

3.30 A random sample of $n = 100$ high school students in a particular region had an average GPA of $\overline{y} = 2.84$.

  a Explain why it is reasonable to assume that the observed mean GPA of 2.84 can be regarded as a random sample of size $n = 1$ from a normal distribution. [Hint: What is the sampling distribution of $\overline{Y}$?]

  b Assume, for the purposes of this problem, that the standard deviation of the population of GPAs of all high school students is 0.7 and construct a 99% confidence interval for μ, the population mean GPA.

  c State the conclusions that can be drawn from the interval constructed in (b).

d   On the basis of the interval computed in (b), would you conclude that the population mean GPA is not equal to 2.75? Explain.

3.31   In an experiment to evaluate the toxic effect of a chemical, a toxicologist determined 15% as the 99% confidence upper bound on the fraction of the exposed population who suffer from the toxic effects of the chemical.

a   What does the 99% confidence upper bound mean in relation to the toxicity?

b   On the basis of the calculated upper bound, is it reasonable to conclude that the actual fraction suffering from toxic side effects is more than 10%? Less than 20%? Explain.

3.32   In Example 3.17, we discussed the derivation of the confidence intervals for the population mean μ when the population standard deviation is known and the sampling distribution of $\overline{Y}$ is normal. We asserted that the lower and upper endpoints of a 90% two-sided confidence interval for μ are the same as the 95% lower and upper bounds for μ, respectively.

a   Give an intuitive explanation of why this is the case.

b   Does the same relationship hold between the endpoints of a $100(1 - \alpha)\%$ two-sided confidence interval and the $100(1 - \alpha/2)\%$ upper and lower bounds? Explain.

3.9
Testing hypotheses

The following example describes a statistical hypothesis-testing problem.

EXAMPLE **3.18**   Forced expiratory volume (FEV) is a standard measure of the capacity of lungs to expel air in breathing. It is claimed that a new dietary and exercise regimen will increase the average FEV of healthy males in a particular adult age group by at least 1. The following are the changes in FEV of $n = 8$ healthy males who were put on the new regimen:

$$1.40, 1.10, 0.82, 1.50, 1.00, 0.90, 1.62, 1.52$$

Do these data support the claim?

The question can be answered by performing a statistical test of a hypothesis about μ, the mean change in FEV in the population of those following the exercise-diet regimen. In such a procedure, the objective is to see if the observed data provide sufficient evidence to conclude that the hypothesis $\mu > 1$ is true.   ■

Research and null hypotheses

Statistical hypothesis tests deal with two hypotheses: a research hypothesis, denoted by H_1; and a null hypothesis, denoted by H_0. A hypothesis that needs confirmation on the basis of observed data is called a *research hypothesis*. A research hypothesis is a hypothesis that an investigator would like to prove. In Example 3.18, the research hypothesis is H_1: the true mean increase in FEV is more than 1. In symbols, this research hypothesis can be expressed as H_1: $\mu > 1$.

A hypothesis that is true if and only if a research hypothesis is false is called a *null hypothesis*. Thus, a hypothesis that an investigator wants to refute is a null hypothesis. In Example 3.18, the null hypothesis is H_0: the true mean increase in FEV is less than or equal to 1. In symbols, the null hypothesis is H_0: $\mu \le 1$.

A test of a null hypothesis H_0 is a statistical inferential procedure that can be used to see if the observed data provide sufficient evidence to reject H_0 in favor of H_1. Just as a confidence interval has an associated confidence coefficient, a hypothesis test has its *significance level* (or simply level). The level of a hypothesis test is a number α such that the probability that the test will falsely support the research hypothesis does not exceed α. In symbols, α is a number such that

$$\Pr\left\{\text{test will reject } H_0 \text{ when } H_0 \text{ is true}\right\} \le \alpha. \tag{3.16}$$

A test whose level is α is called an α-level test.

The objective in a hypothesis test is to see if the data provide sufficient evidence to reject the null hypothesis. We regard the evidence provided by the data as sufficient if the null hypothesis can be rejected by a test for which the probability of falsely rejecting H_0 is less than a value α that we select. Thus, the level of a test can be thought of as the strength of the evidence required by the test (or by the researcher who uses the test) to reject H_0 in favor of H_1.

EXAMPLE **3.19**   For the hypothesis-testing problem in Example 3.18, we have H_0: $\mu \le 1$ and H_1: $\mu > 1$. Suppose that we decide to use a 0.05-level test of H_0. Then the probability that the test will falsely support H_1 is less than or equal to 0.05. In other words, if the true mean change in FEV is actually less than or equal to 1, there is less than 5% chance that we will conclude that the true mean change in FEV is more than 1.   ∎

A research hypothesis about a parameter θ usually has one of the three forms:

1. H_1: $\theta > \theta_0$,

2. H_1: $\theta < \theta_0$,

3. H_1: $\theta \ne \theta_0$,

where θ_0 is a specified value of the parameter. The first two hypotheses are called *one-sided hypotheses*, while the third is a *two-sided hypothesis*. The research hypothesis in Example 3.19 is a one-sided hypothesis; it is of the first type, with $\theta = \mu$ and $\theta_0 = \mu_0 = 1$.

We now describe two common approaches for testing statistical hypotheses: the confidence-interval approach and the test-statistic approach.

Hypothesis testing with confidence intervals

Suppose we want to perform an α-level test of a null hypothesis about a parameter θ. To do so, we could construct an appropriate confidence interval for θ and then reject H_0 in favor of H_1 if the set of acceptable values defined by the confidence interval supports H_1. This is known as the confidence-interval (CI) approach. As an example, consider testing H_0: $\theta \leq \theta_0$ against H_1: $\theta > \theta_0$. Since H_1 implies that θ_0 is a lower bound for θ, the support for H_1 can be evaluated by examining the acceptable values of θ as determined by the lower confidence interval for θ. If the lower confidence bound $\hat{\theta}_L$ is larger than θ_0—that is, if θ_0 is outside the lower confidence interval—then every acceptable value of θ supports H_1, as we see in Figure 3.14a. In that case, we can reject H_0 and conclude that H_1 is true. However, if $\hat{\theta}_L$ is less than θ_0—that is, if θ_0 is within the lower confidence interval, then some of the acceptable values of θ support H_1 and others support H_0, as in Figure 3.14b. Consequently, on the basis of the confidence interval we have constructed, all we can say is that the data do not provide sufficient evidence to conclude in favor of H_1.

F I G U R E   **3.14**

Relationship of lower confidence bound (LCB) to H_1

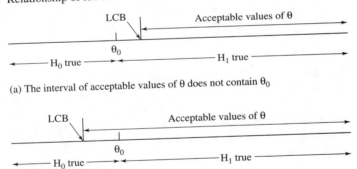

(a) The interval of acceptable values of θ does not contain θ_0

(b) The interval of acceptable values of θ does contain θ_0

It can be shown that, if the lower confidence bound has confidence coefficient α, the level of the hypothesis test based on this lower bound will also be α. In other words, the probability that θ_0 will be outside the $100(1 - \alpha)\%$ lower confidence interval for θ will be less than α for all values of θ for which H_0 is true.

As an illustration, suppose that we want to test the null hypothesis H_0: $\mu \leq \mu_0$, where μ is the mean of a normally distributed population with a known standard deviation σ. Let $\overline{Y}$ be the mean of a random sample from the population, and consider the test procedure that rejects H_0 if μ_0 is less than the $100(1 - \alpha)\%$ lower confidence bound in Equation (3.13)

$$\hat{\mu}_L = \overline{Y} - z(\alpha)\sigma_{\overline{Y}}.$$

In this case, the test will reject H_0 if

$$\mu_0 < \overline{Y} - z(\alpha)\sigma_{\overline{Y}}$$

or equivalently if

$$\mu_0 + z(\alpha)\sigma_{\overline{Y}} < \overline{Y}.$$

The probability that this test will reject H_0 is the same as the probability that the observed value of $\overline{Y}$ is larger than $\mu_0 + z(\alpha)\sigma_{\overline{Y}}$. Since the sampling distribution of $\overline{Y}$ is a normal distribution, the probability that the test will reject H_0 can be expressed as an area under a normal distribution with mean μ and standard deviation $\sigma_{\overline{Y}} = \sigma/\sqrt{n}$. Figure 3.15 shows this probability for two values of μ: $\mu = \mu_0$ and $\mu = \mu_1$ ($\mu_1 < \mu_0$).

FIGURE   3.15

Probabilities of rejecting H_0

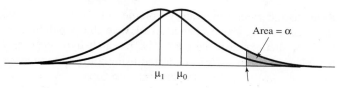

In each case, the shaded area under the corresponding distribution is the required probability. The following conclusions can be drawn from the figure.

1   If the true mean of the population is equal to μ_0, the probability of rejecting H_0 is α. This probability, which equals the shaded area under the normal distribution with mean μ_0, is the area under a normal distribution to the right of $z(\alpha)$ standard deviations above the population mean.

2   If the true population mean is μ_1, where $\mu_1 < \mu_0$, then a comparison of the shaded area under the normal distribution with mean μ_0 to the corresponding area under the normal distribution with mean μ_1 shows that the probability of rejecting H_0 is less than α.

It follows that, if the population mean μ is less than or equal to μ_0—that is, if H_0 is true—then the probability of rejecting H_0 is less than α, so that the test has level α.

The confidence interval procedure for testing the one-sided null hypothesis H_0: $\theta \le \theta_0$ can be modified in a straightforward manner to derive procedures for testing H_0: $\theta \ge \theta_0$ or H_0: $\theta = \theta_0$. These procedures are described in Box 3.4.

BOX 3.4

Confidence-interval approach for testing hypotheses

The following criteria may be formulated for level-α tests of null hypotheses about θ based on confidence intervals.

H_0	H_1	Reject H_0 and accept H_1 if
$\theta \leq \theta_0$	$\theta > \theta_0$	θ_0 is outside the $100(1-\alpha)\%$ lower confidence interval for θ
$\theta \geq \theta_0$	$\theta < \theta_0$	θ_0 is outside the $100(1-\alpha)\%$ upper confidence interval for θ
$\theta = \theta_0$	$\theta \neq \theta_0$	θ_0 is outside the $100(1-\alpha)\%$ confidence interval for θ

The next example illustrates how the confidence-interval approach can be used to test hypotheses.

EXAMPLE 3.20   Refer to the forced expiratory volume (FEV) data in Example 3.18. As before, let μ denote the mean change in FEV in the population of subjects who follow the new diet-exercise regimen. The research hypothesis is H_1: $\mu > 1$. The null hypothesis H_0: $\mu \leq 1$ can be tested using a lower confidence interval for μ. From the results in Box 3.4 (in the present case, $\mu = \theta$, $\theta_0 = 1$), we see that a 0.05-level test will reject H_0 in favor of H_1 if 1 is outside the 95% lower confidence interval for μ.

Let's assume that the FEV data are a random sample from a normally distributed population with a known standard deviation $\sigma = 0.30$; the case of unknown σ will be considered in the next chapter. Equations (3.13)–(3.15) can be used to construct $100(1-\alpha)\%$ confidence intervals for μ. From the given data, we have $n = 8$, and $\bar{y} = 1.2325$, and $\sigma_{\bar{y}} = 0.30/\sqrt{8} = 0.1061$, so that a 95% lower bound for μ is

$$\bar{y} - z(0.05)\sigma_{\bar{y}} = 1.2325 - (1.645)(0.1061) = 1.058.$$

Therefore, a 95% lower confidence interval for μ is $1.058 \leq \mu < +\infty$, so that all acceptable (at 95% confidence level) values are at least as large as 1.058. The value of $\mu = 1$ is outside the 95% lower confidence interval for μ, so that the computed lower confidence interval supports the conclusion that the true mean change in FEV is larger than 1. In other words, on the basis of the observed data, we can reject H_0: $\mu \leq 1$ and accept H_1: $\mu > 1$ at the level of 0.05. The level $\alpha = 0.05$ implies that there is less than 5% probability that the procedure we have used will lead us to accept the research hypothesis H_1: $\mu > 1$ when in fact the null hypothesis H_0: $\mu \leq 1$ is true.

Before concluding this example, note that the computed 95% lower confidence bound of 1.058 implies that the observed data will support, at the 0.05 level, any research hypothesis of the form H_1: $\mu > \mu_0$ provided $\mu_0 < 1.058$. For example, at the 0.05 level, the data support the conclusion that the true mean change in FEV is at least as large as 1.02. ∎

Hypothesis testing with test statistics

In the test-statistic (TS) approach, the hypothesis test is performed by comparing the calculated value of an appropriately chosen statistic, called a *test statistic*, to an appropriate critical value.

For example, consider the problem of testing H_0: $\mu \leq \mu_0$ on the basis of a random sample of size n from a normal distribution with mean μ and a known standard deviation σ. Let $\overline{Y}$ denote the sample mean. As a possible test statistic, consider $\overline{Y} - \mu_0$, the difference between an estimate of μ and the cutoff value μ_0 that differentiates between the null and research hypotheses. Notice that a large positive observed value for this difference would suggest that H_0 should be rejected, so that an intuitively reasonable hypothesis-testing procedure is to reject H_0 if the observed value of $\overline{Y} - \mu_0$ is large. In other words, the procedure would reject the null hypothesis if the observed value of $\overline{Y} - \mu_0$ is larger than a critical value. The critical value is selected in such a way that the resulting test procedure has the desired level. In practice, it is convenient to express the test statistic as a z-score

$$Z = \frac{\overline{Y} - \mu_0}{\sigma_{\overline{Y}}}. \tag{3.17}$$

Since large values of $\overline{Y} - \mu_0$ correspond to large values of Z, the hypothesis test will reject H_0 if the observed value of Z is larger than a preset critical value.

In Exercise 3.35, you will be asked to show that the test procedure that rejects H_0 if the observed value of Z is larger than $z(\alpha)$ is the same as an α-level test of H_0 based on a lower confidence bound for μ. Consequently, the test that rejects H_0 if the calculated value of Z is larger than $z(\alpha)$ is a level-α test of H_0.

Let z_c be the observed value of Z. An α-level test of H_0: $\mu \leq \mu_0$ vs. H_1: $\mu > \mu_0$ will reject H_0 if

$$z_c > z(\alpha).$$

Otherwise, the test will conclude that the data do not provide sufficient evidence to reject the null hypothesis.

EXAMPLE 3.21 In Example 3.20, we used the data on FEV change in $n = 8$ subjects and the confidence-interval approach to test the null hypothesis H_0: $\mu \leq 1$, where μ is the mean change in FEV in the study population. Let's use the test-statistic approach to test H_0: $\mu \leq 1$ against H_1: $\mu > 1$ at $\alpha = 0.05$.

Recall that, for the given data, $n = 8$, $\overline{y} = 1.2325$, and $\sigma_{\overline{y}} = 0.1061$. Therefore, the observed value of the test statistic Z is

$$z_c = \frac{1.2325 - 1}{0.1061} = 2.193.$$

The test will reject H_0 at level $\alpha = 0.05$ if the observed value 2.193 exceeds the critical value $z(0.05) = 1.645$. Since 2.193 is larger than 1.645, we can reject H_0 and conclude that $\mu > 1$. ∎

Figure 3.16a shows the observed values of the test statistic Z that will lead to rejection of H_0: $\mu \leq \mu_0$ by an α-level test. Just as a large observed value of Z supports the rejection of H_0: $\mu \leq \mu_0$, a small observed value of Z supports the rejection H_0: $\mu \geq \mu_0$. Consequently, a hypothesis test based on Z will reject H_0: $\mu \geq \mu_0$ in favor of H_1: $\mu < \mu_0$ if the observed value of Z is less than a critical value. The appropriate critical value for rejecting H_0: $\mu \geq \mu_0$ is $-z(\alpha)$, as shown in Figure 3.16b.

FIGURE   3.16

Observed Z that will reject H_0 at level α

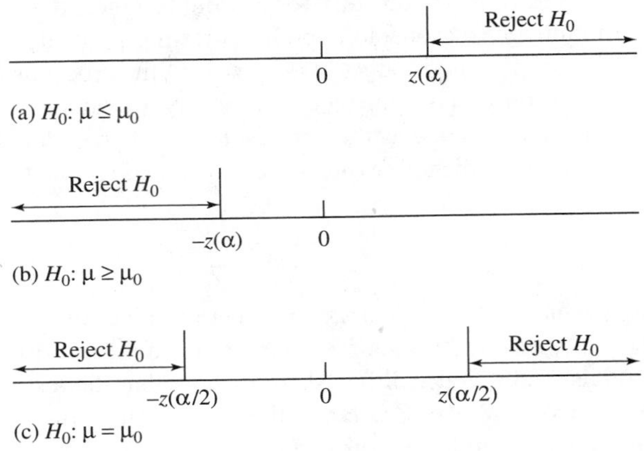

(a) H_0: $\mu \leq \mu_0$

(b) H_0: $\mu \geq \mu_0$

(c) H_0: $\mu = \mu_0$

For a two-sided research hypothesis, both a large positive value and a large (in magnitude) negative value of Z will support H_1. Figure 3.16c shows the values of Z that will lead to rejection of H_0: $\mu = \mu_0$ in favor of H_1: $\mu \neq \mu_0$ at level α. As might be expected, H_0: $\mu = \mu_0$ can be rejected at level α if the calculated value z_c of Z is less than $-z(\alpha/2)$—that is, if Z has a small observed value—or if z_c is larger than $z(\alpha/2)$.

The test-statistic approach to testing hypotheses regarding the mean of a normally distributed population (with a known standard deviation) can be used to test hypotheses about an arbitrary parameter θ if it is possible to find an estimator $\hat{\theta}$ whose sampling distribution is a normal distribution with a mean θ and a known variance $\sigma_{\hat{\theta}}^2$. The test procedure is summarized in Box 3.5.

BOX **3.5**

Test-statistic approach for testing hypotheses about θ on the basis of a normally distributed estimate $\hat{\theta}$

Let $\hat{\theta}$ be an estimator of θ such that the sampling distribution of $\hat{\theta}$ is a normal distribution with a mean θ and a known variance $\sigma_{\hat{\theta}}^2$. Then α-level tests of hypotheses about θ can be performed as follows.

---

1 Calculate the test statistic

$$z_c = \frac{\hat{\theta} - \theta_0}{\sigma_{\hat{\theta}}}.$$

2 a Reject H_0: $\theta \le \theta_0$ in favor of H_1: $\theta > \theta_0$ at level α if $z_c > z(\alpha)$.
    b Reject H_0: $\theta \ge \theta_0$ in favor of H_1: $\theta < \theta_0$ at level α if $z_c < -z(\alpha)$.
    c Reject H_0: $\theta = \theta_0$ in favor of H_1: $\theta \ne \theta_0$ at level α if $z_c > z(\alpha/2)$ or $z_c < -z(\alpha/2)$.

Note that the tests of hypotheses regarding the mean of a normal distribution in Example 3.21 and Figure 3.16 can be obtained by setting $\theta = \mu$, $\hat{\theta} = \overline{Y}$, and $\sigma_{\hat{\theta}}^2 = \sigma^2/n$ in the procedures described in Box 3.5.

Confidence interval vs. test statistic approach to hypothesis testing

The advantages and disadvantages associated with the test-statistic approach and the confidence-interval approach are different. An advantage of the test-statistic approach is that, once the value of the test statistic has been calculated, it is simple to perform tests at different levels. The researcher simply compares the calculated value of the test statistic to an appropriate critical value; no new calculations will be necessary. Thus, in Example 3.21, if we wish to perform the test at the 0.01 level instead of the 0.05 level, we compare the calculated value of the test statistic, $z_c = 2.193$, to the 0.01-level critical value, $z(0.01) = 2.326$. The calculated value is less than the critical value, and we conclude that the data do not support (at 0.01 level) the research hypothesis that $\mu > 1$. The confidence-interval approach, on the other hand, requires a new interval every time the significance level of the test is changed.

A disadvantage of the test-statistic approach can be seen by comparing the results in Examples 3.20 and 3.21. In both examples, we tested the same hypotheses— H_0: $\mu \le 1$ and H_1: $\mu > 1$—at the same level of significance: $\alpha = 0.05$. As we might expect, both the confidence-interval approach and the test-statistic approach indicated that, at the 0.05 level, the data support the conclusion that $\mu > 1$. However, the confidence-interval approach also allowed us to conclude, with 95% confidence, that μ is larger than 1.058. Thus, the confidence-interval approach provided us with a better summary of the information contained in the observed sample. Generally speaking, the test-statistic approach for hypothesis testing concerns itself with whether the null hypothesis can be rejected on the basis of the observed data; it does not address the broader issue of estimating the true value of the parameter(s) of interest.

The attained level (*p*-value) of a hypothesis test

The use of a hypothesis test as an inferential procedure requires us to select the significance level at which the test is to be performed. The choice of the level in a particular test is a subjective matter and depends mainly on the desired strength of evidence needed to reject the null hypothesis. A useful method of summarizing the results of a statistical test without actually committing to a specific significance level is to calculate a quantity known as the attained level of the hypothesis test.

The *attained level* of a hypothesis test is the smallest significance level (value of α) at which the test will reject the null hypothesis. The attained level of a test is also called the *p-value* of the test.

EXAMPLE 3.22   Let's determine the *p*-value of the test of the hypotheses H_0: $\mu \leq 1$ vs. H_1: $\mu > 1$ described in Example 3.21. Recall that the calculated value of the test statistic was $z_c = 2.193$, so that the test will reject the null hypothesis at level α if $2.193 \geq z(\alpha)$.

From Table C.1 (Appendix C), the area to the right of 2.193 under a $N(0, 1)$ distribution is 0.0143; this is the shaded area in Figure 3.17a. Also, it can be seen from Figure 3.17a that, if $\alpha \geq 0.0143$, then $z(\alpha) \leq 2.193$. Thus, the null hypothesis will be rejected by an α-level test if $\alpha \geq 0.0143$. Similarly, the null hypothesis will not be rejected by an α-level test if $\alpha < 0.0143$. Consequently, the smallest α at which the test will reject H_0 is $\alpha = 0.0143$, so that the *p*-value of the test is $p = 0.0143$. It follows that an α-level test will conclude in favor of H_1 if $\alpha \geq 0.0143$ and conclude that there is not enough evidence to support H_1 if $\alpha < 0.0143$.

The *p*-value of the two-sided research hypothesis H_1: $\mu \neq 1$ is shown as the shaded area in Figure 3.17b. Note that both large positive values and large (in magnitude) negative values of z_c favor H_1 in this case. Reasoning similar to that in the one-sided case shows that the *p*-value equals twice the *p*-value obtained for testing H_0: $\mu \leq 1$ against H_1: $\mu > 1$. Thus, the *p*-value is $2 \times 0.0143 = 0.0286$. ∎

In certain situations (see the binomial test in Box 6.2 and Fisher's exact test in Box 6.6), the calculation of a *p*-value is not as straightforward as in Example 3.22. In such cases, it is helpful to note that a *p*-value is the probability, under the null hypothesis, of selecting a sample that is at least as favorable to the research hypothesis as the observed sample. Consider Example 3.23.

EXAMPLE 3.23   Let's look again at the one-sided test of H_0: $\mu \leq 1$ and H_1: $\mu > 1$ in Example 3.22. The observed sample gave a value of $z_c = 2.193$ to the test statistic

$$Z = \frac{(\bar{Y} - \mu_0)}{\sigma_{\bar{Y}}} = \frac{(\bar{Y} - 1)}{0.1061}. \tag{3.18}$$

Recall that large values of the test statistic favor H_1. Therefore, any observed value of Z that is larger than 2.193 corresponds to a sample that is at least as favorable to the research hypothesis as the observed sample. Also, the results in Box 3.2 imply that, if $\mu = 1$, the sampling distribution of Z is a $N(0, 1)$-distribution. Consequently, the area to the right of 2.193 under a $N(0, 1)$-distribution equals the probability, if H_0 is true, of observing a sample at least as favorable to H_1 as the actually observed sample. Recall that, in Example 3.22, we determined the *p*-value of the test to be equal to this area. ∎

FIGURE 3.17

Relationship between the p-value and z_c

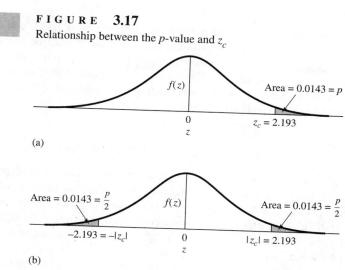

(a)

(b)

There are two ways to interpret a p-value. In the first interpretation, the p-value of a test is the smallest significance level at which the null hypothesis will be rejected by the test. For example, a p-value of 0.04 would mean that any α-level test for which $\alpha \geq 0.04$ will conclude in favor of the research hypothesis. An α-level test with $\alpha < 0.04$ will conclude that the observed data do not support the research hypothesis. Thus, a 0.03-level test will conclude that the data do not support the research hypothesis, whereas a 0.05-level test will reject the null hypothesis in favor of the research hypothesis.

In the second interpretation, the number $1 - p$ is regarded as an index of the strength of the evidence against the null hypothesis provided by the data. The smallest and the largest possible values of this index are 0 and 1, respectively. Thus, a p-value of 0.01, which corresponds to an index of 0.99, strongly supports rejection of the null hypothesis, while a p-value of 0.60, which corresponds to an index of 0.40, indicates very little support for rejecting H_0.

The power of a hypothesis test

The significance level of a hypothesis test determines the degree of protection we have against the possibility that the test will declare incorrectly that the research hypothesis is true. By keeping the significance level low, we make sure that the probability of concluding in favor of a false research hypothesis is small. The conclusion of a test that incorrectly declares in favor of the research hypothesis is called a *Type I error*. Thus, the probability of a Type I error in a level-α test is at most α.

$$\Pr\{\text{Type I error in a level-}\alpha\text{ test}\} \leq \alpha. \tag{3.19}$$

In practice, it is critical to control the Type I error rate because hypothesis testing procedures are designed with emphasis on proving a research hypothesis beyond reasonable doubt. On the other hand, practitioners would also want to ensure that the probability of a test procedure concluding in favor of a true research hypothesis must also be high. The probability that a test will accept a true research hypothesis is called the *power of the test*. The failure of a statistical test to accept a true research hypothesis is called a *Type II error*. Traditionally, the symbol β is used to denote the probability of a Type II error. Thus

$$\Pr\{\text{Type II error}\} = \beta, \tag{3.20}$$

$$\text{Power of a test} = 1 - \Pr\{\text{Type II error}\} = 1 - \beta. \tag{3.21}$$

Of course, in an ideal hypothesis-testing procedure, both Type I and Type II error probabilities are zero: $\alpha = 0$ and $\beta = 0$. Unfortunately, as long as statistical inferences are based on samples, zero error probabilities cannot be guaranteed for any statistical test procedure. In practice, proper selection of the significance level α will allow the Type I error rate of a statistical test procedure to be kept below any desired value. In contrast, as illustrated in Example 3.24, the Type II error rate and, therefore, the power of the test depend on the true value of the parameter under test.

EXAMPLE **3.24**   In Example 3.20, we used data on forced expiratory volume (FEV) for a random sample of $n = 8$ subjects to test the null hypothesis H_0: $\mu \leq 1$ regarding μ, the mean FEV in the population of all who follow the new diet-exercise regimen. Let $\overline{Y}$ be the mean of the observed sample. Then the test procedure will reject H_0 in favor of H_1: $\mu > 1$ at significance level $\alpha = 0.05$ if the observed value of the 95% lower bound

$$\hat{\mu}_L = \overline{Y} - z(0.05)\sigma_{\overline{Y}} = \overline{Y} - (1.645)(0.1061)$$

$$= \overline{Y} - 0.1745$$

exceeds 1. Therefore, the test will reject H_0 if $\overline{Y} - 0.1745 \geq 1$ or, equivalently, if $\overline{Y} > 1.1745$. Thus, the probability that this test will reject the null hypothesis is the same as the probability that, in a random sample of $n = 8$ FEV values, the sample mean $\overline{Y}$ exceeds 1.1745. For a μ such that H_0 is true ($\mu \leq 1$), this probability equals the probability of a Type I error; for a μ such that H_1 is true ($\mu > 1$), this probability equals the power of the test.

Now, the probability of observing a sample with a mean greater than 1.1745 can be calculated on the basis that the sampling distribution of $\overline{Y}$ is a normal distribution, with mean μ and standard deviation $\sigma_{\overline{Y}} = 0.1061$. Indeed, for any given μ, the power of the test equals the area to the right of 1.1745 under a $N(\mu, 0.1061)$-distribution. As seen in Section 2.6, Table C.1 (Appendix C) can be used to determine this probability for any given value of μ. For instance, it can be verified that, for $\mu = 1.25$, the power of the test equals 0.7616. Thus, if the true population mean is 1.25, the probability that the test will conclude in favor of the hypothesis $\mu > 1$ is 0.76 (76%).

Table 3.6 gives the probability that the test will reject H_0 for selected values of μ. These probabilities were determined using the SAS procedure called PROBNORM. Two key properties of the test procedure can be gleaned from the probabilities in Table 3.6. First, the probability of rejecting the null hypothesis is less than 0.05 if

TABLE   3.6

Probability of rejecting H_0

μ	Probability
0.80	0.00021
0.90	0.00484
1.00	0.05002
1.10	0.24129
1.20	0.59497
1.25	0.76164
1.30	0.88156
1.40	0.98322
1.50	0.99892

the null hypothesis is true. Second, the power of the test increases with increase in μ. In other words, the larger the true mean FEV, the more likely it is that the test will reject H_0: $\mu \leq 1$. Figure 3.18 illustrates why the probability that the test will reject H_0 increases with μ. The figure shows the powers of the test as shaded areas under normal distributions for two values μ_1, μ_2 of μ. The shaded areas equal the probabilities of $\overline{Y}$ exceeding 1.1745 in the two cases.   ■

FIGURE   3.18

Probability of rejecting H_0 at two distinct values of μ

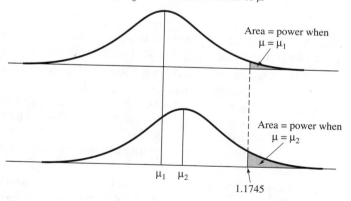

In practice, two criteria are used when selecting a statistical test procedure. First and foremost, a statistical test must have a Type I error rate equal to or less than a specified value (for instance, $\alpha = 0.05$). Second, if the true value of the parameter differs sufficiently from the values specified under the null hypothesis, then the power of the test must be higher than a specified value; correspondingly, the Type II error rate must be lower than a specified value. Example 3.25 illustrates how the criteria for controlling Type I and Type II errors are applied in practice.

EXAMPLE **3.25**   Refer to Example 3.14. Let μ denote the true average percentage N-loss over a 16-week period when Urea + N-Serve is used as fertilizer for sugarcane plants. Suppose that we want to test the research hypothesis H_1: $\mu < 13\%$. In designing a study with this goal, we must ensure that the data collected will enable us to perform a statistical test of the null hypothesis H_0: $\mu \geq 13\%$ against the research hypothesis H_1: $\mu < 13\%$. The test procedure should be at a preset significance level and should have sufficient power to declare in favor of the research hypothesis if the actual value of μ is markedly lower than 13%. For example, we might require the test to have the following properties.

1   The probability of Type I error is 0.05; that is, $\alpha = 0.05$.

2   If the true value of μ is sufficiently distinct from the null values ($\mu \geq 13\%$)—say, $\mu < 10\%$—then the Type II error of the test is less than 0.20; that is, $\beta < 0.20$ if $\mu < 10\%$.

The first property ensures that the test has less than 5% chance of incorrectly concluding that the true mean percentage N-loss is less than 13%. The second property guarantees that, if the true mean percentage N-loss is less than 10%, the test has at least 80% (power $= 1 - \beta$) chance of concluding that H_1: $\mu < 13\%$ is true.   ∎

Some methods for designing studies that provide tests of adequate power will be described in Chapter 7.

Exercises

3.33   Refer to Exercise 3.29, where you were asked to construct a 95% lower confidence bound for μ, the average height of a population of Douglas fir trees.

    a   Explain why the calculated lower bound can be used to test the null hypothesis H_{01} that the population mean height is no larger than 190 cm but not the null hypothesis H_{02} that the population mean height is at least 100 cm.

    b   State, giving reasons, the conclusion about H_{01} that can be drawn from the lower bound calculated in Exercise 3.29.

    c   Use the confidence interval approach to test the null hypothesis H_{02} at a level $\alpha = 0.01$.

3.34   Refer to Exercise 3.30, where you were asked to construct a 99% confidence interval for μ, the population mean GPA of high school students in a particular region.

    a   State the null and research hypotheses that can be tested using the confidence interval.

    b   State the conclusions that can be drawn about the null hypothesis in (a).

3.35   As we have seen, in the confidence-interval approach to testing hypotheses regarding the mean μ of a normal population, the null hypothesis H_0: $\mu \leq \mu_0$ is rejected in favor of the research hypothesis H_1: $\mu > \mu_0$ if the lower bound $\hat{\mu}_L$ is at least as large as the cutoff value μ_0.

a   Verify that

$$\mu_0 \leq \hat{\mu}_L \quad \text{if and only if} \quad Z = \frac{\overline{Y} - \mu_0}{\sigma_{\overline{Y}}} \geq z(\alpha).$$

Hence, argue that, for given data and significance level, the conclusions based on the confidence-interval approach to testing H_0 will be the same as the conclusions based on the test-statistic approach.

b   Following a line of reasoning similar to that in (a), verify that an α-level test will reject the null hypothesis $H_0: \mu \geq \mu_0$ in favor of $H_1: \mu < \mu_0$ if the observed value of Z in (a) is less than $-z(\alpha)$.

3.36   Refer to the nitrogen loss (N-loss) data in Example 3.14. Assume that the data are a random sample from a normal distribution with variance $\sigma^2 = 4$.

a   Use the test-statistic approach to see if the observed sample supports the conclusion that the population mean N-loss is less than 13%. Perform the test at significance level $\alpha = 0.05$.

b   Calculate the p-value of the test in (a) and interpret it.

3.37   Refer to the hypothesis-testing procedure in Exercise 3.36.

a   Calculate the power of a 0.05-level test when the true mean N-loss μ is equal to 12%, 11%, 10%, and 9%.

b   Interpret the numbers you calculated in (a).

c   Plot the numbers you calculated in (a) against the values of μ and comment on the pattern of changes in power as a function of μ.

d   Repeat the calculation in (a) for a 0.01-level test, and plot the power as in (c). Compare the power of the 0.05-level test to that of the 0.01-level test. Which test has higher power? Explain, giving intuitive reasons, why the power of the test at 0.05 level is consistently higher than that at 0.01 level.

3.38   Refer to Exercise 3.34.

a   Determine the p-value associated with the test of the null hypothesis $H_0: \mu = 2.75$.

b   On the basis of that p-value, what conclusion can be drawn at significance level $\alpha = 0.01$? At $\alpha = 0.0001$?

3.10

Predicting future values

Instead of making an inference about a population parameter, we might be interested in predicting the actual value of a future observation from the population. Let's illustrate that distinction by means of an example. A farmer evaluating a new fertilizer is interested in the average yield that it produces. The farmer wants to make an inference about μ, the average of the population of yields from all plants treated with the fertilizer. The yield from any single plant is of no concern, as long as the average for a large number of plants is sufficiently high. By contrast, a gardener who plans to use the fertilizer on a single plant doesn't care about the average yield of the population of all plants. The gardener wants to know what yield the fertilizer would provide for that one plant. In other words, whereas the farmer is interested in making an inference about the population parameter μ, the gardener would like to make an inference about the observed value of a single future observation from the population of yields. Similarly, a laboratory manager will be interested in the average quality of the chemicals supplied by a manufacturer, whereas a chemist in the lab is concerned about the single sample of the chemical he or she is about to use.

A $100(1 - \alpha)\%$ *prediction interval* for a future observation Y is an interval (Y_L, Y_U) such that it will contain Y with probability $1 - \alpha$. The $100(1 - \alpha)\%$ one-sided prediction intervals $(-\infty, Y_U)$ and $(Y_L, +\infty)$, with the associated upper and lower prediction bounds Y_U and Y_L, respectively, are defined in a similar manner.

Formulas for computing prediction intervals can be derived analagously to formulas for confidence intervals. As an example, suppose we have a random sample of size n from a normally distributed population with a known variance σ^2 and that we are interested in constructing a prediction interval for a future observation Y from the same population. Prediction intervals for Y can be constructed on the basis that (see Example B.7 in Appendix B) the sampling distribution of the difference $Y - \overline{Y}$, where $\overline{Y}$ is the sample mean, is a normal distribution with a mean of zero and a variance $\sigma^2 + \sigma_{\overline{Y}}^2 = \sigma^2 + \sigma^2/n$. In Exercise 3.39, you are asked to argue that, since $100(1 - \alpha)\%$ of the observed values of $Y - \overline{Y}$ will be within the interval

$$\left(0 - z(\alpha/2)\sqrt{\sigma^2 + \frac{\sigma^2}{n}}, \ \ 0 + z(\alpha/2)\sqrt{\sigma^2 + \frac{\sigma^2}{n}} \right),$$

a $100(1 - \alpha)\%$ prediction interval for Y has the endpoints

$$Y_L = \overline{Y} - z(\alpha/2)\sqrt{\left(1 + \frac{1}{n}\right)\sigma^2},$$

$$Y_U = \overline{Y} + z(\alpha/2)\sqrt{\left(1 + \frac{1}{n}\right)\sigma^2}. \tag{3.22}$$

Similar reasoning can be used to show that the $100(1 - \alpha)\%$ lower and upper prediction bounds for Y are

$$Y_L = \overline{Y} - z(\alpha)\sqrt{\left(1 + \frac{1}{n}\right)\sigma^2}, \tag{3.23}$$

$$Y_U = \overline{Y} + z(\alpha)\sqrt{\left(1 + \frac{1}{n}\right)\sigma^2}. \tag{3.24}$$

EXAMPLE 3.26   For a random sample of $n = 15$ subjects in the age group 20–30 in a particular population, the systolic blood pressures had a mean of $\overline{y} = 135$ mm Hg. We are interested in predicting the systolic blood pressure of a single subject selected at random from this particular population. Suppose that, in the population from which the sample was selected, the systolic blood pressures have a normal distribution with standard deviation $\sigma = 22$. Then, according to Equation (3.22), the $100(1 - \alpha)\%$

prediction interval for a future observation has the endpoints

$$Y_L = \overline{Y} - z(\alpha/2)\sqrt{\left(1 + \frac{1}{n}\right)\sigma^2},$$

$$Y_U = \overline{Y} + z(\alpha/2)\sqrt{\left(1 + \frac{1}{n}\right)\sigma^2}.$$

Using the values $n = 15$, $\overline{y} = 135$, $\sigma = 22$, and $\alpha = 0.05$, we get

$$Y_L = 135 - 1.96\sqrt{\left(1 + \frac{1}{15}\right)(22)^2} = 90.5 \text{ mm Hg}$$

and similarly

$$Y_U = 135 + 44.53 = 179.5 \text{ mm Hg}.$$

Therefore we can predict with 95% confidence that a randomly selected subject from the population will have a systolic blood pressure between 90.5 mm Hg and 179.5 mm Hg. ∎

Comparing the confidence and prediction intervals

Let's look at the similarities and differences between a confidence interval for a parameter and a prediction interval for a future observation. Both types of intervals provide bounds for unknown quantities. Both, being calculated from sample values, are subject to error, but we can ensure their validity (by appropriate selection of α) at any prescribed confidence level. A confidence interval provides bounds for an unknown population parameter, whereas a prediction interval gives bounds for a future observation from the population. Thus, a confidence interval is an interval for an unknown fixed quantity, while a prediction interval is an interval for an unknown random entity.

Exercises

3.39 Let $\overline{Y}$ and Y denote, respectively, the mean of a random sample of size n and a future observation from a normal distribution with known variance σ^2. On the basis that the sampling distribution of the statistic $Y - \overline{Y}$ is a normal distribution with zero mean and variance $(1 + 1/n)\sigma^2$, derive Equations (3.22)–(3.24) for the endpoints of the prediction intervals for Y.

3.40 Refer to Exercise 3.29, where you were asked to calculate a 95% lower bound for the average height of the population of Douglas fir trees in a region.

a Use the given data to determine a 95% lower bound for a tree that has been randomly selected from the region.

b Interpret the bound calculated in (a).

c Notice that the 95% lower prediction bound in (a) is less than the 95% lower confidence bound you calculated in Exercise 3.29. Examine the formulas for computing the two lower bounds and argue that, for a given data and given confidence level, the lower prediction bound will never be larger than the lower confidence bound. Give an intuitive explanation for this relationship.

3.41   Refer to Exercise 3.30, where you were asked to calculate a 99% confidence interval for the mean GPA of high school students in a particular region.

a   Construct a 99% prediction interval for the GPA of a high school student selected at random from the population of high school students in the region. Interpret the interval you construct.

b   Compare the prediction interval with the corresponding confidence interval and explain (on the basis of the computing formulas and intuitive considerations) why the prediction interval is wider than the confidence interval.

3.11
The role of normal distributions in statistical inference

As you may have surmised, the normal distribution plays a key role in many statistical inference-making procedures. The normal and the log-normal distributions serve as good models for the distributions of many target populations commonly encountered in practice. In addition, the normal distribution can also be used to approximate the sampling distributions of a variety of frequently used statistics. In Section 3.8, we considered the use of normal distributions to describe the sampling distributions of sample means. In Chapter 9, we will see how normal distributions can be used to approximate the sampling distributions of statistics that are linear combinations of sample means. An important example of a linear combination of sample means is $\overline{Y}_1 - \overline{Y}_2$, the difference between the means of two independent samples. As an illustration, the sampling distribution of $\overline{Y}_1 - \overline{Y}_2$ is described in Box 3.6.

BOX **3.6**   ***The sampling distribution of $\overline{Y}_1 - \overline{Y}_2$***

Let $\overline{Y}_1$ and $\overline{Y}_2$ be the means of independent random samples of sizes n_1 and n_2 from populations with means and variances μ_1, μ_2 and σ_1^2, σ_2^2, respectively. Then we can state the following properties of the sampling distribution of $\overline{Y}_1 - \overline{Y}_2$.

I   The mean and variance of the sampling distribution of $\overline{Y}_1 - \overline{Y}_2$ are, respectively

$$\mu_{\overline{Y}_1 - \overline{Y}_2} = \mu_1 - \mu_2$$

$$\sigma_{\overline{Y}_1 - \overline{Y}_2}^2 = \frac{\sigma_1^2}{n_1} + \frac{\sigma_2^2}{n_2}.$$

II   If both n_1 and n_2 are large (as a rule of thumb, $n_1, n_2 \geq 30$), the sampling distribution of $\overline{Y}_1 - \overline{Y}_2$ is approximately a normal distribution with mean $\mu_{\overline{Y}_1 - \overline{Y}_2}$ and variance $\sigma^2_{\overline{Y}_1 - \overline{Y}_2}$. The normal approximation improves as the sample sizes increase.

III If both population distributions are normal, the sampling distribution of $\overline{Y}_1 - \overline{Y}_2$ is also normal.

A comparison of the results in Box 3.6 with those in Box 3.2 reveals many similarities between the sampling distribution of the mean of a single random sample and that of the difference between the means of two independent random samples. Both sampling distributions are normal under very similar conditions (for instance, when populations are normal or sample sizes are large) and both have means and variances that are determined by sample sizes and the means and variances of the target populations. As we shall see in Chapter 9, results similar to those in Boxes 3.2 and 3.6 are applicable under more general conditions.

We already know that the sampling distributions of statistics play a key role in the development and implementation of statistical inferential procedures and that normal distributions can be used (at least approximately) to describe the sampling distributions of a variety of statistics. On that basis, a natural first step in discussing statistical inferential procedures is to consider the case where the inferences can be based on a normally distributed statistic. As we will see, many of the commonly used statistical inferential procedures correspond to intuitively obvious modifications in the procedures based on normally distributed statistics.

In Sections 3.8 and 3.9, we considered the case of a sample mean $\overline{Y}$ that has a normal distribution with a known variance $\sigma_{\overline{Y}}$ and described how to construct confidence intervals and hypothesis tests for the population mean μ. In Exercise 3.42, you will be asked to argue that the methods of Sections 3.8 and 3.9 can be used to make inferences about an arbitrary parameter θ, provided an estimator $\hat{\theta}$ whose sampling distribution is a normal distribution with mean θ and known variance $\sigma^2_{\hat{\theta}}$ can be found. Box 3.7 summarizes the procedures for making inferences about θ on the basis of a normally distributed estimator $\hat{\theta}$.

BOX **3.7**

Inferences about θ when the sampling distribution of $\hat{\theta}$ is normal with a known standard error

Let $\hat{\theta}$ be an estimator of θ, and suppose that the sampling distribution of $\hat{\theta}$ is a normal distribution with mean θ and standard error (standard deviation) $\sigma_{\hat{\theta}}$. Then confidence intervals and hypotheses tests for θ can be constructed as follows.

Confidence intervals

The $100(1 - \alpha)\%$ lower confidence interval (LCI) is

$$\hat{\theta} - z(\alpha)\sigma_{\hat{\theta}} \leq \theta \leq +\infty.$$

The $100(1 - \alpha)\%$ upper confidence interval (UCI) is

$$-\infty \leq \theta \leq \hat{\theta} + z(\alpha)\sigma_{\hat{\theta}}.$$

The $100(1 - \alpha)\%$ confidence interval (CI) is

$$\hat{\theta} - z(\alpha/2)\sigma_{\hat{\theta}} \leq \theta \leq \hat{\theta} + z(\alpha/2)\sigma_{\hat{\theta}}.$$

Hypothesis tests

Let z_c denote the calculated value of the test statistic

$$Z = \frac{\hat{\theta} - \theta_0}{\sigma_{\hat{\theta}}}.$$

Then the criteria for α-level hypothesis tests may be tabulated as follows.

Hypothesis H_0	H_1	Confidence-interval (CI) method Reject H_0 at level α if	Test-statistic (TS) method Reject H_0 at level α if
$\theta \leq \theta_0$	$\theta > \theta_0$	θ_0 is not in LCI	$z_c > z(\alpha)$
$\theta \geq \theta_0$	$\theta < \theta_0$	θ_0 is not in UCI	$z_c < -z(\alpha)$
$\theta = \theta_0$	$\theta \neq \theta_0$	θ_0 is not in CI	$z_c > z(\alpha/2)$ or $z_c < -z(\alpha/2)$

Application of the procedures in Box 3.7 to specific inferential problems involves identification of the parameter of interest θ, choice of its estimator $\hat{\theta}$, and verification that the sampling distribution of $\hat{\theta}$ is normal with mean θ and a known standard error $\sigma_{\hat{\theta}}$. It can be verified that the confidence intervals and hypothesis tests for the population mean μ in Sections 3.9 and 3.10 are special cases of those in Box 3.7 with $\theta = \mu$, $\hat{\theta} = \overline{Y}$, and $\sigma_{\hat{\theta}} = \sigma_{\overline{Y}} = \sigma/\sqrt{n}$.

Example 3.27 illustrates how the procedures in Box 3.7 can be used when the parameter of interest is $\theta = \mu_1 - \mu_2$, the difference between two population means.

EXAMPLE 3.27   In Example 3.14, we presented data on the percentage nitrogen loss (N-loss) over a 16-week period when Urea + N-Serve (UN) was used as the fertilizer for sugarcane. In the following table, we add data for N-loss when Urea alone (U) was used as the fertilizer.

Fertilizer	Percentage N-loss							
Urea + N-Serve (UN)	10.8	10.5	14.0	13.5	8.0	9.5	11.8	10.0
	8.7	9.0	9.8	13.8	14.7	10.3	12.8	
Urea (U)	8.0	7.3	14.1	9.8	7.1	6.3	10.0	7.1
	7.9	6.1	6.9	11.0	10.0			

These data are independent random samples of $n_1 = 15$ plots under fertilizer UN and $n_2 = 13$ plots under fertilizer U. Suppose that the N-loss populations for these two fertilizers have normal distributions with standard deviations $\sigma_1 = \sigma_2 = 2.0$, and that a 95% lower confidence interval for the true differential mean N-loss, $\theta = \mu_1 - \mu_2$, is desired.

The first task in constructing a lower confidence interval is to select a point estimator $\hat{\theta}$ that is normally distributed with mean θ and a known standard error $\sigma_{\hat{\theta}}$. When estimating the difference between population means, an obvious choice is the difference between sample means: $\hat{\theta} = \overline{Y}_1 - \overline{Y}_2$.

We know that the populations of N-losses have normal distributions. In that case, properties I and III in Box 3.7 imply that the sampling distribution of $\hat{\theta} = \overline{Y}_1 - \overline{Y}_2$ is a normal distribution with mean

$$\mu_{\hat{\theta}} = \mu_{\overline{Y}_1 - \overline{Y}_2} = \mu_1 - \mu_2 = \theta$$

and standard error

$$\sigma_{\hat{\theta}} = \sigma_{\overline{Y}_1 - \overline{Y}_2} = \sqrt{\frac{\sigma_1^2}{n_1} + \frac{\sigma_2^2}{n_2}}. \tag{3.25}$$

Therefore, the results in Box 3.7 can be used to make inferences about θ.

Calculations using the N-loss data yield $\overline{y}_1 = 11.15$ and $\overline{y}_2 = 8.58$. Hence, $\hat{\theta} = 11.15 - 8.58 = 2.57$ is the estimated value of θ. By substituting $\sigma_1 = \sigma_2 = 2.0$, $n_1 = 15$, and $n_2 = 13$ in Equation (3.25), we get

$$\sigma_{\hat{\theta}} = \sqrt{\frac{4}{15} + \frac{4}{13}} = \sqrt{0.5744} = 0.76.$$

A 95% lower confidence interval for θ is calculated with $\alpha = 0.05$ from the formula in Box 3.7. The required interval is

$$\hat{\theta} - z(0.05)\sigma_{\hat{\theta}} \leq \theta \leq +\infty \Leftrightarrow 2.57 - (1.645)(0.76) \leq \theta \leq +\infty$$

$$\Leftrightarrow 1.32 \leq \theta \leq +\infty.$$

Therefore, we may conclude with 95% confidence that the mean N-loss for fertilizer UN is at least 1.32% more than that for fertilizer U. We have 95% confidence in this calculated lower bound because 95% of all lower bounds constructed in this fashion from repeatedly drawn samples will be less than the true value of θ. ∎

Example 3.27 illustrated how the information in Box 3.7 can be used to construct confidence intervals for a parameter θ. Application of the results to tests of hypotheses regarding θ is considered in Example 3.28.

EXAMPLE 3.28   Continuing with Example 3.27, let's consider the problem of testing the null hypothesis H_0: $\theta \leq 1.00$ against the research hypothesis H_1: $\theta > 1.00$. In other words, let's examine whether the data provide sufficient evidence to claim that the mean N-loss for fertilizer UN is at least 1% point more than that for fertilizer U.

Clearly, the hypothesis-testing procedure in Box 3.7 is applicable in this situation, and we can use either the confidence-interval method or the test-statistic method. The confidence-interval method with $\alpha = 0.05$ will reject the null hypothesis if the 95% lower confidence interval does not contain $\theta_0 = 1.00$. Since the 95% lower confidence interval $(1.32, +\infty)$ constructed in Example 3.27 does not contain 1.00, it follows that a test with $\alpha = 0.05$ provides sufficient evidence in favor of the claim that the N-loss for fertilizer UN is at least 1% more than the N-loss for fertilizer U. Indeed, any value of θ_0 less than 1.32 will be outside the 95% LCI, and so it follows that a 0.05 level test of H_0: $\theta \leq \theta_0$ against H_1: $\theta > \theta_0$ will lead to the acceptance of H_1 if $\theta_0 < 1.32$. For example, the data provide sufficient evidence to accept H_1: $\theta > 1.25$ at the $\alpha = 0.05$ level. On the other hand, the data do not provide sufficient evidence (at $\alpha = 0.05$) to conclude that $\theta > 1.40$.

When using the test-statistic approach to test H_0: $\theta \leq 1$ against H_1: $\theta > 1$, the calculated value of the test statistic is needed. We have

$$z_c = \frac{\hat{\theta} - \theta_0}{\hat{\sigma}_{\hat{\theta}}} = \frac{2.57 - 1.00}{0.76} = 2.07.$$

If a test at the $\alpha = 0.05$ level of significance is desired, the calculated value 2.07 should be compared with the critical value $z(0.05) = 1.645$ obtained from Table C.1 (Appendix C). In this example, the rejection rule for the one-sided test of H_0: $\theta < \theta_0$ in Box 3.7 leads us to reject H_0 and accept H_1 at the 0.05 level. We conclude, at the 0.05 level of significance, that the data provide evidence to support the conclusion that the mean N-loss for fertilizer UN is at least 1% more than that for fertilizer U.

In this hypothesis test, we have focused on a specific level of significance: $\alpha = 0.05$. Alternatively, we could determine the p-value of the test by calculating the smallest level at which the null hypothesis will be rejected. The required value equals the area of the shaded region in Figure 3.19. We see from Figure 3.19 that, if we decide to perform the test at a level α such that α is less than the area of the shaded region, the corresponding critical value $z(\alpha)$ for the test will be larger than z_c. Therefore, if α is less than the area of the shaded region in Figure 3.19, a hypothesis

F I G U R E   3.19

The p-value of the test of the difference in mean N-losses

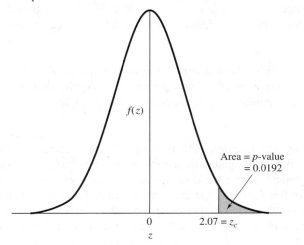

test at level α will not reject the null hypothesis. It follows that the area of the shaded region is equal to the smallest level at which the null hypothesis can be rejected. We can verify from Table C.1 that the area to the right of 2.07 under a standard normal distribution is 0.0192, and so the *p*-value of this test is $p = 0.0192$. The test would conclude in favor of H_1 if the test were done at any level larger than 0.0192. For example, a 0.01-level test will not reject the null hypothesis, but a 0.02-level test will find enough evidence in favor of the research hypothesis. ■

Exercises

3.42 Follow the line of reasoning in Sections 3.8 and 3.9 to argue that the procedures in Box 3.7 lead to valid confidence intervals and hypothesis tests.

3.43 Refer to the N-loss data in Example 3.27.

    a Construct a 90% confidence interval for the true mean differential N-losses for the two fertilizers.

    b Use the confidence interval in (a) to see if the data support the conclusion that the true mean N-losses are not the same for the two fertilizers. Be sure to state the null hypothesis and the level of the test that form the basis for your answer.

    c Determine the *p*-value for the test in (b) and interpret it.

3.12
Overview

The principal objective in this chapter is to bring together some of the key concepts and definitions underlying the inferential procedures of statistics discussed in the remainder of this book.

Two important concepts are the random sample and the sampling distribution of a statistic. By selecting random samples, we can use the observed characteristics of the samples to draw conclusions about the unknown characteristics of the target populations from which the samples are drawn. Probability theory enables us to determine the sampling distributions of the statistics calculated from random samples. The sampling distributions can be used to assess the likelihood that an observed sample has arisen from a specific population. Three of the most frequently used sampling distributions—the *t*-distribution, the χ^2-distribution and the *F*-distribution—are briefly described in this chapter.

Three types of inference problems are described in this chapter: (1) estimating population parameters; (2) testing hypotheses about populations; and (3) predicting future observations. The performance characteristics of a point estimator can be evaluated using criteria such as bias, standard error, and mean squared error; a confidence interval can be evaluated on the basis of its confidence coefficient and/or width. Finally, a hypothesis-testing procedure is evaluated on the basis of the probabilities of two types of errors: the error of accepting a false research hypothesis (Type I error) and the error of not accepting a true research hypothesis (Type II error).

4

Inferences about One or Two Populations: Interval Data

4.1
Introduction

In choosing the statistical techniques that are most appropriate for inferences based on a particular random sample, we must pay attention to the scales of measurement of the underlying random variables. In the next section, we describe three commonly used scales of measurement—the nominal, ordinal, and interval scales. The remainder of the chapter will be devoted to describing some simple but effective statistical procedures used to analyze data consisting of measurements based on interval scales. Among these are procedures that use random sample from a target population for making inferences about the mean and variance of the population, procedures that use independent random samples from two target populations for comparing the means and variances of the two populations, and procedures that use a random sample of pairs of measurements, one from each of two target populations, for making inferences about the difference between the population means. Procedures for analyzing data measured on nominal and ordinal scales will be described in Chapters 5 and 6.

4.2
Scales of measurement

The nominal scale

A *nominal variable* is a variable whose values contain no quantifiable information. They identify some quality of the measurements and, as such, are just names useful for distinguishing between distinct categories. Nominal variables are also called qualitative or categorical variables.

EXAMPLE **4.1**   In Example 2.7, we considered a population consisting of five pest types. The variable Y = pest type is nominal, because its values are just names of different types of pests; these values do not contain any quantifiable information. For instance, the difference of 2 between the values $Y = 1$ and $Y = 3$ (difference = $3 - 1 = 2$) does not give any information about the difference between pest type 3 and pest type 1. Rather, the numerical values 1 and 3 serve as proxies for the names of two distinct types of pests.   ■

The ordinal scale

An *ordinal variable* is a variable whose values are quantifiable to the extent that they can be arranged in order of magnitude. Consider the following example.

EXAMPLE **4.2**   On diagnosis of cancer, histology—the microscopic study of tissue structure—is often used to determine a grade that indicates the patient's survival potential. In one system, the survival potential is graded as 1, 2, 3, or 4; 1 indicates the most favorable outcome (complete cure) and 4 indicates the most unfavorable outcome (death within six weeks). The variable Y = histological grade is ordinal, because a lower value of Y corresponds to a lower risk of unfavorable outcome. However, the difference $3 - 1 = 2$ between grades 3 and 1 does not quantify the degree of difference in survival potential. The values 1, 2, 3, and 4 for Y represent its possible values arranged in order of increasing risk of an unfavorable outcome.   ■

The interval scale

An *interval variable* is a variable whose values can be regarded as points on a number line. It is *quantitative*; the difference between two values of an interval variable provides a numerical measure of the amount by which the values differ. An example of an interval variable is $Y =$ plant height (cm) seven days after planting. A height $Y = 12$ cm for one plant and $Y = 8$ cm for another plant implies that, seven days after planting, the first plant was 4 cm taller than the second.

EXAMPLE **4.3**   In Example 2.2, where we considered the number of bacterial colonies in a growth medium, we can identify an instance of an interval variable: $Y =$ number of colonies in a specimen sample. The difference between two values of Y is a measure of the amount by which one value differs from the other. For instance, the values $Y = 10$ and $Y = 3$ for two specimen samples imply that one specimen has seven more colonies than the other.   ■

The nominal and the interval scales correspond to, respectively, the lowest and the highest levels of the measurement scale in terms of the amount of quantifiable information. Any variable measured at a given level can be measured at all lower levels. In other words, an interval variable can be treated either as an ordinal variable or as a nominal variable. For example, weight is an interval variable but is also an ordinal variable, because weights can be arranged in order by magnitude. Weight is also a nominal variable, because a value of 20 g can be regarded as a measurement in the 20-g weight category. Similarly, a variable that is measured on an ordinal scale can be treated as a nominal variable.

EXAMPLE **4.4**   The histological grade of a cancer patient described in Example 4.2, which is an ordinal variable, can also be regarded as a nominal variable. There are four nominal categories for the values of Y: grade 1, grade 2, grade 3, and grade 4.   ■

As already noted, this chapter is devoted to some common statistical procedures that use data measured on interval scales to make inferences about one or two target populations. The corresponding methods for ordinal and nominal data are presented in Chapters 5 and 6, respectively.

Exercises

4.1   For each of the following variables, determine the highest level of the scale of its measurements.

a $Y =$ the percentage change in tumor weight in a treated rat.

b Y = the socioeconomic status (low, medium, or high) of an experimental subject.

c Y = the type of irrigation method (automatic sprinkler system, manual sprinkler system, underground irrigation system).

d Y = the type of stimulus given to an experimental subject (audio, video, audio and video).

e Y = the yield per plot of a variety of corn.

f Y = the acceptance rating of an ice cream flavor (1, 2, 3, 4, or 5, where 5 and 1 correspond, respectively, to the lowest and the highest ratings).

g Y = the type of vegetation in a plot of land (mostly pine, mostly oak, pine and oak mixed, other).

4.3
Inferences about means

Methods for constructing confidence intervals and hypothesis tests for a parameter θ were described in Sections 3.8 and 3.9. In this section, we discuss how those procedures can be adapted for two important special cases:

1 The case where θ is a population mean ($\theta = \mu$)

2 The case where θ is the difference between two population means ($\theta = \mu_1 - \mu_2$)

Recall that the procedures in Sections 3.8 and 3.9 require an estimator $\hat{\theta}$ such that the sampling distribution of $\hat{\theta}$ is a normal distribution with mean θ. In the two cases considered in this section, such an estimator is provided by the sample mean ($\hat{\theta} = \overline{Y}$) in case 1 and the difference between the sample means ($\hat{\theta} = \overline{Y}_1 - \overline{Y}_2$) in case 2. In symbols, this intuitive estimator of θ can be expressed as

$$\hat{\theta} = \begin{cases} \overline{Y} & \text{if } \theta = \mu \\ \overline{Y}_1 - \overline{Y}_2 & \text{if } \theta = \mu_1 - \mu_2. \end{cases} \tag{4.1}$$

The results in Box 3.2 and Box 3.6 imply that the above estimators will have normal distributions if either the target populations are normal or the samples are large. The standard error of $\hat{\theta}$ is

$$\sigma_{\hat{\theta}} = \begin{cases} \dfrac{\sigma}{\sqrt{n}} & \text{if } \hat{\theta} = \overline{Y} \\[3mm] \sqrt{\dfrac{\sigma_1^2}{n_1} + \dfrac{\sigma_2^2}{n_2}} & \text{if } \hat{\theta} = \overline{Y}_1 - \overline{Y}_2, \end{cases} \tag{4.2}$$

where, σ, σ_1, σ_2, n, n_1, and n_2 are the corresponding population standard deviations and sample sizes. Clearly, if the population standard deviations (σ in case 1 and σ_1 and σ_2 in case 2) are known, we can make inferences about θ by using the value in Equation (4.2) as $\sigma_{\hat{\theta}}$ in Box 3.7. Examples 3.27 and 3.28 illustrated this technique. In practice, however, it is unlikely that the population standard deviations are known.

An appealing solution to this problem is to estimate the population standard deviations using sample standard deviations. Then Equation (4.2) can be used to arrive at the following estimators for the standard error of $\hat{\theta}$ in the cases of one and two populations.

$$\hat{\sigma}_{\hat{\theta}} = \begin{cases} \frac{S}{\sqrt{n}} & \text{if } \hat{\theta} = \overline{Y} \\[3mm] \sqrt{\frac{S_1^2}{n_1} + \frac{S_2^2}{n_2}} & \text{if } \hat{\theta} = \overline{Y}_1 - \overline{Y}_2. \end{cases} \qquad (4.3)$$

In cases with one or two populations, the effect of replacing $\sigma_{\hat{\theta}}$ with its estimated value on the validity of the procedures in Box 3.5 has been much studied. If the samples are large (as a rule of thumb, $n \geq 30$), estimating $\sigma_{\hat{\theta}}$ will have little effect on the validity of the confidence intervals and hypothesis tests in Box 3.7 regardless of the distributions of the target populations. Procedures for making inferences about population means when large sample sizes are available are presented in Box 4.1.

BOX **4.1**

Inferences about population means with unknown population standard deviations and large sample sizes

Suppose $\overline{Y}_1$, $\overline{Y}_2$ and S_1, S_2 are, respectively, the means and standard deviations of independent random samples of sizes n_1 and n_2 from two (not necessarily normal) populations with means μ_1 and μ_2. Then confidence intervals and hypothesis tests for $\theta = \mu_1$ when n_1 is large, and for $\theta = \mu_1 - \mu_2$ when both n_1 and n_2 are large, can be constructed as follows. In the following, we let

$$\hat{\theta} = \overline{Y}_1 \text{ and } \hat{\sigma}_{\hat{\theta}} = \frac{S_1}{n_1} \quad \text{if } \theta = \mu_1,$$

$$\hat{\theta} = \overline{Y}_1 - \overline{Y}_2 \text{ and } \hat{\sigma}_{\hat{\theta}} = \sqrt{\frac{S_1^2}{n_1} + \frac{S_2^2}{n_2}} \quad \text{if } \theta = \mu_1 - \mu_2,$$

Confidence intervals

The $100(1 - \alpha)\%$ LCI is

$$\hat{\theta} - z(\alpha)\hat{\sigma}_{\hat{\theta}} \leq \theta \leq +\infty.$$

The $100(1 - \alpha)\%$ UCI is

$$-\infty \leq \theta \leq \hat{\theta} + z(\alpha)\hat{\sigma}_{\hat{\theta}}.$$

The $100(1 - \alpha)\%$ CI is

$$\hat{\theta} - z(\alpha/2)\hat{\sigma}_{\hat{\theta}} \leq \theta \leq \hat{\theta} + z(\alpha/2)\hat{\sigma}_{\hat{\theta}}.$$

Hypothesis Tests

Let z_c denote the calculated value of the test statistic

$$Z = \frac{\hat{\theta} - \theta_0}{\hat{\sigma}_{\hat{\theta}}}.$$

Then the criteria for α-level hypothesis tests are as follows.

Hypothesis H_0	H_1	CI method Reject H_0 at level α if	TS method Reject H_0 at level α if
$\theta \leq \theta_0$	$\theta > \theta_0$	θ_0 is not in LCI	$z_c > z(\alpha)$
$\theta \geq \theta_0$	$\theta < \theta_0$	θ_0 is not in UCI	$z_c < -z(\alpha)$
$\theta = \theta_0$	$\theta \neq \theta_0$	θ_0 is not in CI	$z_c > z(\alpha/2)$ or $z_c < -z(\alpha/2)$

The next example demonstrates the use of the procedures in Box 4.1 to make inferences about the difference between the means of two populations when the population distributions and population variances are unknown but the sample sizes are large.

EXAMPLE **4.5** Cholesterol level (mg/ml) measurements on random samples of children selected from two ethnic populations yielded the following data.

Population (i)	Sample size (n_i)	Sample mean $(\bar{y}_i)$	Sample variance (s_i^2)
1	48	208.4	948.1
2	40	170.6	902.5

We now use these data for the following purposes:

1 To construct a 95% confidence interval for the difference between the mean population cholesterol levels of the children in the two ethnic groups

2 To compute and interpret the p-value associated with the research hypothesis that, in population 1, the mean cholesterol level is higher than 200 mg/ml

Let μ_1 and μ_2 denote the mean cholesterol levels in populations 1 and 2. Then the parameter of interest is $\theta = \mu_1 - \mu_2$ in calculation 1 and $\theta = \mu_1$ in calculation 2. Since the samples are large, the results in Box 4.1 can be used to make inferences about θ without making any assumptions about the actual distributions of the measurements in the two target populations.

The estimated mean cholesterol levels in the two populations are $\bar{y}_2 = 170.6$ and $\bar{y}_1 = 208.4$. Therefore, the estimated difference between the mean cholesterol levels is $\hat{\theta} = \bar{y}_1 - \bar{y}_2 = 208.4 - 170.6 = 37.8$. From Box 4.1, the estimated standard error of $\hat{\theta}$ is

$$\hat{\sigma}_{\hat{\theta}} = \sqrt{\frac{948.1}{48} + \frac{902.5}{40}} = 6.51.$$

A 95% confidence interval for $\theta = \mu_1 - \mu_2$ is

$$\hat{\theta} \pm z(0.025)\hat{\sigma}_{\hat{\theta}} = 37.8 \pm (1.96)(6.51)$$

$$= 37.8 \pm 12.7$$

or

$$25.1 \leq \theta \leq 50.5.$$

Thus we can conclude, with 95% confidence, that the mean cholesterol level in population 1 is at least 25.1 mg/ml and at most 50.5 mg/ml more than that in population 2.

The parameter of interest in calculation 2 is $\theta = \mu_1$, and the objective is to compute the p-value for the test of the null hypothesis H_0: $\theta \leq 200$ against the research hypothesis H_1: $\theta > 200$. From Box 4.1, the standard error of $\hat{\theta}$ and the calculated value of the test statistic are

$$\hat{\sigma}_{\hat{\theta}} = \sqrt{\frac{948.1}{48}} = 4.44$$

and

$$z_c = \frac{208.4 - 200}{4.44} = 1.892,$$

respectively. Following a reasoning similar to that in Example 3.28, we see that the p-value of the test is the area to the right of 1.892 under a standard normal distribution. We see from Table C.1 in Appendix C that the p-value is 0.0294 and, thus, at any significance level $\alpha \geq 0.0294$, we can conclude that the true mean cholesterol level in population 1 is more than 200 mg/ml. ∎

When the samples are large, the procedures described in Box 4.1 can be used to make inferences about the means of one and two populations. No assumptions about the target population distributions are necessary. When the target populations have normal distributions, these procedures can be modified for use with small sample sizes. Different modifications are required in the following three cases.

1   The parameter of interest is the population mean, the population has a normal distribution, and the population variance is unknown.

2   The parameter of interest is the difference between two population means, the populations have normal distributions, and the population variances are equal but unknown.

3   The parameter of interest is the difference between two population means, the populations have normal distributions, and the population variances are unknown and possibly unequal.

Box 4.2 contains procedures for inferences about the mean of a normal population when the population variance is unknown.

BOX 4.2

Inferences about the mean of a $N(\mu, \sigma^2)$ population with unknown σ^2

Suppose $\overline{Y}$ and S are the mean and standard deviation of a random sample of size n from a normal population with mean μ and standard deviation σ.

Let $\sigma_{\overline{Y}} = S/\sqrt{n}$. The following procedures can be used to make inferences about μ.

Confidence intervals

The $100(1 - \alpha)\%$ LCI is

$$\overline{Y} - t(n - 1, \alpha)\hat{\sigma}_{\overline{Y}} \le \mu \le +\infty.$$

The $100(1 - \alpha)\%$ UCI is

$$-\infty \le \mu \le \overline{Y} + t(n - 1, \alpha)\hat{\sigma}_{\overline{Y}}.$$

The $100(1 - \alpha)\%$ CI is

$$\overline{Y} - t(n - 1, \alpha/2)\hat{\sigma}_{\overline{Y}} \le \mu \le \overline{Y} + t(n - 1, \alpha/2)\hat{\sigma}_{\overline{Y}}.$$

Hypothesis tests

Let t_c denote the calculated value of the test statistic

$$t = \frac{\overline{Y} - \mu_0}{\hat{\sigma}_{\overline{Y}}} = \frac{\sqrt{n}(\overline{Y} - \mu_0)}{S}.$$

Then the criteria for α-level hypothesis tests are as follows.

Hypothesis		CI method	TS method
H_0	H_1	Reject H_0 at level α if	Reject H_0 at level α if
$\mu \le \mu_0$	$\mu > \mu_0$	μ_0 is not in LCI	$t_c > t(n - 1, \alpha)$
$\mu \ge \mu_0$	$\mu < \mu_0$	μ_0 is not in UCI	$t_c < -t(n - 1, \alpha)$
$\mu = \mu_0$	$\mu \ne \mu_0$	μ_0 is not in CI	$t_c > t(n - 1, \alpha/2)$ or $t_c < -t(n - 1, \alpha/2)$

Note that the modifications of the procedures in Box 3.7 needed to obtain the inferential methods in Boxes 4.1 and 4.2 are identical except that the critical values of a standard normal distribution are used to construct confidence intervals and hypothesis tests in Box 4.1, whereas the critical values of t-distributions are used for the same purposes in Box 4.2. In both cases, the population mean is estimated by the sample mean ($\hat{\theta} = \overline{Y}$); and the true standard error $\sigma_{\hat{\theta}} = \sigma/\sqrt{n}$ is replaced by the estimated standard error $\hat{\sigma}_{\hat{\theta}} = S/\sqrt{n}$.

The inferential procedures in Box 4.2 will be referred to as procedures based on one-sample t-tests or simply as t-test procedures. Example B.6 in Appendix B contains a brief theoretical justification of the one-sample t-test procedures. Examples 4.6 and 4.7 illustrate their practical use.

EXAMPLE 4.6 Consider the N-loss data for fertilizer UN in Example 3.27. Assume that the percentage N-losses have a normal distribution, but the population standard deviation σ is not known. Let's compute a 95% upper confidence bound for μ, the mean percentage N-loss for fertilizer UN.

The problem is to construct a (one-sided) confidence interval for the mean of a normal population using a random sample of size $n = 15$. We can use the results in Box 4.2 with $n = 15$ and $\alpha = 0.05$. It can be verified that the sample mean and standard deviation are, respectively, $\bar{y} = 11.15$ and $s = 2.1397$. The estimated standard error of $\bar{Y}$ is

$$\hat{\sigma}_{\bar{Y}} = \frac{s}{\sqrt{n}} = \frac{2.1397}{\sqrt{15}} = 0.5525.$$

The required upper bound for μ can be computed using the formula for UCI in Box 4.2

$$\bar{Y} + t(14, 0.05)\hat{\sigma}_{\bar{Y}} = 11.15 + (1.761)(0.5525) = 12.12.$$

Thus, we can conclude with 95% confidence that, on average, no more than 12.12% N-loss will occur for fertilizer UN. ■

EXAMPLE **4.7**   Refer to the N-loss data for fertilizer UN in Example 3.27. The results in Box 4.2 can be used to test a hypothesis about μ, the population mean N-loss. Suppose we wish to test H_0: $\mu \geq 13$ against H_1: $\mu < 13$. If a test at the $\alpha = 0.05$ level is desired, the upper confidence interval $(-\infty, 12.12)$ computed in Example 4.6 can be used as described in Box 4.2. Since $\mu_0 = 13\%$ is outside the 95% UCI, we can reject the null hypothesis at the $\alpha = 0.05$ level and conclude that the true mean percentage N-loss is less than 13%.

Alternatively, the test-statistic approach can be used. With $\mu_0 = 13$, the calculated value of the test statistic is

$$t_c = \frac{\bar{y} - \mu_0}{\hat{\sigma}_{\bar{Y}}} = \frac{11.15 - 13}{0.5525} = -3.35.$$

The research hypothesis can be accepted at the level $\alpha = 0.05$, because the calculated value of -3.35 for the test statistic is less than $-t(14, 0.05) = -1.761$. Thus, we conclude that the true mean N-loss is less than 13%. The probability of an error when arriving at such a conclusion is less than 5%.

Alternatively, the p-value of this test can be computed. For 14 degrees of freedom in Table C.2 (Appendix C), we see that the p-value is less than 0.005. Therefore, the null hypothesis can be rejected at any level ≥ 0.005. ■

Sections 4.2 and 4.3 of the companion text by Younger (1997) show how to use the UNIVARIATE and MEANS procedures of SAS for performing the calculations in Examples 4.6 and 4.7. The following display shows the codes and the resulting printout when the MEANS procedure is used with the N-loss data.

The first part of the output gives the 95% confidence intervals for the means of the two populations. The second part gives the calculated value of the test statistic and the associated two-sided p-value. The one-sided p-value associated with the one-sided test in Example 4.7 is $0.0047/2 = 0.0023$.

```
---------------------------------------------------------------
   INPUT CODES
---------------------------------------------------------------
options ls=70 ps=60;
data nloss;
input fert $ nloss @@;
nlossdif = nloss-13;
lines;
UN 10.8 UN 10.5 UN 14.0 UN 13.5  UN 8.0 UN  9.5
UN 11.8 UN 10.0 UN  8.7 UN  9.0  UN 9.8 UN 13.8
UN 14.7 UN 10.3 UN 12.8
U 8.0 U 7.3 U 14.1 U  9.8 U  7.1 U 6.3 U 10.0 U 7.1
U 7.9 U 6.1 U  6.9 U 11.0 U 10.0
;
proc sort;
by fert;
proc means clm;
var nloss;
by fert;
proc means t prt;
var nlossdif;
where fert = 'UN';
run;
```

```
      ---------------------------------------------------------
   SAS OUPUT
      ---------------------------------------------------------
```

Analysis Variable : NLOSS

```
         ----------- FERT=U -----------
      Lower 95.0% CLM Upper 95.0% CLM
      -------------------------------
            7.2017972       9.9674336
      -------------------------------
         ----------- FERT=UN -----------
      Lower 95.0% CLM Upper 95.0% CLM
      -------------------------------
            9.9617319      12.3316014
      -------------------------------
```

```
   -------------------------------------------------------
```

Analysis Variable : NLOSSDIF

```
               T  Prob>|T|
      ---------------------
         -3.3546190    0.0047
      ---------------------
```

Next, suppose we want to compare the means of two normally distributed populations when the samples are small and the population standard deviations are equal but unknown. Box 4.3 summarizes the appropriate procedures for this case.

BOX **4.3**   ***Inferences about $\mu_1 - \mu_2$ for normal populations with unknown but equal variances***

Suppose $\overline{Y}_1, \overline{Y}_2$ and S_1, S_2 are the means and standard deviations of independent random samples of sizes n_1 and n_2 from two normal populations with means μ_1, μ_2 and a common standard deviation σ. Let $N = n_1 + n_2$ denote the total sample size and define

$$\hat{\sigma}_{\overline{Y}_1 - \overline{Y}_2} = S_p\sqrt{\frac{1}{n_1} + \frac{1}{n_2}},$$

where

$$S_p^2 = \frac{(n_1 - 1)S_1^2 + (n_2 - 1)S_2^2}{N - 2}.$$

Inferences about $\theta = \mu_1 - \mu_2$, the difference between the population means, can be made as follows.

Confidence intervals

The $100(1 - \alpha)\%$ LCI is

$$\overline{Y}_1 - \overline{Y}_2 - t(N - 2, \alpha)\hat{\sigma}_{\overline{Y}_1 - \overline{Y}_2} \leq \theta \leq +\infty.$$

The $100(1 - \alpha)\%$ UCI is

$$-\infty \leq \theta \leq \overline{Y}_1 - \overline{Y}_2 + t(N - 2, \alpha)\hat{\sigma}_{\overline{Y}_1 - \overline{Y}_2}.$$

The $100(1 - \alpha)\%$ CI is

$$\overline{Y}_1 - \overline{Y}_2 - t(N - 2, \alpha/2)\hat{\sigma}_{\overline{Y}_1 - \overline{Y}_2} \leq \theta \leq \overline{Y}_1 - \overline{Y}_2 + t(N - 2, \alpha/2)\hat{\sigma}_{\overline{Y}_1 - \overline{Y}_2}.$$

Hypothesis tests

Let t_c denote the calculated value of the test statistic

$$t = \frac{\overline{Y}_1 - \overline{Y}_2 - \theta_0}{\hat{\sigma}_{\overline{Y}_1 - \overline{Y}_2}}.$$

Then the criteria for α-level hypothesis tests are as follows.

Hypothesis		CI method	TS method
H_0	H_1	Reject H_0 at level α if	Reject H_0 at level α if
$\theta \leq \theta_0$	$\theta > \theta_0$	θ_0 is not in LCI	$t_c > t(N - 2, \alpha)$
$\theta \geq \theta_0$	$\theta < \theta_0$	θ_0 is not in UCI	$t_c < -t(N - 2, \alpha)$
$\theta = \theta_0$	$\theta \neq \theta_0$	θ_0 is not in CI	$t_c < -t(N - 2, \alpha/2)$ or
			$t_c > t(N - 2, \alpha/2)$

The quantity S_p^2 in Box 4.3, which is a weighted average of the two sample variances S_1^2 and S_2^2, is called the *pooled estimate*. Under the assumption that the population variances are equal, the pooled estimate is a reasonable estimate of the common population variance σ^2.

To obtain the estimated standard error $\hat{\sigma}_{\bar{Y}_1 - \bar{Y}_2}$ in Box 4.3, the population variances σ_1^2 and σ_2^2 in the expression

$$\sigma_{\bar{Y}_1 - \bar{Y}_2} = \sqrt{\frac{\sigma_1^2}{n_1} + \frac{\sigma_2^2}{n_2}}$$

are replaced by their pooled estimate S_p^2. Since $N - 2 = (n_1 - 1) + (n_2 - 1)$, the degrees of freedom for the *t*-test based on the pooled estimate may be calculated as the sum of the degrees of freedom of the *t*-tests when only a single sample is used for estimation of the population variance. In subsequent chapters, the procedures in Box 4.3 will be called procedures based on an independent-sample *t*-test.

EXAMPLE 4.8   To determine whether waste discharged by a chemical plant is polluting the local river, the river water was sampled at two locations—one upstream and one downstream from the discharge site. Independent water samples of sizes $n_1 = 10$ and $n_2 = 15$, respectively, were selected from the upstream and downstream locations. The concentration level (ppm) of a suspected chemical pollutant was determined in each water sample, with the following results:

Upstream	24.5, 29.7, 20.4, 28.5, 25.3, 21.8, 20.2, 21.0, 21.9, 22.2
Downstream	32.8, 30.4, 32.3, 26.4, 27.8, 26.9, 29.0, 31.5, 31.2, 26.7, 25.6, 25.1, 32.8, 34.3, 35.4

Shown below is a side-by-side box-plot of the pollution data. This plot was generated by using the UNIVARIATE procedure of SAS. The computer codes for generating the box-plots as well as other calculations illustrated in this example are shown in Figure 4.6 of the companion text by Younger (1997).

A visual inspection of the box-plots indicates that the downstream pollution level is much higher than the upstream level but the variability in downstream pollution levels seems to be about the same as that found upstream. In this example we will analyze these data assuming that the data are independent random samples from two normal distributions with a common unknown variance.

The means and variances of the samples are as follows:

Sample	Sample size	Sample mean	Sample variance
Downstream	15	29.88	11.066
Upstream	10	23.55	11.283

Let μ_1 and μ_2 denote the true mean chemical level in the downstream and upstream locations. Then, under the assumption that the populations of the

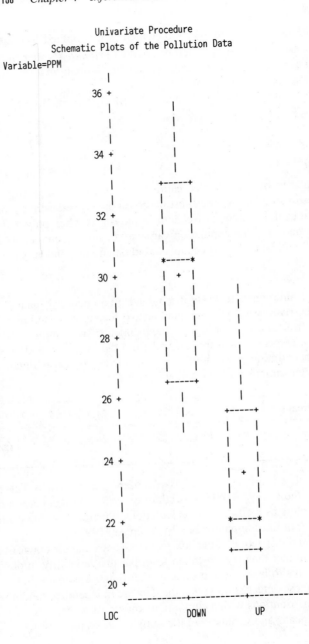

Univariate Procedure
Schematic Plots of the Pollution Data

chemical levels at the two locations have normal distributions with a common variance σ^2, the procedures in Box 4.3 can be used to make inferences about $\theta = \mu_1 - \mu_2$. We now construct a 95% confidence interval for θ under such an assumption.

First, we note that the estimated value of θ is $\bar{y}_1 - \bar{y}_2 = 29.88 - 23.55 = 6.33$ and the pooled estimate of the common population standard deviation σ is

$$s_p = \sqrt{\frac{(14)(11.066) + (9)(11.283)}{25 - 2}} = 3.34.$$

Also, the estimated standard error of $\hat{\theta} = \bar{Y}_1 - \bar{Y}_2$ is

$$\hat{\sigma}_{\bar{Y}_1 - \bar{Y}_2} = 3.34 \sqrt{\frac{1}{15} + \frac{1}{10}} = 1.3633,$$

so that the endpoints of a 95% confidence interval for $\mu_1 - \mu_2$ are

$$\bar{Y}_1 - \bar{Y}_2 \pm t(23, 0.025)\hat{\sigma}_{\bar{Y}_1 - \bar{Y}_2} = 6.33 \pm (2.069)(1.3633).$$

Thus, a 95% confidence interval for $\theta = \mu_1 - \mu_2$ is

$$3.51 \leq \mu_1 - \mu_2 \leq 9.15.$$

We can conclude with 95% confidence that the true difference between the mean chemical levels at the downstream and upstream locations is at least as large as 3.51 ppm but no larger than 9.15 ppm. In other words, the mean downstream pollution level is at least 3.51 and at most 9.15 ppm more than that prevailing in the upstream waters. ■

Finally, consider the case where the two target populations have normal distributions but the assumption that their variances are equal is questionable. Theoretical investigations have shown that, if the population distributions are symmetric and sample sizes are nearly equal, the independent-sample t-test procedures in Box 4.3 are satisfactory for most practical purposes. However, the use of S_p as recommended in Box 4.3 is not justified if the population variances are not equal and there is a considerable difference between the two sample sizes. In such cases, the use of Equation (4.3) to estimate the standard error of $\hat{\theta} = \bar{Y}_1 - \bar{Y}_2$ is recommended. The degrees of freedom needed to determine the critical values of the t-distribution may be approximated using a method suggested by Satterthwaite (1946). Box 4.4 describes how Satterthwaite's approximation is used to make inferences about $\theta = \mu_1 - \mu_2$.

BOX 4.4

Inferences about $\mu_1 - \mu_2$ for normal populations with unknown variances

Suppose $\bar{Y}_1$, $\bar{Y}_2$ and S_1, S_2 are the means and standard deviations of independent random samples of sizes n_1 and n_2 from two normal populations with means μ_1, μ_2 and standard deviations σ_1, σ_2. Let the standard error of $\hat{\theta} = \bar{Y}_1 - \bar{Y}_2$ be estimated by

$$\hat{\sigma}_{\hat{\theta}} = \sqrt{\frac{S_1^2}{n_1} + \frac{S_2^2}{n_2}}$$

and ν be the largest integer less than or equal to k, where

$$\frac{1}{k} = \frac{1}{n_1 - 1}\left[\frac{S_1^2/n_1}{S_1^2/n_1 + S_2^2/n_2}\right]^2 + \frac{1}{n_2 - 1}\left[\frac{S_2^2/n_2}{S_1^2/n_1 + S_2^2/n_2}\right]^2.$$

Approximate tests and confidence intervals for θ can then be constructed as follows.

Confidence intervals

The $100(1 - \alpha)\%$ LCI is

$$\hat{\theta} - t(\nu, \alpha)\hat{\sigma}_{\hat{\theta}} \leq \theta \leq +\infty.$$

The $100(1 - \alpha)\%$ UCI is

$$-\infty \leq \theta \leq \hat{\theta} + t(\nu, \alpha)\hat{\sigma}_{\hat{\theta}}.$$

The $100(1 - \alpha)\%$ CI is

$$\hat{\theta} - t(\nu, \alpha/2)\hat{\sigma}_{\hat{\theta}} \leq \theta \leq \hat{\theta} + t(\nu, \alpha/2)\hat{\sigma}_{\hat{\theta}}.$$

Hypothesis tests

Let t_c denote the calculated value of the test statistic

$$t_c = \frac{\hat{\theta} - \theta_0}{\hat{\sigma}_{\hat{\theta}}}.$$

Then the criteria for α-level hypothesis tests are as follows.

Hypothesis		CI method	TS method
H_0	H_1	Reject H_0 at level α if	Reject H_0 at level α if
$\theta \leq \theta_0$	$\theta > \theta_0$	θ_0 is not in LCI	$t_c > t(\nu, \alpha)$
$\theta \geq \theta_0$	$\theta < \theta_0$	θ_0 is not in UCI	$t_c < -t(\nu, \alpha)$
$\theta = \theta_0$	$\theta \neq \theta_0$	θ_0 is not in CI	$t_c > t(\nu, \alpha/2)$ or $t_c < -t(\nu, \alpha/2)$

Let's restate the circumstances in which the procedures in Box 4.1 and Box 4.4 are appropriate. If the populations are approximately normally distributed and the population variances are unequal and unknown, the procedures in Box 4.4 should be used to make inferences about the difference between the population means. If the samples are large, the population variances are unknown, and the population distributions are nonnormal, then the procedures in Box 4.1 are appropriate.

EXAMPLE **4.9**   In Example 4.8, we analyzed pollution data for a river under the assumption that the populations of pollution levels have normal distributions with a common variance.

In this example, we analyze the same data under the milder assumption that the populations have normal distributions with unknown variances. Equality of the variances is not assumed. We use the method in Box 4.4 to see if the data justify the conclusion that the true mean pollution level is higher at the downstream location than at the upstream location.

As before, let μ_1 and μ_2 denote, respectively, the true mean pollution levels at the downstream and upstream locations. Then the hypotheses to be tested are $H_0: \mu_1 - \mu_2 \leq 0$ and $H_1: \mu_1 - \mu_2 > 0$. From the sample data, we find that $n_1 = 15, n_2 = 10; \bar{y}_1 = 29.88, \bar{y}_2 = 23.55; s_1^2 = 11.066, s_2^2 = 11.283$.

We can use the methods in Box 4.4 with $\theta = \mu_1 - \mu_2$ and $\hat{\theta} = \bar{Y}_1 - \bar{Y}_2$ to get

$$\hat{\theta} = 29.88 - 23.55 = 6.33,$$

$$\hat{\sigma}_{\hat{\theta}} = \sqrt{\frac{s_1^2}{n_1} + \frac{s_2^2}{n_2}} = \sqrt{\frac{11.066}{15} + \frac{11.283}{10}} = 1.366.$$

The calculated value of the test statistic is

$$t_c = \frac{\hat{\theta} - \theta_0}{\hat{\sigma}_{\hat{\theta}}} = \frac{6.33 - 0}{1.366} = 4.634.$$

Hypothesis H_1 can be accepted at level α if $t_c > t(\nu, \alpha)$, where ν is the largest integer that does not exceed k and

$$\frac{1}{k} = \frac{1}{15-1}\left[\frac{11.066/15}{11.066/15 + 11.283/10}\right]^2 + \frac{1}{10-1}\left[\frac{11.283/10}{11.066/15 + 11.283/10}\right]^2$$

$$= 0.0517,$$

so that

$$k = \frac{1}{0.0511} = 19.34.$$

Therefore, $\nu = 19$ and, from Table C.2 (Appendix C), we have $t(19, 0.05) = 1.729$. We can reject the null hypothesis at the 0.05 level and conclude that the true mean pollution level downstream is higher than that upstream.

Finally, the $100(1 - \alpha)\%$ confidence interval for the difference between mean pollution levels is found to be $3.47 \leq \mu_1 - \mu_2 \leq 9.19$. Comparison with Example 4.8 shows that the interval constructed on the assumption of equal population variances is narrower than that constructed without such an assumption. ∎

Comparing means in the case of paired data

When comparing the means of two populations, the methods in Boxes 4.1, 4.3, and 4.4 require that we have independent samples from the populations of interest. In many studies, either by design or by necessity, the assumption of independent samples is violated, because observations are grouped according to shared characteristics that influence response. We'll discuss the benefits of grouping observations according to shared characteristics in Chapter 7. For the present, let's consider an example. In a study of the mechanism by which ethanol affects cardiovascular function, a physiologist measured the effects of an infusion of ethanol to lambs both with and without a pretreatment to deactivate histamine receptors. Each lamb was observed under two conditions. In one, the lamb received ethanol after pretreatment with a histamine blocker; in the other, ethanol was infused after pretreatment with saline solution alone. The measured effects of ethanol resulting from the two conditions could not be treated as two independent samples, because the pair of measurements from the same lamb will share characteristics peculiar to that lamb. Similarly, observations from litter mates, subdivided experimental units, or experimental units matched on meaningful characteristics should not be regarded as independent. Such cases call for special methods in which pairs of observations may be related. Appropriate methods are easily derived from the procedures described in Box 4.1 or their modifications in Box 4.2.

Let Y_1 denote an observation obtained under a first treatment or condition and Y_2 denote the paired observation under a second treatment or condition. Then we have a set of n independent pairs of observations (Y_1, Y_2). If μ_1 and μ_2 are the means of the populations of Y_1 and Y_2, respectively, then the set of n differences

$$D = Y_1 - Y_2$$

is a random sample of size n from the population of all such differences. The mean of the population of differences is equal to the difference between the two population means; that is, $\mu_D = \mu_1 - \mu_2$. Thus, inferences concerning the parameter μ_D can be made by applying the one-sample procedures in Box 4.2 to the sample of differences. Consider the following example.

EXAMPLE **4.10**   In a study, measurements of cell fluidity were made in 20 dishes of pulmonary artery cells from ten dogs. Cell samples from each dog were randomly divided into two dishes; a randomly selected dish from each pair was randomly assigned to an oxygen (O_2) treatment, and the other to a control treatment consisting of no O_2 exposure. The following are the cell fluidity measurements in the ten pairs of dishes; the variables Y_1 and Y_2 denote, respectively, the cell fluidities in the untreated and the oxygen-treated dishes in a pair.

Pair	Y_1	Y_2	$D = Y_1 - Y_2$	Pair	Y_1	Y_2	$D = Y_1 - Y_2$
1	0.308	0.308	0.000	6	0.278	0.293	−0.015
2	0.304	0.309	−0.005	7	0.296	0.302	−0.006
3	0.305	0.305	0.000	8	0.301	0.300	0.001
4	0.304	0.311	−0.007	9	0.302	0.308	−0.006
5	0.301	0.303	−0.002	10	0.237	0.250	−0.013

Suppose we are interested in comparing μ_2, the mean fluidity of cells exposed to O_2, with μ_1, the mean for untreated cells. Methods for comparing the means of two populations that use data from independent samples are not appropriate here, because the measurements from dishes within a pair are unlikely to be independent. However, the sample of ten differences can be regarded as a random sample of size $n = 10$ from a population with mean $\mu_D = \mu_1 - \mu_2$. ■

Appropriate methods of inference for paired observations are summarized in Box 4.5.

BOX **4.5**   *Inferences about μ_D, the mean of the differences between paired observations*

Suppose that $\overline{D}$ and S_D are, respectively, the mean and standard deviation of a random sample of n differences between paired observations. Let μ_D denote the mean of the population of differences and let the estimated standard error $S_D/\sqrt{n}$ of $\overline{D}$ be denoted by $\hat{\sigma}_{\overline{D}}$. If the population of differences has a normal distribution, or if the sample size n is large, then the following procedures can be used to make inferences about μ_D.

Confidence intervals

The $100(1-\alpha)\%$ LCI is

$$\overline{D} - t(n-1, \alpha)\hat{\sigma}_{\overline{D}} \leq \mu_D \leq +\infty.$$

The $100(1-\alpha)\%$ UCI is

$$-\infty \leq \mu_D \leq \overline{D} + t(n-1, \alpha)\hat{\sigma}_{\overline{D}}.$$

The $100(1-\alpha)\%$ CI is

$$\overline{D} - t(n-1, \alpha/2)\hat{\sigma}_{\overline{D}} \leq \mu_D \leq \overline{D} + t(n-1, \alpha/2)\hat{\sigma}_{\overline{D}}.$$

Hypothesis tests

Let t_c denote the calculated value of the test statistic

$$t = \frac{\overline{D} - \mu_{D0}}{\hat{\sigma}_{\overline{D}}} = \frac{\sqrt{n}(\overline{D} - \mu_{D0})}{S_D}.$$

Hypothesis		CI method	TS method
H_0	H_1	Reject H_0 at level α if	Reject H_0 at level α if
$\mu_D \leq \mu_{D0}$	$\mu > \mu_{D0}$	μ_{D0} is not in LCI	$t_c > t(n-1, \alpha)$
$\mu_D \geq \mu_{D0}$	$\mu < \mu_{D0}$	μ_{D0} is not in UCI	$t_c < -t(n-1, \alpha)$
$\mu_D = \mu_{D0}$	$\mu \neq \mu_{D0}$	μ_{D0} is not in CI	$t_c < -t(n-1, \alpha/2)$ or $t_c > t(n-1, \alpha/2)$

The procedures in Box 4.5 are called procedures based on paired *t*-tests. The next example illustrates the calculations and interpretations when paired *t*-test procedures are used to make inferences about μ_D, the difference between two population means.

EXAMPLE **4.11**   On the basis of the data in Example 4.10, let's construct a 95% confidence interval for μ_D, the difference between the true mean cell fluidity of treated and untreated cells. It can be verified from the data that $\overline{D}$, the mean of the sample of differences, and $\hat{\sigma}_{\overline{D}}$, the standard error of $\overline{D}$, are -0.0053 and 0.0017, respectively. From Table C.2 (Appendix C), we get $t(9, 0.025) = 2.262$, so that the endpoints of the 95% confidence interval for μ_D are $-0.0053 \pm 2.262(0.0017)$. The required interval is

$$-0.0091 \le \mu_D \le -0.0015.$$

Thus, with 95% confidence, we can assert that the difference between the mean cell fluidity of treated and untreated cells is within the range from -0.0091 to -0.0015. (The negative sign implies that the treated dishes have higher mean fluidity than do the untreated dishes.) Zero is outside the 95% confidence interval for μ_D, and so we can reject H_0: $\mu_D = 0$ at the 0.05 level and conclude that the mean fluidity of the cells treated with O_2 is different from that for untreated cells.

In Exercise 3.32, you showed that the upper endpoint of a 95% confidence interval is also the upper endpoint of a 97.5% upper confidence interval. Since the upper bound for the 95% confidence interval for μ_D is negative, the value $\mu_{D0} = 0$ is outside the 97.5% upper confidence interval for μ_D. Therefore, at the 0.025 level, the one-sided null hypothesis H_0: $\mu_D \ge 0$ can be rejected in favor of the research hypothesis H_1: $\mu_D < 0$. We can conclude, at the 97.5% confidence level, that the treated cells have higher fluidity (the difference is at least $0.0015 \approx 0.002$) than the untreated cells.

The hypotheses H_0: $\mu_D = 0$ against H_1: $\mu_D \ne 0$ can also be tested by the test-statistic method. With $\mu_{D0} = 0$, the calculated value of the test statistic is

$$t_c = \frac{\overline{D} - \mu_{D0}}{\hat{\sigma}_{\hat{\theta}}} = \frac{-0.0053 - 0}{0.0017} = -3.12.$$

For a 0.05-level test, the lower critical value is $-t(9, 0.025) = -2.262$. Since the calculated value t_c is less than the lower critical value, we can reject the null hypothesis at the 0.05 level. Figure 4.9 in the companion text by Younger (1997) shows how to use the MEANS procedure of SAS for analyzing paired data. ■

Because standard deviations and, hence, standard errors are virtually never known in practice, Boxes 4.2–4.5 contain the procedures that are most commonly used to make inferences about one or two population means. In the next section, some procedures for making inferences about the variances of normally distributed populations are presented.

Exercises

4.2   A mental health clinic in a local health department conducted a study of mothers' understanding of instructions regarding care for their children, who had been clients of the clinic. Following a consultation in which the instructions were communicated, each mother was interviewed by a team of professionals who assigned her a score designed to measure her level of understanding of the instructions. Two methods of communicating the instructions were employed. The following data summarize the scores assigned to two samples of mothers.

Method	Number of mothers	Mean score	Standard deviation of scores
1	38	56.8	6.2
2	42	43.2	3.9

Let μ_1, σ_1 and μ_2, σ_2 denote, respectively, the mean and standard deviation of the scores for the conceptual population of mothers receiving instructions using methods 1 and 2.

a   What can you say about the sampling distribution of $\overline{Y}_1 - \overline{Y}_2$, the difference between the mean scores of independent random samples of $n_1 = 38$ and $n_2 = 42$ mothers selected from the two populations?

b   Using the data, conduct an appropriate statistical test to see if the true mean score for method 1 is higher than that for method 2.

c   Using the data, construct a 99% lower confidence bound for $\mu_1 - \mu_2$ and interpret the result you obtain.

4.3   In an experiment with rats, a behavioral scientist used an auditory signal to indicate that food was available through an open door in the cage. The scientist counted the number of trials needed by each of ten rats to learn to recognize the signal. Answer the following questions on the assumption that the conceptual population of numbers of trials needed by the rats has a normal distribution with mean μ and standard deviation $\sigma = 4$.

a   For the purpose of statistical analysis, the data can be regarded as a random sample from a target population. Give a graphical description of the target population.

b   Let $\overline{Y}$ denote the mean number of trials needed by the experimental rats. What does it mean to say that a proportion π of the values in the sampling distribution of $\overline{Y}$ is within 2 units of the mean of the target population?

c   Determine the value of π and interpret its meaning.

d   Suppose that the following data were obtained by the behavioral scientist.

$$18, 19, 13, 14, 18, 15, 14, 21, 14, 11$$

   i   Calculate a 95% confidence interval for μ and interpret the interval you obtain.

   ii   Perform a statistical test to see if the data support the conclusion that, on average, a rat needs less than 18 trials to learn to recognize the signal.

4.4   The main cause of illness and death in patients suffering from hairy cell leukemia (HCL) is their susceptibility to infection from uncommon infectious agents. The following data were obtained in a study designed to determine the infection rate

$$IR = \frac{\text{Number of infections}}{\text{Number of months}} \times 1000$$

in HCL patients who responded (as determined by some typical blood values) to a new therapy:

$$2.95, 2.11, 1.83, 2.06, 1.90, 1.95, 1.71, 1.98, 2.24, 1.59, 1.87, 1.93,$$
$$2.11, 1.96, 1.94, 1.17, 1.77, 1.98, 2.34, 2.18, 2.26, 1.69, 2.12, 2.18,$$
$$1.91, 1.28, 2.27, 2.13, 1.38, 2.03, 2.29, 1.74, 2.07, 1.25, 2.37$$

For these data, we find that

$$\bar{y} = 1.96; \qquad\qquad\qquad s = 0.3517$$

a   Construct a 90% confidence interval for μ, the true mean infection rate among responding patients, and interpret the interval you obtain.

b   Perform a statistical test to see if the observed data support the conclusion that the infection rate for the responding treated patients is less than 2.5 per month.

c   What assumptions, if any, did you make in (a) and (b)?

4.5   A psychologist has designed an index to measure the social perceptiveness of elementary school children. The index is based on ratings of a child's responses to questions about a set of photographs showing different social situations. The following are the indices measured on a random sample of 16 children attending elementary schools in a particular school district:

$$48, 75, 69, 58, 60, 68, 59, 66, 71, 52, 49, 60, 54, 55, 70, 57$$

a   State the assumptions under which one-sample t-test procedures can be used with these data.

b   Making the assumptions asked for in (a), construct a 90% confidence interval for the mean μ of the population from which the sample was selected.

c   Explain how the interval constructed in (b) should be interpreted by the psychologist.

d   Calculate the p-value of the test of H_0: $\mu \leq 50$ and H_1: $\mu > 50$.

e   Explain how the p-value in (d) should be interpreted by the psychologist.

f   Suppose that the psychologist wishes to use the interval in (b) to test the null hypothesis in (d). Explain, giving reasons, the level at which the test can be performed. What is the conclusion drawn from such a test?

4.6   The following data are the lengths of ears (cm) collected in a study of size inheritance in a hybrid variety of corn:

$$18.1, 15.3, 14.4, 10.5, 18.2, 18.7, 16.4, 15.8, 17.1, 12.4, 14.8, 16.2, 11.5, 14.7, 13.3$$

a   State the assumptions under which these data can be used to construct confidence intervals and perform hypothesis tests about the mean length μ of ears in plants of this particular variety.

b   Construct a 90% confidence interval for the mean length μ of ears in plants of this variety. Interpret the interval you construct.

c   Explain, giving reasons, how the interval in (b) can be used to conduct a hypothesis test at the $\alpha = 0.10$ level if the null hypothesis is H_0: $\mu = 17$ cm, against the research hypothesis H_1: $\mu \neq 17$ cm.

d   Write the conclusions that can be drawn from the test in (c).

e   Calculate the *p*-value for the test in (c) and interpret this number.

f   Perform a statistical test to see if it is reasonable to conclude that, on average, the ear length for this variety of corn is larger than 14 cm. State your conclusions.

4.7   In Examples 3.27 and 3.28, independent random samples of nitrogen losses were used to make inferences about $\theta = \mu_1 - \mu_2$, the true mean differential N-loss between two fertilizer regimens UN and U. The inferences were made under the assumption that the populations of measured N-losses had normal distributions with a common variance $\sigma^2 = 4$.

a   Use the data in Example 3.27 to construct a 95% lower bound for θ under the assumption that the populations have normal distributions with unknown but equal variances.

b   Interpret the bound you calculated in (a).

c   Repeat (a) and (b) under the assumption that the populations have normal distributions with unknown (and not necessarily equal) variances.

d   The lower bound in Example 3.27 and those in (a) and (c) are three lower bounds for θ calculated from the same set of data. Briefly discuss what factors we should consider when making a choice between the three lower bounds.

4.8   The following are data reported by Potoff and Roy (1964) on the dental measurements (distance, mm, from the center of the pituitary to the pterygomaxillary fissure) of 11 girls and 16 boys, all of whom were 16 years old:

Girls	23.0, 25.5, 26.0, 26.5, 23.5, 22.5, 25.0, 24.0, 21.5, 19.5, 28.0
Boys	31.0, 26.5, 27.5, 27.0, 26.0, 28.5, 26.5, 25.5, 26.0, 31.5, 25.0, 28.0, 29.5, 26.0, 30.0, 25.0

a   State a set of assumptions under which the *t*-tests and the associated confidence intervals can be used to make inferences about the difference between the mean dental measurements in the populations of 16-year-old boys and 16-year-old girls.

b   Set up null and alternative hypotheses (clearly explaining the meanings of the symbols) for a statistical test to see if these data support the conclusion that the mean dental measurements in the population of 16-year-old boys is higher than that in the population of 16-year-old girls.

c   Determine the *p*-value of the *t*-test of the hypotheses in (b). Write your conclusions.

d   Construct a 95% confidence interval for the difference between the means of dental measurements in the two populations. Interpret the interval you construct.

4.9   In a study, the following lymphocyte counts were obtained in 2-year-old Holstein cows and 2-year-old Guernseys:

Holsteins	5166, 6080, 7290, 7031, 6700, 8908, 4214, 5135, 5002, 4900, 8043, 6205, 3800
Guernseys	6310, 6295, 4497, 5182, 4273, 6591, 6425, 4600, 5407, 5509

a   State the assumptions under which the independent-sample t-test procedures can be used to analyze these data.

b   Under the assumptions in (a), obtain a 95% confidence lower bound for the differential mean lymphocyte counts between Holstein cows and Guernsey cows.

c   Interpret the bound calculated in (b).

d   Obtain the bound calculated in (b) without assuming that the variances of the two lymphocyte count populations are equal.

e   Compare the variances of the two samples to determine whether the result in (b) or (d) is the appropriate lower bound for the differential mean lymphocyte counts between Holstein cows and Guernsey cows.

4.10   In order to determine the effectiveness of a potential anticancer drug, seven matched pairs of mice were used; the animals within each pair were similar in their biological characteristics. At the start of the experiment, each animal was implanted with a type of cancer cell. A randomly selected animal from each pair was treated with the experimental drug for a specified period of time. The other animal within each pair served as the control. At the end of the experiment, the tumor was extracted from each animal and weighed. The following data are the tumor weights (g):

Pair	1	2	3	4	5	6	7
Control	1.321	1.423	2.682	0.934	1.230	1.670	3.201
Treatment	0.841	0.932	2.011	0.762	0.991	1.120	2.312

a   Analyze the data to see if the expected tumor weight for treated animals is lower than that for the control animals. Write the appropriate conclusions.

b   Construct a 95% confidence interval for the average difference in tumor weights for the treated and untreated animals. Interpret the interval.

c   State the assumptions about the data needed to ensure that the statistical methods used to answer (a) and (b) are valid.

4.11   The following data are measures of pulmonary vascular resistance (PVR) in eight lambs before and after infusion of the drug histamine:

Lamb	PVR before histamine	PVR after histamine
1	0.095	0.176
2	0.106	0.142
3	0.082	0.194
4	0.152	0.136
5	0.090	0.115
6	0.086	0.084
7	0.137	0.103
8	0.121	0.189

a   Perform a statistical test to determine whether histamine increases PVR on average. Be sure to state the null and alternative hypotheses and the conclusions that can be drawn from the test.

    b Calculate lower and upper bounds that bracket the mean effect of histamine with 95% confidence. What does it mean to say that these bounds are calculated with 95% confidence?

4.4
Inferences about variances

The need to make inferences about the variances of normally distributed populations arises in many contexts. Perhaps the most important application of inferences about population variances is a technique known as analysis of variance (ANOVA). As will be explained in Chapter 8, analysis of variance is a statistical procedure for comparing several population means; it is based on the idea that increase in the difference in population means is associated with increase in the variability of sample means. Inferences about population variances are also employed when there is some doubt about the assumption of equal population variances underlying the independent-sample t-test procedures described in Box 4.3. If σ_1^2 and σ_2^2 are the variances of two normally distributed populations, we might be interested in determining whether the observed data indicate that $\sigma_1^2 \neq \sigma_2^2$. Another situation in which an inference about a population variance is helpful is described in Example 4.12.

EXAMPLE **4.12**   A laboratory manager wants to ensure that 95% of the carbon analysis measurements made in the laboratory are within 0.1 ppm of the true value. On the basis of experience, the manager is willing to assume that the carbon content measurements made on identical soil samples are (approximately) normally distributed with mean μ and standard deviation σ. Recall that 95% of the measurements in a normal population will lie within 1.96 ($\sim$2) standard deviations of the mean. Therefore, the laboratory manager's requirement will be met by the carbon content measurements in the lab if $2\sigma < 0.10$ (that is, $\sigma < 0.05$).

    A technician presented with ten identical soil samples for carbon analysis produced the following results (ppm):

$$0.560, 0.842, 0.731, 0.782, 0.673, 0.718, 0.791, 0.726, 0.760, 0.798$$

Do these data indicate that the technician's measurements meet the lab manager's requirements? In other words, can the observed sample be regarded as a random sample from a population with standard deviation less than 0.05? To answer this question, the manager can perform a hypothesis test in which the null and alternative hypotheses are, respectively, H_0: $\sigma \geq 0.05$ and H_1: $\sigma < 0.05$.   ■

    Inferences about the variance of a normal population utilize the sample variance S^2 and critical values of χ^2-distributions. We summarize the relevant inferential procedures in Box 4.6. Recall that the α-level critical value of a $\chi^2(v)$-distribution is denoted by $\chi^2(v, \alpha)$.

BOX **4.6**

Inferences about the variance of a normal population

Let S^2 denote the variance calculated on the basis of a random sample of size n taken from a normal population with variance σ^2. Confidence intervals and hypothesis tests for σ^2 can be constructed as follows.

Confidence intervals for σ^2

The $100(1 - \alpha)\%$ UCI is

$$0 \leq \sigma^2 \leq \frac{(n - 1)S^2}{\chi^2(n - 1, 1 - \alpha)}.$$

The $100(1 - \alpha)\%$ LCI is

$$\frac{(n - 1)S^2}{\chi^2(n - 1, \alpha)} \leq \sigma^2 \leq +\infty.$$

The $100(1 - \alpha)\%$ CI is

$$\frac{(n - 1)S^2}{\chi^2(n - 1, \alpha/2)} \leq \sigma^2 \leq \frac{(n - 1)S^2}{\chi^2(n - 1, 1 - \alpha/2)}.$$

Hypothesis tests

Let χ_c^2 denote the calculated value of the test statistic

$$\chi^2 = \frac{(n - 1)S^2}{\sigma_0^2}.$$

Then the criteria for α-level hypothesis tests for σ^2 are as follows.

Hypothesis		CI method	TS method
H_0	H_1	Reject H_0 at level α if	Reject H_0 at level α if
$\sigma^2 \leq \sigma_0^2$	$\sigma^2 > \sigma_0^2$	σ_0^2 is not in LCI	$\chi_c^2 > \chi^2(n - 1, \alpha)$
$\sigma^2 \geq \sigma_0^2$	$\sigma^2 < \sigma_0^2$	σ_0^2 is not in UCI	$\chi_c^2 < \chi^2(n - 1, 1 - \alpha)$
$\sigma^2 = \sigma_0^2$	$\sigma^2 \neq \sigma_0^2$	σ_0^2 is not in CI	$\chi_c^2 > \chi^2(n - 1, \alpha/2)$ or $\chi_c^2 < \chi^2(n - 1, 1 - \alpha/2)$

Intuitive explanations for the formulas in Box 4.6 are similar to those in Box 4.1. In other words, the endpoints of the confidence intervals for σ^2 are obtained by adjusting its estimated value S^2 to account for random fluctuations. In this case, however, the adjustment factor is multiplicative. For example, the upper bound for the UCI is S^2 multiplied by the factor $(n - 1)/\chi^2(n - 1, 1 - \alpha)$. Of course, the adjustment factor depends upon the sample size and the desired confidence coefficient. In the test-statistic approach, the estimate of σ^2 is compared with the hypothetical value σ_0^2 by looking at χ_c^2, which equals a constant multiple of their ratio;

the multiplication factor is $(n - 1)$. If χ_c^2 is large, as determined by an appropriate critical value, then we conclude that $\sigma^2 > \sigma_0^2$.

EXAMPLE 4.13 Now suppose that the lab manager in Example 4.12 wants to send the technician for further training if the standard deviation for the population of all measurements by this technician exceeds 0.05. On the basis of the given data what would be your recommendation to the lab manager?

We need to test the hypothesis H_0: $\sigma \leq 0.05$ against H_1: $\sigma > 0.05$. If we decide to use the CI method, we calculate a LCI for σ. From the data we have, $n = 10$ and $s^2 = 0.0062$, so that the 95% lower confidence bound for σ^2 calculated from the LCI formula in Box 4.6 is

$$\frac{(n - 1)s^2}{\chi^2(n - 1, \alpha)} = \frac{(10 - 1)(0.0062)}{\chi^2(9, 0.05)} = \frac{(9)(0.0062)}{16.9190} = 0.0033,$$

where 16.9190 is the 0.05-level critical value for a $\chi^2(9)$-distribution. Therefore, a 95% lower confidence interval for the population variance is

$$0.0033 \leq \sigma^2 \leq +\infty,$$

so that a 95% confidence interval for the population standard deviation is

$$\sqrt{0.0033} \leq \sigma \leq +\infty \quad \Rightarrow \quad 0.058 \leq \sigma \leq +\infty.$$

Since $\sigma_0 = 0.05$ falls outside the lower confidence interval, we recommend (at the $\alpha = 0.05$ level) that the technician be sent for further training.

If we adopt the test-statistic approach, we rewrite the hypotheses as H_0: $\sigma^2 \leq (0.05)^2 = 0.0025$, H_1: $\sigma^2 > 0.0025$. The calculated value of the test statistic is

$$\chi_c^2 = \frac{(n - 1)s^2}{\sigma_0^2} = \frac{(9)(0.0062)}{0.0025} = 22.32.$$

From Table C.3 (Appendix C), $\chi^2(9, 0.01) = 21.6660$, so that the *p*-value is less than 0.01. Thus, if testing at any α greater than or equal to 0.01, further training for the technician should be recommended.

Let's compare the information provided by the point estimate, interval estimate, and the hypothesis test in this example. A point estimate of the technician's standard deviation is $s = \sqrt{0.0062} = 0.079$, which is 58% higher than the specified upper bound of 0.05. The 95% lower confidence interval implies that the smallest plausible value for the true standard deviation is 0.058; hence, σ is not likely to be less than 0.05 as claimed in the null hypothesis. We are led to the conclusion that $\sigma > 0.05$ at the 0.05 level of significance.

The test statistic does not give us the smallest plausible value for true σ but tells us that the claim $\sigma > 0.05$ is valid at a confidence level higher than 0.99 ($\alpha \leq 0.01$). If we desire information about the extent by which the true standard deviation exceeds the specified upper bound of 0.05, we should construct a confidence interval for σ. Calculations by the CI formula in Box 4.6 show that a 95% confidence interval for σ is $0.0542 \leq \sigma \leq 0.1437$. Therefore, the true standard deviation of the technician's measurements is at least 108% ($0.054/0.05 = 1.08$) and no more than 288% ($0.144/0.05 = 2.88$) of the specified value of 0.05. Figure 4.11 in the companion text by Younger (1997) shows an SAS program that can be used to perform the calculations associated with inferences about σ^2. ■

The test procedures in Box 4.6 are based on the assumption that the distribution of the target population is normal. In contrast to the procedures for comparing population means, the procedures in Box 4.6 may perform very poorly if that assumption is incorrect. For nonnormal populations, approximate inferences about the population variance will require large sample sizes. Such procedures fall outside the scope of our present discussion.

Box 4.7 summarizes formulas for comparing the variances of two normally distributed populations. Recall that $F(m, n, \alpha)$ is the α-level critical value of an F-distribution with m degrees of freedom for the numerator and n degrees of freedom for the denominator. These values are found in Table C.4 (Appendix C).

BOX 4.7

Inferences about variances of two normal populations

Let S_1^2 and S_2^2 denote the variances of two independent random samples of sizes n_1 and n_2 from two normal populations with variances σ_1^2 and σ_2^2, respectively. Confidence intervals and hypothesis tests for the ratio σ_1^2/σ_2^2 can be constructed as follows.

Confidence intervals for σ_1^2/σ_2^2

The $100(1 - \alpha)\%$ UCI is

$$0 \leq \frac{\sigma_1^2}{\sigma_2^2} \leq \frac{S_1^2}{S_2^2} \frac{1}{F(n_1 - 1, \, n_2 - 1, \, 1 - \alpha)}.$$

The $100(1 - \alpha)\%$ LCI is

$$\frac{S_1^2}{S_2^2} \frac{1}{F(n_1 - 1, \, n_2 - 1, \, \alpha)} \leq \frac{\sigma_1^2}{\sigma_2^2} \leq +\infty.$$

The $100(1 - \alpha)\%$ CI is

$$\frac{S_1^2}{S_2^2} \frac{1}{F(n_1 - 1, \, n_2 - 1, \, \alpha/2)} \leq \frac{\sigma_1^2}{\sigma_2^2} \leq \frac{S_1^2}{S_2^2} \frac{1}{F(n_1 - 1, \, n_2 - 1, \, 1 - \alpha/2)}.$$

Hypothesis tests

Let δ_0 be a given nonnegative number and let F_c be the calculated value of the test statistic

$$F = \frac{S_1^2}{S_2^2}.$$

Then the criteria for the α-level hypothesis tests are as follows.

Hypothesis		CI method	TS method
H_0	H_1	Accept H_1 if	Accept H_1 if
$\frac{\sigma_1^2}{\sigma_2^2} \leq \delta_0$	$\frac{\sigma_1^2}{\sigma_2^2} > \delta_0$	$\delta_0 \notin \text{LCI}$	$F_c > \delta_0 F(n_1 - 1, \, n_2 - 1, \, \alpha)$
$\frac{\sigma_1^2}{\sigma_2^2} \geq \delta_0$	$\frac{\sigma_1^2}{\sigma_2^2} < \delta_0$	$\delta_0 \notin \text{UCI}$	$F_c < \delta_0 F(n_1 - 1, \, n_2 - 1, \, 1 - \alpha)$
$\frac{\sigma_1^2}{\sigma_2^2} = \delta_0$	$\frac{\sigma_1^2}{\sigma_2^2} \neq \delta_0$	$\delta_0 \notin \text{CI}$	$F_c > \delta_0 F(n_1 - 1, \, n_2 - 1, \, \alpha/2)$ or
			$F_c < \delta_0 F(n_1 - 1, \, n_2 - 1, \, 1 - \alpha/2)$

Example 4.14 illustrates the use of the results in Box 4.7.

EXAMPLE 4.14 After a period of training, the technician in Example 4.12 was asked to analyze 15 more identical samples with the following results:

$$0.721, 0.731, 0.732, 0.741, 0.762, 0.713, 0.752, 0.738,$$
$$0.700, 0.728, 0.736, 0.746, 0.730, 0.739, 0.733$$

Compute a 95% confidence interval for σ_1/σ_2, the ratio of true standard deviations of the technician's measurements before and after training. Would you conclude that there is a demonstrable improvement after training?

Assuming that the measurements before and after training follow normal distributions with standard deviations σ_1 and σ_2, respectively, we can compute the required confidence interval using the CI formula in Box 4.7. We have $n_1 = 10$, $n_2 = 15$, $s_1^2 = 0.0062$, and $s_2^2 = 0.0002$. A 90% confidence interval for the ratio σ_1^2/σ_2^2 is

$$\left(\frac{0.0062}{0.0002}\right) \frac{1}{F(9, 14, 0.05)} \leq \frac{\sigma_1^2}{\sigma_2^2} \leq \left(\frac{0.0062}{0.0002}\right) \frac{1}{F(9, 14, 0.95)}$$

From Table C.4 (Appendix C), we find that $F(9, 14, 0.05) = 2.65$. To find $F(9, 14, 0.95)$, we use Equation (3.8)

$$F(9, 14, 0.95) = \frac{1}{F(14, 9, 0.05)} = \frac{1}{3.025} = 0.3305.$$

Therefore, the confidence interval for the ratio of variances is

$$\left(\frac{0.0062}{0.0002}\right) \frac{1}{2.65} \leq \frac{\sigma_1^2}{\sigma_2^2} \leq \left(\frac{0.0062}{0.0002}\right) \frac{1}{0.3305},$$

that is

$$11.6981 \leq \frac{\sigma_1^2}{\sigma_2^2} \leq 93.7973.$$

The 90% confidence interval for the ratio of standard deviations before and after training is then obtained by taking the square roots of the endpoints

$$\sqrt{11.6981} \le \frac{\sigma_1}{\sigma_2} \le \sqrt{93.7973} \qquad \text{or} \qquad 3.4202 \le \frac{\sigma_1}{\sigma_2} \le 9.6848.$$

A 90% confidence interval for the ratio of standard deviations after and before training is calculated by taking the inverses of the endpoints

$$\frac{1}{9.6848} \le \frac{\sigma_2}{\sigma_1} \le \frac{1}{3.4202} \qquad \text{or} \qquad 0.103 \le \frac{\sigma_2}{\sigma_1} \le 0.292.$$

With 90% confidence, we conclude that the standard deviation after training is at least 10.3% and at most 29.2% of the standard deviation before training. Furthermore, as noted in Exercise 3.31, the upper bound of 0.292 in the 90% confidence interval is also a 95% upper confidence interval for σ_2/σ_1. Thus, with 95% confidence, we can conclude that $\sigma_2/\sigma_1 < 0.292$; that is, that the standard deviation after training is less than 29.2% of the standard deviation before training.   ∎

EXAMPLE 4.15   For the N-loss data in Example 3.27, assume normal populations and check whether it is reasonable to assume equal population variances, as t-test procedures require.

We have $n_1 = 15$, $n_2 = 13$. Straightforward calculations yield $s_1^2 = 4.5784$ and $s_2^2 = 5.2364$. The assumption of equal variances is equivalent to the assumption that $\sigma_1^2/\sigma_2^2 = 1$. Taking $\delta_0 = 1$ in Box 4.7, we can test H_0: $\sigma_1^2/\sigma_2^2 = 1$ against H_1: $\sigma_1^2/\sigma_2^2 \ne 1$. The calculated value of the test statistic is

$$F_c = \frac{s_1^2}{s_2^2} = \frac{4.5784}{5.2364} = 0.8743$$

and from Table C.4 (Appendix C) we see that $F(14, 12, 0.10) = 2.117$ and $F(14, 12, 0.90) = 1/F(12, 14, 0.10) = 0.487$. Hence, the p-value exceeds 0.20. The hypothesis of equal population variances cannot be rejected even at a level of $\alpha = 0.20$. Thus the assumption of equal population variances is not contradicted by the observed data.   ∎

As already noted, a wide range of software can be used to implement the statistical procedures described in this book. Sections 4.4 and 4.7 of the companion text by Younger (1997) describe in detail how the TTEST procedure of SAS can be used to compare the means and variances of two populations. The following is a SAS printout generated when the TTEST procedure was used to analyze the pollution data in Example 4.8.

```
----------------------------------------------------------------
                        TTEST PROCEDURE
                           PART 1

   Variable: PPM
   LOC      N      Mean            Std Dev          Std Error
----------------------------------------------------------------
```

dwn	15	29.88000000	3.32655978	0.85891404
up	10	23.55000000	3.35898463	1.06220421

PART 2

Variances	T	DF	Prob>|T|
Unequal	4.6339	19.3	0.0002
Equal	4.6433	23.0	0.0001

PART 3

For HO: Variances are equal, F´ = 1.02   DF = (9,14)

Prob>F´ = 0.9396

This printout is divided into three parts. The first shows the sample sizes, the sample means, the sample standard deviations, and the estimated standard errors $(s/\sqrt{n})$ of the sample means.

The second part shows the calculated values of the test statistics first without assuming equal variances ($t_c = 4.6339$) and next under the assumption that the variances are equal ($t_c = 4.6433$). The numbers listed in the right-hand column (headed Prob>|T|) in the second table are the p-values for the corresponding two-sided t-tests. These p-values indicate a significant difference between the means regardless of whether the hypothesis is tested under the assumption that the population variances are equal.

Finally, the third part of the display shows the result of the F-test for comparing the variances of the two populations. The p-value of 0.9396 indicates that the observed data do not give credence to the hypothesis that the population variances are different.

Like the χ^2-procedures of Box 4.6, the F-test procedures in Box 4.7 may perform poorly if the requirement that the target populations are normally distributed is not met. When large sample sizes are available, it is possible to construct approximate confidence intervals and tests for the ratio of two variances. The details of such procedures fall outside the scope of the present discussion.

In Sections 4.3 and 4.4, we discussed procedures for making inferences about means and variances of one or two populations. These procedures are examples of the methods in Sections 3.8 and 3.9 where we considered estimation and hypothesis testing in the context of inferences about population parameters. It remains for us to describe how the information in samples may be used to predict future observations. Such methods are most easily described in terms of statistical models, which we consider next.

Exercises

4.12 The data below are the times (min) to complete two variations of a standardized aptitude test. The completion times were observed in an experiment in which the tests were given to two independent random samples of ten subjects each.

Test	Completion times	Mean	Variance
1	14.8, 11.4, 13.2, 16.0, 16.6, 13.3, 13.8, 14.5, 14.1, 12.7	14.04	2.3627
2	11.1, 11.7, 10.5, 11.9, 10.5, 8.4, 10.2, 11.0, 11.6, 7.8	10.47	1.8890

    a Perform a statistical test to see if the data indicate a difference between the variances of the completion times for the two aptitude tests. Write your conclusion.

    b State the assumptions on which your test in (a) depends.

    c Construct a 95% confidence interval for the difference $\mu_1 - \mu_2$ between the mean completion times in the populations of all subjects who take the two tests. Interpret the interval you obtain.

    d State the assumptions on which the confidence interval in (c) depends.

4.13 Refer to the auditory signal data in Exercise 4.3.

    a Describe the parameter that is estimated by the standard deviation of the sample.

    b Use the χ^2-test procedure to construct a 95% confidence interval for the parameter described in (a).

    c Interpret the interval constructed in (b).

    d State the assumptions needed to justify the procedures used to construct the interval in (b).

4.14 In the study described in Exercise 4.3, the behavioral scientist also determined the number of trials needed by rats to learn to recognize a visual signal. The following data were obtained for an independent sample of 15 rats:

$$15, 12, 18, 16, 14, 16, 17, 12, 13, 28, 22, 19, 21, 22, 26$$

    a Perform a hypothesis test to see if the data contradict the null hypothesis that the variability in the auditory measurements is the same as the variability in the visual measurements.

    b Based on the test in (a), suggest an appropriate method of constructing a confidence interval for the difference between the mean numbers of trials needed to learn to recognize the two types of signals. (Assume that the populations of interest have normal distributions.)

    c Use the procedure recommended in (b) to construct a 99% lower confidence bound for $\mu_V - \mu_A$, the difference between means of the populations from which the responses for the visual and auditory trials were measured.

    d State the conclusions that can be drawn from the bound calculated in (c).

4.15 Let σ_B and σ_G denote, respectively, the standard deviations of the dental measurements of the boys and girls, respectively, in Exercise 4.8.

    a Making suitable assumptions, construct a 95% confidence interval for σ_B/σ_G.

    b Interpret the interval constructed in (a).

    c State the assumptions that were made in (a).

4.16 In Exercise 4.9b, you were asked to construct a lower confidence bound for the differential mean lymphocyte counts under the assumption that the relevant populations have normal distributions with a common variance. Perform an appropriate test to see if the observed data contradict the assumption of equal population variances.

4.17 Work Exercise 4.4 using a suitable statistical computing software.

4.18 Work Exercise 4.8 using a suitable statistical computing software.

4.19 Work Exercise 4.11 using a suitable statistical computing software.

4.5
Statistical models

As we have seen, the assumptions that can be made about our data will determine which inferential procedure is most appropriate for answering the research question. In choosing the best technique, we need to ask questions such as the following: Can we assume that the data are measured on an interval scale? Can we assume that the data are independent random samples from normal populations? Is it reasonable to assume that the population variances are equal? A *statistical model* is a convenient mathematical tool for summarizing the assumptions about a set of measurements. Box 4.8 shows a statistical model for a random sample of size n from a normal population with mean μ and variance σ^2.

BOX **4.8**

> ### A statistical model for a random sample from a $N(\mu, \sigma^2)$ population
>
> Let $Y_j, j = 1, 2, \ldots, n$, denote the j-th observation in the sample. Then
>
> $$Y_j = \mu + E_j, \qquad\qquad j = 1, 2, \ldots, n,$$
>
> where μ is the population mean and $E_j = Y_j - \mu$ is the amount by which the j-th observation differs from the population mean. The values $E_1, E_2, \ldots, E_n$ are assumed to constitute a random sample from a $N(0, \sigma^2)$ population; we can express this more concisely by saying that the E_j are independent $N(0, \sigma^2)$.

The statistical model in Box 4.8 is one of the simplest encountered in practice. Notice that it represents each observation as a sum of two components, one fixed and one random. The fixed component μ, which remains the same for all responses in the sample, is the mean of the population from which the sample will be selected. The random component E_j, which varies from observation to observation, is called the *error component* of the observed value. The error component of an observed value is simply the amount by which the observed value deviates from the population mean. There is no implication that an error has been made in the measurement process. The error term of the model in Box 4.8 is illustrated in Figure 4.1 for two responses, one above and one below the population mean.

Even though the model represents each Y_j as the sum of μ and E_j, neither μ nor E_j are observable. Because the observations are selected from a normal population with mean μ and standard deviation σ, the errors could be thought of as a random sample

FIGURE   4.1

Errors in two observations from a normally distributed population

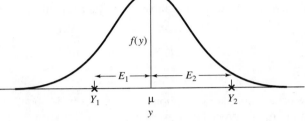

from a conceptual population of errors obtained when each measurement Y in the target population is replaced by $E = Y - \mu$, its deviation from the population mean. The Y population has mean μ, and so the mean of the E population will be zero. Also, measurements in the population of errors are the same as measurements in the Y population shifted by a constant amount μ, and so both populations will have the same standard deviation (because, as we know, the standard deviation remains unaltered when a constant is subtracted from every measurement) and their distributions are of the same shape. Thus, the requirement that the errors constitute a random sample from a $N(0, \sigma^2)$ population is the same as the requirement that the responses are a random sample from a $N(\mu, \sigma^2)$ population.

The relationship between the distribution of the Y values and the distribution of the E values is shown in Figure 4.2. Notice that the Y and E populations are two normally distributed populations with equal variances and the means shifted by an amount μ.

FIGURE   4.2

Relationship between the target population and the population of errors in Box 4.8

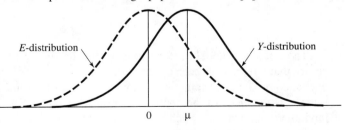

EXAMPLE 4.16   In Example 4.6, we analyzed 15 N-loss measurements under the assumption that these measurements can be regarded as a random sample of size $n = 15$ measurements from a $N(\mu, \sigma^2)$ population. The model in Box 4.8 can be adapted to summarize the assumptions about the N-loss data as follows

$$Y_j = \mu + E_j, \qquad j = 1, 2, \ldots, 15,$$

where μ is the population mean N-loss for fertilizer UN and E_j is the random error in Y_j; the E_j are independent $N(0, \sigma^2)$. ■

All statistical models considered in this book have the same basic structure as that in Box 4.8. Each observed response is regarded as the sum of two components: a fixed quantity representing a location parameter of the population; and a random quantity representing the amount by which the particular observation deviates from the location parameter. In most traditional models, such as the ANOVA models introduced in Chapter 8 and the regression models introduced in Chapter 10, the fixed quantity is the mean of the corresponding population (as in Box 4.8). In Chapter 5, we will see examples of statistical models in which the fixed components are the population medians. In this book, we refer to the fixed component as the *expected response*. When the fixed component is the population mean, the expected response is the response we expect to observe on average. In that case we will use the notation $\mathcal{E}(Y_j)$ introduced in Chapter 3, to denote the expected response. Thus in Box 4.8, $\mu = \mathcal{E}(Y_j)$.

The statistical model in Box 4.9 summarizes the assumptions underlying the use of the independent-sample t-test in Box 4.3.

BOX **4.9**

A model for two independent samples from normal populations with a common variance

Let Y_{ij} denote the j-th observation in the i-th sample, $j = 1, 2, \ldots, n_i$; $i = 1, 2$. Then

$$Y_{ij} = \mu_i + E_{ij}$$

where μ_i is the mean of the i-th population and E_{ij} is the error in Y_{ij}; the E_{1j} and the E_{2j} are independent random samples of sizes n_1 and n_2, respectively, from a $N(0, \sigma^2)$ population; that is, the E_{ij} are independent $N(0, \sigma^2)$.

The model in Box 4.9 uses two subscripts. The first subscript i identifies the population number and takes the two values 1 and 2. The second subscript j, which identifies the particular observation within a sample, takes n_i values for population number i; that is, for the sample of size n_1 from population 1, the subscript j takes the values $1, 2, \ldots, n_1$, and, for the sample of size n_2 from population 2, j takes the values $1, 2, \ldots, n_2$. As in the one-sample model in Box 4.8, the expected response for observations from population i is μ_i. That is, $\mu_i = \mathcal{E}(Y_{ij})$. The error E_{ij} in the j-th response in the i-th sample can be interpreted as the difference between the j-th observed response and its expected value. Furthermore, the errors E_{1j} and E_{2j} in the j-th responses in samples from populations 1 and 2 are independent $N(0, \sigma^2)$. Thus, the model in Box 4.8 is just another way of stating that the independent-sample t-test procedures rest on the following assumptions:

1   The data are independent random samples from two populations.

2   Both populations have normal distributions.

3   The two populations have equal but unknown variances.

Figure 4.3 displays the relationship between the distributions of the two target populations (assumed to be normal with means μ_1 and μ_2 and a common standard deviation σ) and the population of errors.

FIGURE   4.3
Relationship between the target and error populations in Box 4.9

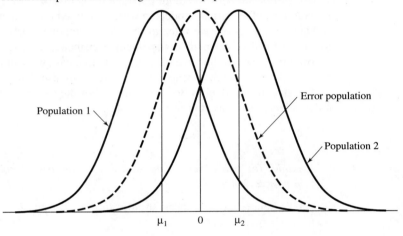

The population of errors can be conceptualized as the measurements in population 1 minus μ_1, the mean of population 1, or the measurements in population 2 minus μ_2, the mean of population 2. Because μ_1 is negative in Figure 4.3, the error population is the same as population 1 shifted to the right by μ_1 units. Similarly, μ_2 is positive, and so the error population can also be obtained by shifting population 2 to the left by μ_2 units.

EXAMPLE 4.17   The statistical model in Box 4.9 may be adapted as follows to summarize the assumptions in Example 4.8 about the measured chemical concentration levels

$$Y_{ij} = \mu_i + E_{ij}, \qquad i = 1, 2;$$
$$j = 1, 2, \ldots, 10 \text{ if } i = 1 \text{ and } j = 1, 2, \ldots, 15 \text{ if } i = 2,$$

where μ_1 is the upstream true mean chemical concentration; μ_2 is the downstream true mean chemical concentration; and E_{ij} is the random error in Y_{ij}. The E_{ij} are independent $N(0, \sigma^2)$.   ■

Exercises

4.20   Write a statistical model that summarizes the assumptions necessary to use the *t*-test procedures on the auditory stimulus data in Exercise 4.3.

4.21   Write a statistical model that summarizes the assumptions needed to justify the one-sample *t*-test procedures for analyzing the corn ear length data in Exercise 4.6.

4.22   Write a statistical model to summarize the assumption that the N-loss data in Example 3.27 are independent random samples from normal populations with a common variance of 4.

4.23   Write a statistical model to summarize the assumption that the dental data in Exercise 4.8 are random samples from normal populations with a common variance.

4.24   Write a statistical model summarizing the assumption that the lymphocyte count data in Exercise 4.9 are independent random samples from normally distributed populations with unknown and possibly unequal variances.

4.6
Predicting future observations

As we saw in Section 3.10, a $100(1 - \alpha)\%$ prediction interval for a future observation Y is an interval $(Y_L,\ Y_U)$ that contains Y with probability $1 - \alpha$.

A $100(1 - \alpha)\%$ prediction interval for a future observation from a normal distribution with a known variance σ^2 was given in Equation (3.22). When the population variance is not known, as is often the case, the prediction interval in Equation (3.22) has to be modified: The population standard deviation σ is replaced by its sample estimate, and the critical value $z(\alpha)$ of the standard normal distribution is replaced by an appropriate critical value of a t-distribution. Box 4.10 contains the $100(1 - \alpha)\%$ prediction intervals for the mean of a future sample of m observations taken from a normally distributed population with an unknown variance. The prediction intervals for a single future observation are obtained by setting $m = 1$.

BOX 4.10

$100(1 - \alpha)\%$ prediction interval for the mean of a future sample of m observations from a normal distribution

Let $\overline{Y}$ and S^2 denote the mean of a random sample of size n from a $N(\mu,\ \sigma^2)$ population. The prediction bounds for the mean of m future observations from the same population are as follows.

The $100(1 - \alpha)\%$ lower prediction bound (LPB) is

$$\overline{Y} - t(n - 1,\ \alpha)\sqrt{\left(\frac{1}{m} + \frac{1}{n}\right) S^2}.$$

The $100(1 - \alpha)\%$ upper prediction bound (UPB) is

$$\overline{Y} + t(n - 1, \ \alpha)\sqrt{\left(\frac{1}{m} + \frac{1}{n}\right)S^2}.$$

The $100(1 - \alpha)\%$ prediction interval (PI) is

$$\overline{Y} - t(n - 1, \ \alpha/2)\sqrt{\left(\frac{1}{m} + \frac{1}{n}\right)S^2};$$

$$\overline{Y} + t(n - 1, \ \alpha/2)\sqrt{\left(\frac{1}{m} + \frac{1}{n}\right)S^2}.$$

An intuitive explanation of the formulas for these intervals can be obtained by exploring their connection to the confidence intervals for the population mean. To do so, let $\overline{Y}_f$ denote the mean of a future sample of size m. Then, from property III in Box 3.2, the sampling distribution of $\overline{Y}_f$ is a normal distribution with mean μ and variance $\sigma_{\overline{Y}_f}^2 = \sigma^2/m$. In other words, $\overline{Y}_f$ can be regarded as an observed response from a $N(\mu, \ \sigma^2/m)$ population. Consequently, $\overline{Y}_f$ can be expressed as

$$\overline{Y}_f = \mu + E, \tag{4.4}$$

where E has a $N(0, \ \sigma^2/m)$ distribution. The model in Equation (4.4) suggests that an interval for predicting $\overline{Y}_f = \mu + E$ can be obtained by first constructing a confidence interval for μ and then expanding it in such a way that the expanded interval has probability $1 - \alpha$ of containing $\mu + E$.

The prediction bounds in Box 4.10 can be interpreted as intervals obtained in precisely this manner. The intervals are the expanded versions of $100(1 - \alpha)\%$ confidence intervals for μ; the expansion results from the addition of the term S^2/m, the estimated variance of E, to the term S^2/n, the estimated variance used in the construction of a confidence interval for μ.

EXAMPLE **4.18**   The survival times (hours) of ten mice inoculated with a particular dose of an antibiotic are as follows:

24.5, 25.0, 26.3, 26.7, 27.3, 28.4, 28.6, 28.6, 29.0, 30.1

Assume that the logarithm of the survival time has a normal distribution and use the given data to construct a 95% prediction interval for the survival time of a new mouse inoculated with the same dose of the antibiotic.

Log-survival times can be regarded as a random sample from a normal distribution, and so we work with the transformed variable $Y = \log_e(\text{survival time})$. The transformed data are

3.199, 3.219, 3.270, 3.285, 3.307, 3.346, 3.353, 3.353, 3.367, 3.405

The transformed data yield $\bar{y} = 3.3104$, $s^2 = 0.00446$, and $s = 0.067$. A 95% prediction interval for a future observation of $Y = \log_e(\text{survival time})$ of a new mouse can be constructed by setting $n = 10$ and $m = 1$ in Box 4.10. The resulting prediction interval is

$$\bar{y} \pm t(9,\ 0.02)\sqrt{\left(\frac{1}{1} + \frac{1}{10}\right)s^2} = 3.3104 \pm 2.262\sqrt{(1 + 0.01)0.00446}$$

$$= (3.1586, 3.4622).$$

A 95% prediction interval for the survival time can be obtained by raising to the power e the endpoints of the interval for $\log_e(\text{survival time})$. Thus, a 95% prediction interval for a future observation of the survival time is $(e^{3.1586},\ e^{3.4622}) = (23.54, 31.89)$. Therefore, we can predict with 95% confidence that a mouse injected with the given dose of the antibiotic will survive between 23.54 and 31.89 hours. ∎

Prediction intervals as normal limits

Prediction intervals are useful when screening large populations to identify individuals who are at risk for abnormalities. In such applications, each subject is classified as low-risk (normal) or high-risk (abnormal), and the endpoints of the $100(1 - \alpha)\%$ prediction interval for a response from the low-risk population are called the $100(1 - \alpha)\%$ *normal limits*. The $100(1 - \alpha)\%$ normal limits provide an interval containing $100(1 - \alpha)\%$ of the values in a normal (low-risk) population. Such limits are calculated using large samples from populations known to be normal (low-risk). If we select a small value for α (say, $\alpha = 0.05$) and a large sample size, then an individual with a value outside the calculated interval will be regarded as abnormal, because the probability that the value for a normal individual will be outside this interval is small (approximately α). If, for example, a 6-month-old calf has more antibodies to bovine leukemia virus (BLV) than the upper bound of 95% normal limits calculated from a large sample of calves of similar age, then BLV infection of that calf may be suspected. This conclusion is justified in that only about 2.5% of all noninfected calves have more BLV antibodies than this upper bound. The use of the results in Box 4.10 to construct normal limits is illustrated in the next example.

EXAMPLE 4.19   Some programs to eradicate bovine leukemia virus from dairy herds rely on separation of infected animals from the rest of the herd. However, the identification of infected animals may be difficult, because calves are often fed on pooled colostrum, the so-called first milk, with high antibody levels. Such calves yield abnormally high titers (amount of antibody) during their first few months of life, even though they may not be infected. To permit early identification of infected calves, a researcher measured BLV antibody titers in $n = 202$ 3-month-old uninfected calves who had been fed colostrum containing BLV antibodies.

The average logarithm of BLV antibody titers in these calves was $\bar{y} = 0.30$, with a standard deviation $s = 0.16$. Assume that the logarithm of the BLV values for uninfected calves is normally distributed. Then a 95% upper prediction bound on log titers is obtained from Box 4.10 as

$$0.30 + t(201, 0.05)\sqrt{(1 + 1/202)(0.16)^2} = 0.57.$$

Because the population sampled consisted of normal, uninfected 3-month-old calves and the sample size is large, the 95% UPB is an approximate upper normal limit for the log of BLV antibody titers. Thus, for example, a 3-month-old calf with a log titer larger than 0.57 (say, 0.70) is likely to be truly infected and should be separated from noninfected animals.　■

Exercises

4.25 Refer to the infection rate data in Exercise 4.4.

  a Construct a 95% lower prediction bound for the average infection rate of a sample of five HCL patients.

  b Interpret the bound in (a).

  c Explain why you would expect this bound to be less than the corresponding 95% lower confidence bound for the mean infection rate.

4.26 Refer to the social perceptiveness data in Exercise 4.5.

  a Construct a 99% prediction interval for the social perceptiveness index.

  b Interpret the interval constructed in (a).

4.27 Elevated serum creatinine level is an indicator of renal (kidney) damage. A sample of 1000 asymptomatic subjects with no renal disease yielded a mean creatinine level $\bar{y} = 1.10$ with a standard deviation $s = 0.35$.

  a Use these data to construct 95% normal limits for the serum creatinine level of a subject with no renal damage.

  b Explain the meaning of the limits calculated in (a).

  c What conclusions can you draw about a subject whose serum creatinine level is 2.0?

4.28 The data in Example 4.9 were obtained from normal (nonlymphocytotic) cows. Assume that the square root of the lymphocyte counts is normally distributed.

  a Use these data to estimate 95% upper normal limits on lymphocyte counts.

  b What conclusions would you draw about a Holstein cow with a lymphocyte count of 9548 and a Guernsey cow with a count of 6500?

4.7
The assumptions for t-, χ^2-, and F-test procedures

Because standard deviations—and hence standard errors—are virtually never known in practice, the procedures based on t-tests are very commonly used to make inferences about population means. Similarly, the procedures based on the χ^2- and F-tests are frequently used to make inferences about population variances. As we have already seen, these procedures are designed for interval data and depend for their validity on one or more of the following assumptions.

1   The populations have normal distributions.

2   The population variances are equal but unknown.

3   The data are independent random samples from the populations.

In practice, it is too much to expect that there will be no violations of these assumptions. The best we can hope for is that the assumptions are reasonable—that any violations will not seriously affect the quality of the inferences made using the procedures. Thus, it is important to know how sensitive a given procedure is to violations of its underlying assumptions. Is it possible that small departures from assumptions will cause serious problems for the procedure? For instance, if the population distribution is not exactly normal, will the actual level of a 0.05-level one-sample *t*-test equal 0.05? If the population variances are not equal, will the actual confidence coefficient of a 95% confidence interval for the difference between two population means be 0.95? Answers to such questions are necessary to assess the appropriateness of a statistical procedure for a particular set of data. The effects of violating the assumptions of one- and two-sample tests will not be considered in depth here. In the following, we present a brief account of this topic, which is considered in more detail in Chapter 10 of Scheffe (1959) and Chapters 1 and 2 of Bradley (1968).

First, consider assumption 1. In the present context, the assumption that a target population has a normal distribution can be violated in two important ways: because the population distribution is lacking in symmetry; or because it is less or more flat than a normal distribution. A population whose distribution is not symmetric is said to be *skewed*. Figure 4.4 shows two skewed distributions. Recall that we came across skewed distributions when discussing population medians in Section 2.4. The distribution in Figure 4.4a is said to be skewed to the right because the range of the values in the populations is larger to the right of the median than to the left of the median. For similar reasons, the distribution in Figure 4.4b is said to be skewed to the left.

FIGURE   4.4
Skewed distributions

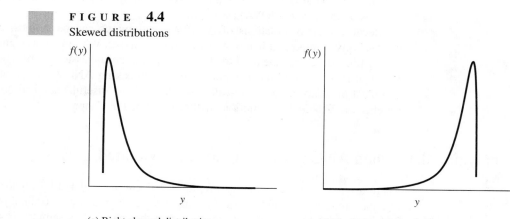

(a) Right-skewed distribution   (b) Left-skewed distribution

Figure 4.5 shows three distributions—a normal distribution, a distribution that is flatter than the normal, and a distribution that has a higher concentration of measurements near the median than the normal. The distribution that is flatter than normal has a greater proportion of measurements far from the median. Such a distribution is said to be *heavy-tailed*. Recall that we saw an example of a heavy-tailed distribution in Figure 2.4. For a similar reason, the distribution with a higher concentration of measurements in the center is said to be *light-tailed*.

Generally speaking, the procedures based on *t*-tests (especially one-sided procedures) are more sensitive to skewing of the populations than to heaviness or lightness of their tails. If the populations are reasonably symmetric, the *t*-test procedures usually perform relatively well, regardless of whether the population distribution is truly normal. For populations with heavy tails, however, some of the nonparametric alternatives discussed in Chapter 5 are preferable to *t*-tests. The χ^2- and *F*-tests for inferences about population variances are more sensitive to violations of assumption 1 than are the procedures based on *t*-tests. Consequently, the results of these tests should be viewed with caution if there is reason to believe that the populations may be nonnormal.

Violations of assumption 2—that the population variances are equal—can have serious consequences on the independent-sample *t*-test, especially if the sample sizes are not equal. However, when the sample sizes are nearly equal, the performance of procedures based on independent-sample *t*-tests tends to be satisfactory even when the variance of one population is as much as three times the variance of the other population.

Finally, assumption 3—that the samples are independent—concerns the way that the responses are measured. Violation of this assumption calls for appropriate modification of the procedures. For example, it might be necessary to replace the independent-sample *t*-test by the paired-sample *t*-test.

In Section 8.3, we will see that the assumptions needed for the validity of the independent-sample *t*-tests are also the assumptions underlying analysis of variance (ANOVA). ANOVA is a statistical procedure that extends the independent-sample and paired-sample *t*-tests to situations where comparisons of the means of more than two populations are desired. For this reason, assumptions 1–3 are often called the analysis of variance (ANOVA) assumptions.

Because certain violations of ANOVA assumptions can be serious, checking their validity is important in any statistical data analyses. While various significance-testing procedures may be used to check many of the usual assumptions (such as the *F*-test in Box 4.7), the most popular tools for checking ANOVA assumptions are graphical methods based on so-called residuals. Some useful graphical procedures for checking ANOVA assumptions will be described in Chapter 8.

Analyzing data when ANOVA assumptions are violated

There are two ways in which assumptions 1–3 can be violated by a particular set of data. First, the data may consist of measurements made on an ordinal or a nominal scale. Clearly, such measurements cannot be regarded as samples from a normally distributed population and therefore do not satisfy assumption 1. Methods for

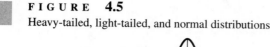

FIGURE   **4.5**

Heavy-tailed, light-tailed, and normal distributions

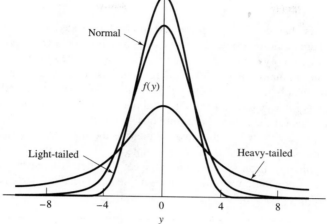

making inferences about one or two populations when the data are measured on ordinal and nominal scales are described in Chapters 5 and 6, respectively. Second, the
data, though measured on an interval scale, may not satisfy one or more of the three
assumptions. In such cases, there are two options when selecting inference procedures. One is to take advantage of the fact that an interval variable can be treated as
an ordinal or nominal variable. Frequently, the required inferences can be made using
the methods described in Chapters 5 and 6. The other option is to see if the measurements could be transformed in such a way as to satisfy assumptions 1–3. If so, it is
very often possible to use the *t*-test, χ^2-test, and *F*-test procedures to make the required inferences. The analysis of transformed data will be considered in Section 8.7.

Exercises

4.29   Using a box-plot of the ear length data in Exercise 4.6, determine whether the sample
indicates:

a   that the population is skewed;

b   that the population has heavy tails, as indicated by the presence of extreme
values.

4.30   Using box-plots of the dental data for boys and girls in Exercise 4.8, compare the
following characteristics of the two populations:

a   the variances of the populations;   **b**   the skewness of the populations;

c   the heaviness of the tails of the populations.

4.31   Use a box-plot to see if the data in Exercise 4.12 indicate:

a   asymmetry in the populations of completion times for the two tests.

b   heaviness of tails in the populations of completion times for the two tests.

4.8
Overview

In this chapter, we have presented inferential procedures for use with interval data that are random samples from one or two populations. These procedures are based on normal theory: Either the target populations sampled must be normally distributed, or the estimator on which the inference is based must have a normal—or at least approximately normal—sampling distribution. We have discussed confidence intervals and hypothesis tests for inferences about means, and variances, as well as prediction intervals for the mean of m future observations.

In the following table, we summarize the methods described in this chapter in terms of the type of data (one sample vs. two sample), the parameter of interest (population mean, population variance, difference between population means and ratio of population variances), the sample size (small vs. large) and assumptions that are appropriate to the target population(s).

Data type	Parameter of interest	Sample size(s)	Inference procedure	Assumptions
One sample	μ	large	z-test	random sample
		small	t-test	random sample from a normal population
	σ	small	χ^2-test	random sample from a normal population
Two samples	$\mu_1 - \mu_2$	large	z-test	independent random samples
		small	t-test	independent random samples from normal populations with common σ^2
			approximate t-test	independent random samples from normal populations
	σ_1^2/σ_2^2	small	F-test	independent random samples from normal populations
Paired sample	$\mu_1 - \mu_2$	large	z-test	random sample of differences
		small	t-test	random sample of differences from a normal population

The concept of a statistical model was introduced for the first time in this chapter. As we will see, statistical models play a key role when the techniques described in this chapter are extended to more complex situations involving more than just two target populations.

5

Inferences about One or Two Populations: Ordinal Data

5.1
Introduction

In Chapter 4, we presented several procedures for using interval data to make inferences about one or two populations. In the absence of large sample sizes, the efffectiveness of these procedures depends on the validity of certain assumptions about the data. For example, the independent sample t-test for comparing two population means requires the assumption that the data are independent random samples from normal distributions with a common variance.

In this chapter, we describe three common procedures that can be used with ordinal data to make inferences about one or two populations. As noted in Chapter 4, these methods are also appropriate for interval data, especially when the assumptions underlying t-test procedures are of uncertain validity.

The ordinal data procedures discussed in this chapter are known as distribution-free methods based on ranks. They are said to be *distribution-free* because their validity does not depend on the form of the population distributions. It is often enough to assume that, for instance, the population distribution is continuous or is continuous and symmetric.

Three distribution-free procedures are described in this chapter. The sign test procedures in Section 5.3 can be used for making inferences about the median of a continuous population assuming only that the population distribution is continuous. Under the additional assumption that the population distribution is symmetric, the procedures based on the Wilcoxon signed rank test, described in Section 5.4, can be used for drawing conclusions about the population median. Finally, the Wilcoxon rank sum test procedures of Section 5.5 are designed for comparing the locations (medians or means) of two target populations on the basis of independent samples of measurements that can be ordered using an ordinal scale.

5.2
Ranks of ordinal measurements

The *rank* of a particular measurement in a set of ordinal data specifies its position when the measurements are arranged in order of increasing ordinal levels. For example, in a taste-testing trial, tasters may be asked to rate the flavors of several brands of ice cream. Suppose four brands, A, B, C, and D, are rated in appeal by a taster as $D < A < C < B$. Brand A occupies the second position when the brands are arranged in increasing order of their ratings, and so we assign it the rank 2. Similarly, the ranks for B, C, and D are, respectively, 4, 3, and 1. Unfortunately, it is not always possible to give a unique rank to each value in a set of measurements. For example, if the taster is not able to distinguish between brands A and C, the rating of the four brands will be $D < A = C < B$. In that case, there is no unique rank for A, because there are two ways in which the ratings can be arranged: $D < A = C < B$; and $D < C = A < B$. Both 2 and 3 are equally valid as the rank for Brand A or Brand C. We say that the ratings for brands A and C are *tied* and assign each of them the average of the two possible ranks: $\frac{1}{2}(2 + 3) = 2.5$. The resulting ranks for A, B, C, and D are, respectively, 2.5, 4, 2.5, and 1.

This rule for breaking ties allows us to determine unique ranks for each measurement in a set of ordinal measurements; the resulting ranks are called *mid-ranks*. While there are other rules for breaking ties, mid-ranks will always be used in this book.

The ranks of a set of values measured on an interval scale can be determined in the same manner. For example, the set of eight measurements {2, 12, 38, 12, 6, 12, 4, 2} can be arranged in increasing order as follows

$$2 = 2 < 4 < 6 < 12 = 12 = 12 < 38.$$

Accordingly, they may be assigned the following ranks:

Measurement	Rank
2	$\frac{1}{2}(1 + 2) = 1.5$
4	3
6	4
12	$\frac{1}{3}(5 + 6 + 7) = 6$
38	8

To rank the measurements in a particular data set, it is convenient to look for groups of measurements with a common value. Thus, for the set {2, 12, 38, 12, 6, 12, 4, 2}, we can identify five groups: {2, 2}, {4}, {6}, {12, 12, 12}, and {38}. Each group is called a *tied group*, and the number of elements in it is called the *tie size*. In this example, we have five tied groups of sizes 2, 1, 1, 3, and 1. Clearly, the number of groups in a data set that has no tied ranks is the same as the number of measurements in the data set. The size of each tied group will be 1 in that case.

Exercises

5.1   The following are data on socioeconomic status (SES)—low (L), medium (M), or hight (H)—for a random sample of $n = 8$ females:

Subject	1	2	3	4	5	6	7	8
SES	L	L	H	M	L	M	L	M

a   Determine the tied groups and tie sizes in the SES data.
b   Determine the ranks for the SES data.

5.2   Refer to the infection rate data in Exercise 4.4.

a   Determine the tie groups and tie sizes.
b   Determine the ranks.

5.3   Refer to the lymphocyte count data in Exercise 4.9.

a   Determine the tie sizes and tie groups associated with the 23 lymphocyte counts obtained by combining the data for Holstein and Guernsey cows.
b   Determine the ranks associated with the combined data in (a).
c   Compare the average of the ranks for the Holstein cows with that for the Guernsey cows and state the conclusions that can be drawn from this comparison.

5.3
Procedures based on the sign test

Let's begin with an example in which the assumptions underlying the t-test and χ^2-test procedures may be violated by the data.

EXAMPLE 5.1   Schrier and Junga (1981) studied the entry and distribution of the drugs chloropromazine (CPZ) and vinblastine (VBL) into human red blood cells during endocytosis (absorption). The following data are the percentages of CPZ found in membrane fractions of red blood cells suspended in Hanks' solution or sucrose for samples from eight donors; the data for donors 7 and 8 are fictitious and have been included to facilitate illustration of some specific data analysis techniques.

				Donor				
Medium	1	2	3	4	5	6	7	8
Hanks' solution	5	6	7	15	17	14	14	27
Sucrose	9	4	7	13	16	22	21	26

Note that the two CPZ values for cells from the same donor are likely to be related to each other, and so these data should be regarded as a sample of $n = 8$ pairs of measurements. The data can be used to draw conclusions about either of the two populations: the values obtained with Hanks' solution and the values for sucrose. If we can assume normal distributions for either of the two populations, then the one-sample t-test procedures of Box 4.2 and the χ^2-procedures of Box 4.6 can be used to make inferences about the means and variances of the corresponding population. If the population of the differences between the measurements within pairs can be assumed to be normally distributed, then the paired t-test procedures in Box 4.5 can be used to make inferences about the difference between the means of the two populations.

However, the procedures based on normally distributed populations may not be appropriate for these data. Practical experience indicates that instrument errors associated with the detection of low levels may cause the distribution of such measurements to be skewed to the right. The reader can see the tendency towards right-skewed distributions in the following computer printout (obtained using SAS software), which shows stem-leaf plots and box-plots of the CPZ values observed in the two mediums.

```
----------------------------------------------------------------
                         The SAS System
                      Univariate Procedure

Variable=HANKs´

          Stem Leaf                    #          Boxplot
            2 7                        1             |
            2                                        |
            1 57                       2          +-----+
            1 44                       2          *--+--*
            0 567                      3          +-----+
              ----+----+----+----+
          Multiply Stem.Leaf by 10**+1

Variable=SUCROSE

          Stem Leaf                    #          Boxplot
            2 6                        1             |
            2 12                       2          +-----+
            1 6                        1          |  +  |
            1 3                        1          *-----*
            0 79                       2          +-----+
            0 4                        1             |
              ----+----+----+----+
          Multiply Stem.Leaf by 10**+1

----------------------------------------------------------------
```

Of course, plots based on eight measurements do not provide reliable information about the population distributions but, if supported by experience with such data, they suggest that it may be preferable in this case to use distribution-free inferential techniques, which do not require the populations to have normal distributions.   ∎

Various distribution-free procedures may be employed to make inferences about the location parameter of a population distribution. The sign test and the Wilcoxon signed rank test procedures are two of the most commonly used. Both are designed for inferences about the median of a continuous but not necessarily normal population. The mean and median are the same for a normal distribution, and so the sign test and the Wilcoxon signed rank test can also be used to make inferences about the mean of a normal population. Therefore, these two procedures are alternatives to the procedures based on the one-sample t-test and are useful in situations where violations of the normal-distribution assumption may be serious enough to invalidate one-sample t-test procedures.

In this section, we look at distribution-free procedures based on the sign test. Procedures based on the Wilcoxon signed rank test will be described in the next section.

Sign test procedures

Let η be the median of a continuous population. Sign test procedures are based on the *sign statistic,* whose value equals the number of measurements in the sample that exceed a hypothesized population median η_0. More precisely, if $\{y_1, y_2, \ldots, y_n\}$ is a random sample from a continuous population with median $\eta = \eta_0$, then the sign statistic is defined as

$$B = \text{the number of } y_i > \eta_0. \tag{5.1}$$

As its name suggests, the sign statistic is also equal to the number of values in the modified sample $\{y_1 - \eta_0, y_2 - \eta_0, \ldots, y_n - \eta_0\}$ that have positive signs. The sign test procedures are described in Box 5.1.

BOX **5.1**

The sign test procedure for inferences about a population median

Let $y_{(1)} < y_{(2)} < \cdots < y_{(n)}$ denote the order statistics of a random sample of size n from a continuous population with median η. Then the sign test procedure for inferences about η is as follows.

Point Estimation

A point estimate for the population median is the sample median

$$\hat{\eta} = \begin{cases} \frac{1}{2}(y_{(m)} + y_{(m+1)}) & \text{if } n = 2m \\ y_{(m+1)} & \text{if } n = 2m + 1. \end{cases}$$

Interval Estimation

The $100(1 - \alpha)\%$ confidence interval for η is

$$y_{(L)} < \eta < y_{(U)},$$

where $U = b(n, \alpha/2)$, $L = n + 1 - b(n, \alpha/2)$, and $b(n, \alpha)$ is the α-level critical value for the sign test. This critical value can be determined using statistical software such as SAS or StatXact or from Table C.5 in Appendix C. If n is outside the range of Table C.5, the following large sample approximation can be used

$$b(n, \alpha/2) \simeq \frac{n}{2} + z(\alpha/2)\sqrt{\frac{n}{4}},$$

where $z(\alpha/2)$ is the $\alpha/2$-level critical value of a standard normal distribution.

Hypothesis Test

Let η_0 be a specified value of η and let B_c and n^* denote, respectively, the calculated value of the statistic B in Equation (5.1) and the number of nonzero differences in the modified sample $y_1 - \eta_0, y_2 - \eta_0, \ldots, y_n - \eta_0$. Consider three sets of hypotheses: (1) H_0: $\eta \leq \eta_0$ against H_1: $\eta > \eta_0$; (2) H_0: $\eta \geq \eta_0$ against H_1: $\eta < \eta_0$; and (3) H_0: $\eta = \eta_0$ against H_1: $\eta \neq \eta_0$. The rejection criteria for the three null hypotheses are as follows:

1   Reject H_0: $\eta \leq \eta_0$ at level α if $B_c \geq b(n^*, \alpha)$.

2   Reject H_0: $\eta \geq \eta_0$ at level α if $B_c \leq n^* - b(n^*, \alpha)$.

3   Reject H_0: $\eta = \eta_0$ at level α if $B_c \geq b(n^*, \alpha/2)$ or if $B_c \leq n^* - b(n^*, \alpha/2)$.

Connection between sign test and binomial experiment

In Box 2.3, a binomial experiment, also called a $B(n, \pi)$ experiment, was defined as an experiment for counting the number of successes in n independent trials, each of which can result in a success with probability π. The probability of observing y successes in a binomial experiment with n trials can be calculated using the binomial probability function in Equation (2.10). If observing $y_i > \eta_0$ is regarded as a success in the i-th trial, and π denotes the probability that y_i will exceed η_0 (that is, the probability of a success), then B in Equation (5.1) equals the total number of successes in a $B(n, \pi)$ experiment.

Now, if η_0 is the true population median—that is, if $\eta = \eta_0$—then the probability that y_i will be larger than η_0 is $\pi = 1/2$. Consequently, if the hypothesis H_0: $\eta = \eta_0$ is true, the sampling distribution of B is a binomial distribution with n trials and $\pi = 1/2$. The critical values of the $B(n, 0.5)$ distributions listed in Table C.5 in Appendix C can be used for testing H_0: $\eta = \eta_0$.

The form of the rejection region for a sign test can be determined as follows. If $\eta > \eta_0$, then more than half $(n/2)$ of the measurements in the sample may be expected to exceed η_0 (Figure 5.1a). Indeed, increase in the population median will

FIGURE 5.1

Relationship of the population median η to the expected value of B

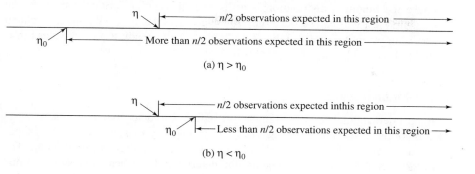

(a) $\eta > \eta_0$

(b) $\eta < \eta_0$

be accompanied by increase in the number of sample values that are greater than η_0. In other words, larger values of B favor the research hypothesis that the true median is larger than η_0. In the sign test, large values of B favor the conclusion that the true population median is greater than η_0. Similarly, small values of B support the research hypothesis that the population median is less than η_0 (Figure 5.1b).

Point and interval estimates based on sign test

The use of the sample median to estimate the population median is intuitively reasonable, but the formula for the confidence interval may look odd at first sight. It can be shown that the $100(1 - \alpha)\%$ confidence interval in Box 5.1 consists of all values of η that will not be rejected by an α-level sign test. In other words, the $100(1 - \alpha)\%$ confidence interval for η is the same as the set of acceptable (based on an α-level sign test) values of η. Readers interested in the details of how this property can be verified should refer to Section B.7 in Appendix B.

Experimenting with different values of n and α will show that the endpoints of a sign test confidence interval are two measurements located a fixed number (depending on the values of n and α) of observations above and below the sample median. For example, if $n = 9$ and $\alpha/2 = 0.0195$, we get $b(9, 0.0195) = 8$ from Table C.5 (Appendix C). Therefore, a $100(1 - \alpha)\% = 100[1 - (2 \times 0.0195)]\% = 96.1\%$ confidence interval for the population median is $(y_{(L)}, y_{(U)})$, where $U = 8$ and $L = 9 + 1 - 8 = 2$. Thus, the endpoints of the confidence interval are the sample values located three $(5 - 2 = 8 - 5 = 3)$ observations above and below the sample median

$y_{(1)}$	$y_{(2)}$	$y_{(3)}$	$y_{(4)}$	$y_{(5)}$	$y_{(6)}$	$y_{(7)}$	$y_{(8)}$	$y_{(9)}$
	↑			↑			↑	
	Lower			Point			Upper	
	endpoint			estimate			endpoint	

In contrast, the confidence interval based on the t-test will have its upper and lower endpoints located a fixed distance (depending upon n, α, and the estimated standard error of $\overline{Y}$) above and below the sample mean.

Availability of test levels and confidence coefficients

Because the binomial distribution is a discrete distribution, critical values for an α-level sign test will be available only for selected values of α. (We made a related observation when discussing Table 2.3.) Table C.5 (Appendix C) gives values of α such that $\Pr\{B \geq y\} = \alpha$ for selected values of y and n.

Computational aspects

The most demanding step in a sign test procedure is to determine the critical values of a binomial distribution with n trials and probability of success $\pi = 0.5$. Once the critical values are known, implementing sign test procedures is not difficult. The SAS PROBBNML function can be used to determine the critical values of binomial distributions. The SAS UNIVARIATE procedure can be used to determine the approximate p-values when the sample size is large. Section 5.2 of the companion text by Younger (1997) shows how to use SAS for these purposes.

An example of the printout from StatXact for sign test procedures will be illustrated in Section 6.2.

EXAMPLE **5.2**   Let's analyze the data in Example 5.1 to estimate the median percentage of CPZ in a membrane fraction of red blood cells suspended in Hanks' solution. As already noted, our analysis will depend on which assumptions seem reasonable for the data.

Under the assumption that the CPZ values have a continuous distribution with median η, we can use the sign test procedure to make inferences about η.

A point estimate for η is the median of the sample. Arranging the sample values in increasing order, we get: $y_{(1)} = 5$, $y_{(2)} = 6$, $y_{(3)} = 7$, $y_{(4)} = 14$, $y_{(5)} = 14$, $y_{(6)} = 15$, $y_{(7)} = 17$, and $y_{(8)} = 27$.

The sample size is even ($n = 2m$, where $m = 4$), and so the median of the sample is the average of the middle two values. Thus, an estimate of the population median is

$$\hat{\eta} = \frac{1}{2}(y_{(4)} + y_{(5)}) = \frac{1}{2}(14 + 14) = 14.$$

In order to construct a confidence interval for η, we refer to Table C.5 and note that, for $n = 8$, the two smallest available α are $\alpha = 0.035$ and $\alpha = 0.004$.

The two smallest α correspond to the critical values $b(8, 0.035) = 7$ and $b(8, 0.004) = 8$. If we take $\alpha/2 = 0.035$ ($\alpha = 2 \times 0.035 = 0.07$) in the formula for the confidence interval in Box 5.1, we get

$$U = b(n, \alpha/2) = b(8, 0.035) = 7,$$

$$L = n + 1 - b(n, \alpha/2) = 8 + 1 - 7 = 2,$$

so that a $100(1 - 0.07)\% = 93\%$ confidence interval for η is

$$y_{(L)} < \eta < y_{(U)} \Leftrightarrow y_{(2)} < \eta < y_{(7)} \Leftrightarrow 6 < \eta < 17.$$

Thus, we have 93% confidence that the median percentage CPZ in the membrane fraction is between 6% and 17%.

It can readily be verified that the one-sample t-test procedure in Box 4.2 will yield a 93% confidence interval of (7.89, 18.36) for the population mean. As noted in Section 4.7, such an interval should work well even if the population is not normal, provided the population distribution is not too skewed or heavy-tailed, but the interval based on the sign test is valid regardless of the shape of the population distribution. ∎

Like the one-sample t-test for paired data, the sign test can be used to make inferences about the population of differences between paired observations.

EXAMPLE 5.3 Do the data in Example 5.1 indicate a difference in the percentage of CPZ in membrane fractions of red blood cells suspended in the two solutions?

Because the data are paired (by donor), the differences between the paired values can be regarded as a single sample from a population of differences. The sign test procedure permits analysis of the data without the need to assume that the population of differences has a normal distribution. The only assumption needed for the sign test is that the population of differences must have a continuous distribution.

Let η_D denote the population median of the difference D between the pairs of CPZ values obtained for the two mediums. Point and interval estimates for the population median of the difference can be calculated as in Example 5.2. The following are the calculated values of the differences d_i = Hanks' solution − Sucrose for the i-th pair and the order statistics of the differences $d_{(i)}$.

Observation number i	1	2	3	4	5	6	7	8
Difference d_i	−4	+2	0	+2	+1	−8	−7	+1
Order statistic $d_{(i)}$	−8	−7	−4	0	+1	+1	+2	+2
Point estimate	$\hat{\eta}_D = \frac{1}{2}(0 + 1) = 0.5$							
98.4% confidence interval	$-7 < \eta_D < +2$							

Even though the point estimate indicates that the value for Hanks' solution will be 0.5% more than the corresponding value for sucrose in about 50% of the instances, the confidence interval indicates that the true median difference could be anywhere between −7% and 2%.

Let's use the test-statistic approach to verify the null hypothesis that the true median difference is zero; that is, under the null hypothesis, we are asserting that the difference is as likely to be positive as negative. We use the test based on the statistic B_c in Box 5.1. Note that, with $\eta_0 = 0$, there is no need to modify the sample. If we discard the zero value from the sample of differences, we get a reduced sample of size $n^* = 7$. There are four positive values in the reduced sample, and so the value of the test statistic is $B_c = 4$. From Table C.5 (Appendix C), we see that $b(7, 0.062) = 6$. Hence, the null hypothesis $H_0: \eta = 0$ can be rejected at level $\alpha = 2 \times 0.062 = 0.124$ if $B_c \geq 6$ or $B_c \leq 7 + 1 - 6 = 2$. Therefore, the null hypothesis of zero median cannot be rejected even at a 12.4% significance level. We do not have sufficient evidence to reject the null hypothesis that the true median difference is zero.

If there are no serious violations of the normal-distribution assumption for the population of differences, the null hypothesis of zero median can be tested using the one-sample t-test. It is easy to verify that the calculated value of the test statistic is $t_c = -1.12$, which is not significant ($p = 0.2996$). Thus, in this particular instance, similar conclusions emerge from the sign test and the paired-sample t-test.   ■

As already noted, the sign test is an alternative to the t-test for inferences about a population mean. The advantage of the sign test over the t-test is that its results are valid even if the population distribution is not normal. A disadvantage is that, if the population does have a normal distribution, the power of the sign test (its ability to detect a true research hypothesis) will be less, and the width of its confidence interval is likely to be more, than those of the corresponding t-tests.

Exercises

5.4   Refer to the infection rate data in Exercise 4.4.

    a   Determine the possible confidence levels for which these data may be used to construct sign test confidence intervals for the population median infection rate.

    b   Construct a confidence interval for the median infection rate at the confidence level that is closest to 95%.

    c   Compare the confidence interval in (b) with the corresponding confidence interval based on the one-sample t-test.

    d   Describe the assumptions underlying the intervals constructed in (b) and (c).

5.5   Refer to the N-loss data for fertilizer UN in Example 3.27.

    a   Use a 0.05-level sign test to see if the data provide sufficient evidence to conclude that the true median N-loss is less than 13%.

    b   Use Table C.5 (Appendix C) to determine the p-value of the test in (a) and write your conclusions on the basis of the p-value.

    c   Compare the p-value in (b) with that for the corresponding t-test. Describe the considerations involved in choosing between the two p-values.

5.6   In a study to evaluate a drug for the relief of eye irritation, a light puff of air was blown into one eye of each of 20 experimental subjects at a low but fixed velocity. The time (min) elapsed before the redness of the exposed (irritated) eye returned to the color of the unexposed eye was recorded. The time-to-relief data obtained are as follows:

0.4, 4.6, 2.2, 1.2, 4.5, 5.7, 8.0, 2.1, 4.8, 3.0, 8.8, 11.4, 1.3, 1.4, 2.1, 1.3, 12.5, 2.4, 4.6, 2.8

    a   Draw a histogram and a box-plot of the data to see if it is reasonable to assume that the time-to-relief data can be regarded as a random sample from a normal distribution.

    b   On the basis of your conclusions in (a), perform a statistical test to determine whether the data support the conclusion that more than 50% of those treated by the drug will obtain relief in less than 5 minutes.

   c Use the sign test procedure to construct a 99% confidence interval for the population median. Interpret the interval you construct.

5.4
Procedures based on the Wilcoxon signed rank test

Often, we encounter data for which the population distribution is not normal but may reasonably be assumed to be symmetric. In that case, a distribution-free compromise between the sign test procedures and the t-test procedures is to use the *Wilcoxon signed rank test*.

The Wilcoxon signed rank statistic

The statistic on which the sign test is based equals the number of measurements in the sample that exceed a specified value η_0 of the population median. The statistic for the Wilcoxon signed rank test depends not only on the number of sample values that exceed η_0, but also on ranks of the absolute differences $|y_i - \eta_0|$ between the sample values and η_0. The Wilcoxon signed rank statistic is defined as follows.

Let $\{y_1, y_2, \ldots, y_n\}$ be a random sample from a continuous population. Rank the absolute values in the modified sample $\{y_1 - \eta_0, y_2 - \eta_0, \ldots, y_n - \eta_0\}$, and denote these ranks by $R_1^{(abs)}, R_2^{(abs)}, \ldots, R_n^{(abs)}$, where $R_i^{(abs)}$ is called the *absolute rank* of $y_i - \eta_0$. The *signed rank* of $y_i - \eta_0$ is defined as

$$R_i^+ = \begin{cases} R_i^{(abs)} & \text{if } y_i - \eta_0 \text{ is positive} \\ 0 & \text{otherwise.} \end{cases}$$

The Wilcoxon signed rank statistic is the sum of the signed ranks of the modified sample

$$W^+ = \sum_i R_i^+. \tag{5.2}$$

As we see in Box 5.2, the point and interval estimators in the Wilcoxon signed rank test procedures are best described in terms of statistics called the Walsh averages.

Walsh averages

The Walsh averages of a sample are the collection of means of all possible pairs of measurements in the sample. In a sample of size n, there are $N = \frac{1}{2}n(n+1)$ Walsh averages. Let w_{ij} denote the Walsh average of y_i and y_j; that is, $w_{ij} = \frac{1}{2}(y_i + y_j)$. As an example, consider the sample $\{2, 6, 6, 9\}$. The sample size is $n = 4$, and so there are $N = \frac{1}{2}(4)(5) = 10$ Walsh averages: $w_{11} = \frac{1}{2}(2+2) = 2$; $w_{12} = \frac{1}{2}(2+6) = 4$; $w_{13} = \frac{1}{2}(2+6) = 4$; $w_{14} = \frac{1}{2}(2+9) = 5.5$; $w_{22} = \frac{1}{2}(6+6) = 6$, $w_{23} = \frac{1}{2}(6+6) = 6$; $w_{24} = \frac{1}{2}(6+9) = 7.5$; $w_{33} = \frac{1}{2}(6+6) = 6$; $w_{34} = \frac{1}{2}(6+9) = 7.5$; and $w_{44} = \frac{1}{2}(9+9) = 9$.

Wilcoxon signed rank procedures

The Wilcoxon signed rank procedure for inferences about a population median are summarized in Box 5.2.

BOX **5.2**   ***The Wilcoxon signed rank procedures for inferences about a population median***

Suppose $\{y_1, y_2, \ldots, y_n\}$ is a random sample of size n from a continuous symmetric population with median η. Let $W_{(i)}$ denote the i-th order statistic of the set of Walsh averages $\{W_{ij}: 1 \le i \le j \le n\}$. The Wilcoxon signed rank test procedure for inferences about η is as follows.

Point Estimation

A point estimate for the population median is the median of the $N = n(n+1)/2$ Walsh averages

$$\hat{\eta} = \begin{cases} \frac{1}{2}(W_{(k)} + W_{(k+1)}) & \text{if } N = 2k \\ W_{(k+1)} & \text{if } 2k+1. \end{cases}$$

Interval Estimation

A $100(1 - \alpha)\%$ confidence interval for η is

$$W_{(L)} < \eta < W_{(U)},$$

where $U = w(n, \alpha/2), L = N + 1 - w(n, \alpha/2); w(n, \alpha)$ is the α-level critical value of the sampling distribution of W^+ in Equation (5.2). These critical values are tabulated in Table C.6 (Appendix C). If n is outside the range in Table C.6, use the large-sample approximation

$$w(n, \alpha/2) \simeq \mu_{W^+} + z(\alpha/2)\sigma_{W^+},$$

where

$$\mu_{W^+} = \frac{N}{2}; \quad \sigma_{W^+} = \sqrt{\frac{1}{12}\left[N(2n+1) - \frac{1}{4}\sum_{j=1}^{g} t_j(t_j^2 - 1)\right]},$$

$t_1, t_2, \ldots, t_g$ are the tie sizes in the modified sample, and $z(\alpha/2)$ is the $\alpha/2$-level critical value of a standard normal distribution.

Hypothesis Test

Consider three sets of hypotheses: (1) $H_0: \eta \le \eta_0$ against $H_1: \eta > \eta_0$; (2) $H_0: \eta \ge \eta_0$ against $H_1: \eta < \eta_0$; and (3) $H_0: \eta = \eta_0$ against $H_1: \eta \ne \eta_0$.

Delete the zero differences (that is, $y_i - \eta_0 = 0$) from the modified sample and adjust the sample size accordingly. In other words, calculate W^+

using nonzero differences where n is the number of nonzero values in the modified sample.

Let W_c^+ be the calculated value of W^+. Then the rejection criteria for three null hypotheses about η are as follows.

1   Reject H_0: $\eta \leq \eta_0$ at level α if $W_c^+ \geq w(n, \alpha)$.

2   Reject H_0: $\eta \geq \eta_0$ at level α if $W_c^+ \leq N - w(n, \alpha)$.

3   Reject H_0: $\eta = \eta_0$ at level α if $W_c^+ \geq w(n, \alpha/2)$ or if $W_c^+ \leq N - w(n, \alpha/2)$.

Rationale for signed rank test procedures

The rationale for the Wilcoxon signed rank test is more difficult to explain than that for the sign test. The test utilizes the fact that, when the population is symmetric, the distribution of the absolute ranks of the negative measurements in the modified sample will be the same as the distribution of the absolute ranks of the positive measurements. It can be verified that the $100(1 - \alpha)\%$ confidence interval for η in Box 5.2 consists of the set of all acceptable (based on an α-level signed rank test) values of the population median η. For more details on the rationale for the Wilcoxon signed rank test procedures, see Hollander and Wolfe (1973), Lehmann (1975), or Sprent (1993).

Critical values

The α-level critical values of W^+—denoted by $w^+(n, \alpha)$—are presented in Table C.6 (Appendix C) for selected values of n and α. Under the assumption that the data constitute a random sample from a continuous distribution symmetric about η_0, Table C.6 gives the values of n, α, and $w^+(n, \alpha)$ for which

$$\Pr\{W^+ \geq w^+(n, \alpha)\} = \alpha.$$

For example, $w^+(7, 0.023) = 26$.

As with the sign test statistic B, the sampling distribution of W^+ is a discrete distribution, so that critical values for W^+ are not available for all combinations of n and α.

The large-sample approximation for $w(n, \alpha)$ is based on the theoretical finding that, when the samples are large and the null hypothesis is true, the sampling distribution of W^+ can be approximated by a normal distribution with mean μ_{W^+} and standard deviation σ_{W^+} in Box 5.2.

Effect of ties

The assumption that the sample is a random sample from a continuous symmetric population implies that there will be no tied observations. When used with samples containing tied values, *p*-values and significance levels obtained from the critical values in Table C.6 (Appendix C) tend to be higher than their true values. Consequently, the resulting test procedures will be conservative, in the sense that acceptance of research hypotheses will become less likely.

Sign rank test and sign test confidence intervals

Comparison of the point and interval estimates obtained in the signed rank test and the sign test shows that the main difference between the two procedures is that the sign test uses the raw order statistics of the sample, whereas the signed rank procedure uses the order statistics of the means of all possible pairs of measurements in the observed sample. When the population distribution is symmetric, the confidence interval based on the signed rank test can be expected to be narrower than the corresponding interval based on the sign test.

Computational aspects

The UNIVARIATE procedure in SAS can be used to calculate the Wilcoxon signed rank test statistic. StatXact software can be used to compute the test statistic and determine the associated *p*-value. When computing point and interval estimates, determining the order statistics of the Walsh averages can be a tedious task for moderate to large samples. As we see in Example 5.4, some of the computing effort can be reduced by adopting a systematic approach. The companion text by Younger (1997) contains a SAS macro that can be used to perform this task.

Application of signed rank procedures

Let's look at an example.

EXAMPLE **5.4**   We now use the Wilcoxon signed rank procedures to analyze the CPZ data in Example 5.3. In addition to the assumption that the population of differential CPZ values has a continuous distribution, we need to assume that the population is symmetric.

There are $N = \frac{1}{2}(8)(8 + 1) = 36$ Walsh averages. Table 5.1 shows a convenient method of displaying the 36 Walsh averages. The entry at the intersection of row i and column j—cell (i, j)—in Table 5.1 is the Walsh average of the i-th and j-th order statistics. Thus, the entry -3.5 in the $(2, 4)$ cell (row 2 and column 4) is w_{24}, the Walsh average of the second and fourth order statistics

$$w_{24} = \frac{y_{(2)} + y_{(4)}}{2} = \frac{-7 + 0}{2} = -3.5.$$

As already noted, determining the order statistics of the Walsh averages can become tedious for large data sets. However, the fact that the values in Table 5.1 increase from left to right within any row and from top to bottom within any column implies certain order relationships. With some practice, the researcher can take advantage of these relationships to reduce the effort needed for the signed rank test procedures. It is also helpful to note that, in Table 5.1, the entry in row $i + 1$ can be

TABLE 5.1

The Walsh averages of the sample

| | | Order statistics $y_{(j)}$ | | | | | | |
		−8	−7	−4	0	1	1	2	2
	−8	−8.0	−7.5	−6.0	−4.0	−3.5	−3.5	−3.0	−3.0
	−7		−7.0	−5.5	−3.5	−3.0	−3.0	−2.5	−2.5
Order	−4			−4.0	−2.0	−1.5	−1.5	−1.0	−1.0
statistics	0				0.0	0.5	0.5	1.0	1.0
$y_{(i)}$	1					1.0	1.0	1.5	1.5
	1						1.0	1.5	1.5
	2							2.0	2.0
	2								2.0

obtained by adding the fixed quantity $\frac{1}{2}(y_{(i+1)} - y_{(i)})$ to the entry directly above in row i. For example, the entries in the third row ($i = 3$) can be obtained by adding $\frac{1}{2}[-4 - (-7)] = 1.5$ to the entries in the second row.

A point estimate of the median differential CPZ is the median of the Walsh averages

$$\hat{\eta} = \frac{w_{(18)} + w_{(19)}}{2} = \frac{-1.5 + (-1.5)}{2} = -1.5.$$

Referring to Table C.6 (Appendix C) with $n = 8$, we see that if we use the value α in this table as the value of $\alpha/2$ for constructing a confidence interval, a $100(1 - \alpha)\%$ confidence interval for the median difference can be constructed using the value $\alpha/2 = 0.027$. A $100(1 - 0.054)\% = 94.6\%$ confidence interval for the median difference can be constructed as follows.

From Table C.6 (Appendix C), we get $w(8, 0.027) = 32$, so that a 94.6% confidence interval for η_D is $(W_{(L)}, W_{(U)})$, where $L = N + 1 - w(8, 0.027) = 36 + 1 - 32 = 5$ and $U = w(8, 0.027) = 32$. The required confidence interval is

$$w_{(5)} \le \eta_D \le w_{(32)} \Leftrightarrow -5.5 \le \eta_D \le +1.5.$$

For the purpose of illustration, let's use the large-sample approximation to construct a 94.6% confidence interval. We first note that the tie sizes in the modified sample are $t_1 = 1$, $t_2 = 2$, $t_3 = 2$, $t_4 = 1$, $t_5 = 1$, and $t_6 = 1$. Also, we find from Table C.1 that $z(0.027) = 1.92$, so that we use the approximation

$$w(8, 0.054) \simeq \frac{36}{2} + (1.92)\sqrt{\frac{36(2 \times 8 + 1) - 3}{12}} = 18.0 + 13.6779 \simeq 32,$$

which is the same as the exact value obtained from Table C.6. Thus, even for a sample size as small as 8, the large-sample approximation is quite accurate.

The 95% confidence interval contains zero, and so the null hypothesis—that the differential percentage CPZ values are symmetrically distributed about zero—cannot be rejected at the $\alpha = 0.05$ level.

The same null hypothesis can be tested (with $\eta_0 = 0$ in Box 5.2) using the test-statistic approach. The sample of differences contains one value equal to $\eta_0 = 0$;

discarding this value, we rank the absolute values of the remaining $n = 7$ nonzero differences with the following results:

Nonzero differences $d_j - 0$	-8	-7	-4	1	1	2	2
Absolute values $\|d_j - 0\|$	8	7	4	1	1	2	2
Absolute ranks $R_j^{(abs)}$	7	6	5	1.5	1.5	3.5	3.5
Signed ranks R_j^+	0	0	0	1.5	1.5	3.5	3.5

The value of the Wilcoxon signed rank statistic is the sum of the signed ranks of the modified sample

$$W_c^+ = 0 + 0 + 0 + 1.5 + 1.5 + 3.5 + 3.5 = 10.$$

From Table C.6, we see that $w(7, 0.023) = 26$, so that a test at $\alpha = 0.046$ will reject H_0 if $W_c^+ \geq 26$ or if $W_c^+ \leq 7(7 + 1)/2 - 26 = 2$. Consequently, there is not enough evidence to reject the null hypothesis that the population of differences is symmetrically distributed about zero. Thus, as expected, the test-statistic and confidence-interval approaches lead to the same conclusion.

The following printout was produced by using the Wilcoxon signed rank test to analyze the CPZ data in StatXact.

```
                StatXact printout for CPZ data
-------------------------------------------------------------------

WILCOXON SIGNED RANK TEST


Summary of Exact distribution of WILCOXON SIGNED RANK statistic:
    Min      Max     Mean    Std-dev   Observed     Standardized
    0.     28.00    14.00    5.895     10.00          -.6786

-------------------------------------------------------------------

Asymptotic Inference:
    One-sided p-value: Pr { Test Statistic .LE. Observed } =  0.2487
    Two-sided p-value: 2 * One-sided                       =  0.4974
-------------------------------------------------------------------

Exact Inference:
    One-sided p-value: Pr { Test Statistic .LE. Observed } =  0.2813
    Point probability: Pr { Test Statistic .EQ. Observed } =  0.0469
    Two-sided p-value: 2 * One-sided                       =  0.5625

-------------------------------------------------------------------
```

The display is divided into three parts. The first shows the observed value $W^+ = 10$, along with the mean (μ_{W^+}) and standard deviation (σ_{W^+}) of its sampling distribution. The second part shows the large-sample one-sided and two-sided p-values of the signed rank test. The third part contains the p-values for the one- and two-sided tests. The exact one-sided p-value is

$$\Pr\left\{W^+ \leq 10\right\} = \Pr\left\{\text{test statistic} \leq \text{observed value}\right\} = 0.2813,$$

which means that there is insufficient evidence to conclude that the population of differential CPZ values is symmetrically distributed about zero. ∎

Models for inferences about a population median: A comparison

As already noted, inferential procedures based on the sign test, the signed rank test, and the *t*-test can be used to make inferences about a population median under three different sets of assumptions. These assumptions are summarized in Box 5.3.

BOX **5.3**

Statistical models associated with the sign, signed rank, and t-tests for inferences about a population median

Let $Y_j, j = 1, 2, \ldots, n$, denote the *j*-th observation in a sample. Suppose that Y_j satisfies the model

$$Y_j = \eta + E_j, \qquad j = 1, 2, \ldots, n,$$

where the following assumptions are made:

I   The term η is the population median.

II The term $E_j = Y_j - \eta$ is the error in the *j*-th observation.

Assumptions for the sign test

The sign test procedures can be used to make inferences about η provided the following assumption is added to assumptions I and II.

IIIa The E_j are a random sample from a continuous population with zero median.

Assumptions for the signed rank test

The Wilcoxon signed rank test procedures can be used to make inferences about η provided the following assumption is added to assumptions I and II:

IIIb The E_j are a random sample from a continuous symmetric population with zero median.

Assumptions for the one-sample t-test

The one-sample *t*-test procedures can be used to make inferences about η provided the following assumption is added to assumptions I and II:

IIIc The E_j are a random sample from a normal population with zero median (mean).

Notice in Box 5.3 that the basic structure of the model remains the same for the three procedures. The difference is in the assumptions about the random errors. The sign test procedures require the least restrictive assumption—that the errors have a continuous distribution with zero median. The assumption for the Wilcoxon signed rank test procedures is more restrictive—not only that the errors are continuously distributed with zero median but also that the error distributions are symmetric.

Finally, the t-test procedures require the most restrictive assumption—that the errors have a normal distribution with zero median—because the assumption of normal distribution with zero mean implies that the error distribution is continuous and symmetric about a zero median.

Wilcoxon signed rank test of symmetry

The model for the Wilcoxon signed rank test in Box 5.2 assumes that the population distribution is symmetric about an unknown median η. When it is reasonable to assume that the population is symmetric, the Wilcoxon signed rank procedures provide methods for making inferences about η, the unknown population median. The Wilcoxon signed rank test can also be used to test the null hypothesis that the population distribution is symmetric about a particular median. In such an application, the test is usually called the Wilcoxon signed rank test of symmetry. The corresponding null hypothesis H_0 is that the population is continuous and symmetric with median η_0, where η_0 is a given number. The alternative hypothesis H_1 states that either the population median is not equal to η_0 or the population distribution is skewed.

The test procedure is the same as the two-sided Wilcoxon signed rank test in Box 5.2. Example 5.5 describes an application of the Wilcoxon signed rank test of symmetry.

EXAMPLE **5.5**   In Example 4.11, we used paired t-procedures to compare the mean cell fluidities of dog pulmonary artery cells exposed to oxygen treatment and unexposed cells. The paired t-test rests on the assumption that the difference in cell fluidities has a normal distribution. If that assumption is thought to be too stringent, we might want to compare the two treatments using the Wilcoxon signed rank test of symmetry. The null hypothesis H_0 for the test is that the distribution of the differential cell fluidities is symmetric about the median $\eta = 0$. No difference between the cell fluidities of the exposed and unexposed cells would imply that an observed differential cell fluidity is as likely to be larger than a given positive value a as it is to be less than the corresponding negative value $-a$; in other words, areas A_1 and A_2 will be equal in Figure 5.2. Thus, rejection of the null hypothesis can be interpreted as the demonstration of a difference between the exposed and unexposed cells. Accordingly, the signed rank test of symmetry with $\eta_0 = 0$ is an alternative to the two-sided paired t-test. Actual calculations for performing the test are left as an exercise.   ■

Exercises

5.7   In Example 5.4, we used Wilcoxon signed rank procedures to make inferences about η_D, the median differential percentage CPZ values.

    a   Write an appropriate statistical model under which the Wilcoxon signed rank procedures can be used to analyze the CPZ data.

FIGURE 5.2

The distribution of the differential cell fluidities if H_0 is true

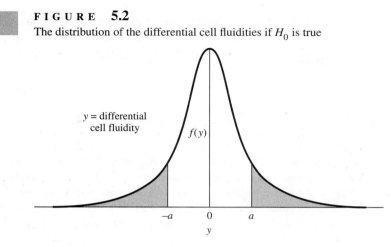

b Use the Wilcoxon signed rank test to see if the null hypothesis $H_0: \eta_D \geq 1.5$ can be rejected in favor of $\eta_D < 1.5$ at a level $\alpha = 0.05$.

c On the basis of the critical values in Table C.6, what can you say about the p-value of the test in (b)?

5.8 The following are the temperatures (°C) recorded at a weather station (daily high) and at an experimental site in a pasture in South Florida on eight days in June and July:

| Weather station | 34.5 | 35.4 | 35.9 | 33.4 | 36.9 | 35.8 | 36.0 | 37.4 |
| Pasture | 33.2 | 32.9 | 33.4 | 32.3 | 34.1 | 33.0 | 33.3 | 34.6 |

a Use the sign test, the Wilcoxon signed rank test, and the t-test procedures to construct approximate 95% confidence intervals for the median difference between the daily high temperature recorded at the weather station and the temperature recorded at the experimental site.

b Using suitable statistical models, describe the assumptions under which each of the three intervals will be a valid confidence interval.

c If all three intervals are valid, which would you prefer? Why?

5.9 Refer to the eye irritation data in Exercise 5.6.

a Use the Wilcoxon signed rank test procedure to construct an approximate 90% confidence interval for the median time elapsed before the redness in the irritated eye returned to the color of the unirritated eye.

b Use the histogram and the box-plots calculated in Exercise 5.6b to see if there is any indication that some of the assumptions for the Wilcoxon signed rank procedures may be unreasonable for these data.

5.10 Refer to the tumor weight data in Exercise 4.10.

a Explain why it is reasonable to characterize the null hypothesis that the experimental drug has no effect on the treatment as the null hypothesis that the observed differences between tumor weights within pairs are a random sample from a continuous symmetric population with zero median.

b   Apply an appropriate distribution-free test to the null hypothesis in (a).

5.5
Procedures based on the Wilcoxon rank sum test

In this section, we consider a distribution-free alternative to the independent-sample t-test procedures. These distribution-free procedures are useful for comparing two populations when there is reason to doubt the assumptions underlying the independent-sample t-test (Box 4.3) or the approximate t-test (Box 4.4).

The most commonly used distribution-free procedures for comparing two continuous populations are based on a statistical hypothesis-testing procedure known as the *Wilcoxon rank sum test*. The Wilcoxon rank sum test can be used to test the null hypothesis that two continuous populations have identical distributions. The statistical model in Box 5.4 summarizes a set of assumptions under which the Wilcoxon rank sum test may be used.

BOX **5.4**   *A statistical model for the Wilcoxon rank sum test*

Let Y_{ij} denote the j-th observation in the i-th sample; $j = 1, 2, \ldots, n_i$; $i = 1, 2$. Then

$$Y_{ij} = \eta_i + E_{ij},$$

where η_i is the mean (or median) of the i-th population and E_{ij} is the error in Y_{ij}; the E_{1j} and the E_{2j} are independent random samples of sizes n_1 and n_2, respectively, from a continuous population with zero mean (median).

Compare the model for the Wilcoxon rank sum test in Box 5.4 with the model for the independent-sample t-test in Box 4.9. Both tests require independent samples from continuous populations having the same shape. In both instances, the assumptions imply that the populations may have different location parameters (medians, means), so that inferences concerning the differences between the two populations can be formulated as inferences concerning the parameter $\theta = \eta_1 - \eta_2$. The t-test requires the further assumption that the shapes of the populations correspond to a normal distribution, whereas the Wilcoxon rank sum test is applicable for any population shape. The relationship between the distributions of the two populations corresponding to the model in Box 5.4 is shown in Figure 5.3.

Comparison of Figure 5.3 and Figure 4.3 helps illustrate the differences and similarities between the conditions under which the independent-sample t-test and the Wilcoxon rank sum test are appropriate. The most important common feature of the two tests is the assumption that the distributions of the two populations are

FIGURE 5.3

Relationship between the target populations and the population of errors in Box 5.4

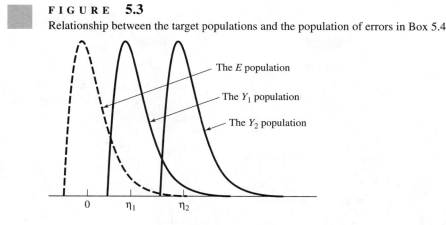

The E population

The Y_1 population

The Y_2 population

0 η_1 η_2

identical, except for a possible shift (implying $\theta \neq 0$) in their locations. For this reason, the model in Box 5.4 is often known as a *two-sample shift model*.

The Wilcoxon rank sum statistic

Let y_{11}, y_{12}, $\ldots$, y_{1n_1} and y_{21}, y_{22}, $\ldots$, y_{2n_2} be independent random samples from two populations with respective medians η_1 and η_2; let θ_0 be a specified value of the difference $\theta = \eta_1 - \eta_2$. Rank the $N = n_1 + n_2$ values $\{y_{11}, y_{12}, \ldots, y_{1n_1}, y_{21} - \theta_0, y_{22} - \theta_0, \ldots, y_{2n_2} - \theta_0\}$ obtained by combining the first sample with modified values from the second sample. Let $R_1, R_2, \ldots, R_{n_2}$ denote the ranks of the modified second sample values in the combined sample. The Wilcoxon rank sum statistic is defined as the sum of the ranks of the modified second sample values

$$W = R_1 + R_2 + \cdots + R_{n_2}. \tag{5.3}$$

The Wilcoxon rank sum procedures

The Wilcoxon rank sum procedures for inferences about two populations are summarized in Box 5.5.

BOX 5.5

> **Wilcoxon rank sum procedures for inferences about the difference between two population medians**
>
> Suppose $\{y_{11}, y_{12}, \ldots, y_{1n_1}\}$ and $\{y_{21}, y_{22}, \ldots, y_{2n_2}\}$ are independent random samples from continuous populations with medians η_1 and η_2, respectively.

The Wilcoxon rank sum test procedure for inferences about $\theta = \eta_2 - \eta_1$ is as follows.

Point estimation

Let $D_{ij} = y_{2j} - y_{1i}$ be the $n_1 n_2$ differences between the observations in the two samples, and $D_{(1)}, D_{(2)}, \ldots, D_{(n_1 n_2)}$ denote their order statistics. A point estimate of θ is given by the median of the D_{ij}

$$\hat{\theta} = \begin{cases} \frac{1}{2}(D_{(k)} + D_{(k+1)}) & \text{if } n_1 n_2 = 2k \\ D_{(k+1)} & \text{if } n_1 n_2 = 2k + 1. \end{cases}$$

Interval estimation

A $100(1 - \alpha)\%$ confidence interval for θ is

$$D_{(L)} < \theta < D_{(U)},$$

where

$$U = w(n_1, n_2, \alpha/2) - \frac{n_2(n_2 + 1)}{2};$$

$$L = \frac{n_2(N + n_1 + 1)}{2} + 1 - w(n_1, n_2, \alpha/2),$$

$N = n_1 + n_2$, and $w(n_1, n_2, \alpha)$ is the α-level critical value of the Wilcoxon rank sum test. Values of $w(n_1, n_2, \alpha)$ are given in Table C.7 (Appendix C) for selected values of $n_1, n_2,$ and α.

For values of n_1 and n_2 outside the range of Table C.7, use the large-sample approximation

$$w(n_1, n_2, \alpha) \simeq \mu_W + z(\alpha)\sigma_W,$$

where

$$\mu_W = \frac{n_2(N + 1)}{2},$$

$$\sigma_W = \sqrt{\frac{n_1 n_2}{12}\left[N + 1 - \frac{\sum_{j=1}^{g} t_j(t_j^2 - 1)}{N(N + 1)}\right]},$$

and $t_1, t_2, \ldots, t_g$ are the sizes of the tied groups in the modified combined sample $\{y_{11}, y_{12}, \ldots, y_{1n_1}, y_{21} - \theta_0, y_{22} - \theta_0, \ldots, y_{2n_2} - \theta_0\}$.

Hypothesis test

Let W_c denote the calculated value of the Wilcoxon rank sum statistic. Consider three sets of hypotheses: (1) H_0: $\theta \leq \theta_0$ against H_1: $\theta > \theta_0$;

(2) $H_0\colon \theta \geq \theta_0$ against $H_1\colon \theta < \theta_0$; (3) $H_0\colon \theta = \theta_0$ against $H_1\colon \theta \neq \theta_0$. The following are the corresponding rejection criteria:

1 Reject $H_0\colon \theta \leq \theta_0$ at level α if $W_c \leq n_2(N + 1) - w(n_1, n_2, \alpha)$.
2 Reject $H_0\colon \theta \geq \theta_0$ at level α if $W_c \geq w(n_1, n_2, \alpha)$.
3 Reject $H_0\colon \theta = \theta_0$ at level α if $W_c \geq w(n_1, n_2, \alpha/2)$ or $W_c \leq n_2(N + 1) - w(n_1, n_2, \alpha/2)$.

Wilcoxon rank sum test

Why do the Wilcoxon rank sum procedures work in practice? Consider a simple example, in which we have two independent random samples, each of size $n = 2$, from two continuous populations. Assume that the two populations are identical except for a possible shift in their locations. Let $\theta = \eta_2 - \eta_1$, where η_1 and η_2 are the population medians, and suppose that we are interested in testing the null hypothesis $H_0\colon \theta = \theta_0$ against the research hypothesis $H_1\colon \theta > \theta_0$.

Let $\{R_1, R_2\}$ denote the ranks of the two measurements in the modified second sample when the four measurements in the combined sample are ranked as a single group. The following are the six possible values for $\{R_1, R_2\}$:

$$\{1, 2\}, \ \{1, 3\}, \ \{1, 4\}, \ \{2, 3\}, \ \{2, 4\}, \ \{3, 4\}$$

Each of these six values corresponds to a calculated value for W, the Wilcoxon rank sum statistic. For example, the ranks $\{1, 2\}$ will result in the calculated value $W_c = 1 + 2 = 3$. It can be verified that the six sets of values for the ranks will yield five distinct values of W_c: 3, 4, 5, 6, and 7. Thus, the distribution of the population of all possible values for W is a discrete distribution.

Figure 5.4 shows three populations differing only in their medians. Population 1 and Population 2 represent the two target populations. The medians of these two populations are η and $\eta + \theta$ so that the difference between the medians equals θ. The third population, called the modified population, is the population obtained by subtracting θ_0 from each value in the second population. The median of the modified population is $\eta + \theta - \theta_0$. From Figure 5.4, it can be seen that the effect of modifying the second population by subtracting a constant amount from each value is to shift the second population toward the first population. The amount of the shift is equal to the amount subtracted. If the amount subtracted is equal to the difference θ between the two medians, the distribution of the shifted second population will be identical to that of the first population. Consequently, if the null hypothesis is true—that is, if $\theta = \theta_0$—the modified second sample can be regarded as a random sample from the first population. If $\theta > \theta_0$, then the modified second sample is a random sample from a population shifted to the right of the first population (Figure 5.4).

Consequently, if H_1 is true, the values in the modified second sample will tend to be larger than the values in the first sample, and the ranks associated with the values in the modified second sample will tend to be the larger values in the set $\{1, 2, 3, 4\}$.

FIGURE   5.4

Effect of shifting the second population by an amount less than the true difference between the population medians

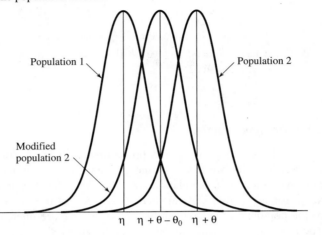

Thus, large values of the Wilcoxon rank sum test will favor the research hypothesis $H_1: \theta > \theta_0$.

Recall that the rejection region for an α-level test is determined by the sampling distribution of the test statistic when the null hypothesis is true. The required sampling distribution of W can be obtained as follows. As already noted, if the null hypothesis is true, then the combined sample $\{y_{11}, y_{12}, y_{21} - \theta_0, y_{22} - \theta_0\}$ can be regarded as a random sample of size $n = 4$ from the first population. Therefore, under $H_0: \theta = \theta_0$, the six possible sets of values given earlier are equally likely (with probability 1/6) as the ranks of the modified second sample. The possible calculated values of W, the ranks generating these values, and the probability of obtaining these values of W when H_0 is true are shown in Table 5.2.

Table 5.2 displays the sampling distribution of the Wilcoxon rank sum statistic under the null hypothesis that both samples are random samples from a continuous

TABLE   5.2

Probabilities of the Wilcoxon rank sum statistic if H_0 is true

Possible ranks $\{R_1, R_2\}$	Value of W w_c	Probability $p(w_c)$
$\{1, 2\}$	3	1/6
$\{1, 3\}$	4	1/6
$\{1, 4\}; \{2, 3\}$	5	2/6
$\{2, 4\}$	6	1/6
$\{3, 4\}$	7	1/6

population. Therefore, Table 5.2 can be used to determine the critical values for the Wilcoxon rank sum test when the sample sizes are $n_1 = 2$ and $n_2 = 2$. For example, from Table 5.2 we see that

$$P\{W > 6\} = p(7) = 1/6 = 0.17,$$

so that $w(2, 2, 0.17) = 6$.

Confidence intervals

The confidence intervals based on the Wilcoxon rank sum test for the difference between population medians might look odd at first sight. However, it can be shown that the interval generated by the formula in Box 5.5 can be interpreted as the set of all possible values θ_0 of the difference between the population medians that will be accepted by a two-sided α-level rank sum test of the null hypothesis H_0: $\theta = \theta_0$.

Critical values

As in the case of the sign and signed rank statistics, the sampling distribution of the Wilcoxon rank sum statistic is discrete, so that the critical values are not available for all combinations of the values of n_1, n_2, and α. Also, if the critical values in Table C.7 (Appendix C) are used with samples in which tied observations can occur, the resulting intervals and tests will be conservative, in the sense that true confidence levels will be higher and true significance levels (p-values) will be lower than the values obtained by the test.

Computational aspects

The SAS NPAR1WAY procedure can be used to calculate the Wilcoxon rank sum statistic and perform the large-sample test. For small sample sizes, where the large-sample approximation is not appropriate, the test can be performed using the WI option in the linear ranks tests in StatXact. StatXact can also be used to construct confidence intervals based on the Wilcoxon rank sum test.

Application of rank sum procedures

Let's look at an example.

EXAMPLE **5.6**   During an in vivo comparison of two antitumor drugs (conducted in the laboratory of Dr. Norman Reed, Department of Microbiology, Montana State University), Collings and Hamilton (1988) collected the following data on negative (natural) log-transformed tumor weights for two groups of 12 mice, each exposed to a different drug:

| Standard drug | | Test drug | |
Logarithm of tumor weight	Rank in combined sample	Logarithm of tumor weight	Rank in combined sample
−1.56444	1	−0.95089	6
−1.52562	2	−0.94663	7
−1.33500	3	−0.69315	10
−1.11711	4	−0.59222	11
−1.05222	5	−0.55962	12
−0.94507	8	−0.52117	14
−0.80648	9	−0.00995	16
−0.54349	13	0.07472	17
−0.01489	15	0.14156	18
0.16487	19	0.19601	20
0.46681	21	0.65887	22
1.80181	23	2.93746	24

Collings and Hamilton (1988) note that the negative log-transformed data conform relatively well to the two-sample shift model in Box 5.4 and that both distributions appear to be positively skewed; that is, the right tail is longer. Thus, we should look for an alternative to the *t*-test procedure for inferences about the shift between two population distributions, which is the same as the difference between the medians.

Let's assume that the distributions of the negative log-transformed tumor weights are of the same shape for two drugs and use the Wilcoxon rank sum procedure to test the null hypothesis H_0 that the distributions of the tumor weights are the same for both drugs against the research hypothesis H_1 that the distribution of the tumor weight measurements for the test drug is shifted to the right of the distribution of the tumor weight measurements for the standard drug. If we identify the values for the standard and test drugs with the first and second samples in Box 5.5, respectively, then we want to test the null hypothesis H_0: $\theta = 0$ against the research hypothesis H_1: $\theta > 0$.

We apply the procedures in Box 5.5 with $\theta_0 = 0$. The calculated value of the Wilcoxon rank sum statistic is the sum of the ranks of the second sample values (because $\theta_0 = 0$, there is no need to shift the second sample values)

$$W_c = 6 + 7 + 10 + 11 + 12 + 14 + 16 + 17 + 18 + 20 + 22 + 24$$

$$= 177.$$

An α-level test will reject the null hypothesis of equal distributions in favor of the research hypothesis that the distribution of test drug values is shifted to the right if

$$W_c \geq w(n_1, n_2, \alpha),$$

where $n_1 = n_2 = 12$ are the sample sizes and $w(n_1, n_2, \alpha)$ is the α-level critical value obtained from Table C.7 (Appendix C) or calculated from the large-sample approximation in Box 5.5. In the present case, the values of n_1 and n_2 are outside the range in Table C.7, and so we use the large-sample approximation. There are no ties in the combined sample, and hence $g = N = 24$ and $t_j = 1$, for $j = 1, 2, \ldots, 24$. Thus

$$\mu_W = \frac{12(24 + 1)}{2} = 150$$

and

$$\sigma_W = \sqrt{\frac{12 \times 12}{12} \left[24 + 1 - \frac{\sum_{j=1}^{24}(1)(1^2 - 1)}{24(24 + 1)} \right]} = 17.32.$$

For $\alpha = 0.05$, the approximation in Box 5.5 yields

$$w(12, 12, 0.05) \simeq 150 + 1.645(17.32) = 179.$$

The calculated value of the test statistic $W_c = 177$, which is less than 179, and we conclude at the 0.05 level that there is not enough evidence that the median negative log of the tumor weights for the test drug is greater than the median of the corresponding values for the standard drug. The calculated value of the test statistic

$$Z = \frac{W_c - \mu_W}{\sigma_W}$$

can be used with Table C.1 to determine the *p*-value associated with the large-sample Wilcoxon rank sum test. For the observed value $W_c = 177$, the calculated value of Z is $Z_c = 1.559$, so that the *p*-value of the one-sided test is 0.0595.

The following is the StatXact printout for the Wilcoxon rank sum test procedures with the tumor weight data.

```
                 StatXact printout for tumor weight data
-----------------------------------------------------------------------
WILCOXON RANK SUM TEST

Summary of Exact distribution of WILCOXON-MANN-WHITNEY statistic:
    Min      Max      Mean     Std-dev    Observed   Standardized
   78.00    222.0    150.0      17.32      177.0         1.559


Mann-Whitney Statistic =      99.00


Asymptotic Inference:
  One-sided p-value: Pr {   Test Statistic .GE. Observed } =      0.0595
  Two-sided p-value: 2 * One-sided                         =      0.1190
Exact Inference:

  One-sided p-value: Pr {   Test Statistic .GE. Observed } =      0.0638
  Point probability: Pr {   Test Statistic .EQ. Observed } =      0.0071
  Two-sided p-value: Pr { | Test Statistic - Mean |
                      .GE. | Observed - Mean |             =      0.1277
  Two-sided p-value: 2*One-sided                           =      0.1277
```

```
-----------------------------------------------------------------

HODGES-LEHMANN ESTIMATES OF SHIFT PARAMETER

Point Estimate of Shift : Theta = POP_1 - POP_2 =    0.5280

95.00% Confidence Interval for Theta :
         Asymptotic : (-0.1402  ,   1.127)
         Exact      : (-0.1444  ,   1.136)
-----------------------------------------------------------------
```

As we see from the printout, the difference between the *p*-values computed using the exact method (0.0638) and the large-sample approximation (0.0595) is very small. ∎

Confidence intervals for the difference between the median negative log tumor weights can be constructed using the procedure in Box 5.5. The StatXact printout in Example 5.6 shows that with 95% confidence, this difference is contained in the interval $(-0.1444, 1.136)$. Example 5.7 illustrates the construction of a confidence interval another data set.

EXAMPLE **5.7**   In a study of serum albumin levels in children with clinical evidence of protein-energy malnutrition (PEM) the following serum albumin levels (g/dliter) were obtained for four control children and three children with a form of PEM known as marasmus (Smith et al., 1981).

Group	Serum albumin levels			
Marasmus	3.0	3.2	3.6	
Control	5.1	4.7	5.1	4.8

Let's use these data to verify the research hypothesis that the median serum albumin level for control children is 1.5 g/dliter more than the median level for children with the marasmus form of PEM. Let η_1 and η_2 denote, respectively, the median serum albumin levels for the control and marasmus children, and let $\theta = \eta_1 - \eta_2$. Then we want to test the null hypothesis $H_0: \theta \leq 1.5$ against the research hypothesis $H_1: \theta > 1.5$. We can use the Wilcoxon rank sum test in Box 5.5 with $\theta_0 = 1.5$.

If we treat the control sample as the second sample, we can use rejection criterion 2 in Box 5.5. The modified samples are obtained by subtracting 1.5 from the values in the second sample. The modified samples may be arranged in increasing order of magnitude as follows, along with their ranks:

Samples	3.0 (1)	3.2 (1)	3.2 (2)	3.3 (2)	3.6 (1)	3.6 (2)	3.6 (2)
Ranks	1	2.5	2.5	4	6	6	6

The numbers in parentheses indicate whether the value is from the sample of control children (2) or the sample of children with marasmus (1).

The calculated value of the Wilcoxon rank sum statistic is the sum of the modified second-sample ranks

$$W_c = 2.5 + 4 + 6 + 6 = 18.5.$$

From Table C.7 (Appendix C), when $n_1 = 3$ and $n_2 = 4$, we get $w(3, 4, 0.029) = 18$, so that the null hypothesis can be rejected at the level $\alpha = 0.029$. Therefore, we can conclude, at the $\alpha = 0.029$ level, that the median serum albumin value for control subjects is at least 1.5 g/dliter more than the median value for children with the marasmus form of PEM. Because the combined samples had ties, the significance level at which H_0 can be rejected is less than the value of 0.029 obtained from Table C.7.

We now construct a $100(1 - 2 \times 0.029)\% = 94.2\%$ confidence interval for θ, the difference between the medians, with $\alpha/2 = 0.029$ in Box 5.5. First, we need to calculate the values of D_{ij}. There are $3 \times 4 = 12$ such values. These values are shown in the two-way table below, in which the entry common to row i and column j is the difference $D_{ij} = y_{2j} - y_{1i}$:

		y_{2j}		
y_{1i}	4.7	4.8	5.1	5.1
3.6	1.1	1.2	1.5	1.5
3.2	1.5	1.6	1.9	1.9
3.0	1.7	1.8	2.1	2.1

Using the values $n_1 = 3, n_2 = 4, N = 7$, and $w(3, 4, 0.029) = 18$ in the formulas for U and L in Box 5.5, we get

$$U = 18 - \frac{4(4 + 1)}{2} = 8 \quad \text{and} \quad L = \frac{4(7 + 3 + 1)}{2} + 1 - 18 = 5.$$

Therefore, a 94.2% confidence interval for θ is

$$D_{(5)} < \theta < D_{(8)} \Leftrightarrow 1.5 < \theta < 1.8.$$

We conclude, at the 94.2% confidence level, that the median serum albumin value for the control subjects is at least 1.5 g/dliter and at most 1.8 g/dliter more than the corresponding median value for children with the marasmus form of PEM.   ∎

The Wilcoxon rank sum test of stochastic order

Under the assumptions of the model in Box 5.4, the Wilcoxon rank sum procedures may be used to make inferences about the difference between the means or medians of two populations that have identical shapes. The hypothesis tests based on the Wilcoxon rank sum tests may also be used under the less restrictive assumption that the populations are *stochastically ordered*.

Given two populations, called the Y_1 population and the Y_2 population, the Y_1 population is said to be *stochastically larger* than the Y_2 population if the following conditions are satisfied:

1   For any y, the proportion of measurements greater than y in the Y_1 population is at least as large as the corresponding proportion in the Y_2 population. In symbols

$$\Pr\{Y_1 > y\} \geq \Pr\{Y_2 > y\} \qquad \text{for any } y. \tag{5.4}$$

2   There is at least one value y_0 such that

$$\Pr\{Y_1 > y_0\} > \Pr\{Y_2 > y_0\}. \tag{5.5}$$

The symbol $Y_1 \gg Y_2$ will be used to indicate that the Y_1 population is stochastically larger than the Y_2 population. The conditions in Equations (5.4) and (5.5) have the following practical implications. Equation (5.4) implies that a measurement selected at random from the Y_1 population is at least as likely to exceed any specified value as a measurement selected from the Y_2 population. Equation (5.5) implies that there is at least one value y_0 such that a randomly selected value from the Y_1 population has a higher chance of exceeding y_0 than does a randomly selected value from the Y_2 population. For example, the assumption that the population of yields for variety A is stochastically larger than the population of yields for variety B implies that:

1   for any y ($y = 10$ bushels, $y = 20$ bushels, $y = 550$ bushels, and so on), the probability that the yield for variety A exceeds y bushels is at least as large as the probability that the yield for variety B exceeds y bushels;

2   there is at least one value—say, $y_0 = 35$ bushels—such that the probability of a yield for variety A exceeding 35 bushels is higher than the corresponding probability for variety B.

Figure 5.5 shows two stochastically ordered populations. Notice that, in contrast to Figure 5.3, the populations need not be of the same shapes.

The area with horizontal shading equals the probability that a randomly selected value from the Y_1 population will exceed the value $y = 10$. The corresponding probability for the Y_2 population is shown by the area with vertical shading. It can be verified, by moving the vertical line along the y-axis and comparing the corresponding areas, that the Y_1 population is stochastically larger than the Y_2 population.

Comparison of Figures 5.3 and 5.5 shows the differences between two location-shifted populations and two stochastically ordered populations. Location-shifted populations are of the same shape and are stochastically ordered, but stochastically ordered populations need not be of the same shape. Thus, the assumption that populations are stochastically ordered is less restrictive than the assumption that they are location-shifted.

Note that some populations are not stochastically ordered; an example is shown in Figure 5.6, where the probability that an observation from the Y_1 population exceeds 10 is at least as large as the probability that an observation from the Y_2 population will exceed 10 but the opposite is true for the probabilities that observations will exceed 20.

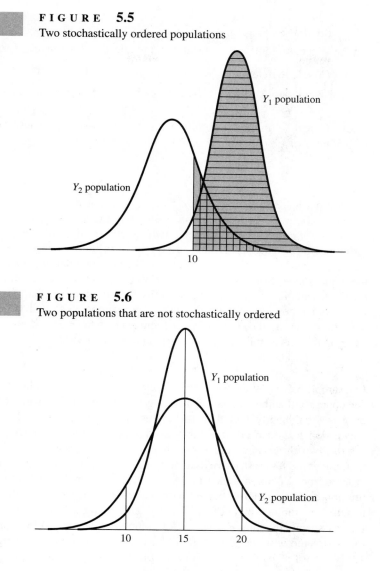

FIGURE   5.5
Two stochastically ordered populations

Y_1 population

Y_2 population

10

FIGURE   5.6
Two populations that are not stochastically ordered

Y_1 population

Y_2 population

10   15   20

Box 5.6 describes the Wilcoxon rank sum test for stochastically ordered populations.

BOX **5.6**

Wilcoxon rank sum test of stochastic order

Let $y_{11}, y_{12}, \ldots, y_{1n_1}$ and $y_{21}, y_{22}, \ldots, y_{2n_2}$ denote independent random samples from two stochastically ordered continuous populations. Three sets of hypotheses about the difference between two populations can be tested using the Wilcoxon rank sum test statistic; the rejection criteria are as follows:

1   Conclude in favor of $H_1: Y_1 \gg Y_2$ at level α if $w_c \geq w(n_1, n_2, \alpha)$.
2   Conclude in favor of $H_1: Y_1 \ll Y_2$ at level α if $w_c \leq n_2(N + 1) - w(n_1, n_2, \alpha)$.
3   Conclude in favor of $H_1: Y_1 \gg Y_2$ or $Y_1 \ll Y_2$ at level α if $w_c \geq w(n_1, n_2, \alpha/2)$ or $w_c \leq n_2(N + 1) - w(n_1, n_2, \alpha/2)$.

Comparing Box 5.5 with Box 5.6, we see that the procedure for testing the difference between two population medians is exactly the same as the procedure for testing the difference between two stochastically ordered populations. The difference is in what we assume to be known in the two cases. Comparing medians is appropriate if the distributions of the two populations are known to be of the same shape. Stochastic ordering of the populations, by contrast, implies that measurements in one population generally tend to be larger than the measurements in the other population, in the sense of Equations (5.4) and (5.5). We rule out situations such as those depicted in Figure 5.6, where the difference between the two populations is mainly due to a difference in their variability.

EXAMPLE **5.8**   In Example 5.6 the Wilcoxon rank sum test indicated that at the level $\alpha = 0.0638$, we can conclude that the median negative log-transformed tumor weight for the standard drug is significantly less than the corresponding median for the test drug. This conclusion was reached under the assumption that the tumor weight distributions for the two drugs have the same shape. The test procedure in Box 5.6 implies that, under the less restrictive assumption that the two populations are stochastically ordered, the Wilcoxon rank sum test leads to the conclusion that the negative log tumor weight of a mouse treated with the test drug is stochastically larger than the negative log tumor weight of a mouse treated with the standard drug. In other words, for any specified value of negative log tumor weight—say, 1.00—the chance that a mouse treated with the test drug will have a negative log tumor weight greater than 1.0 is at least as large as the chance that the negative log tumor weight for a mouse treated with the standard drug will exceed 1.00.   ■

Exercises

5.11   Refer to the dental data in Exercise 4.8.

a   Describe a statistical model that will justify using the Wilcoxon rank sum test to compare the mean dental measurement of boys with that for girls.
b   Assuming the model in (a), perform a test to see if the mean dental measurements for boys is higher than that for girls.

c   Construct an approximate 95% confidence interval for the difference between the mean dental measurements of the boys and girls.

d   Compare the confidence interval in (c) with the corresponding interval based on the two-sample t-test. What conclusions can you draw from this comparison?

5.12   In a study to compare the consumer ratings of two brands of red wine, 20 adult shoppers were randomly selected from those entering a big shopping mall on a weekend and divided into two groups of ten people. The two brands were assigned at random to the two groups, and each subject was asked to taste and rate the wine assigned to his or her group. The rating was based on a five-point scale (-2, very poor; -1, poor; 0, neutral; 1, good; and 2, very good). The following ratings were observed.

Brand	Ratings									
A	1	0	-2	-1	1	0	-1	0	1	0
B	2	1	1	0	2	0	1	2	1	0

a   In the context of this problem, explain what is meant by the claim that the rating for brand B is stochastically larger than that for brand A.

b   State appropriate null and research hypotheses for a test to see if the data support the conclusion that the ratings for brand B tend to be higher than those for brand A.

c   Explain why the two-sample t-test may not be appropriate to test the null hypothesis in (b).

d   Use the Wilcoxon rank sum procedure to test the null hypothesis in (b) and write your conclusions.

5.13   An advantage of using a procedure based on ranks rather than a procedure that uses actual measured values is that the rank-based procedures are less sensitive to extreme values. For example, gross errors in measured values will have less impact on Wilcoxon rank sum procedures than on the corresponding procedures based on a two-sample t-test. To illustrate this property, refer to Example 5.7.

a   Change the control value of 5.1 to 25.1 and test $H_0: \theta \leq 1.5$ against $H_1: \theta > 1.5$ using the Wilcoxon rank sum test.

b   Test the hypothesis in (a) using the two-sample t-test.

c   Compare the results in (a) and (b) with the results when the same tests are applied to the original data.

5.6
Overview

The ordinal data procedures presented in this chapter can be used in two ways. First, the procedures are useful to analyze data measured on an ordinal scale, for which the procedures of Chapter 4 are not applicable. Second, the procedures are useful for interval data when there is reasonable doubt about the validity of the assumptions on which the procedures in Chapter 4 depend (equal population variances, normally distributed populations, and so on).

With interval data, one option is to use the methods in Chapter 4 after a suitable transformation of the data. This approach will be discussed in Chapter 8. For the present, note that this option opens up a rich variety of well-documented techniques for analyzing our data, but inferences based on transformed data may not accurately correspond to inferences about populations of interest. Also, as we will see in Chapter 8, determining the appropriate transformation may be a difficult task, and even the transformed data may violate the needed assumptions in some cases.

The second option is to apply the rank based methods discussed in this chapter. The main advantage of these methods is that they require less restrictive assumptions about data than the methods described in Chapter 4; they can be used with interval data as well as ordinal data. An obvious disadvantage of the rank based procedures is that, when used with interval data, they replace the observed values by their ranks and/or signs, resulting in a certain loss of information. Of course, the loss of information is only a problem when rank test procedures are used for data for which the methods in Chapter 4 are appropriate. In such cases, inferences based on the methods of Chapter 4 are bound to be more efficient, simply because these procedures incorporate more information about the populations by way of additional assumptions. However, extensive theoretical and practical investigations have shown that, in general, the use of rank based procedures when the data meet the assumptions in Chapter 4 does not result in a serious loss of efficiency. On the other hand, with certain violations of these assumptions, the efficiency gain resulting from rank test procedures can be considerable. For instance, when the assumptions needed for the t-test procedures are satisfied by the data, and the sample sizes are not too small, the use of rank procedures instead of the corresponding t-test procedures will result in a loss of efficiency equivalent to losing about 5% of the data. For nonnormal symmetric populations with heavy tails, the use of rank test procedures will have a large sample efficiency, equivalent to the efficiency of the corresponding t-test procedures applied to a sample containing 50% more observations. For more details on the efficiency of rank test procedures relative to procedures based on ANOVA assumptions, see Hollander and Wolfe (1973), Lehmann (1975), and Sprent (1993).

The main disadvantage of rank test procedures is that there are not enough of them. Despite the rapid development of such procedures in recent years, nonparametric procedures for complicated inference problems are still not available.

6

Inferences about One or Two Populations: Categorical Data

6.1
Introduction

In this chapter, we present some categorical data methods that are frequently used to make inferences about one or two populations. The conditions under which the methods described in this chapter are useful parallel the conditions in Chapters 4 and 5. In Section 6.2, we describe methods that use a single random sample for making inferences about population proportions. Section 6.3 is devoted to describing methods for comparing population proportions (a) using two independent samples and (b) using a single paired sample. The binomial test for a population proportion, the Fisher's exact test for comparing two population proportions, the Chi-square goodness-of-fit test, and the McNemar test are among the procedures described in this chapter.

Categorical data are measurements of nominal variables. Because the values assumed by nominal variables have no quantitative interpretations, categorical data are best described using counts and proportions. Consider the following example.

EXAMPLE 6.1   To assess the effectiveness of a new treatment for aphid infestation of tobacco plants, 100 treated plants were classified into three infestation categories. It was found that 67 plants had no aphid infestation, 24 had only stem infestation, and 9 had only leaf infestation.

Let Y be a variable that can take the following values

$$Y = \begin{cases} 1 & \text{if the plant has only leaf infestation} \\ 2 & \text{if the plant has only stem infestation} \\ 3 & \text{if the plant has no infestation.} \end{cases}$$

Then Y is a nominal variable whose observed value indicates the type of infestation of a plant. The data on 100 plants can be regarded as a random sample of size $n = 100$ from a conceptual population of values of Y corresponding to the population of all treated plants. This population can be described using three parameters π_1, π_2, and π_3 ($\pi_1 + \pi_2 + \pi_3 = 1$), which represent the population relative frequencies of $Y = 1$, $Y = 2$, and $Y = 3$, respectively. As noted in Chapter 2, the proportion π_i ($i = 1, 2, 3$) can be interpreted as the probability that a randomly selected treated tobacco plant will have infestation of type i.

The information in the sample can be summarized using the observed counts of the values of Y in the three categories

$$f_1 = 9, f_2 = 24, f_3 = 67. \quad \blacksquare$$

In Example 6.1, the measurements fall into three categories. More generally, suppose a random sample of size n is selected from a population with C mutually exclusive and exhaustive categories. Let $f_1, f_2, \ldots, f_C$ denote the number of responses in categories $1, 2, \ldots, C$, respectively. We shall call f_k the *observed frequency* of responses in the k-th category. The sample data can be displayed in a table with one row and C columns, as in Table 6.1.

TABLE   6.1
Frequencies in a sample of size n

Category	1	2	$\cdots$	k	$\cdots$	C	Total
Frequency	f_1	f_2	$\cdots$	f_k	$\cdots$	f_C	$f_+ (= n)$

Notice that we have used the symbol f_+ to denote the sum of all frequencies. Thus f_+ is another symbol for the sample size n. The convention of using subscript $+$ will prove convenient when we have to deal with more complex sampling schemes. Subscript $+$ means that the corresponding quantity can be obtained by adding over all possible values of the subscript replaced by $+$. Thus, f_+ is the sum of f_k over all values of k; that is, $f_+ = f_1 + f_2 + \cdots + f_C$.

Table 6.2 shows the probability distribution of the population from which the sample in Table 6.1 was selected. The proportion of values in the k-th category is π_k ($k = 1, 2, \ldots, C$), so that the probability of a randomly selected value belonging

TABLE 6.2

Probability distribution of a categorical population

Category	1	2	$\cdots$	k	$\cdots$	C	Total
Proportion	π_1	π_2	$\cdots$	π_k	$\cdots$	π_C	$\pi_+ (= 1)$

to the k-th category is π_k. Clearly, $\pi_1 + \pi_2 + \cdots + \pi_C = 1$. In Chapter 2, we have seen several examples of probability distributions of categorical populations. For instance, the probability distribution of the types of pests in a region in Table 2.2 (reproduced in Table 6.3) is the probability distribution of a categorical population with $C = 5$ categories.

TABLE 6.3

Probability distribution of pest types

Pest type	1	2	3	4	5	Total
Proportion	0.16	0.18	0.10	0.46	0.10	1

Inferences about a categorical population are inferences about the population proportions.

Dichotomous populations

A 0–1 population in which the proportion of 1s equals π can be viewed as a nominal or categorical population with two categories ($C = 2$). Categorical populations with two categories are called *dichotomous populations*. If we let $Y = 1$ for category 1 and $Y = 0$ for category 2, then $\pi_1 = \pi$ and $\pi_2 = 1 - \pi$ are the population relative frequencies of $Y = 1$ and $Y = 0$, respectively. The observed frequencies f_1 and f_2 are the number of 1s and 0s, respectively, in the sample.

As we see, only one proportion, $\pi_1 = \pi$, is needed to specify a dichotomous population, and so inferences about dichotomous populations are inferences about a single population proportion.

Exercises

6.1 Display the probability distribution of each of the following populations in the form of Table 6.3, and determine the values of C and $\pi_1, \ldots, \pi_C$:

  a the 0–1 population in Example 1.4.

  b the litter-size population in Exercise 2.10.

  c the tobacco leaf population in Exercise 2.12.

  d the germinating seed population in Exercise 2.14.

   e   the age and blood pressure population in Exercise 2.15.

   f   the population of errors in Exercise 2.19.

   g   the caterpillar population in Example 2.12.

6.2
Inferences about a single categorical population

In this book, we consider two common types of inferences about a single categorical population: inferences about a single population proportion; and inferences about a specified set of values for the population proportions. For information on techniques useful for other types of inferences, see Agresti (1990) and Fleiss (1981).

Inferences about a single proportion

We begin with an example of a situation in which inference about a population proportion is one of the objectives of data analysis.

EXAMPLE **6.2**   Suppose that in Example 6.1 we want to see if the observed data justify the claim that less than 15% of the population of treated plants will have only leaf infestation. Let π denote the proportion of plants with only leaf infestation. Then we want to test the null hypothesis H_0 *that* $\pi \geq 0.15$ against the research hypothesis H_1 *that* $\pi < 0.15$. Alternatively, if we are interested in evaluating the overall effectiveness of the treatment, we might want to use the data to construct a confidence interval for π. ■

In Box 6.1, we present procedures that can be used to test hypotheses and construct confidence intervals for a population proportion when large sample sizes are available.

BOX **6.1**

   Large-sample inferences about a population proportion

Let $\hat{\pi}$ be the observed proportion of responses belonging to a particular category in a random sample of size n. Let π denote the corresponding population proportion. Then, if the sample size is large ($n \geq 30$), hypothesis tests and confidence intervals for π can be constructed as follows.

1   A test of the null hypothesis H_0: $\pi = \pi_0$ can be performed using the calculated value z_c of the test statistic

$$Z = \frac{\hat{\pi} - \pi_0}{\hat{\sigma}_{\hat{\pi}}},$$

where

$$\hat{\sigma}_{\hat{\pi}} = \sqrt{\frac{\hat{\pi}(1 - \hat{\pi})}{n}}$$

is the estimated standard error of $\hat{\pi}$.

2   A $100(1 - \alpha)\%$ confidence interval for π is calculated as

$$\hat{\pi} \pm z(\alpha/2)\hat{\sigma}_{\hat{\pi}},$$

where $z(\alpha)$ is the α-level critical value of a standard normal distribution.

The procedures in Box 6.1 can be derived from the procedures in Box 4.1 for making inferences about μ, the population mean. If we regard the target population as a 0–1 population in which the proportion of 1s is π, then, as noted in Equation (2.8), the mean and variance of the population are $\mu = \pi$ and $\sigma^2 = \pi(1 - \pi)$. For large sample sizes, if we let $\hat{\theta} = \overline{Y}$ and $\theta = \mu$, the results in Box 4.1 imply that inferences about μ can be based on a confidence interval of the type

$$\overline{Y} \pm z(\alpha/2)\frac{S}{\sqrt{n}}$$

and a test statistic of the type

$$Z = \frac{\overline{Y} - \mu_0}{\frac{S}{\sqrt{n}}},$$

where $\overline{Y}$ and S are the sample mean and sample standard deviation, respectively. Using Equations (3.1) and (3.3), we can verify that, when the sample consists only of 0s and 1s, the mean and standard deviation of the sample can be expressed as

$$\overline{Y} = \hat{\pi}, \quad S = \sqrt{\frac{n}{n - 1}\hat{\pi}(1 - \hat{\pi})}, \tag{6.1}$$

where $\hat{\pi}$ denotes the proportion of 1s in the sample. When n is large, the fraction $n/(n - 1)$ will be close to 1, so that the expression for $\sigma_{\hat{\pi}}$ in Box 6.1 is a large sample approximation for $S/\sqrt{n}$. Thus, when sample sizes are large, the conclusions about the population proportion based on confidence intervals and test statistics in Box 4.1 will be similar to those based on the procedures described in Box 6.1.

The use of the procedures in Box 6.1 is illustrated in the next example.

EXAMPLE **6.3**   Suppose that, in Example 6.1, we want to determine whether the observed data justify the claim that less than 15% of the population of treated plants will have only leaf infestation. In other words, if π denotes the population proportion of treated plants with only leaf infestation, we want to test the null hypothesis H_0: $\pi \geq 0.15$ against the research hypothesis H_1: $\pi < 0.15$.

The frequencies of infestation types in Example 6.1 can be tabulated as follows:

Type of infestation	Leaf	Stem	None	Total
Number of plants	9	24	67	100

We see that $\hat{\pi} = 9/100 = 0.09$ is the sample proportion of plants with leaf infestation. We have a large sample size ($n = 100$), and so H_0 can be tested as in Box 6.1. The test statistic has the value

$$z_c = \frac{0.09 - 0.15}{\sqrt{\dfrac{0.09(1 - 0.09)}{100}}} = \frac{-0.06}{0.0286} = -2.10.$$

From Table C.1 (Appendix C), the p-value of the test is 0.0179. Therefore, at the 0.05 level of significance (in fact, at any level greater than 0.0179), we can conclude that the new treatment for aphids keeps the leaf infestation rate below 15%.

The endpoints of a 95% confidence interval for the proportion of leaf-infested treated tobacco plants are

$$0.09 \pm 1.96\sqrt{\frac{0.09(1 - 0.09)}{100}} = 0.09 \pm (1.96)(0.0286),$$

so that the required confidence interval is

$$0.034 \leq \pi \leq 0.146.$$

At the 95% confidence level, we can conclude that between 3.4% and 14.6% of the treated tobacco plants will have leaf infestation.   ■

The procedures in Box 6.1 are appropriate for inferences about a population proportion provided the sample is large. When the sample is small, we can make inferences about π on the basis that the number of 1s in a random sample of size n from a 0–1 population can be regarded as the number of successes in a binomial experiment with n trials and probability of success π. The procedure in Box 6.2, which is called the *binomial test*, can be used to calculate the p-value for testing hypotheses about π.

BOX **6.2**

The binomial test for a population proportion

Suppose we have a random sample of size n from a categorical population. Let r denote the number of responses in a particular category, and let π denote the population proportion in that category. Also let

$$f(y|n, \pi) = \frac{n!}{y!(n - y)!}\pi^y(1 - \pi)^{n-y}$$

denote the probability function of a binomial distribution with n trials and probability of success π; see Equation (2.10). Then the p-values of the tests of hypotheses about π can be computed as follows.

a To test H_0: $\pi \le \pi_0$ against H_1: $\pi > \pi_0$

$$p = f(r|n, \pi_0) + f(r + 1|n, \pi_0) + \cdots + f(n|n, \pi_0).$$

b To test H_0: $\pi \ge \pi_0$ against H_1: $\pi < \pi_0$

$$p = f(0|n, \pi_0) + f(1|n, \pi_0) + \cdots + f(r|n, \pi_0).$$

c To test H_0: $\pi = \pi_0$ against H_1: $\pi \ne \pi_0$

$$p = p_- + p_+$$

where

$$p_- = f(0|n, \pi_0) + f(1|n, \pi_0) + \cdots + f(k|n, \pi_0),$$

$$p_+ = f(n - k|n, \pi_0) + f(n - k + 1|n, \pi_0) + \cdots + f(n|n, \pi_0),$$

$$k = \min(r, n - r).$$

Rationale for the binomial test

Suppose that we have a random sample of size n from a categorical population in which a proportion π of measurements belong to a particular category A. Let Y denote the number of sample measurements belonging to A. Then formulas for the p-values in a binomial test are based on the fact that the random variable Y has a binomial distribution with parameters n and π. The three p-values in Box 6.2 are the probabilities of observing samples that are at least as favorable to the research hypothesis as the corresponding observed samples. For example, consider testing H_0: $\pi \ge \pi_0$ against H_1: $\pi < \pi_0$. The probability of observing a response in the category of interest is less if H_1 is true than if H_0 is true, and so we would expect fewer responses in that category for H_1 than for H_0. Thus, a smaller number of observed responses in the category of interest is more favorable to the research hypothesis H_1. The expression for the p-value in case (b) in Box 6.2 gives the probability that, under the null hypothesis H_0: $\pi = \pi_0$, the observed number of responses in the category of interest is at least as favorable to H_1 as the observed value r. A similar interpretation can be given to the other two p-values in Box 6.2.

Calculating the *p*-value

In the comments following Example 3.22, we noted that the p-value of a test can be calculated as the probability under the null hypothesis of observing a sample that is at least as favorable to the research hypothesis as the observed sample. The expressions for the p-values in Box 6.2 can be derived using this principle. For example, a large value for r, the observed number of responses, favors the research hypothesis H_1: $\pi > \pi_0$, and so the p-value for testing H_0: $\pi \le \pi_0$ against H_1: $\pi > \pi_0$ can be calculated by adding the probabilities—under the null hypothesis—of observing $r, r + 1, \cdots$, and n responses in category 1.

Computational aspects

Computation of the p-values associated with a binomial test can be laborious, even for moderate sample sizes. For instance, when $n = 15$, $r = 7$, and $\pi_0 = 0.25$, we need to add eight terms—$f(0|15, 0.25)$, $f(1|15, 0.25)$, ..., $f(7|15, 0.25)$—in order to obtain the p-value for testing the hypotheses in case (b) in Box 6.2. Many of the standard statistical software packages have programs that can be used for rapid computation of the binomial probability $f(y\,|n, \pi)$. For example, the PROBBNML function in SAS can be used to compute the sum

$$f(0|n, \pi) + f(1|n, \pi) + \cdots + f(r|n, \pi)$$

for any combination of values of r, n, and π. See Figure 6.3 in the companion text by Younger (1997) for an example. A procedure in StatXact for computing confidence intervals for a population proportion can be used to determine the p-value associated with the binomial test; for more information, see Sprent (1993).

Example 6.4 describes an application of the binomial test.

EXAMPLE **6.4**   A plant breeder observed two dwarf plants among $n = 10$ F_2-generation plants. Do these data indicate that less than 25% of all F_2-generation plants will be dwarfs?

We have a sample of size $n = 10$ from a categorical population with two categories: tall and dwarf. The observed frequency in the category of interest (dwarf) is $f = 2$. Let π denote the population proportion of dwarf plants. We need a test of the hypothesis H_0: $\pi \geq 0.25$ against H_1: $\pi < 0.25$. The p-value of the observed sample is computed using the formula for case (b) in Box 6.2

$$p = f(0|10, 0.25) + f(1|10, 0.25) + f(2|10, 0.25)$$

$$= \frac{10!}{0!10!}(0.25)^0(0.75)^{10} + \frac{10!}{1!9!}(0.25)^1(0.75)^9 + \frac{10!}{2!8!}(0.25)^2(0.75)^8$$

$$= 0.0563 + 0.1877 + 0.2816 = 0.5256.$$

The p-value is relatively high, and so it is not reasonable to claim that less than 25% of the F_2-generation plants will be dwarfs.

The p-value of a test of H_0: $\pi = 0.25$ against H_1: $\pi \neq 0.25$ is

$$k = \min(2, 10 - 2) = 2$$

$$p = f(0|10, 0.25) + f(1|10, 0.25) + f(2|10, 0.25)$$

$$\qquad + f(10 - 2|10, 0.25) + f(10 - 2 + 1|10, 0.25) + f(10|10, 0.25)$$

$$= f(0|10, 0.25) + f(1|10, 0.25) + f(2|10, 0.25)$$

$$\qquad + f(8|10, 0.25) + f(9|10, 0.25) + f(10|10, 0.25) = 0.526. \quad \blacksquare$$

Box 6.2 presented a procedure for testing hypotheses about a population proportion π when the sample size is small. The corresponding interval estimates of π can be constructed using Table C.8 in Appendix C. For sample sizes $n = 5$–15, Table C.8 gives the 95% and 99% confidence intervals for π as functions of f, the observed frequency. Confidence intervals for other sample sizes are available in Crow (1956). Alternatively, option BI in the one-sample procedures of StatXact can be used to compute these confidence intervals.

EXAMPLE **6.5** A 95% confidence interval for the population proportion π of dwarf plants in the F_2-generation in Example 6.4 can be obtained from Table C.8. For $n = 10$, $f = 2$, and $1 - \alpha = 0.95$, we see from Table C.8 that the required confidence interval is $0.0252 \leq \pi \leq 0.5561$. Thus, with 95% confidence, we can assert that between 2.52% and 55.61% of F_2-generation plants will be dwarfs. This is a wide confidence interval because the sample size is relatively small.

When option BI in StatXact is used to construct a 95% confidence interval for the proportion of dwarfs in the F_2-generation plants, the following printout is obtained.

```
-------------------------------------------------------------------
                        StatXact Printout

ESTIMATION OF BINOMIAL PARAMETER (PI)

                    NUMBER OF TRIALS      =         10

                    NUMBER OF SUCCESSES   =         2

        Point Estimate of PI = 0.2000

        95.00% Confidence Interval for PI = (0.2521E-01*, 0.5561)
```
*The symbol E-01 means that the decimal should be moved one place to the left of its current position. Thus 0.2521E-01 should be read as .02521.
```
-------------------------------------------------------------------
```

As we see from the printout, the confidence interval provided by StatXact is the same as that obtained using Table C.8. ∎

Inferences about a specified set of values for population proportions

In addition to making inferences about the proportion of values in a specific category, a researcher may also be interested in seeing whether the population proportions differ from a specified set of values. Here are two examples.

EXAMPLE **6.6** In Example 6.1, suppose that the population relative frequencies of the three types of aphid infestation in untreated plants are known to be 20% for leaf infestation only, 30% for stem infestation only, and 50% for no infestation. It is of practical interest to determine whether the observed data indicate that the population proportions for the treated plants are different from the known proportions for the untreated plants. ∎

EXAMPLE **6.7**   In plant and animal breeding and in human genetics, investigators often conduct statistical hypothesis tests to see if a set of observed data contradict a theory about the inheritance of genetic traits (such as dark hair or wrinkled skin) from generation to generation. For instance, let π_A and π_B be the proportions of traits A and B in the F_1 (present) generation. Then, it can be shown that, if A and B are independently inherited, the proportions of phenotypes (combinations of A and B) in the F_2 (next) generation will be as in Table 6.4.

TABLE 6.4

Phenotype proportions under independent inheritance

Phenotype	Proportion in F_2
AB: Both A and B present	$\pi_{AB} = \pi_A \pi_B$
Ab: A is present, B is absent	$\pi_{Ab} = \pi_A(1 - \pi_B)$
aB: A is absent, B is present	$\pi_{aB} = (1 - \pi_A)\pi_B$
ab: Both A and B are absent	$\pi_{ab} = (1 - \pi_A)(1 - \pi_B)$

In practice, the problem is to use the observed data to test the theory of independent inheritance of given traits under one of the following conditions.

1   Both π_A and π_B are known, so that the population proportions are completely specified. For example, suppose that each of the traits occurs in 75% of the F_1-generation—that is, $\pi_A = \pi_B = 3/4$. Then, if the theory of independent inheritance holds for A and B, the phenotype proportions in the F_2-generation can be determined by setting $\pi_A = \pi_B = 3/4$ in Table 6.4. Table 6.5 shows the phenotype proportions under the hypothesis of independent inheritance. Notice in Table 6.5 that, when two traits are independently inherited, the phenotypes in the F_2-generation occur in the ratio 9:3:3:1.

TABLE 6.5

$$\pi_{AB} = \tfrac{3}{4} \times \tfrac{3}{4} = \tfrac{9}{16}$$
$$\pi_{Ab} = \tfrac{3}{4} \times \tfrac{1}{4} = \tfrac{3}{16}$$
$$\pi_{aB} = \tfrac{1}{4} \times \tfrac{3}{4} = \tfrac{3}{16}$$
$$\pi_{ab} = \tfrac{1}{4} \times \tfrac{1}{4} = \tfrac{1}{16}$$

2   It is known that π_A and π_B are equal to a common but unknown value π; that is, $\pi_A = \pi_B = \pi$, where π is not known. Under this condition, testing the theory of independent inheritance is equivalent to testing the null hypothesis that the proportions of phenotypes in F_2 can be expressed as in Table 6.6.

TABLE   6.6

$$\pi_{AB} = \pi^2$$
$$\pi_{Ab} = \pi(1 - \pi)$$
$$\pi_{aB} = (1 - \pi)\pi$$
$$\pi_{ab} = (1 - \pi)(1 - \pi)$$

3   Both π_A and π_B are unknown. Under this condition, testing the theory of independent inheritance is equivalent to a test in which the null hypothesis states that the phenotype proportions are as in Table 6.4.

Under condition 1, the hypothesis-testing problem is the same as that in Example 6.6; that is, we are interested in making inferences about a completely specified set of values for the population proportions. Under conditions 2 and 3, however, the inference is about a partially specified set of population proportions.   ■

The inferential problems in Examples 6.6 and 6.7 are hypothesis-testing problems in which the objective is to see if the distribution of the population is different from that specified under the null hypothesis.

In Box 6.3, we describe a large-sample procedure for testing the null hypothesis H_0 that the population has a specified distribution against the research hypothesis H_1 that the population distribution is not as specified under H_0.

As we shall see, the test in Box 6.3 utilizes a statistic that measures how well the observed frequencies fit the theoretical frequencies under the null hypothesis. The test is commonly known as a χ^2 goodness-of-fit test.

BOX 6.3

The χ^2 goodness-of-fit test of a completely specified set of proportions

Let f_k denote the frequency of Category k in a random sample of size $n = f_+$ from a categorical population with C categories. Let π_k and π_{k0} denote, respectively, the true proportion and its hypothesized value, in the k-th ($k = 1, 2, \ldots, C$) category. A large-sample test of

$$H_0\colon \pi_1 = \pi_{10}, \pi_2 = \pi_{20}, \ldots, \pi_k = \pi_{k0}$$

against the research hypothesis H_1 that the population proportions are not as specified by H_0, can be performed using the test statistic

$$\chi_c^2 = \sum_{k=1}^{C} \frac{(f_k - \hat{f}_k)^2}{\hat{f}_k},$$

where

$$\hat{f}_k = f_+ \pi_{k0}.$$

The null hypothesis is rejected at level α if

$$\chi_c^2 \geq \chi^2(C - 1, \alpha),$$

where $\chi^2(C - 1, \alpha)$ is the α-level critical value of a χ^2-distribution with $C - 1$ degrees of freedom.

Criteria for large samples

The large-sample requirement means that the $\hat{f}_k$ are not all small (as a rule of thumb, $\hat{f}_k \geq 5$ in at least 75% of the categories and no $\hat{f}_k$ equals zero).

Interpreting $\hat{f}_k$ as an expected frequency

If H_0: $\pi_k = \pi_{k0}$ were true, we would expect that, in a sample of size f_+, a proportion π_{k0} of measurements will be in the k-th category. In other words, under H_0, we should expect $\hat{f}_k = (f_+)(\pi_{k0})$ measurements in the k-th category. For this reason, $\hat{f}_k = f_+\pi_{k0}$ is called the expected frequency in the k-th category under the null hypothesis.

A measure of goodness of fit

The test statistic in Box 6.3 is a weighted sum of the squares of the differences, $f_k - \hat{f}_k$, between the observed and expected frequencies. A weight of $1/\hat{f}_k$ is assigned to the k-th squared difference $(f_k - \hat{f}_k)^2$. Under this weighting scheme, a deviation in a category where we expect a large number of observations is assigned a lower weight than a deviation of the same magnitude in a category where we expect few observations. This is intuitively reasonable because, for instance, a deviation of 5 in a category where we expect 100 observations (weight 0.01) is more likely (and provides less evidence in favor of H_1) under the null hypothesis than a deviation of 5 in a category where we expect only 10 observations (weight 0.10). The test statistic is an overall measure of the deviation of the observed frequencies from what would be expected if the null hypothesis were true. As its value increases, the test statistic provides more evidence in favor of the alternative hypothesis that the population proportions are not as specified under H_0.

A simple computing formula

Using straightforward algebraic manipulations, it can be shown that

$$\chi_c^2 = \sum_{k=1}^{C} \frac{f_k^2}{\hat{f}_k} - f_+.$$

This expression for χ_c^2 is more convenient for computing purposes than that in Box 6.3.

Here's an example of the χ^2 goodness-of-fit test in Box 6.3.

EXAMPLE 6.8   Over a specified period, observers sighted 200 birds at a particular location. The birds may be classified into four species categories as follows:

Species	1	2	3	4	Total
Number of birds	40	80	65	15	200

Do these data provide evidence that the composition of the species in the location has changed from a previously known proportion of 3:3:3:1?

Let π_k denote the proportion of the birds belonging to species k ($k = 1, 2, 3, 4$). We need a test of the null hypothesis H_0: $\pi_1 = 0.30, \pi_2 = 0.30, \pi_3 = 0.30$, and $\pi_4 = 0.10$ against the alternative hypothesis H_1 that the population proportions are not as specified by the null hypothesis.

The test can be carried out by the procedure in Box 6.3. We have a sample of $f_+ = 200$ observations from a categorical population of $C = 4$ categories. The expected frequencies under the null hypothesis are

$$\hat{f}_1 = (200)(0.30) = 60, \qquad \hat{f}_2 = (200)(0.30) = 60,$$

$$\hat{f}_3 = (200)(0.30) = 60, \qquad \hat{f}_4 = (200)(0.10) = 20.$$

The calculated value of the test statistic is

$$\chi_c^2 = \frac{(40)^2}{60} + \frac{(80)^2}{60} + \frac{(65)^2}{60} + \frac{(15)^2}{20} - 200 = 15.$$

From Table C.3 (Appendix C), the 0.05-level critical value of a χ^2-distribution with $C - 1 = 4 - 1 = 3$ degrees of freedom is 7.8147. Therefore, at a significance level of 0.05, the null hypothesis can be rejected, and we conclude that the species composition has changed from the previously known proportion of 3:3:3:1.

The p-value of the test is equal to the probability that an observed value of χ^2 with three degrees of freedom will exceed the computed value of 15; that is, the p-value is equal to the area to the right of 15 under a χ^2-distribution with three degrees of freedom. From Table C.3, it can be seen that the p-value is between 0.001 and 0.005. The p-value is relatively small, and so we conclude that the data provide strong support for the research hypothesis that the population proportions are not 3:3:3:1. Younger (1977) in Figure 6.4 shows how to use the PROBCHI function of SAS to find the p-value associated with a χ^2-test.   ■

The χ^2-test in Box 6.3 is appropriate when the distribution is completely specified by the null hypothesis. However, as seen in Example 6.7, there are interesting practical situations in which the population distributions are only partially specified by the null hypotheses. In such cases it might be possible to express the population distributions in terms of q parameters whose values are unknown. For example, under condition 2 in Example 6.7, the population distribution is specified in terms of $q = 1$ parameter, namely, π. The null hypothesis under condition 3 specifies the population distribution in terms of $q = 2$ parameters, π_A and π_B.

Generally speaking, there are various practical settings in which researchers are interested in testing a null hypothesis that the population proportions are known functions of q unknown parameters. In symbols, such null hypotheses can be expressed as

$$\pi_{k0} = \phi_k(\theta_1, \ldots, \theta_q), \qquad k = 1, 2, \ldots, C, \qquad \text{(6.2)}$$

where π_{k0} is the proportion in the k-th category, the ϕ_k are known functions, and the θ_j are unknown parameters. For instance, under condition 2 of Example 6.6, the null hypothesis is specified in terms of $q = 1$ parameter $\theta = \pi$, and the functions that specify the π_k in terms of θ are

$$\pi_{10} = \phi_1(\theta) = \theta^2,$$

$$\pi_{20} = \phi_2(\theta) = \theta(1 - \theta),$$

$$\pi_{30} = \phi_3(\theta) = (1 - \theta)\theta,$$

$$\pi_{40} = \phi_4(\theta) = (1 - \theta)(1 - \theta).$$

Under condition 3 of Example 6.6, the null hypothesis is specified in terms of $q = 2$ parameters $\theta_1 = \pi_A$ and $\theta_2 = \pi_B$, with functions

$$\pi_{10} = \phi_1(\theta_1, \theta_2) = \theta_1\theta_2$$

$$\pi_{20} = \phi_2(\theta_1, \theta_2) = \theta_1(1 - \theta_2)$$

$$\pi_{30} = \phi_3(\theta_1, \theta_2) = (1 - \theta_1)\theta_2$$

$$\pi_{40} = \phi_4(\theta_1, \theta_2) = (1 - \theta_1)(1 - \theta_2).$$

In the case of a null hypothesis under which the population distribution is partially specified, the χ^2 goodness-of-fit test in Box 6.3 is modified as in Box 6.4, where the null hypothesis is that the population proportions are specified by Equation (6.2).

BOX **6.4**   ***The χ^2 goodness-of-fit test of a partially specified set of proportions***

Let f_k and π_k, $k = 1, \ldots, C$ be as in Box 6.3 and assume that the null hypothesis H_0 states that, for each k, $\pi_k = \pi_{k0}$, where

$$\pi_{k0} = \phi_k(\theta_1, \ldots, \theta_q)$$

is a known function of q parameters $\theta_1, \ldots, \theta_q$. The research hypothesis H_1 is that H_0 is not true.

If the sample is large, the null hypothesis H_0 can be tested using the test statistic

$$\chi_c^2 = \sum_{k=1}^{C} \frac{(f_k - \hat{f}_k)^2}{\hat{f}_k},$$

where

$$\hat{f}_k = f_+ \phi_k(\hat{\theta}_1, \ldots, \hat{\theta}_q)$$

and $\hat{\theta}_1, \hat{\theta}_2, \ldots, \hat{\theta}_q$ are the maximum-likelihood estimators of $\theta_1, \theta_2, \ldots, \theta_q$, respectively. The null hypothesis is rejected at level α if

$$\chi_c^2 \geq \chi^2(C - 1 - q, \alpha),$$

where $\chi^2(C - 1 - q, \alpha)$ is the α-level critical value of a χ^2-distribution with $C - 1 - q$ degrees of freedom.

Criteria for large sample sizes

As in Box 6.3, the large-sample requirement means that not all $\hat{f}_k$ are small (as a rule of thumb, $\hat{f}_k \geq 5$ in at least 75% of the categories and no $\hat{f}_k$ equals zero).

The method of maximum likelihood

The maximum-likelihood method is a widely used statistical method of estimating unknown parameters. In the standard goodness-of-fit problems considered in this book, the maximum-likelihood estimators turn out to be as we would expect on the basis of intuition; for instance, the sample means and sample proportions are the maximum-likelihood estimators of the corresponding population means and population proportions, respectively.

Interpreting $\hat{f}_k$ as an expected frequency

The expected frequencies $\hat{f}_k$ in Box 6.4 are calculated in exactly the same manner as those in Box 6.3. The only difference is that, while the proportions π_{k0} are completely known in Box 6.3, they are estimated from the data in Box 6.4. The estimation procedure involves obtaining the maximum-likelihood estimators of $\theta_1, \ldots, \theta_q$ and then substituting them into Equation (6.2) to get the estimated values $\hat{\pi}_{k0} = \phi(\hat{\theta}_1, \ldots, \hat{\theta}_q)$.

EXAMPLE **6.9**   A random sample of 300 subjects was selected from a population. Each subject was classified according to gender and color blindness, as follows:

Color-blind	Gender Female	Male	Total
No	130	119	249
Yes	41	10	51
Total	171	129	300

Let θ_1 and θ_2 denote, respectively, the proportion of subjects with normal eyesight and the proportion of females in the population. Then it can be shown that, under the hypothesis that gender and color blindness are independently inherited traits, the theoretical proportions of individuals possessing the four possible combinations of gender and color blindness will be as follows:

Color-blind	Gender	
	Female	Male
No	$\theta_1\theta_2$	$\theta_1(1-\theta_2)$
Yes	$(1-\theta_1)\theta_2$	$(1-\theta_1)(1-\theta_2)$

For the purpose of illustration, we will perform two statistical tests to see if the observed data contradict the null hypothesis of independent inheritance for two different settings. In the first setting, we assume that the female-to-male ratio in the population is known to be 1:1. In the second, we assume that the female-to-male ratio in the population is unknown.

If the female-to-male ratio is 1:1, 50% of the population are females, so that $\theta_2 = 0.50$. The null hypothesis in this case specifies the population proportions as known functions of $q = 1$ parameter: the unknown population proportion θ_1 of normal subjects. It turns out that the maximum-likelihood estimate of θ_1 is the intuitive estimate provided by the sample proportion of normal subjects; that is, the maximum likelihood estimate of θ_1 is

$$\hat{\theta}_1 = \frac{249}{300} = 0.83.$$

To calculate the expected frequencies under the null hypothesis, we estimate the proportions in the four categories by setting $\theta_1 = 0.83$, $\theta_2 = 0.50$ in the table of theoretical proportions. For example, the estimated proportion of normal males is

$$\hat{\theta}_1 \times 0.50 = 0.83 \times 0.50 = 0.4150,$$

so that the expected frequency in this category is $300 \times 0.4150 = 124.5$. The estimated proportions and the corresponding observed and expected frequencies in each category are as follows:

Category		Estimated	Frequencies	
Color-blind	Gender	proportion	Observed	Expected
No	Female	0.415	130	124.5
No	Male	0.415	119	124.5
Yes	Female	0.085	41	25.5
Yes	Male	0.085	10	25.5
Total		1.0	300	300

The calculated value of the test statistic is

$$\chi_c^2 = \frac{(130 - 124.5)^2}{124.5} + \cdots + \frac{(10 - 25.5)^2}{25.5} = 19.3291.$$

Since we have estimated $q = 1$ parameter, the calculated value should be compared with the critical value of a χ^2-distribution with $C - 1 - q = 4 - 1 - 1 = 2$

degrees of freedom. From Table C.3 (Appendix C), the p-value associated with the calculated value is seen to be less than 0.001. Thus, under the assumption that the female-to-male ratio in the population is 1:1, these data seem to contradict the theory that colorblindness and gender are independently inherited.

To test the null hypothesis without assuming a 1:1 female-to-male ratio will require maximum-likelihood estimation of both θ_1 and θ_2. As before, the estimate of θ_1 is 0.83. The maximum-likelihood estimate of θ_2 turns out to be the same as the observed proportion of females, so that $\hat{\theta}_2 = 171/300 = 0.57$. The expected frequencies are calculated as in the previous case, except that we use $\hat{\theta}_2 = 0.57$ instead of $\hat{\theta}_2 = 0.50$. The resulting calculated value of the test statistic is $\chi_c^2 = 13.72$. We are estimating $q = 2$ parameters, and so the calculated value should be compared to the critical value of a χ^2-distribution with $C - 1 - q = 4 - 1 - 2 = 1$ degree of freedom. The associated p-value is small, so that it is reasonable to reject the null hypothesis of independent inheritance. ∎

The χ^2-tests in Boxes 6.3 and 6.4 are large-sample tests in that they require expected frequencies of at least 5 in 75% of the categories and at least 1 in every category. When the expected frequencies are small, exact tests similar to the binomial test for a single proportion are used.

Exercises

6.2 A large seed company wants to estimate a 95% lower bound to the germination rate (proportion of planted seeds that germinate) for a new hybrid variety of corn. In an experiment in which 50 randomly selected seeds were planted, 44 seeds germinated.

   a Construct the required lower bound for the seed germination rate and interpret it.

   b On the basis of the lower bound calculated in (a), what conclusions can you draw about the company's claim that the germination rate exceeds 90%?

   c Use the p-value of a suitable hypothesis test to draw conclusions about the claim that the germination rate exceeds 90%.

6.3 Out of 100 patients treated by a new antibiotic, 86 were cured of a particular type of bacterial infection.

   a Construct a 90% confidence interval for the probability that a patient treated with the antibiotic will be cured of the infection. Interpret the interval that you construct.

   b Determine the p-value of an appropriate hypothesis test to see if the observed data support the conclusion that less than 85% of the patients will be cured by the new antibiotic.

6.4 The sensitivity, specificity, and predictive power of a diagnostic test for a disease are defined as follows:

- *Sensitivity* is the probability that the test will give a positive result (indicating the presence of the disease) in a subject who has the disease.

- *Specificity* is the probability that the test will give a negative result (indicating the absence of the disease) in a subject who does not have the disease.

- *Predictive power* is the probability that the test will make the correct diagnosis.

The following data show the results of performing a radiologic diagnostic test for coronary artery disease (CAD) in 200 subjects, of whom 123 actually had the disease:

Test result	CAD		Total
	Present	Absent	
Positive	95	25	120
Negative	28	52	80
Total	123	77	200

a   Estimate the sensitivity, specificity, and predictive power of the diagnostic test.

b   Construct 95% confidence intervals for the parameters estimated in (a) and interpret them.

c   Perform a hypothesis test to verify the claim that the test will detect more than 70% of the cases who have the disease.

d   Construct a 95% lower confidence bound to the predictive power of the test and interpret it.

6.5   A new diagnostic blood test for detection of a particular type of cancer showed positive results for 13 of the 15 patients who actually had the disease.

a   Estimate the sensitivity of the diagnostic test and interpret it.

b   Construct a 95% confidence interval for the sensitivity of the diagnostic test. Interpret the interval you obtain.

c   The manufacturer of the test kit that is used to perform the test claims that it will diagnose the disease in more than 80% of those who actually have the disease. Calculate the *p*-value of an appropriate hypothesis test to verify the manufacturer's claim using:

   i   the results in Box 6.2;

   ii   suitable statistical computing software.

6.6   Refer to the infestation data in Example 6.1.

a   Construct a 95% confidence interval for the population proportion of treated plants that will be free of infestation. Interpret the interval you construct.

b   Perform a suitable hypothesis test to see if it is reasonable to conclude that more than 60% of the treated plants will be free of infestation.

6.7   In a wine-tasting experiment, two judges were asked to rate 15 samples of a brand of red wine as acceptable or unacceptable. The samples were presented to the judges in such a way that they were unaware that the same brand of wine was presented to them each time. The results obtained are as follows:

Judge 1	Judge 2 Acceptable	Unacceptable	Total
Acceptable	8	3	11
Unacceptable	1	3	4
Total	9	6	15

a Let π_1 denote the probability that, in a given sitting, judge 1 will rate the wine as acceptable. Construct a 99% confidence interval for π_1 and interpret it.

b Compute the *p*-value of a hypothesis test to see if the data support the conclusion that there is more than a 60% chance that judge 1 will find a sample of the wine acceptable.

c Let π denote the probability that, in a given sitting, both judges will rate the wine as acceptable. Construct a 95% confidence interval for π and interpret it.

6.8 The following blood-type frequencies were obtained from a sample of 1000 subjects screened at a shopping mall over a period of one month:

Blood type	O	A	B	AB	Total
Frequency	465	394	96	45	1000

a Do these data support the conclusion that less than 5% of the population screened has blood type AB?

b Perform a statistical test to see if these data contradict the claim that, in this population, blood types O, A, B, and AB are in the ratio 9:8:2:1.

6.9 According to Mendelian theory, the probability distribution of the color and shape of a variety of pea is as follows:

y	Round and yellow	Round and green	Angular and yellow	Angular and green
f(y)	$\frac{9}{16}$	$\frac{3}{16}$	$\frac{3}{16}$	$\frac{1}{16}$

A random sample of 200 peas showed the following frequency distribution:

y	Round and yellow	Round and green	Angular and yellow	Angular and green
Frequency	110	40	42	8

a Construct a 95% confidence interval for the population proportion of peas that are round and yellow. Interpret the interval you obtain.

b Perform a hypothesis test to determine whether the given data contradict the Mendelian theory.

6.10 Refer to the pest-type distribution in Table 6.3. Suppose that, on a particular occasion, the following frequencies were observed for the pest types:

Pest type	1	2	3	4	5
Frequency	19	24	8	23	36

a On the basis of these data, is it reasonable to conclude that the frequency distribution in Table 6.3 overestimates the frequency of pest type 4?

b On the whole, is there reason to doubt that the frequency distribution in Table 6.3 is a good model of the pest-type distribution in this case?

6.11 In a microbial mutagenesis assay, as we noted in Exercise 2.43, a plate of bacteria is exposed to a test compound, and the number of revertants is counted after incubation. The data from 50 independent assays of a new test compound are as follows:

Number of revertants	5	6	7	8	9	10	11	12	13
Frequency	3	5	6	6	12	8	5	3	2

a It is known that the mean of a random sample from a population with a Poisson distribution is the maximum-likelihood estimator of λ, the rate parameter. Assuming that the number of revertants has a Poisson distribution, calculate the maximum-likelihood estimator of the rate parameter.

b Use the maximum-likelihood estimator calculated in (a) to estimate the expected frequencies in eight categories defined according to the number of revertants, as follows:

Category Number	1	2	3	4	5	6	7	8
Number of revertants	Less than 6	6	7	8	9	10	11	More than 12

c Use the χ^2 goodness-of-fit test to see if the data contradict the claim that the number of revertants has a Poisson distribution.

d Is the large-sample requirement satisfied for the test in (c)?

6.12 Suppose you want to test the null hypothesis that the number of revertants in Exercise 6.11 has a Poisson distribution with the parameter $\lambda = 9$.

a How will you modify the test procedure in Part (c) of Exercise 6.11 in this case?

b Perform the test in (a) and write your conclusions.

6.3
Comparing proportions in two populations

Because the distribution of a categorical population is specified by the proportions in its categories, comparing proportions in two categorical populations is the same as comparing two categorical distributions. In this section, we describe some statistical methods for such comparisons. Many features of categorical populations could be compared, but we restrict our attention to two of the most frequent types of inferences:

1 Inferences about the difference between the proportions in a specific category.
2 Inferences about the overall difference between two population distributions.

As in Chapter 4, each of these inference problems will be considered for two settings:

1 The data consist of two independent random samples from two categorical populations.
2 The data consist of a single random sample of paired responses, where every pair contains one observation from each of the two populations.

In the following example, the data consist of two independent samples from two categorical populations.

EXAMPLE **6.10** In Example 6.1, we described the frequency distribution of types of aphid infestation of treated tobacco plants. The observed frequency distribution of aphid infestation in 100 treated tobacco plants (from Example 6.1) and 100 untreated plants is as follows:

| Treatment | Infestation type | | | Total |
	Leaf infestation	Stem infestation	No infestation	
Treated	9	24	67	100
Untreated	39	44	17	100
Total	48	68	84	200

We have two independent random samples of sizes $n_1 = 100$ and $n_2 = 100$ from two populations. As in Example 6.1, the samples are measurements of a nominal variable $Y =$ Infestation type, with three nominal categories: leaf infestation, stem infestation, and no infestation. The following questions might be asked about these data.

1 Is the proportion of leaf infestation in untreated plants higher than that in treated plants? What is a 95% confidence interval for the difference between these two proportions?
2 Are the proportions of plants in the various infestation categories different in the populations of treated and untreated plants? In other words, are the population distributions across the three infestation categories the same for the treated and untreated plants? ■

The notation used for the observed frequencies in a sample from a categorical population (Table 6.1) can be extended in a straightforward manner to the case of two independent samples from two categorical populations. As before, C denotes the number of categories, and the random variable Y_{ij} denotes the j-th observation from the i-th population. Thus, Y_{ij} is a nominal variable that can take one of C possible values. The number of responses from the i-th population in the k-th category will be denoted by f_{ik}. Thus, for each i, f_{ik} is the number of Y_{ij} in category k and is called the observed frequency of category k in sample i. It is convenient to display f_{ik} as in Table 6.7.

T A B L E   6.7

Frequencies of categories in two samples, each of size n

Sample number i	Category number k						Total
	1	2	$\cdots$	k	$\cdots$	C	
1	f_{11}	f_{12}	$\cdots$	f_{1k}	$\cdots$	f_{1C}	$f_{1+} (= n_1)$
2	f_{21}	f_{22}	$\cdots$	f_{2k}	$\cdots$	f_{2C}	$f_{2+} (= n_2)$
Total	f_{+1}	f_{+2}	$\cdots$	f_{+k}	$\cdots$	f_{+C}	

The convention used in Table 6.7 is a straightforward generalization of that in Table 6.1. The $+$ subscript here means that the corresponding quantity is obtained by adding the terms with all the possible values of the subscript replaced by $+$ and with the other subscript fixed at the level indicated. For example, f_{2+} is the sum of $f_{21}, f_{22}, \ldots$, and f_{2C} and thus equals n_2, the size of the second sample. Similarly, f_{+2} is the sum of f_{12} and f_{22}, the total number of observations in category 2.

A notation similar to that in Table 6.2 can be used to describe the proportions in two categorical populations. Let π_{ik} ($i = 1, 2; k = 1, 2, \ldots, C$) denote the true proportion in category k from population i. Then π_{ik} is the probability that a randomly selected measurement in population i belongs to category k. Observe that $\pi_{i1}, \pi_{i2}, \ldots, \pi_{ik}$ determine the probability distribution of the i-th population ($i = 1, 2$) and that they satisfy the constraint $\pi_{i1} + \pi_{i2} + \cdots + \pi_{ik} = 1$. Table 6.8 is the two-population analog of Table 6.2.

In the next example, the data in Example 6.10 are described in terms of the notation just introduced.

T A B L E   6.8

Probability distributions of two categorical populations

Population number i	Category number k						Total
	1	2	$\cdots$	k	$\cdots$	C	
1	π_{11}	π_{12}	$\cdots$	π_{1k}	$\cdots$	π_{1C}	$\pi_{1+} (= 1)$
2	π_{21}	π_{22}	$\cdots$	π_{2k}	$\cdots$	π_{2C}	$\pi_{2+} (= 1)$
Total	π_{+1}	π_{+2}	$\cdots$	π_{+k}	$\cdots$	π_{+C}	

EXAMPLE 6.11   The data in Example 6.10 are the observed frequencies of the categories in two samples of sizes $n_1 = n_2 = 100$ from two categorical populations. There are $C = 3$ categories of the nominal variable $Y = $ type of infestation. The j-th observed response from the i-th population is Y_{ij}, $i = 1, 2; j = 1, 2, \ldots, 100$. Let $k = 1$ for leaf infestation, $k = 2$ for stem infestation, and $k = 3$ for no infestation. Then

$$Y_{ij} = \begin{cases} 1 & \text{if the } j\text{-th plant in the } i\text{-th sample has leaf infestation} \\ 2 & \text{if the } j\text{-th plant in the } i\text{-th sample has stem infestation} \\ 3 & \text{if the } j\text{-th plant in the } i\text{-th sample has no infestation.} \end{cases}$$

The observed number of responses that fall into the k-th category is f_{ik}. Thus, for example, the observed frequency of category 2 (stem infestation) in the first sample is $f_{12} = 24$ and the observed frequency of category 3 (no infestation) in the second sample $(i = 2)$ is $f_{23} = 17$. The total number of responses in the second sample is $f_{2+} = 100$, and the total number of responses in the second category is $f_{+2} = 24 + 44 = 68$.   ∎

The data in Example 6.10 consist of two independent random samples from the two populations. In some situations, the samples from two populations may be related, because each response from a particular population has something in common with a corresponding response from the other population, as in the following example.

EXAMPLE 6.12   In a controlled clinical trial to determine the efficacy of an experimental drug for treating migraine headache, each of 90 patients was treated with two drugs—the experimental drug and a placebo—in a random order. Each treatment lasted 12 weeks. At the end of the treatment period, the effect of the drug was classified into three categories: completely effective (CE); somewhat effective (SE); and not effective (NE). The migraine headache data obtained are as follows:

	Response to placebo			
Response to drug	CE	SE	NE	Total
CE	12	14	19	45
SE	6	8	14	28
NE	4	6	7	17
Total	22	28	40	90

The migraine headache data consist of $n = 90$ pairs of responses from 90 patients. Let Y_{1j} and Y_{2j} denote, respectively, the responses of the j-th patient to drug and placebo. The table shows the frequencies of the categories of the pairs (Y_{1j}, Y_{2j}). For example, in 12 pairs the response category is (CE, CE); that is, $(Y_{1j}, Y_{2j}) = $ (CE, CE). In 19 pairs, the response category is (CE, NE). Clearly, because the responses to the two treatments are measured on the same subject, it is not reasonable to assume that the drug responses (the Y_{1j}) and the placebo responses (the Y_{2j}) are independent samples.

Note that the table of migraine headache data shows the observed frequencies of pairs of responses, one from each population, whereas the numbers in Table 6.7 are the observed frequencies of the responses from each of the two samples. In the table of frequencies of paired responses, the frequencies of responses from individual populations appear as the row and column totals. For comparison, the migraine data may be tabulated in a form analagous to Table 6.7 as follows; notice that the rows now represent the drug and placebo populations, and the columns represent the categories.

Treatment	Effect			Total
	CE	SE	NE	
Drug	45	28	17	90
Placebo	22	28	40	90
Total	69	56	55	

The questions that are relevant for independent-sample data are also relevant for paired samples. The following are two examples.

1 What is a 95% confidence interval for the difference between the population proportions of subjects for whom the two treatments are completely effective? More specifically, let π_{ij} denote the population proportion of subjects in the i-th treatment ($i = 1$ for drug treatment and $i = 2$ for placebo) corresponding to response category k ($k = 1$ for CE, $k = 2$ for SE, and $k = 3$ for NE). What is a 95% confidence interval for $\theta = \pi_{11} - \pi_{21}$?

2 On the whole, is there a difference between the patterns of responses to the two treatments? In other words, is there a difference between the probability distributions of the responses to the experimental and placebo treatments? ∎

However, as with interval data in Chapter 4, the inferential methods for comparing population proportions that use independent samples are different from those that use paired samples.

Comparing two population proportions: Independent samples

In Section 6.2, we described how the large-sample procedures in Box 4.1 for inferences about a population mean can be used to derive large-sample procedures for making inferences about a population proportion. The key consideration that allows procedures for making inferences about means to be adapted so as to make inferences about proportions is that the mean of a collection of measurements that contains only 0s and 1s is the same as the proportion of 1s in the collection. In Box 6.5, the procedures for comparing two population means given in Box 4.1 are adapted for the comparison of two population proportions.

BOX **6.5**

Large-sample inferences about the difference between two proportions: Independent samples

Let $\hat{\pi}_1$ and $\hat{\pi}_2$ be the proportions in a particular category observed for two independent random samples of sizes n_1 and n_2, respectively. Let π_1 and π_2 be the corresponding population proportions. If n_1 and n_2 are both large (≥ 30), hypothesis tests and confidence intervals for $\theta = \pi_1 - \pi_2$ can be constructed as follows.

Hypothesis test

The null hypothesis H_0: $\theta = \theta_0$ can be tested using the calculated value z_c of the test statistic

$$Z = \frac{\hat{\theta} - \theta_0}{\hat{\sigma}_{\hat{\theta}}},$$

where $\hat{\theta} = \hat{\pi}_1 - \hat{\pi}_2$ is the estimated difference between population proportions and $\hat{\sigma}_{\hat{\theta}}$ is the estimated standard error of $\hat{\theta}$

$$\hat{\sigma}_{\hat{\theta}} = \sqrt{\frac{\hat{\pi}_1(1 - \hat{\pi}_1)}{n_1} + \frac{\hat{\pi}_2(1 - \hat{\pi}_2)}{n_2}}.$$

Confidence interval

A $100(1 - \alpha)\%$ confidence interval for θ is

$$\hat{\theta} \pm z(\alpha/2)\hat{\sigma}_{\hat{\theta}},$$

where $z(\alpha)$ is the α-level critical value of a standard normal distribution.

Example 6.13 shows an application of the procedures in Box 6.5.

EXAMPLE **6.13**   In Example 6.11, the category representing no infestation corresponded to $k = 3$. Let π_{13} and π_{23} denote, respectively, the proportion of uninfested plants in the treated and control populations. Then, on the basis of the data in Example 6.10, the estimated proportions are

$$\hat{\pi}_{13} = \frac{67}{100} = 0.67 \quad \text{and} \quad \hat{\pi}_{23} = \frac{17}{100} = 0.17.$$

Suppose we want to scrutinize the claim that the proportion of treated plants that remain free of infestation is larger. Then, we should test the null hypothesis H_0: $\pi_{13} - \pi_{23} \leq 0$ against the research hypothesis H_1: $\pi_{13} - \pi_{23} > 0$. Setting $\pi_1 = \pi_{13}, \pi_2 = \pi_{23}$, and $\theta = \pi_1 - \pi_2$ in Box 6.5, we can calculate the estimated standard error of $\hat{\theta}$

$$\hat{\sigma}_{\hat{\theta}} = \sqrt{\frac{0.67(1 - 0.67)}{100} + \frac{0.17(1 - 0.17)}{100}} = 0.0602,$$

and so the calculated value of the test statistic is

$$z_c = \frac{0.67 - 0.17}{0.0602} = 8.31.$$

From Table C.1 (Appendix C), the p-value for a one-sided test is less than 0.001, so that the null hypothesis can be rejected in favor of the conclusion that the proportion of treated plants that are uninfested is larger.

A 95% confidence interval for the difference between proportions is

$$(0.67 - 0.17) \pm (1.96)(0.0602) \qquad \Rightarrow \qquad 0.38 \leq \pi_{13} - \pi_{23} \leq 0.62$$

so that, compared to untreated plants, between 38% and 62% more of the treated plants will be free of infestation. ∎

The procedures in Box 6.5 are meant to be used when sample sizes are large and the proportions being compared are estimated from independent samples.

Fisher's exact test

For small independent samples, H_0: $\pi_1 = \pi_2$ can be tested using a procedure known as Fisher's exact test. Whereas the binomial test for testing a hypothesis about a single population proportion uses the binomial distribution, Fisher's exact test for comparing two proportions uses the hypergeometric distribution. The probability functions of hypergeometric distributions are useful in a variety of statistical applications; for more details, see Johnson and Kotz (1970).

Suppose that independent random samples of sizes n_1 and n_2 from two categorical populations contain a total of m_1 values in a specific category—say, the first category. The observed frequencies of the categories in the two samples are as in Table 6.9, where we have departed from the notation in Table 6.7 in order to simplify the expressions for Fisher's exact test.

TABLE 6.9

Frequencies of categories in two samples

Sample	Category 1	Not 1	Total
1	a	b	n_1
2	c	d	n_2
Total	m_1	m_2	n

Let π_1 and π_2 denote the true proportions in the first category of the two populations and let A be the random variable that denotes the frequency in the first sample of the observed responses in the first category. Assume the following conditions:

1   The null hypothesis H_0: $\pi_1 = \pi_2$ is true.

2   Of the n observed responses, m_1 belong to the first category.

Then the probability that the random variable A will take the value a may be shown to be

$$\Pr\{A = a\} = f(a|m_1, n_1, n) = \frac{m_1! m_2! n_1! n_2!}{n! a! b! c! d!}. \tag{6.3}$$

The function $f(a|m_1, n_1, n)$ is called the *hypergeometric probability function* with parameters m_1, n_1, and n. Under the null hypothesis H_0: $\pi_1 = \pi_2$, the hypergeometric probability function can be used to calculate the probability of observing a given set of cell frequencies for Table 6.9.

As we noted in our comments following Example 3.22, the *p*-value of a test can be interpreted as the probability, under the null hypothesis, of observing data that are at least as favorable to the research hypothesis as the observed sample. As we see in Box 6.6, Fisher's exact test uses this interpretation and Equation (6.3) to calculate the *p*-value of a test of H_0.

BOX 6.6

> ### Fisher's exact test for comparing two proportions
>
> Suppose independent random samples of sizes n_1 and n_2 taken from two populations are categorized as in Table 6.9. Let π_1 and π_2 be the proportions in the first category in the two populations. Consider the following hypotheses:
>
> **a** H_0: $\pi_1 \le \pi_2$ against H_1: $\pi_1 > \pi_2$;
> **b** H_0: $\pi_1 \ge \pi_2$ against H_1: $\pi_1 < \pi_2$;
> **c** H_0: $\pi_1 = \pi_2$ against H_1: $\pi_1 \ne \pi_2$.
>
> The corresponding *p*-values of Fisher's exact test can be calculated as follows
>
> $$p = \sum f(y|r, n_1, n),$$
>
> where $f(y|r, n_1, n)$ is the hypergeometric probability function in Equation (6.3) and the sum is taken over all values of y such that the frequencies in the following 2×2 table are at least as favorable to H_1 as are the observed frequencies:
>
> | Sample | Category 1 | Not 1 | Total |
> |--------|------------|-------|-------|
> | 1 | y | b | n_1 |
> | 2 | c | d | n_2 |
> | Total | m_1 | m_2 | n |

The calculations needed for Fisher's exact test are illustrated in the next example.

EXAMPLE 6.14   A random sample of six sites in each of two North Florida counties is classified into three categories: predominantly pine; predominantly hardwood; and other. The results obtained are as follows:

County	Terrain type Pine	Hardwood	Other	Total
1	1	2	3	6
2	4	1	1	6
Total	5	3	4	12

Let π_1 and π_2 denote, respectively, the proportions of predominantly pine forest sites in the two counties. Fisher's exact test can be used to test the null hypothesis H_0: $\pi_1 \geq \pi_2$ against the alternative hypothesis H_1: $\pi_1 < \pi_2$. The 2×2 table for this test is obtained by combining the frequencies of the hardwood and other categories into a single category: not pine. The resulting table is as follows:

County	Terrain type Pine	Not pine	Total
1	1	5	6
2	4	2	6
Total	5	7	12

In the notation of Table 6.9 we have $n = 12$, $n_1 = 6$, $n_2 = 6$, $a = 1$, $b = 5$, $c = 4$, $d = 2$, $m_1 = 5$ and $m_2 = 7$. To calculate the p-value, we first note that, out of $n = 12$ sites, $m_1 = 5$ sites were in the first category (predominantly pine). Of these five sites, one was in county 1. Any sample in which fewer than one (out of five) category 1 sites are in county 1 would be at least as favorable to H_1 as the observed sample. Thus, there are two 2×2 tables at least as favorable to H_1 as the observed table. One of them, corresponding to $y = 1$, is the observed table itself; the other, corresponding to $y = 0$, is as follows:

County	Terrain type Pine	Not pine	Total
1	0	6	6
2	5	1	6
Total	5	7	12

The required p-value is the sum of the probabilities of observing the frequencies in the two tables

$$p = p(1|5, 6, 12) + p(0|5, 6, 12)$$

$$= \frac{5!7!6!6!}{12!1!5!4!2!} + \frac{5!7!6!6!}{12!0!6!5!1!}$$

$$= 0.1136 + 0.0076$$

$$= 0.1212,$$

which is relatively large. Therefore, the sample data do not support the conclusion that the proportion of predominantly pine forest sites in county 2 is larger than that in county 1.

As is evident from these computations, Fisher's exact test can be laborious. Fortunately, researchers can use appropriate statistical computing software, including the FREQ procedure in SAS and the option FI/EX in StatXact. The following is a portion of the printout obtained using the SAS FREQ procedure on the terrain frequency data. For details of how to implement the FREQ procedure in SAS, see Figure 6.6 in the companion text by Younger (1997).

```
---------------------------------------------------------------
       PARTIAL PRINTOUT FROM  THE FREQ PROCEDURE IN SAS
   FISHER'S EXACT TEST FOR COMPARING TERRAIN FREQUENCIES.
             TABLE OF COUNTY BY CATEGORY
       COUNTY          CATEGORY

          Frequency|    1 |    2 | Total
          ---------+--------+--------+
              1 |    1 |    5 |   6
          ---------+--------+--------+
              2 |    4 |    2 |   6
          ---------+--------+--------+
          Total        5      7     12


        STATISTICS FOR TABLE OF COUNTY BY CATEGORY
```

Statistic	DF	Value	Prob
Chi-Square	1	3.086	0.079
Likelihood Ratio Chi-Square	1	3.256	0.071
Continuity Adj. Chi-Square	1	1.371	0.242
Mantel-Haenszel Chi-Square	1	2.829	0.093
Fisher's Exact Test (Left)			0.121
(Right)			0.992
(2-Tail)			0.242
Phi Coefficient		-0.507	
Contingency Coefficient		0.452	
Cramer's V		-0.507	
Sample Size = 12			

```
        WARNING: 100% of the cells have expected counts less
                than 5. Chi-Square may not be a valid test.
```

Notice that the SAS procedure gives three p-values corresponding to the three research hypotheses in Box 6.6. The left p-value in the printout is appropriate for this particular problem, because it corresponds to the null hypothesis in which the difference $\pi_1 - \pi_2$ is to the left of zero—that is, $\pi_1 - \pi_2 < 0$.   ∎

Caution is required when using small-sample exact tests, such as the binomial test for a single proportion and Fisher's exact test for difference between two proportions. When used with very small samples, these tests may not have adequate power to detect differences that are of practical importance. Consequently, nonsignificant results obtained using exact tests should be interpreted with care when the samples are small.

Comparing proportions using paired samples

The McNemar test

The McNemar test is very commonly used to compare proportions based on paired categorical data. The test statistic has the same structure as that for independent samples. As in Box 6.5, let $\theta = \pi_1 - \pi_2$ be the difference between population proportions. Then the McNemar statistic for testing H_0: $\theta = \theta_0$ takes the form

$$Z = \frac{\hat{\theta} - \theta_0}{\hat{\sigma}_{\hat{\theta}}},$$

where $\hat{\theta}$ is the estimate of θ and $\hat{\sigma}_{\hat{\theta}}$ is the estimated standard error of $\hat{\theta}$. As in the case of independent samples, θ is estimated as the difference between the sample proportions but the formula for the estimated standard error differs from that in Box 6.5.

In Box 6.7, we present the large-sample procedures based on the McNemar test for inferences about the difference between the proportions of two populations in a particular category.

BOX **6.7**

Large-sample McNemar test for inferences about the difference between population proportions

Suppose that the frequencies of paired categories in a random sample of N independent paired responses from two categorical populations are arranged in a 2 × 2 table as follows:

Category of population 1	Category of population 2		Total
	1	Not 1	
1	A	B	N_1
Not 1	C	D	N_2
Total	M_1	M_2	N

Let π_1 and π_2 denote the proportions of populations 1 and 2, respectively, in category 1 and let $\theta = \pi_1 - \pi_2$.

Hypothesis test

The null hypothesis H_0: $\theta = \theta_0$ can be tested using the test statistic

$$Z = \frac{\hat{\theta} - \theta_0}{\hat{\sigma}_{\hat{\theta}}},$$

where

$$\hat{\theta} = \frac{B - C}{N}$$

and

$$\hat{\sigma}_{\hat{\theta}} = \frac{\sqrt{B + C}}{N}.$$

The hypothesis test is performed by comparing the calculated value of Z with the critical values of the standard normal distribution in Table C.1 (Appendix C).

Confidence interval

A $100(1 - \alpha)\%$ confidence interval for θ is

$$\hat{\theta} \pm z(\alpha/2)\hat{\sigma}_{\hat{\theta}},$$

where $z(\alpha)$ is the α-level critical value of a standard normal distribution.

The following comments on the McNemar test are relevant here:

1   Let $\hat{\pi}_1 = N_1/N$ and $\hat{\pi}_2 = M_1/N$ denote the sample proportions in category 1 for populations 1 and 2 respectively. Then $\theta = \pi_1 - \pi_2$, the difference between the two population proportions, can be estimated by

$$\hat{\theta} = \hat{\pi}_1 - \hat{\pi}_2 = \frac{N_1}{N} - \frac{M_1}{N} = \frac{A + B}{N} - \frac{A + C}{N} = \frac{B - C}{N}.$$

2   When $\theta_0 = 0$, the expression for the McNemar statistic can be simplified as

$$Z = \frac{B - C}{\sqrt{B + C}}.$$

EXAMPLE **6.15**   Let's see if the migraine headache data in Example 6.12 indicate that the proportion of patients for whom the treatment is at least somewhat effective—those whose response is SE or CE—is higher for the experimental drug than for the placebo. If we classify a patient for whom the treatment is at least somewhat effective as category 1, the observed data can be arranged in a 2×2 table as follows:

| | Category for placebo treatment | | Total |
Category for drug treatment	1 (SE or CE)	Not 1 (NE)	
1 (SE or CE)	40	33	73
Not 1 (NE)	10	7	17
Total	50	40	90

Let the population proportion of responses in category 1 be π_1 for the drug treatment and π_2 for the placebo treatment. We can use the McNemar procedure in Box 6.7 to make inferences about $\theta = \pi_1 - \pi_2$. From the data, we see that

$$\hat{\theta} = \frac{33 - 10}{90} = 0.2556, \qquad \hat{\sigma}_{\hat{\theta}} = \frac{\sqrt{33 + 10}}{90} = 0.0729$$

are the estimates of θ and its standard error, respectively. To test the null hypothesis $H_0: \theta \leq 0$ against $H_1: \theta > 0$, the calculated value of the test statistic z is

$$z_c = \frac{0.2556 - 0}{0.0729} = 3.51.$$

The p-value is small ($p < 0.001$), so that it is reasonable to conclude that the effect of the new treatment is greater than that of the placebo. A 95% confidence interval for the difference between the proportions is

$$0.2556 \pm (1.96)(0.0729) = 0.2556 \pm 0.1429 \qquad \Leftrightarrow \qquad 0.11 \leq \theta \leq 0.40,$$

showing that, at the 95% confidence level, the difference between the proportions can be anywhere between 0.11 and 0.40.   ■

The McNemar test in Box 6.7 is a large-sample test. When the sample size n is small, exact tests can be performed using the MC option in the StatXact tests on $R \times C$ tables.

Comparing two categorical distributions: Independent samples

The problem of comparing two categorical populations based on independent samples was encountered in Example 6.10. In typical applications, we have data as in Table 6.7, and the problem is to test the null hypothesis that the two populations have identical distributions—that is, have same proportions in the k categories—against the alternative hypothesis that the two distributions are not identical. In the notation of Table 6.8, the null hypothesis can be stated as follows

$$H_0: \pi_{11} = \pi_{21}, \pi_{12} = \pi_{22}, \ldots, \pi_{1C} = \pi_{2C}.$$

The corresponding research hypothesis H_1 is that at least one of the equalities in H_0 does not hold.

A large-sample test of the equality of two categorical populations is presented in Box 6.8.

BOX 6.8

The χ^2-test of the equality of distribution of two categorical populations: Large samples

Suppose that we have two independent random samples of sizes $n_1 = f_{1+}$ and $n_2 = f_{2+}$ from two categorical populations, each with C categories. Suppose further that the frequencies of the responses are as in Table 6.7. Let π_{ik} denote the proportion of values from population i in category k ($i = 1, 2$; $k = 1, 2, \ldots, C$). Then the maximum-likelihood estimates of the expected frequencies under the hypothesis H_0: $\pi_{1k} = \pi_{2k}$, $k = 1, \ldots, C$, are

$$\hat{f}_{ik} = \frac{f_{i+}f_{+k}}{f_{++}}.$$

If $\hat{f}_{ik}$ are not too small, the following test statistic can be used to test H_0 against the research hypothesis H_1 that the population distributions are not equal

$$\chi_c^2 = \sum_{i=1}^{2} \sum_{k=1}^{C} \frac{(f_{ik} - \hat{f}_{ik})^2}{\hat{f}_{ik}}$$

$$= \sum_{i=1}^{2} \sum_{k=1}^{C} \frac{f_{ik}^2}{\hat{f}_{ik}} - f_{++}.$$

The null hypothesis is rejected at level α if

$$\chi_c^2 \geq \chi^2(C - 1, \alpha),$$

where $\chi^2(C - 1, \alpha)$ is the α-level critical value of a $\chi^2(C - 1)$-distribution.

Criteria for large sample

The requirement of large sample size is similar to that in Boxes 6.3 and 6.4: that $\hat{f}_{ik} \geq 5$ in at least 75% of the cells and no $\hat{f}_{ik} = 0$.

Interpreting $\hat{f}_{ik}$ as an expected frequency

The formula for the expected frequency $\hat{f}_{ik}$ in Box 6.8 can be justified intuitively for the example of the infestation frequency data in Example 6.10. Out of a total of 200 plants, $39 + 9 = 48$ plants have leaf infestation. Thus, a proportion equal to 48/200 of the plants are leaf-infested. If the proportion of leaf-infested plants is the same in both populations, approximately $(100)(48/200)$ leaf-infested plants would be expected among the 100 treated plants. Therefore, the expected frequency of leaf infestation in category 1 is

$$\hat{f}_{11} = \frac{100 \times 48}{200} = \frac{f_{1+}f_{+1}}{f_{++}}.$$

The expected frequencies in the six cells are shown in parentheses in the following 2×3 table of observed frequencies:

Treatment	Leaf infestation	Stem infestation	No infestation	Total
Treated	9 (24)	24 (34)	67 (42)	100
Control	39 (24)	44 (34)	17 (42)	100
Total	48	68	84	200

Calculating with computers

The χ^2-test in Box 6.8 is a large-sample test. The SAS FREQ procedure can be used to perform this test. If the samples are small, an exact test can be performed using the option CH/EX in the tests for $R \times T$ tables available in StatXact.

EXAMPLE **6.16**   Let's see if the data in Example 6.10 support the claim that the distributions of the types of aphid infestation are different for treated and untreated tobacco plants. From the expected frequencies calculated earlier, we see that, if the rates of infestation are the same in the two populations, then leaf, stem, and no infestation will occur, respectively, in 24%, 34%, and 42% of the plants in each sample. Since the expected frequencies are not too small, the large-sample χ^2-test in Box 6.8 can be used to test the hypotheses. Using the second expression for χ_c^2 in Box 6.8, we get

$$\chi_c^2 = \frac{9^2}{24} + \frac{24^2}{34} + \frac{67^2}{42} + \frac{39^2}{24} + \frac{44^2}{34} + \frac{17^2}{42} - 200 = 254.39 - 200$$
$$= 54.39.$$

The calculated value of the test statistic should be compared with the critical value of a χ^2-distribution with $3 - 1 = 2$ degrees of freedom. From Table C.3 (Appendix C), we see that the calculated value is larger than $\chi^2(2, 0.001) = 13.8155$, so that the p-value is less than 0.001. The null hypothesis can be rejected at any level $\alpha \geq 0.001$. We conclude that the distributions of infestation rates are different for treated and untreated plants.   ∎

The case of two dichotomous populations ($C = 2$)

When $C = 2$, the frequencies of observed categories can be arranged in a 2×2 table such as Table 6.9. In that case, the calculated value χ_c^2 in Box 6.8 can be calculated using a somewhat simpler expression

$$\chi_c^2 = \frac{n(ad - bc)^2}{m_1 m_2 n_1 n_2}.$$ (6.4)

Furthermore, testing the equality of two dichotomous population distributions is equivalent to testing the equality of population proportions in a specific category. To see this, let π_{11} and π_{21} denote the proportion of values in the first category of populations 1 and 2, respectively. Since there are only two categories in each population, the proportion of values in the second category will be $\pi_{12} = 1 - \pi_{11}$ for population 1 and $\pi_{22} = 1 - \pi_{21}$ for population 2. Thus, the null hypothesis H_0: $\pi_{11} = \pi_{21}$ will be true if and only if the null hypothesis H_0: $\pi_{12} = \pi_{22}$ is also true. This equivalence means that the z-test in Box 6.5 and the χ^2-test in Box 6.8 serve the same purpose. At first sight, this might seem confusing; we might wonder which is more appropriate in a particular situation. However, the two tests lead to the same conclusion from the same data, as the next example demonstrates.

EXAMPLE **6.17**   In a study to determine the possible association between the HLA phenotype of premature infants and factors leading to chronic lung disease, 44 HLA-A2-type and 33 non-HLA-A2-type premature newborns were followed (Clark et al., 1982). It was found that 25 of the HLA-A2 types and 11 of the non-HLA-A2 types developed hyaline membrane disease. Do these data indicate that the rate of incidence of hyaline membrane disease is different in the two groups?

The disease frequency data can be arranged in a 2×2 table as follows:

	Disease status		
Type	Yes	No	Total
HLA-A2	25	19	44
Non-HLA-A2	11	22	33

We have two samples of sizes $n_1 = f_{1+} = 44$ and $n_2 = f_{2+} = 33$ from two dichotomous populations. The frequencies of responses in the two categories are: $f_{11} = 25, f_{12} = 19$ for sample 1; and $f_{21} = 11, f_{22} = 22$ for sample 2. Let π_{11} and π_{21} denote, respectively, the proportion of HLA-A2-type and non-HLA-A2-type premature infants who develop hyaline membrane disease. In other words, π_{11} and π_{21} are the proportion of values in the yes category in the two populations.

To see if the incidence rates of the disease in the two populations are different, we should test the null hypothesis H_0: $\pi_{11} = \pi_{21}$ against the research hypothesis H_1: $\pi_{11} \neq \pi_{21}$. This test can be performed using the z-test in Box 6.5 or the χ^2-test in Box 6.8. We perform the z-test first.

Let $\theta = \pi_{11} - \pi_{21}$ and $\theta_0 = 0$. Then, $\hat{\pi}_{11} = 25/44 = 0.5682$, $\hat{\pi}_{21} = 11/33 = 0.3333$, $\hat{\theta} = 0.5682 - 0.3333 = 0.2349$, and

$$\hat{\sigma}_{\hat{\theta}} = \sqrt{\frac{(0.5682)(0.4318)}{44} + \frac{(0.3333)(0.6667)}{33}} = 0.1109,$$

so that the test statistic is

$$z_c = \frac{\hat{\theta} - \theta_0}{\hat{\sigma}_{\hat{\theta}}} = \frac{0.2349 - 0}{0.1109} = 2.117.$$

From Table C.1 (Appendix C), the p-value for a two-sided test is $2(0.0166) = 0.0332$. The null hypothesis can be rejected at any level $\alpha \geq 0.0332$. For instance, the null hypothesis can be rejected at the $\alpha = 0.05$ level but not at the $\alpha = 0.01$ level.

To perform the χ^2-test in Box 6.8, we shall use Equation (6.4). It can be verified that the same χ^2_c value will result if the formula in Box 6.8 is used. With $a = 25$, $b = 11$, $c = 19$, $d = 22$, $n_1 = 25 + 19 = 44$, $n_2 = 11 + 22 = 33$, $m_1 = 25 + 11 = 36$, $m_2 = 19 + 22 = 41$, and $n = 25 + 19 + 11 + 22 = 77$, we get

$$\chi^2_c = \frac{77(25 \times 22 - 11 \times 19)^2}{44 \times 33 \times 36 \times 41} = 4.178.$$

From Table C.3 (Appendix C), it can be verified that the p-value of the test is between 0.05 and 0.025. This is consistent with the p-value of 0.0332 obtained for the z-test. ∎

The following is a StatXact printout showing the results of asymptotic (large sample) and exact methods for comparing proportions of hyaline membrane disease in HLA and non-HLA groups. The large-sample method is the same as that in Box 6.5 and requires large sample sizes for its validity. The exact method can be used with any sample size, but is hard to implement without good computing software.

```
                        StatXact Output
---------------------------------------------------------------------
DIFFERENCE OF TWO BINOMIAL PROPORTIONS


Statistic based on the observed 2 by 2 table :

   Binomial proportion for column <col1   > :     pi_1     = 0.3333
   Binomial proportion for column <col2   > :     pi_2     = 0.5682
   Difference of binomial proportions :    Delta = pi_2 - pi_1  = 0.2348
   Standardized difference of binomial proportions : Delta/Stdev = 2.117


Results:
   ------------------------------------------------------------------
   Method          P-value(2-sided)       95.00% Conf. Interval of Delta
   ------------------------------------------------------------------

   Asymp              0.0343              (0.01739,     0.4523)
   Exact              0.0465              (0.003689,    0.4618)

   ------------------------------------------------------------------
```

In this particular case, the large-sample (asymptotic) and exact methods give similar results because the sample sizes are relatively large.

If the z-test and the χ^2-test are equivalent for this purpose, why do we need them both? The χ^2-test is a two-sided test designed to test the null hypothesis that the population distributions are equal against the research hypothesis that the population distributions are not equal and so is equivalent to the two-sided z-test. If we are interested in a one-sided comparison—for example, $H_0: \pi_{11} \leq \pi_{21}$ against $H_1: \pi_{11} > \pi_{21}$—then the χ^2-test is not appropriate.

Comparing two categorical distributions: Paired samples

As with independent samples, the McNemar test for comparing two population proportions can be used to compare the distributions of two dichotomous populations. A test for comparing two populations that include more than two categories can be found in Fliess (1981).

Exercises

6.13   Refer to Example 6.10.

a   Construct a 95% lower bound to the difference between the proportions of leaf- and stem-infested plants in the treated and untreated populations.

b   On the basis of the interval constructed in (a), what conclusions can be drawn about the claim that treated tobacco plants have less infection than untreated plants?

c   Suppose it is claimed that, in the untreated plants, the ratio of infected (in the leaf or stem) plants to uninfected plants is 80:20. Perform an appropriate statistical test to see if the data for untreated plants contradict this claim.

6.14   In a study, two groups of patients with a particular infection are given two different antibiotic treatments. The numbers of patients cured within various periods are as follows:

Treatment	Time to cure (days)				
	1–4	5–6	7–8	9–10	> 10
Oral	28	32	24	10	6
By injection	44	20	12	11	10

a   Perform a statistical test to see if the data indicate a difference between the frequency distributions of the times to cure under the two treatments. Write your conclusion.

b   Construct a 95% confidence interval for the difference between the proportions of patients under the two treatments who are cured within six days. Interpret the interval you obtain.

6.15   The following data on the self-esteem of male and female children in a particular socioeconomic group were collected by administering a questionnaire:

Gender	Self-esteem level		
	Low	Medium	High
Male	30	56	19
Female	42	50	8

a Perform a test to see if the data indicate that the proportion of females with low self-esteem is higher than the corresponding proportion of males.

b Do the data support the conclusion that the distributions of self-esteem levels are different for males and females?

6.16 Random samples of oranges grown under two different conditions were classified into three categories (1, discard; 2, ship for sale; 3, ship for juice) as follows:

Growth condition	Category		
	1	2	3
1	18	56	76
2	29	41	80

a Construct a 95% upper bound to the proportion of oranges grown under condition 1 that have to be discarded. Interpret the bound you obtain.

b On the basis of these data, is it reasonable to conclude that the overall quality of the oranges grown under the two conditions is different?

6.17 In a study, six out of ten seeds of variety 1 germinated, whereas three out of ten seeds of variety 2 germinated. Perform a statistical test to see if the data support the conclusion that the germination rate for variety 1 is higher than that for variety 2.

6.18 Refer to Example 6.17.

a Verify that the same value of χ_c^2 will result if we use the second formula in Box 6.8.

b Do these data indicate that the proportion of HLA-A2 types with hyaline membrane disease is higher than the proportion of non-HLA-A2 types?

c Construct a 95% confidence interval for the difference between the proportions of individuals with hyaline membrane disease in the two HLA types. Interpret your result.

6.19 Each person in a random sample of 200 customers arriving at a grocery store was asked to rate two brands of vanilla ice cream according to two categories: like and dislike. The following data were obtained:

Brand A rating	Brand B rating		Total
	Like	Dislike	
Like	38	92	130
Dislike	13	57	70
Total	51	149	200

Construct a 99% confidence interval for the difference between the proportions of consumers who like Brand A and the proportion who like Brand B ice creams. Interpret the interval you construct.

6.20   In order to see the effectiveness of a job training program, each of 93 trainees was tested on basic skills immediately before and immediately after undergoing the training program, with the following results:

Pretraining test result	Posttraining test result		Total
	Pass	Fail	
Pass	14	5	19
Fail	49	25	74
Total	63	30	93

a   Do the data indicate that the training program is effective?

b   Construct a 95% confidence interval for the proportion of trainees who fail the posttraining test after having passed the pretraining test. Interpret the interval you calculate.

6.21   Case-control studies provide an efficient means of investigating the possibility of an association between exposure to a risk factor and subsequent development of a disease. In a typical case-control study, each subject in a random sample of n subjects who have the disease (each case) is matched with a subject who does not have the disease (a control), on the basis of one or more shared characteristics that might influence the likelihood of the disease. The prior exposure characteristics of each member of the n case-control pairs are determined for comparison purposes.

For example, in a study of association between endometrial cancer (disease) and estrogen use (risk factor), each of 451 patients with endometrial cancer was matched with a patient without endometrial cancer, on the basis of several factors such as race, age, hospital, and date of admission (Antunes et al., 1979). Each patient was classified into one of two categories: $+$ if the patient had used estrogen in the past; and $-$ otherwise. Thus, each case-control pair fell into one of four categories: $(+, +)$, in which both patients used estrogen; $(+, -)$, in which only the case used estrogen; and so on. The frequencies of case-control pairs in a subgroup in which the cases had stage-I cancer are as follows:

Classification of case	Classification of control		Total
	$+$	$-$	
$+$	11	65	76
$-$	9	102	111
Total	20	167	187

Construct a 99% lower bound for the difference between the proportions of estrogen users among the cases and controls. Interpret the bound you calculate.

6.4
Overview

This chapter is devoted to methods for making inferences about one and two categorical populations. Because the distribution of a categorical population is completely determined by the population proportions of values in the various categories, inferences about categorical populations reduce to inferences about population proportions. In the one-population setting, the procedures in this chapter may be used to make inferences about the proportion in a single specific category or to test the null hypothesis that the population proportions have a specified set of values. In the two-sample setting, the methods in this chapter may be used to make inferences about the difference between two population proportions and to see if the distributions of two categorical populations are identical.

Categorical methods have a wide variety of applications in biology and life sciences, where nominal data are frequently encountered. Furthermore, they can be used for inferences about ordinal and interval data, after appropriate reduction, and so they provide an alternative to the interval data methods of Chapter 4, and rank data methods of Chapter 5.

7

Designing Research Studies

7.1
Introduction

The need for careful design of any research study was emphasized in Section 1.3. In this chapter, we focus on two key questions faced by a researcher embarking on a new study:

1. What steps are possible to ensure that the study conclusions are not distorted by the influence of extraneous sources on the measured values?

2. How many measurements are needed to make sure that the study conclusions meet appropriate reliability criteria?

Section 7.2 introduces some useful terminology. In Section 7.3, we address the influence of uncontrollable sources on the measurement process. Two methods of eliminating their effects on the study conclusions are described: blocking at the design stage; and statistically adjusting at the analysis stage. In Section 7.4, two common methods of determining sample sizes are considered; their use in designing studies to make inferences about one or two population means is demonstrated.

7.2
Some useful terminology

Types of variables

A research study can be viewed as a protocol for measuring the values of a set of variables—called *response variables*—under a set of conditions called *study conditions*. Consider the following example.

EXAMPLE **7.1**   In a study to compare the absorption rates of three drugs, A, B, and C, a predetermined dose of each was administered to six subjects (18 subjects in all). Three subjects within each group were given the drug orally; the other three received it intravenously. For each test subject, the amount of drug absorbed into the blood within four hours of its administration was measured. Each subject's body weight and age were also measured, in case these variables also influence drug absorption.

The response variable in this study was $Y =$ the amount of drug absorbed within four hours of its administration. The conditions under which the measurements are made could be determined by the type of drug administered (A, B, or C), the method of administering the drug (intravenous or oral), the subject's body weight, and the subject's age. All other variables—such as gender and blood pressure—were considered unimportant as indicators of the rate of drug absorption.   ■

Study conditions are best described in terms of *explanatory variables*. We think of the measured values of a response variable as capable of being explained by the values of a number of explanatory variables. For example, in the drug absorption study in Example 7.1, some of the explanatory variables were $X_1 =$ type of drug, $X_2 =$ age of the recipient, $X_3 =$ weight of the subject, and $X_4 =$ method of administration. Since no record of blood pressure or gender was kept, we assume that the investigators felt that these two variables were not useful for predicting drug absorption rate.

The condition under which a response is measured is determined by the corresponding values of the explanatory variables. For instance, the response from a subject who is 50 years of age, weighs 160 lbs, and receives drug A orally corresponds to the study condition: $X_1 =$ A, $X_2 = 50$, $X_3 = 160$, and $X_4 =$ oral.

EXAMPLE **7.2**   To compare the weight gains of chickens fed four experimental diets, 100 chickens were assigned to 20 pens; each pen received five birds. The pens were divided into four groups of five pens each, and the four diets were assigned to the four groups at random. The birds in a pen were fed a common diet for 12 weeks. At the end of the study period, the weight gain of each of the 100 birds was measured. The measured values in this study—the weight gains of the experimental chickens—may depend to various degrees on several variables, including the type of diet, the bird's initial weight, age, and gender, the type of pen in which the animal was housed, and the amount of water available during the study period. Thus, the response variable is $Y =$ weight gain of the animal, while some of the explanatory variables are: $X_1 =$ type of diet, $X_2 =$ initial weight, $X_3 =$ age, and $X_4 =$ daily allowance of water. The study conditions can be described by specifying the values of the explanatory variables for every animal in the study.   ∎

The level of an explanatory variable

The value of an explanatory variable is called its *level*. In Example 7.2, there are four levels of the explanatory variable $X_1 =$ type of diet, corresponding to the four diets, whereas the actual initial weight of a chicken is the level of the explanatory variable $X_2 =$ initial weight. The number of distinct values of an explanatory variable can be finite or infinite. For example, X_1 has a finite number of possible levels (four), whereas X_2 has infinitely many levels.

The level of an explanatory variable can be measured on an interval, ordinal, or nominal scale. Following the terminology in Section 4.2, we say that an explanatory variable is *quantitative* if its levels are measured on an interval scale; otherwise, it is *qualitative* or *nominal*. Thus, initial weight is a quantitative variable, whereas type of diet is a qualitative variable.

Study material

The smallest units of the study material for which responses are measured are called *observational units*. Because many statistical procedures are based on the premise that the measured data are samples from statistical populations, the observational units are also called *sampling units*. Thus, in Example 7.1, the subjects who were treated with drugs were the observational units; in Example 7.2, the individual chickens on which measurements were made were the observational units. In different contexts, an observational unit might be an animal, a plot of land, a specimen sample, a human subject, or any of a wide variety of other objects.

Types of explanatory variables

An explanatory variable whose effect on the response is a primary objective of the study is called a *factor*.

EXAMPLE **7.3**   A study was conducted to see how soil type and soil moisture affect the root growth of a variety of plant. Two soil types—sandy and clay—and three soil moisture levels—10%, 20%, and 30%—were used in the study. Plants were grown in nine pots of sandy soil, of which three had 10% moisture, three had 20% moisture, and three had 30% moisture, and in nine pots of clay soil, with the same moisture distribution. In each pot, the root volume was measured when the plants were 10 weeks old.

The response variable is Y = 10-week root volume. The objective of the study is to collect information about the effects on Y of two explanatory variables: X_1 = soil type and X_2 = soil moisture content. Thus both explanatory variables are factors.

In Example 7.2, the objective of the study was to compare the effects of the diets on weight gain. Thus, the only factor in that study is the explanatory variable X_1 = type of diet. ■

An explanatory variable whose effect has the potential to mask (or enhance) the effect of a factor is called a *confounding variable*. In the study in Example 7.3, potential confounding variables include X_3 = amount of light, X_4 = amount of fertilizer, and X_5 = location of the pot in the greenhouse. In Example 7.4, we see why the amount of light should be regarded as a confounding variable.

EXAMPLE **7.4**   Let's assume, for the sake of simplicity, that the investigators in Example 7.3 wanted to study the effect of moisture content in sandy soil only. Suppose they used nine pots: three with sandy soil at 10% moisture, three with sandy soil at 20% moisture, and three with sandy soil at 30% moisture. Suppose, further, that the study was conducted in a greenhouse with three benches: one for the three pots with 10% moisture, one for the 3 pots with 20% moisture, and one for the three pots with 30% moisture. Assume that the bench locations in the greenhouse were such that the bench with 10% moisture received the most light and the bench with 30% moisture received the least light. Display 7.1 shows the study design. The entries in the display are the soil moisture contents of three pots—numbered 1, 2, and 3—for each of the three blocks (benches).

A careful examination of Display 7.1 reveals a major flaw in the study design. The design does not take account of the possible confounding effect of the light condition. It is well known that plants receiving more light are likely to produce a higher root volume. Therefore, if we measure higher root volume for soil with 10% moisture content than for soil with, say, 30% moisture content, we cannot know whether the increased root volume is due to a difference in the effects of 10% and

DISPLAY **7.1**

A study design for comparing the effects of moisture contents on root volume

	Pot number		
Bench number	1	2	3
1 (high light)	10	10	10
2 (medium light)	20	20	20
3 (low light)	30	30	30

30% moisture contents or due to the difference in light exposure of the plants in soil with 10% and 30% moisture contents. The amount of light is a confounding variable, because its effect may mask (or enhance) the differences in measured responses at different levels of the experimental factor.   ■

Types of studies

Research studies can be classified into one of two types depending on the manner in which the levels of the explanatory variables associated with the observed responses are selected. A study in which the investigator selects the levels of at least one factor is called an *experimental study.* A factor whose levels are determined by the investigator is called an *experimental factor*. A design in which the levels of all the explanatory variables are determined as part of the observational process is called an *observational study*.

EXAMPLE **7.5**   In Example 7.3, the 18 pots in which root volumes were measured are the observational units. By controlling the moisture content and soil type (levels of the explanatory variables) in each pot, the investigator specified a priori the levels associated with the 18 responses that will be measured in the study. Thus, this is an experimental study.

Now consider a research study in which an investigator wanted to assess the effect of mothers' alcohol consumption during pregnancy on the birth-weights of their babies. Mothers of 76 babies born in a group of hospitals during a specified period were classified into three categories according to their alcohol consumption during pregnancy: none, moderate, and excessive. Also, the birth-weights of the 76 babies were recorded.

In this study, the observational units were the 76 newborn babies and the response variable was $Y =$ birth-weight. The value of one qualitative explanatory variable $X =$ mother's alcohol consumption (with three levels: none, moderate, and excessive) was measured for each observational unit. This is an observational study because the investigator did not control the assignment of levels of the explanatory variable X. Rather, the levels of X were measured along with the responses.   ■

In an experimental study, the number of responses at each level of an explanatory variable is determined as a part of the study design. For example, in the study of the effects of soil type and soil moisture on root volume, the design called for three measured responses at each of the six combinations of the levels of soil type and soil moisture content. In an observational study, in contrast, the number of responses at various levels of an explanatory variable is not under the direct control of the investigator. In the study of the effect of pregnant women's alcohol consumption in Example 7.5, the number of mothers in the various alcohol consumption groups (levels of the explanatory variable) depended upon the characteristics of the mothers of the 76 babies born during the study period.

As we will see in Section 7.4, if we are able to control the number of measured responses at each level of an explanatory variable, we can design the study so as to permit analysis of the results using tests of known power and estimates of known precision. This is a definite advantage of an experimental study over an observational study.

In many instances, unfortunately, an experimental study is either impossible or prohibitively expensive. For example, in an experimental study of birth-weight and maternal alcohol consumption, we would have to select a specified number (say, 25) of pregnant women in each of the three alcohol consumption groups. Clearly it is a lot easier and less expensive to locate 75 newborn babies and determine the alcohol consumption groups of their mothers (an observational study) than to identify 25 pregnant women in each of the alcohol consumption groups who are willing to participate in the study and then determine the birth-weights of their babies.

Our primary focus in the remainder of this book will be the design and analysis of experimental studies. However, many of the analytical methods used for experimental studies can also be applied to observational studies. The strict validity of the conclusions drawn from observational studies will depend on the extent to which the assignment of levels of the study factors was controlled by the investigator.

Treatments and experimental units

The methodology for the design and analysis of experimental studies was first developed at the Rothamsted Agricultural Experiment Station in London, England, under the direction of the late Sir R. A. Fisher. As a result, much of the terminology—such as treatments and plots—reflects the agricultural heritage of the discipline. A combination of the levels of one or more experimental factors is called a *treatment*.

EXAMPLE **7.6**   In Example 7.2, the objective was to compare the weight gain in chickens fed four different diets. There was only one experimental factor, and so the treatments are combinations of levels of a single factor—the type of diet fed to the chicken. The four treatments in this experiment could be identified by the levels 1, 2, 3, and 4. In Example 7.3, six treatments result from the possible combinations of the two levels of the soil type with the three levels of the soil moisture content. The six treatments can be identified by the factor levels that are combined to form the treatments. For instance, sandy soil with 10% moisture may be denoted as treatment (S, 10). The other five treatments are (C, 10), (S, 20), (C, 20), (S, 30), and (C, 30).   ■

The smallest unit of the study material sharing a common treatment is called an *experimental unit*. Like an observational unit, an experimental unit can be an animal, a plot of land, a specimen sample, a human subject, and so on. However, it is important to distinguish between an observational unit and an experimental unit. For instance, in Example 7.2, the observational units are the chickens, because the smallest units on which responses were measured were the individual chickens. On the other hand, the birds in a single pen received a common treatment, because the chickens within a pen were fed from a common source containing their assigned diet. Therefore, the experimental units in this study are the pens. In contrast, the observational units and experimental units are identical in the drug absorption study in Example 7.1 and the root volume study in Example 7.3.

Sources of variation in the observed responses

The primary objective of a research study is to make inferences about the effects of experimental factors on the observed responses. In an experimental study, such inferences are obtained by comparing the observed responses of the experimental units to the various treatments. For instance, in the experiment in Example 7.3, the effects of soil type and soil water content can be examined by comparing the measured root volume of plants subjected to the six treatments—that is, the six combinations of levels of soil type and soil moisture content.

A characteristic of all measured responses is their variability. Good research designs take into account that variability in the measured responses may be attributed to one or more of the following sources:

1   Variation due to the effects of experimental factors. The experimental units are subjected to different treatments, which produce different responses.

2   Variation due to the effects of identified confounding variables. Different levels of one or more confounding variables cause variation in the measured responses of the experimental units.

3   Variation due to unidentified sources. This is often called the *error variation*. Because it is impossible to repeat measurements under constant experimental conditions, measured values made under apparently identical conditions will not always be the same. The error variation can be thought of as the result of a number of unidentified confounding variables.

EXAMPLE **7.7**   Tarla (1988) conducted an experiment to compare the utilization of selenium from calcium selenite and sodium selenite by sheep fed dietary selenium over different feeding periods. He used six treatments resulting from combining the levels of two experimental factors:

1   Selenium source at three levels:
    a   basal diet supplemented with alfalfa meal and urea;
    b   basal diet containing 0.3 μg/g added selenium as sodium selenite;
    c   basal diet containing 0.3 μg/g added selenium as calcium selenite.

2   Feeding period at two levels:
    a   40 days;   **b**   80 days.

In the study, 42 crossbred wethers (male sheep castrated before sexual maturity), each with an approximate initial body weight of 34 kg, were randomly divided into six groups of seven animals. The six treatments (combinations of three types of diet and two feeding periods) were randomly assigned to the six groups. At the end of the study period (40 days for three treatments and 80 days for three treatments), samples of kidney tissue, liver tissue, and serum from each of the 42 animals (experimental units) were analyzed for selenium content. Data on feed intake and weight gain (g/day) for each animal were also collected.

The primary response variables in this study were: $Y_1 =$ serum selenium content; $Y_2 =$ liver tissue selenium content; and $Y_3 =$ kidney tissue selenium content. Two other response variables that may also depend on the diet and feeding period— $Y_4 =$ amount of feed intake (g/day) and $Y_5 =$ weight gain (g/day)—were measured to see how they differed between treatment groups.

The measured responses from two experimental animals may differ because of the following possible sources of variation:

1 Effects of experimental factors (treatments): Responses could differ because the animals were on different diets over different feeding periods.

2 Effects of identified confounding variables: Confounding variables such as the initial weight and the amount of available water may affect the observed responses. In other words, responses could vary because, for example, the actual initial weights of the animals were not the same or the amount of available water differed from pen to pen.

3 Effects of unidentified sources of variation: Responses could vary because of differences, however small, in the conditions under which tissue samples were analyzed; small undetectable differences in the exact composition of the diet fed to different animals; and so on.   ■

Exercises

7.1 The presence of bacteria in urine (bacteriuria) is known to be associated with kidney disease. The following are three possible designs for a study in which the primary objective is to see if users of oral contraceptives (OC) are more likely to have bacteriuria than are nonusers.

1 The experimenters select 25 female OC users and 25 female OC nonusers in each of four age groups: 16–19 years; 20–29 years; 30–39 years; and 40–49 years. Each subject is tested; a 1 is recorded if bacteriuria is detected and a 0 is recorded otherwise.

2 The experimenters select 100 female OC users and 100 female OC nonusers between the ages of 16 and 49. Each subject is classified into one of the four age groups described in the preceding case. Each subject is tested and a 1 or 0 is recorded to denote the presence or absence of bacteriuria.

3 The experimenters select 200 female subjects between the ages of 19 and 49. Each subject is classified into one of eight groups: OC user, age 16–19; OC nonuser, age 16–19; and so on. Each subject is tested, and a 1 or 0 is recorded to denote the presence or absence of bacteriuria.

a Classify the three designs as experimental or observational studies, giving reasons.

b Identify, giving reasons, the response and explanatory variables.

   c Identify, giving reasons, the levels of measurement (nominal, ordinal, or interval) of the response and explanatory variables.

   d Identify, giving reasons, the factors and confounding variables.

   e Identify, giving reasons, the observational units.

7.2 To compare the yield of a variety of wheat under four fertilizer treatments, experimenters prepared 20 pots, each containing two wheat plants (40 plants in all). Each fertilizer treatment was used on five pots, and the yield of wheat was measured for each of the 40 plants.

   a Identify, giving reasons, the response and explanatory variables in this study.

   b Identify, giving reasons, the levels of measurements (nominal, ordinal, or interval) of the response and explanatory variables.

   c Explain why this is an experimental study.

   d Identify the treatments, the observational units, and the experimental units.

   e For each plant, let $X_1 =$ the amount of daily light exposure and $X_2 =$ the amount of daily water received. Given that yield may depend on the amounts of light and water, explain why X_1 and X_2 are potential confounding variables.

7.3 In a study to evaluate how a subject's weight depends on gender, daily calorie intake, and daily amount of exercise, researchers examined four combinations of two levels of daily calorie intake (low, high) and two levels of daily exercise (low, high) for males and females. A group of 12 male subjects was divided into four subgroups of three subjects; the subgroups were assigned at random to the four combinations of food intake and exercise levels. A group of 12 female subjects was assigned in the same way to food intake and exercise levels. Over a one-year period, each subject was asked to follow a regimen corresponding to his or her assigned combination of calorie intake and exercise levels. During the study period, the weight of each subject was recorded four times: at the beginning of the study and then after 4, 8, and 12 months. Because it was suspected that a subject's weight may depend on age and bone structure (small, medium, large), the values of these two variables were recorded for each subject.

   a Explain why this is an experimental study.

   b Identify the response variable(s).

   c Identify the experimental factors and indicate the levels (quantitative, qualitative) at which they are measured.

   d Identify the potential confounding variables and indicate the levels (quantitative, qualitative) at which they are measured.

   e Identify the treatments.

   f Identify the observational and experimental units.

7.3
Tools for developing experimental designs

A good study design makes efficient use of available resources to collect the data needed to meet the study objectives. In this section, we discuss three useful techniques for developing good experimental designs: randomization, replication, and blocking.

Randomization

A basic rule in any experimental study is that, to the extent possible, treatments should be allocated to experimental units at random. For instance, in the study to compare the weight gain of chickens on four diets in Example 7.2, the diets should be assigned to the pens at random, under the condition that five pens are allocated to each of the four diets. Example 7.8 describes a method by which the treatments can be randomly assigned to the experimental units.

EXAMPLE **7.8**   In the study described in Example 7.2, there are many ways of randomizing the diets to animals subject to the condition that each diet is assigned to exactly five animals. In one method, we number the animals from 1 to 100 and the pens from 1 to 20. Then we select a random sample of five numbers between 1 and 100 and assign the animals with those numbers to pen 1; we select a random sample of five numbers from the remaining 95 numbers and assign the animals with those numbers to pen 2; and so on. When all the animals have been assigned to the 20 pens, we select five random numbers between 1 and 20 and assign the pens with the selected numbers to diet 1, and so on. Section 7.2 of the companion text by Younger (1997) describes how SAS can be used to do this randomization.   ■

Essentially, there are two reasons for advocating random allocation of treatments to experimental units.

1   Randomization allows the observed responses to be regarded as random samples from appropriate populations. For instance, in the weight gain study in Example 7.2, randomly selecting five chickens to receive a particular diet—say, diet 1—allows us to assume that the five measured values for diet 1 constitute a random sample of size $n = 5$ from the population of weight gains of all chickens fed diet 1. Since many statistical inferential techniques are based on the assumption that the observed data are random samples from the relevant populations, randomization justifies the use of these techniques to draw conclusions from the observed data.

2   Randomization eliminates the influence of systematic biases on the measured values. Such biases, if undetected, might act as confounding variables. Thus, in Example 7.2, randomization ensures that every pen has the same chance of receiving each of the four experimental diets, so that the initial weight of the bird, a potential confounding variable, will have the same chance of affecting the 20 responses.

Replication

The number of experimental units for which responses to a particular treatment are observed is called the number of *replications* of that treatment. In the weight gain study in Example 7.2, each diet was fed to birds in five pens, and so there were five replications of each treatment. In the root volume study in Example 7.3, there were three replications per treatment.

The technique of replication can be used for two purposes. First, observing more than one response to a treatment enables us to estimate the variability in responses that is not associated with treatment differences. For instance, in Example 7.2, the variance of the weight gains of animals receiving the same diet measures variability that cannot be ascribed to the different effects of diets. Second, increasing the number of replications increases the reliability of the conclusions drawn from the observed data. For example, let $\overline{Y}$ and S^2 denote, respectively, the mean and variance of the responses to n independent replications of a treatment. Then $\overline{Y}$ is an estimate of the expected response to the treatment and, on the basis of Equation (4.3), the standard error of $\overline{Y}$ can be estimated as

$$\hat{\sigma}_{\overline{Y}} = \frac{S}{\sqrt{n}}.$$

Thus, the standard error of the estimate of the expected response will decrease as the number of replications increases. In general, increasing the number of replications means collecting more information about a treatment, so that we can make more precise inferences about its effects.

Blocking

Blocking is a technique used to eliminate the effects of selected confounding variables when comparing the treatments. Under suitable conditions, blocking allows us to compare the expected responses for different treatments without regard to the levels of the associated confounding variables.

Blocks are groups of experimental units sharing a common level of a confounding variable. For instance, consider Example 7.3, where we discussed a greenhouse experiment to compare the effects of three levels of moisture content on the root volume of plants. The experimental units were nine pots placed on three benches exposed to different amounts of lights. The pots in a single bench could be regarded as a block, because they were affected by the same level of the confounding variable—light exposure.

The simplest—and probably the most common—use of blocking is to design the experiment so that every treatment is applied to exactly one randomly selected experimental unit within a block. This is a *randomized complete block design* and will be discussed in Chapter 15. Note that the experiment in Example 7.3 is not a randomized complete block design because all units within a block (bench) received the same treatment. A randomized complete block design would have randomly allocated the three moisture content levels to the three pots within each bench. A possible randomized complete block design for the root volume study is shown in Display 7.2. The figures in the display are the moisture contents of the soil in three pots—numbered 1, 2, and 3—in each of the three blocks (benches).

The main difference between the experimental designs in Displays 7.1 and 7.2 is the way in which the treatments are assigned to the experimental units. For the design in Display 7.2, there will be a response for every treatment (moisture level) under

DISPLAY 7.2

A randomized complete block design for comparing the effect of soil moisture content on root volume

Bench number	Pot number		
	1	2	3
1 (high light)	20	10	30
2 (medium light)	10	20	30
3 (low light)	10	30	20

each light condition, so that the influence of light condition on the mean of the three responses for any one treatment will be the same as its influence on the mean response for any other treatment. Consequently, we may expect that, under some conditions, the difference between mean responses for any two treatments will be free of the confounding effect of light condition. Indeed, as we will see in Chapter 15, Display 7.2 represents a good design for comparing the effects of soil moisture levels, provided it is reasonable to assume that the combined effect of moisture content and light exposure satisfy a property called *additivity*. If that assumption is reasonable, the responses from the randomized complete block design will enable us to eliminate the effects of light conditions when comparing the effects of moisture levels.

If blocking is to be effective in eliminating the influence of confounding variables, more than one experimental unit must be exposed to the effect of each level of the confounding variable. Thus, in Display 7.2, blocking on the basis of the confounding variable—amount of light—was possible only because three pots (experimental units) were exposed to each light condition. On many occasions, unfortunately, because of cost or feasibility considerations, it is not possible to control the number of experimental units that will be exposed to each level of the confounding variable.

EXAMPLE **7.9**   The selenium study in Example 7.7 is a good example of a situation where blocking on the basis of a suspected confounding variable is not a practical proposition. Recall that the experiment involved six treatments corresponding to combinations of three diets and two feeding periods. The initial weight of the animal was regarded as a confounding variable. A randomized complete block design in which experimental units were blocked on the basis of initial weight would have required the investigators to divide the 42 animals into seven groups of six animals in such a way that animals within a group had equal (or nearly equal) initial weights. Obviously, finding 42 crossbred animals with such specific weight requirements could be quite expensive, if not impossible. Consequently, blocking on the basis of initial weight is not a practical solution in this case.   ■

When blocking is not feasible, statistical adjustments can often be used to allow for the effects of confounding variables. Such adjustments are made at the time of data analysis using a technique known as *analysis of covariance,* which will be discussed in Chapter 12.

Exercises

7.4   Refer to the drug absorption study in Example 7.1. Suppose that the investigators wanted to use the observed responses to see: (1) if the absorption rate of any one of the three drugs depended on the way it was administered; and (2) if there were any differences between the absorption rates of the three drugs in either of the two modes of application.

   a   Argue that this is an experimental study and identify the experimental factors.
   b   What are the treatments and experimental units?
   c   How many replications are there for each treatment?
   d   Are there any confounding variables? Explain.

7.5   Refer to the study in Example 7.1.

   a   Give the layout of a randomized complete block design in which the study subjects are divided into the following four blocks according to their age and body weight:

Block	Age group	Body weight group
1	Young	Low
2	Young	High
3	Old	Low
4	Old	High

   b   How many replications are there for the treatments in the design in (a)?
   c   For the design in (a), suggest a method of randomizing the treatments to the experimental units.

7.6   Refer to the study of weight gains in chickens on four experimental diets in Example 7.2.

   a   Suppose that the pens housing the chickens are in five rows of four pens, as follows:

Pen 1	Pen 2	Pen 3	Pen 4
Pen 5	Pen 6	Pen 7	Pen 8
Pen 9	Pen 10	Pen 11	Pen 12
Pen 13	Pen 14	Pen 15	Pen 16
Pen 17	Pen 18	Pen 19	Pen 20

      Give the layout of a design that uses the rows as blocks.
   b   How many replications of treatments are there in the design in (a)?

c For the design in (a), explain how you would randomize the treatments to the experimental units.

d Describe a set of conditions under which you would recommend the use of the design in (a).

7.4
Statistical models and experimental designs

As we saw in Section 4.5, statistical models provide a convenient means of summarizing the assumptions about a set of measured responses. Example 7.10 shows how statistical models can be used when selecting the design for a research study.

EXAMPLE **7.10**   An animal nutritionist was interested in comparing the average 16-week weight gain of animals on two diets—diet 1 and diet 2. The nutritionist decided to observe the weight gains of eight animals housed in four pens (two animals per pen) over a 16-week period. Which of the following two designs for this study would be better? Why?

1 Assign animals at random to the four pens in such a way that each pen has two animals. Select two pens at random and call them pen 1 and pen 2; the other two pens are designated pen 3 and pen 4. Assign diet 1 to the four animals in pens 1 and 2. The remaining four animals in pens 3 and 4 receive diet 2. The layout of design 1 is shown in the first three rows of Display 7.3. Let Y_{ij} denote the j-th response (weight gain) for treatment (diet) i ($i = 1, 2; j = 1, 2, 3, 4$). The last row of Display 7.3 shows the symbols denoting the eight responses.

DISPLAY **7.3**

Assignment of diets and pens to animals: Design 1

Pen	1		2		3		4	
Animal	1	2	3	4	5	6	7	8
Diet	1	1	1	1	2	2	2	2
Response	Y_{11}	Y_{12}	Y_{13}	Y_{14}	Y_{21}	Y_{22}	Y_{23}	Y_{24}

2 Assign animals at random to the four pens in such a way that each pen has two animals. Within each pen, randomly assign the two diets to the two animals. A layout of design 2 is shown in Display 7.4, along with the symbols denoting the responses.

To decide which design is better suited to the nutritionist's research objectives, we need to determine the statistical models that are appropriate for the observed responses in the two designs. These models can be used to see if the responses yield sufficient information to provide answers to the research questions.

DISPLAY   7.4

Assignment of diets and pens to animals: Design 2

Pen	1		2		3		4	
Animal	1	2	3	4	5	6	7	8
Diet	1	2	1	2	1	2	1	2
Response	Y_{11}	Y_{21}	Y_{12}	Y_{22}	Y_{13}	Y_{23}	Y_{14}	Y_{24}

Let μ_1 and μ_2 denote, respectively, the true (population) mean weight gains for animals on diets 1 and 2; that is, μ_i is the mean of the conceptual population of weight gains of all animals on diet i ($i = 1, 2$). The researcher's objective is to use the data from eight animals to make inferences about the differential mean weight gain $\theta = \mu_1 - \mu_2$.

Let's begin by writing a suitable model for the responses from design 1. We must first determine whether there are variables that have confounding effects on the responses—that is, variables whose effect on the responses may mask (or enhance) the observed difference between the mean weight gains for the two diets.

Let's assume that each animal is fed its own diet individually; in that case, the experimental units are the animals. Suppose that the experimental animals are cross-bred so that they may be expected to respond similarly to the same diet. Thus, there is no reason why we should expect a specific animal (say, animal 1) to gain more weight than another animal (say, animal 2) on the same diet. Thus, the experimental units can be regarded as similar in terms of their ability to respond to the diets.

Could pen effects act as a confounding variable? In other words, is there reason to believe that, on average, animals in different pens will respond differently to the same diet? Perhaps, because of the location of pen 1, the animals in pen 1 will do better than those in pen 2. If so, the effect of the pens should be regarded as a potential confounding variable.

Let's begin by assuming that the pen effect is not a confounding variable—in other words, that the pen effect on the expected weight gain is the same for all pens. Then the expected response (weight gain) for an animal in a particular pen is the same as that for an animal on the same diet in any other pen. Consequently, the observed weight gain for the j-th animal receiving the i-th diet can be written as

$$Y_{ij} = \mu_i + E_{ij}, \qquad i = 1, 2; \quad j = 1, 2, 3, 4. \tag{7.1}$$

Here μ_i is the expected weight gain for diet i and E_{ij} is the amount by which the observed weight gain will differ from its expected value. The term E_{ij} in Equation (7.1) accounts for the error variation due to sources other than the diets.

The model in Equation (7.1) implies that an observed response depends only on the diet and not on the pen in which the animal is housed. Display 7.5 shows the expected responses in design 1 under the assumption that the pen effect is not a confounding source of variation.

Notice that the model in Equation (7.1) is the same as the model for two independent samples in Box 4.9. The data comprising the responses from the eight animals can be regarded as two independent samples: a sample of size $n_1 = 4$ from

DISPLAY   7.5

Expected responses for design 1 when pens do not have confounding effects

Pen	Animal	Diet	Expected response
1	1	1	μ_1
1	2	1	μ_1
2	3	1	μ_1
2	4	1	μ_1
3	5	2	μ_2
3	6	2	μ_2
4	7	2	μ_2
4	8	2	μ_2

a population with mean μ_1; and a sample of size $n_2 = 4$ from a population with mean μ_2. It follows that the independent-sample t-test procedures in Box 4.3 can be used to make inferences about $\theta = \mu_1 - \mu_2$ provided the E_{ij} in Equation (7.1) satisfy the assumptions in Box 4.9—that is, provided the weight gains under the two diets have independent normal distributions with a common variance. The Wilcoxon rank sum procedure in Section 5.5 can be used to make inferences about θ if the assumptions in Box 4.9 are too stringent for these data.

If the expected response to a diet may vary from pen to pen, the pens could have a confounding effect on the comparison of the effects of diets. To see this, suppose that the diet and pen effects are *additive*—that is, that the expected response for an animal on a particular diet is the sum of two components, one due to the effect of the diet and the other due to the effect of the pen. In symbols, assume that the expected response for diet i in pen j is $\mu_i + \rho_j$, where the parameter ρ_j accounts for the change in the expected response due to the fact that the animal is housed in pen j. Then, in design 1, the expected responses are $\mu_1 + \rho_1, \mu_1 + \rho_2, \mu_2 + \rho_3$, and $\mu_2 + \rho_4$ in pens 1, 2, 3, and 4 respectively. Display 7.6 shows the expected responses in design 1 under the assumption that pen and diet effects are additive.

Since the expected responses are the same within any given pen, the four sets of responses in the four pens could be regarded as four random samples, each of size $n = 2$, from four populations with means $\mu_1 + \rho_1, \mu_1 + \rho_2, \mu_2 + \rho_3$, and $\mu_2 + \rho_4$. For example, the two responses from pen 1 form a random sample of size $n = 2$ from the conceptual population of all responses from animals fed diet 1 in pen 2. While the means of these four populations may be estimated by the corresponding sample means—for example, $\mu_1 + \rho_1$ may be estimated by the mean weight gain for animals in pen 1—it is not possible to find a simple way to estimate the parameter of interest: $\theta = \mu_1 - \mu_2$. For example, the difference between the mean weight gains for pens 1 and 2 estimates the difference between the corresponding population means: $(\mu_1 + \rho_1) - (\mu_1 + \rho_2) = \rho_1 - \rho_2$. Similarly, the difference between mean responses for pens 1 and 3 estimates $(\mu_1 + \rho_1) - (\mu_2 + \rho_3) = (\mu_1 - \mu_2) + (\rho_1 - \rho_3)$. Therefore, if pen is a confounding variable, design 1 will not permit the estimation of $\theta = \mu_1 - \mu_2$ in a simple manner.

DISPLAY   7.6

Expected responses for design 1 when pens could have a confounding effect

Pen	Animal	Diet	Expected response
1	1	1	$\mu_1 + \rho_1$
1	2	1	$\mu_1 + \rho_1$
2	3	1	$\mu_1 + \rho_2$
2	4	1	$\mu_1 + \rho_2$
3	5	2	$\mu_2 + \rho_3$
3	6	2	$\mu_2 + \rho_3$
4	7	2	$\mu_2 + \rho_4$
4	8	2	$\mu_2 + \rho_4$

Now consider design 2. First, if the pen effects are the same for all pens, Equation (7.1) is appropriate as a model, because the expected response from an experimental unit (animal) for a given diet will be the same for all pens. Display 7.7 shows the expected responses for design 2 under the assumption that pens do not have a confounding effect.

As for Display 7.5, an examination of Display 7.7 shows that the eight responses could be regarded as two samples each of size $n = 4$: (1) a sample consisting of four responses for diet 1 (the responses from animals 1, 3, 5, and 7); (2) a sample consisting of four responses for diet 2 (the responses from animals 2, 4, 6, and 8). Thus, the two-sample model in Equation (7.1) is appropriate in this case also. It follows that the independent-sample t-test in Section 4.3 or the Wilcoxon rank sum test in Section 5.5 can be used to make inferences about $\theta = \mu_1 - \mu_2$. However, design 1 would be the recommended design in this case, because it is usually easier to implement than design 2.

DISPLAY   7.7

Expected responses for design 2 when pens do not have a confounding effect

Pen	Animal	Diet	Expected response
1	1	1	μ_1
1	2	2	μ_2
2	3	1	μ_1
2	4	2	μ_2
3	5	1	μ_1
3	6	2	μ_2
4	7	1	μ_1
4	8	2	μ_2

Finally, if the pens are a potential source of confounding effects, and we assume that the expected response for diet i in pen j is $\mu_i + \rho_j$, then the model for Y_{ij}, the weight gain for the animal receiving diet i in pen j, can be expressed as

$$Y_{ij} = \mu_i + \rho_j + E_{ij}, \qquad i = 1, 2; \quad j = 1, 2, 3, 4. \tag{7.2}$$

Display 7.8 shows the expected responses for design 2 under the assumption that Equation (7.2) is the appropriate model.

DISPLAY 7.8

Expected responses for design 2 when pens could have a confounding effect

Pen	Animal	Diet	Expected response
1	1	1	$\mu_1 + \rho_1$
1	2	2	$\mu_2 + \rho_1$
2	3	1	$\mu_1 + \rho_2$
2	4	2	$\mu_2 + \rho_2$
3	5	1	$\mu_1 + \rho_3$
3	6	2	$\mu_2 + \rho_3$
4	7	1	$\mu_1 + \rho_4$
4	8	2	$\mu_2 + \rho_4$

The expected responses in Display 7.8 imply that, if the model in Equation (7.2) is appropriate—that is, if the pens have an additive effect—then the responses from design 2 should be regarded as eight samples, each of size $n = 1$, from eight populations. The population means are $\mu_1 + \rho_1$, $\mu_2 + \rho_1$, $\mu_1 + \rho_2$, $\mu_2 + \rho_2$, $\mu_1 + \rho_3$, $\mu_2 + \rho_3$, $\mu_1 + \rho_4$, and $\mu_2 + \rho_4$. Thus, design 2 will produce eight samples from eight populations, instead of the four samples from four populations obtained in design 1. The fact that both diets appear together in every pen gives design 2 a decided advantage over design 1. To see this, let's look at the within-pen differences, $D_j = Y_{1j} - Y_{2j}$, $j = 1, 2, 3, 4$, between the responses for the two diets. Consider D_1, the observed differential weight gain for the two diets in pen 1. The expected differential weight gain for pen 1 can be determined by subtracting the expected response for diet 2 in pen 1 from the expected response for diet 1 in pen 1. Thus, the expected differential weight gain in pen 1 is $(\mu_1 + \rho_1) - (\mu_2 + \rho_1) = \mu_1 - \mu_2 = \theta$. We see that the expected differential weight gain is $\mu_D = \mu_1 - \mu_2$ in every pen. Thus, a model for the observed differential weight gain in pen i can be written as

$$D_i = \mu_D + E_i, \qquad i = 1, 2, 3, 4, \tag{7.3}$$

where E_i is the error that accounts for the difference between the observed and expected differential weight gain in pen i.

The model in Equation (7.3) has the same structure as the one-sample model in Box 4.8. If it is reasonable to assume that the differential responses in the four pens are a random sample from a normally distributed population, procedures based on the one-sample t-test can be used to make inferences about $\mu_D = \mu_1 - \mu_2$. Alternatively, if the assumption of a normal distribution for the population of differential weight gains is too stringent, inferences based on the sign test (Box 5.1) or the signed rank test (Box 5.2) are also possible. Therefore, in the presence of pen-to-pen variation in the expected responses, design 2 is preferable to design 1.   ■

Example 7.10 shows how, under some conditions, the principle of blocking may be used to develop an experimental design that eliminates the effect of a confounding variable (pens) when comparing the expected responses for two treatments (diets). By randomizing the two treatments for the two animals within each pen, we make sure that experimental units within a block (pen) are exposed to a common level of the confounding variable. If the effects of diets and pens are additive, the data obtained from design 2 will eliminate pen effects when the mean responses for the diets are compared.

The lesson of Example 7.10 is that study designs must be planned carefully and statistical models can be used to that end. If the experiment is not designed properly, the results could have undesirable characteristics, as we saw when design 1 was adopted in the presence of a confounding effect of the pens.

Typically, when planning a new study, the research scientist or the statistician needs to ask questions such as the following:

1   What is the research question?

2   What is being measured?

3   What experimental material is available?

4   Are there any potential confounding variables?

5   What assumptions are reasonable for the measured values?

Answers to these questions can be used to develop a model for the data, which in turn can be used to select an appropriate study design.

The missing data problem

Even in a very carefully planned and conducted study, we can expect that responses to some treatments may be missing either because some experimental units failed to respond, or because of circumstances beyond the control of the experimenter. For instance, in a survey of yearly incomes of farmers, data for some farmers in the sample may be missing either because these farmers refused to divulge their incomes or because the investigators were not able to contact them. In a study of 4-week heights of plants treated with different treatments, missing data may result because some plants were destroyed by insects during the first week after planting. When missing data occur in a research study, the causes and pattern of the missing data should be taken into consideration for making appropriate changes in the planned analysis of the data.

Exercises

7.7 The following are two designs for a study in which the primary objective is to test the efficacy of a new treatment for relieving minor eye irritation.

1 Irritate one eye of each of 20 subjects by blowing a puff of air at sufficient pressure to cause minor irritation. Divide the subjects into two groups of ten. Treat each subject in group 1 with the experimental treatment and record the time taken for relief from eye irritation. The subjects in group 2 are not treated. This design will result in 20 responses with ten replications of each of the two methods (treatment, no treatment).

2 Irritate both eyes of each of ten subjects by blowing puffs of air at sufficient pressure to cause minor irritation. Treat one eye of each subject with the experimental treatment and observe the times to relief in the treated and untreated eyes. This design will result in 20 responses with ten replications of each of the two methods (treatment, no treatment).

Using statistical models as appropriate, discuss the advantages and disadvantages of the two designs, under suitable assumptions about the way the factor(s) and confounding variables are associated with the measured responses.

7.8 To compare the heights of three varieties of an ornamental plant at 4, 8, and 16 weeks, a horticulturist considers the following study designs.

1 Grow 24 plants in 24 pots. Divide the pots into three groups of eight—groups 1, 2, and 3. Record the 4-week heights of the plants in group-1 pots, the 8-week heights in group-2 pots, and the 16-week heights in group-3 pots. This study design will produce 24 responses with eight replications of each time period.

2 Grow eight plants in eight pots. For each plant, record the 4-, 8-, and 16-week heights. This study design will also produce 24 responses with eight replications of each time period.

Using statistical models as appropriate, discuss the advantages and disadvantages of the two designs, under suitable assumptions about the way the factor(s) and confounding variables are associated with the measured responses.

7.5
Determining sample sizes

As we saw in Section 7.3, determining the number of replications (the sample sizes) is an important step in designing a research study. Ideally, the procedure used for sample size determination should involve a cost-benefit analysis that balances the cost of increased sample size against the corresponding increase in available information. Unfortunately, this can be complicated, not only because it is difficult to specify realistic cost functions in a complex experiment, but also because there is no universally accepted criterion that can be used to quantify the available information.

In this section, we look at two common approaches to calculating sample sizes: the hypothesis-testing approach and the confidence-interval approach. As an illustration of these approaches, we consider experiments in which the primary objective is to compare two population means or two population proportions. The application of these two approaches to more complex experimental settings will be described in subsequent chapters.

In the hypothesis-testing approach, the sample sizes are chosen so as to ensure the required power at a specified parameter value for a particular test of a null hypothesis. In the confidence-interval approach, the sample sizes are chosen so as to ensure a desired bound on the confidence interval for a selected population parameter.

EXAMPLE 7.11   Let's consider the sample sizes for a study to compare the mean light frequencies emitted by two species of fireflies.

Obviously, the appropriate sample sizes will depend on the study objective. Let μ_A and μ_B, respectively, denote the mean frequencies (Hz) of the light emitted by the two species A and B. Given that the purpose of the experiment is to compare μ_A and μ_B, let's focus our attention on inferences about the parameter $\theta = \mu_A - \mu_B$. We could test an appropriate null hypothesis about θ, or we could estimate the value of θ using a confidence interval, in order to verify the research hypothesis that, say, the mean frequency for species A is higher than that for species B—in symbols, $H_1: \theta > 0$.

First, let's consider the hypothesis-testing approach. We want to test the null hypothesis $H_0: \theta = 0$ against the research hypothesis $H_1: \theta > 0$, at a particular level α. As we saw in Section 3.9, α is the probability of wrongly concluding in favor of H_1 and its value (say, $\alpha = 0.05$) is selected so as to minimize the likelihood of concluding in favor of a false research hypothesis.

On the other hand, a good study design should guarantee a reasonable success rate in the identification of a true research hypothesis, especially when there is substantial departure from the null hypothesis. Of course, what constitutes a reasonable success rate and substantial departure will vary from problem to problem. In the present example, we may want to design the study so that, if the difference between μ_A and μ_B were larger than, say, 1 Hz, the probability that the study would conclude in favor of the research hypothesis is, say, 0.80.

As we saw in Section 3.9, the probability of concluding in favor of a true research hypothesis—the power of the test—will typically increase with increasing sample size. In the hypothesis-testing approach to sample size determination, we select sample sizes to guarantee that our study has a specified power (say, 0.80) when the degree of departure from the null hypothesis (the amount by which μ_A exceeds μ_B) is more than the specified amount (say, $\Delta = 1$ Hz).

Next, consider the confidence-interval approach. If our primary concern is to estimate the difference $\theta = \mu_A - \mu_B$, we might start with the intuitive consideration that, if we have two confidence intervals for θ at the same confidence level, the narrower interval will provide more information about θ. Accordingly, in the confidence-interval approach to sample size calculation, we select a value for α—say, $\alpha = 0.05$—and choose the sample sizes so as to ensure that the width of a $100(1 - \alpha)\%$ confidence interval for the difference between μ_A and μ_B is less than a specified value W. As we'll see later in this section, specifying a bound on the width of a confidence interval can often be interpreted as specifying a bound on the error of estimation. ■

In Example 7.11, the parameter of interest was the difference between two population means. Clearly, however, the hypothesis-testing and confidence-interval approaches are applicable for any parameter θ.

The hypothesis-testing approach

In general, in the hypothesis-testing approach to sample size determination, we are interested in testing a hypothesis about a parameter θ at a particular level α. We want to be sure that, if the value of θ specified by the null hypothesis deviates from its true value by an amount larger than some *threshold value* Δ, the test will reject the null hypothesis with a probability of at least $1 - \beta$. In other words, the power of the test must be at least $1 - \beta$ if the value of θ under the null hypothesis deviates from its true value by more than Δ. In practice, the level α, the threshold value Δ, and the power $1 - \beta$ are specified by the investigators. Figure 7.1 illustrates how a threshold value Δ can be used to quantify the notion of substantial departures from the null hypothesis in three cases: H_{01}: $\theta \leq \theta_0$; H_{02}: $\theta \geq \theta_0$; and H_{03}: $\theta \neq \theta_0$. The θ values within regions A and B in Figure 7.1 indicate substantial departures from H_{01} and H_{02}, respectively. Important violations of H_{03} are indicated by values of θ in either A or B.

FIGURE   7.1

Rejection criteria based on a threshold value

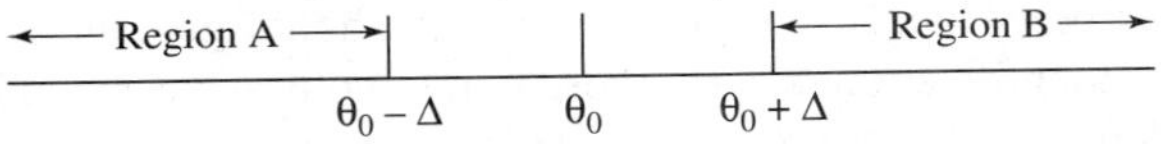

$\longleftarrow$ Region A $\longrightarrow$ | $\;$ | $\longleftarrow$ Region B $\longrightarrow$

$\theta_0 - \Delta \qquad \theta_0 \qquad \theta_0 + \Delta$

The actual method for computing sample sizes varies from problem to problem, depending on the study design, the type of measured responses, the interpretation of the parameter θ, and the test procedure. Sample size calculation methods when using a t-test of a hypothesis about a population mean or the difference between two population means are summarized in Box 7.1; the corresponding calculation methods when using a z-test to compare two population proportions are given in Box 7.2.

BOX **7.1**

Sample size calculation when using a t-test of a hypothesis about one or two population means

If θ is a population mean or the difference between two population means, consider the problem of testing a hypothesis about θ using a t-test. Let n be the number of measurements per sample needed to guarantee a probability of at least $1 - \beta$ that the test will detect a deviation of magnitude Δ from the null value, in the direction of the research hypothesis. Then n is the smallest integer such that

$$n \geq a\,[t(v,\ b\alpha) + t(v,\ \beta)]^2 \left(\frac{\sigma}{\Delta}\right)^2,$$

where σ is the population standard deviation; v denotes the degrees of freedom for the t-test; and

$$a = \begin{cases} 1 & \text{for a one-sample test} \\ 2 & \text{for a two-sample test;} \end{cases}$$

$$b = \begin{cases} 1 & \text{for a one-sided test} \\ \frac{1}{2} & \text{for a two-sided test.} \end{cases}$$

Comments on the sample size method in Box 7.1

1   A practical problem with the sample size procedure in Box 7.1 is that it calls for a good estimate of σ. In studies of variables that have been investigated by others, good estimates of σ can be based on past experience with similar data. When in doubt, it is best to be conservative and overestimate σ; overestimation of σ results in sample sizes that are larger than those actually needed. Estimating σ can be difficult in new studies, where very little information is available about the variability of measurements within the populations. In such cases, the need to specify σ can be eliminated by measuring the extent of departures from the null hypothesis in units of σ. We return to this topic later in the present section.

2   The degrees of freedom v depend on the sample size n: $v = n - 1$ for the one-sample t-test and $v = 2(n - 1)$ for the two-sample t-test.

3   This relationship between v and n implies that large sample sizes are associated with large degrees of freedom for t. On the basis that a t-distribution with large degrees of freedom can be approximated by a standard normal distribution, we can replace $t(v,\ b\alpha)$ and $t(v,\ \beta)$ by $z(b\alpha)$ and $z(\beta)$, respectively, for studies involving large sample sizes. In that case, the required sample size is the smallest n such that

$$n \geq a\,[z(b\alpha) + z(\beta)]^2 \left(\frac{\sigma}{\Delta}\right)^2. \tag{7.4}$$

Unlike the sample size inequality in Box 7.1, the expression on the right-hand side of Equation (7.4) does not involve n. Consequently, the smallest value of n satisfying Equation (7.4) is

$$n = a\,[z(b\alpha) + z(\beta)]^2 \left(\frac{\sigma}{\Delta}\right)^2. \tag{7.5}$$

4   A useful property of the sample size method in Box 7.1 is that, for given values of α and β, the sample size depends only on the ratio σ/Δ. This ratio is a measure of the minimum amount, in units of σ, by which research hypotheses of interest deviate from the null hypothesis. Thus, the sample sizes associated with $\Delta = 8$ and $\sigma = 2$ are the same as the sample sizes associated with $\Delta = 4$ and $\sigma = 1$.

Application of the method in Box 7.1

Let's look at an example.

EXAMPLE **7.12**   We'll use the hypothesis-testing approach to calculate the sample sizes for the study in Example 7.11. Recall that the hypothesis-testing objective of the study was to use a 0.05-level test of H_0: $\theta = 0$ against H_1: $\theta > 0$, where $\theta = \mu_A - \mu_B$ is the difference between the mean light frequencies emitted by two species of fireflies. We want to determine the sample sizes such that, if μ_A is 1 Hz more than μ_B, the probability of concluding that $\mu_A > \mu_B$ is at least 0.80.

If the independent-sample t-test in Box 4.3 is to be used in this study, the sample sizes can be calculated using the results in Box 7.1, with $\alpha = 0.05$, $1 - \beta = 0.80$, $\Delta = 1$, and an estimated value for σ. We take $a = 2$ because a two-sample t-test will be used and $b = 1$ because it will be a one-sided test. The two-sample t-test will have sample sizes $n_1 = n_2 = n$, and so the degrees of freedom for t will be $\nu = (n - 1) + (n - 1) = 2(n - 1)$.

It remains to determine a value for σ. Selection of σ is usually based on past experience or on preliminary data collected for this purpose. In the present case, experience with similar data suggests a standard deviation of the measured light frequencies in the range 0.75–0.80 Hz. We will use the conservative value $\sigma = 0.80$.

The determination of n in Box 7.1 is by trial and error. We start with a trial value of n and calculate the value of the expression

$$2 \left[t(2(n - 1),\ 0.05) + t(2(n - 1),\ 0.20) \right]^2 \left(\frac{0.80}{1} \right)^2$$

obtained by setting $\nu = 2(n - 1)$ in the right-hand side (RHS) of the inequality in Box 7.1. If n is larger than RHS, we try a smaller value and compare the new value of RHS with n. If n is smaller than RHS, we try a larger value of n. We continue this process until we find the smallest value of n satisfying the inequality in Box 7.1. Display 7.9 shows the required calculations if we start with a trial value of $n = 8$. As is clear from Display 7.9, the smallest n for which $n >$ RHS is $n = 9$. Therefore, 9 replications per species are required to attain the hypothesis-testing objectives of Example 7.11. ■

DISPLAY **7.9**
Sample size calculation

n	$\nu = 2(n-1)$	$t(\nu,\ 0.05)$	$t(\nu,\ 0.20)$	RHS	Comments
8	14	1.761	0.868	8.85	$n <$ RHS; try larger n.
10	18	1.734	0.862	8.63	$n <$ RHS; try smaller n.
9	16	1.746	0.865	8.73	$n >$ RHS; stop

Clearly, the trial-and-error procedure in Example 7.11 is best performed on a computer. Section 7.3 in the companion text by Younger (1997) describes an SAS program for the sample size calculations in Display 7.9.

In the following example, Box 7.1 is used to plan a study with large samples on the basis of a z-test of a hypothesis about a population proportion.

EXAMPLE 7.13   An opinion poll of hunters who had a license in previous years is to be conducted to see if they are in favor of extending the hunting season by two weeks, with an accompanying $5 increase in the license fee. Rather than conducting an expensive and time-consuming census of the total population, the poll takers will use a random sample of n hunters. Let π denote the proportion of hunters who favor the proposition, and consider testing H_0: $\pi \leq 0.50$ against H_1: $\pi > 0.50$. Determine n such that, if the true proportion in favor of extending the hunting season is more than 60%, then the probability that a 0.05-level test will conclude in favor of H_1 is at least 0.90.

As in Example 6.13 where we compared the proportions of uninfested plants, we have a discrete population of 0s and 1s; in this case, 0 represents a hunter who does not favor the proposition, and 1 represents a hunter who does favor the proposition. The population mean π is the parameter of interest. The population is not normal, and so we consider using a z-test based on a large sample.

The sample size calculation can be carried out using Equation (7.5). We have $\theta = \pi$, $\theta_0 = 0.50$, $\Delta = 0.10$, $\alpha = 0.05$, $1 - \beta = 0.90$, $a = 1$, and $b = 1$. Here $\Delta = 0.10$, because we would like to detect a difference of magnitude 0.10 from the null value 0.50. Also, because we have a one-sided, one-sample test, we set $a = b = 1$.

To determine an estimate of σ, we have two options. In both cases we use Equation 2.8 to obtain the expression $\sigma^2 = \pi(1 - \pi)$.

The first option is to guess a value $\hat{\pi}$ for π and use the estimate $\hat{\sigma}^2 = \hat{\pi}(1 - \hat{\pi})$. For example, if we guessed that approximately 70% favored the proposition, we would set $\hat{\pi} = 0.70$ and get $\hat{\sigma}^2 = 0.70(1 - 0.70) = 0.21$.

The second option is to adopt a conservative approach and choose the largest possible value for σ^2. As already noted, choosing a value of σ^2 in Box 7.1 that is larger than is necessary will increase the required sample size. Display 7.10 shows values of $\sigma^2 = \pi(1 - \pi)$ for several values of π.

As is clear from Display 7.10, the largest possible value of the variance $\sigma^2 = \pi(1 - \pi)$ of a 0–1 population is obtained when $\pi = 0.50$. In other words, the largest variance is obtained when exactly half of the population measurements equal 1. In that case, $\sigma^2 = 0.50(1 - 0.50) = 0.25$.

DISPLAY   7.10
Values of π and $\sigma^2 = \pi(1 - \pi)$

π	0.00	0.10	0.20	0.30	0.40	0.50	0.60	0.70	0.80	0.90	1.00
$\pi(1 - \pi)$	0.00	0.09	0.16	0.21	0.24	0.25	0.24	0.21	0.16	0.09	0.00

Using a conservative approach, we take $\sigma = \sqrt{0.25} = 0.50$ and find from Equation (7.5) that

$$n = [z(\alpha) + z(\beta)]^2 \left(\frac{\sigma}{\Delta}\right)^2 = (1.64 + 1.28)^2 \left(\frac{0.50}{0.10}\right)^2 = 213.16.$$

The required sample size is 214.   ∎

Example 7.13 shows how the results in Box 7.1 can be used to calculate sample sizes for large-sample inferences about a single population proportion. Although a similar approach can be used to determine sample sizes for large-sample inferences about the difference between two population proportions, such an approach is not recommended in practice, mainly because of the difficulties associated with estimating the variances of two 0–1 populations.

Other methods of sample size calculation

In Box 7.2, we present an alternative to the method in Box 7.1 for calculating the sample sizes when testing a hypothesis about the difference between two population proportions. The sample size formula in Box 7.2, which was derived by Casagrande, Pike, and Smith (1978), calls on the investigator to specify the values of π_1 and π_2 at which the desired power should be attained. Table A.3 in Fleiss (1981) lists sample sizes for various values of α, β, π_1, and π_2; for details on the derivation of the formula in Box 7.2, see Chapter 3 in Fleiss (1981).

In certain settings, the researcher can specify only the order of magnitude of the difference between the two proportions at which a specified power is desired, rather than the actual values of the proportions. In that case, Box 7.2 is not appropriate for sample size calculation. Instead, a set of tables in Chapter 6 of Cohen (1977) can be used.

BOX **7.2**   *Sample size calculation for specified power when using a z-test to compare two population proportions*

If $\theta = \pi_1 - \pi_2$ denotes the difference between two population proportions, consider the problem of testing the null hypothesis H_0: $\theta = 0$ at level α. Suppose we want to design a study in which the probability of rejecting the null hypothesis is at least $1 - \beta$ when π_1 and π_2 have specified values π_{10} and π_{20}, respectively. Let n be the number of replications per group needed for such a study. Then an approximate value for n is given by

$$n = A^2 \left[\frac{K}{2\Delta}\right]^2,$$

where

$$A = z(b\alpha)\sqrt{2\pi(1 - \pi)} + z(\beta)\sqrt{\pi_{10}(1 - \pi_{10}) + \pi_{20}(1 - \pi_{20})},$$

$$\Delta = \left| \pi_{10} - \pi_{20} \right|,$$

$$K = 1 + \sqrt{1 + \frac{4\Delta}{A^2}},$$

$$\pi = \frac{1}{2}(\pi_{10} + \pi_{20}),$$

and

$$b = \begin{cases} 1 & \text{for a one-sided test} \\ \frac{1}{2} & \text{for a two-sided test.} \end{cases}$$

Example 7.14 illustrates the use of the procedure in Box 7.2.

EXAMPLE **7.14**   Chemical control of undesirable plant life often results in environmentally hazardous side effects. A less harmful option is biological control. For instance, several researchers feel that the most effective method of biological control of water hyacinths in Florida is an integrated method using both insects and pathogens. The theory behind combining treatment with pathogens and exposure to insects is that insect infestation increases the probability that the weed will contract the disease caused by the pathogens. Thus, in integrated weed management programs, researchers want to know whether π_D, the proportion of insect-infested plants among the diseased plants, is higher than π_{DF}, the proportion of insect-infested plants among the disease-free plants.

A plant pathologist is interested in combining a pathogen A with an insect type B for the integrated biological management of a type of weed. To determine whether there is any advantage in combining A with B, the pathologist would like to test the null hypothesis H_0: $\pi_D = \pi_{DF}$ against the research hypothesis H_1: $\pi_D > \pi_{DF}$ at the $\alpha = 0.01$ level. He would like to design a study that requires two independent random samples of n plants, one from the population of diseased plants and the other from the population of disease-free plants. Obviously, the measured responses in such a study are nominal (1, infested plant; 0, uninfested plant) and produce counts of infested plants in each sample. Assume that the investigator wishes to test the null hypothesis H_0: $\pi_D = \pi_{DF}$ using the z-test in Box 6.5.

How many plants should the plant pathologist study from each of the two populations? To answer the question, the pathologist needs to specify values for π_D and π_{DF} that, if true, would be considered a practically important departure from the null hypothesis. The pathologist also needs to specify the desired probability of rejecting the null hypothesis if the specified values are actually true. Suppose that the researcher conjectures that, for the particular types of pathogen and insect in the study, π_D and π_{DF} would be in the neighborhood of 80% and 30%, respectively, and chooses a probability of at least 80% that the test will conclude in favor of H_1 if $\pi_D = 0.80$ and $\pi_{DF} = 0.30$.

The desired sample size can be calculated using the formula in Box 7.2 with $\alpha = 0.01$, $1 - \beta = 0.80$, $\pi_{10} = 0.80$, $\pi_{20} = 0.30$.

From Table C.1 (Appendix C), we get $z(\alpha) = 2.33$; $z(\beta) = 0.84$. Also, $\Delta = 0.80 - 0.30 = 0.50$, and $\pi = \frac{1}{2}(0.30 + 0.80) = 0.55$. Thus, from Box 7.2

$$A = 2.33\sqrt{2(0.55)(1 - 0.55)} + 0.84\sqrt{0.80(1 - 0.80) + 0.30(1 - 0.30)}$$

$$= 2.1503$$

$$K = 1 + \sqrt{1 + \frac{4(0.50)}{(2.1503)^2}} = 2.1969$$

and finally

$$n = (2.1503)^2 \left[\frac{2.1969}{2(0.50)}\right]^2 = 22.3161 \approx 23.$$

Therefore, 23 plants per sample are needed to attain the desired objectives.　■

The use of Box 7.1 to calculate the sample size in Example 7.12 depended on the fact that we planned to use the independent-sample t-test to compare the two populations. Implicit in the use of this test is the requirement that the data from the study satisfy the assumptions in Box 4.3. If these assumptions are deemed inappropriate, then alternative two-sample procedures—such as the Wilcoxon rank sum test for comparing two population medians in Box 5.5 or the χ^2-test in Box 6.8 for comparing two categorical populations—would be used to determine the sample sizes.

An approximate formula for determining sample sizes associated with the Wilcoxon rank sum test is given by Noether (1989), who also presents approximate sample size formulas associated with the sign and signed rank tests in Boxes 5.1 and 5.2, respectively.

Power analysis

The sample size calculation based on the inequality in Box 7.1 uses a direct method, in that the sample sizes are calculated for a single set of specifications for the type of test (for example, a two-sided independent-sample t-test), the level of the test (α), the power of the test ($1 - \beta$), and a quantitative measure of practically important difference between the null and alternative hypotheses (Δ). Precise specification of α, β, and Δ is difficult, and so it is helpful to analyze how the power of the test will vary for different choices of α, Δ, and n. On the basis of such an analysis, the researcher can determine whether a particular design is of adequate power and can also judge whether the cost and inconvenience of increasing the sample sizes are justified by the resulting decrease in α or increase in power. Thus, in *power analysis*, the power of a test is analyzed for a variety of sample sizes (n), a variety of violations of the null hypothesis (Δ), and a variety of Type I error rates (α).

Power analysis is best done using statistical computing software. Such software is becoming increasingly sophisticated. Currently, an attractive option is a set of SAS macros (OneWyPow, PowSetUp, FPowTab1, and so on) developed by O'Brien (O'Brien & Muller, 1993). These modules are easy to use and, at the time of writing, are available as shareware from their developer. Some relevant examples of their use can be found in the Appendix to Chapter 7 of the companion text by Younger (1997).

If we use the O'Brien SAS modules to determine the power of the independent-sample t-test when $\alpha = 0.01, 0.05$, $\sigma = 0.75, 0.80$, $\Delta = 1$, and $N = 14, 16, 20, 24$, where N is the total sample size ($N = 2n$), we obtain the following output:

```
****************************************************
        OneWyPow: Power Analysis Software for SAS
                     Ralph G. O'Brien
       Statistics/Biostatistics, University of Florida
****************************************************
```

ALPHA 0.05

		Std Dev							
		0.75				0.8			
		Total N				Total N			
		14	16	20	24	14	16	20	24
		Pow-er	Pow-er	Pow-er	Pow-er	Pow-er	Pow-er	Pow-er	Pow-er
Two-Group Test	2-tailed t	.630	.699	.805	.877	.575	.643	.753	.833
	1-tailed t	.760	.813	.889	.936	.712	.768	.851	.907

ALPHA 0.01

		Std Dev							
		0.75				0.8			
		Total N				Total N			
		14	16	20	24	14	16	20	24
		Pow-er	Pow-er	Pow-er	Pow-er	Pow-er	Pow-er	Pow-er	Pow-er
Two-Group Test	2-tailed t	.338	.412	.552	.671	.290	.357	.485	.601
	1-tailed t	.456	.533	.666	.769	.402	.473	.602	.707

In this printout, we see that a 0.05-level two-sided, independent-sample t-test applied to two samples, each of size 10 ($N = 20$), will have 80.5% power to detect a difference of $\Delta = 1$ unit in the population means if the population standard deviation is $\sigma = 0.75$ and 75.3% power if $\sigma = 0.80$.

Using Δ/σ to specify departures from H_0

One of the difficult tasks in the sample size procedure in Box 7.1 is to specify σ, the population standard deviation. However, if practically important deviations from the null hypothesis are specified in terms of Δ/σ, the sample size can be calculated from the formula in Box 7.1 without having to specify σ and Δ separately.

A probabilistic criterion for specifying departures from H_0 in terms of Δ/σ can be based on the result in Box 7.3, which is derived in Example B.8 (Appendix B).

BOX **7.3**

The probability that a randomly selected value from a $N(\mu_0 + \Delta, \sigma^2)$-distribution will exceed a randomly selected value from a $N(\mu_0, \sigma^2)$-distribution

Let Y_1 and Y_2 denote independently and randomly selected values from two normal populations, with respective means $\mu_0 + \Delta$ and μ_0 and a common variance σ^2. Then the probability that Y_1 will be at least as large as Y_2 is

$$\Pr\left(Y_1 \geq Y_2\right) = A\left(-\frac{\Delta}{\sigma\sqrt{2}}\right),$$

where $A(z)$ is the area to the right of z under a standard normal distribution.

To see how the result in Box 7.3 can be used to specify Δ/σ in sample size calculations, let's first suppose that we want to use an independent-sample t-test of H_0: $\mu_1 - \mu_2 \leq 0$ against H_1: $\mu_1 - \mu_2 > 0$. By letting $\mu_0 = \mu_2$ and $\mu_0 + \Delta = \mu_1$, the hypotheses H_0 and H_1, which then take the form H_0: $\Delta \leq 0$ and H_1: $\Delta > 0$, can be cast as hypotheses about the populations described in Box 7.3. For any specific value of Δ/σ, the result in Box 7.3 can be used to determine the probability that an observation from population 1 will be at least as large as an independent observation from population 2.

Table 7.1 shows values of $A(-\Delta/\sigma\sqrt{2})$ for selected Δ/σ. Let P denote the probability that a response from population 1 is at least as large as a response from population 2. Then Table 7.1 shows that the values of Δ/σ corresponding to H_0 imply that $P \leq 0.50$, whereas the values of Δ/σ corresponding to H_1 imply that $P > 0.50$. We see that P increases with increasing Δ, and so P can be used as the criterion for specifying practically important departures from H_0. For example, an investigator might decide to reject H_0: $\mu_1 \leq \mu_2$ if the probability that a response from population 1 exceeds a response from population 2 is sufficiently high—say, 90%. Now, Table C.1 (Appendix C) shows that $A(z) = 0.90$

TABLE 7.1
Probability that $Y_1 \geq Y_2$

Δ/σ	If H_0 is true $\Pr(Y_1 \geq Y_2)$	Δ/σ	If H_1 is true $\Pr(Y_1 \geq Y_2)$
−3.0	0.017		
−2.5	0.039	0.5	0.638
−2.0	0.079	1.0	0.760
−1.5	0.144	1.5	0.856
−1.0	0.240	2.0	0.921
−0.5	0.362	2.5	0.961
0.0	0.500	3.0	0.983

corresponds to $z = -1.28$, so that

$$\frac{\Delta}{\sigma\sqrt{2}} = 1.28$$

corresponds to the desired 90% probability. Consequently, the investigator's objective will be attained if the sample sizes are determined using the value $\Delta/\sigma = 1.28\sqrt{2} = 1.81$.

Similar reasoning can be used in the one-sample setting to specify departures from the null hypothesis in terms of Δ/σ. For example, consider the sample size problem when testing H_0: $\mu \leq 0$ against H_1: $\mu > 0$ using a one-sample t-test. If we let $\mu_0 = 0$ in Box 7.3, the probability that an observation from the population specified by the null hypothesis will exceed an observation from the population specified by the alternative hypothesis is found to be $A(-\Delta/\sigma\sqrt{2})$. Also, as the research hypothesis differs more from the null hypothesis, this probability will increase.

EXAMPLE 7.15 In Example 7.12, we determined the sample sizes such that, with probability 0.80, a 0.05-level independent-sample t-test will detect a difference of 1 Hz between population means. In the calculation, we assumed a value $\sigma = 0.80$ for the common standard deviation of the two populations.

Alternatively, we can calculate the sample sizes without assuming a specific σ. To do so, we need to quantify meaningful departures from the null hypothesis in terms of the probability that the light frequency emitted by a randomly selected firefly of species A will be higher than that emitted by a randomly selected firefly of species B. Thus, if Y_A and Y_B denote, respectively, the frequencies of light emitted by randomly selected fireflies of species A and B, we measure the degree of departure from the null hypothesis in terms of the value of $\Pr(Y_A \geq Y_B)$. Suppose we decide that we want a 0.05-level t-test to reject the null hypothesis H_0: $\mu_A \leq \mu_B$ if there is 85% chance that Y_A will be at least as large as Y_B. In other words, we want to reject H_0 if

$$\Pr(Y_A \geq Y_B) = A\left(-\frac{\Delta}{\sigma\sqrt{2}}\right) = 0.85.$$

From Table C.1, $A(z) = 0.85$ corresponds to $z = -1.04$, and so the required sample size can be determined from Box 7.1 with $\Delta/\sigma = (\sqrt{2})(1.04) = 1.47$.   ∎

The confidence-interval approach

In this approach, the sample sizes are determined so that, at a specified confidence level, the expected width of the confidence interval for θ is less than a specified value W. The procedure in Box 7.4 uses the confidence-interval approach to calculate sample sizes associated with confidence intervals based on t-distributions.

BOX **7.4**

A sample size procedure based on the confidence-interval approach

Let θ be a population mean or the difference between two population means, and assume that a confidence interval for θ can be constructed using a t-statistic. Let n be the number of measurements per sample needed to ensure that the width of a $100(1-\alpha)\%$ confidence interval for θ is less than W. Then n is the smallest integer such that

$$ n \geq b\,[t(\nu,\ \alpha/2)]^2 \left(\frac{\sigma}{B}\right)^2, $$

where σ^2 is the population variance; ν denotes the degrees of freedom for t; $B = W/2$; $1 - \alpha$ is the desired confidence coefficient; and

$$ b = \begin{cases} 1 & \text{if } \theta \text{ is a population mean} \\ 2 & \text{if } \theta \text{ is the difference between two population means.} \end{cases} $$

Comments on the confidence-interval procedure in Box 7.4

1 A $100(1 - \alpha)\%$ confidence interval for θ takes the form $(\hat{\theta} - \hat{B},\ \hat{\theta} + \hat{B})$, where $\hat{\theta}$ is the sample mean or the difference between two sample means and

$$ \hat{B} = t(\nu, \alpha/2)\hat{\sigma}_{\hat{\theta}}; $$

$\hat{\sigma}_{\hat{\theta}}$ is the estimate of the standard deviation of $\hat{\theta}$. The width of the confidence interval, which equals $\hat{W} = 2\hat{B}$, is a random variable, because the standard deviation of $\hat{\theta}$ is estimated from the observed data. The sample size formula in Box 7.4 is designed to ensure that the expected value of this random variable— that is, the average of the widths of the population of all $100(1 - \alpha)\%$ confidence intervals—is less than W.

2 The quantity B in the sample size formula may be interpreted as a bound on the error of estimation or the *margin of error* of estimation. We see from Figure 7.2 that, for each θ within the interval $(\hat{\theta} - \hat{B},\ \hat{\theta} + \hat{B})$, the quantity $|\theta - \hat{\theta}|$—that is, the magnitude of the difference between θ and $\hat{\theta}$—will be less than $\hat{B}$. We already know that a confidence coefficient of $(1 - \alpha)$ means that $100(1 - \alpha)\%$

FIGURE 7.2

The estimate $\hat{B}$ as a bound on the error of estimation

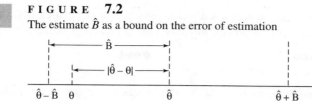

of the intervals contain the true value of θ. Consequently, ensuring that the expected halfwidth of the $100(1 - \alpha)\%$ confidence intervals is less than $\hat{B}$ is the same as ensuring that the expected error of estimation will be less than B with probability $1 - \alpha$.

3 As in Box 7.1, the population standard deviation is usually unknown and has to be estimated.

4 The degrees of freedom for t will depend on the sample size. If θ is a single mean, then $\nu = n - 1$; otherwise, $\nu = 2(n - 1)$.

5 When large sample sizes are acceptable, the sample size can be based on a standard normal distribution. The corresponding formula is

$$n = b\,[z(\alpha/2)]^2 \left(\frac{\sigma}{B}\right)^2. \tag{7.6}$$

Applications of the procedure in Box 7.4

Consider the following example.

EXAMPLE 7.16 For the light emission study in Example 7.11, let's calculate the sample sizes needed to ensure that the length of a 95% confidence interval for $\theta = \mu_A - \mu_B$ will be less than 2 Hz.

As in Example 7.12, we choose the conservative estimate $\sigma = 0.80$. Then the sample sizes can be calculated as in Box 7.4, with $\sigma = 0.80$, $b = 2$, $\nu = 2(n - 1)$, $W = 2$, and $1 - \alpha = 0.95$. The required sample size can be calculated by trial and error. The calculations are summarized in Display 7.11, where RHS $= 2[t(\nu,\ 0.025)]^2(\frac{0.80}{1})^2$; the first trial value is $n = 14$. We see from Display 7.11 that the required sample size is $n = 7$ replications per species. ∎

DISPLAY 7.11

Sample size calculation using the confidence-interval method

n	ν	$t(\nu,\ 0.025)$	RHS	Comments
14	26	2.056	5.41	$n >$ RHS; try smaller n
6	10	2.228	6.35	$n >$ RHS; try larger n
7	12	2.179	6.08	$n <$ RHS; stop

The next example shows how the sample size calculation is simplified when the confidence intervals can be constructed using procedures based on a standard normal distribution.

EXAMPLE 7.17   In Example 7.13, we used the hypothesis-testing approach to determine the sample size for a survey of hunters. Let's now determine the sample size needed to estimate the true proportion of hunters favoring the proposition with a 90% confidence interval of length less than 0.10.

Because we plan to use a large sample, Equation (7.6) can be used for the calculation. Taking the conservative estimate $\sigma = 0.50$ and the values $b = 1$, $W = 0.10$, and $1 - \alpha = 0.90$ in Equation (7.6), we obtain

$$n = [z(0.10/2)]^2 \left(\frac{2 \times 0.50}{0.10} \right)^2 = (1.64)^2 \left(\frac{2 \times 0.50}{0.10} \right)^2 = 268.96 \approx 269.$$

Thus, we need to survey 269 hunters to obtain a 90% confidence interval of length less than 0.10. In other words, if we use a sample of 269 hunters to estimate π, the true proportion of hunters who favor the proposition, we will have 90% confidence that our estimate is within 0.05 of the true value of π.   ∎

Exercises

7.9   For the pollution data in Example 4.8, use the pooled estimate of the common population standard deviation to determine sample sizes ensuring with 90% confidence that, if the actual difference between the true mean chemical levels in the downstream and upstream waters is more than 5 ppm, then a 0.05-level independent-sample t-test will conclude that the mean pollution level downstream is higher than the mean pollution level upstream.

7.10   By means of a statistical computing package or otherwise, conduct a power analysis of the study in Exercise 7.9. Consider values of $\sigma = 10, 11, 12$; $\alpha = 0.01, 0.05$; and $\beta = 0.10, 0.20$. Briefly summarize the results of your analysis.

7.11   For the pollution data in Example 4.8, use the pooled estimate of the common population standard deviation to determine the sample sizes ensuring with 95% probability that the difference between the true mean chemical levels in the downstream and upstream waters can be estimated with a margin of error less than 2 ppm.

7.12   Refer again to the pollution study in Example 4.8. Let P be the probability that the chemical level in a randomly selected water sample from downstream exceeds the level in a sample selected from upstream, and suppose that the state environmental regulatory agency requires $P \leq 0.60$.

    Calculate the sample sizes necessary to ensure that a 0.01-level independent-sample t-test has 80% chance of rejecting $H_0 : \mu_1 - \mu_2 \leq 0$ if P is actually higher than 0.60.

7.13   An experiment to compare the germination rates of seeds cultivated under two growth media gave the following results. 180 seeds out of 210 cultivated in medium I and 140 out of 196 cultivated in medium II germinated. Determine the sample size for a new study in which the objective is to estimate the germination rate for medium I with a margin of error less than 5%.

7.14 Refer to the seed germination experiment in Exercise 7.13. Suppose it is desired to plan a new study to compare the germination rate for medium I with that for a third medium (medium III). Determine the sample sizes such that, in a 0.01-level z-test, the null hypothesis—that there is no difference between the germination rates for the two media—will be rejected with 90% probability if, in fact, the two rates differ by more than 10%.

7.15 Refer to the infection rate data in Exercise 4.4. Using the estimated population standard deviation, determine the appropriate sample sizes to ensure 99% confidence that the true mean infection rate can be estimated with an error less than 0.5.

7.16 Refer to the infection rate data in Exercise 4.4. Determine the sample size such that, in a 0.05-level t-test, the null hypothesis H_0 that the true mean infection rate is less than 1 will be rejected with 90% probability if the actual infection rate is more than 2.

7.17 Refer again to the infection rate study in Exercise 4.4. Let μ denote the true mean infection rate. Suppose that we plan to use a 0.05-level t-test of H_0: $\mu \leq 1$ against H_1: $\mu > 1$. If there is a probability of at least 65% that the infection rate of a person selected at random from the study population exceeds the rate of a person randomly selected from a population with mean 1, we want to reject H_0 with a probability of at least 80%. Determine the corresponding sample size.

7.18 Explain why the sample size requirements will be the same for the following three specifications of the null hypothesis and the corresponding research hypothesis. Assume that an independent-sample t-test will be used in each case.

Specification	H_0	H_1	α	β	σ
1	$\mu_1 = 30, \mu_2 = 40$	$\mu_1 = 30, \mu_2 = 45$	0.05	0.20	2
2	$\mu_1 = 20, \mu_2 = 30$	$\mu_1 = 25, \mu_2 = 30$	0.05	0.20	2
3	$\mu_1 = 20, \mu_2 = 30$	$\mu_1 = 20, \mu_2 = 40$	0.05	0.20	4

7.6
Overview

In this chapter, we discuss some of the basic issues related to planning a research study. The primary focus in designing a study is to develop a measurement protocol such that the results will enable the investigator to answer the research questions in an efficient manner. There are two aspects to a good study design. First, it should take into consideration the influence of potential confounding variables. Ideally, a study design should enable the investigator to estimate the effects of experimental factors after eliminating the effects of confounding variables. Blocking is an effective technique for achieving this goal. In many instances, blocking is not feasible, however, and adjustments for the effects of confounding variables must be made in the statistical analysis of the data.

Second, a good study design will include a carefully chosen sample size. The idea is to minimize the sample size, while ensuring that it is large enough to guarantee a specified accuracy of selected estimators or to permit the detection of important departures from the null hypothesis with high probability. In the first case, the investigator determines the sample size needed to ensure a specific upper bound on the length of a confidence interval. In the second case, the investigator determines the sample size that will provide the hypothesis tests with sufficient power to detect practically important departures from the null hypothesis.

8 Single-Factor Studies: One-Way ANOVA

8.1
Introduction

The studies involving samples from one or two populations in Chapters 4–6 investigate the effects of a single experimental factor. The one-sample problem can be thought of as a problem involving only one level of a factor; the objective of the study is to make inferences about the population of responses for that particular level. Similarly, in studies involving samples from two populations, two levels of a single factor are considered. Often, the primary objective in such studies is to compare the populations of responses at the two levels under consideration.

EXAMPLE **8.1**   In Example 4.8, investigators obtained independent water samples from upstream and downstream locations in a river to study pollution by the waste discharged from a chemical plant. The data can be treated as responses at two levels of one experimental factor: sampling location in the river. The primary objective of the study was to compare the two populations of responses at the two selected locations (levels): upstream and downstream.   ■

In this chapter, some commonly used statistical procedures for comparing the populations of responses at t ($t \geq 2$) levels of a single factor will be presented, for the cases when the responses are measured on interval, ordinal, or nominal scales. As we'll see, the methods described in this chapter are natural generalizations of those in Chapters 4–6.

The simplest studies designed for the comparison of t populations are based on completely randomized designs. Although the methods of analyzing data from completely randomized designs can be used in more general settings, it is instructive to consider them first in this context.

8.2
Completely randomized designs

Let's begin with an example.

EXAMPLE **8.2**   Photomorphogenetic studies of plants are often conducted under a light condition called safelight, whose effect on certain plant properties is the same as the effect of darkness. The effect of two sources of safelight (A and B), each at two intensities (low and high), was compared with the effect of darkness (D) in the following experiment.

The experimenters chose 20 identical seedlings. Four randomly selected seedlings were grown under each of $t = 5$ treatments: D, AL (source A at low intensity), AH (source A at high intensity), BL (source B at low intensity), and BH (source B at high intensity). The heights (cm) of the 20 plants were measured after four weeks.

This is an experimental study involving 20 experimental units (seedlings) and $t = 5$ treatments. Each treatment is a level of a single factor: type of light condition. There are $n = 4$ replications per treatment. The plant height data, shown in Table 8.1, can be regarded as five random samples, each of size $n = 4$, from five populations.   ■

Because the treatments are assigned to experimental units completely at random, the design of the study in Example 8.2 is called a *completely randomized design*. In

TABLE   8.1

Four-week heights of plants grown in five light conditions

Treat-ment	Factor level	Response: four-week height (cm)				Number of replications	Total	Mean	Variance
1	D	32.94	35.98	34.76	32.40	4	136.08	34.02	2.7267
2	AL	30.55	32.64	32.37	32.04	4	127.60	31.90	0.8702
3	AH	31.23	31.09	30.62	30.42	4	123.36	30.84	0.1465
4	BL	34.41	34.88	34.07	33.87	4	137.23	34.31	0.1954
5	BH	35.61	35.00	33.65	32.91	4	137.17	34.29	1.5202
					Total	20	661.44		

a general completely randomized design (CRD), a fixed number t of treatments are randomly assigned to N experimental units in such a way that the i-th treatment is assigned to exactly n_i experimental units. Thus, n_i is the number of replications of the i-th treatment and

$$n_1 + n_2 + \cdots + n_t = N.$$

An experiment that uses a completely randomized design is called a *completely randomized experiment*.

Each treatment in the completely randomized design in Example 8.2 is replicated $n = 4$ ($n_1 = n_2 = \cdots = n_t = 4$) times. For obvious reasons, this is called a *balanced design*. Notice that, in a balanced CRD with n replications per treatment, the total number of experimental units will be $N = nt$.

Notation

Consider a balanced CRD with n replications per treatment. Let y_{ij} denote the j-th observed response for treatment i, $i = 1, 2, \ldots, t$; $j = 1, 2, \ldots, n$. Display 8.1 describes a convenient notation for representing some key statistics calculated using the y_{ij}.

The convention adopted in Display 8.1 is the same as that in Section 6.1. Subscript $+$ indicates that the quantity is the sum of the quantities with all possible values of the subscript that $+$ replaces. The corresponding mean is indicated by a bar over the symbol. Thus, y_{i+} and $\bar{y}_{i+}$ denote, respectively, the total and the mean of all responses for treatment i: $y_{i1}, \ldots, y_{ij}, \ldots, y_{in}$. The symbols y_{++} and $\bar{y}_{++}$ stand for the total and mean of all of the N responses. The variance calculated from the n responses for treatment i is denoted by s_i^2.

The lowercase letters y and s in Display 8.1 denote the observed values of certain random variables. For example, $\bar{y}_{1+}$ and s_1^2 are, respectively, the observed values of the random variables representing the mean ($\bar{Y}_{1+}$) and the variance (S_1^2) of the n responses to treatment 1. For instance, the data in Table 8.1 are the actually observed responses in a completely randomized design with $t = 5$, $n = 4$, and $N = 20$. Also, $y_{2+} = 127.60$, $\bar{y}_{2+} = 31.90$, $y_{++} = 661.44$, $s_2^2 = 0.8702$, and so on.

DISPLAY 8.1

Some statistics based on the responses in a balanced completely randomized experiment

Treatment number	1	...	j	...	n	Number of replications	Total	Mean	Variance
1	y_{11}	...	y_{1j}	...	y_{1n}	n	y_{1+}	$\bar{y}_{1+}$	s_1^2
2	y_{21}	...	y_{2j}	...	y_{2n}	n	y_{2+}	$\bar{y}_{2+}$	s_2^2
.	...	...	...	...	...	.	.	.	.
.	...	...	...	...	...	.	.	.	.
i	y_{i1}	...	y_{ij}	...	y_{in}	n	y_{i+}	$\bar{y}_{i+}$	s_i^2
.	...	...	...	...	...	.	.	.	.
.	...	...	...	...	...	.	.	.	.
t	y_{t1}	...	y_{tj}	...	y_{tn}	n	y_{t+}	$\bar{y}_{t+}$	s_t^2
All observations						$N = nt$	y_{++}	$\bar{y}_{++}$	s^2

Exercises

8.1   Which of the two designs in Exercise 7.7 corresponds to a completely randomized experiment? How many treatments are there in this design? How many replications per treatment are there?

8.2   Which of the two designs in Exercise 7.8 is completely randomized? Identify the treatments and the number of replications per treatment in this design.

8.3   In a study of the effect of glucose on insulin release, 12 identical specimens of pancreatic tissue were divided into three groups of four specimens each. Three levels (low, medium, high) of glucose concentration were randomly assigned to the three groups, and each specimen within a group was treated with its assigned concentration of glucose. The amounts of insulin released by the tissue samples are as follows:

	Concentration	
Low	Medium	High
1.59	3.36	3.92
1.73	4.01	4.82
3.64	3.49	3.87
1.97	2.89	5.39

Designate the low, medium, and high levels as treatment 1, treatment 2, and treatment 3, respectively. Adopting the notation in Display 8.1, calculate the following:

a   the treatment totals y_{i+}, $i = 1, 2, 3$;

b   the overall total y_{++} and the overall mean $\bar{y}_{++}$;

c   the treatment means $\bar{y}_{i+}$, $i = 1, 2, 3$;

d   the treatment variances s_i^2, $i = 1, 2, 3$.

8.4   The effects of three grazing practices on the bulk density of a particular type of soil were compared in the following study. Three roughly identical tracts of land, each containing the type of soil under investigation, were randomly assigned to the three grazing practices. After subjecting each tract to its assigned grazing practice for 12 weeks, the experimenters measured the bulk density (g/cm^3) of four randomly selected core samples from each tract, with the following results:

Treatment number	Treatment	Soil bulk density (g/cm^3)
1	Continuous grazing	1.83 2.01 1.94 1.79
2	Two-week grazing with a one-week rest	1.74 1.68 1.85 1.72
3	Two-week grazing with a two-week rest	1.53 1.60 1.56 1.62

a   Determine n, t, and N.

b   Calculate y_{i+}, $\bar{y}_{i+}$ for $i = 1, 2, 3$.

c   Calculate y_{++} and $\bar{y}_{++}$.

d   Calculate s_1^2, s_2^2, s_3^2, and s^2.

8.3
Analysis of variance (ANOVA)

Suppose that we have interval data from a completely randomized design with t treatments and n replications per treatment. Let $\mu_1, \mu_2, \ldots, \mu_t$ denote the expected responses for the t treatments. Then the t population means can be compared by testing the null hypothesis H_0: $\mu_1 = \mu_2 = \cdots = \mu_t$ (that is, the expected responses are the same for all treatments) against the research hypothesis H_1: $\mu_i \neq \mu_j$ for some $1 \leq i \neq j \leq t$ (that is, there is at least one pair of means μ_i, μ_j such that $\mu_i \neq \mu_j$). In the safelight experiment in Example 8.2, μ_i is the expected four-week height of a plant grown under the i-th type of light and the null hypothesis H_0 states that the expected heights are the same for all five types of light.

In the special case of $t = 2$ treatments, the problem of comparing t expected responses reduces to testing H_0: $\mu_1 = \mu_2$ against H_1: $\mu_1 \neq \mu_2$. If the two populations have normal distributions with a common variance, these hypotheses can be tested using a two-sided independent-sample t-test (Box 4.3). The calculated value of the test statistic for the independent-sample t-test can be expressed as

$$t_c = \frac{1}{\sqrt{\frac{1}{n_1} + \frac{1}{n_2}}} \frac{\bar{y}_{1+} - \bar{y}_{2+}}{s_{(p)}},$$

and a large absolute value $|t_c|$ provides evidence in favor of H_1. Hence, the independent-sample t-test compares $s_{(p)}$, a measure of the magnitude of the observed variability within samples, with $|\bar{y}_{1+} - \bar{y}_{2+}|$, a measure of the magnitude of the observed variability (difference) between the two sample means.

The *analysis of variance (ANOVA) method* for testing the equality of the expected responses for t treatments can be viewed as a direct extension of the independent-sample t-test. Like the independent-sample t-test, the ANOVA method compares a measure of the magnitude of the observed variability within the *t* samples with a measure of the observed variability between the means of *t* samples. Also like the independent-sample *t*-test, the ANOVA method for analyzing data from a completely randomized experiment requires some assumptions, known as the *ANOVA assumptions* (Box 8.1).

BOX **8.1** ***The ANOVA assumptions for a CRD with fixed treatments***

The ANOVA assumptions about the observed responses in a completely randomized experiment are as follows:

1 The populations have normal distributions.

2 The population variances are equal.

3 The observed responses are independent random samples from *t* populations.

As you see, the title of Box 8.1 refers to fixed treatments. The meaning of this expression will be explained in Chapter 14. In the present context, it indicates that the levels of the factor under study are selected (fixed) as a part of the study design. For instance, the design of the photomorphogenetic study in Example 8.2 has fixed treatment effects, because the five levels of the factor were selected by the experimenter as a part of the experimental design. Completely randomized experiments in which the levels of the factors being studied are selected as a part of the observational process will be described in Chapter 14. In such studies, the levels of the factor under investigation are regarded as a random sample from a larger population of levels.

Note that the ANOVA assumptions for a CRD with $t = 2$ treatments are the same as the assumptions for the independent-sample *t*-test.

The ANOVA *F*-test of H_0: $\mu_1 = \mu_2 = \cdots = \mu_t$

Assume that the ANOVA assumptions of Box 8.1 are satisfied by the observed responses in a completely randomized experiment with *t* treatments and *n* replications per treatment. How can we use the data to test the null hypothesis H_0: $\mu_1 = \mu_2 = \cdots = \mu_t$ against the research hypothesis H_1 that at least one pair of μ_i and μ_j are different?

Under the ANOVA assumptions, the observed responses can be regarded as *t* independent random samples from *t* normal populations with means $\mu_1, \mu_2, \ldots, \mu_t$ and a common variance σ^2. Let $\overline{Y}_{i+}$ denote the mean of the responses that will be observed for the *i*-th ($i = 1, 2, \ldots, t$) treatment. Then the results listed in Box 3.2 imply that the sampling distribution of $\overline{Y}_{i+}$ is a normal distribution with mean μ_i and variance σ^2/n. Thus, under the ANOVA assumptions, the mean of the responses that will be observed for the i-th treatment can be regarded as a measurement selected at random from a $N(\mu_i, \sigma^2/n)$-distribution.

Now suppose that the null hypothesis H_0 is true and assume that the expected responses for each of the t treatments is μ—that is, that $\mu_1 = \mu_2 = \cdots = \mu_t = \mu$. Then $\overline{Y}_{1+}, \overline{Y}_{2+}, \ldots, \overline{Y}_{t+}$, the means of the observed samples, can be regarded as a random sample of size t from a population with a $N(\mu, \sigma^2/n)$-distribution. Recall from Chapter 3 that the variance of a random sample is an unbiased estimate of the population variance. Therefore, the variance of the sample means

$$S_T^2 = \frac{(\overline{Y}_{1+} - \overline{Y}_{++})^2 + (\overline{Y}_{2+} - \overline{Y}_{++})^2 + \cdots + (\overline{Y}_{t+} - \overline{Y}_{++})^2}{t - 1}$$

is an unbiased estimate of σ^2/n, the variance of the population. Consequently, nS_T^2 is an unbiased estimate of σ^2. In other words, if H_0: $\mu_1 = \mu_2 = \cdots = \mu_t$ is true, then

$$\text{MS[T]} = nS_T^2 \tag{8.1}$$

is an unbiased estimate of the common population variance σ^2. The symbol MS[T] in Equation (8.1) is an abbreviation of *mean square for treatments*, a term commonly used to describe nS_T^2.

What happens if H_0 is not true? In Example B.4 (Appendix B), it is shown that, if H_1 is true, the mean square for treatments will be an unbiased estimate of a quantity larger than σ^2. This makes sense intuitively because, under H_1, the variability between sample means will be increased on account of differences between the corresponding population means. For a graphical illustration of why nS_T^2 would be larger under H_1 than under H_0, see Figure 8.1. In Figure 8.1a, $n = 5$ responses ($\times$) and their means ($\oplus$) are shown for each of $t = 3$ treatments when H_0 is true. The vertical line at μ marks the location of the common population mean relative to the three samples. The variance of the responses within each sample is an estimate of the common population variance σ^2. The variance of the sample means is less than any of the sample variances and is an estimate of σ^2/n, where $n = 5$.

In Figure 8.1b, we show the responses and the sample means if the population means are unequal. Notice that the sample variances are the same as in Figure 8.1a, but the variability among the sample means is larger. Clearly, differences between population means caused increased variation among sample means.

Suppose we can calculate, from the data, a quantity that is an unbiased estimate of σ^2 regardless of whether the population means are equal. Denote this quantity by MS[E] (an abbreviation of *mean square for error*). Then, if H_0 is true, both MS[T] and MS[E] are estimates of the same value σ^2. However, if H_1 is true, only MS[E] will estimate σ^2, and MS[T] will estimate a quantity larger than σ^2. In other words, MS[E] estimates σ^2 regardless of whether H_0 or H_1 is true, whereas the quantity estimated by MS[T] will be larger than σ^2 except when H_0 is true. Therefore, we would expect that the ratio

$$F = \frac{\text{MS[T]}}{\text{MS[E]}}$$

F I G U R E   8.1

Plot of the data from a CRD under H_0 and H_1

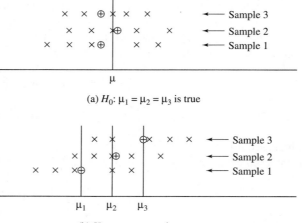

(a) H_0: $\mu_1 = \mu_2 = \mu_3$ is true

(b) H_0: $\mu_1 = \mu_2 = \mu_3$ is not true

will be close to 1 when H_0 is true and larger than 1 when H_1 is true. A test of H_0 against H_1 can be based on the test statistic F. The test will reject H_0 in favor of H_1 if F_c, the calculated value of F, is large. Of course, to implement the test procedure just described, we need:

a a method for calculating MS[E], the estimate of σ^2 irrespective of the validity of H_0;

b a criterion to determine whether the calculated value F_c provides sufficient evidence in favor of H_1 at the prescribed level.

To calculate MS[E], recall from the results in Box 4.3 that, when $t = 2$, the common population variance σ^2 can be estimated by pooling the two sample variances. Analogously, when $t > 2$, we can estimate σ^2 by pooling the t sample variances

$$MS[E] = \frac{(n-1)S_1^2 + (n-1)S_2^2 + \cdots + (n-1)S_t^2}{(n-1) + (n-1) + \cdots + (n-1)}$$

$$= \frac{(n-1)S_1^2 + (n-1)S_2^2 + \cdots + (n-1)S_t^2}{N-t}. \tag{8.2}$$

Now let's consider the test criterion. It can be shown that, when the ANOVA assumptions are satisfied, H_0 may be rejected at the α level if

$$F = \frac{MS[T]}{MS[E]} > F(t-1, N-t, \alpha), \tag{8.3}$$

where $F(t-1, N-t, \alpha)$ is the α-level critical value of the F-distribution with $t-1$ degrees of freedom for the numerator and $N-t$ degrees of freedom for the denominator (See Example B.16 in Appendix B for a proof when $t = 3$).

Exercises

8.5   Refer to Exercise 8.3.

a   Plot the data and the means as in Figure 8.1.

b   On the basis of your plot in (a), would it be reasonable to assume that the variances are the same for the three populations of insulin release measurements?

c   Assume that the variances of populations of insulin release measurements have a common value σ^2. Calculate an estimate for σ^2.

d   Let μ_1, μ_2, and μ_3 denote the true mean insulin releases at glucose concentrations 1, 2, and 3, respectively. On the basis of the plot in (a), do you think that the null hypothesis H_0: $\mu_1 = \mu_2 = \mu_3$ can be rejected?

e   Describe in words the appropriate conclusion when null hypothesis H_0 in (d) is rejected.

f   Describe the ANOVA assumptions (Box 8.1) as they pertain to these data.

g   Suppose that the ANOVA assumptions are satisfied by these data. Calculate the values of MS[T] and MS[E] and answer (d) on the basis of these calculated values.

8.6   Repeat Exercise 8.5 for the data in Exercise 8.4.

8.7   To evaluate the effects of high levels of copper in their feed, six chicks were fed a standard basal diet to which three levels of copper (0, 400, and 800 ppm) were added. The following data show the feed efficiency ratio (g feed/g weight gain) at the end of three weeks:

Copper level	Chick					
	1	2	3	4	5	6
0	1.57	1.54	1.65	1.57	1.59	1.58
400	1.91	1.71	1.55	1.67	1.64	1.67
800	1.88	1.62	1.75	1.97	1.78	2.20

Repeat Exercise 8.5 for these data.

8.4
Analysis of a CRD: Computational details

While useful for interpreting MS[T] and MS[E], Equations (8.1) and (8.2) are not ideally suited for computations, which are more easily performed using the following five steps.

Step 1.   Calculate the *correction for the mean*

$$\text{CM} = \frac{Y_{++}^2}{N}.$$

Step 2.   Calculate the *total sum of squares*

$$\text{SS[TOT]} = \sum_{i=1}^{t}\sum_{j=1}^{n} Y_{ij}^2 - \text{CM}.$$

Step 3.   Calculate the *treatment sum of squares*

$$SS[T] = \sum_{i=1}^{t} \frac{Y_{i+}^2}{n_i} - CM.$$

Step 4.   Calculate the *error sum of squares*

$$SS[E] = SS[TOT] - SS[T].$$

Step 5.   Calculate the *mean square for treatments* and the *mean square for error*

$$MS[T] = \frac{SS[T]}{t-1}, \qquad MS[E] = \frac{SS[E]}{N-t}.$$

The quantities $t - 1$ and $N - t$ are called the degrees of freedom for SS[T] and SS[E], respectively. Note that these computational formulas assume that there are n_i replications for the i-th treatment; consequently, for a balanced experiment with n replications per treatment, n_i should be replaced by n.

The various sums of squares calculated in these five steps are usually summarized as in Table 8.2, the *analysis of variance (ANOVA) table* for a CRD, which is constructed as follows:

1   To obtain the entries in the MS column, the SS column is divided by the *df* column; that is, MS = SS/*df*.

2   The first two entries in the Error row are obtained by subtracting the entries in the Treatment row from the corresponding entries in the Total row.

T A B L E   8.2
The ANOVA table for a CRD

Source of variation	Degrees of freedom (*df*)	Sums of squares (SS)	Mean squares (MS)
Treatment	$t - 1$	SS[T]	MS[T]
Error	$N - t$	SS[E]	MS[E]
Total	$N - 1$	SS[TOT]	

A comparison of the formula for SS[TOT] in step 2 with the computing formula for sample variance in Equation (3.3) shows that $SS[TOT]/(N - 1)$ is the variance of the N responses. Therefore, SS[TOT] is a measure of the overall variability of the observed responses. In Example B.4 (Appendix B), we see that

$$SS[TOT] = SS[T] + SS[E],$$

and consequently SS[TOT] is the sum of SS[T], a measure of variability between the observed treatment means, and SS[E], a measure of variability between the observed responses within treatments.

EXAMPLE 8.3 On the assumption that the populations of plant responses for the five light types have normal distributions with a common variance, do the data in Example 8.2 indicate that the expected plant heights for the five light sources are different? Let's use the ANOVA method to answer this question.

We need to test the null hypothesis $H_0: \mu_1 = \mu_2 = \cdots = \mu_5$ against the research hypothesis $H_1: \mu_i \neq \mu_j$ for at least one pair $i \neq j$. The calculations to construct the ANOVA table are as follows. We have $t = 5$, $n = 4$, and $N = 4 \times 5 = 20$. Thus

$$CM = \frac{(661.44)^2}{20} = 21{,}875.14,$$

$$SS[TOT] = (32.94)^2 + (35.98)^2 + \cdots + (32.91)^2 - CM = 57.46,$$

$$SS[T] = \frac{(136.08)^2}{4} + \cdots + \frac{(137.17)^2}{4} - CM = 41.08,$$

$$SS[E] = 57.46 - 41.08 = 16.38.$$

An examination of these calculations reveals a pattern that recurs in a variety of settings where the ANOVA principles are used. As a general rule, whenever a sum of measurements is squared, the resultant value has to be divided by the number of measurements in the sum. For instance, in the calculation of CM, $(661.44)^2$ is divided by 20, the corresponding number of measurements. Similarly, in the calculation of SS[T], the value $(136.08)^2$ is divided by 4, because four values were added to obtain 136.08. Finally, in the computation of SS[TOT], individual measurements (such as 32.94) were squared, so that the divisor for each squared value is 1.

The ANOVA table for the plant height data is as follows:

Source of variation	Degrees of freedom (*df*)	Sums of squares (SS)	Mean squares (MS)	F_c
Treatment	4	41.08	10.27	9.41
Error	15	16.38	1.09	
Total	19	57.46		

As we see from Table C.4 (Appendix C), the 0.01-level critical value of an F-distribution with four degrees of freedom for the numerator and 15 degrees of freedom for the denominator is $F(4, 15, 0.01) = 4.89$; this is smaller than $F_c = 9.41$, and so we reject H_0 at the $\alpha = 0.01$ level. We can conclude, at the $\alpha = 0.01$ level, that the expected heights of the plants grown under the five light types are not equal. ∎

The ANOVA F-test for the equality of treatment means in a completely randomized experiment can be performed using either the ANOVA or the GLM procedure in the SAS software package. The output resulting from the ANOVA procedure, shown in Section 8.5, states a p-value for the ANOVA F-test of 0.0005; this is the tail area to the right of 9.41 under an $F(4, 15)$-distribution. Thus, the null hypothesis can be rejected at a level as low as $\alpha = 0.0005$.

The next example illustrates the equivalence of the ANOVA F-test and the two-sided independent sample t-test in the case of a completely randomized experiment with two treatments.

EXAMPLE **8.4**   In Example 4.8, we analyzed the chemical concentration data from two locations in a river into which waste was discharged from a chemical plant. We constructed a 95% confidence interval for $\mu_1 - \mu_2$, where μ_1 and μ_2 were, respectively, the true mean concentrations at the downstream and upstream locations. Suppose that we now wish to test the null hypothesis H_0: $\mu_1 = \mu_2$ against H_1: $\mu_1 \neq \mu_2$ using the two-sample t-test as well as the ANOVA F-test.

Most of the calculations needed for the t-test are already available in Example 4.8. Let $\theta = \mu_1 - \mu_2$. Then we want to test H_0: $\theta = 0$ against H_1: $\theta \neq 0$. The test statistic for the t-test is

$$t = \frac{\hat{\theta} - 0}{\hat{\sigma}_{\hat{\theta}}},$$

where $\hat{\theta} = \bar{Y}_{1+} - \bar{Y}_{2+}$ and the formula for $\hat{\sigma}_{\hat{\theta}}$ is given in Box 4.3. From Example 4.8, we see that $\hat{\theta} = 29.88 - 23.55 = 6.33$ and $\hat{\sigma}_{\hat{\theta}} = 1.3633$, so that the calculated value of the test statistic is

$$t_c = \frac{6.33 - 0}{1.3633} = 4.643. \tag{8.4}$$

We reject the null hypothesis in favor of the research hypothesis at level α if $|t_c| \geq t(\nu, \alpha/2)$, where $\nu = n_1 + n_2 - 2 = 15 + 10 - 2 = 23$. Since 4.643 is larger than $t(23, 0.025) = 2.069$, we reject the null hypothesis at the $\alpha = 0.05$ level. In other words, a 0.05-level two-sample t-test leads us to the conclusion that $\mu_1 \neq \mu_2$.

To use the ANOVA F-test, we regard the data as arising from an unbalanced CRD with $t = 2, n_1 = 15, n_2 = 10$, and $N = 15 + 10 = 25$. The calculations are carried out using steps 1–5. The appropriate ANOVA table for the pollution data is as follows:

Source of variation	Degrees of freedom (df)	Sums of squares (SS)	Mean squares (MS)	F_c
Treatment	1	240.41	240.41	21.56
Error	23	256.47	11.15	
Total	24	496.88		

Since F_c is larger than the 0.05-level critical value of the $F(1, 23)$-distribution, we reject the null hypothesis H_0: $\mu_1 = \mu_2$ at the 0.05 level of significance and conclude that $\mu_1 \neq \mu_2$.

As expected, the two-sample t-test and the ANOVA F-test lead to the same conclusion. Indeed, when there are only two treatments, it can be shown that the two-sided t-test and the ANOVA F-test are equivalent.

Also, the mean square error MS[E] and the associated degrees of freedom are exactly the same as $S^2_{(p)}$ and its degrees of freedom in Example 4.8. In fact, both $S^2_{(p)}$

and MS[E] are the same quantity—the pooled estimate of σ^2, the common population variance. ■

The ANOVA F-test for a completely randomized experiment is based on the assumption that the number of replications of the factor levels (treatments) is fixed by the experimenter. Unfortunately, this ideal setting may not be possible in some studies, for the following reasons. First, measurements on some experimental units may be lost during the study period, on account of circumstances beyond the researcher's control. In the pollution study analyzed in Example 8.4, for instance, the researchers may have started with 15 water samples at each location, but five samples from downstream may have been accidentally lost or rendered useless by contamination during testing. Second, in some studies, the number of replications of the factor levels must be determined as part of the observational process. Example 8.5 describes a study in which the numbers of treatment replications are not under the investigator's control.

EXAMPLE **8.5** To assess the relationship between soil drainage class and high seasonal water table depth (HSWT), 99 well locations were monitored in eight counties of Florida (Sawka, Collins, & Rao, 1988). The soil drainage class and HSWT (in.) over the period of monitoring were recorded at each well location. The results, including the mean and the standard deviation (S.D.), are summarized in the following table:

Drainage class	Number of wells	Mean HSWT	S.D. of HSWT
Poorly drained	36	11.5	15.7
Somewhat poorly drained	37	55.0	31.4
Moderately well drained	26	96.4	41.8

The HWST data can be regarded as the result of a single-factor study with three levels of the factor, drainage class. In contrast to a completely randomized experiment, the numbers of replications of the factor levels are not fixed by the investigators. The levels at which the responses were measured and the associated number of replications were determined as a part of the observational process, because the actual levels investigated in the study depended upon the classifications of the 99 wells. ■

The research project in Example 8.5 is an example of an observational study. Observational studies of a single factor such as that in Example 8.5 have structures and objectives similar to experimental studies with a CRD, and the ANOVA F-test can be used in either case. A subtle difference in interpretation between ANOVA results from observational studies and those from completely randomized experiments is noted in Section 8.10.

Exercises

8.8 The following ANOVA table was obtained from a balanced completely randomized design:

Source	df	SS	MS	F_c
Treatments		126		
Error	20		16	
Total	23			

a Fill in the blanks in this table.
b Determine the number of treatments.
c Determine the number of replications per treatment.
d Perform a statistical test to see if there is a difference between the true mean responses to the treatment.

8.9 Suppose that the ANOVA assumptions in Box 8.1 are reasonable for the insulin release data in Exercise 8.3.

a Perform an analysis of variance to see if the data support the conclusion that the expected insulin release varies with the concentration of glucose.
b Write appropriate conclusions.

8.10 Suppose that the ANOVA assumptions in Box 8.1 are reasonable for the bulk density data in Exercise 8.4.

a Perform an analysis of variance to test the null hypothesis that the average bulk density of soil under the three grazing practices is the same.
b Write appropriate conclusions.

8.11 A study to investigate possible causes for high serum phosphate levels in adults with sickle cell anemia included nine patients with sickle cell anemia and seven healthy African American hospital employees with no sickle cell disease (Smith, Valika, et al., 1981). Among the nine sickle cell patients, five had high serum phosphate levels (> 4.5 mg/dliter); the other four had normal levels (3–4.5 mg/dliter).

For the 16 patients, the measured values of the tubular reabsorptive capacity for phosphate (TRCP), which is a measure of the phosphate reabsorption rate in the kidney, are as follows:

Group	Subjects	Number of subjects	TRCP values
1	Patients, high serum phosphate	5	5.36 4.56 4.66 4.51 4.12
2	Patients, normal serum phosphate	4	3.89 3.69 3.18 3.54
3	Normal controls	7	3.79 3.24 4.08 2.63 3.43 2.60 3.38

a Argue that these results should be regarded as data from an observational study in which the objective is to compare the population mean responses at three levels of a single factor. Identify the factors, their levels, and the number of replications for each level.

b Perform an analysis of variance to determine if the data indicate differences between expected TRCP values for the three subject groups. Write your conclusions.

8.12 Refer to Example 8.5.

   a Define symbols to denote the relevant population means, and use them to express the null hypothesis that there is no relationship between soil drainage class and mean HWST (high seasonal water table depth).

   b Use the given data for an ANOVA test of the null hypothesis. Note that MS[E] can be computed using Equation (8.2) modified for unbalanced CRD.

8.13 The common bean is often grown in soils with low phosphorus (P) availability. It is known that P deficiency may cause reduction in plant growth by inhibiting leaf expansion and photosynthesis. Accordingly, to determine how P applied to root medium or leaves influences vegetative growth, a study was conducted with bean plants growing in pots in a greenhouse (Lynch, Läuchili, & Epstein, 1991). Three methods of P application—P released to pots through alumina granules, P sprayed to shoots, and P applied by inoculation of pots—were studied. The study investigated the effects of 12 treatments resulting from combining levels of three experimental factors. Factor 1, the concentration at which P is released to the pots, had three levels: low, 0.4 μM; medium, 1.0 μM; and high, 27 μM. Factor 2, inoculation of the pots (adding to the pots the root debris of bean plants that had been cultivated with glomus macrocarpum), had two levels: yes and no. Finally, factor 3, foliar P application (spraying of the shoot to saturation) had two levels: yes and no. There were three replications for each treatment combination.

The mean and standard deviation (in parentheses) of the leaf area ratio (m^2/kg^{-1}) after 33 days of plant growth in the pots are as follows:

P release	Inoculation	Foliar P	
		No	Yes
Low	No	11.6 (1.04)	10.7 (0.69)
	Yes	12.8 (0.69)	13.5 (0.87)
Medium	No	13.0 (0.52)	12.7 (0.35)
	Yes	14.8 (0.87)	14.9 (0.69)
High	No	14.5 (0.69)	14.2 (0.35)
	Yes	15.7 (0.87)	15.9 (1.04)

   a State the null and research hypotheses about the treatment effects that can be tested using an analysis of variance of these data. State the hypotheses both in words and symbolically using appropriate notation to denote the population means of interest.

   b State the assumptions under which the ANOVA referred to in (a) can be justified.

   c Suppose that the assumptions in (b) are satisfied. Perform the analysis of variance and write your conclusions.

8.14 In a study of the mechanical and photosynthetic properties of sea-palm, a kelp that grows on the wave-swept rocky shores of the western United States, researchers divided the plants into three types depending on their surroundings and their height (Holbrook, Denny, & Koehl, 1991): understory plants, which are less than 25 cm tall and are located at least 30 cm (approximately) inward from the edge of a dense stand; canopy plants, which resemble the understory plants, except that they are more than 25 cm tall; and isolated plants, which are located at least 30 cm from their nearest neighbors. The values in the following table of treatment means are based on measurements of the blade tissue concentration of chlorophyll for a single healthy blade of each of the three plants selected from each of the three types of plants. The values of the chlorophyll concentration in this table are expressed as the ratio of the amount of chlorophyll (mg) to the fresh weight (g) of the blade.

Treatment Treatment number i	Understory 1	Canopy 2	Isolated 3
Number of replications n_i	3	3	3
Mean $\bar{y}_i$	0.466	0.353	0.140
Standard deviation s_i	0.1060	0.0270	0.0270

The corresponding (partial) ANOVA table is as follows:

Source	Sum of squares
Treatment	0.1644
Error	0.0042

a   Perform a test of the null hypothesis that the mean chlorophyll concentration of the blades does not differ between the three types of plants. State your assumptions.

b   Write your conclusions.

8.15   Refer to the data for the feed efficiency ratio in Exercise 8.7.

a   Perform an ANOVA F-test of the null hypothesis that the true mean feed efficiency ratio varies with the amount of copper in the diet. State your assumptions.

b   State the appropriate research hypothesis as well as your conclusions.

8.16   The following data are the crisis frequency rates (obtained by dividing the total number of painful crises by the number of months of observation) in 24 sickle cell anemia patients, 10 of whom were designated as having high serum phosphate levels (Smith, Valika, et al., 1981):

High phosphate	0.22	0.19	0.29	0.09	0.44	0.21	0.15	0.17	0.16	0.26				
Normal phosphate	0.18	0.06	0.10	0.04	0.07	0.07	0.06	0.02	0.03	0.06	0.02	0.09	0.04	0.01

a   State a set of assumptions under which the independent-sample t-test can be used to compare the mean crisis frequency of the high-P patients with that for the normal-P patients. Compare the assumptions for the t-test with the corresponding assumptions for an ANOVA F-test.

b Calculate the pooled estimate of the common population variance using two methods: (1) by pooling the two sample variances; and (2) by constructing an ANOVA table.

c Calculate the t- and F-statistics and verify the relationship $t^2 = F$.

d Discuss the implications of the relationship in (c) for tests of hypotheses about the difference between the true mean crisis frequencies of the two groups of patients.

8.17 In an observational study to assess the effects of menopause on cholesterol levels, a researcher measured the cholesterol levels of women in seven menopausal groups. The sample means, standard deviations, and the numbers of women in each group are as follows:

	Recent treatment	No recent treatment
Surgical menopause (> 40 years old)	$\bar{y}_1 = 225$ $s_1 = 45$ $n_1 = 11$	$\bar{y}_2 = 211$ $s_2 = 19$ $n_2 = 9$
Surgical Menopause (< 40 years old)	$\bar{y}_3 = 195$ $s_3 = 31$ $n_3 = 23$	$\bar{y}_4 = 248$ $s_4 = 133$ $n_4 = 3$
Natural menopause	$\bar{y}_5 = 232$ $s_5 = 30$ $n_5 = 13$	$\bar{y}_6 = 210$ $s_6 = 53$ $n_6 = 23$
Premenopause		$\bar{y}_7 = 162$ $s_7 = 24$ $n_7 = 21$

Note that the women were not randomly assigned to groups; they were simply observed without experimental manipulation. Accordingly, this is an observational study. The objectives and hypotheses of interest, however, were similar to those of an experimental study with a one-way completely randomized design. For example, the investigator wanted to know whether the average of the means of the first six groups was equal to the mean of the last group; methods of testing such contrasts will be presented in Chapter 9. Using the summary statistics in the table, perform an ANOVA test to see if there are any differences between the mean cholesterol levels of the seven groups.

8.5

The one-way classification model

Because data collected in single-factor studies can be conveniently classified into groups (populations) according to the levels of the factor being studied, the models used to describe such data are known as *one-way classification models*. In this section, we describe a one-way classification model that summarizes the ANOVA assumptions in Box 8.1.

The key point to remember when constructing a model for data collected in a completely randomized experiment is that such data can be viewed as independent random samples from t populations. The ANOVA assumptions in Box 8.1 imply that these populations are normally distributed with a common variance, and so the required model can be obtained by extending the model for two independent samples in Box 4.9 to the case of t independent samples. Thus, in a straightforward manner, the responses at the i-th level of the factor can be regarded as constituting a random sample from the i-th population, where the subscript i takes the values $1, 2, \ldots, t$.

Let Y_{ij} and μ_i denote, respectively, the j-th response and its expected value at the i-th level. In other words, μ_i is the mean of the population of all responses for the i-th population. Then, as in Box 4.9, the model for Y_{ij} can be expressed as

$$Y_{ij} = \mu_i + E_{ij}, \qquad j = 1, 2, \cdots, n_i; \quad i = 1, 2, \cdots, t. \tag{8.5}$$

Here E_{ij} is the random error in Y_{ij} (the deviation of the measured value from its expected value). The E_{ij} are assumed to be a random sample from a $N(0, \sigma^2)$ population; that is, the E_{ij} are independent $N(0, \sigma^2)$.

While Equation (8.5) is a natural way of writing a one-way classification model, an equivalent formulation that focuses on the differences between the expected responses μ_i, rather than on the μ_i themselves, is frequently used in practice. This alternative formulation is the basis of the language that users need in order to implement the procedures of popular statistical software packages such as SAS and SPSS and to understand their output.

Alternative formulation of the one-way ANOVA model

A model that emphasizes the differences between the population means can be obtained by expressing Equation (8.5) in terms of the deviation of μ_i from some baseline value μ. Of course, there are many ways of defining the baseline value, but the weighted average of the t population means is one of the most commonly used baseline values. Let

$$\mu = \frac{n_1 \mu_1 + n_2 \mu_2 + \cdots + n_t \mu_t}{N}, \tag{8.6}$$

where $N = n_1 + n_2 + \cdots + n_t$ is the total number of observations. Note that, if the sample sizes are all equal to n, then $N = nt$, and μ equals the average of the μ_i

$$\mu = \frac{\mu_1 + \mu_2 + \cdots + \mu_t}{t}. \tag{8.7}$$

The parameter μ in Equation (8.6) is the average expected value of the N observed responses and is known as the *overall mean*. The deviation of the expected response at level i from the overall mean is called the *effect of treatment i* and will be denoted by τ_i; that is,

$$\tau_i = \mu_i - \mu. \tag{8.8}$$

Note that, despite its name, τ_i is simply the effect of the i-th level of the single factor under investigation, whether the study is experimental or not; the factor levels do not have to be selected in advance. The following discussion is appropriate for any single-factor study.

By writing

$$\mu_i = \mu + \tau_i, \tag{8.9}$$

we see that Equation (8.8) is equivalent to expressing μ_i, the expected response for the i-th treatment, as the sum of two terms: μ, the overall mean expected response for the whole experiment; and τ_i, the deviation of the expected response for treatment i from the overall mean. If we substitute μ_i from Equation (8.9) into Equation (8.5), we get the equivalent model

$$Y_{ij} = \mu + \tau_i + E_{ij} \tag{8.10}$$

Box 8.2 contains a formal description of the one-way classification model in Equation (8.10) under the ANOVA assumptions of Box 8.1.

BOX **8.2**

> **The one-way ANOVA model with fixed treatment effects**
>
> Let Y_{ij} denote the j-th response for treatment i $(j = 1, 2, \ldots, n_i;\ i = 1, 2, \ldots, t)$. Then
>
> $$Y_{ij} = \mu + \tau_i + E_{ij},$$
>
> where μ is the overall mean expected response; τ_i is the effect of treatment i; and E_{ij} is random error in Y_{ij}. The treatment effects satisfy the constraint $n_1\tau_1 + \cdots + n_t\tau_t = 0$, and the E_{ij} are independent $N(0,\ \sigma^2)$.

Comments on the one-way ANOVA model in Box 8.2

1   The condition, $n_1\tau_1 + \cdots + n_t\tau_t = 0$ may seem rather strange at first. Without exploring the mathematical implications of this condition, we simply note here that it is implied by the definition of τ_i in Equation (8.8). If the sample sizes are equal—that is, if $n_1 = n_2 = \cdots = n_t = n$—this condition simplifies to $\tau_1 + \cdots + \tau_t = 0$.

2   From Equation (8.9), it can be shown that the null hypothesis that the true (population) treatment means are equal—H_0: $\mu_1 = \mu_2 = \cdots = \mu_t$—is equivalent to the null hypothesis that the treatment effects are all zero—H_0: $\tau_1 = \tau_2 = \cdots = \tau_t = 0$. Unfortunately, this equivalence has led to occasional misinterpretation of the conclusions from ANOVA F-tests. The null hypothesis H_0 that the treatment effects are all zero does not imply that no treatment has an effect on the response. Rather, H_0 implies that all treatments have the same effect on the response.

3   An important aspect of the model in Box 8.2 is the relationship of the differences between treatment effects to the differences between expected responses. Indeed, it can be verified from Equation (8.9) that for any pair of treatments—say, treatment i and treatment j—the following relationship holds

$$\mu_i - \mu_j = (\mu + \tau_i) - (\mu + \tau_j) = \tau_i - \tau_j. \qquad (8.11)$$

Thus, the difference between any two expected responses equals the difference between the corresponding pair of treatment effects. Consequently, inferences about the differences between the expected responses at different levels of the study factor are equivalent to inferences about the corresponding differences between treatment effects.

EXAMPLE **8.6**   The following is the one-way ANOVA model for the data from the photomorphogenetic study in Example 8.2

$$Y_{ij} = \mu + \tau_i + E_{ij} \qquad i = 1, \ldots, 5; j = 1, \ldots, 4. \qquad (8.12)$$

Here Y_{ij} is the measured four-week height of the j-th plant exposed to the i-th light source; μ is the overall mean (the expected average four-week height of the 20 experimental seedlings); τ_i is the effect of the i-th light source (the amount by which μ_i, the expected four-week height of a seedling exposed to the i-th light source, deviates from μ); E_{ij} is the error in the j-th response for the i-th light source (the difference between the measured and expected four-week heights of the j-th seedling exposed to the i-th light source).

The τ_i satisfy the condition: $\tau_1 + \tau_2 + \tau_3 + \tau_4 + \tau_5 = 0$, and the E_{ij} are independent $N(0, \sigma^2)$.   ■

ANOVA using SAS

As noted earlier, either the SAS ANOVA procedure or the SAS GLM procedure can be used to perform a one-way analysis of variance. In order to implement these procedures, we input the data with names for all the variables, plus three pieces of information: the name of the procedure to be used; the names of the factors (classification variables) whose levels are to be used to classify the data into groups; and the model that summarizes the underlying assumptions. For example, suppose we wanted to perform a one-way ANOVA of the plant height data in Example 8.2. Let HGT and LTSOURCE be the variables that denote the observed plant height and the type of light source. Then the procedure can be invoked by the codes

```
PROC ANOVA;
CLASS LTSOURCE;
MODEL HGT = LTSOURCE;
```

These codes imply that we want to implement the procedure called ANOVA; that our data contains one classification variable—namely, LTSOURCE; and that the model we want to use is HGT = LTSOURCE, which is a compact way of stating that, in our model, the plant height and light sources should be regarded as the response (or *dependent*) and factor (or *independent*) variables, respectively. For details on how to analyze the plant height data using SAS, consult Chapter 8 in the companion text by Younger (1997). A part of the output obtained from the SAS ANOVA procedure is as follows:

Analysis of Variance of Plant Height Data

```
             Analysis of Variance Procedure
                 Class Level Information
            Class     Levels    Values
            LTSOURCE     5      AH AL BH BL D


      Number of observations in data set = 20
```

Dependent Variable: HGT

Source	DF	Sum of Squares	Mean Square	F Value	Pr > F
Model	4	41.08077000	10.27019250	9.41	0.0005
Error	15	16.37655000	1.09177000		
Corrected Total	19	57.45732000			

The contents of the output are easily interpreted. For instance, the name of the classification variable is given as LTSOURCE, with five levels named AH, AL, BH, BL, and D. The variable HGT is the dependent (response) variable in this analysis. The output also shows the ANOVA table, including the *p*-value (in the column headed Pr > F) for the ANOVA *F*-test. It is clear that the null hypothesis of equal treatment effects can be rejected at the $\alpha = 0.0005$ level.

The output obtained from statistical software packages such as SPSS and BMDP will be similar. For information on how to enter and manipulate the computing codes, refer to the relevant user's manuals.

The assumptions underlying the model in Box 8.2 imply that the data satisfying the model can be classified into groups based on one possible source of variation—namely, the treatments. For this reason, the model in Box 8.2 is often called the *one-way analysis of variance model*. In the following chapters, we will see how the one-way ANOVA model can be extended to two-way and multiway models.

Exercises

8.18   Refer to Exercise 8.1. Write the appropriate model for the eye irritation measurements resulting from the CRD study. Interpret the parameters in the model you suggest.

8.19   Refer to Exercise 8.2. Write the appropriate models for the measured four-week heights resulting from the CRD study. Interpret the parameters in the models you suggest.

8.20   Write an appropriate model for the HSWT data in Example 8.5. Interpret the parameters in your model.

8.21   Write an appropriate model for the leaf area data in Exercise 8.13. Interpret the parameters in your model.

8.22   Write a model for the chlorophyll concentration data in Exercise 8.14. Interpret the parameters in your model.

8.23   Write a model for the feed efficiency data in Exercise 8.7. Interpret the parameters in your model.

8.24   Write a model for the data in Exercise 8.17 on the cholesterol levels of seven menopausal groups.

8.6
Checking for violations of assumptions

As noted in Chapter 4, the ANOVA assumptions for analyzing one-way classification data are a straightforward extension of the assumptions needed for the independent-sample t-tests. The effects of violations of ANOVA assumptions on the inferences based on ANOVA F-tests are very similar to the effects of similar violations on the independent-sample t-tests.

Generally speaking, the ANOVA F-test is not very sensitive to violations of the assumption of a normal distribution. The procedure is also insensitive to moderate violations of the equal-variance assumption, provided the sample sizes per group are equal and not too small. However, unequal variances can have a marked effect on the level of the test, especially if smaller sample sizes are associated with groups having larger variances.

There are a variety of procedures for testing whether the ANOVA assumptions are violated. Among these are the Shapiro-Wilk test and Lilliefors-Kolmgorov-Smirnov test of the normality of populations. The SAS package can be used to perform a combined version of these two tests of normality. Equality of variances can be evaluated by Bartlett's test or Hartley's F_{max}-test, among others. Unfortunately, both of these tests, like the F-test for comparing two population variances, assume that the populations are normal and can give misleading results when that is not the case.

Graphical methods based on residuals

With the expanded graphics capabilities of modern computers, graphical methods are increasingly popular as tools to check for violations of assumptions. Some of the most commonly used graphical methods for checking the ANOVA assumptions are based on graphs called *residual plots*.

Residuals

Let Y be a random variable denoting an observed response and $\hat{Y}$ be an estimate of its expected value; that is, $\hat{Y}$ is an estimate of the mean of the population from which

Y was selected. We refer to $\hat{Y}$ as the *predicted value* of Y. The difference between Y and its predicted value $\hat{Y}$ is called the *residual* of Y. In symbols, the residual of Y is e, where

$$e = Y - \hat{Y}. \tag{8.13}$$

Suppose that Y_{ij} satisfies the one-way ANOVA model in Box 8.2. Then

$$Y_{ij} = \mu + \tau_i + E_{ij}, \quad j = 1, \ldots, n_i; \ i = 1, \ldots, t. \tag{8.14}$$

Here $\mu_i = \mu + \tau_i$ is the expected value of Y_{ij}, as in Equation (8.9); E_{ij} is the random error. The residual of Y_{ij} is $e_{ij} = Y_{ij} - \hat{Y}_{ij}$, where $\hat{Y}_{ij}$ is an estimate of μ_i. Because μ_i is the mean of the population of all responses for the i-th treatment, an unbiased estimate for μ_i is $\overline{Y}_{i+}$, the observed mean response for the i-th treatment. Indeed, it can be shown that $\overline{Y}_{i+}$ is the best estimate of $\mu_i = \mu + \tau_i$ according to the theory of least squares. Therefore, the residual e_{ij} of the j-th observation for the i-th treatment is

$$e_{ij} = Y_{ij} - \overline{Y}_{i+}. \tag{8.15}$$

In Exercise 8.25, you will be asked to verify that, under the one-way ANOVA model in Box 8.2, the observed mean response for treatment i can be expressed as the sum of μ, τ_i, and $\overline{E}_{i+}$, where $\overline{E}_{i+}$ is the mean of the random errors associated with the n_i responses to the i-th treatment; that is

$$\overline{Y}_{i+} = \mu + \tau_i + \overline{E}_{i+}. \tag{8.16}$$

By subtracting Equation (8.16) from Equation (8.14), the residual of Y_{ij} can be expressed as

$$e_{ij} = E_{ij} - \overline{E}_{i+}. \tag{8.17}$$

Residual plots for testing normality

If the E_{ij} are a random sample from a population with zero mean, then $\overline{E}_{i+}$, the mean of the errors in the i-th sample, should be close to zero, especially if n_i, the number of replications for the i-th treatment, is large. Thus, we would expect e_{ij} in Equation (8.17) to be close to E_{ij}, so that, if the E_{ij} are a random sample from a $N(0, \sigma^2)$ population, then the e_{ij} should behave like a random sample from a $N(0, \sigma^2)$ population.

Graphical techniques to check whether a sample of residuals looks like a random sample from a normal distribution include the histogram, the stem-and-leaf plot, and the box-plot. Another important technique is the *normal q-q plot*. In a normal q-q plot, the residuals are plotted against a set of suitably chosen percentiles of the standard normal distribution. Under the assumption of normality, the normal q-q plot should approximate a straight line. A sigmoid plot is an indication that the population is heavy-tailed or light-tailed. Skewing of the population is indicated by concave (left skew) and convex (right skew) plots. To interpret a normal q-q plot, look at Figure 8.2. Both SAS and Splus have programs for the easy construction of normal q-q plots.

FIGURE   8.2

Interpretation of normal q-q plots

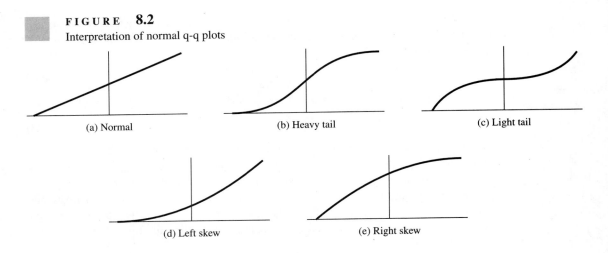

(a) Normal

(b) Heavy tail

(c) Light tail

(d) Left skew

(e) Right skew

The first step in the construction of a normal q-q plot is the computation of the *cumulative empirical probability* associated with every residual. The cumulative empirical probability associated with e_{ij} is defined as

$$p_{ij} = \frac{\text{Number of residuals} \le e_{ij}}{N + 1}.$$

As we saw in Section 5.2, the number of residuals that are less than or equal to e_{ij} is the same as the rank of e_{ij}, so that

$$p_{ij} = \frac{\text{Rank of } e_{ij}}{N + 1}.$$

For example, the cumulative empirical probability associated with a residual that is the sixth largest value (is of rank 6) in a set of $N = 10$ residuals is $p = 6/11 = 0.545$.

The normal q-q plot of a set of residuals is obtained by plotting the residuals e_{ij} against

$$q_{ij} = z(1 - p_{ij}), \tag{8.18}$$

where $z(\alpha)$ is the α-level critical value of a standard normal distribution. In Exercise 8.26, you will be asked to verify that q_{ij} in Equation (8.18) is the $100p_{ij}$ percentile of the standard normal distribution.

EXAMPLE **8.7**   Let's construct the box-plot and the normal q-q plot to see if the normality assumption appears reasonable for the plant height data in Example 8.2. Table 8.3 presents the plant height data, their residuals, and the associated percentiles. The box-plot and the normal q-q plot for the plant height data in Table 8.3 are shown in Figure 8.3. The two residual plots appear to suggest that the assumption of normality is reasonable for the plant height data. The corresponding plots drawn using SAS are given in Figure 8.7 of the companion text by Younger (1997). ■

TABLE   8.3

Residuals and the associated percentiles for the plant height data

i	j	Y_{ij}	$\hat{Y}_{ij}$	e_{ij}	Rank	p_{ij}	q_{ij}
1	1	32.94	34.02	−1.08	04	.1905	−0.876
1	2	35.98	34.02	+1.96	20	.9523	+1.668
1	3	34.76	34.02	+0.74	18	.8571	+1.067
1	4	32.40	34.02	−1.62	01	.0476	−1.669
2	1	30.55	31.90	−1.35	03	.1429	−1.067
2	2	32.64	31.90	+0.74	17	.8095	+0.876
2	3	32.37	31.90	+0.47	14	.6667	+0.431
2	4	32.04	31.90	+0.14	11	.5238	+0.060
3	1	31.23	30.84	+0.39	13	.6190	+0.303
3	2	31.09	30.84	+0.25	12	.5714	+0.180
3	3	30.62	30.84	−0.22	09	.4286	−0.180
3	4	30.42	30.84	−0.42	07	.3333	−0.431
4	1	34.41	34.31	+0.10	10	.4762	−0.060
4	2	34.88	34.31	+0.57	15	.7143	+0.566
4	3	34.07	34.31	−0.24	08	.3810	−0.303
4	4	33.87	34.31	−0.44	06	.2857	−0.566
5	1	35.61	34.29	+1.32	19	.9048	+1.309
5	2	35.00	34.29	+0.71	16	.7619	+0.712
5	3	33.65	34.29	−0.64	05	.2381	−0.712
5	4	32.91	34.29	−1.38	02	.0952	−1.309

FIGURE   8.3

Box-plot (a) and normal q-q plot (b) of plant height data

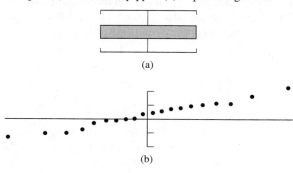

(a)

(b)

Residual plots for checking the equal-variance assumption

The residuals e_{ij} are expected to be close to the errors E_{ij}, and so it is reasonable to expect that the variance of a set of residuals will be close to the variance of the corresponding errors. Thus, the assumption of equality of population variances in the one-way ANOVA model could be checked by visual comparison of the variability of the residuals in the t treatment groups. One approach would be to plot the residuals e_{ij} against the corresponding group means $\bar{y}_{i+}$; in other words, we could plot the residuals against the corresponding predicted values. In what follows, the plot of the residuals against the predicted values will be called an $e \vee p$ plot. Figure 8.4 shows the $e \vee p$ plot for the plant height data.

F·I·G·U·R·E   8.4

$e \vee p$ plot for the plant height data

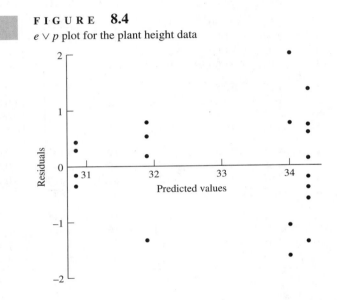

On the basis of the $e \vee p$ plot in Figure 8.4, it appears that the equal-variance assumption is somewhat suspect for the plant height data. Indeed, the plot suggests that the error variance increases with the expected response.

Exercises

8.25   Refer to the one-way ANOVA model in Box 8.2. Let $n_1 = n_2 = \cdots = n_t = n$. Show that, under the assumptions of the model, the mean response for the i-th treatment can be expressed as in Equation (8.16). [Hint: First verify that the sum of $Y_{i1}, Y_{i2}, \ldots, Y_{in}$ equals $n\mu + n\tau_i + E_{i+}$, where $E_{i+} = E_{i1} + E_{i2} + \cdots + E_{in}$.]

8.26   Use the definition of the $100p$ percentile of a population to show that q_{ij} in Equation (8.18) is the $100p_{ij}$ percentile of the standard normal distribution.

8.27   Use the residual plots for the insulin release data in Exercise 8.3 to check the ANOVA assumptions. Comment on the appropriateness of the ANOVA assumptions.

8.28   Refer to the bulk density data in Exercise 8.4. Use residual plots to check the ANOVA assumptions.

8.29 Refer to the TRCP data in Exercise 8.11. Use residual plots to check the validity of the ANOVA assumptions.

8.30 In a study of the effect of added dietary molybdenum (Mo) on the molybdenum concentrations in sheep kidneys, 20 sheep were randomly divided into four groups of five animals each (Pott, 1992). Diets supplemented with four levels of Mo (0, 15, 30, and 45 ppm) were randomly assigned to the four groups. The measured kidney Mo concentrations after 14 days of feeding are as follows:

Added molybdenum (ppm)	Kidney molybdenum concentration (ppm)				
0	2.8126	3.2084	3.5771	3.4228	4.0606
15	4.5645	2.7061	4.3560	3.3012	5.1091
30	4.4732	8.6097	4.9598	4.8945	3.6786
45	7.7524	5.9198	9.1892	9.7367	15.7416

Use residual plots to see if the ANOVA assumptions appear reasonable for these data.

8.7
Analysis of transformed data

There are two ways in which the ANOVA assumptions can be violated. First, the data may consist of measurements on an ordinal or a nominal scale; in that case, methods suitable for ordinal and nominal data are needed. Second, the data, although measured on an interval scale, may not satisfy at least one of the three requirements listed in Box 8.1. In that case, two options are available when selecting a data analysis procedure. One is to reduce the interval data to data measured on an appropriately chosen ordinal or nominal scale and to perform the analysis using methods designed for such data. The other possibility is to see whether the data could be transformed in such a way as to satisfy the ANOVA assumptions. If such a transformation can be found, the transformed data can often be analyzed by the ANOVA method.

Methods of analyzing ordinal and nominal data in the two-population ($t = 2$) setting were described in Chapters 5 and 6. The corresponding methods for the one-way ANOVA setting are presented in Sections 8.8 and 8.9. In the remainder of this section, we describe some common transformations that are useful for handling violations of ANOVA assumptions.

Let y denote a typical measurement in a set of data that do not satisfy the ANOVA assumptions. Let $g(y)$ be a transformation of y such that the data obtained by transforming every y to $g(y)$ satisfy the ANOVA assumptions. Then, provided inferences based on the transformed data can be translated into useful inferences about the population(s) of interest, analysis of the transformed data by ANOVA procedures is possible.

We know that the assumption of equal variances is essential to the performance of ANOVA procedures. Accordingly, a promising approach is to look for a transformation such that the transformed data satisfy that assumption. Fortunately, in many cases, a transformation that brings the population variances closer to each other also makes the data more closely resemble samples from normal populations.

A general method of determining a transformation suitable for a particular data set is presented in Section B.8 of Appendix B. According to this method, the transformation suitable for a particular data set is determined by the relationship between the mean and variance of the populations generating the data. Table 8.4 lists some useful transformations derived using this method.

TABLE 8.4
Some common transformations

Name	Relationship between σ^2 and μ	$g(y)$
Logarithmic	$\sigma^2 \simeq k\mu^2$	$g(y) = \log y$ or $g(y) = \log(1 + y)$
arcsin	$\sigma^2 \simeq k\mu(1 - \mu),$ $0 < \mu < 1$	$g(y) = \arcsin \sqrt{y}$
Square root	$\sigma^2 \simeq k\mu$	$g(y) = \sqrt{y}$ or $g(y) = \sqrt{y + 0.375}$
Power	$\sigma^2 \simeq k\mu^{2(1-\lambda)}$	$g(y) = \begin{cases} \log y & \text{if } \lambda = 0 \\ y^\lambda & \text{if } \lambda > 0 \end{cases}$

To select a transformation from Table 8.4, we need to know the relationship between the mean and variance. That relationship may be diagnosed in different ways. In some instances, a reasonable (for example, binomial, Poisson, or log-normal) population model may be assumed in order to deduce the required relationships. For example, if it is reasonable to assume Poisson distributions of the populations from which the samples are selected, then according to Box 2.6, the relationship between the means and variances of the populations takes the form $\sigma^2 = \mu$, which corresponds to $\sigma^2 = k\mu$ with $k = 1$. Thus, when the populations have Poisson distributions, the square root transformation $g(y) = \sqrt{y}$ is appropriate. If it is difficult to specify reasonable population distributions, as is usually the case when several populations are involved, residual plots are often used to determine suitable transformations.

The log transformation

The *log transformation* $g(y) = \log y$ is appropriate when the population variance is proportional to the square of the population mean

$$\sigma^2 = k\mu^2. \tag{8.19}$$

Notice that, according to Equation (8.19), the standard deviation of a measured value is proportional to the magnitude of its expected value. This is often the case in biology and the life sciences, where the standard deviation increases with the magnitude of the measurement. Because $\log 0$ is an undefined quantity, the transformation $\log(y + 1)$ is usually used instead of $\log y$ when some measurements in the data set are equal to zero.

The log transformation is indicated when the responses take only positive values and the associated random errors have a multiplicative effect, rather than an additive effect. In the context of the one-way ANOVA, the distinction between additive and multiplicative error effects can be illustrated as follows. The one-way ANOVA model

$$Y_{ij} = \mu + \tau_i + E_{ij}$$

in Box 8.2 assumes an additive error effect, because the model implies that an observed response Y_{ij} is the *sum* of the expected response $\mu_i = \mu + \tau_i$ and the error E_{ij}. The corresponding model in which the error effect is multiplicative is

$$Y_{ij} = \mu_i E_{ij}, \qquad \mu_i \neq 0. \tag{8.20}$$

Writing Equation (8.20) in the form

$$E_{ij} = \left(\frac{1}{\mu_i}\right) Y_{ij},$$

we see that, when the error effect is multiplicative, the error in a measurement is proportional to the measured value Y_{ij} (with proportionality factor $1/\mu_i$). In practice, such an assumption is often reasonable for measurements that are comparable in relative terms (for instance, concentrations and rates).

An important property of the log transformation is that the log of a ratio is equal to the difference between logs: $\log(a/b) = \log a - \log b$. Thus, measurements that are comparable in relative terms are ideally suited for ANOVA-based procedures after log transformation, because ANOVA procedures are designed for the comparison of means based on their differences.

Let σ^2 denote the variance of the E_{ij}. Then, the variance of the i-th population will be equal to the variance of $\mu_i E_{ij}$, which in turn will equal μ_i^2(variance of $E_{ij}) = \mu_i^2 \sigma^2$. Thus, if the population means μ_i are different, the population variances will also be different. Furthermore, if the observed responses are nonnegative and have multiplicative error effects, the log of the observed responses will have additive error effects

$$\log Y_{ij} = \log(\mu_i E_{ij}) = \log \mu_i + \log E_{ij}.$$

Thus, if the data satisfy a model with multiplicative error effects such that the errors have a log-normal distribution, then log-transformed data will satisfy the ANOVA assumptions.

The effectiveness of the log transformation as a means of meeting ANOVA assumptions does not depend on the base of the logarithm. However, a natural logarithm (to the base e) is often preferred over logarithms with respect to other bases, because small percentage increments on the relative scale are equivalent to approximately the same difference in the $\log_e$ scale. For example, if $a/b = 1.10$, then $\log_e a - \log_e b = 0.0953 \simeq 0.10$, which implies that, if a is 10% more than b, then $\log_e a$ exceeds $\log_e b$ by an amount approximately equal to 0.10. In the remainder of this text, unless otherwise stated, all logarithms will be to the base e.

EXAMPLE **8.8**   In Example 4.18, we constructed a 95% prediction interval for the survival time of mice inoculated with a predetermined dose of an antibiotic. The data were analyzed after a logarithmic transformation, because practical experience suggests that survival times are more likely to be log-normal than normal; that is, the log of the survival times is likely to be distributed as a normal distribution.   ∎

The arcsin transformation

The *arcsin transformation*—also called the inverse sine transformation—is a trigonometric function that can be used to transform a proportion between 0 and 1 into an angle expressed in degrees or radians (1 radian $= 57.2958$ degrees). Standard hand calculators and statistical computing packages can be used to implement the arcsin transformation for a given data set.

In Example B.2 (Appendix B), the mean and variance of the proportion of successes in a binomial experiment with n trials and probability of success π are shown to be $\mu = \pi$ and $\sigma^2 = (1/n)\pi(1 - \pi)$. Thus, if the observed responses can be expressed as the proportions of successes in t binomial experiments, each with n trials and probability of success $\pi_1, \ldots, \pi_t$, respectively, the data can be regarded as having arisen from populations in which the mean and variance of the populations satisfy the relationship $\sigma^2 = k\mu(1 - \mu)$, where $k = 1/n$ and μ is the population proportion. Consequently, it follows from Table 8.4 that, after the arcsin transformation $g(y) = \arcsin \sqrt{y}$, a data set consisting of the proportions of successes in binomial experiments will better satisfy the ANOVA assumptions. Because the arcsin is undefined for 0 and 1, Bartlett (1947) suggested that $0.25/n$ and $1 - 0.25/n$ be substituted for proportions of 0 and 1, respectively.

EXAMPLE **8.9**   In a study to compare the effects of three diets on the quality of chicken eggs, 24 birds were randomly divided into three groups of eight and assigned to one of the three diets: diet 1, diet 2, and diet 3. The numbers of eggs (out of a total of 25) classified as grade A for each bird are as follows:

Diet 1	21 18 21 24 22 20 21 20
Diet 2	20 17 19 14 19 12 16 11
Diet 3	22 24 21 22 24 23 21 23

These data can be regarded as the numbers of successes in binomial experiments with $n = 25$ trials (grading of the 25 eggs laid by each chicken); the probability of success is equal to the probability of laying a grade A egg. Let π_1, π_2, and π_3 denote, respectively, the probability of Grade A eggs under diets 1, 2, and 3. Then, to see if the data confirm the claim that the rates of grade A egg production are not the same under the three diets, we test the null hypothesis H_0: $\pi_1 = \pi_2 = \pi_3$ against the research hypothesis H_1: $\pi_i \neq \pi_j$ for some $i \neq j$.

Clearly, a one-way ANOVA of the raw data is inappropriate, because the populations are not normal (the samples are from binomial populations) and the population variances may not be equal (see Exercise 8.32). However, if we apply the arcsin transformation to the data, then a one-way ANOVA can be used to compare the means of the transformed populations. Let μ_1, μ_2, and μ_3 denote the means of the populations of transformed proportions. Then testing the null hypothesis H_0: $\pi_1 = \pi_2 = \pi_3$ is equivalent to testing H_0: $\mu_1 = \mu_2 = \mu_3$. Therefore, the required hypothesis tests can be performed using the transformed data.

The arcsin-transformed data are obtained by replacing each observed number of grade A eggs by its arcsin value. For example, the observed number 21 yields the proportion $21/25 = 0.84$, and so the transformed value is arcsin $\sqrt{0.84} = 66.4218°$. The complete set of transformed data is as follows:

Diet 1	66.4218	58.0519	66.4218	78.4630	69.7321	63.4349	66.4218	63.4349
Diet 2	63.4349	55.5501	60.6661	48.4461	60.6661	43.8538	53.1301	41.5539
Diet 3	69.7321	78.4630	66.4218	69.7321	78.4630	73.5701	66.4218	73.5701

The ANOVA table obtained when the SAS ANOVA procedure is applied to the transformed data is as follows:

```
-------------------------------------------------------------------

                    Analysis of Variance Procedure
             ANOVA for the arcsin-transformed egg data


Dependent Variable: Y

                            Sum of        Mean
Source              DF      Squares      Square     F Value   Pr > F
Model                2    1466.6688    733.3344      17.67    0.0001
Error               21     871.6818     41.5087
Corrected Total     23    2338.3506

-------------------------------------------------------------------
```

The p-value of 0.0001 indicates a significant difference between the transformed proportions; this suggests that the proportion of grade A eggs laid by the chickens on the three diets is not the same. ∎

Comments on the arcsin transformation

1   The arcsin transformation is particularly recommended for proportions in the range 0–0.20 or 0.80–1.00. No transformation is needed when the proportions are in the range 0.20–0.80.

2   The arcsin transformation is not appropriate for proportions that are not based on counts. For instance, if Y denotes the tissue water content, calculated as the ratio of the weight or volume of water to the total (tissue + water) weight or volume, then arcsin transformation is not recommended.

3   The arcsin transformation should not be used (without appropriate adjustments) to compare binomial proportions based on an unequal number of trials.

4   In any analysis of transformed data, we should remember that transformation is only useful if conclusions from the transformed data will lead to meaningful conclusions about the original data. For example, in a hypothesis-testing problem, we must make sure that rejection of the null hypothesis for the transformed data is equivalent to rejection of the null hypothesis for the original data. Similarly, point and interval estimates constructed from transformed data will be of practical use only if they can be transformed back to the original scale. For instance, it is not easy to interpret the estimate of the difference between population means of arcsin-transformed proportions (μ_1 and μ_2, in our notation) in terms of the difference between π_1 and π_2, the true population proportions. In such cases, it is better to look for methods of analysis that do not involve transformation.

The square-root transformation

The *square-root transformation* $g(y) = \sqrt{y}$ is appropriate when the means of the populations are proportional to the variances. A square-root transformation is commonly used to analyze count data from Poisson-distributed populations. If the data have many zero or near-zero counts, then $\sqrt{y + 0.375}$ transformation is recommended, rather than $\sqrt{y}$ transformation.

EXAMPLE **8.10**   The numbers of nematodes per soil sample taken from around citrus trees treated with four nematocides are as follows:

Nematocide 1	63 58 67 54 56 64 56 38
Nematocide 2	72 71 64 65 62 64 68
Nematocide 3	64 52 59 46 36 55 52 65
Nematocide 4	53 45 54 66 59 65 61 65

Assume that these are independent random samples from four Poisson populations with parameters λ_1, λ_2, λ_3, and λ_4, where λ_i is the mean number of nematodes per soil sample when nematocide i is applied to a population of soil samples.

Do these data indicate a difference in the effectiveness of the nematocides?

In Box 2.6, we saw that the variance of a Poisson population equals the population mean. Consequently, if the λ_i are unequal, the population variances will also be unequal. In that case, accordingly, the populations are not normally distributed and their variances are unequal.

We would like to test the null hypothesis H_0 that there is no difference between the four populations. Because the populations are discrete and their variances may not be equal, the one-way ANOVA is not recommended for testing H_0. However, it is reasonable to test H_0 by applying a one-way ANOVA to the data after square-root transformation, because we know that samples from Poisson populations are more likely to have equal population variances and to satisfy the normal-distribution assumption after such a transformation. Since no difference between the populations of nematode counts implies no difference between the corresponding populations of the square root of the counts, a one-way ANOVA can be used to test H_0 using transformed samples. You are asked to perform this analysis in Exercise 8.36. ■

The power transformation

The *power transformation*

$$g(y) = \begin{cases} \log y & \text{if } \lambda = 0 \\ y^\lambda & \text{if } \lambda > 0 \end{cases}$$

was developed by Box and Cox (1964), who described a method of estimating a value $\lambda \geq 0$ such that the transformed data correspond more closely to the ANOVA assumptions than the original data do. The Box-Cox power transformation includes the log and square-root transformations as special cases with $\lambda = 0$ and $\lambda = \frac{1}{2}$, respectively. For more details about the power transformation, see Draper and Smith (1981, Section 4.10).

Exercises

8.31 Use residual plots to see if the plant height data in Example 8.2 will satisfy the ANOVA assumptions more fully after log transformation.

8.32 Explain why the egg count data in Example 8.9 violate the equal-variance assumption.

8.33 Refer to the kidney Mo data in Exercise 8.30.

   a Examine the relationship between the mean and standard deviations to determine whether log transformation of the data is called for.

   b Compare the residual plots of the log-transformed data with the corresponding residual plots of the untransformed data.

8.34 Refer to the insulin release data in Example 8.3.

   a Examine the relationship between the sample mean and sample standard deviation to determine whether log transformation of the data will improve the validity of ANOVA assumptions.

   b Compare the residual plots of the log-transformed data with the corresponding plots in Exercise 8.27 and comment.

8.35 Refer to the bulk density data in Exercise 8.4.

   a Examine the relationship between the sample mean and sample standard deviation to determine whether log transformation of the bulk density data will improve the validity of the ANOVA assumptions.

   b Compare the residual plots of the log-transformed data with the corresponding plots in Exercise 8.28 and comment.

8.36 Use the square-root transformation to perform a one-way ANOVA of the nematode count data in Example 8.10 and write your conclusions.

8.8

One-way classified ordinal data

When one-way classified data can be regarded as independent samples from normal populations with equal variances, the test of the null hypothesis H_0 that the means of the t populations are equal can be carried out using the ANOVA F-test. Under the assumption that the populations have normal distributions with equal variances, the null hypothesis that the population means are equal is equivalent to the null hypothesis that the population distributions are identical. Therefore, the one-way ANOVA F-test is a test of the null hypothesis that the populations have identical normal distributions.

When the data are measured on an interval scale but the ANOVA assumptions are not satisfied, it may be possible to compare the populations by performing one-way ANOVA on suitably transformed data, as we saw in Section 8.7. In some circumstances, however, it may be preferable to employ specially designed analytical procedures that can be used on ordinal and nominal data.

Some one- and two-sample procedures for ordinal or nominal data were discussed in Chapters 5 and 6. In this section, we describe a procedure that utilizes one-way classified ordinal data to test the null hypothesis that the distributions of t ($t \geq 2$) populations are equal. In Section 8.9, we present the corresponding procedure for one-way classified nominal data. Unsurprisingly, the procedures we describe here are straightforward generalizations of the Wilcoxon rank sum test and the χ^2-test for comparing two populations described in Chapters 5 and 6.

The Kruskal-Wallis test

In the case of one-way classified interval data for which the ANOVA assumptions do not look reasonable, the null hypothesis that t populations have identical distributions can be tested using the *Kruskal-Wallis test*. The Kruskal-Wallis test, which is based on the ranks of the measured values, is a generalization of the two-sided Wilcoxon rank sum test in Box 5.5.

EXAMPLE **8.11**   To compare the resistance of a group of bacteria to an antibiotic, investigators tested five specimens of each of three strains against concentrations of the antibiotic and determined the lowest concentration that inhibited visible bacteria growth (the minimal inhibitory concentration). The following values were obtained for the minimal inhibitory concentration (mg/liter) of the antibiotic:

Strain	Specimen				
	1	2	3	4	5
1	20	25	30	28	26
2	8	10	2	8	11
3	40	35	30	20	30

We can regard these data as independent random samples, each of size $n = 5$, from three populations. If the populations are normally distributed with a common variance, then we can apply the ANOVA F-test to the null hypothesis that the populations of minimum inhibitory concentration levels of the antibiotic are the same for the three bacteria strains. However, because of problems associated with measuring low and/or high concentrations, ANOVA assumptions are often not satisfied for this type of data. One alternative is to look for a transformation that makes the ANOVA assumptions reasonable for the transformed data. Another is to proceed as in Chapter 5 and perform an analysis based on the ranks. ∎

As already noted, the Kruskal-Wallis test is an extension of the two-sided, two-sample Wilcoxon rank sum test to the case where we have t ($t \geq 2$) independent samples from t continuous populations. As in the two-sample Wilcoxon rank sum test, the assumed model can be quite general; we can simply require that, under the null hypothesis, the responses for the t treatments have a common continuous distribution.

To work in the setting of the one-way ANOVA model, the corresponding model for the Kruskal-Wallis test is similar, except that the assumption of a normal distribution for the errors is replaced by the assumption that the errors have a common continuous distribution. Box 8.3 summarizes the assumptions for the Kruskal-Wallis test of equal treatment effects under a one-way ANOVA type model.

BOX **8.3**   *Assumptions for the Kruskal-Wallis test*

Let Y_{ij} denote the j-th response from population i ($j = 1, 2, \ldots, n_i$; $i = 1, 2, \ldots, t$). Then

$$Y_{ij} = \mu + \tau_i + E_{ij}$$

where μ is the overall mean expected response; τ_i is the effect of treatment i; and E_{ij} is the random error in Y_{ij}.

The treatment effects satisfy the constraint $n_1\tau_1 + \cdots + n_t\tau_t = 0$, and the E_{ij} are independent and have a common continuous distribution.

Comments on the Kruskal-Wallis model in Box 8.3

1   A comparison of Box 8.3 with Box 8.2 shows that the main difference between the assumptions for the ANOVA F-test and the Kruskal-Wallis test is that the F-test assumes a normal distribution with a common variance for the errors, whereas the Kruskal-Wallis test only assumes identical continuous error distributions.

2   For nonnormal populations, as we know, the median is often a better indicator of location than the mean. Accordingly, the assumptions for the Kruskal-Wallis test and other rank tests are often stated in terms of population medians. Minor

modifications in Box 8.3 will yield such a model. All we need to do is to replace $\mu + \tau_i$, the true mean of population i, by the median of the i-th population. Then τ_i, the effect of treatment i, should be interpreted as the difference between the median of the i-th population and the overall median. When $\mu + \tau_i$ is the median of the i-th population, concluding that there is a difference between the populations is the same as concluding that the population medians are different.

3   Comparison with the two-sample model in Box 5.4 shows that the model in Box 8.3 is a straightforward generalization of the model underlying the Wilcoxon rank sum test.

EXAMPLE 8.12   A model for the minimal inhibitory concentration (MIC) data in Example 8.11 can be written as follows. Let Y_{ij} denote the j-th measured MIC for the i-th strain. Then

$$Y_{ij} = \mu + \tau_i + E_{ij},$$

where μ is a baseline MIC value (for instance, the overall median); τ_i is the effect of the i-th strain (the difference between μ and the true median MIC for the i-th strain); and E_{ij} is the random error (the difference between the j-th measured and true MIC for the i-th strain). The E_{ij} are a random sample from a continuous population. We do not assume that this continuous population is normally distributed.   ■

Like the Wilcoxon rank sum test, the Kruskal-Wallis test uses the sum of the treatment ranks when the combined sample of N observations is arranged in increasing order of magnitude. If there are ties in the data, the mid-ranks are used. The details of the Kruskal-Wallis test are presented in Box 8.4.

BOX 8.4

The Kruskal-Wallis Rank Test

Suppose we have one-way classified data satisfying the model in Box 8.3 and we wish to test the null hypothesis H_0 that the population distributions (treatment effects) are all equal against the alternative hypothesis H_1 that the population distributions are not all equal.

Let R_{ij} and R_{i+} denote, respectively, the rank of the j-th response for the i-th treatment in the combined sample of N responses and the sum of the ranks for the i-th treatment. Suppose there are g tied groups (see Section 5.2) with tie sizes $t_1, t_2, \ldots, t_g$. The Kruskal-Wallis test statistic for testing H_0 against H_1 is

$$K = \frac{K_1}{C},$$

where

$$K_1 = \frac{12}{N(N+1)} \sum_{i=1}^{t} \frac{R_{i+}^2}{n_i} - 3(N+1),$$

$$C = 1 - \frac{\sum\limits_{j=1}^{g} t_j (t_j^2 - 1)}{N^3 - N}.$$

The null hypothesis is rejected at level α if the calculated value K_c of the test statistic exceeds the α-level critical value listed in Table C.9 in Appendix C. If either t or $n_1, n_2, \ldots, n_t$ are outside the range in Table C.9, we reject H_0 at approximate level α if K_c exceeds $\chi^2(t-1, \alpha)$.

Comments on the Kruskal-Wallis test in Box 8.4

1 As in the case of the Wilcoxon rank sum test, the critical values in Table C.9 are computed on the assumption that there will be no ties among the observations. For data that can have ties, the test tends to be conservative; that is, the true critical value will be smaller than that in Table C.9. Thus, for tied data, a nominal 0.05-level test based on Table C.9 will have less than a 0.05 chance of rejecting a true H_0. The net effect of using a conservative test is that, when we reject the null hypothesis at a chosen level, we will have more confidence in our decision than the level indicates. On the other hand, if we find a lack of significance at a chosen level, we need to remember that the actual level of the test is lower than the chosen level.

2 The critical values in Table C.9 are given for $t = 3$ treatments with no more than five replications per treatment. The case of only two treatments is covered by the Wilcoxon rank sum test. For more than three treatments and/or more than five replications per treatment, we can use appropriate statistical software to perform an exact test or employ the $\chi^2(t-1)$ approximation suggested in Box 8.4.

3 Exact p-values associated with the Kruskal-Wallis test or good approximations can be obtained using the the KW option available in the StatXact procedures for tests on $R \times C$ contingency tables.

EXAMPLE 8.13 Let's use the Kruskal-Wallis test to see if the data in Example 8.11 indicate a significant difference between the distributions of minimum inhibitory concentrations for the three bacteria strains. We have $t = 3$ treatments, and $n_1 = n_2 = n_3 = 5$ replications per treatment. Arranging the data in increasing order of magnitude as

$$2 < 8 = 8 < 10 < 11 < 20 = 20 < 25 < 26 < 28 < 30 = 30 = 30 < 35 < 40,$$

we find that there are $g = 11$ groups

$$2;\ (8, 8);\ 10;\ 11;\ (20, 20);\ 25;\ 26;\ 28;\ (30, 30, 30);\ 35;\ 40,$$

with tie sizes

$$t_1 = 1;\ t_2 = 2;\ t_3 = t_4 = 1;\ t_5 = 2;\ t_6 = t_7 = t_8 = 1;\ t_9 = 3;\ t_{10} = t_{11} = 1.$$

The ranks of the $N = 15$ MIC values for the three strains of bacteria are as follows (tied observations are assigned their mid-ranks):

Strain	Specimen					Total
i	1	2	3	4	5	R_{i+}
1	6.5	8.0	12	10	9.0	45.5
2	2.5	4.0	1.0	2.5	5.0	15.0
3	15	14	12	6.5	12	59.5

Calculation of the test statistic in Box 8.4 is straightforward. First we calculate K_1

$$K_1 = \frac{12}{15(15+1)} \left[\frac{45.5^2}{5} + \frac{15.0^2}{5} + \frac{59.5^2}{5} \right] - 3(15+1) = 10.355.$$

Next we calculate the correction for ties. If $t_j = 1$, then $t_j(t_j^2 - 1) = (1)(1^2 - 1) = 0$, and so we have

$$C = 1 - \left[\frac{2(2^2 - 1) + 2(2^2 - 1) + 3(3^2 - 1)}{15^3 - 15} \right] = 0.9893.$$

The calculated value of the test statistic is

$$K_c = \frac{K_1}{C} = \frac{10.355}{0.9893} = 10.467.$$

Referring to Table C.9 with $n_1 = n_2 = n_3 = 5$, we see that the critical value at the $\alpha = 0.01$ level is 7.98, so that the p-value associated with the calculated value of 10.467 is less than 0.01. Such a small p-value leads us to conclude that there is a difference in the MIC distributions of the three strains. Under the assumptions implied by our model in Example 8.12, we may conclude that the median inhibitory concentrations for the three strains are not equal.

Now let's see what happens if we use the χ^2-approximation. From Table C.3, we see that the 0.01-level critical value of a χ^2-distribution with $k - 1 = 3 - 1 = 2$ degrees of freedom is 9.21, leading us to the same conclusion as before.

The following is a portion of the StatXact output obtained using the KW option in tests on $R \times C$ tables.

```
                    KRUSKAL-WALLIS TEST


               Minimal Inhibitory Concentration Data


Statistic based on the observed   3 by  11 table (x):
          KW(x) = Kruskal-Wallis statistic =   10.47


Asymptotic p-value: (based on Chi-squared distribution with    2 df)
          Pr { KW(X) .GE.  10.47    } =      0.0053


Exact p-value and point probability:
```

```
Pr { KW(X) .GE.  10.47   } =   0.0005
Pr { KW(X) .EQ.  10.47   } =   0.0002
```

As we see from the StatXact output, the exact p-value is 0.0005, whereas the approximate p-value based on the large-sample χ^2-approximation is 0.0053.   ∎

More information about the rank analysis of one-way classified ordinal data is available in texts on nonparametric methods, such as Hollander and Wolfe (1973) or Sprent (1993).

Exercises

8.37   Refer to the data in Exercise 8.3 concerning the effect of glucose concentration on insulin release in pancreatic tissues.

   a   Perform the Kruskal-Wallis test to see if the true median insulin releases differ between concentrations.

   b   Write a model that forms a basis for the test in (a). Interpret the parameters in your model.

   c   On the basis of your model in (b) and the Kruskal-Wallis test in (a), what can you say about the differences between the true mean MIC values for the three strains?

   d   Compare the result in (c) with the result of the ANOVA test in Exercise 8.9. What can you say about the validity of these two results?

8.38   Refer to the data in Exercise 8.4 concerning the effect of grazing pressure on soil bulk density.

   a   Perform the Kruskal-Wallis test to see if the true mean soil bulk density depends on the grazing pressure.

   b   Write a model that forms a basis for the Kruskal-Wallis test to compare the population means.

   c   Compare the result in (a) with the result of the ANOVA F-test in Exercise 8.10.

8.39   Refer to Exercise 8.11, concerning the effect of serum phosphate level on renal phosphate reabsorptive capacity (TRCP).

   a   Perform the Kruskal-Wallis test to see if the true median TRCP values vary depending on the serum phosphate level.

   b   Write a model that forms a basis for the Kruskal-Wallis test to compare the population means.

   c   Compare the result in (a) with the result of the ANOVA F-test in Exercise 8.11.

8.40   Refer to Exercise 8.30, concerning kidney Mo levels in sheep.

   a   Perform the Kruskal-Wallis test to see if the kidney Mo concentrations change with the level of added dietary Mo.

   b   Describe conditions under which the test procedure in (a) would be preferable to an ANOVA F-test.

8.41 In an experiment to see if the color of ice cream has any effect on its palatability, a judge was asked to rank nine samples of vanilla ice cream in increasing order of acceptability, with 1 as the most favorable rating and 9 as the least favorable rating. Three of the samples were red (R), three were white (W), and three were yellow (Y). The following are the judge's ratings and the color of the nine samples:

Sample	1	2	3	4	5	6	7	8	9
Color	R	W	W	Y	R	R	Y	W	Y
Rating	3	7	8	6	4	2	1	9	5

a Perform an appropriate test to see if the color of ice cream affected its acceptability.

b State the null and research hypotheses that you are testing, as well as the assumptions that you are making.

8.42 In a study to compare four methods of treating a type of bone fracture, each method was used on five subjects. Four weeks after the initial treatment, a radiologist was asked to examine each patient's X-ray and to rate the degree of healing on a five-point scale:

-2 = no healing at all; alternative procedures should be used right away;

-1 = some progress; let's monitor the subject for one more week;

0 = the fracture is healing, but slowly;

1 = good progress; should be able to discontinue treatment soon;

2 = complete healing; discontinue the treatment right away.

The following results were obtained:

Treatment method	Degree of healing				
1	-1	0	-2	1	2
2	2	2	-1	1	2
3	2	-1	1	0	1
4	0	0	-1	-1	-2

a Use the Kruskal-Wallis test to test the null hypothesis H_0 that the four methods of treating fracture are no different from each other against the research hypothesis H_1 that the four methods do not give the same results. Write your conclusions.

b Explain why it is not appropriate to use the ANOVA F-test for the null hypothesis in (a).

8.43 Lethal doses of the toxic drug adriamycin were given to 64 rats. These 64 rats were then randomly assigned to four treatment groups. The survival time of each rat was recorded, with the following results:

Treatment	Survival times (hours)
1	66 72 72 95 95 95 95 95 98 116 116 116 116 116 116 138
2	72 95 95 95 95 95 95 95 116 116 116 138 143 163 163 192
3	66 72 72 72 95 95 95 95 95 95 95 116 116 116 138 360
4	72 95 95 95 95 95 116 116 116 138 360 360 360 368 360 274

Perform the most appropriate analysis for comparing the four survival distributions. Justify your choice of methods.

8.44 Refer to the MIC data in Example 8.11.

a Use residual plots to see if log transformation of the data improves the validity of the ANOVA assumptions.

b On the basis of your conclusions in (a), perform an appropriate test on the transformed data to see if the mean minimum inhibitory concentrations differ for the three groups.

8.9
One-way classified nominal data

Let's begin with an example of a one-way classified nominal data set.

EXAMPLE **8.14** To determine whether the distribution of blood types is different in three ethnic subgroups of a population, blood samples from 100 subjects within each subgroup were classified according to blood type. The following results were obtained:

Group	Blood type				Total
	O	A	B	AB	
1	44	31	14	11	100
2	38	39	10	13	100
3	56	26	12	6	100
Total	138	96	36	30	300

We wish to use the data to test the null hypothesis that the distributions of subjects across the blood type categories are identical for the three ethnic groups.

The response from each subject is nominal; it takes the value O, A, B, or AB. Formally, we can write the j-th response for treatment i ($i = 1, 2, 3; j = 1, 2, \ldots, 100$) as follows

$$Y_{ij} = \begin{cases} 1 & \text{if the } j\text{-th individual in the sample from the } i\text{-th group had blood type O} \\ 2 & \text{if the } j\text{-th individual in the sample from the } i\text{-th group had blood type A} \\ 3 & \text{if the } j\text{-th individual in the sample from the } i\text{-th group had blood type B} \\ 4 & \text{if the } j\text{-th individual in the sample from the } i\text{-th group had blood type AB.} \end{cases}$$

The responses are nominal (categorical), with four categories. The observed responses can be classified according to the levels of one factor: the individual's ethnic subgroup. There are three levels of the factor being studied; each level is replicated 100 times in the observed data.

The given data are the observed frequencies (number of responses) in each blood type category. Let π_{ik} denote the proportion of the population of individuals in ethnic group i whose blood type is in category k. Then the blood type distribution in population i is a discrete distribution; the measurements take values 1, 2, 3, and 4 with probabilities π_{i1}, π_{i2}, π_{i3}, and π_{i4}, respectively. The null hypothesis that the three blood type distributions are identical is equivalent to the claim that the proportion of individuals in each category is the same for the three populations. For instance, under the null hypothesis, $\pi_{11} = \pi_{21} = \pi_{31}$; that is, the proportion of blood type O is the same for all three populations.

Clearly, the assumption of normally distributed populations is not justified in this case. An alternative to the ANOVA F-test is needed to test the null hypothesis of equal population distributions. ∎

Nominal data cannot be compared on the basis of their numerical values, and so statistical analyses of nominal data such as those in Example 8.14 are based on the observed frequencies of responses in the various categories.

Notation

Suppose that we have independent random samples from R categorical populations, each consisting of measurements that can be classified into C categories. In other words, we have one-way classified nominal data at R levels of the study factor. Each measurement in the data set falls into one of C categories.

Let f_{ik} denote the number of responses from population i in category k. Then the observed responses can be summarized as in Display 8.2.

DISPLAY 8.2

Observed frequencies of one-way classified data

| Population number i | \multicolumn{6}{c}{Category number k} | Total |
	1	2	$\cdots$	k	$\cdots$	C	
1	f_{11}	f_{12}	$\cdots$	f_{1k}	$\cdots$	f_{1C}	f_{1+}
2	f_{21}	f_{22}	$\cdots$	f_{2k}	$\cdots$	f_{2C}	f_{2+}
$\cdots$	$\cdots$	$\cdots$	$\cdots$	$\cdots$	$\cdots$		
i	f_{i1}	f_{i2}	$\cdots$	f_{ik}	$\cdots$	f_{iC}	f_{i+}
$\cdots$	$\cdots$	$\cdots$	$\cdots$	$\cdots$	$\cdots$		
R	f_{R1}	f_{R2}	$\cdots$		$\cdots$	f_{RC}	f_{R+}
Total	f_{+1}	f_{+2}	$\cdots$	f_{+k}	$\cdots$	f_{+C}	f_{++}

Compare the entries in Display 8.2 with those in Display 8.1. The entries in Display 8.1 are the actual observed responses; the margins show the corresponding totals. Display 8.2, on the other hand, is an extension of Table 6.7; the main entries give the observed frequencies in the corresponding categories. In both displays, the row subscript i denotes the population number. However, the column subscript in Display 8.1 refers to the replication number, whereas in Display 8.2 it refers to the identification number of the category to which the response belongs. The row totals are the number of replications in each sample, while the column totals denote the total number of observed responses in the different categories. Thus, f_{1+} is the number of observations from population 1, whereas f_{+1} is the total number of observations in category 1.

Large-sample χ^2-test

Let π_{ik} denote the probability that a response from population i falls into category k. If, for every category, the probability of a response in that category is the same for all populations, then the populations will have identical distributions across response categories. Consequently, the null hypothesis that there are no differences between populations can be formulated as

$$H_0: \pi_{1k} = \cdots = \pi_{ik} = \cdots = \pi_{Rk}, \ k = 1, 2, \ldots, C. \tag{8.21}$$

In Box 8.5, we present a large-sample procedure for testing H_0. The logic behind the test in Box 8.5 is the same as that in Chapter 6, where we considered the analysis of nominal data from one and two samples. The test is performed by assuming that H_0 is true and comparing the observed and the expected frequencies. The test rejects H_0 if the observed frequencies depart significantly from the frequencies expected under the null hypothesis.

BOX **8.5**

A χ^2-test for equality of R categorical populations: Large samples

Suppose that we have one-way classified nominal data at R levels of a factor, with C categories of responses. Suppose further that the frequencies of the responses are as in Display 8.2. Let π_{ik} denote the proportion of measurements in category k of population i ($i = 1, 2, \ldots, R$; $k = 1, 2, \ldots, C$). Then the maximum-likelihood estimates of the expected frequencies under the hypothesis $H_0: \pi_{1k} = \pi_{2k} = \cdots = \pi_{Rk}, k = 1, \ldots, C$, are

$$\hat{f}_{ik} = \frac{f_{i+}f_{+k}}{f_{++}}.$$

If the $\hat{f}_{ik}$ are not too small, a test of H_0 against the research hypothesis H_1 that the population distributions are not equal can be performed using the test statistic

$$\chi_c^2 = \sum_{i=1}^{R} \sum_{k=1}^{C} \frac{(f_{ik} - \hat{f}_{ik})^2}{\hat{f}_{ik}}$$

$$= \sum_{i=1}^{R} \sum_{k=1}^{C} \frac{f_{ik}^2}{\hat{f}_{ik}} - f_{++}.$$

The null hypothesis is rejected at level α if

$$\chi_c^2 \geq \chi^2((R-1)(C-1), \alpha).$$

Comments on the large-sample χ^2-test procedures in Box 8.5

1 When $R = 2$, the test procedure in Box 8.5 reduces to the two-sample procedure in Box 6.8.

2 Analagously to Boxes 6.3, 6.4, and 6.8, the requirement of large sample size is that $\hat{f}_{ik} \geq 5$ in at least 75% of the cells and no $\hat{f}_{ik} = 0$.

3 The formula for $\hat{f}_{ik}$ can be justified intuitively by reasoning similar to that used for justifying $\hat{f}_{ik}$ in Box 6.8. In the blood type frequency table in Example 8.14, notice that the proportion with blood type O is 138/300. If the null hypothesis is true, then the proportions of individuals with blood type O in the three populations are equal. An estimate of this common proportion is 138/300, the sample proportion of blood type O. Therefore, if H_0 is true, we should expect approximately 138/300 of the 100 subjects in sample 1 to have blood type O, and so the estimated expected frequency is

$$\hat{f}_{11} = \frac{138}{300} \times 100 = \frac{f_{1+}f_{+1}}{f_{++}}.$$

Expected frequencies for other cells can be estimated in the same fashion. The following table shows the observed and expected (in parentheses) frequencies of the blood types for the three ethnic groups:

Group	Blood type				Total
	O	A	B	AB	
1	44 (46)	31 (32)	14 (12)	11 (10)	100
2	38 (46)	39 (32)	10 (12)	13 (10)	100
3	56 (46)	26 (32)	12 (12)	6 (10)	100
Total	138	96	36	30	300

4 As already noted, the χ^2-test in Box 8.5 is a large-sample test. When the expected cell frequencies are small, exact tests similar to Fisher's exact test for the two-sample problem should be used. Thanks to recent advances in computational speed and computing algorithms, implementation of exact tests is possible in most cases. The FI option in the StatXact tests on $R \times C$ contingency tables can be used for this purpose. The FREQ procedure of SAS can also be used if R and C are not too large.

EXAMPLE 8.15   Let's apply the χ^2-test to the null hypothesis that the distributions of blood types are the same for the three ethnic groups in Example 8.14. We have

$$\chi_c^2 = \frac{44^2}{46} + \frac{38^2}{46} + \frac{56^2}{46} + \cdots + \frac{11^2}{10} + \frac{13^2}{10} + \frac{6^2}{10} - 300 = 9.61.$$

The calculated value of the test statistic must be compared with the critical values of a χ^2-distribution with $(R - 1)(C - 1) = (3 - 1)(4 - 1) = 6$ degrees of freedom. From Table C.3 (Appendix C), we see that $\chi^2(6, 0.15) < 9.61 < \chi^2(6, 0.10)$. Therefore, the p-value is between 0.10 and 0.15, and the hypothesis of equal blood type distributions can be rejected at the $\alpha = 0.15$ level but not at the $\alpha = 0.10$ level.

The StatXact output for the blood type data is as follows.

```
FISHER'S EXACT TEST

Statistics based on the observed   3 by   4 table (x):
        P(x)  = Hypergeometric Prob. of the table   = 0.9914E-07
        FI(x) = Fisher statistic                    = 9.569


Asymptotic p-value: (based on Chi-squared distribution with   6 df)
        Pr { FI(X)  .GE.  9.569 } =      0.1440


Exact p-value and point probability:
    Pr { FI(X) .GE.  9.569 } = Pr { P(X) .LE. 0.9914E-07 } = 0.1413
    Pr { FI(X) .EQ.  9.569 } = Pr { P(X) .EQ. 0.9914E-07 } = 0.0000
```

From this output we see that the exact p-value of 0.1413 is very close to the asymptotic (large-sample) p-value of 0.1440. In this particular case, accordingly, the large-sample test in Box 8.5 is as good as Fisher's exact test, as we would expect, because the expected cell frequencies meet the large-sample criterion in comment 2.   ■

For more information on the analysis of categorical data, see Agresti (1990).

Exercises

8.45   In a study, 50 randomly selected sites in each of four counties in North Florida were classified according to their type of terrain, with the following results:

County	Type of terrain			Total
	Predominantly pine	Predominantly hardwood	Predominantly swamp	
1	9	16	25	50
2	25	9	16	50
3	14	8	28	50
4	25	11	14	50
Total	73	44	83	200

a   Let Y_{ij} denote the measured value for the j-th site in the i-th county. As for the blood type data in Example 8.14, give a symbolic description of the possible values that Y_{ij} can take.

b   Using the notation in Display 8.2, determine the values of $R, C, f_{12}, f_{33}, f_{2+}, f_{+3}$, and f_{++}.

c   Let π_{ik} denote the actual proportion of sites in county i that have terrain of type k. Describe, in words, the null hypothesis H_0: $\pi_{1k} = \pi_{2k} = \pi_{3k} = \pi_{4k}, k = 1, 2, 3$.

d   State and interpret the usual alternative hypothesis associated with H_0 in (c).

e   Perform a test of H_0 and write your conclusions.

8.46   In a study, the relation between the incidence of cancer among phosphate mine workers and their exposure to radiation was investigated. Cancer incidence was monitored among 150 phosphate mine workers, of whom 50 had 0–29 working level months (WLM) of exposure, 50 had 30–89 WLM of exposure, and 50 had 90–120 WLM of exposure. The results obtained for the frequency of incidence are as follows:

Exposure level	Cancer incidence		Total
	Yes	No	
0–29 WLM	19	31	50
30–89 WLM	26	24	50
90–120 WLM	37	13	50

Perform a test of the null hypothesis that the cancer risk is the same for each of the three exposure levels. Write your conclusions.

8.47   In a study to assess the protective effects of a new agent A against paraquat, 50 rats were randomly assigned to three groups: I, paraquat alone; II, paraquat plus a low dose of A; III, paraquat plus a high dose of A. The following survival rates were recorded for the rats:

Treatment	Survivors after eight days	Deaths in the first eight days
I	20	30
II	25	25
III	35	15

Perform an appropriate statistical test to see if the administration of paraquat with A affects the survival rate.

8.10
Overview

In this chapter, we consider the comparison of t populations using information contained in independent samples from each population. Although the estimation of differences between population parameters is of major concern in practice, we focus attention here on testing the null hypothesis H_0 that the population distributions are equal, in order to point out similarities between one-, two-, and k-population procedures. Methods of estimating differences and ordering subsets of populations will be presented in Chapter 9.

An observational process that generates independent samples from normal populations with a common variance leads to one of the most important tools of data analysis: the analysis of variance (ANOVA). In later chapters, the ANOVA technique will be applied in a wide variety of experimental situations.

Like the two-sample t-test, the ANOVA method is useful because minor deviations from the underlying assumptions do not seriously affect its validity. The α-level ANOVA F-test will have an approximate level of α even if the population distributions are not exactly normal. However, the power of the test to detect true differences may decline.

When the failure of ANOVA assumptions is serious—for instance, when we have nominal data or asymmetry and/or unequal variances in the populations—we need to resort to alternative methods of analysis. For nominal data, proportions (frequencies) in different categories of responses are compared, while quantitative nonnormal data can be analyzed using ranks. Both of these methods are briefly described in this chapter.

Methods in this chapter are appropriate for one-way classifications arising from either CRD or observational studies. A subtle difference is that, whereas the sample sizes are fixed by the experimenter in CRD studies, that is not always so in observational studies. The methods described in this chapter require fixed sample sizes. Thus, when they are used in observational studies with random n, the results are conditional on the observed sample sizes. This distinction should be kept in mind when applying ANOVA methods to data where sample sizes are not determined by the investigator.

9

Single Factor Studies: Comparing Means and Determining Sample Sizes

9.1 Introduction

In Chapter 8, we saw how the ANOVA F-test can be used to determine whether the expected responses at the t levels of an experimental factor differ from each other. Often, in practice, it is not of primary interest to test the null hypothesis that the population means are equal. Instead, investigators want to make specific comparisons of the means of the populations under consideration. In most instances, such comparisons fall into one of the following categories:

1   Comparisons specified prior to performing the experiment

2   Comparisons specified after observing the outcome of the experiment

3   All possible pairwise comparisons

The next example illustrates the differences between these three types of comparisons.

EXAMPLE 9.1   The reduction in systolic blood pressure (BP) after a drug for hypertension is administered is one of the key indicators of how well the patient is responding to the drug. When treating for hypertension, the side effects associated with the drug are of particular concern. In a study, two drugs A and B for reducing the side effects of a standard hypertension drug S were evaluated. Drugs A and B were administered concurrently with drug S. The study was conducted using a completely randomized design, with five treatments, as follows:

Treatment	Drug
1	Standard drug (S)
2	S combined with a low dose of A (S + AL)
3	S combined with a high dose of A (S + AH)
4	S combined with a low dose of B (S + BL)
5	S combined with a high dose of B (S + BH)

There were four replications of each treatment. The reduction in blood pressure (mm Hg) over a period of four weeks observed for the experimental subjects may be tabulated as follows:

Treatment	Responses (mm Hg)	Total	Mean
1	27, 26, 21, 26	100	25.00
2	19, 13, 15, 16	63	15.75
3	15, 10, 10, 11	46	11.50
4	22, 15, 21, 18	76	19.00
5	20, 18, 17, 16	71	17.75

The following are three of the questions that the investigators asked about the treatment effects.

1   Is there a difference between the effects of low and high doses of A? In other words, on average, is the reduction in systolic BP the same for S + AL and for S + AH? In practical terms, are the expected responses for treatment 2 and treatment 3 different?

2   Is there a difference between the effects of low and high doses of B? In other words, are the expected responses for treatment 4 and treatment 5 different?

3   Is there a difference between the average of the expected responses for the two doses of A (S + AL and S + AH) and the average of the expected responses for the two doses of B (S + BL and S + BH)?

Each of these three questions calls for a comparison of subsets of treatment means. To see this, let μ_i denote the expected response for treatment i $(i = 1, \ldots, 5)$. Then the first question requires a comparison of μ_2 and μ_3, whereas the second question requires a comparison of μ_4 and μ_5. The third question asks about the difference between the average of μ_2 and μ_3 and the average of μ_4 and μ_5; in other words, it involves a comparison of $\{\mu_2, \mu_3\}$ with $\{\mu_4, \mu_5\}$. Because these three comparisons were specified at the planning stages of the study, they fall into our first category: comparisons specified prior to performing the study.

Sometimes, the outcome of a study suggests a research question that did not occur to the investigators in planning the experiment. For instance, on the basis of the observed mean responses in the blood pressure study, it appears that S is the best treatment and S + BL is the second best. It is natural to wonder whether the observed difference between the mean responses for these two treatments indicates that there is a difference between the corresponding population means. An answer to the question requires a comparison of μ_1 and μ_4, the expected mean reductions in systolic BP for S and S + BL respectively. This comparison was specified after observing the outcome of the experiment, and the corresponding statistical techniques are different from those for comparisons specified before observing the outcome of the study.

Finally, in studies that are exploratory in nature, the researcher may not have any preplanned comparisons in mind. In such cases, the investigator would like to arrive at an overall summary of the significance of the observed differences between the sample means. Such comparisons are helpful in deciding which population means are different. The desired information can be obtained by making all possible pairwise comparisons of the treatment means. ∎

In Section 9.2, we look at methods for comparisons of population means specified during the planning stages of the study. In Section 9.3, we turn to comparisons specified after the study has been performed. In Section 9.4, we consider methods specifically designed for simultaneous inferences about several comparisons. In Section 9.5, we examine methods for comparing all possible pairs of treatment means and, in Section 9.6, we explore methods for constructing simultaneous confidence intervals. Finally, the question of predicting linear functions of the means of future samples is considered in Section 9.7 and determining sample sizes for single-factor studies is considered in Section 9.8.

9.2
Comparisons specified before performing the experiment

As already noted, comparisons of treatment means that are specified before the experiment is performed are useful because they allow the investigators to answer the questions that motivated the research study. When the ANOVA assumptions are satisfied by the responses in a single-factor study, the procedures for making such comparisons are direct extensions of the methods for comparing two population means in Chapter 4.

As in the comparison of two population means, we begin by formulating the comparison as an inference about a parameter θ. For instance, in Example 9.1, the first question can be cast in terms of the parameter $\theta_1 = \mu_2 - \mu_3$. To determine if there is a difference between the effects of low and high doses of A, we test H_0: $\theta_1 = 0$ against H_1: $\theta_1 \neq 0$. Similarly, the second and third questions can be answered by testing suitable hypotheses about the parameters $\theta_2 = \mu_4 - \mu_5$ and $\theta_3 = \frac{1}{2}(\mu_2 + \mu_3) - \frac{1}{2}(\mu_4 + \mu_5)$.

Having identified the parameter of interest θ, we need to obtain an estimate $\hat{\theta}$ such that the sampling distribution of $\hat{\theta}$ is a normal distribution with mean θ and standard error $\hat{\sigma}_{\hat{\theta}}$. In this chapter, we'll see how, under the ANOVA assumptions, t-test procedures similar to those in Box 4.4 can be used to make inferences about θ.

Linear combinations of means

Each of the three parameters θ_1, θ_2, and θ_3 just defined has the general form

$$\theta = c_1\mu_1 + c_2\mu_2 + \cdots + c_t\mu_t, \tag{9.1}$$

where $c_1, c_2, \ldots, c_t$ are known constants and $\mu_1, \mu_2, \ldots, \mu_t$ are population means. To see this, let $t = 5, c_1 = 0, c_2 = 1, c_3 = -1,$ and $c_4 = c_5 = 0$. Then Equation (9.1) reduces to

$$\theta = (0)\mu_1 + (1)\mu_2 + (-1)\mu_3 + (0)\mu_4 + (0)\mu_5 = \mu_2 - \mu_3,$$

which is the same as θ_1. Similarly, setting $t = 5, c_1 = c_2 = c_3 = 0, c_4 = 1,$ and $c_5 = -1$ in Equation (9.1) will result in $\theta = \theta_2$. Finally, when $t = 5, c_1 = 0, c_2 = c_3 = \frac{1}{2},$ $c_4 = c_5 = -\frac{1}{2}$ in Equation (9.1), $\theta = \theta_3$.

Expressions of the form in Equation (9.1) are called *linear combinations of population means*. The constants $c_1, c_2, \ldots, c_t$ are called the *coefficients* in the linear combination.

It is intuitively reasonable to estimate a linear combination of population means by the corresponding linear combination of sample means. In other words, if $\overline{Y}_i$ denotes the mean of the sample from the i-th population $(i = 1, \ldots, t)$, then an intuitive estimator of θ in Equation (9.1) is

$$\hat{\theta} = c_1\overline{Y}_1 + c_2\overline{Y}_2 + \cdots + c_t\overline{Y}_t. \tag{9.2}$$

In Box 9.1, we summarize some key results about the sampling distribution of statistics of the form

$$\hat{\theta} = c_0 + c_1\overline{Y}_1 + c_2\overline{Y}_2 + \cdots + c_t\overline{Y}_t. \tag{9.3}$$

Notice that the statistic in Equation (9.3) is of the same form as the linear combination in Equation (9.2), except for the extra constant c_0. We refer to Equation (9.3) as a *linear function of sample means*. The constant c_0 is called the *intercept*. Notice that a linear combination is a linear function with zero intercept.

BOX **9.1**

Sampling distributions of linear functions of sample means

Let $\overline{Y}_1, \overline{Y}_2, \ldots, \overline{Y}_t$ be the means of independent random samples of sizes $n_1, n_2, \ldots, n_t$ from populations with means $\mu_1, \mu_2, \ldots, \mu_t$ and variances $\sigma_1^2, \sigma_2^2, \ldots, \sigma_t^2$. Let

$$\theta = c_0 + c_1\mu_1 + c_2\mu_2 + \cdots + c_t\mu_t,$$

where $c_0, c_1, c_2, \ldots, c_t$ are known constants, and consider the linear function of sample means

$$\hat{\theta} = c_0 + c_1\overline{Y}_1 + c_2\overline{Y}_2 + \cdots + c_t\overline{Y}_t.$$

Then we can state the following conclusions:

1   The sampling distribution of $\hat{\theta}$ is a distribution with mean

$$\mu_{\hat{\theta}} = \theta$$

and variance

$$\sigma_{\hat{\theta}}^2 = \frac{c_1^2\sigma_1^2}{n_1} + \frac{c_2^2\sigma_2^2}{n_2} + \cdots + \frac{c_t^2\sigma_t^2}{n_t}.$$

2   If the target populations have normal distributions, then the sampling distribution of $\hat{\theta}$ is a normal distribution with mean θ and variance $\sigma_{\hat{\theta}}^2$.

Comments on the sampling distributions in Box 9.1

1   We have already encountered two special cases of the results in Box 9.1. The first case is property III in Box 3.2, which states that the sampling distribution the mean of a random sample from a population with a $N(\mu, \sigma^2)$ distribution has a normal distribution with mean μ and standard deviation $\sigma/\sqrt{n}$. The same result can be obtained by letting $t = 1$, $c_0 = 0$, and $c_1 = 1$ in Box 9.1. Then $\theta = \mu_1$, $\hat{\theta} = \overline{Y}_1$, and

$$\sigma_{\hat{\theta}} = \sqrt{\frac{(1)^2\sigma_1^2}{n_1}} = \frac{\sigma_1}{\sqrt{n_1}},$$

so that the results in Box 9.1 imply property III of Box 3.2. In Exercise 9.1, you are asked to show that, by a suitable choice of values for t and c_i, the results in Box 9.1 can be used to derive the sampling distribution of the difference between sample means described in Box 3.6.

2   Recall from Section 3.8 that, if the mean of the sampling distribution of an estimator $\hat{\theta}$ equals θ, then $\hat{\theta}$ is said to be an unbiased estimator of θ. From conclusion 1 in Box 9.1, it follows that, regardless of whether the populations have normal distributions, a linear function of sample means is an unbiased estimator of the corresponding linear function of the population means.

3 If the population variances are all equal to σ^2, the standard error of $\hat{\theta}$ can be expressed as

$$\sigma_{\hat{\theta}} = \sqrt{\left\{ \frac{c_1^2}{n_1} + \frac{c_2^2}{n_2} + \cdots + \frac{c_t^2}{n_t} \right\} \sigma^2}. \qquad (9.4)$$

Suppose that we have independent random samples from normal populations with a common variance σ^2 and that we want to make inferences about a linear combination θ of the form in Equation (9.1). A procedure similar to that in Chapter 4 for making inferences about the difference between population means can be used for this purpose. Box 9.2 summarizes this procedure under a slightly more general setting, when the parameter of interest is a linear function rather than a linear combination.

BOX **9.2**

Inferences about linear functions of population means

Let $\overline{Y}_1, \overline{Y}_2, \ldots, \overline{Y}_t$ be the means of independent random samples of sizes $n_1, n_2, \ldots, n_t$ from normal distributions with means $\mu_1, \mu_2, \ldots, \mu_t$ and a common variance σ^2. Inferences about the linear function

$$\theta = c_0 + c_1\mu_1 + c_2\mu_2 + \cdots + c_t\mu_t,$$

where $c_0, c_1, c_2, \ldots, c_t$ are known constants, can be made as follows.

Confidence interval

A $100(1 - \alpha)\%$ confidence interval for θ is

$$\hat{\theta} - t(N - t, \ \alpha/2)\hat{\sigma}_{\hat{\theta}} \leq \theta \leq \hat{\theta} + t(N - t, \ \alpha/2)\hat{\sigma}_{\hat{\theta}},$$

where

$$\hat{\theta} = c_0 + c_1\overline{Y}_1 + \cdots + c_t\overline{Y}_t,$$

$$\hat{\sigma}_{\hat{\theta}} = \sqrt{\left\{ \frac{c_1^2}{n_1} + \frac{c_2^2}{n_2} + \cdots + \frac{c_t^2}{n_t} \right\} \text{MS[E]}},$$

and MS[E] is the mean square error in a one-way ANOVA of the data.

Hypothesis testing

The null hypothesis H_0: $\theta = \theta_0$ can be tested using the test statistic

$$t = \frac{\hat{\theta} - \theta_0}{\hat{\sigma}_{\hat{\theta}}}.$$

The calculated value of the test statistic should be compared to an appropriate critical value of a $t(N - t)$-distribution.

Comments on the inferences in Box 9.2

1   The standard error of the estimate of θ is estimated by replacing σ^2 with MS[E], the pooled estimate of the common population variance. In Section 8.3, we noted that MS[E] is the extension of the pooled estimate s_p^2 defined in Box 4.3 to the case of t samples.

2   By setting $c_0 = 0$ and $t = 1$ or 2, it can be verified that the one-sample and independent-sample procedures in Boxes 4.2 and 4.3, respectively, are special cases of the procedures in Box 9.2.

The next three examples illustrate how the results in Box 9.2 can be used to make inferences about linear combinations of the means of normally distributed populations.

EXAMPLE **9.2**   The blood pressure data in Example 9.1 were obtained in a completely randomized experiment with $t = 5$ treatments and $n = 4$ replications of each treatment. The ANOVA table obtained when the SAS GLM procedure is used to conduct a one-way ANOVA of the data is as follows.

```
-----------------------------------------------------------------
Dependent Variable: BP
                            Sum of        Mean
Source              DF      Squares     Square   F Value   Pr > F
Model                4    388.70000    97.17500    15.10   0.0001
Error               15     96.50000     6.43333
Corrected Total     19    485.20000
-----------------------------------------------------------------
```

Let's first use the information in Box 9.2 to construct a 95% confidence interval for the expected drop in blood pressure (BP) for a patient treated with the standard drug (treatment 1). The parameter of interest $\theta = \mu_1$, the expected response for treatment 1, can be expressed as the linear function

$$\theta = (0) + (1)\mu_1 + (0)\mu_2 + (0)\mu_3 + (0)\mu_4 + (0)\mu_5.$$

Thus, with $c_1 = 1$ and $c_0 = c_2 = c_3 = c_4 = c_5 = 0$ in Box 9.1, we see that an unbiased estimate of $\theta = \mu_1$ is $\hat{\theta} = \bar{Y}_1$. From Box 9.2, the standard deviation of $\hat{\theta}$ is

$$\sigma_{\hat{\theta}} = \sqrt{\left(\frac{1^2}{4} + \frac{0^2}{4} + \frac{0^2}{4} + \frac{0^2}{4} + \frac{0^2}{4}\right)\sigma^2} = \sqrt{\frac{\sigma^2}{4}}.$$

Because σ^2 is unknown, we replace it with its pooled estimate MS[E]. From the ANOVA table, we have MS[E] = 6.43, so that the estimated standard error of $\hat{\theta}$ is

$$\hat{\sigma}_{\hat{\theta}} = \sqrt{\frac{\text{MS[E]}}{4}} = \sqrt{\frac{6.43}{4}} = 1.27.$$

A 95% confidence interval for μ_1 can be constructed using the formula in Box 9.2

$$\overline{Y}_1 \pm t(\nu,\ 0.025)\sqrt{\frac{MS[E]}{4}},$$

where $\nu = N - t = 20 - 5 = 15$ is the degrees of freedom for MS[E]. Thus, the required confidence interval is

$$25.00 \pm (2.131)(1.27) = 25.00 \pm 2.71 = (22.29,\ 27.71).$$

With 95% confidence, we can expect a mean reduction in BP between 22.29 mm Hg and 27.71 mm Hg for the standard drug. ∎

EXAMPLE 9.3   A principal objective of the BP reduction study in Example 9.1 was to see how the effects of the four combination therapies—S + AL, S + AH, S + BL, and S + BH—compare with the effect of the standard therapy. In this example, let's use the data to see if there is reason to believe that the expected BP reduction for S + AH is significantly lower than that for the standard therapy.

Let $\theta = \mu_1 - \mu_3$. Then we want to test H_0: $\theta \le 0$ against H_1: $\theta > 0$. The estimate of θ is $\bar{y}_1 - \bar{y}_3 = 25 - 11.50 = 13.50$. With $c_0 = 0, c_1 = 1, c_2 = 0, c_3 = -1, c_4 = c_5 = 0$ in Box 9.2, we get

$$\hat{\sigma}_{\hat{\theta}} = \sqrt{\left(\frac{1}{4} + \frac{1}{4}\right) MS[E]} = \sqrt{\left(\frac{1}{4} + \frac{1}{4}\right) 6.43} = 1.793.$$

The calculated value of the test statistic is

$$t_c = \frac{\hat{\theta} - 0}{\hat{\sigma}_{\hat{\theta}}} = \frac{13.50 - 0}{1.793} = 7.529.$$

From Table C.2 (Appendix C) with $df = 15$, we see that the p-value is less than 0.001, so that H_0 can be rejected at a level as low as $\alpha = 0.001$. We conclude that the expected drop in BP resulting from S + AH is significantly lower than that for the standard drug. The desirability of replacing S by S + AH will depend on how significant the difference between μ_1 and μ_3 is in practice. A confidence interval for this difference can be constructed using the formula $\hat{\theta} \pm t(\nu,\ \alpha/2)\hat{\sigma}_{\hat{\theta}}$. As an exercise, construct this confidence interval and interpret it. ∎

EXAMPLE 9.4   Do the BP reduction data in Example 9.1 indicate that there is a difference between the average of expected BP drops for S + AL and S + AH and the average of expected drops for S + BL and S + BH?

Let $\theta = \frac{1}{2}(\mu_2 + \mu_3) - \frac{1}{2}(\mu_4 + \mu_5)$. Then this question can be answered by testing H_0: $\theta = 0$ against H_1: $\theta \ne 0$. The method in Box 9.2 can be used for this purpose.

The estimate of θ is

$$\hat{\theta} = \frac{1}{2}(\bar{y}_2 + \bar{y}_3) - \frac{1}{2}(\bar{y}_4 + \bar{y}_5) = \frac{1}{2}(15.75 + 11.50) - \frac{1}{2}(19.00 + 17.75) = -4.75.$$

With $c_0 = c_1 = 0$, $c_2 = c_3 = \frac{1}{2}$, $c_4 = c_5 = -\frac{1}{2}$, and MS[E] $= 6.43$ in Box 9.2, we get $\hat{\sigma}_{\hat{\theta}} = 1.27$. The calculated value of the test statistic is

$$t_c = \frac{\hat{\theta} - 0}{\hat{\sigma}_{\hat{\theta}}} = \frac{-4.75 - 0}{1.27} = -3.74.$$

Consulting Table C.2 (Appendix C) with $\nu = 15$ degrees of freedom, we see that the p-value for a two-sided test is less than 0.002. The p-value is small, and so it is reasonable to reject the null hypothesis and conclude that, on the basis of the data, there is a difference between the average of μ_2 and μ_3 and that of μ_4 and μ_5.   ∎

Orthogonal comparisons

Examples 9.3 and 9.4 illustrate how linear combinations of sample means can be used to make inferences about the corresponding linear combinations of population means. In each case, the linear combination considered was an example of a contrast of sample means.

Contrasts of sample means

Let $\overline{Y}_1, \overline{Y}_2, \ldots, \overline{Y}_t$ denote the means of t samples. A linear combination

$$\hat{\theta} = c_1 \overline{Y}_1 + c_2 \overline{Y}_2 + \cdots + c_t \overline{Y}_t$$

is called a *contrast of sample means* if the coefficients satisfy the condition

$$c_1 + c_2 + \cdots + c_t = 0. \tag{9.5}$$

A contrast in which only two coefficients are nonzero is called a *simple contrast*. Contrasts with more than two nonzero coefficients are called *complex contrasts*. Thus, $\hat{\theta} = \overline{Y}_1 - \overline{Y}_3$ is a simple contrast, whereas $\hat{\theta} = \frac{1}{2}(\overline{Y}_2 + \overline{Y}_3) - \frac{1}{2}(\overline{Y}_4 + \overline{Y}_5)$ is a complex contrast.

Linear combinations of sample means that are contrasts are often used to compare one subset of sample means with another. For instance, the contrast $\hat{\theta} = \frac{1}{2}(\overline{Y}_2 + \overline{Y}_3) - \frac{1}{2}(\overline{Y}_4 + \overline{Y}_5)$ compares the means of $\{\overline{Y}_2, \overline{Y}_3\}$ and $\{\overline{Y}_4, \overline{Y}_5\}$. If there is no difference between these two means, then $\hat{\theta} = 0$. With increase in $\hat{\theta}$, the difference between the means of the two subsets increases, and hence the magnitude of $\hat{\theta}$ is a measure of the variation (disparity) between the means of the subsets being compared. Similarly, the contrast $\hat{\theta} = \overline{Y}_1 - \frac{1}{4}(\overline{Y}_2 + \overline{Y}_3 + \overline{Y}_4 + \overline{Y}_5)$ is a measure of the disparity between the means of the subsets $\{\overline{Y}_1\}$ and $\{\overline{Y}_2, \overline{Y}_3, \overline{Y}_4, \overline{Y}_5\}$. In general, the magnitude of a contrast between t sample means can be interpreted as a measure of the disparity between the means of a specific pair of subsets of sample means.

In Chapter 8, we noted that the treatment sum of squares SS[T] in a one-way ANOVA provides a measure of the variability of t sample means. It turns out that this measure of the overall variability between t sample means can be subdivided into parts, such that each part is a measure of the variability between the means of a specific pair of subsets of sample means. Such subdivisions of the treatment sum of squares allow us to explain the observed differences between the treatment means in terms of the differences between selected subsets of treatment means. The notion of orthogonal contrasts is needed to accomplish this type of subdivision.

Mutually orthogonal contrasts

Let $\overline{Y}_1, \overline{Y}_2, \ldots, \overline{Y}_t$ denote the means of random samples of sizes $n_1, n_2, \ldots, n_t$ from t populations with a common variance σ^2. Two contrasts

$$\hat{\theta}_1 = c_1\overline{Y}_1 + \cdots + c_t\overline{Y}_t$$

and

$$\hat{\theta}_2 = d_1\overline{Y}_1 + \cdots + d_t\overline{Y}_t$$

are said to be mutually orthogonal if

$$\frac{c_1 d_1}{n_1} + \frac{c_2 d_2}{n_2} + \cdots + \frac{c_t d_t}{n_t} = 0. \tag{9.6}$$

If the sample sizes are equal—that is, if $n_1 = n_2 = \cdots = n_t$—Equation (9.6) reduces to the form

$$c_1 d_1 + c_2 d_2 + \cdots + c_t d_t = 0. \tag{9.7}$$

A set of k contrasts is said to be a *mutually orthogonal set* if all the pairs in the set are mutually orthogonal. Consider the following example.

EXAMPLE **9.5**   Let $\overline{Y}_1, \ldots, \overline{Y}_5$ denote the observed mean responses for the five treatments in Example 9.1. Consider the linear combinations

$$\hat{\theta}_1 = \overline{Y}_1 - \frac{1}{4}(\overline{Y}_2 + \overline{Y}_3 + \overline{Y}_4 + \overline{Y}_5),$$

$$\hat{\theta}_2 = \overline{Y}_2 - \overline{Y}_3,$$

$$\hat{\theta}_3 = \overline{Y}_4 - \overline{Y}_5,$$

$$\hat{\theta}_4 = \frac{1}{2}(\overline{Y}_2 + \overline{Y}_3) - \frac{1}{2}(\overline{Y}_4 + \overline{Y}_5).$$

Is $\{\hat{\theta}_1, \hat{\theta}_2, \hat{\theta}_3, \hat{\theta}_4\}$ a mutually orthogonal set of comparisons?

Consider the pair $\{\hat{\theta}_1, \hat{\theta}_2\}$. Writing

$$\hat{\theta}_1 = (1)\overline{Y}_1 + \left(-\frac{1}{4}\right)\overline{Y}_2 + \left(-\frac{1}{4}\right)\overline{Y}_3 + \left(-\frac{1}{4}\right)\overline{Y}_4 + \left(-\frac{1}{4}\right)\overline{Y}_5,$$

$$\hat{\theta}_2 = (0)\overline{Y}_1 + (+1)\overline{Y}_2 + (-1)\overline{Y}_3 + (0)\overline{Y}_4 + (0)\overline{Y}_5,$$

we determine the coefficients of the linear combinations as

$$c_1 = 1, \ c_2 = c_3 = c_4 = c_5 = -\frac{1}{4}$$

for $\hat{\theta}_1$ and

$$d_1 = 0, \; d_2 = 1, \; d_3 = -1, \; d_4 = d_5 = 0$$

for $\hat{\theta}_2$.

Both linear combinations are contrasts, because the coefficients in $\hat{\theta}_1$ and $\hat{\theta}_2$ satisfy Equation (9.5)

$$c_1 + \cdots + c_5 = d_1 + \cdots + d_5 = 0.$$

The sample sizes are equal, and so we can use Equation (9.7) to see if $\hat{\theta}_1$ and $\hat{\theta}_2$ are orthogonal. We have

$$c_1 d_1 + \cdots + c_5 d_5$$
$$= (1)(0) + \left(-\frac{1}{4}\right)(+1) + \left(-\frac{1}{4}\right)(-1) + \left(-\frac{1}{4}\right)(0) + \left(-\frac{1}{4}\right)(0)$$
$$= -\frac{1}{4} + \frac{1}{4} = 0.$$

Accordingly, $\hat{\theta}_1$ and $\hat{\theta}_2$ are mutually orthogonal. The mutual orthogonality of the pairs $\{\hat{\theta}_1, \hat{\theta}_3\}$, $\{\hat{\theta}_1, \hat{\theta}_4\}$, $\{\hat{\theta}_2, \hat{\theta}_3\}$, $\{\hat{\theta}_2, \hat{\theta}_4\}$, and $\{\hat{\theta}_3, \hat{\theta}_4\}$ can be verified similarly. All possible pairs in the set are mutually orthogonal, and so we conclude that this is a mutually orthogonal set.   ∎

Each of the four linear combinations in Example 9.5 can be used to compare the means of a pair of subsets of the five mean responses. For instance, the linear combination $\hat{\theta}_1$ compares the observed mean response to treatment 1 (the standard drug) with the average of the mean responses to the other four treatments. Similarly, $\hat{\theta}_2$ provides a comparison of the mean responses to the drug combination $S + A$ at low and high doses of A.

From comment 2 in Box 9.1, it follows that every linear combination of sample means is an unbiased estimate of the corresponding linear combination of population means; that is, $\hat{\theta} = c_1 \overline{Y}_1 + \cdots + c_t \overline{Y}_t$ is an unbiased estimate of $\theta = c_1 \mu_1 + \cdots + c_t \mu_t$. Contrasts between sample means can be used to test whether the data indicate differences between specific subsets of population means. A contrast $\hat{\theta}$ is said to be significant at level α if the null hypothesis H_0: $\theta = 0$ can be rejected in favor of H_1: $\theta \neq 0$ at level α.

The significance of a contrast can be evaluated using the two-sided t-test in Box 9.2. Since a two-sided t-test is equivalent to a one-sided F-test, there are two equivalent ways of performing such a test.

Method 1.   Declare $\hat{\theta}$ significant at level α if

$$|t_c| = \left| \frac{\hat{\theta}}{\hat{\sigma}_{\hat{\theta}}} \right| > t(\nu, \; \alpha/2),$$

where ν and $\hat{\sigma}_{\hat{\theta}}$ are as defined in Box 9.2.

Method 2. Declare $\hat{\theta}$ significant at level α if

$$F_c = t_c^2 = \left(\frac{\hat{\theta}}{\hat{\sigma}_{\hat{\theta}}}\right)^2 > F(1, \nu, \alpha).$$

If this F-statistic can be interpreted as a ratio of mean squares, we can establish a direct connection to ANOVA concepts. To construct such an interpretation, note that, on the basis of the results in Box 9.2, the test statistic in method 2 can be expressed as

$$F_c = \frac{\hat{\theta}^2}{\left(\dfrac{c_1^2}{n_1} + \cdots + \dfrac{c_t^2}{n_t}\right) \text{MS}[E]}. \tag{9.8}$$

For reasons that will soon be clear, we define the sum of squares for $\hat{\theta}$ as

$$\text{SS}[\hat{\theta}] = \frac{\hat{\theta}^2}{\left(\dfrac{c_1^2}{n_1} + \cdots + \dfrac{c_t^2}{n_t}\right)}. \tag{9.9}$$

If the sample sizes are equal—that is, if $n_1 = \cdots = n_t = n$—the sum of squares in Equation (9.9) can be expressed as

$$\text{SS}[\hat{\theta}] = \frac{n\hat{\theta}^2}{\left(c_1^2 + \cdots + c_t^2\right)}. \tag{9.10}$$

The statistic for testing the significance of $\hat{\theta}$ can be expressed as

$$F_c = \frac{\text{SS}[\hat{\theta}]}{\text{MS}[E]}. \tag{9.11}$$

We want to compare F_c in Equation (9.11) with the critical value of an F-distribution with one degree of freedom for the numerator and ν degrees of freedom for the denominator, and we know that the quantity $\text{MS}[E]$ in the denominator has ν degrees of freedom. Accordingly, let's think of $\text{SS}[\hat{\theta}]$ as a sum of squares with one degree of freedom. We then define the mean square for $\hat{\theta}$ as

$$\text{MS}[\hat{\theta}] = \frac{\text{SS}[\hat{\theta}]}{1},$$

so that the test statistic in Equation (9.11) can be expressed as

$$F_c = \frac{\text{MS}[\hat{\theta}]}{\text{MS}[E]}. \tag{9.12}$$

This expression can be interpreted as the ratio of two mean squares, one for testing $\hat{\theta}$ and the other for the error. Note the analogy between the calculation of F_c in Equation (9.12) and the calculation of the ANOVA F-statistic $F_c = \text{MS}[T]/\text{MS}[E]$ in Equation (8.3). Another useful property of the statistic in Equation (9.12) is that multiplying the coefficients of $\hat{\theta}$ by a constant does not change the value of $\text{SS}[\hat{\theta}]$

and hence $M\dot{S}[\hat{\theta}]$. For example, the sum of squares for testing the contrast

$$\hat{\theta} = \frac{1}{2}(\overline{Y}_1 + \overline{Y}_2) - \frac{1}{2}(\overline{Y}_3 + \overline{Y}_4)$$

is the same as the sum of squares for testing the contrast

$$\hat{\theta}^* = (\overline{Y}_1 + \overline{Y}_2) - (\overline{Y}_3 + \overline{Y}_4).$$

Notice that the contrast $\hat{\theta}^*$ can be obtained by multiplying the coefficients of $\hat{\theta}$ by 2.

EXAMPLE **9.6**   Let's determine the significance of the comparison $\hat{\theta}_1$ in Example 9.5 and interpret the result.

From the data in Example 9.1, we have $n_1 = n_2 = \cdots = n_5 = 4$, $\overline{Y}_1 = 25.00$, $\overline{Y}_2 = 15.75$, $\overline{Y}_3 = 11.50$, $\overline{Y}_4 = 19.00$, and $\overline{Y}_5 = 17.75$, so that

$$\hat{\theta}_1 = 25.00 - \frac{1}{4}(15.75 + 11.50 + 19.00 + 17.75) = 9.00.$$

Also, from Example 9.2, the mean square error $MS[E] = 6.43$ with $\nu = 15$ degrees of freedom, and so from Equation (9.10)

$$SS[\hat{\theta}_1] = MS[\hat{\theta}_1] = \frac{4(9.00)^2}{\left((1)^2 + (-\frac{1}{4})^2 + (-\frac{1}{4})^2 + (-\frac{1}{4})^2 + (-\frac{1}{4})^2\right)} = 259.20.$$

As an exercise, verify that this sum of squares will remain the same if the coefficients of $\hat{\theta}_1$ are multiplied by a constant—say, 4.

The calculated value of the test statistic is

$$F_c = \frac{MS[\hat{\theta}_1]}{MS[E]} = \frac{259.20}{6.43} = 40.31.$$

If we look at Table C.4 (Appendix C) with 1 and 15 degrees of freedom for the numerator and denominator, respectively, we find that the calculated value F_c is significant at the $\alpha = 0.01$ level. It follows that, even at the level $\alpha = 0.01$, the data support the conclusion that the expected drop in blood pressure for the standard treatment is different from the average of those expected for the other four treatments.   ∎

As already noted, the concept of mutually orthogonal contrasts allows us to explain the variability between observed treatment means in terms of differences between selected treatment subsets. It can be shown that, for t treatments, there can be at most $t - 1$ contrasts in a mutually orthogonal set and, for any given set of $t - 1$ mutually orthogonal contrasts, the sum of the sums of squares due to the contrasts will equal the treatment sum of squares. These results are summarized in Box 9.3.

BOX **9.3**   *Subdivision of treatment sum of squares*

1   If $\{\hat{\theta}_1, \hat{\theta}_2, \ldots, \hat{\theta}_k\}$ is a set of k mutually orthogonal contrasts between t sample means, then $k \leq t - 1$.

> **2** Let $\{\hat{\theta}_1, \hat{\theta}_2, \ldots, \hat{\theta}_{t-1}\}$ be a set of $t-1$ mutually orthogonal comparisons between t sample means and let SS[T] be the treatment sum of squares. Then
>
> $$SS[T] = SS[\hat{\theta}_1] + SS[\hat{\theta}_2] + \cdots + SS[\hat{\theta}_{t-1}].$$

It follows from Box 9.3 that the treatment sum of squares, which provides information about the observed differences between t sample means, can be expressed as the sum of $t-1$ quantities, each of which provides information about the observed difference between two specific subgroups of treatment means. Of course, the subgroups must be chosen in such a way that the contrasts for comparing the subgroups are mutually orthogonal.

In practice, the contrasts are selected on the basis of the objectives of the study. If these contrasts are mutually orthogonal, the contrast sum of squares provides a useful summary of the contributions of the contrasts to the treatment differences.

EXAMPLE **9.7** In Example 9.2, we presented the ANOVA table for a one-way analysis of BP reduction data. In this example, let's subdivide the treatment sum of squares using the set of orthogonal contrasts in Example 9.5.

Proceeding as in Example 9.6, we find that

$$\hat{\theta}_1 = 9.00, \quad SS[\hat{\theta}_1] = 259.20,$$

$$\hat{\theta}_2 = 4.25, \quad SS[\hat{\theta}_2] = 36.13$$

$$\hat{\theta}_3 = -1.25, \quad SS[\hat{\theta}_3] = 3.13$$

$$\hat{\theta}_4 = -4.75, \quad SS[\hat{\theta}_4] = 90.25.$$

Adding the sums of squares for the four comparisons, we get

$$SS[\hat{\theta}_1] + SS[\hat{\theta}_2] + SS[\hat{\theta}_3] + SS[\hat{\theta}_4] = 259.20 + 36.13 + 3.13 + 90.25 = 388.71,$$

which equals the treatment sum of squares given in Example 9.2, if we allow for rounding error. The sums of squares for the four orthogonal comparisons and the calculated values of the associated test statistics are most conveniently represented in an ANOVA table, as follows.

Source	df	SS	MS	F_c
Drugs	4	388.7	97.175	15.10
S versus other treatments ($\hat{\theta}_1$)	1	259.200	259.200	40.29
S + AH versus S + AL ($\hat{\theta}_2$)	1	36.125	36.125	5.62
S + BH versus S + BL ($\hat{\theta}_3$)	1	3.125	3.125	< 1
A versus B ($\hat{\theta}_4$)	1	90.250	90.250	14.03
Error	15	96.5	6.433	
Total	19	485.2		

The values of F_c are calculated using Equation (9.11). From Table C.4 (Appendix C), the critical value of the $F(1, 15)$-distribution is 3.07 for $\alpha = 0.10$, 4.54 for $\alpha = 0.05$, and 8.68 for $\alpha = 0.01$. We conclude that the expected drop in blood pressure is not significantly different for the high and low doses of drug B and is significant at the 0.05 level for the high and low doses of drug A. The difference between the expected BP reduction for the standard drug alone and the average of the expected BP reduction for the other four treatments is highly significant ($\alpha < 0.01$), as is the difference between the average results for A and B. It appears, therefore, that the observed differences between the sample means are mainly due to a difference between the response to the standard treatment and the average response to the other four treatments and a difference between the response to treatment with drug A and the response to treatment with drug B. However, this overall conclusion was based on the significance levels (or p-values) of four separate hypothesis tests. At this point, what significance level can be associated with this conclusion remains an open question. The significance levels associated with conclusions resulting from multiple hypothesis tests will be discussed in Section 9.4. ∎

Computational aspects

Calculations for testing the significance of contrasts can be performed using the GLM procedure in SAS. For details, see Section 9.2 in the companion text by Younger (1997). The SAS input codes for subdividing the treatment sum of squares using the four orthogonal comparisons considered in the previous example are as follows.

```
------------------------------------------------------------------
data bp;
  input trt rep bp @@;
  cards;
   1 1 27 1 2 26 1 3 21 1 4 26 2 1 19 2 2 13 2 3 15 2 4 16 3 1 15 3 2 10
   3 3 10 3 4 11 4 1 22 4 2 15 4 3 21 4 4 18 5 1 20 5 2 18 5 3 17 5 4 16
   ;
proc glm;
  class trt;
  model bp = trt;
  means trt / tukey;
  contrast 'S vs. rest'      trt 1 -.25 -.25 -.25 -.25;
  contrast 'S + AH vs. S + AL' trt 0 -.50  .50  0    0 ;
  contrast 'S + BH vs. S + BL' trt 0  0    0  -.50  .50;
  contrast 'A vs. B'         trt 0 -.50 -.50  .50  .50;
------------------------------------------------------------------
```

Portions of the corresponding SAS output are as follows.

```
------------------------------------------------------------------
              General Linear Models Procedure
Dependent Variable: BP
```

Source	DF	Sum of Squares	Mean Square	F Value	Pr > F
Model	4	388.70000	97.17500	15.10	0.0001
Error	15	96.50000	6.43333		
Corrected Total	19	485.20000			

General Linear Models Procedure

Dependent Variable: BP

Contrast	DF	Contrast SS	Mean Square	F Value	Pr > F
S vs. rest	1	259.20000	259.20000	40.29	0.0001
S + AH vs. S + AL	1	36.12500	36.12500	5.62	0.0316
S + BH vs. S + BL	1	3.12500	3.12500	0.49	0.4965
A vs. B	1	90.25000	90.25000	14.03	0.0019

Exercises

9.1 Let $\overline{Y}_1, \overline{Y}_2, \ldots, \overline{Y}_t$ be t treatment means and $c_1, c_2, \ldots, c_t$ be given constants such that $c_1 + c_2 + \cdots + c_t = 0$.

a Show that the linear combinations

$$\hat{\theta}_1 = c_1\overline{Y}_1 + c_2\overline{Y}_2 + \cdots + c_t\overline{Y}_t,$$
$$\hat{\theta}_2 = dc_1\overline{Y}_1 + dc_2\overline{Y}_2 + \cdots + dc_t\overline{Y}_t,$$

where d is some nonzero constant, are both contrasts.

b Verify that, for a given set of data, the statistics for testing the significance of $\hat{\theta}_1$ and $\hat{\theta}_2$ will have identical values.

c Use the result in (b) to conclude that multiplying a contrast by a constant or changing its sign will have no effect on the p-value of the associated significance test.

d Outline intuitive considerations that support the conclusion in (c).

9.2 Refer to the insulin release data in Exercise 8.3.

a Construct a 95% confidence interval for the expected insulin release at each of the three glucose concentrations. Interpret the confidence intervals you construct.

b Perform a hypothesis test to see if the data support the claim that the mean insulin release at high glucose concentration is at least one unit more than that at medium glucose concentration.

c Perform the hypothesis test in (b) using the confidence interval approach. Interpret the interval you construct.

d Let μ_1, μ_2, and μ_3 denote, respectively, the expected amount of insulin release at low, medium, and high glucose concentrations. Construct two contrasts such that one measures the difference between the expected releases at low and medium concentrations, while the other measures the difference between the expected release at high concentration and the average of the expected releases at low and medium concentrations.

e Check whether the contrasts you constructed in (d) are orthogonal.

f Test the significance of the contrasts in (d). Write the corresponding conclusions.

g Do the sums of squares for the contrasts in (d) subdivide the treatment sum of squares into single degree of freedom sum of squares? Explain.

9.3 In Exercise 8.4, data on soil bulk density for three grazing practices were presented. Let $(\bar{y}_1, \mu_1), (\bar{y}_2, \mu_2)$, and $(\bar{y}_3, \mu_3)$ denote sample and population mean bulk densities for treatment 1 (continuous grazing), treatment 2 (two-week grazing with a one-week rest), and treatment 3 (two-week grazing with a two-week rest), respectively.

a Construct a 95% confidence interval for $\mu_2 - \mu_3$. Use the confidence interval you construct to make an appropriate inference about the difference between soil bulk densities when two weeks of grazing is followed by one-week and two-week rest periods.

b Describe in words the quantity estimated by each of the following four contrasts:

  i $\hat{\theta}_1 = \bar{y}_1 - \frac{1}{2}(\bar{y}_2 + \bar{y}_3)$;   **ii** $\hat{\theta}_2 = \bar{y}_1 - \bar{y}_2$;

  iii $\hat{\theta}_3 = \bar{y}_2 - \bar{y}_3$;   **iv** $\hat{\theta}_4 = \frac{1}{2}(\bar{y}_1 + \bar{y}_2) - \bar{y}_3$.

c Without performing any calculations, argue that the four contrasts in (b) are not mutually orthogonal.

d Divide the four contrasts in (b) into two sets such that the two contrasts within a set are orthogonal. Which of these two sets would give you a meaningful subdivision of the treatment sum of squares?

e Test the significance of the two orthogonal contrasts you selected in (d) and write an overall summary of your conclusions.

9.4 Refer to the TRCP (tubular reabsorptive capacity for phosphate) data in Exercise 8.11.

a Construct a set of two orthogonal contrasts between treatment means such that one contrast measures the difference between the expected TRCP values for high and normal serum phosphate patients and the other measures the difference between the expected TRCP values of patients and normal controls.

b Use the contrasts in (a) to decompose the treatment sum of squares into two single degree of freedom components. Test the significance of these contrasts and write your conclusions.

9.5 Refer to the data on leaf area ratios in Exercise 8.13. Denote the 12 treatment means by $\bar{y}_1, \bar{y}_2, \ldots, \bar{y}_{12}$.

a Construct a set of four mutually orthogonal contrasts useful for estimating the following quantities:

  i the difference between the expected leaf area ratios for treatments with and without foliar P;

  ii the difference between the expected leaf area ratios for treatments with and without inoculation;

  iii the difference between the expected leaf area ratios for high and low release of P;

  iv the difference between the expected leaf area ratios for high P-release treatments with and without inoculation.

b Construct 95% confidence intervals for the quantities estimated by the four contrasts in (a). Interpret the resulting intervals.

9.3

Testing contrasts suggested by data

The Fisher critical contrast value

The significance of a contrast can be determined using a two-sided t-test. Such a test declares a contrast $\hat{\theta}$ significant—that is, rejects the null hypothesis H_0: $\theta = 0$—at level α if

$$|t_c| = \left| \frac{\hat{\theta}}{\hat{\sigma}_{\hat{\theta}}} \right| > t_{\nu,\,\alpha/2}. \tag{9.13}$$

Multiplying both sides of Equation (9.13) by $\hat{\sigma}_{\hat{\theta}}$, we obtain the equivalent inequality

$$|\hat{\theta}| > t(\nu,\,\alpha/2)\hat{\sigma}_{\hat{\theta}}. \tag{9.14}$$

Thus, the two-sided t-test declares $\hat{\theta}$ significant at level α if $|\hat{\theta}|$, the absolute value of the contrast, exceeds $t(\nu,\,\alpha/2)\hat{\sigma}_{\hat{\theta}}$. We refer to $t(\nu,\,\alpha/2)\hat{\sigma}_{\hat{\theta}}$ as the *Fisher critical contrast value* (Fisher CCV) for an α-level test of $\hat{\theta}$. Thus, according to Equation (9.14), we declare $\hat{\theta}$ significant at level α if its absolute value is larger than the α-level Fisher CCV. If the decision to test the significance of $\hat{\theta}$ is made without regard to the observed outcome of the experiment, then the test in Equation (9.14) guarantees that the probability of falsely declaring significance is α.

The Scheffe critical contrast value

If the decision to test was based on the outcome of the study, Equation (9.14) is no longer appropriate for testing the significance of $\hat{\theta}$. To understand why this is so, consider two possible reasons for deciding to compare two particular means selected from a set of four means. The first possibility is that the objectives of the study call for a comparison of the expected responses to two specific treatments—say, treatment 1 and treatment 2. Then the decision to make the comparison is made prior to observing the outcome, and the comparison can be made by testing the simple contrast

$$\hat{\theta} = \overline{Y}_1 - \overline{Y}_2 \tag{9.15}$$

on the basis of the Fisher CCV in Equation (9.14). A second possibility is that the sample means to be compared are selected because they correspond to the treatments having the smallest and the largest observed mean responses. Such a comparison can be made by testing the significance of the contrast

$$\hat{\theta}^* = \overline{Y}_{(4)} - \overline{Y}_{(1)}, \tag{9.16}$$

where $\overline{Y}_{(1)}$ and $\overline{Y}_{(4)}$ denote, respectively, the smallest and the largest observed means. It is inappropriate to determine the significance of $\hat{\theta}^*$ using the criterion for $\hat{\theta}$, because the difference between the largest and the smallest means will always be at least as large as the difference between any two means; that is, the observed value of $\hat{\theta}^*$ in Equation (9.16) will be at least as large as the observed value of $\hat{\theta}$ in Equation (9.15). Consequently, for any α, the cutoff value of $\hat{\theta}^*$ for rejecting the null hypothesis that there is no difference between the corresponding population means may be expected to be larger than the cutoff value of $\hat{\theta}$ needed to reject the corresponding null hypothesis. Indeed, generally speaking, the cutoff value for testing the significance of a contrast specified on the basis of the observed outcome of an experiment will differ from the cutoff value for a contrast specified without regard to the observed outcome. Box 9.4 summarizes a method of determining the cutoff value for testing a contrast suggested by the data. The cutoff value in Box 9.4 is called the *Scheffe critical contrast value* (Scheffe CCV).

BOX **9.4**

> ### The Scheffe method of testing contrasts suggested by data
>
> Let $\hat{\theta}$ be a contrast of the means of independent random samples from t normal populations with a common variance σ^2. Let $\hat{\sigma}_{\hat{\theta}}$ denote the estimated standard error of $\hat{\theta}$ and ν denote the degrees of freedom for the mean square for estimating σ^2. Then the test that declares $\hat{\theta}$ significant if
>
> $$|\hat{\theta}| > \left[\sqrt{(t-1)F(t-1, \nu, \alpha)}\right]\hat{\sigma}_{\hat{\theta}}$$
>
> has level no higher than α.

EXAMPLE **9.8**

Suppose that, on examining the data in Example 9.1, the investigator noticed that the mean drop in blood pressure for the two treatments involving drug B are higher than those for two treatments involving drug A. Surprised by this unexpected result, the investigator decided to see whether the data indicate a significant difference, at the $\alpha = 0.05$ level, between the expected BP reductions for treatments $S + A$ and $S + B$.

The comparison to be tested is the same as comparison $\hat{\theta}_4$ in Example 9.5. Because this comparison was suggested by the data, we use the Scheffe CCV to test its significance. We have $t = 5$, $n_1 = \cdots = n_5 = 4$, MS[E] $= 6.43$, $\nu = 15$, $\hat{\theta}_4 = -4.75$, and

$$\hat{\sigma}_{\hat{\theta}_4} = \sqrt{\left(\frac{(\frac{1}{2})^2}{4} + \frac{(\frac{1}{2})^2}{4} + \frac{(-\frac{1}{2})^2}{4} + \frac{(-\frac{1}{2})^2}{4}\right)(6.43)} = 1.268.$$

Therefore, the Scheffe CCV for $\alpha = 0.05$ is

$$\left(\sqrt{(5-1)F(4, 15, 0.05)}\right)(1.268) = (\sqrt{4 \times 3.06})(1.268) = 4.43.$$

Since $|\hat{\theta}_4| = 4.75$ is larger than 4.43, we conclude that $\hat{\theta}_4$ is significant at the 0.05 level; that is, there is a difference between the expected BP reductions associated with drugs A and B.   ■

Let's compare the conclusion in Example 9.8 with the corresponding result in Example 9.7. The significance test in Example 9.7 was based on the assumption that the decision to test $\hat{\theta}_4$ was made before the outcome of the study was observed. In other words, it was the same as the two-sided t-test that uses the Fisher CCV in Equation (9.14). A simple calculation will show that the Fisher CCV for testing $\hat{\theta}_4$ equals $(2.131)(1.268) = 2.70$, which is smaller than the corresponding Scheffe CCV of 4.43. Thus, the test in Example 9.8 is more conservative than the test in Example 9.7; that is, the test in Example 9.7 is more likely to declare significance. Indeed, this is as it should be, because in Example 9.8 two groups that look different on the basis of the observed outcome of the experiment are compared, whereas in Example 9.7 two groups selected prior to observing the data are compared.

Exercises

9.6   Refer to the TRCP data in Exercise 8.11.

   a   An examination of the data reveals that the TRCP values for patients with high serum phosphate (treatment 1) were consistently larger than the values for patients with normal serum phosphate (treatment 2), as well as the values for normal controls (treatment 3). Furthermore, the TRCP values for treatments 2 and 3 look about the same. Suggest a contrast that can be used to test the research hypothesis that the expected TRCP value for treatment 1 is larger than the average of the expected responses for treatments 2 and 3.

   b   Use the contrast in (a) to see if the data support the research hypothesis. Write your conclusions.

9.7   Refer to the data for the leaf area ratio in Exercise 8.13. Consider the four means for treatments in which P release was low. Notice that the differential mean response for the two foliar P levels is -0.9 ($10.7 - 11.6 = -0.9$) when there is no inoculation and $+0.7$ ($13.5 - 12.8 = +0.7$) when there is inoculation. Thus, it appears that, with low P release, foliar P causes an increase in the leaf area ratio for inoculated plants and a decrease in the leaf area ratio for noninoculated plants.

   a   Select an appropriate contrast to test the null hypothesis that, with low P release, the differential mean responses for the two foliar P levels are the same for inoculated and noninoculated plants.

   b   Use the contrast in (a) to see if the data support the conclusion that, with low P release, the difference between the effects of foliar P application depends on whether the plants have been inoculated.

9.4
Multiple comparison of means

In this section, we consider methods appropriate for simultaneously testing more than one contrast between t means. We saw such a situation in Example 9.7, where we tested $k = 4$ contrasts between $t = 5$ treatment means. A procedure for simultaneously testing more than one contrast will be called a *multiple comparison test*.

An obvious method of conducting a multiple comparison test of k contrasts is to perform k individual significance tests, one for each of the k contrasts. This is precisely what we did in Example 9.7, where we performed four separate F-tests of four contrasts. Each of the four tests was evaluated on the basis of its attained p-value without regard to the results of other tests.

When the final answer to a research question depends on the results from more than one significance test, the investigator needs to understand how the error rates of individual tests influence the error rates applicable to the overall conclusions from a consideration of several tests. This topic has attracted much attention from theoretical statisticians. From a practitioner's point of view, it is important to know what can be said about the joint performance of a set of k significance tests, each of which has a significance level α.

There are two popular criteria for evaluating the joint performance of multiple significance tests: the comparisonwise error rate and the experimentwise error rate. Our next task is to formally define these two types of error rates and consider multiple comparison procedures for which these error rates can be controlled.

The comparisonwise error rate

In a multiple comparison procedure, the expected proportion of the contrasts that will be declared significant when there are no differences between the population means is called the *comparisonwise error rate*. More precisely, if k_s denotes the number of contrasts declared significant in a multiple comparison procedure involving k contrasts, then the expected value of the proportion k_s/k when none of the contrasts tested is truly significant is the comparisonwise error rate of the procedure.

A multiple comparison procedure in which each contrast is tested at level α will have a comparisonwise error rate of α, because the chance of obtaining significance in a single α-level test of a contrast that is not truly significant—that is, a contrast where H_0 is true—is α. Consequently, in k individual α-level tests of k contrasts, none of which are truly significant, the expected proportion of significant results—that is, the proportion of times H_0 is rejected—will be α. One method of constructing a multiple comparison procedure with comparisonwise error rate α is to use the α-level Fisher CCV as the cutoff point for testing each of the k comparisons. Such a procedure is called the *Fisher multiple comparison method* and is described in Box 9.5.

BOX 9.5

> ### The Fisher multiple comparison method
>
> Let $\hat{\theta}_1, \hat{\theta}_2, \ldots, \hat{\theta}_k$ denote k contrasts between the means of t independent random samples from normal populations with a common variance σ^2. The multiple comparison procedure that declares $\hat{\theta}_i$ significant if
>
> $$|\hat{\theta}_i| > t(\nu, \alpha/2)\hat{\sigma}_{\hat{\theta}_i},$$
>
> where ν denotes the degrees of freedom for the estimate of σ^2 and $\hat{\sigma}_{\hat{\theta}_i}$ is the estimated standard error of $\hat{\theta}_i$, has comparisonwise error rate α.

The experimentwise error rate

The comparisonwise error rate of a multiple comparison procedure has the major drawback that the expected number of false significances will depend on the actual number of contrasts tested. In the long run, when used to test the significance of k contrasts, none of which is truly significant, a multiple comparison procedure with comparisonwise error rate α will declare $k\alpha$ false significances, on average. Thus, the expected number of false significances will increase as the number of contrasts increases. For example, a procedure using a comparisonwise error rate of $\alpha = 0.05$ can be expected to declare one false significance when testing 20 contrasts and five false significances when testing 100 contrasts.

The experimentwise error rate of a multiple comparison procedure is a natural generalization of the Type I error rate associated with the significance test of a single contrast. A multiple comparison procedure is said to have an *experimentwise error rate* α if the probability of declaring at least one (false) significance when testing k contrasts that are not truly significant is α.

The experimentwise error rate can be interpreted as the average number of experiments in which the procedure will declare one or more false significances when none of the contrasts is truly significant. Let's look at two multiple comparison methods that can be designed so as to have an experimentwise error rate less than a specified value α.

The first is the Scheffe multiple comparison method (Box 9.6).

BOX 9.6

> ### The Scheffe multiple comparison method
>
> Let $\hat{\theta}_1, \hat{\theta}_2, \ldots, \hat{\theta}_k$ denote k contrasts between the means of independent samples from t normal populations with a common variance σ^2. The multiple comparison procedure that declares $\hat{\theta}_i$ significant if
>
> $$|\hat{\theta}_i| > \left(\sqrt{(t-1)F(t-1, \nu, \alpha)} \right) \hat{\sigma}_{\hat{\theta}_i},$$

where ν denotes the degrees of freedom for the estimate of σ^2 and $\hat{\sigma}_{\hat{\theta}_i}$ is the estimated standard error of $\hat{\theta}_i$, has an experimentwise error rate no higher than α.

An important characteristic of the Scheffe method is that it does not actually depend on k, the number of contrasts that will be tested. As such, the α-level Scheffe method will guarantee that the experimentwise error rate does not exceed α, regardless of the number of contrasts tested. Indeed, the Scheffe multiple comparison procedure guarantees the experimentwise error rate even if we test all possible contrasts, including those suggested by the data. For this reason, the cutoff value in Box 9.6 is exactly the same as the cutoff value in Box 9.4 for testing a contrast suggested by the data.

Since the Scheffe procedure is designed to protect against declaring false significances of arbitrarily many contrasts (including those that are not under test), the associated cutoff values tend to be rather large. Consequently, when the number of contrasts to be tested is not very large, the Scheffe method will be too conservative; that is, it will mask too many true significances. For small to moderate values of k and contrasts specified in advance, a method based on a probabilistic result due to Bonferroni is preferable (Box 9.7).

BOX **9.7**   ***The Bonferroni multiple comparison method***

Let $\hat{\theta}_1$, $\hat{\theta}_2$, ... , $\hat{\theta}_k$ denote k contrasts between the means of independent samples from t normal populations with a common variance σ^2. The multiple comparison procedure that declares $\hat{\theta}_i$ significant if

$$|\hat{\theta}_i| > t(\nu, \alpha, k)\hat{\sigma}_{\hat{\theta}_i},$$

where ν denotes the degrees of freedom for the estimate of σ^2, $t(\nu, \alpha, k)$ is the $\alpha/2k$-level critical value of a $t(\nu)$-distribution, and $\hat{\sigma}_{\hat{\theta}_i}$ is the estimated standard error of $\hat{\theta}_i$, has an experimentwise error rate no greater than α.

The Bonferroni method at level α is equivalent to the Fisher method with level α/k. For instance, the 0.05-level Bonferroni cutoff value for testing $k = 4$ contrasts utilizes the critical value of a t-distribution corresponding to the level $\alpha = (0.05)/(2 \times 4) = 0.0063$. You'll notice that Table C.2 (Appendix C) does not provide critical values at levels such as $\alpha = 0.0063$, but Table C.10 lists the Bonferroni critical values $t(\nu, \alpha, k)$ for $\alpha = 0.05$ and some typical values of k and ν.

Section 9.4 in the companion text by Younger (1997) describes how to use SAS to determine the Bonferroni critical values for any combination of α, v, and k.

EXAMPLE **9.9**   Let's use the multiple comparison methods described in this section to test the significance of the four comparisons in Example 9.5, first at the comparisonwise error rate $\alpha = 0.05$ and then at the experimentwise error rate $\alpha = 0.05$.

In Example 9.5 there were $t = 5$ treatments, each replicated $n = 4$ times. The estimated value of σ^2 was MS[E] = 6.43, with $v = 15$ degrees of freedom. The four contrasts of interest were

$$\hat{\theta}_1 = \overline{Y}_1 - \frac{1}{4}(\overline{Y}_2 + \overline{Y}_3 + \overline{Y}_4 + \overline{Y}_5),$$

$$\hat{\theta}_2 = \overline{Y}_2 - \overline{Y}_3,$$

$$\hat{\theta}_3 = \overline{Y}_4 - \overline{Y}_5,$$

$$\hat{\theta}_4 = \frac{1}{2}(\overline{Y}_2 + \overline{Y}_3) - \frac{1}{2}(\overline{Y}_4 + \overline{Y}_5).$$

The calculated values of the contrasts were given in Example 9.7. The estimated standard errors of these contrast estimates can be calculated as in Box 9.2

$$\hat{\sigma}_{\hat{\theta}} = \sqrt{\left(\frac{c_1^2}{n_1} + \frac{c_2^2}{n_2} + \cdots + \frac{c_t^2}{n_t}\right) \text{MS[E]}}. \tag{9.17}$$

The contrast values and their estimated standard errors are listed below for ready reference.

i	1	2	3	4
$\hat{\theta}_i$	9.00	4.25	-1.25	-4.75
$\hat{\sigma}_{\hat{\theta}_i}$	1.42	1.80	1.80	1.27

To attain a comparisonwise error rate of $\alpha = 0.05$, we use the Fisher method at level $\alpha = 0.05$. The Fisher CCV for $\hat{\theta}_i$ is

$$t(15, \ 0.025)\hat{\sigma}_{\hat{\theta}_i} = (2.131)\hat{\sigma}_{\hat{\theta}_i},$$

where $\hat{\sigma}_{\hat{\theta}_i}$ is the estimated standard error calculated using Equation (9.17). The Scheffe CCV at an experimentwise error rate $\alpha = 0.05$ is

$$\left(\sqrt{4F(4, \ 15, \ 0.05)}\right)\hat{\sigma}_{\hat{\theta}_i} = (3.50)\hat{\sigma}_{\hat{\theta}_i}$$

and the corresponding Bonferroni CCV is

$$t(15, 0.05, \ 4)\hat{\sigma}_{\hat{\theta}_i} = (2.837)\hat{\sigma}_{\hat{\theta}_i}.$$

The absolute values of the four contrasts and the corresponding 0.05-level cutoff values for the three multiple comparison procedures are shown in the following table:

Contrast	Absolute value of contrast	Standard Error	Fisher CCV	Bonferroni CCV	Scheffe CCV
$\hat{\theta}_1$	9.00	1.42	3.02	4.03	4.97
$\hat{\theta}_2$	4.25	1.80	3.84	5.11	6.30
$\hat{\theta}_3$	1.25	1.80	3.80	5.11	6.30
$\hat{\theta}_4$	4.75	1.27	2.71	3.60	4.45

Any contrast with an absolute value larger than the cutoff value will be declared significant by the corresponding method. From this table, it can be verified that the 0.05-level Fisher test, which has a comparisonwise error rate $\alpha = 0.05$, will declare $\hat{\theta}_1$, $\hat{\theta}_2$, and $\hat{\theta}_4$ significant. The more conservative 0.05-level experimentwise error rate tests will declare only $\hat{\theta}_1$ and $\hat{\theta}_4$ significant. An examination of the cutoff values associated with the three methods shows that the Scheffe method is the most conservative (largest cutoff values) and the Fisher method is the least conservative.   ∎

Which multiple comparison method is most appropriate depends on the error rate that is to be controlled. The problem with the Fisher method is that, when simultaneously testing many contrasts, the likelihood of finding one or more false significance can be quite large. For instance, suppose that we use the 0.05-level Fisher method to test $k = 4$ contrasts, none of which is truly significant. Suppose further that the four individual tests are statistically independent—that is, that the outcome of one test in no way depends on the outcome of any of the other three. Then the probability that the Fisher procedure will declare at least one false significance is 0.185. The following table displays the probabilities of one or more false significances in the Fisher method, for some typical values of k.

	α	
k	0.05	0.01
2	0.098	0.020
3	0.143	0.030
4	0.185	0.039
5	0.226	0.049
10	0.401	0.096
20	0.642	0.182

As we see from this table, the experimentwise error rate of the 0.05-level Fisher method can be as high as 22% even when testing only five contrasts. The problem with the Fisher method is that it tends to declare too many false significances, especially when k is large.

Now let's turn to experimentwise error rates. The Scheffe method controls error rates associated with all possible contrasts, even those that are not under test. Thus, unlike the Bonferroni method, the Scheffe method controls the experimentwise error rate when testing multiple contrasts specified on the basis of the observed outcome. If we are testing all simple contrasts—that is, differences between two sample means—the Tukey method (Section 9.5) is preferable to the Scheffe method.

To test contrasts specified in advance, we can choose between the Scheffe and Bonferroni methods. Smaller cutoff values result in increased probability of declaring significance, and hence tests with smaller cutoff values will have higher power to detect true significances. Consequently, of the two methods for testing at a given significance level, the one with the smaller cutoff value is preferred.

An examination of the results in Boxes 9.6 and 9.7 shows that, if we know the values of α, ν, k, and t, we can compare the cutoff values by comparing the magnitudes of the corresponding coefficients of $\hat{\sigma}_{\hat{\theta}_i}$: $\sqrt{(t-1)F(t-1, \nu, \alpha)}$ for the Scheffe method; and $t(\nu, \alpha, k)$ for the Bonferroni method.

EXAMPLE **9.10**   Suppose that we want to test $k = 6$ contrasts among $t = 10$ means. Assume that the available estimate of σ^2 has $\nu = 20$ degrees of freedom and that we desire an experimentwise error rate $\alpha \leq 0.05$. Which multiple comparison method should we use?

To find out, we can compare the coefficients of $\hat{\sigma}_{\hat{\theta}_i}$ in the cutoff values associated with the two procedures. The Scheffe coefficient is

$$\sqrt{(t-1)F(t-1, \nu, \alpha)} = \sqrt{(10-1)F(10-1, 20, 0.05)} = \sqrt{(9)(2.393)} = 4.64$$

and the Bonferroni coefficient is

$$t(\nu, \alpha, k) = t(20, 0.05, 4) = 2.93.$$

The smaller cutoff value is associated with the Bonferroni method, which will consequently have higher power than the Scheffe method.

Now suppose that we have $t = 8$ treatments and the estimate of σ^2 has $\nu = 5$ degrees of freedom. Suppose further that we want an overall summary of the significances of the differences between the observed mean responses for the eight treatments. Such a summary can be obtained by testing all possible simple contrasts between the eight treatment means. As we see in Section 9.5, testing all possible simple contrasts between eight means entails testing 28 simple contrasts of the form $\hat{\theta} = \bar{Y}_i - \bar{Y}_j$. When $t = 8$, $\nu = 5$, $k = 28$, and $\alpha = 0.05$, the coefficients of $\hat{\sigma}_{\hat{\theta}}$ in the cutoff values for the Scheffe and Bonferroni methods are, respectively, 5.842 and 6.045. Thus, in this case, the Scheffe method is preferable to the Bonferroni method.   ∎

Exercises

9.8   Refer to Exercise 9.2.

a   Use the Fisher, Scheffe, and Bonferroni multiple comparison methods, each at the 0.05 level, to test the significance of the two contrasts in Exercise 9.2d.

b   Summarize the results of the three tests in (a).

9.9   Refer to Exercise 9.3.

a   Use the 0.05-level Bonferroni and Scheffe multiple comparison methods to test the significance of the four contrasts in Exercise 9.3b and write your conclusions.

    b   Use the Fisher method at the 0.05 level to test the significances of the two comparisons in Exercise 9.3e. Explain why your results are the same as the result from the corresponding single degree of freedom *F*-tests at the 0.05 level.

    c   Perform the significance tests of the two comparisons in Exercise 9.3e using the Bonferroni and Scheffe methods at the 0.05 level. Explain why the results for the Scheffe method will be the same as the corresponding results in (a).

9.10   Refer to Exercise 9.4.

    a   Use the Bonferroni and Scheffe methods, each at the 0.05 level, to test the significance of the contrasts in Exercise 9.4a.

    b   Compare the results in (a) with the results in Exercise 9.4b. Comment on the differences and similarities of the results from the three procedures.

9.5
Multiple pairwise comparison of means

In experiments that are exploratory in nature, the researcher may not have any comparisons in mind at first, but rather may be interested in an overall summary of the significances of the differences between the mean responses. As noted in Example 9.10, such a summary can be obtained by making all possible pairwise comparisons of the sample means. In this section, we look at some multiple comparison procedures for this task.

To perform all possible pairwise comparisons between t means, we need to test $k = t(t - 1)/2$ simple contrasts, each of the form $\hat{\theta} = \overline{Y}_i - \overline{Y}_j$. For instance, with $t = 3$ means, there will be $k = 3(3 - 1)/2 = 3$ pairwise comparisons involving the pairs $\{\overline{Y}_1, \overline{Y}_2\}$, $\{\overline{Y}_1, \overline{Y}_3\}$, and $\{\overline{Y}_2, \overline{Y}_3\}$. Similarly, pairwise comparison of four means entails testing $k = 4(4 - 1)/2 = 6$ contrasts.

Clearly, the methods of multiple comparison in Section 9.4 can be used to test all possible pairwise comparisons. However, $k = t(t - 1)/2$ increases rapidly with increasing t (for instance, when $t = 6$, $k = 15$; when $t = 8$, $k = 28$; when $t = 10$, $k = 45$), and so the number of pairwise comparisons that need to be tested may be large. When used to test a large number of comparisons, the Fisher method tends to be too liberal (it declares too many false significances), whereas the Scheffe and Bonferroni methods tend to be too conservative (they declare too few true significances).

Nevertheless, multiple pairwise comparison is one of the most frequently used techniques of data summarization, and there are several methods specifically designed for this task. As for the multiple comparison methods in Section 9.4, each multiple pairwise comparison method is associated with a set of critical values called *least significant differences* (LSD). The procedure is implemented by comparing each of the $k = t(t - 1)/2$ pairwise differences between the t means with the corresponding LSD value. If a difference between a pair of means is larger than the corresponding LSD value, it is regarded as significant.

When a pairwise multiple comparison procedure requires a large number of significance tests, it is best to use a computer; the companion text by Younger (1997) shows how to use SAS for this purpose. The systematic scheme in Box 9.8 can be used for pairwise comparisons involving a large number of significance tests if no computer is available.

BOX **9.8**

A scheme for multiple pairwise comparisons

Multiple pairwise comparison of t means $\overline{Y}_1, \overline{Y}_2, \ldots, \overline{Y}_t$ can be implemented using the following four steps.

Step 1.   Arrange the means in increasing order. Let $\overline{Y}_{(i)}$ denote the i-th largest mean; that is, $\overline{Y}_{(1)}$ is the smallest mean, and $\overline{Y}_{(t)}$ is the largest mean.

Step 2.   Determine the order in which the differences will be tested. Start with the largest mean minus the smallest mean $(\overline{Y}_{(t)} - \overline{Y}_{(1)})$ and end with the second smallest mean minus the smallest mean $(\overline{Y}_{(2)} - \overline{Y}_{(1)})$. The order of testing for the case when $t = 4$ is shown in the first two columns of the table in Example 9.11.

Step 3.   Test the significance of a specific difference $\overline{Y}_{(i)} - \overline{Y}_{(j)}$, where $i > j$, as follows. If $\overline{Y}_{(i)}$ and $\overline{Y}_{(j)}$ are within the range of two means that have previously been declared not significantly different, then the difference between $\overline{Y}_{(i)}$ and $\overline{Y}_{(j)}$ is declared nonsignificant. Otherwise, the difference is compared to the appropriate LSD value, LSD_{ij}. The means are declared significantly different if

$$\overline{Y}_{(i)} - \overline{Y}_{(j)} > \text{LSD}_{ij}.$$

Step 4.   Group together the means that are not significantly different from each other. The groups may overlap. They are identified by underlining the ordered means belonging to the same group.

The next example illustrates the use of the scheme in Box 9.8.

EXAMPLE **9.11**   Suppose we wish to perform a pairwise multiple comparison of the $t = 4$ means $\overline{Y}_1 = 13.6$, $\overline{Y}_2 = 12.0$, $\overline{Y}_3 = 16.2$, and $\overline{Y}_4 = 9.3$. The first step is to order the means in increasing order. The ordered means are

$$\overline{Y}_{(1)} = 9.3, \ \overline{Y}_{(2)} = 12.0, \ \overline{Y}_{(3)} = 13.6, \ \text{and} \ \overline{Y}_{(4)} = 16.2.$$

In the second step, we decide on the order in which the $k = 4(4 - 1)/2 = 6$ pairwise comparisons will be tested. The first comparison to be tested is the difference between $\overline{Y}_{(4)}$ and $\overline{Y}_{(1)}$. This corresponds to $i = 4$ and $j = 1$ in Box 9.8. The pairs of means to be compared and their differences are listed in the first two columns of the following table. The third column contains a hypothetical set of LSD values for implementing the third step in Box 9.8. The LSD values needed in a specific application will depend on the multiple comparison procedure chosen and the desired error rate. The fourth column contains the results (S for significant; NS for not significant) of the significance tests in the third step in Box 9.8.

i	j	$\overline{Y}_{(i)} - \overline{Y}_{(j)}$	LSD_{ij}	Result
4	1	$16.2 - 9.3 = 6.9$	5.8	S
4	2	$16.2 - 12.0 = 4.2$	5.3	NS
4	3	—	—	NS
3	1	$13.6 - 9.3 = 4.3$	3.8	S
3	2	—	—	NS
2	1	$12.0 - 9.3 = 2.7$	3.6	NS

The difference between $\overline{Y}_{(4)}$ and $\overline{Y}_{(1)}$ is significant, because the value of the contrast $\overline{Y}_{(4)} - \overline{Y}_{(1)}$ is 6.9, which is larger than the corresponding LSD value, $LSD_{41} = 5.8$. However, the difference between $\overline{Y}_{(4)}$ and $\overline{Y}_{(2)}$ is not significant. Also, the difference between the means in the pairs $\{\overline{Y}_{(4)}, \overline{Y}_{(3)}\}$ and $\{\overline{Y}_{(3)}, \overline{Y}_{(2)}\}$ is not significant, because both pairs lie within the range of the pair $\{\overline{Y}_{(4)}, \overline{Y}_{(2)}\}$, which have already been declared not significantly different. On the basis of the fourth step in Box 9.8, the final groupings corresponding to the significances in the fourth column of the table can be displayed as follows:

Treatment	4	2	1	3
Treatment mean	9.3	12.0	13.6	16.2

In what follows, we describe six multiple pairwise comparison methods. Three are direct adaptations of the multiple comparison procedures discussed in Section 9.4. The other three are specially designed for multiple pairwise comparison. Each of the six procedures can be used with any selected value of α $(0 < \alpha < 1)$. The choice of α will determine the error rates associated with each procedure. As for the multiple comparison procedures in Section 9.4, the relationship of α to the comparisonwise or experimentwise error rates will determine the desirability of using the procedure in any particular application.

In the following, $\overline{Y}_{(i)}$ is the i-th largest in a set of t means and LSD_{ij} will denote the LSD value for testing the simple contrast $\hat{\theta} = \overline{Y}_{(i)} - \overline{Y}_{(j)}$. Also, $n_{(i)}$, MS[E], and ν will denote, respectively, the number of replications in $\overline{Y}_{(i)}$, the mean square error for estimating σ^2, and the degrees of freedom for MS[E].

The Fisher multiple pairwise comparison method

To test a simple contrast $\hat{\theta}$, the α-level Fisher multiple pairwise comparison method uses the LSD value

$$\text{LSD}_{ij}(\text{F}) = t(v, \alpha/2) \sqrt{\left(\frac{1}{n_{(i)}} + \frac{1}{n_{(j)}}\right) \text{MS[E]}}. \tag{9.18}$$

The LSD value in Equation (9.18) is the Fisher critical contrast value in Box 9.5 appropriate for testing the simple contrast $\hat{\theta} = \overline{Y}_{(i)} - \overline{Y}_{(j)}$. Therefore, the α-level Fisher multiple pairwise comparison method will have a comparisonwise error rate of α. The corresponding experimentwise error rate will be between α and $k\alpha$, where $k = t(t-1)/2$. Thus, for large values of t, the experimentwise error rate can be fairly large; the test will declare an excessive number of false significances. For example, with $t = 6$ treatments requiring $k = 15$ pairwise comparisons, the experimentwise error rate for a 0.05-level Fisher multiple pairwise comparison procedure can be as large as $15 \times 0.05 = 0.75$.

The Scheffe multiple pairwise comparison method

The LSD values for the Scheffe multiple pairwise comparison method are the same as the Scheffe critical contrast values in Box 9.6 for testing simple contrasts

$$\text{LSD}_{ij}(\text{S}) = \sqrt{(t-1)F(t-1, v, \alpha) \left(\frac{1}{n_{(i)}} + \frac{1}{n_{(j)}}\right) \text{MS[E]}}. \tag{9.19}$$

Unlike the Fisher method, the Scheffe multiple pairwise comparison method will control the experimentwise error rate. The experimentwise error rate of an α-level Scheffe method will not exceed α.

The Bonferroni multiple pairwise comparison method

To control the experimentwise error rate, the Bonferroni modification of the Fisher method is an alternative to the Scheffe multiple pairwise comparison method. The LSD value for the α-level Bonferroni multiple pairwise comparison procedure can be obtained by setting $k = t(t-1)/2$ in Box 9.7

$$\text{LSD}_{ij}(\text{B}) = t(v, \alpha, k) \sqrt{\left(\frac{1}{n_{(i)}} + \frac{1}{n_{(j)}}\right) \text{MS[E]}}, \tag{9.20}$$

where $t(v, \alpha, k)$ can be obtained from Table C.10 (Appendix C).

The Tukey multiple pairwise comparison method

If the sample sizes are equal—that is, if $n_1 = n_2 = \cdots = n_t = n$—a procedure less conservative than the Scheffe method is the *Tukey multiple pairwise comparison method.* The LSD value for the Tukey method is given by

$$LSD_{ij}(T) = q(t, \nu, \alpha)\sqrt{\frac{MS[E]}{n}}, \tag{9.21}$$

where $q(t, \nu, \alpha)$ is called the α-level *studentized range* for t means and ν degrees of freedom. These values are tabulated in Table C.11 (Appendix C). Just like the Scheffe and Bonferroni methods, the α-level Tukey multiple pairwise comparison method has an experimentwise error rate of less than α.

Alternatives to the Tukey method

Two alternatives to the Tukey method are often used for multiple pairwise comparisons. These methods are less conservative than the Tukey procedure; they are compromise procedures that have a higher chance of detecting true significances but do not have a guaranteed upper bound on the experimentwise error rate.

One such compromise is the α-level *Student-Neuman-Keuls multiple pairwise comparison method.* The LSD values needed by the Student-Neuman-Keuls (SNK) method are defined as

$$LSD_{ij}(SNK) = q(p, \nu, \alpha)\sqrt{\frac{MS[E]}{n}}, \qquad i > j, \tag{9.22}$$

where $p = i - j + 1$; $q(p, \nu, \alpha)$ and n are as in Equation (9.21). Note that, for given i and j, $i - j + 1$ is the number of means contained in the range of values between the i-th and j-th largest (both inclusive) sample means.

The performance characteristics of the SNK method fall somewhere between the Fisher and Tukey methods. There is no definite relationship between α and the comparisonwise or experimentwise error rates. When there is no difference between the population means, the probability of declaring a significant difference between the i-th and j-th largest means will be α for all $i \leq j$.

Another alternative to the Tukey method is the α-level *Duncan multiple pairwise comparison method.* The LSD value in this case is given by

$$LSD_{ij}(D) = R(p, \nu, \alpha)\sqrt{\frac{MS[E]}{n}}, \qquad j < i, \tag{9.23}$$

where $p = i - j + 1$ and $R(p, \nu, \alpha)$ is the α-level *Duncan's studentized range* for p means and ν degrees of freedom. The values of $R(p, \nu, \alpha)$ are listed in Table C.12 (Appendix C). As for the SNK method, the value of α in Equation (9.23) does not have any specific relationship to the experimentwise or comparisonwise error rates of the method. If the population means are equal, then, for any $1 \leq j < i \leq t$, the probability is $1 - (1 - \alpha)^{i-j}$ that the α-level Duncan method will declare the i-th and the j-th largest means to be significantly different. For example, if the means of four populations are equal, the probability that a 0.05-level Duncan's method will declare the smallest ($j = 1$) and the largest ($i = 4$) sample means to be significantly different is $1 - (1 - 0.05)^{4-1} = 1 - (0.95)^3 = 0.14$.

The next example illustrates the calculations and conclusions corresponding to the six multiple pairwise comparison procedures discussed thus far.

EXAMPLE 9.12   Caceres (1990) used a completely randomized design with six replications per treatment to compare the digestibility, in sheep, of four types of hay: June-harvested Mott elephant grass (JM); September-harvested Mott elephant grass (SM); June-harvested Pensacola Bahia grass (JP); and September-harvested Pensacola Bahia grass (SP). The experimental animals—24 crossbred mature wethers, each of approximately 60-kg body weight—were randomly allotted to individual metabolism pens. The four types of hay were randomly assigned to six pens each. The measured organic matter content (expressed as a percentage of total dry matter) in the feces of animals fed the four types of hay is as follows:

Hay	JP	SP	JM	SM
Treatment	1	2	3	4
Sample size	6	6	6	6
Mean	60.3	63.8	67.2	67.1

The corresponding ANOVA table is as follows:

Source	df	SS	MS
Hay	3	192.84	64.28
Error	20	63.95	3.20
Total	23	256.79	

Let's use the six pairwise comparison methods (with $\alpha = 0.05$) to compare the four means. In accordance with step 1 in Box 9.8, the treatment means may be arranged in order of increasing magnitude as follows:

Hay	JP	SP	SM	JM
Rank	1	2	3	4
Symbol	$\overline{Y}_{(1)}$	$\overline{Y}_{(2)}$	$\overline{Y}_{(3)}$	$\overline{Y}_{(4)}$
Sample size	6	6	6	6
Mean	60.3	63.8	67.1	67.2

First, let's use the 0.05-level Fisher method. Substituting $\alpha = 0.05$, $t = 4$, $n_i = n_j = 6$, and MS[E] $= 3.20$ into Equation (9.18), we get

$$\text{LSD}_{ij}(\text{F}) = t(20, 0.025)\sqrt{\left(\frac{1}{6} + \frac{1}{6}\right)(3.20)} = 2.15,$$

so that any difference that is at least as large as 2.15 will be declared significant by the 0.05-level Fisher method. In accordance with step 2 in Box 9.8, testing starts with the comparison of JM and JP (the largest mean and the smallest mean) and ends with the comparison of SP and JP (the second smallest mean and the smallest mean). The result of these comparisons is in the following table, where S denotes a significant difference and NS denotes a nonsignificant difference:

Comparison	Difference	Conclusion
JM versus JP	$67.2 - 60.3 = 6.9 > 2.15$	S
JM versus SP	$67.2 - 63.8 = 3.4 > 2.15$	S
JM versus SM	$67.2 - 67.1 = 0.1 < 2.15$	NS
SM versus JP	$67.1 - 60.3 = 6.8 > 2.15$	S
SM versus SP	$67.1 - 63.8 = 3.3 > 2.15$	S
SP versus JP	$63.8 - 60.3 = 3.5 > 2.15$	S

The groupings resulting from these tests may be indicated by joining with a common underline the group means that are not significantly different from each other

$$JP \quad SP \quad \underline{SM \quad JM} \qquad \text{(9.24)}$$

On the basis of the 0.05-level Fisher multiple pairwise comparison method, we can conclude that the difference observed between the mean organic matter contents in the feces from animals fed SM and JM is not significant. All other pairs of means are significantly different from each other.

The groupings in Equation (9.24) are based on a method that has a 5% comparisonwise error rate. To obtain an experimentwise error rate of less than 5%, we have two options: the 0.05-level Scheffe method or the 0.05-level Bonferroni method. Let's look at the Bonferroni method first.

We are making $k = 4(4 - 1)/2 = 6$ pairwise comparisons, and so the LSD value is computed with $\alpha = 0.05$ and $k = 6$ in Equation (9.20)

$$\text{LSD}_{ij}(\text{B}) = t(20, 0.05, 6)\sqrt{\left(\frac{1}{6} + \frac{1}{6}\right)(3.20)} = 3.02.$$

The Bonferroni LSD value of 3.02 is larger than the Fisher LSD value of 2.15, and so the Bonferroni method is less likely than the LSD method to detect differences. However, in this instance, the conclusions resulting from the 0.05-level Bonferroni procedure will be identical to those obtained on the basis of the 0.05-level Fisher method.

The Scheffe method with $\alpha = 0.05$ in Equation (9.19) will also guarantee an experimentwise error rate of less than 5%. The LSD value is calculated by taking $t = 4$, $\nu = 20$, and $\text{MS}[E] = 3.20$ in Equation (9.19)

$$\text{LSD}_{ij}(\text{S}) = \sqrt{(4 - 1)F(4 - 1, 20, 0.05)\left(\frac{1}{6} + \frac{1}{6}\right)(3.20)} = 3.15.$$

The LSD value for the Scheffe method is larger than the value for the Bonferroni method, and so the Scheffe test will be more conservative. However, for these data, the groupings given by the Scheffe method will be exactly the same as those from the Bonferroni method, because none of the observed differences between pairs of treatment means has a value between 3.02 and 3.15.

The Tukey method is another pairwise comparison procedure that guarantees a preset experimentwise error rate. From Table C.11 (Appendix C), the value of $q(4, 20, 0.05)$—the 0.05-level studentized range for four means and 20 degrees of freedom—is 3.96. Using this value in Equation (9.21), we get the following

Tukey LSD value

$$LSD_{ij}(T) = (3.96)\sqrt{\frac{3.20}{6}} = 2.89. \tag{9.25}$$

The resulting groupings are

$$\text{JP} \quad \text{SP} \quad \underline{\text{SM} \quad \text{JM}}. \tag{9.26}$$

Thus, for these data, the Tukey groupings show the same number of significances as the Scheffe or Bonferroni groupings. Note, however, that the LSD value for the Tukey method lies between the LSD values for the Bonferroni and Fisher methods.

The SNK modification of the Tukey method involves different studentized ranges for different comparisons. The appropriate LSD values will depend on the value of $p = i - j + 1$, the number of means within the range of means being compared. For example, when comparing SM with JP, we have $i = 3$, $j = 1$, and $p = 3 - 1 + 1 = 3$; three means (JP, SP, and SM) are contained in the range between JP and SM. The corresponding studentized range is $q(3, \ 20, \ 0.05) = 3.58$, and from Equation (9.22) the LSD value is

$$LSD_{31}(SNK) = (3.58)\sqrt{\frac{3.20}{6}} = 2.61.$$

The observed difference between the means for JP and SM is $67.1 - 60.3 = 6.8$ and thus is larger than 2.61. Consequently, this difference will be declared significant by the SNK method. The calculations for all the differences may be summarized as follows:

Comparison	Difference	p	LSD	Conclusion
JM versus JP	$67.2 - 60.3 = 6.9$	4	2.89	S
JM versus SP	$67.2 - 63.8 = 3.4$	3	2.61	S
JM versus SM	$67.2 - 67.1 = 0.1$	2	2.15	NS
SM versus JP	$67.1 - 60.3 = 6.8$	3	2.61	S
SM versus SP	$67.1 - 63.8 = 3.3$	2	2.15	S
SP versus JP	$63.8 - 60.3 = 3.5$	2	2.15	S

The groupings of the treatments are exactly the same as in the previous cases, but note that these groupings are derived using a different set of LSD values.

As we know, the SNK method does not have a guaranteed experimentwise error rate and can be regarded as a compromise between the very liberal Fisher method and the rather conservative Tukey method. Another such compromise is the Duncan method, which also changes the LSD value depending on the number of means within the range of means being compared. The LSD values for the Duncan method are calculated by replacing $q(p, \ \nu, \ \alpha)$ in the SNK method with $R(p, \ \nu, \ \alpha)$, the α-level Duncan's studentized range for p treatments and ν degrees of freedom. For example, with $n = 4$, $i - j + 1 = p = 3$, $\alpha = 0.05$, $\nu = 20$, and $MS[E] = 3.20$, we have

$$LSD_{ij}(D) = R(3, 20, 0.05)\sqrt{\frac{3.20}{6}}.$$

From Table C.12 (Appendix C), we get $R(3, 20, 0.05) = 3.10$. Hence, if $p = i - j + 1 = 3$, then

$$\text{LSD}_{ij}(D) = 3.10\sqrt{0.53} = 2.26.$$

Similar calculations show that the LSD values for $p = 4$ and $p = 2$ are, respectively, 2.32 and 2.15. Again, the final groupings will be the same as those resulting from the Fisher method. ∎

When selecting an appropriate multiple pairwise comparison method, as when selecting a procedure for testing multiple contrasts among treatment means (see Example 9.10), we want a procedure that, ideally, not only controls the experimentwise error rate but also has high power to detect true significances. Each of the three procedures—the Scheffe, Bonferroni, and Tukey methods—controls the experimentwise error rate. Among these, the procedure that uses the smallest LSD values will have the highest power. Of the other three methods, the Fisher method is not recommended, because of the risk of too many false significances (Type I errors). The Duncan and SNK methods generally have higher power than do the Scheffe, Bonferroni, and Tukey procedures, but it is hard to be sure of the error rates they control.

EXAMPLE 9.13   Suppose that we want to perform all possible pairwise comparisons of the observed treatment means in a completely randomized experiment, in which six treatments are each replicated four times. If we desire an experimentwise error rate of 0.05, should we use the Scheffe method, the Bonferroni method, or the Tukey method?

We have $t = 6$ treatments and $v = 18$ degrees of freedom for MS[E]. Also, $n_1 = n_2 = \cdots = n_6 = n = 4$. To answer the question, we compare the three LSD values and select the procedure that corresponds to the lowest value of LSD. From Equations (9.19)–(9.21), we get

$$\text{LSD}_{ij}(S) = \sqrt{(6-1)F(6-1, 18, 0.05)\left(\frac{1}{4} + \frac{1}{4}\right)\text{MS[E]}} = 2.63\sqrt{\text{MS[E]}},$$

$$\text{LSD}_{ij}(B) = t(18, 0.05, 15)\sqrt{\left(\frac{1}{4} + \frac{1}{4}\right)\text{MS[E]}} = 2.39\sqrt{\text{MS[E]}},$$

$$\text{LSD}_{ij}(T) = q(6, 18, 0.05)\sqrt{\frac{\text{MS[E]}}{4}} = 2.25\sqrt{\text{MS[E]}}.$$

Since the coefficient of $\sqrt{\text{MS[E]}}$ is smallest for the Tukey LSD value, the Tukey procedure is the most powerful of the three. ∎

Exercises

9.11   Refer to the insulin release data in Exercise 8.3. Use the Bonferroni, Scheffe, Tukey, Student-Neuman-Keuls, and Duncan pairwise multiple comparison methods, each at the 0.05 level, to compare the mean insulin releases at the three glucose concentrations. Write your conclusions on the basis of the results from each of the five pairwise multiple comparison methods. Compare and contrast your conclusions, with special reference to the error rates and powers associated with these techniques.

9.12 Repeat Exercise 9.11 for the soil bulk density data in Exercise 8.4.

9.13 Repeat Exercise 9.11 for the TRCP data in Exercise 8.11.

9.14 Repeat Exercise 9.11 for the leaf area data in Exercise 8.13.

9.15 The conclusions obtained on applying Fisher, Scheffe, Duncan, and Tukey multiple pairwise comparison procedures to the same set of six sample means may be summarized as follows:

a $\overline{y}_{(1)} \; \overline{y}_{(2)} \; \overline{y}_{(3)} \; \overline{y}_{(4)} \; \overline{y}_{(5)} \; \overline{y}_{(6)}$

b $\overline{y}_{(1)} \; \overline{y}_{(2)} \; \overline{y}_{(3)} \; \overline{y}_{(4)} \; \overline{y}_{(5)} \; \overline{y}_{(6)}$

c $\overline{y}_{(1)} \; \overline{y}_{(2)} \; \overline{y}_{(3)} \; \overline{y}_{(4)} \; \overline{y}_{(5)} \; \overline{y}_{(6)}$

d $\overline{y}_{(1)} \; \overline{y}_{(2)} \; \overline{y}_{(3)} \; \overline{y}_{(4)} \; \overline{y}_{(5)} \; \overline{y}_{(6)}$

Identify, giving reasons, the procedure that was responsible for the conclusion in each case.

9.6
Simultaneous confidence intervals for contrasts

The multiple comparison methods in Sections 9.4 and 9.5 are procedures for simultaneously testing several hypotheses about contrasts among population means. In this section, we describe how these multiple comparison methods can be adapted to construct simultaneous confidence intervals for several contrasts among population means.

EXAMPLE **9.14** To determine the effects of four environmental conditions on the growth of a certain species of fish, 12 aquariums in which the water temperature and water movement could be controlled were divided into four groups of three. The groups were randomly assigned to the following four conditions:

1 Still cold water (SC)   **2** Still warm water (SW)

3 Flowing cold water (FC)   **4** Flowing warm water (FW)

The tanks were stocked with fish of uniform size and fed identical diets over a period of 12 weeks. The mean weight gains (lbs) for a sample of 12 fish obtained by selecting one fish at random from each of the 12 tanks are as follows:

Treatment	SC	SW	FC	FW
Treatment number i	1	2	3	4
Sample size n_i	3	3	3	3
Treatment mean $\overline{y}_i$	1.55	1.59	1.08	2.01

The corresponding ANOVA table looks like this:

Source	df	SS	MS
Treatments	3	1.302	0.434
Error	8	0.347	0.043

The investigators were mainly interested in estimating the following four expected differential weight gains:

a between fish grown in still cold water (SC) and fish grown in flowing cold water (FC);

b between fish grown in still warm water (SW) and fish grown in flowing warm water (FW);

c between fish grown in still cold water (SC) and fish grown in still warm water (SW);

d between fish grown in flowing cold water (FC) and fish grown in flowing warm water (FW).

Let μ_1, μ_2, μ_3, and μ_4 denote, respectively, the expected mean weight gains for fish grown under experimental conditions SC, SW, FC, and FW. Our goal is to estimate four simple contrasts between the population means

$$\theta_1 = \mu_1 - \mu_3,$$
$$\theta_2 = \mu_2 - \mu_4,$$
$$\theta_3 = \mu_1 - \mu_2,$$
$$\theta_4 = \mu_3 - \mu_4.$$

Unbiased estimators of these contrasts are given by the corresponding linear combinations of the sample means. Thus, estimates of the four contrasts are

$$\hat{\theta}_1 = \bar{y}_1 - \bar{y}_3 = 1.55 - 1.08 = \quad 0.47,$$
$$\hat{\theta}_2 = \bar{y}_2 - \bar{y}_4 = 1.59 - 2.01 = -0.42,$$
$$\hat{\theta}_3 = \bar{y}_1 - \bar{y}_2 = 1.55 - 1.59 = -0.04,$$
$$\hat{\theta}_4 = \bar{y}_3 - \bar{y}_4 = 1.08 - 2.01 = -0.93.$$

More useful information about the contrasts can be obtained by constructing confidence intervals. A $100(1-\alpha)\%$ confidence interval for θ_i is given by $\hat{\theta}_i \pm t(\nu, \alpha/2)\hat{\sigma}_{\hat{\theta}_i}$, where $\nu = 8$ is the degrees of freedom for MS[E] (see Example 9.2). Because all the treatments are replicated the same number of times, the standard errors of the four contrast estimators are identical and equal to 0.17. Thus, a 95% confidence interval for θ_i is $\hat{\theta}_i \pm (2.306)(0.17) = \hat{\theta}_i \pm 0.39$, and the set of four intervals

$$0.08 \leq \theta_1 \leq \quad 0.86,$$
$$-0.81 \leq \theta_2 \leq -0.03,$$
$$-0.43 \leq \theta_3 \leq \quad 0.35,$$
$$-1.32 \leq \theta_4 \leq -0.54$$

(9.27)

is a set of 95% confidence intervals for the four contrasts. The method used to construct these intervals is such that in repeated application of the method, 95% of the intervals for any given contrast will contain the true value of that contrast. In other words, we have 95% confidence in the performance of each interval in this set.

From a practical viewpoint, it is natural to ask what confidence level can be assigned to the joint performance of the four intervals. In other words, in repeated application of the method used to construct this set of intervals, what is the proportion of the cases in which all of the four intervals simultaneously contain the true values of their corresponding contrasts?   ■

As in the case of the α-level simultaneous hypothesis tests in Sections 9.4 and 9.5, the $100(1-\alpha)\%$ confidence coefficient associated with a collection of k confidence intervals can be defined in two ways.

Individual and simultaneous confidence coefficients

A set of k confidence intervals is said to have an *individual confidence coefficient* of $1 - \alpha$ if each of the k intervals in the set is a $100(1-\alpha)\%$ confidence interval. An individual confidence coefficient of $1 - \alpha$ implies that the probability that the interval for any single contrast will contain the true value of that contrast is $1 - \alpha$. Thus, an individual confidence coefficient relates to the performance of each interval in the set and says nothing about the joint performance of the k intervals. The set of four confidence intervals in Example 9.14 has an individual confidence coefficient of 0.95.

A set of k confidence intervals is said to have a *simultaneous confidence coefficient* of $1-\alpha$ if the probability that all of the k intervals will simultaneously contain the true values of their corresponding contrasts is $1 - \alpha$. Three methods of constructing $100(1 - \alpha)\%$ simultaneous confidence intervals for a set of k contrasts are summarized in Box 9.9. As can easily be verified, the expressions in Box 9.9 follow directly from the formulas for the critical contrast values associated with the Scheffe, Bonferroni, and Tukey multiple comparison methods.

BOX **9.9**   *Simultaneous confidence intervals for k contrasts*

Let $\hat{\theta}_1, \hat{\theta}_2, \ldots, \hat{\theta}_k$ denote k contrasts of the means of independent samples from t normal populations with a common variance σ^2. If ν denotes the degrees of freedom for estimating σ^2, and $\hat{\sigma}_{\hat{\theta}_i}$ is the estimated standard error of $\hat{\theta}_i$ (Box 9.2), then we can derive the following expressions:

1   The Scheffe $100(1 - \alpha)\%$ simultaneous confidence intervals for the k contrasts between population means are

$$\hat{\theta}_i \pm \left[\sqrt{(t-1)F(t-1, \nu, \alpha)} \right] \hat{\sigma}_{\hat{\theta}_i}. \quad (i = 1, 2, \cdots, k)$$

2 The Bonferroni $100(1-\alpha)\%$ simultaneous confidence intervals for the k contrasts between population means are

$$\hat{\theta}_i \pm t(\nu,\ \alpha,\ k)\hat{\sigma}_{\hat{\theta}_i}, \quad (i = 1, 2, \cdots, k)$$

where $t(\nu,\ \alpha,\ k)$ is the $(\alpha/2k)$-level critical value of a t distribution with ν degrees of freedom (Table C.10 in Appendix C).

3 The Tukey $100(1-\alpha)\%$ simultaneous confidence intervals for k simple contrasts between population means are

$$\hat{\theta}_i \pm q(t,\ \nu,\ \alpha)(\sqrt{2})^{-1}\hat{\sigma}_{\hat{\theta}_i}, \quad (i = 1, 2, \ldots, k)$$

where $q(t,\ \nu,\ \alpha)$ is the α-level studentized range for t means and ν degrees of freedom (see Table C.11 in Appendix C).

The next example illustrates the use of the procedures in Box 9.9 to construct simultaneous confidence intervals.

EXAMPLE **9.15**   Let's construct a set of 95% simultaneous confidence intervals for the $k = 4$ contrasts in Example 9.14. First, as noted in Example 9.14, the estimated standard errors of the four contrast estimates are all equal to 0.17. Therefore, the Scheffe 95% simultaneous confidence intervals are obtained by taking $t = 4$, $\nu = 8$, $\alpha = 0.05$, and $\hat{\sigma}_{\hat{\theta}_i} = 0.17$ in Box 9.9. The required intervals are $\hat{\theta}_i \pm \left[\sqrt{(4-1)(4.066)} \right] (0.17) = \hat{\theta}_i \pm 0.59$, where the $\hat{\theta}_i$ are the estimated values of the contrasts calculated in Example 9.14.

Entering Table C.10 (Appendix C) with $\alpha = 0.05$, $k = 4$, $\nu = 8$, we obtain $t(8,\ 0.05,\ 4) = 3.206$. Using this in Box 9.9, we see that the Bonferroni intervals are $\hat{\theta}_i \pm (3.206)(0.17) = \hat{\theta}_i \pm 0.55$.

Finally, knowing that the four contrasts are simple contrasts and the sample sizes are equal, we can use the Tukey method to construct a set of 95% simultaneous confidence intervals. From Table C.11, the 0.05-level studentized range for $t = 4$ means and $\nu = 8$ degrees of freedom is seen to be 4.53. The Tukey simultaneous intervals are computed as $\hat{\theta}_i \pm (4.53)(\sqrt{2})^{-1}(0.17) = \hat{\theta}_i \pm 0.54$.

In this instance, the Tukey method is recommended, because it yields the narrowest 95% simultaneous confidence intervals. The resulting intervals are

$$-0.07 \leq \theta_1 \leq 1.01,$$
$$-0.96 \leq \theta_2 \leq 0.12,$$
$$-0.58 \leq \theta_3 \leq 0.50,$$
$$-1.47 \leq \theta_4 \leq -0.39.$$

(9.28)

Comparison of the 95% individual intervals in Equation (9.27) with the 95% simultaneous confidence intervals in Equation (9.28) reveals that the simultaneous confidence intervals are wider than their individual counterparts. This is to be expected, because the confidence level associated with the simultaneous intervals refers to their joint performance, whereas the individual confidence level refers to the performance of individual intervals.   ■

Exercises

9.16   Use the Bonferroni method to construct a set of 99% simultaneous confidence intervals for the two contrasts in Exercise 9.2d. Compare the conclusions that can be drawn from these intervals and from the corresponding set of 99% individual confidence intervals.

9.17   Refer to Exercise 9.3.

a   Construct the 90% Bonferroni simultaneous confidence intervals for the set of four contrasts in Exercise 9.3b. Interpret the results.

b   Construct the 90% Bonferroni simultaneous confidence intervals for the two meaningful contrasts you selected in Exercise 9.3d. Compare the intervals with the corresponding intervals obtained in (a). Explain the difference between the conclusions that can be drawn from the intervals constructed here and their counterparts in (a).

c   Construct a set of 90% Tukey simultaneous confidence intervals for all possible pairwise differences between the three population means. Interpret your results.

d   Explain why the confidence interval for $\theta_2 = \mu_1 - \mu_2$ obtained by the Tukey method is different from that obtained by the Bonferroni method in (a).

9.18   Use the Scheffe and Bonferroni methods to construct simultaneous 95% confidence intervals for the two contrasts in Exercise 9.4a. Compare the two sets of intervals and comment on the differences between them.

9.19   Suppose that we use the Bonferroni and Tukey methods to construct 95% simultaneous confidence intervals for a set of ten contrasts. Which method will give narrower intervals? Why?

9.7
Predicting linear functions of sample means

In certain circumstances, it is useful to construct prediction intervals for linear functions of the means of independent random samples taken from normally distributed populations. To do so, we can use a direct extension of the method in Box 4.10 for constructing prediction intervals for the mean of a future sample. The method is described in Box 9.10.

BOX **9.10**

Prediction intervals for linear functions of means of independent random samples from normal distributions

Let $\overline{Y}_{1+}, \overline{Y}_{2+}, \ldots, \overline{Y}_{t+}$ be the means of independent random samples of sizes $n_1, n_2, \ldots, n_t$ from t normally distributed populations with a common variance σ^2. Let MS[E] denote the mean square error in the associated one-way ANOVA. Furthermore, let $\overline{Y}_{if}$ denote the mean of a future random sample of size m_i from the i-th normal population. Assume that the future samples are independent.

A $100(1 - \alpha)\%$ prediction interval for the linear function

$$\hat{\theta}_f = c_0 + c_1\overline{Y}_{1f} + c_2\overline{Y}_{2f} + \cdots + c_t\overline{Y}_{tf},$$

where $c_0, c_1, \ldots, c_t$ are known constants, is given by

$$\hat{\theta} \pm t(N - t, \alpha/2)\hat{\sigma}_{\hat{\theta}_f},$$

where

$$\hat{\theta} = c_0 + c_1\overline{Y}_{1+} + c_2\overline{Y}_{2+} + \cdots + c_t\overline{Y}_{t+},$$

$$\hat{\sigma}_{\hat{\theta}_f} = \sqrt{\left[c_1^2\left(\frac{1}{m_1} + \frac{1}{n_1}\right) + c_2^2\left(\frac{1}{m_2} + \frac{1}{n_2}\right) + \cdots + c_t^2\left(\frac{1}{m_t} + \frac{1}{n_t}\right)\right]\text{MS[E]}}.$$

Note that, if we set $t = 1$, $c_0 = 0$, $c_1 = 1$, $n_1 = n$, and $m_1 = m$, the prediction interval in Box 9.10 is the same as the prediction interval in Box 4.10.

The expression in Box 9.10 can be easily adapted when using the Bonferroni method to construct a set of $100(1 - \alpha)\%$ simultaneous prediction intervals for a set of k future linear functions. All that is necessary is to replace $t(N - t, \alpha/2)$ with $t(N - t, \alpha, k)$, which is given in Table C.10 (Appendix C).

EXAMPLE **9.16**   In Example 9.14, we considered a study designed to compare the mean weight gains of fish grown in four environmental conditions: SC (still cold water), SW (still warm water), FC (flowing cold water), and FW (flowing warm water). Let Y_{SC}, Y_{SW}, Y_{FC}, and Y_{FW} denote the gains in weight of four fish grown under the four conditions SC, SW, FC, and FW, respectively. Let's construct a set of 95% simultaneous prediction intervals for the following four comparisons of the weight gains

$$\hat{\theta}_{1f} = Y_{SC} - Y_{FC},$$

$$\hat{\theta}_{2f} = Y_{SW} - Y_{FW},$$

$$\hat{\theta}_{3f} = Y_{SC} - Y_{SW},$$

$$\hat{\theta}_{4f} = Y_{FC} - Y_{FW}.$$

We can regard Y_{SC}, Y_{SW}, Y_{FC}, and Y_{FW} as the means of four future random samples, each of size $m = 1$, taken from the normally distributed populations from which the data in Example 9.14 have been collected. Then the problem can be stated as the problem of constructing simultaneous prediction intervals for $k = 4$ linear functions of the means of four future random samples of sizes $m_1 = m_2 = m_3 = m_4 = 1$. The coefficients of the linear functions are easily determined by the definition of the contrasts of interest. For example, for the linear function $\hat{\theta}_{1f}$, the coefficients are $c_0 = 0, c_1 = 1, c_2 = 0, c_3 = 1$, and $c_4 = 0$. For a 95% simultaneous confidence level, we replace $t(N - t, 0.05/2)$ in Box 9.10 with $t(N - t, 0.05, 4)$. From Table C.10, we see that $t(8, 0.05, 4) = 3.206$. Direct substitution into the formula in Box 9.10 shows that the prediction interval for $\hat{\theta}_{if}$ is of the form

$$\hat{\theta}_i \pm 3.206 \sqrt{\left[(1^2) \left(\frac{1}{1} + \frac{1}{3} \right) + (-1)^2 \left(\frac{1}{1} + \frac{1}{3} \right) \right] \text{MS[E]}}$$

$$= \hat{\theta}_i \pm (3.206)(0.34) = \hat{\theta}_i \pm 1.09,$$

where $\hat{\theta}_i$ is the estimate of the contrast θ_i in Example 9.14. The resulting set of simultaneous prediction intervals is

$$
\begin{aligned}
-0.62 &\leq Y_{SC} - Y_{FC} \leq 1.56, \\
-1.51 &\leq Y_{SW} - Y_{FW} \leq 0.67, \\
-1.13 &\leq Y_{SC} - Y_{SW} \leq 1.05, \\
-2.02 &\leq Y_{FC} - Y_{FW} \leq 0.16.
\end{aligned}
$$

∎

Exercises

9.20 Refer to Example 9.16.

   a Construct a set of 95% individual prediction intervals for Y_{SC}, Y_{SW}, Y_{FC}, and Y_{FW} and interpret them.

   b Construct a set of 95% simultaneous prediction intervals for Y_{SC}, Y_{SW}, Y_{FC}, and Y_{FW} and interpret them.

9.21 Refer to the insulin release data Example 8.3.

   a Construct a set of 90% prediction intervals for the amount of insulin released by three future tissue samples treated with the three levels (low, medium, high) of glucose concentration. Interpret the results.

   b Construct a 95% prediction interval for the difference between the mean amount released from a future sample of three tissue specimens treated with a medium glucose concentration and a future sample of five tissue specimens treated with a high glucose concentration. Interpret the result.

9.8
Determining sample sizes

In Section 7.5, we considered sample size calculations for studies in which the primary objective is to compare the means or proportions of two populations. We examined two approaches: the hypothesis-testing approach and the confidence interval approach. In this section, we use these two approaches to calculate sample sizes for completely randomized experiments. As before, we let $\mu_1, \mu_2, \ldots, \mu_t$ denote the population means and assume that the populations have normal distributions with a common variance σ^2.

The hypothesis-testing approach

In this approach, the objective is to determine sample sizes such that an α-level test has power $1 - \beta$ to detect practically important departures from the null hypothesis $H_0: \mu_1 = \mu_2 = \cdots = \mu_t$. Let

$$\sigma_\mu^2 = \frac{1}{t} \Sigma (\mu_i - \bar{\mu})^2 \tag{9.29}$$

be the variance of the population means. We know that the null hypothesis $H_0: \mu_1 = \mu_2 = \cdots = \mu_t$ is equivalent to the null hypothesis $H_0: \sigma_\mu^2 = 0$, and large values of σ_μ^2 can be interpreted as indicating large departures from the null hypothesis. Consequently, the variance σ_μ^2 provides a quantitative criterion to measure the extent of departures from the null hypothesis. A common approach to the sample size problem is to select a critical value $\sigma_{\mu 0}^2$ for σ_μ^2 and to determine the sample sizes such that, if the actual value of σ_μ^2 exceeds the critical value, then the probability of an α-level test rejecting $H_0: \sigma_\mu^2 = 0$ is at least $1 - \beta$.

The next example illustrates how the critical value of σ_μ^2 may be specified.

EXAMPLE **9.17**   On the basis of a careful literature review and some preliminary studies, an investigator expects that the weight gains (kg) of sheep fed three experimental diets will be in the neighborhood of $\mu_1 = 13$, $\mu_2 = 16$, and $\mu_3 = 22$. To confirm this hypothesis, the investigator wants to conduct a completely randomized experiment in which n animals will be allocated to each of the three diets and the weight gain after 12 weeks on the diet will be measured for each animal. The experimenter wants to ensure that, if the research hypothesis is true, there will be at least an 80% chance that a 0.05-level F-test will reject the null hypothesis $H_0: \mu_1 = \mu_2 = \mu_3$. What should the value of n be?

The means $\mu_1 = 13$, $\mu_2 = 16$, and $\mu_3 = 22$ yield $\sigma_\mu^2 = 14$, and so we will regard any set of means for which σ_μ^2 is larger than 14 as indicating an important departure from the null hypothesis. Then the number of replications per treatment that meets the investigator's requirements is the value of n such that the probability of rejecting the null hypothesis $H_0: \sigma_\mu^2 = 0$ by the 0.05-level ANOVA F-test is at least 80% when the actual value of σ_μ^2 exceeds the critical value 14.   ∎

The sample size problem in Example 9.17 is a special case of the following general problem. We want to find n, the number of replications per treatment, such that, if σ_μ^2 exceeds a critical value $\sigma_{\mu0}^2$, then the probability that an α-level ANOVA F-test will reject H_0: $\sigma_\mu^2 = 0$ is at least $1 - \beta$.

We saw in Chapter 8 that, when data arise from a completely randomized experiment, the null hypothesis H_0: $\sigma_\mu^2 = 0$ can be tested using the ANOVA F-test, in which the test statistic is the ratio of mean squares

$$F = \frac{\text{MS[T]}}{\text{MS[E]}}. \tag{9.30}$$

If there are n replications per treatment, the test will reject H_0 when the calculated value of F in Equation (9.30) exceeds the α-level critical value of the F-distribution with $\nu_1 = t - 1$ and $\nu_2 = nt - t$ degrees of freedom for the numerator and denominator, respectively. The requirement that the test should have power $1 - \beta$ when $\sigma_\mu^2 = \sigma_{\mu0}^2$ is the same as the requirement that, when $\sigma_\mu^2 = \sigma_{\mu0}^2$, the probability that the ratio of mean squares in Equation (9.30) exceeds $F(\nu_1, \nu_2, \alpha)$ should equal $1 - \beta$. In symbols

$$\Pr\left\{ \frac{\text{MS[T]}}{\text{MS[E]}} > F(\nu_1, \nu_2, \alpha) | \sigma_\mu^2 = \sigma_{\mu0}^2 \right\} = 1 - \beta. \tag{9.31}$$

When the ANOVA assumptions are satisfied, the probability specified by Equation (9.31) equals the area to the right of $F(\nu_1, \nu_2, \alpha)$ under a distribution known as a *noncentral F-distribution*. This noncentral F-distribution is characterized by three parameters ν_1, ν_2, and λ, where

$$\nu_1 = t - 1,$$

$$\nu_2 = t(n - 1),$$

$$\lambda = \frac{nt\sigma_\mu^2}{2\sigma^2}. \tag{9.32}$$

The parameters ν_1 and ν_2 are the degrees of freedom for the numerator and the degrees of freedom for the denominator, respectively; λ is the *noncentrality parameter*. The following properties of the noncentral F-distribution needed to calculate the probability in Equation (9.31) are useful in sample size determination:

1   The noncentral F-distribution corresponding to $\lambda = 0$ is the same as the F-distribution described in Section 3.6. For this reason, the F-distributions in Section 3.6 are sometimes called *central F-distributions*.

2   The condition $\sigma_\mu^2 = 0$ implies that $\lambda = 0$. Hence from the previous property, it follows that when $\sigma_\mu^2 = 0$, the probability in Equation (9.31) will equal the area to the right of $F(\nu_1, \nu_2, \alpha)$ under an $F(\nu_1, \nu_2)$-distribution. Thus, when the null hypothesis H_0: $\sigma_\mu^2 = 0$ is true, the probability in Equation (9.31) will be equal to α.

3   It can be shown that for given values of ν_1, ν_2, and α, the probability $1 - \beta$ in Equation (9.31) increases as λ increases. In other words, for fixed values of t, n, and α, $1 - \beta$ will increase with increase in $\sigma_{\mu0}^2$.

Statistical computing packages such as SAS and Splus can be used to calculate the areas under noncentral F-distributions. If t, $\sigma_{\mu0}^2$, σ^2, and α are known, these packages can be used to determine $1 - \beta$ for various selected values of n. The value of n for which $1 - \beta$ is closest to the desired value is the required sample size. The next example illustrates these calculations.

EXAMPLE **9.18**   Suppose that the researcher in Example 9.17 conservatively estimates the standard deviation of the weight gains of animals on any of the experimental diets to be about $\sigma = 5$ kg. Let's determine the sample size such that, if $\sigma_\mu^2 = 14$, the probability that H_0: $\sigma_\mu^2 = 0$ will be rejected at the $\alpha = 0.05$ level is 80%.

We have $t = 3$, $\alpha = 0.05$, $\sigma^2 = 25$, and $\sigma_{\mu 0}^2 = 14$. We want to find n such that $1 - \beta$ in Equation (9.31) is close to 0.80. To do so, we first note that, for a given sample size n, the noncentrality parameter in Equation (9.32) equals

$$\lambda = \frac{n(3)(14)}{2(25)} = 0.84n.$$

If we start with a trial value $n = 10$, we have $\nu_1 = t - 1 = 3 - 1 = 2$, $\nu_2 = nt - t = 10(3) - 3 = 27$, and $\lambda = 0.84(10) = 8.4$. Thus, the value of $1 - \beta$ that corresponds to $n = 10$ is the area to the right of $F(2, 27, 0.05) = 3.354$ under a noncentral F-distribution with $\nu_1 = 2$, $\nu_2 = 27$, and $\lambda = 8.4$. The companion text by Younger (1997) shows how to use the SAS functions FINV and PROBF to determine this area. For the present, we simply note that $1 - \beta$ equals 0.944 for $n = 10$. This is much higher than the target value of 0.80, and so we try a smaller sample size—say, $n = 6$—and continue the process until we find the smallest n for which $1 - \beta$ is at least 0.80. The steps in this trial-and-error process may be summarized as follows:

n	ν_1	ν_2	λ	$1 - \beta$	Comments
10	2	27	8.40	0.944	Too large; try smaller n
6	2	15	5.04	0.729	Too small; try larger n
8	2	21	6.72	0.872	Too large; try smaller n
7	2	18	5.88	0.812	Stop

The required sample size is $n = 7$, and so the experiment should use 21 animals, with seven animals randomized to each diet. ∎

We can make the following comments on the sample size procedure in Example 9.18.

1   In Example 8.4, we saw that a two-sided t-test with ν_2 degrees of freedom is equivalent to the ANOVA F-test with $\nu_1 = 1$ degrees of freedom for the numerator and ν_2 degrees of freedom for the denominator. Consequently, the sample size procedure for the ANOVA F-test with $t = 2$ treatments can be used to determine sample sizes for controlling the power of two-sided t-tests. In Section B.9 of Appendix B, it is shown that, when $t = 2$, the absolute difference $|\mu_1 - \mu_2|$ will exceed the critical value Δ if and only if σ_μ^2 exceeds the critical value $\sigma_{\mu 0}^2 = \Delta^2/4$.

Therefore, the sample sizes needed by a two-sided t-test to detect a difference of magnitude Δ with probability $1 - \beta$ can be calculated using the areas under a noncentral F-distribution with parameters $\nu_1 = 1$, $\nu_2 = 2(n - 1)$, and

$$\lambda = \frac{(n)(2)(\Delta^2/4)}{2\sigma^2} = \frac{n\Delta^2}{4\sigma^2}.$$

2   As with the sample size procedures in Box 7.1, the sample size procedure for comparing the means of t normal populations requires an estimate of the common population standard deviation σ.

3   In Example 9.17, the critical value of σ_μ^2 that corresponded to practically important departures from the null hypothesis was determined on the basis of a set of specified values for the population means. In practice, researchers may find it unrealistic to select a critical value for σ_μ^2 on the basis of a single set of population mean values. An alternative is to specify a value Δ such that violations of H_0 are regarded as practically important if the difference between at least one pair of population means is equal to or larger than Δ. For instance, in Example 9.17, the researcher might consider it important to conclude that the expected weight gains for the three diets are different if the expected differential weight gain between at least two of the experimental diets exceeds $\Delta = 8$ kg.

In Section B.9 of Appendix B, it is shown that, if the difference between at least one pair of population means is at least Δ, then the variance of the population means is at least $\Delta^2/2t$. In symbols, if $|\mu_i - \mu_j| \geq \Delta$ for at least one pair i, j, then $\sigma_\mu^2 \geq \Delta^2/2t$. In Exercise 9.25, you will be asked to argue that the sample sizes calculated to ensure a power of $1 - \beta$ when we set $\sigma_{\mu 0}^2 = \Delta^2/2t$ in Equation (9.32) will ensure a power of at least $1 - \beta$ when detecting a minimum difference of Δ between at least one pair of means. In other words, the required sample size can be calculated using a noncentral F-distribution with noncentrality parameter

$$\lambda = \frac{nt(\Delta^2/2t)}{2\sigma^2} = \frac{n\Delta^2}{4\sigma^2}. \tag{9.33}$$

EXAMPLE 9.19   Suppose that the investigator in Example 9.17 wants 80% power for a 0.05-level ANOVA F-test if the differential weight gain between at least one pair of diets exceeds 8 kg. Determine the appropriate sample size if $\sigma^2 = 25.0$ g.

We have $\alpha = 0.05$, $1 - \beta = 0.80$, $t = 3$, $\sigma^2 = 25.0$, and $\Delta = 8.0$. Then the critical value for σ_μ^2 is

$$\sigma_{\mu 0}^2 = \frac{8^2}{2(3)} = 10.67,$$

so that the noncentrality parameter is

$$\lambda = \frac{n(3)(10.67)}{2(25)} = 0.64n.$$

The sample size calculations may be summarized as follows:

n	v_1	v_2	λ	$1 - \beta$	Comments
10	2	27	6.40	0.866	Too large; try smaller n
7	2	18	4.48	0.691	Too small; try larger n
9	2	24	5.76	0.821	Too large; try smaller n
8	2	21	5.12	0.763	Stop

The required sample size is $n = 9$ replications per diet. ■

When there is no reasonable estimate for σ^2, the sample size can be determined on the basis of the approach in Box 7.3. The importance of a specific departure from the null hypothesis can be measured in terms of the probability that, for at least one pair of populations, a randomly selected measurement from one population will exceed a randomly selected measurement from the other population. The next example illustrates how that idea can be put into practice.

EXAMPLE **9.20**   Suppose we are interested in planning a completely randomized experiment to study the differences between the response times to four methods of treating migraine headache. An objective in such a study might be to test the null hypothesis that there is no difference between the mean response times against the research hypothesis that the means do differ. We regard the differences between the four treatments as practically important if, for at least one pair of treatments, the probability that the response time under one of the treatments will be at least as large as that under the other is 90%. Let's determine the sample size necessary to guarantee that the 0.05-level ANOVA F-test will have 80% power to detect practically important differences between the four treatments.

The required sample size can be determined using Equation (9.31) provided we have an appropriate value for the noncentrality parameter λ in Equation (9.33). Notice that λ can be expressed in terms of Δ / σ, where Δ is the minimum difference between the population means that we desire to detect. As we saw in Section 7.5, when two population means differ by Δ, the requirement of a 90% chance that a response from one of the populations will be at least as large as a response for the other is equivalent to the requirement that

$$A\left(-\frac{\Delta}{\sigma\sqrt{2}}\right) = 0.90.$$

From Table C.1 (Appendix C), we see that $A(z) = 0.90$ corresponds to $z = -1.28$, so that

$$-\frac{\Delta}{\sigma\sqrt{2}} = -1.28 \text{ or } \frac{\Delta}{\sigma} = 1.81.$$

The sample size calculated with $\alpha = 0.05$, $1 - \beta = 0.80$, $t = 4$, and noncentrality parameter

$$\lambda = \frac{n}{4}(1.81)^2 = (0.82)n$$

will ensure that a 0.05-level ANOVA F-test will have the required power. In Exercise 9.26, you will be asked to determine the required sample size. ∎

The confidence interval approach

The confidence interval approach for determining sample sizes in experiments to compare the means of two normal populations was described in Section 7.5. In this approach, the sample sizes were determined to ensure that the length of a $100(1 - \alpha)\%$ confidence interval for $\theta = \mu_1 - \mu_2$ is less than $2B$.

In studies involving more than two populations, we usually want to estimate more than one parameter. For instance, in Example 9.14, where we considered an experiment to compare the effects of four aquarium conditions on the growth of a species of fish, the primary interest was in estimating $k = 4$ contrasts between the population means. When planning experiments that involve more than two populations, it may be desirable to determine sample sizes such that the lengths of simultaneous confidence intervals for k ($k \geq 2$) specified parameters are less than $2B$. Usually, these parameters are linear combinations of the population means. Such a procedure is described in Box 9.11.

BOX 9.11

> ### Sample sizes to ensure that the lengths of k $100(1 - \alpha)\%$ simultaneous confidence intervals are less than $2B$
>
> Let $\theta_1, \theta_2, \ldots, \theta_k$ denote k linear combinations of t population means and assume that simultaneous confidence intervals for $\{\theta_1, \theta_2, \ldots, \theta_k\}$ can be constructed using at least one of the procedures described in Box 9.9. Let n be the number of measurements per sample needed to ensure that the lengths of the k simultaneous confidence intervals are less than $2B$. Then n is the smallest integer for which
>
> $$n \geq CT^2 \left(\frac{\sigma}{B}\right)^2,$$
>
> where σ^2 is the estimate of the common population variance; C is the maximum of $\{C_1, C_2, \ldots, C_k\}$; C_k is the sum of squares of the coefficients of θ_k; and
>
> $$T = \begin{cases} \sqrt{(t-1)F(t-1, \nu, \alpha)} & \text{for the Scheffe intervals} \\ t(\nu, \alpha, k) & \text{for the Bonferroni intervals} \\ q(t, \nu, \alpha) & \text{for the Tukey intervals.} \end{cases}$$
>
> and ν denotes the degrees of freedom for the error sum of squares.

The procedure in Box 9.11 may be compared with that in Box 7.4. Whereas the method of determining the sample sizes in Box 7.4 is intended to control the length of the confidence interval for a single parameter $\theta = \mu_1 - \mu_2$, the method in Box 9.11 is intended to control the lengths of simultaneous confidence intervals for k parameters.

In Example 9.21, we use the procedure in Box 9.11 to determine the sample sizes for the diet study in Example 9.17.

EXAMPLE **9.21**   In Example 9.18, we determined the sample sizes needed to compare the mean weight gains from three experimental diets on the basis of the hypothesis-test approach. Specifically, we calculated the sample sizes such that the experiment had a high (80%) probability of detecting departures from the null hypothesis corresponding to the values $\mu_1 = 13$, $\mu_2 = 16$, and $\mu_3 = 22$.

Alternatively, suppose that the primary objective of the experimenter is to make inferences about two contrasts

$$\theta_1 = \mu_1 - \mu_2,$$

$$\theta_2 = \mu_1 - \frac{1}{2}(\mu_2 + \mu_3).$$

Let's determine the sample sizes such that the lengths of the 90% simultaneous confidence intervals for θ_1 and θ_2 will not exceed 12 kg. Assume, as in Example 9.18, that the standard deviation of the weight gains of animals on any of the three diets is approximately 5 kg.

In the notation in Box 9.11, we have $t = 3$, $k = 2$, $2B = 12$, $\sigma = 5$, and $1 - \alpha = 0.90$. The degrees of freedom for error in a completely randomized experiment with t treatments and n replications per treatment are known to be $\nu = nt - t = t(n - 1)$, and so we have $\nu = 3(n - 1)$. To determine C, we must calculate the sum of squares of the coefficients in each of the two contrasts. For θ_1, the coefficients are $c_1 = 1$, $c_2 = -1$, and $c_3 = 0$, so that the required sum of squares is $C_1 = (1)^2 + (-1)^2 + (0)^2 = 2$. For θ_2, the corresponding value is $C_2 = (1)^2 + \left(-\frac{1}{2}\right)^2 + \left(-\frac{1}{2}\right)^2 = 1.5$. Therefore, C is the maximum of $\{2, 1.5\}$, which is 2.

The quantity T in Box 9.11 depends on which simultaneous interval procedure we choose. The Tukey method is not appropriate, because θ_2 is a complex contrast. That leaves two possibilities: the Scheffe method and the Bonferroni method. For the Bonferroni method, $T = t(\nu, 0.10, 2)$, the critical value of a t-distribution with $\nu = 3(n - 1)$ degrees of freedom at the level $\alpha = 0.10/4 = 0.025$. Because T depends on n, the sample size calculations are carried out by a trial-and-error method. For each choice of n, the value of T is determined from Table C.10 (Appendix C). If we start with the trial value $n = 10$, the calculation may be summarized as follows:

n	$\nu = 3(n - 1)$	$T = t(\nu, 0.10, 2)$	RHS	Comments
10	27	2.052	5.890	n too large; try smaller n
5	12	2.179	6.595	n too small; try larger n
6	15	2.131	6.307	n too small; try larger n
7	18	2.101	6.131	n too large. Stop

In this table, RHS denotes the right-hand side of the inequality in Box 9.11

$$\text{RHS} = CT^2 \left(\frac{\sigma}{B}\right)^2 = 2T^2 \left(\frac{5}{6}\right)^2 = 1.389T^2.$$

The calculations in the table show that we must choose $n = 7$ animals per diet to guarantee that the length of the 90% simultaneous confidence intervals for θ_1 and θ_2 will be less than 12 kg. In other words, if seven animals per diet are used in the study, then there is 90% chance that the estimated values of the contrasts will be within 6 kg of their true values. ■

In Example 9.21, the sample sizes were computed on the premise that the primary objective was to estimate two specific linear combinations of population means. In practice, it is unlikely that a researcher will be able to identify a single set of contrasts that are of primary concern at the beginning of the study. The sample size can then be determined to control the lengths of simultaneous confidence intervals for (a) all possible contrasts between population means; or (b) all possible pairwise differences between population means. In the former case, the procedure in Box 9.11 will entail the use of the Scheffe method; in the latter case, either the Tukey method or the Bonferroni method can be used. From a practical viewpoint, the sample sizes designed to control the lengths of all possible contrasts tend to be much too large. For this reason, in the absence of a specific set of contrasts that are of particular interest, the sample sizes are usually determined so as to control the lengths of confidence intervals for all possible pairwise differences between the population means. Of the three methods considered here for determining sample sizes in this case, the Scheffe method tends to be too conservative. Usually the choice is between the Bonferroni and Tukey methods. In Exercises 9.22e and 9.23e, you are asked to determine sample sizes so as to control the lengths of confidence intervals for all possible pairwise differences between population means.

Exercises

9.22 Refer to the insulin release data in Exercise 8.3. Let μ_L, μ_M, and μ_H denote, respectively, the true mean amount of insulin released at low, medium, and high glucose concentration. Estimate the sample sizes needed in future studies designed to meet the following objectives.

a If $\mu_L = 2.0$, $\mu_M = 3.5$, and $\mu_H = 4.5$, we want a 0.05-level ANOVA F-test to reject H_0: $\mu_L = \mu_M = \mu_H$ with probability 0.80.

b If the difference between any pair of true means exceeds 1.0, we want to reject H_0: $\mu_L = \mu_M = \mu_H$ with probability 0.85.

c We want to reject H_0: $\mu_L = \mu_M = \mu_H$ with probability 0.90 if there is at least one pair of levels of glucose concentration such that, with a probability of 90% (or more), the amount of insulin released at one level will exceed the amount released at the other level.

d   We want 90% confidence that the estimates of the two contrasts $\theta_1 = \mu_H - \mu_M$ and $\theta_2 = \frac{1}{2}(\mu_M + \mu_H) - \mu_L$ will be within 1.5 units of their true values.

e   We want 95% confidence that the difference between every pair of true means is estimated within 1.5 units of its true value.

9.23   Refer to the bulk density data of Exercise 8.4. Let μ_1, μ_2, and μ_3 denote, respectively, the true mean soil bulk densities under the three treatments (grazing practices). Estimate the sample sizes needed in future studies designed to meet the following objectives:

a   If $\mu_1 = 1.8$, $\mu_2 = 1.7$, and $\mu_3 = 1.6$, we want a 0.05-level ANOVA F-test to reject H_0: $\mu_1 = \mu_2 = \mu_3$ with probability 0.80.

b   If the difference between any pair of true means exceeds 0.1 g/cm^3, we want to reject H_0: $\mu_1 = \mu_2 = \mu_3$ with probability 0.80.

c   We want to reject H_0: $\mu_1 = \mu_2 = \mu_3$ with probability 0.90 if there is at least one pair of treatments such that, with a probability of 90% (or more), the bulk density for one treatment exceeds that for the other treatment.

d   We want 90% confidence that the estimates of the two contrasts $\theta_1 = \mu_1 - \mu_3$ and $\theta_2 = \mu_1 - 2\mu_2 + \mu_3$ will be within 0.1 g/cm^3 of their true values.

e   We want 95% confidence that the difference between every pair of true means is estimated within 0.2 g/cm^3 of its true value.

9.24   In a study to compare the average indoor radiation levels in three types of homes built on reclaimed phosphate mine lands (see Example 1.6), a random sample of six homes of each type was selected from the population of all homes. Assume that a one-way ANOVA would be appropriate for testing the null hypothesis H_0 that there is no difference between the true mean indoor radiation levels in the three types of homes. For $c = 1.0$, 1.25, 1.50, 1.75, 2.0, 2.25, 2.50, 2.75, and 3.0, calculate the (approximate) probability that the ANOVA F-test will reject the null hypothesis if the difference between any one pair of means is more than $c\sigma$, where σ is the true standard deviation of the radiation levels found in homes of a given type. Interpret the results you obtain.

9.25   We know that, for given values of ν_1, ν_2, and α, the probability $1 - \beta$ in Equation (9.31) increases with increasing values of $\sigma_{\mu 0}^2$. On this basis, argue that the sample sizes calculated by solving Equation (9.31) with $\sigma_{\mu 0}^2 = \Delta^2/2t$ will ensure a power of at least $1 - \beta$ when detecting a difference Δ between at least one pair of treatments.

9.26   Use statistical software such as SAS to determine the sample size in Example 9.20.

9.9
Overview

In this chapter, we focus on two types of problems associated with the design and analysis of single-factor studies. In Sections 9.2–9.6, we consider various methods of comparing the means of t normally distributed populations. We have seen how the problem of comparing population means can be set in the framework of inferences about treatment contrasts. The actual method used for inferences about these contrasts will depend on several factors, the most important of which are: (1) whether the contrasts are specified in advance or after looking at the data; and (2) whether inferences are desired for a single contrast or more than one contrast. When inferences are desired for more than one contrast, the appropriateness of a procedure depends on the type of error rate (comparisonwise, experimentwise, or a compromise between the two) that is to be controlled.

In Section 9.8, we consider methods of determining sample sizes. As in the one- and two-population settings discussed in Section 7.5, sample sizes for studies involving t populations can be determined in two ways: (1) by calculating sample sizes such that the test of a key null hypothesis has specified power to detect stated alternative hypotheses; or (2) by calculating sample sizes such that the lengths of confidence intervals for specific contrasts are less than a specified value. Of course, as we saw in Section 3.7, controlling the length of a confidence interval can be interpreted as controlling the magnitude of the error of estimation. Thus, we can determine sample sizes such that the margin of estimation errors for a specific set of contrasts is within the desired bounds.

The methods in this chapter are appropriate for comparing the means of t normal populations under the assumption that the populations have a common variance and the data are independent random samples from these populations. In other words, these methods can be used to compare population means if the data satisfy ANOVA assumptions for the one-way classification model. As we see in subsequent chapters, these methods can also be used, with minor modifications, for more complex models that satisfy ANOVA assumptions.

10

Simple Linear Regression

10.1
Introduction

In Chapter 7, we discussed how research studies can be formulated in terms of statistical models that describe assumed relationships between explanatory (independent) and response (dependent) variables. As we saw, a dependent variable is a response variable that can be affected by, or associated with, the values of one or more independent (explanatory) variables. The one-way ANOVA model in Box 8.2 can be used to investigate the changes in the expected value of a quantitative response variable corresponding to changes in a nominal explanatory variable. The following example illustrates the idea.

EXAMPLE 10.1   In Example 8.2, a study to compare the effects of five light sources on the heights of four-week-old plants was described. In Example 8.6, a one-way ANOVA model in which the dependent variable Y is the four-week height of a plant (an interval variable), while the independent variable X is the type of light source, was proposed for the data. The independent variable in the model is a nominal variable taking five possible values corresponding to the five light sources in the study. The model assumes that the expected value of a measured response at the i-th level of the experimental factor is

$$\mathcal{E}(Y|i\text{-th factor level}) = \mu_i = \mu + \tau_i,$$

where μ is the overall mean of the population of heights of all plants and τ_i is the effect of the i-th level of the nominal variable X. Recall that τ_i can be interpreted as the difference $\mu_i - \mu$ between the population mean response for the i-th light source μ_i and the overall mean μ. The one-way ANOVA model provided a framework for comparing the expected plant heights (expected responses) corresponding to different light sources (levels of the independent variable). ∎

When the response and explanatory variables are measured on an interval scale, we are often interested in the possibility of a mathematical relationship between the expected response and the levels of a number of independent variables. Let Y and $(X_1, \ldots, X_p)$ denote, respectively, a quantitative dependent variable and a set of p quantitative independent variables. Let $\mu(x_1, \ldots, x_p)$ denote the mean of the population of all responses when $X_1 = x_1, \ldots, X_p = x_p$. *Regression analysis* is a statistical technique that can be used to draw conclusions regarding the mathematical relationship between $\mu(x_1, \ldots, x_p)$ and $x_1, \ldots, x_p$.

Regression analysis may conveniently be conceptualized in terms of the means of statistical populations characterized by conditional distributions. The distribution of the (conceptual) population of all measurements of Y when the independent variables are held fixed at $X_1 = x_1, \ldots, X_p = x_p$ is called the *conditional distribution* of Y given $X_1 = x_1, \ldots, X_p = x_p$. The mean of the conditional distribution of Y given $X_1 = x_1, \ldots, X_p = x_p$ is called the *expected value* of Y given $X_1 = x_1, \ldots, X_p = x_p$. In what follows, we will occasionally use the symbol $\mathcal{E}(Y|x_1, \ldots, x_p)$ to denote the mean of the conditional distribution of Y given $X_1 = x_1, \ldots, X_p = x_p$. Inclusion of the variable names $Y, x_1, \ldots, x_p$ in the symbol for the conditional mean will be helpful when it is important to keep track of the roles of the various variables in a discussion.

EXAMPLE **10.2**   Here are some examples of research settings in which regression analysis is useful:

1   Let Y = the time to relief from pain and X = the dose of a particular drug. Then the question of how the drug dose affects time to relief can be investigated using a regression analysis in which the dependent variable is Y and the independent variable is X. The conditional distribution of interest is the distribution of the conceptual population of all relief times observed at given drug doses. Here $\mu(x)$, the expected relief time given drug dose x, is the mean of the conditional distribution of Y given $X = x$; that is, $\mu(1.5)$ is the mean of the conceptual population of all relief times when the drug dose is 1.5. On certain occasions, $\mu(x)$ will be denoted by $\mathcal{E}(Y|x)$.

2   Let Y = the level of exposure to radiation, X_1 = the distance from the source of radiation, and X_2 = the wind velocity at the time the radiation exposure is measured. The effects of distance and wind velocity on the level of exposure can be studied using a regression analysis in which Y is the dependent variable and X_1 and X_2 are $p = 2$ independent variables. The distributions of the conceptual populations of exposure levels at given distances and wind velocities are the conditional distributions of interest. The mean level of exposure when $X_1 = x_1$ and $X_2 = x_2$ is denoted by $\mu(x_1, x_2)$ or $\mathcal{E}(Y|x_1, x_2)$.

3   Let Y = the weight gain after six months on a diet, and X_1, X_2, and X_3 denote, respectively, the diet consumption rate, percentage of fat in the diet, and percentage of protein in the diet. Then the expected weight gain for given consumption rate, percentage of fat, and percentage of protein can be analyzed using a regression analysis involving $p = 3$ independent variables. ■

The next example illustrates the essential difference between analysis of variance and regression analysis.

EXAMPLE **10.3**   To study the effect of an experimental drug on the influx of a particular chemical in the brains of rats, 16 rats were divided into four groups of four. Each group received one of the four doses of the drug. Table 10.1 shows d, the dose (nM) of the drug; $x = \log d$ (recall that, unless stated otherwise, log is to base e); y, the measured chemical influx (nmol/mg protein); $\bar{y}$, the mean chemical influx for the group of four

T A B L E   **10.1**
Chemical influx y at drug dose d

d	$x = \log d$	Influx y (nmol/mg protein)				$\bar{y}$	s
1	0.00	21	24	26	25	24.00	2.16
10	2.30	36	38	36	35	36.25	1.26
100	4.61	43	47	45	49	46.00	2.58
1000	6.91	54	56	58	53	55.25	2.22

rats receiving dose d; and s, the standard deviation of the measured influxes at each dose. These data could be analyzed in at least two ways: by the ANOVA method in Chapter 8; and by regression analysis. Let's compare these two methods.

In the ANOVA approach, a one-way ANOVA model with doses as treatments (four replications per treatment) is used for data analysis. Some typical questions that can be answered using a one-way ANOVA are as follows.

1   Is there a difference between the expected responses for the four doses (treatments)? In other words, is there a difference between the mean chemical influxes at the four drug doses?

2   What can be said about the mean chemical influx for rats exposed to a dose $d = 100$ nM? What is a 95% confidence interval for the mean influx at $d = 100$ nM?

3   What can be said about a future response at dose $d = 100$ nM? What is a 95% prediction interval for the chemical influx in the brain of a rat receiving a dose $d = 100$ nM?

4   Is the expected chemical influx for rats exposed to dose $d = 1000$ nM higher than that for rats exposed to the low dose $d = 1$ nM? How much higher?

Note that all the questions pertain to four specific populations—the four conceptual populations generated by the responses to the four doses (1, 10, 100, and 1000 nM) used in the study design. As we'll see in this chapter, the ANOVA assumptions do not justify conclusions about responses at levels other than those used in the study design.

The regression approach is not subject to the same limitation. When the experimental factor (independent variable) is quantitative, regression analysis can be used to make inferences about the populations of responses at levels other than those in the study design. For instance, a regression analysis of the influx data will provide answers to questions such as the following:

1   Does the expected chemical influx increase with drug dose?

2   What is the expected chemical influx at dose $d = 250$ nM?

3   What is a 95% prediction interval for the chemical influx in the brain of a rat exposed to a dose $d = 150$ nM?

4   At what dose of the drug will the expected chemical influx be 30 nmol/mg protein?

Let $\mu(d)$ denote the mean chemical influx for the population of rats exposed to dose d. Thus, $\mu(d)$ is the expected response at dose d. The first step in the regression analysis is to make a reasonable assumption about the mathematical relationship between $\mu(d)$ and d. Such an assumption can be based either on past experience or on the pattern observed in the current data.

Past experience with dose-response data suggests that a straight-line relationship between the expected response $\mu(d)$ and $\log d$ is a reasonable assumption in many instances. Figure 10.1 shows a scatter plot of the measured influxes against $x = \log d$. The observed mean responses (based on $n = 4$ responses at each level) are also plotted in the same graph.

F I G U R E   10.1

Plot of influx versus dose

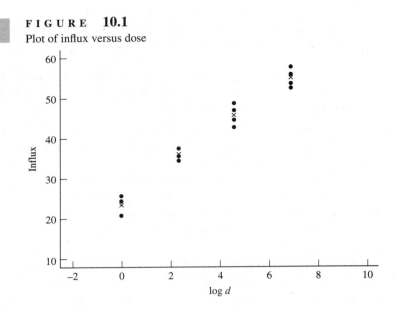

Figure 10.1 shows that the mean responses lie close to a straight line so that the assumption of a straight-line relationship between $x = \log d$ and the population mean response $\mu(d)$ is a reasonable starting point for a regression analysis of these data. The first step in the analysis is to use the observed data to estimate the equation of the straight line on which the population mean responses will lie. Then the resulting straight line can be used to estimate the mean response at any given value of $\log d$. Details of the estimation procedure will be presented later. Let's suppose that the estimated mean influx at dose d, denoted by $\hat{\mu}(d)$, is related to $\log d$ as follows

$$\hat{\mu}(d) = 26 + 4 \log d. \tag{10.1}$$

Then Equation (10.1) can be used to make inferences about the conditional distribution of the chemical influx at any dose. For example, the estimated expected influx at dose $d = 250$ is $\hat{\mu}(250) = 26 + 4 \log 250 = 48.08$, and the estimated dose at which the expected influx is 30 is obtained by solving the equation $30 = 26 + 4 \log d$. Solving the equation, we get

$$d = \exp\left(\frac{30 - 26}{4}\right) = 2.72,$$

so that we can expect an average influx of 30 nmol/mg protein at a dose $d = 2.72$ nM.   ■

The analysis suggested in Example 10.3 is called a simple linear regression analysis and forms the topic of this chapter. Simple linear regression may be used to analyze a straight-line relationship between a single quantitative independent variable X and the expected value (population mean) of a response variable. Regression analysis with multiple independent variables will be discussed in Chapter 11.

Exercises

10.1   Refer to the chemical influx data in Example 10.3.

a   Write a one-way ANOVA model for the data. Explain all the terms in your model.

b   Construct a one-way ANOVA table. What conclusions can be drawn from the results in the ANOVA table?

c   Construct a 95% prediction interval for the chemical influx in the brain of a rat receiving a dose $d = 100$ nM of the experimental drug. Interpret this interval.

d   Construct a 95% lower confidence bound for the change in mean chemical influx when the drug dose is changed from 1 to 10 nM. Interpret the bound.

10.2   The measured soil water content of 16 soil samples at four depths is as follows:

Depth (cm)	Soil water contents (cm^3/cm^3)			
10	0.313	0.299	0.340	0.289
40	0.315	0.310	0.294	0.293
60	0.289	0.284	0.281	0.275
120	0.259	0.259	0.245	0.251

a   Write a one-way ANOVA model for the data. Explain all the terms in your model.

b   Construct a one-way ANOVA table for the data and test the null hypothesis that the mean soil water contents at the four depths are not significantly different.

c   Construct a scatter plot of the data. Does an inspection of the scatter plot suggest a straight-line relationship between the mean soil water content and depth? Explain.

10.3   Weight gain of the mother during pregnancy is known to be a critical factor in determining the birth-weight of the infant. Some data collected in a study of the relationship between average weight gain and mother's age are as follows:

Age (years)	24	22	16	17	19	15	21	23
Weight gain (kg)	17.17	10.34	7.04	6.91	7.56	6.32	13.01	13.8

a   Draw a scatter plot of the data. Do you think a straight line adequately represents the relationship of weight gain to mother's age?

b   Suppose the investigator wanted to perform a one-way ANOVA in which age was the experimental factor, with eight levels, to see if the observed differences between the weight gains can be explained by the differences in the mothers' ages. Do you see any problems with this type of analysis? Explain.

10.2
The simple linear regression model

As already noted, simple linear regression analysis is a special case of a general statistical procedure called regression analysis. Regression analysis is applicable when the dependent and independent variables are all quantitative (measured on interval scales) and the mean of the conditional distribution of the dependent variable can be expressed as a mathematical function of the independent variables. That mathematical function is called a *regression function*. In other words, a regression function specifies a mathematical relationship between $\mu(x_1, \cdots, x_p)$—sometimes written equivalently as $\mathcal{E}(Y|x_1, \ldots, x_p)$—and $x_1, \ldots, x_p$. In a simple linear regression analysis, only one independent variable is involved, and the regression function specifies a straight-line relationship between the dependent and independent variables. The simple linear regression function is defined in Box 10.1.

BOX 10.1

The simple linear regression function

Suppose that the independent variable X and the dependent variable Y are both quantitative. Let $\mu(x)$ denote the mean of the conditional distribution of Y given $X = x$. A regression function of the form

$$\mu(x) = \beta_0 + \beta_1 x,$$

where β_0 and β_1 are constants, is called a simple linear regression function. The constants β_0 and β_1 are called *regression parameters*.

The regression function in Box 10.1 specifies a straight-line relationship between $\mathcal{E}(Y|x)$ and x. Let's consider the interpretation of the regression parameters. First, if $x = 0$, we get

$$\mu(0) = \beta_0 + \beta_1(0) = \beta_0,$$

so that β_0 is the expected response when the independent variable is zero. Thus, β_0 is a population mean; it is the mean of the conditional distribution of Y given $X = 0$. Sometimes it is convenient to refer to β_0 as the *intercept parameter* because the y-intercept of the regression function is β_0 (Figure 10.2).

Second, if we examine the change in the expected response resulting from a one-unit increase (from x to $x + 1$) in the value of the independent variable, we find that

$$\mu(x + 1) - \mu(x) = \left[\beta_0 + \beta_1(x + 1)\right] - (\beta_0 + \beta_1 x)$$
$$= (\beta_0 + \beta_1 x + \beta_1) - (\beta_0 + \beta_1 x)$$
$$= \beta_1.$$

Thus, β_1 is the change in the population mean response corresponding to unit increase in the value of the independent variable. As we see from Figure 10.2, β_1 is the slope of the regression function and is sometimes called the *slope parameter*.

A common mistake when interpreting β_0 and β_1 is to disregard the units of measurement. The intercept parameter β_0 is the mean of a conditional distribution of the response variable Y, so that the units for β_0 are the same as the units for Y.

FIGURE 10.2

Plot of a simple linear regression function

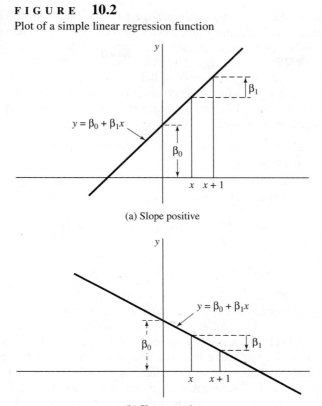

(a) Slope positive

(b) Slope negative

The slope parameter β_1 measures the change in the conditional mean of Y per unit change in X; thus, the units of β_1 will depend on the units of both X and Y. For instance, if Y is measured in centimeters and X is measured in minutes, then β_0 is measured in centimeters and β_1 in centimeters per minute (cm/min).

In the next example, the interpretation of the regression parameters is illustrated for the chemical influx data in Example 10.3.

EXAMPLE 10.4   We saw in Example 10.3 that a straight-line relationship between the expected chemical influx and the logarithm of the drug dose is a reasonable assumption. Therefore, if we let $X = \log(\text{dose})$ and $Y = $ chemical influx, we can assume that

$$\mu(x) = \beta_0 + \beta_1 x, \qquad \text{(10.2)}$$

where $\mu(x)$ is the mean of the population of measured influxes when $\log(\text{dose})$ is equal to x. For instance, when the dose is 1, the value of X is $x = \log 1 = 0$, so that $\mu(0)$ is the mean influx for the population of all rats that receive a dose of 1. Similarly, since $\log 10 \simeq 2.3$, $\mu(2.3)$ is the mean influx for the population of all rats that receive a drug dose of 10.

Equation (10.2) defines a simple linear regression function that specifies how the population mean chemical influx changes with $X = \log d$. The values of β_0 and β_1 are unknown and have to be estimated from the data. Methods for estimating these parameters and making related inferences will be described later in this chapter.

For the present, let's suppose that $\beta_0 = 25$ and $\beta_1 = 4$. The intercept value $\beta_0 = 25$ means that, for the population of rats receiving a dose corresponding to $x = 0$—that is, $d = 1$— the mean chemical influx is 25 nmol/mg protein. The value $\beta_1 = 4$ has the following implications:

- Because the slope is positive, the expected chemical influx increases with increasing dose.

- For every 1-unit increase in x, the average influx will change (increase) by 4 units. Let d_1 and d_2 be doses such that $x = \log d_1$ and $x + 1 = \log d_2$. Then

$$1 = \log d_2 - \log d_1 = \log \left(\frac{d_2}{d_1} \right).$$

Hence

$$\frac{d_2}{d_1} \doteq e^1 \simeq 2.72,$$

so that increasing x by 1 unit is the same as changing the corresponding dose d_1 to $d_2 = 2.72 d_1$. Thus, if $\beta_1 = 4$, every increase by the factor 2.72 in drug dose corresponds to an average increase of 4 nmol/mg protein in chemical influx. Conversely, increasing the drug dose d by a factor of 10—that is, changing d to $10d$—is equivalent to changing $x = \log d$ by the amount $\log 10d - \log d = \log 10 = 2.30$. Consequently, if $\beta_1 = 4$, every increase in drug dose by a factor of 10 corresponds to an average increase in the chemical influx of $4 \times 2.3 = 9.2$ nmol/mg protein. ■

Like the analysis of variance, regression analysis is best described in terms of a model that summarizes the underlying assumptions. In Box 10.2, we describe one of the simplest and most frequently used regression models.

BOX **10.2**   ***The simple linear regression model***

Let X and Y denote, respectively, quantitative independent and dependent variables. For $i = 1, \ldots, n$, let Y_i denote the response that will be observed at the value $X = x_i$. Then

$$Y_i = \beta_0 + \beta_1 x_i + E_i, \qquad i = 1, \cdots, n,$$

where β_0 is the expected response when $X = 0$; β_1 is the expected change in response per unit increase in X; and E_i is the random error in Y_i. The E_i are independent $N(0, \sigma^2)$.

> When it is convenient to do so, the subscript i will be omitted, and the simple linear regression model will be written in the form
>
> $$Y = \beta_0 + \beta_1 x + E,$$
>
> where the errors are independent $N(0, \sigma^2)$.

Comparison of the ANOVA and simple linear regression models

Comparison of the simple linear regression model with the one-way ANOVA model in Box 8.2 reveals the following similarities and differences:

1 The structure of the two models is similar. Both models represent an observed response as the sum of two components: a fixed but unknown component representing the expected response and a random component representing the error

$$\text{Observed response} = \text{Expected response} + \text{Random error}.$$

In symbols, both models have the structure

$$Y_i = \mu(x_i) + E_i,$$

where $\mu(x_i)$ is the expected response when $X = x_i$.

2 In both models, the error components are assumed to be independent $N(0, \sigma^2)$; that is, the responses are assumed to be independent and normal, with a common variance σ^2. The difference between the models is in the relationship of the expected response to the level of the independent variable. In the ANOVA model, no specific relationships are assumed between the expected response and the value of X; the only assumption is that the expected responses may be different at different levels of the independent variable. In the simple linear regression model, on the other hand, the expected response is related to the level of the experimental factor through the simple linear regression function

$$\mu(x) = \beta_0 + \beta_1 x.$$

Thus, the assumptions for a simple linear regression model include the ANOVA assumptions plus an assumption about the form of the regression function.

3 Because no relationships between the expected responses are assumed in a one-way ANOVA model, the model is specified in terms of $t + 1$ parameters, where t is the number of treatments. Of these, t parameters specify the expected responses for the t treatments, and one parameter specifies the error variance. Thus, inferences based on a one-way ANOVA model will require the estimation of $t + 1$ parameters. When there are n responses available for analysis, t degrees of freedom are needed to estimate the t expected responses (population means). The remaining $n - t$ degrees of freedom are available for estimation of the error variance.

By contrast, only two parameters (β_0 and β_1) are needed to specify the expected responses in a simple linear regression model. One additional parameter specifies the error variance, so that the simple linear regression model requires only three parameters for its specification. This difference plays an important role in deciding which model is appropriate for analyzing a given data set. In situations where several models seem to describe the data equally well, the model with the fewest parameters is preferable.

4 The type of information summarized by β_0 and β_1 is quite different. The slope parameter β_1 contains information about the manner in which the expected responses vary with the value of the independent variable X. If $\beta_1 = 0$, then there is no difference between the expected responses at different levels of X. Therefore, the null hypothesis H_0: $\beta_1 = 0$ in the simple linear regression model plays a role similar to that of the null hypothesis H_0 that there is no difference between the treatment effects in a one-way ANOVA model.

The intercept parameter β_0, on the other hand, has nothing to do with the differences between the expected responses; rather, it is the value of the expected response at one specific value $X = 0$. Like the overall mean μ in the one-way ANOVA model, the intercept parameter is a reference point used as a baseline for the comparison of expected responses.

5 The ANOVA and regression models also make different assumptions about the conditional distributions. Under the simple linear regression model, the conditional distribution of Y given $X = x_i$ is a normal distribution with mean

$$\mu(x_i) = \beta_0 + \beta_1 x_i$$

and variance σ^2. Under the one-way ANOVA model, the conditional distribution of Y given $X = x_i$ is a normal distribution with mean

$$\mu(x_i) = \mu + \tau_i$$

and variance σ^2, where μ is the overall mean and τ_i is the effect of the level x_i. The assumptions for a simple linear regression model are more restrictive, in that they specify not only the form of the conditional distribution of Y at each value of X, but also a straight-line relationship between the conditional mean and the value of X. Because of the assumed straight-line relationship, two parameters (β_0 and β_1) are sufficient to determine the means of the conditional distributions at all levels of X. Figure 10.3 illustrates the differences between these assumptions under the two models.

The differences between the assumptions about the expected responses in the two models are illustrated in Figure 10.4, which displays some possible patterns of expected responses at four levels ($X = 5$, 10, 15, 20) of the independent variable. Four patterns of differences between the expected responses $\mu(x_i)$, $i = 1, 2, 3, 4$, are displayed for the ANOVA model. Of these, patterns 1 and 2 are straight lines of positive and zero slope, respectively. These patterns show, respectively, an increasing straight-line trend and no difference among the expected responses. Notice that, in the ANOVA context, a line with zero slope implies no difference between treatment effects ($\tau_i = 0$, $i = 1, \cdots, 4$). Patterns 3 and 4 are more complex. Line 3 shows no particular trend, while line 4 shows a nonlinear decreasing trend.

FIGURE   10.3

Conditional distributions under ANOVA and simple linear regression models

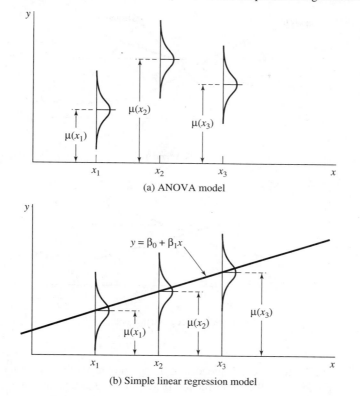

Because the simple linear regression model specifies a linear relationship, only three patterns are permissible. Increasing and decreasing trends (patterns 1 and 3) correspond to positive and negative slopes (β_1), while zero slope (pattern 2) implies that changing the level of X does not have any effect on the expected response.

6 As in a one-way ANOVA model, the assumptions in the simple linear regression model imply that the levels of the experimental factor are fixed by the investigators. In that case, the data can be treated as a collection of independent random samples of fixed sizes from several populations, one for each of the levels of X at which responses were observed. For instance, the chemical influx data can be regarded as four independent random samples (one at each of the four levels of the drug) of size $n = 4$.

7 Like the one-way ANOVA model, the simple linear regression model is often used in observational studies, where the levels of the independent variable X are not under the control of the investigator. The consequences of using a simple linear regression analysis when the levels of X are not fixed in advance are

FIGURE   **10.4**

Relationships between $\mu(x)$ and x

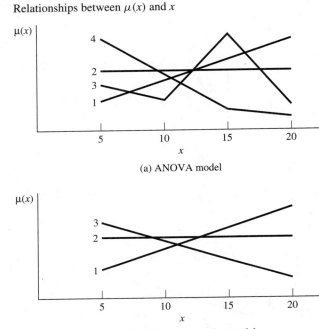

(a) ANOVA model

(b) Simple linear regression model

similar to the consequences of using one-way ANOVA to analyze data in which the levels of the experimental factor are not fixed in advance.

EXAMPLE **10.5**   Under the assumption that a simple linear regression model is appropriate for the chemical influx data in Example 10.3, an intuitively reasonable method of estimating the regression parameters is to examine a scatter plot of the observed data (Figure 10.1) and to visually determine a straight line that fits the data. The y-intercept and slope of the visually fitted line can serve as estimates of the intercept and slope parameters. A plot of the data is shown in Figure 10.5, along with a straight line that appears to describe the data well.

On the basis of the visually fitted line, the estimates of the intercept and slope parameters are $\hat{\beta}_0 = 24$ and $\hat{\beta}_1 = 5$.   ■

Exercises

10.4   Refer to the chemical influx data in Example 10.3

    a   Write a simple linear regression model with $X = \log d$, where d is the drug dose, as the independent variable. Explain all the terms in the model.

FIGURE   10.5

Visually estimated regression line

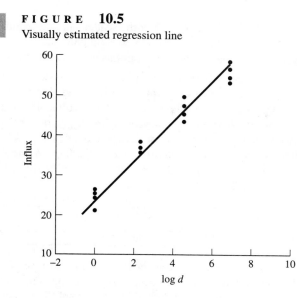

b   Write a simple linear regression model with $X = d$ as the independent variable. Explain all the terms in the model.

c   What are the units of measurement for the regression coefficients in the two models in (a) and (b)?

d   Under the model in (a), what is assumed about the distribution of the observed chemical responses when the drug dose is $d = 150$ nM?

e   Compare and contrast the two models in (a) and (b) with the one-way ANOVA model in Exercise 10.1.

10.5   Refer to the soil water content data in Exercise 10.2.

a   Write a simple linear regression model for the data. Explain all the terms in the model.

b   Compare and contrast the simple linear regression model in (a) with the one-way ANOVA model in Exercise 10.2.

c   Set up the null hypothesis H_0 that the expected soil water content is the same at the four depths, in terms of the parameters in the model in (a).

d   On the basis of an inspection of the scatter plot, draw the straight line that best describes the soil water content data. Graphically determine the intercept and slope of the line.

e   Is it reasonable to treat the soil depths as fixed by the investigator? Explain.

10.6   Refer to the weight gain data in Exercise 10.3.

a   Write a simple linear regression model for the weight gain, with the age of the mother as the independent variable. Explain all the terms in the model.

b   Explain how the interpretations of the regression parameters in (a) will change if the independent variable is

  i X = the mother's age $-$ 10;   **ii** X = the mother's age in months.

c   On the basis of a visual inspection of the scatter plot, determine the intercept and slope of the line that best describes the weight gain data.

d   Is it reasonable to treat the mothers' ages as fixed by the investigator? Explain.

10.7   Peas are known to be a self-intolerant crop, because repeated planting in the same field makes them susceptible to root rot diseases. In the Netherlands, one pea crop in a six-year (1:6) rotation is considered desirable agronomically. In a study to examine how the crop interval (the number of years without pea or other legume) is related to the severity of root rot diseases in pea crops, ten soil samples from different parts of the Netherlands were investigated (Oyarzun, Gerlagh, & Hoogland, 1993). The following data were obtained for the crop interval and the disease index (on a scale from 0, healthy, to 5, dead:

Crop interval	0	4	6	6	6	8	9	9	9	14
Disease index	3.5	3.7	5.0	3.0	3.1	3.1	1.5	3.0	3.3	1.6

a   Write a simple linear regression model relating the disease index to the crop interval (independent variable). Explain all the terms in the model.

b   Construct a scatter plot of the data, draw the straight line that best describes the data (on the basis of visual inspection), and determine its intercept and slope.

c   Can the crop interval values in the data be regarded as fixed by the investigators? Explain.

10.3
Estimating parameters

Inferences based on a simple linear regression model require the estimation of three parameters: the regression parameters β_0 and β_1; and the error variance σ^2. Before describing how to estimate these parameters, we need to revisit the concepts of predicted and residual values discussed in Section 8.6.

Predicted and residual values

Under the simple linear regression model, the *predicted value* of the dependent variable Y when the independent variable X is equal to x_i is

$$\hat{y}_i = \hat{\beta}_0 + \hat{\beta}_1 x_i,$$

where $\hat{\beta}_0$ and $\hat{\beta}_1$ are estimated values of the regression parameters. Since the expected response at $X = x_i$ is $\mu(x_i) = \beta_0 + \beta_1 x_i$, the predicted value $\hat{y}_i$ is an estimate of this expected value. When it is convenient to do so, we shall use the alternative symbol $\hat{\mu}(x_i)$ to denote the predicted value when $X = x_i$.

The equation

$$\hat{y} = \hat{\beta}_0 + \hat{\beta}_1 x$$

is called the *prediction equation* based on the estimates $\hat{\beta}_0$ and $\hat{\beta}_1$. The line represented by the prediction equation is called the *prediction line* or the *estimated regression line*.

The prediction line is an estimate of the *true regression line*, which describes the assumed straight-line relationship between the population mean and the value of the independent variable. The true regression line can be represented by the equation

$$\mathcal{E}(Y|x) = \beta_0 + \beta_1 x,$$

where $\mathcal{E}(Y|x)$ is the expected response when $X = x$.

The difference between the observed and predicted values of a response variable is called the *residual*. If y_i and $\hat{y}_i$ are the observed and predicted values at $X = x_i$, the residual is defined as

$$e_i = y_i - \hat{y}_i.$$

Under the simple linear regression model, the residual at $X = x_i$ is

$$e_i = y_i - \left(\hat{\beta}_0 + \hat{\beta}_1 x_i \right).$$

Note that the predicted and residual values are defined in exactly the same manner in the setting of simple linear regression analysis as in the one-way ANOVA settings (Section 8.6). In both cases, the predicted value is an estimate of the mean of a conditional distribution (expected response), and the corresponding residual is the difference between the observed and predicted values.

EXAMPLE **10.6**   In Example 10.5, we visually estimated a straight line to describe the chemical influx data in Example 10.3. On the basis of the visually fitted line, the estimates of the intercept and slope parameters are $\hat{\beta}_0 = 24$ and $\hat{\beta}_1 = 5$. The estimated regression line is

$$\hat{y} = 24 + 5x. \tag{10.3}$$

Table 10.2 shows the predicted values and residuals based on Equation (10.3).

All the entries in Table 10.2 are calculated using the estimated regression line in Equation (10.3). For instance, at dose $d = 10$, X has the value $x = 2.30$, and the predicted value is $\hat{y} = 24 + 5(2.3) = 35.5$. The residuals for the four responses ($i = 5, 6, 7,$ and 8) at dose $d = 10$ are $e_5 = 36 - 35.5 = 0.5$, $e_6 = 38 - 35.5 = 2.5$, $e_7 = 36 - 35.5 = 0.5$, and $e_8 = 35 - 35.5 = -0.5$. An examination of the residuals reveals that one of the observed responses (y_2) is the same as its predicted value; six (with positive residuals) exceed their predicted values; and nine (with negative residuals) are less than their predicted values. Figure 10.6 shows a plot of the estimated regression line in Equation (10.3), along with the observed responses at x_2, x_7, and x_{11}. Figure 10.6 also shows the magnitudes of the predicted and residual values for these three observations.

Notice from Figure 10.6 that positive, negative, and zero residuals correspond, respectively, to responses that are above, below, and on the estimated line. The value of -14.40 for the sum of the residuals could be interpreted to mean that, on average, an observed value is $14.40/16 = 0.9$ units below (because the sum is negative) the prediction line. However, observing that a large positive residual will cancel the influence on the mean residual of a large negative residual, we conclude that the

TABLE   **10.2**

Predicted values and residuals based on a visually estimated prediction equation

i	x_i	y_i	$\hat{y}_i$	e_i	e_i^2
1	0.00	21	24.00	−3.00	9.00
2	0.00	24	24.00	0.00	0.00
3	0.00	26	24.00	2.00	4.00
4	0.00	25	24.00	1.00	1.00
5	2.30	36	35.50	0.50	0.25
6	2.30	38	35.50	2.50	6.25
7	2.30	36	35.50	0.50	0.25
8	2.30	35	35.50	−0.50	0.25
9	4.61	43	47.05	−4.05	16.40
10	4.61	47	47.05	−0.05	0.00
11	4.61	45	47.05	−2.05	4.20
12	4.61	49	47.05	1.95	3.80
13	6.91	54	58.55	−4.55	20.70
14	6.91	56	58.55	−2.55	6.50
15	6.91	58	58.55	−0.55	0.30
16	6.91	53	58.55	−5.55	30.80
Sum				−14.40	103.70

FIGURE   **10.6**

Visually estimated regression line and residuals

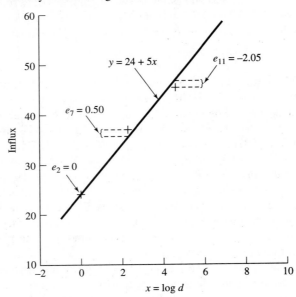

sum of the residuals is not a good measure of how well the estimated line fits the observed data.

An overall measure of the closeness of the observed values to the predicted values is provided by the sum of squares of the residuals. As the residual sum of squares decreases, the overall fit of the line to the data improves. We see from Table 10.2 that the sum of squares of the residuals for the estimated line in Equation (10.3) is 103.70. Any line that produces a residual sum of squares less than 103.70 fits the observed data better. ■

An obvious drawback of estimating β_0 and β_1 on the basis of a visual estimate of the regression line is that the result is subjective. The degree to which an estimated line will describe the data will depend on the individual entrusted with the task. An objective alternative is the *method of least squares*, in which we determine regression coefficients such that the corresponding estimated line has the smallest possible residual sum of squares.

Least squares estimates of regression parameters

For $i = 1, \ldots, n$, let y_i denote the observed value of Y when $X = x_i$. Then $\hat{\beta}_0$ and $\hat{\beta}_1$, the least squares estimates of β_0 and β_1, respectively, are determined such that the sum of squares of the residuals

$$SS[E] = \sum_{i=1}^{n} \left(y_i - \hat{y}_i\right)^2$$

$$= \sum_{i=1}^{n} \left[y_i - \left(\hat{\beta}_0 + \hat{\beta}_1 x_i\right)\right]^2 \tag{10.4}$$

has the smallest possible value.

Recall that in Section 8.4, SS[E] denoted the error sum of squares in the analysis of a completely randomized design. The same symbol is used to denote the sum of squares of the residuals in a simple linear regression analysis, because the error sum of squares in a completely randomized design is the same as the sum of squares of the residuals based on a one-way ANOVA model. In what follows, the error sum of squares and the residual sum of squares are regarded as the same quantity.

The least squares estimates of β_0 and β_1 are obtained by solving a set of two equations, called *normal equations*

$$\sum_{i=1}^{n} y_i = n\hat{\beta}_0 + \left(\sum_{i=1}^{n} x_i\right) \hat{\beta}_1,$$

$$\sum_{i=1}^{n} x_i y_i = \left(\sum_{i=1}^{n} x_i\right) \hat{\beta}_0 + \left(\sum_{i=1}^{n} x_i^2\right) \hat{\beta}_1. \tag{10.5}$$

The solution of Equation (10.5) can be expressed as a set of compact mathematical formulas (Box 10.3). These formulas provide useful insight into the structure of the least squares estimates of regression coefficients.

BOX **10.3**   *Least squares estimates of the coefficients in simple linear regression*

Let (x_i, y_i), $i = 1, \ldots, n$, denote n measured values of the response variable Y at n values of the independent variable X. The least squares estimates of the intercept and slope parameters in the simple linear regression of Y on X are given by

$$\hat{\beta}_1 = \frac{S_{xy}}{S_{xx}},$$

$$\hat{\beta}_0 = \bar{y} - \hat{\beta}_1 \bar{x},$$

where

$$S_{xx} = \sum_{i=1}^{n} (x_i - \bar{x})^2 = \sum_{i=1}^{n} x_i^2 - \frac{\left(\sum_{i=1}^{n} x_i\right)^2}{n}$$

and

$$S_{xy} = \sum_{i=1}^{n} (x_i - \bar{x})(y_i - \bar{y}) = \sum_{i=1}^{n} x_i y_i - \frac{\left(\sum_{i=1}^{n} x_i\right)\left(\sum_{i=1}^{n} y_i\right)}{n}.$$

The residual sum of squares for the least squares fitted line equals

$$SS[E] = S_{yy} - \hat{\beta}_1^2 S_{xx},$$

where

$$S_{yy} = \sum_{i=1}^{n} (y_i - \bar{y})^2 = \sum_{i=1}^{n} y_i^2 - \frac{\left(\sum_{i=1}^{n} y_i\right)^2}{n}.$$

Notice in Box 10.3 that S_{xx} and S_{yy} are the numerators in the expressions for the variances of $\{x_1, \ldots, x_n\}$ and $\{y_1, \ldots, y_n\}$, respectively. As such, they are indicators of how dispersed the values of the independent and dependent variables in the data set are. Indeed, S_{yy} is the same as the total sum of squares SS[TOT] defined in Section 8.4. For this reason, we will use the symbol SS[TOT] instead of S_{yy} whenever it is convenient to do so. The quantity S_{xy} is the numerator in the expression for the sample covariance of the x and y values.

The next example illustrates the calculation and interpretation of least squares estimates of the coefficients in a simple linear regression model.

EXAMPLE **10.7**   To determine whether the age of a rat (dependent variable) can be predicted using independent variables such as body weight, hematocrit (percentage of red blood cells), and protein (mg/ml blood), a study was conducted on a total of 37 rats, ranging in age from 20 to 89 days. The data on two variables—Y = age and X = weight—for $n = 10$ rats are as follows (Brandt, Waters, Muron, & Block, 1982, Figure 2):

Rat number i	1	2	3	4	5	6	7	8	9	10
Weight x (g)	76	89	154	189	180	180	241	320	271	370
Age y (days)	28	33	35	37	38	47	66	67	70	82

Let's suppose that a simple linear regression model is appropriate for these data. Then the data can be modeled as

$$Y = \beta_0 + \beta_1 x + E.$$

The calculations needed for least squares estimation of β_0 and β_1 are

$$\sum x = 2070, \qquad \bar{x} = 207.0;$$

$$\sum y = 503, \qquad \bar{y} = 50.3;$$

$$\sum x^2 = 508,756, \qquad S_{xx} = 508,756 - \frac{2070^2}{10} = 80,266;$$

$$\sum y^2 = 28,589, \qquad S_{yy} = 28,589 - \frac{503^2}{10} = 3288;$$

$$\sum xy = 119,404, \qquad S_{xy} = 119,404 - \frac{(2070)(503)}{10} = 15,283.$$

The least squares estimates are

$$\hat{\beta}_1 = \frac{S_{xy}}{S_{xx}} = \frac{15,283}{80,266} = 0.1904,$$

$$\hat{\beta}_0 = \bar{y} - \hat{\beta}_1 \bar{x} = 50.3 - (0.19)(207) = 10.97,$$

so that the estimated regression line is

$$\hat{y} = 10.97 + 0.19x.$$

Let's interpret the estimated regression parameters. What does the value of 10.97 for the intercept parameter mean? Clearly, it doesn't make sense to interpret this number as the estimated mean age of the population of rats weighing $x = 0$ g! Even if we were to conceptualize a population of rats with zero mean body weight, a mean age of 10.97 days does not make practical sense for this population. We would want the mean age of this idealized population of rats to be zero. Should we conclude that a simple linear regression analysis is not appropriate for these data? It appears from the plot of the data and the least squares line in Figure 10.7a that the estimated line describes the data well. Why is it, then, that the least squares line gives a poor predicted value at $X = 0$?

To answer this question, note that $X = 0$ is far outside the range of values of X used to fit the least squares line. A straight line may not describe the mathematical relationship between $\mathcal{E}(Y|x)$ and x for all values of x. The nonlinear line in Figure 10.7b fits the data as well as the least squares line does, in terms of the residual sum of squares (see Exercise 10.8), although the two lines will obviously give different predicted values in the range $0 \leq X < 76$.

The estimated value $\hat{\beta}_1 = 0.19$ can be interpreted as the estimated difference in the mean ages of rats with a difference of 1 g in their body weights. Again, whether this interpretation is valid for rats with body weights outside the range of the data depends on whether a simple linear regression model is appropriate in this range.

As a general rule, great care is needed when using a regression line to make inferences outside the range of the data. We will address this problem in greater detail later.

FIGURE   10.7

Least squares regression line (a) and nonlinear line (b) for predicting the age of rats

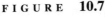

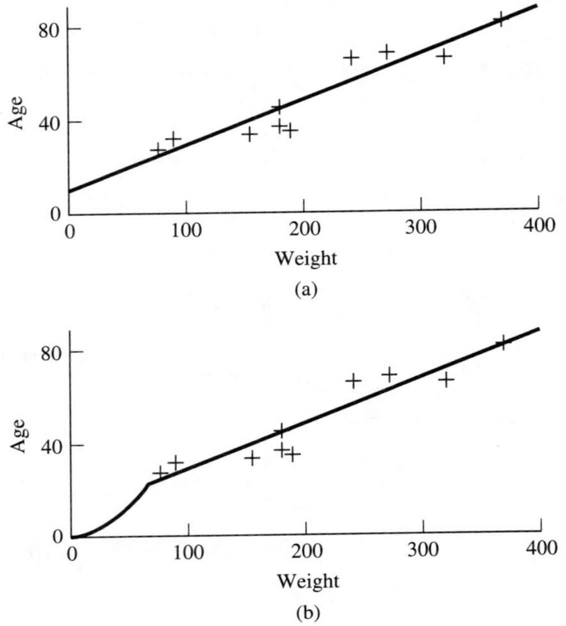

Finally, the residual sum of squares for the least squares estimated line is

$$\text{SS[E]} = S_{yy} - \hat{\beta}_1^2 S_{xx} = 3288 - (0.1904)^2 (80{,}266) = 378.18,$$

which is the smallest possible residual sum of squares that can be obtained by fitting a straight line to these data. In other words, if any other line were used to predict age, the SS[E] would be larger. ∎

As in the one-way ANOVA model, inferences based on a simple linear regression model require a suitable estimate of σ^2, the error variance. Such an estimate is presented in Box 10.4.

BOX **10.4**

Estimate of the error variance in the simple linear regression model

Assume a simple linear regression model, as described in Box 10.2. Then the least squares estimate of σ^2, the error variance, is

$$\text{MS[E]} = \hat{\sigma}^2 = \frac{\text{SS[E]}}{n-2},$$

where SS[E] is the sum of squares of the residuals defined in Box (10.3).

Theorem B6.2 in Appendix B can be used to show that MS[E] in Box 10.4 is an unbiased estimate of σ^2. The denominator of MS[E] is called the *degrees of freedom for SS[E]*. Notice that the structure of MS[E] is very similar to the corresponding quantity in a one-way ANOVA. The only difference is in the degrees of freedom. In a one-way ANOVA, the degrees of freedom may be calculated as $n - t$, where n is the total number of responses and t is the number of treatments (levels of the independent variable). For simple linear regression analysis, the corresponding expression is $n - 2$.

The degrees of freedom for SS[E]

As we will see in Chapter 12, the one-way ANOVA model and the simple linear regression model are just two examples of a *general linear model*. The following rule is often used to determine the degrees of freedom for SS[E] in any model of this type. If n is the number of observations and k is the number of parameters needed to specify the expected response in the linear model, then the degrees of freedom available for estimating the error variance may be calculated as $df_{SS[E]} = n - k$.

Since the one-way ANOVA model requires t parameters to specify the expected responses, t degrees of freedom are needed to estimate these parameters. The remaining $n - t$ degrees of freedom are available for estimating σ^2. In a simple linear regression model, two degrees of freedom are used to estimate the parameters specifying the expected responses, and $n - 2$ degrees of freedom are available for estimating σ^2.

EXAMPLE **10.8**   For the rat weight data in Example 10.7, the error sum of squares was seen to be $SS[E] = 378.28$. There were $n = 10$ responses, and so the degrees of freedom for the SS[E] may be calculated as $n - 2 = 8$. Hence, the least squares estimate of the error variance is

$$\hat{\sigma}^2 = MS[E] = \frac{378.18}{8} = 47.27. \quad \blacksquare$$

Exercises

10.8   Argue that the residual sum of squares for the nonlinear line in Figure 10.7 is the same as the residual sum of squares for the least squares straight line. Hence, conclude that, according to the least squares criterion, both lines fit the data equally well.

10.9   Refer to the chemical influx data in Example 10.3.

a   Determine the least squares estimates of the intercept and slope parameters in the simple linear regression of Y = chemical influx on $X = \log d$.

b   Use the prediction equation based on a simple linear regression model to estimate the expected chemical influx when the drug dose is:

i   100 nM;   **ii**   150 nM.

c   Explain why the residual sum of squares (SS[E]) based on the least squares prediction line will not be larger than that based on the visually estimated line in Example 10.3. Verify that this is so for these data.

d   Explain why the residual sum of squares based on the least squares prediction line will not be smaller than that based on the one-way ANOVA model in Exercise 10.1. Verify that this is so for these data.

10.10   Refer to the soil water content data in Exercise 10.2. Let Y = soil water content and X = depth of soil.

a   Estimate the coefficients in the simple linear regression of Y on X.

b   On the basis of a simple linear regression model, what can you say about the change in soil water content with increasing depth?

c   On the basis of a simple linear regression model, estimate the water content at the surface of the soil.

d   On the basis of a simple linear regression model, estimate the expected water content of the soil at a depth of 50 cm.

e   Estimate the error variance in the simple linear regression of Y on X.

10.11   Refer to the weight gain data in Exercise 10.3. Let Y = mother's weight gain and X = mother's age.

a   Assuming a simple linear regression model, estimate the expected change in mother's gain in weight per 1-month increase in her age.

b   Estimate the expected weight gain of a mother aged 18 years.

c   Estimate the error variance in the simple linear regression of mother's weight on mother's age. Interpret this estimate.

10.12   Refer to the disease index data described in Exercise 10.7.

a   Estimate the least squares straight line for predicting the disease index as a function of the crop interval.

b   Estimate the expected disease index in a field in which no peas or other legumes have been planted during the previous seven years.

10.4
Inferences about expected and predicted responses

In this section, we examine how a simple linear regression model can be used to make inferences about population means and future responses. Let's begin with methods for making inferences about a linear combination of β_0 and β_1, the parameters needed to specify the expected response in a simple linear regression model. If c_0 and c_1 are known constants, consider the linear combination

$$\theta = c_0\beta_0 + c_1\beta_1. \tag{10.6}$$

Some examples of θ that are frequently encountered in practice are as follows.

1   When $c_0 = 1$ and $c_1 = 0$, we have $\theta = \beta_0$. Thus, β_0, the expected response when $x = 0$, is itself a linear combination of β_0 and β_1.

2 If we set $c_0 = 0$ and $c_1 = 1$, we obtain $\theta = \beta_1$. Thus, β_1, the amount of change in the expected response per unit increase in the value of the independent variable, is itself a linear combination of β_0 and β_1.

3 The case $c_0 = 1$ and $c_1 = x$ corresponds to $\theta = \mu(x)$; this implies that $\mu(x)$, the expected response when the level of the independent variable is equal to x, is a linear combination of β_0 and β_1.

An estimate of a linear combination of the regression parameters is provided by the same linear combination of the corresponding least squares estimates. Thus, an estimate of θ in Equation (10.6) is

$$\hat{\theta} = c_0 \hat{\beta}_0 + c_1 \hat{\beta}_1, \tag{10.7}$$

where $\hat{\beta}_0$ and $\hat{\beta}_1$ are the least squares estimates of β_0 and β_1, respectively. Inferences about θ can be based on the sampling distribution of $\hat{\theta}$ summarized in Box 10.5

BOX **10.5**

> **Properties of the least squares estimates of linear combinations of β_0 and β_1**
>
> For $i = 1, \ldots, n$, let Y_i denote an observed value of the dependent variable Y when the independent variable X has the value x_i. Assume the simple linear regression model
>
> $$Y_i = \beta_0 + \beta_1 x_i + E_i, \qquad i = 1, \ldots, n,$$
>
> where the E_i are independent, with zero mean and variance σ^2. (Note that, at this point, a normal distribution is not assumed for the E_i.) For given constants c_0 and c_1, let θ and $\hat{\theta}$ be the linear combination and its estimate given by Equations (10.6) and (10.7), respectively.
>
> **1** Then we may state the following properties:
>   **a** The quantity $\hat{\theta}$ is an unbiased estimate of θ.
>   **b** The standard error of $\hat{\theta}$ is $\sigma_{\hat{\theta}}$, where
>
> $$\sigma_{\hat{\theta}}^2 = \sigma^2 \left\{ s_{00} c_0^2 + 2 s_{01} c_0 c_1 + s_{11} c_1^2 \right\},$$
>
> $$s_{00} = \left[\frac{1}{n} + \frac{\bar{x}^2}{S_{xx}} \right],$$
>
> $$s_{11} = \frac{1}{S_{xx}},$$
>
> $$s_{01} = -\frac{\bar{x}}{S_{xx}}.$$
>
> **2** If the error components are a random sample from a normally distributed population—that is, if the E_i are independent $N(0, \sigma^2)$—then $\hat{\theta}$ has a normal distribution with mean θ and variance $\sigma_{\hat{\theta}}^2$; that is, $\hat{\theta}$ is $N(\theta, \sigma_{\hat{\theta}}^2)$.

3 Let $\hat{\sigma}^2$ be the least squares estimate of the error variance defined in Box 10.4; that is, $\hat{\sigma}^2 = MS[E]$. Then, if

$$\hat{\sigma}_{\hat{\theta}}^2 = \hat{\sigma}^2 \left\{ s_{00}c_0^2 + 2s_{01}c_0c_1 + s_{11}c_1^2 \right\}$$

denotes the estimated variance of $\hat{\theta}$ obtained by replacing σ^2 by its least squares estimate, the null hypothesis H_0: $\theta = \theta_0$ can be tested using the test statistic

$$t_{\hat{\theta}} = \frac{\hat{\theta} - \theta_0}{\hat{\sigma}_{\hat{\theta}}}.$$

The value of the test statistic should be compared to an appropriate critical value of a $t(n-2)$-distribution.

4 A $100(1-\alpha)\%$ confidence interval for θ is given by

$$\hat{\theta} - t(\nu, \alpha/2)\hat{\sigma}_{\hat{\theta}} \leq \theta \leq \hat{\theta} + t(\nu, \alpha/2)\hat{\sigma}_{\hat{\theta}}.$$

The results in Box 10.5 can be derived using results from the theory of least squares summarized in Theorem B6.2 in Appendix B.

The sampling distributions of $\hat{\beta}_0$ and $\hat{\beta}_1$

In the special cases where $c_0 = 1$, $c_1 = 0$ and $c_0 = 0$, $c_1 = 1$, Box 10.5 gives results for the sampling distributions of the least squares estimates of β_0 and β_1. These frequently used results may be summarized as follows.

1 From property 1 in Box 10.5, it follows that, regardless of whether the errors have normal distributions, the following properties may be stated:

a The least squares estimates $\hat{\beta}_0$ and $\hat{\beta}_1$ are unbiased estimates of β_0 and β_1, respectively.

b The standard error of $\hat{\beta}_0$ is (set $c_0 = 1$, $c_1 = 0$ in the formula for $\sigma_{\hat{\theta}}$)

$$SE(\hat{\beta}_0) = \sigma_{\hat{\beta}_0} = \sigma\sqrt{s_{00}} = \sigma\sqrt{\frac{1}{n} + \frac{\bar{x}^2}{S_{xx}}},$$

and the standard error of $\hat{\beta}_1$ is (set $c_0 = 0$, $c_1 = 1$ in the same formula)

$$SE(\hat{\beta}_1) = \sigma_{\hat{\beta}_1} = \sigma\sqrt{s_{11}} = \frac{\sigma}{\sqrt{S_{xx}}}.$$

2 Property 2 in Box 10.5 states that, if the errors are independent $N(0, \sigma^2)$, then the least squares estimates of the regression coefficients have normal distributions. In other words, $\hat{\beta}_0$ is

$$N\left(\beta_0, \ \sigma^2 \left[\frac{1}{n} + \frac{\bar{x}^2}{S_{xx}}\right]\right)$$

and $\hat{\beta}_1$ is

$$N\left(\beta_1, \ \frac{\sigma^2}{S_{xx}}\right).$$

3 Let $\hat{\sigma}_{\hat{\beta}_0}$ and $\hat{\sigma}_{\hat{\beta}_1}$ denote the estimated standard errors of $\hat{\beta}_0$ and $\hat{\beta}_1$, respectively. Then, from property 3 in Box 10.5, we can conclude that tests of the null hypotheses H_0: $\beta_0 = \beta_{0(0)}$ and H_0: $\beta_1 = \beta_{1(0)}$ can be based on the test statistics

$$t_{\hat{\beta}_0} = \frac{\hat{\beta}_0 - \beta_{0(0)}}{\hat{\sigma}_{\hat{\beta}_0}},$$

$$t_{\hat{\beta}_1} = \frac{\hat{\beta}_1 - \beta_{1(0)}}{\hat{\sigma}_{\hat{\beta}_1}},$$

respectively. The observed values of the test statistics should be compared to the critical values of a $t(n-2)$-distribution.

4 Property 4 in Box 10.5 gives an expression for the confidence interval for θ. From this expression, we can deduce the form of the $100(1-\alpha)\%$ confidence intervals for β_0

$$\hat{\beta}_0 - t(\nu, \alpha/2)\hat{\sigma}_{\hat{\beta}_0} \le \beta_0 \le \hat{\beta}_0 + t(\nu, \alpha/2)\hat{\sigma}_{\hat{\beta}_0}$$

and for β_1

$$\hat{\beta}_1 - t(\nu, \alpha/2)\hat{\sigma}_{\hat{\beta}_1} \le \beta_1 \le \hat{\beta}_1 + t(\nu, \alpha/2)\hat{\sigma}_{\hat{\beta}_1},$$

where $\nu = n - 2$.

The next example illustrates the use of the sampling distributions of $\hat{\beta}_0$ and $\hat{\beta}_1$ to make inferences about the slope and intercept parameters in a simple linear regression model.

EXAMPLE **10.9** To understand the pattern of variation in plant species in Mediterranean grasslands, data were collected on $Y =$ density of plant species (the number of species per 0.04 m^2) and $X =$ the altitude of the region (in 1000 m) from 12 experimental plots over a period of five years (1986–1990). The results obtained for the average (over the five-year period) species density are as follows (Montalvo, Casado, Levassor, & Pineda, 1993, Figure 2):

Plot	\multicolumn{6}{c}{Altitude x (1000 m)}					
	0.64	0.86	0.89	1.22	1.45	1.72
1	15	16	13.5	11	11.5	8
2	20	18.5	16	19.5	12	8.5

Figure 10.8 shows a scatter plot of the data.

FIGURE 10.8

Scatter plot of plant species density data

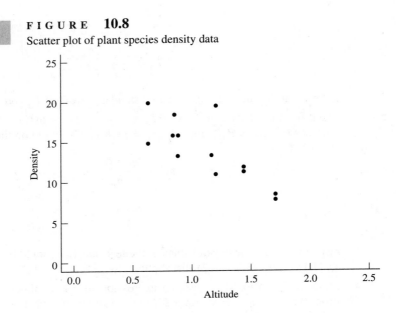

It appears that the relationship between the mean species density and altitude can be well approximated by a straight line. Therefore let's assume the simple linear regression model

$$Y_i = \beta_0 + \beta_1 x_i + E_i, \quad i = 1, \dots, 12,$$

where the notation employed is as follows:

- Y_i is the species density (the number of species per 0.04 m^2) that will be observed in the i-th plot;
- x_i is the altitude (in 1000 m) of the i-th plot;
- β_0 is the expected species density at zero altitude;
- β_1 is the expected change in species density for each unit (1000 m) increase in altitude;
- E_i is the random error in the measured species density for the i-th plot.

The E_i are independent $N(0, \sigma^2)$.

The formulas in Box 10.3 can be used to calculate the least squares estimates of the regression coefficients and the associated error sum of squares. We have

$$\bar{x} = 1.13; \quad \bar{y} = 14.125; \quad S_{xx} = 1.6584; \quad S_{yy} = 181.0625; \quad \text{and } S_{xy} = -13.5450.$$

Thus

$$\hat{\beta}_1 = -8.17, \quad \hat{\beta}_0 = 23.35, \quad \text{and} \quad \text{SS[E]} = 70.4336.$$

First of all, notice that the estimate of the slope parameter is negative; in other words the plant species density decreases as the altitude increases. The estimated rate of decrease is 8.17 species per 0.04 m^2 for every 1000-m increase in altitude.

The estimated value of the intercept parameter is 23.35, so that the estimated plant density at zero altitude is 23.35 species per 0.04 m^2. Of course, because zero altitude is outside the range of the present data set, the practical implication of the estimate of β_0 will depend on whether the simple linear regression model is appropriate near $x = 0$.

The results in Box 10.5 can be used to construct confidence intervals and hypothesis tests for the regression parameters. First, we need to calculate the least squares estimate of the error variance σ^2, which is given by MS[E], the error mean square. We know that SS[E] equals 70.4336 with $\nu = n - 2 = 12 - 2 = 10$ degrees of freedom, and so the required estimate is

$$\hat{\sigma}^2 = \text{MS[E]} = \frac{70.4336}{12 - 2} = 7.0434.$$

The estimated standard errors of $\hat{\beta}_0$ and $\hat{\beta}_1$ are

$$\hat{\sigma}_{\hat{\beta}_0} = \sqrt{\text{MS[E]}\left(\frac{1}{n} + \frac{\bar{x}^2}{S_{xx}}\right)} = \sqrt{7.0434\left[\frac{1}{12} + \frac{(1.13)^2}{1.6584}\right]} = 2.4515$$

and

$$\hat{\sigma}_{\hat{\beta}_1} = \sqrt{\frac{\text{MS[E]}}{S_{xx}}} = \sqrt{\frac{7.0434}{1.6584}} = 2.0609.$$

A 95% confidence interval for the slope parameter is

$$\hat{\beta}_1 \pm t(10, \ 0.025)\hat{\sigma}_{\hat{\beta}_1} = -8.17 \pm (2.228)(2.0609) \quad \Rightarrow \quad -12.76 \le \beta_1 \le -3.58.$$

We can conclude with 95% confidence that, for every 1000-m increase in altitude, there will be a decrease ($\hat{\beta}_1$ is negative) of between 3.58 and 12.76 species per 0.04 m^2, on average.

A confidence interval for β_0 can be constructed in a similar manner. As already noted, however, meaningful interpretation of such an interval is only possible if the simple linear regression model is appropriate in a range of altitudes containing the value $x = 0$.

Tests of hypotheses concerning the regression parameters can be performed using either the confidence-interval approach or the test-statistic approach. The 95% confidence interval of $-12.76 \le \beta_1 \le -3.58$ for the slope parameter means that, for any value of $\beta_{1(0)}$ outside the range from -12.76 to -3.58, the null hypothesis H_0: $\beta_1 = \beta_{1(0)}$ will be rejected ($\alpha = 0.05$) in favor of the two-sided alternative hypothesis H_1: $\beta_1 \neq \beta_{1(0)}$. For instance, at the $\alpha = 0.05$ level, the null hypothesis H_0: $\beta_1 = 0$, which states that the altitude has no effect on the species density, can be rejected in favor of H_1: $\beta_1 \neq 0$. ■

Confidence intervals for the expected responses

By setting $c_0 = 1$ and $c_1 = x$ in Box 10.5, we obtain results for making inferences about $\mu(x) = \beta_0 + \beta_1 x$, the expected response at $X = x$. In particular, we can obtain the expression for a $100(1 - \alpha)\%$ confidence interval for $\mu(x)$ in this manner (Box 10.6).

BOX 10.6

Confidence interval for an expected response in a simple linear regression model

Let $\mu(x_0) = \beta_0 + \beta_1 x_0$ denote the expected response at $X = x_0$, and let $\hat{\mu}(x_0) = \hat{\beta}_0 + \hat{\beta}_1 x_0$ denote its least squares estimate. Recall that $\hat{\mu}(x_0)$ is the same as the predicted value of Y when $X = x_0$. Then we can state the following properties.

1 The standard error of $\hat{\mu}(x_0)$ is

$$\sigma_{(x_0)} = \sqrt{\left[\frac{1}{n} + \frac{(x_0 - \bar{x})^2}{S_{xx}} \right] \sigma^2},$$

where $\bar{x}$ is the mean of the n settings of X at which the responses were measured.

2 The least squares estimate of the standard error of $\hat{\mu}(x_0)$ is

$$\hat{\sigma}_{(x_0)} = \sqrt{\left[\frac{1}{n} + \frac{(x_0 - \bar{x})^2}{S_{xx}} \right] MS[E]},$$

3 A $100(1 - \alpha)\%$ confidence interval for $\mu(x_0)$ is

$$\hat{\mu}(x_0) - t(n - 2, \alpha/2)\hat{\sigma}_{(x_0)} \leq \mu(x_0) \leq \hat{\mu}(x_0) + t(n - 2, \alpha/2)\hat{\sigma}_{(x_0)}.$$

The confidence interval in Box 10.6 can be derived from the confidence interval for θ in Box 10.5, by straightforward algebraic manipulations with $c_0 = 1$, $c_1 = x$.

Some important properties of the confidence interval for $\mu(x_0)$ may be derived from the formula in Box 10.6 for $\hat{\sigma}_{(x_0)}$, the standard error of the estimate of the expected response.

1 The standard error decreases with increase in n; hence, increasing the sample size will decrease the width of the confidence interval.

2 The standard error decreases with increase in S_{xx}; hence, the width of the confidence interval can be decreased by increasing the spread of the values of X at which the responses are measured.

3 The standard error increases with increase in $(x_0 - \bar{x})^2$; hence, the width of the confidence interval for $\mu(x_0)$ can be decreased by decreasing the value of $(x_0 - \bar{x})^2$. In other words, as x_0 moves toward $\bar{x}$, the confidence interval for $\mu(x_0)$ becomes narrower. For a given data set, the shortest confidence interval is obtained when $x_0 = \bar{x}$.

These properties prove useful when designing a study on the basis of a simple linear regression model.

EXAMPLE 10.10   On the basis of the plant species density data in Example 10.9, let's estimate the mean species density at an altitude of 750 m. The altitude is measured in units of 1000 m, and so we want to estimate $\mu(x_0)$, where $x_0 = 0.75$. The required estimate is $\hat{\mu}(0.75) = \hat{\beta}_0 + \hat{\beta}_1(0.75) = 23.35 - (8.17)(0.75) = 17.22$.

The least squares estimate of the standard error of $\hat{\mu}(0.75)$ can be computed using the expression in Box 10.6

$$\hat{\sigma}_{(0.75)} = \sqrt{\left[\frac{1}{12} + \frac{(0.75 - 1.13)^2}{1.6584}\right](7.0434)} = 1.10.$$

A 95% confidence interval for $\mu(0.75)$ may be calculated as

$$\hat{\mu}(0.75) \pm t(10, 0.025)\hat{\sigma}_{(0.75)}$$

and thus

$$14.78 \le \mu(0.75) \le 19.67.$$

Therefore, with 95% confidence, we can conclude that, at an altitude of 750 m, we can expect between 14.78 and 19.67 species per 0.04 m^2.

A 95% lower bound for the number of species at altitude 750 m is

$$\hat{\mu}(0.75) - t(10, 0.05)\hat{\sigma}_{(0.75)} = 17.23 - (1.812)(1.10) = 15.24.$$

We can conclude with 95% confidence that, on average, at least 15.24 species per 0.04 m^2 can be found at an altitude of 750 m.   ■

Confidence bands for expected responses

The expressions in Box 10.6 can be used to construct a confidence interval for $\mu(x)$ at each value of x. If the endpoints of the $100(1 - \alpha)\%$ confidence intervals for the expected responses are plotted against x, the band between the resulting curves is called a $100(1 - \alpha)\%$ *confidence band*.

A confidence band provides a convenient graphical representation of the collection of confidence intervals for the expected response at various values of the independent variable. Figure 10.9 shows the 95% confidence band for the plant species density data. The intersection of the vertical line at x and the confidence band is the confidence interval for the expected response at x.

Prediction intervals and prediction bands

In Chapter 9, we described how prediction intervals can be used in the setting of a one-way ANOVA model. Similar ideas are applicable when working with a simple

FIGURE   10.9

95% confidence band for plant species density

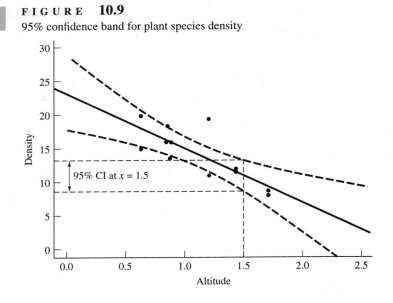

linear regression model. As we see in the next example, the problem of prediction arises in a very natural way in a regression setting.

EXAMPLE **10.11**   For a particular variety of plant, researchers wanted to develop a formula for predicting the quantity of seeds as a function of the density of plants. They conducted a study with four levels of the factor X, the number of plants per plot, in a completely randomized design; there were four replications per treatment. The data obtained for the number of plants per plot and the quantity of seeds per plot (g) are as follows:

| | Plot number | | | |
Plants per plot x	1	2	3	4
10	12.6	11.0	12.1	10.9
20	15.3	16.1	14.9	15.6
30	17.9	18.3	18.6	17.8
40	19.2	19.6	18.9	20.0

To predict the response at one of the four levels used in the study—that is, to predict when $X = 10$, 20, 30, or 40—prediction intervals can be constructed using the methods described in Chapter 9, if the assumptions underlying the one-way ANOVA model are reasonable. However, if we want a prediction formula that is valid for values of X not included in the study design—say, $X = 25$—we need an assumption regarding the mathematical relationship between the expected response and the value of the independent variable. For example, if we can assume that the seed yield data satisfy a simple linear regression model, it is possible to construct formulas for prediction intervals at any given value of X.   ■

Under the simple linear regression model, an observed response at $X = x$ can be represented as

$$Y_x = \mu(x) + E,$$

where $\mu(x) = \beta_0 + \beta_1 x$ is the expected response at $X = x$, and E is the random error. The predicted value of Y_x is $\hat{y}_x = \hat{\beta}_0 + \hat{\beta}_1 x$, because $\hat{y}_x$ is an estimate of the expected response at $X = x$. In other words, we use the same value for estimating the expected response and predicting the observed response at $X = x$. However, the standard error used to construct a prediction interval for Y_x is larger than that used to construct a confidence interval for $\mu(x)$, because a confidence interval is designed to cover $\mu(x)$ alone, whereas a prediction interval should cover $\mu(x) + E$. A prediction interval for the mean of m future values of Y at a given value x of the independent variable X may be constructed on the basis of the formula in Box 10.7. By setting $m = 1$, we can use the formula to construct a prediction interval for a single future value.

BOX **10.7**

> ### Prediction intervals for future observations
>
> Let $\mu(x) = \beta_0 + \beta_1 x$ denote the expected response when $X = x$, and let $\hat{\mu}(x)$ be the least squares estimate (based on n responses) of $\mu(x)$. If $\overline{Y}_{x_0}$ denotes the mean of a random sample of m observed responses at $X = x_0$, then a $100(1 - \alpha)\%$ prediction interval for $\overline{Y}_{x_0}$ is given by
>
> $$\hat{\mu}(x_0) - t(n - 2, \ \alpha/2)\hat{\sigma}_P(x_0) \leq \overline{Y}_{x_0} \leq \hat{\mu}(x_0) + t(n - 2, \ \alpha/2)\hat{\sigma}_P(x_0),$$
>
> where
>
> $$\hat{\sigma}_P^2(x_0) = \left[\frac{1}{m} + \frac{1}{n} + \frac{(x_0 - \bar{x})^2}{S_{xx}} \right] \mathrm{MS[E]}$$
>
> and MS[E] is the least squares estimate of the error variance.

Compare the formula for the prediction interval in Box 10.7 with the formula for the confidence interval in Box 10.6. The only difference between the two intervals is the way in which the standard errors are calculated. The standard error $\hat{\sigma}_P(x)$ in the prediction interval has an extra term $1/m$ ($m = 1$ for a single response) to account for the contribution of random error to the variability of an observed value. It is simple to verify that all the comments about the properties of the confidence interval for an expected response are equally valid for the prediction interval in Box 10.7. Also, the endpoints of the $100(1 - \alpha)\%$ prediction intervals can be plotted against x to generate a $100(1 - \alpha)\%$ *prediction band* for a future observation of the response variable.

EXAMPLE **10.12**   Let's assume that the simple linear regression model is appropriate for the seed yield data in Example 10.11. In that case, we write

$$Y_x = \beta_0 + \beta_1 x + E_x,$$

where x is the number of plants per plot; Y_x is an observed seed yield when there are x plants in a plot; and E_x is the random error in Y_x. The random errors are independent $N(0, \sigma^2)$.

The following statistics can be calculated from the data in Example 10.11

$$\bar{x} = 25, \quad S_{xx} = 2000, \quad \text{MS[E]} = 0.7479, \quad \hat{\beta}_0 = 9.68, \quad \hat{\beta}_1 = 0.26.$$

The least squares prediction equation for predicting an observed seed yield is

$$\hat{y} = 9.68 + 0.26x.$$

The estimated standard error of a predicted seed yield for a single plot with x_0 plants can be calculated using the expression for $\hat{\sigma}_P(x_0)$ with $m = 1$. For a plot with 35 plants, the predicted seed yield is $9.68 + 0.26(35) = 18.78$, and the estimated standard error of the predicted yield is

$$\hat{\sigma}_P(35) = \sqrt{\left[1 + \frac{1}{16} + \frac{(35 - 25)^2}{2000}\right](0.7479)}$$

$$= 0.9122.$$

A 95% prediction interval for the yield of a plot with 35 plants can be calculated as

$$18.78 \pm t(14, \ 0.025)(0.9122).$$

Thus

$$16.82 \leq Y_{35} \leq 20.73.$$

We can conclude with 95% confidence that a plot with 35 plants will yield between 16.82 and 20.73 g of seeds. In contrast, calculations using the formula for a confidence interval for $\mu(x_0)$ show that a 95% confidence interval for the expected (population mean) quantity of seeds produced in plots containing 35 plants is (18.16, 19.40). As we would expect, this interval is narrower than the corresponding prediction interval.

The REG procedure in SAS may be used to fit a simple linear regression model to the seed yield data. Section 10.2 in the companion text by Younger (1997) provides the SAS codes needed to implement the REG procedure.

The following printout shows the results given by the REG procedure for the seed yield data. The printout is divided into two parts. The first part contains the estimates of β_0 and β_1, their estimated standard errors ($\hat{\sigma}_{\hat{\beta}_0}$ and $\hat{\sigma}_{\hat{\beta}_1}$), the values of the t-statistics ($t_{\hat{\beta}_0}$ and $t_{\hat{\beta}_0}$) for testing the null hypothesis that the parameter is zero (H_{00}: $\beta_0 = 0$ and H_{01}: $\beta_1 = 0$), and the two sided p-values for testing H_{00} and H_{01}.

The second part shows the observed values y, the corresponding predicted values $\hat{y}$, the residuals $e = y - \hat{y}$, and the 95% prediction intervals at every value of X at which an observation is available ($X = 10, 20, 30,$ and 40 plants).

PART 1

Parameter Estimates

Variable	DF	Parameter Estimate	Standard Error	T for H0: Parameter=0	Prob > |T|
INTERCEP	1	9.675000	0.52957193	18.269	0.0001
X	1	0.260000	0.01933723	13.446	0.0001

PART 2

X	Dep Var Y	Predict Value	Std Err Predict	Lower95% Predict	Upper95% Predict	Residual
10	12.6000	12.2750	0.362	10.2645	14.2855	0.3250
10	11.0000	12.2750	0.362	10.2645	14.2855	-1.2750
10	12.1000	12.2750	0.362	10.2645	14.2855	-0.1750
10	10.9000	12.2750	0.362	10.2645	14.2855	-1.3750
20	15.3000	14.8750	0.237	12.9519	16.7981	0.4250
20	16.1000	14.8750	0.237	12.9519	16.7981	1.2250
20	14.9000	14.8750	0.237	12.9519	16.7981	0.0250
20	15.6000	14.8750	0.237	12.9519	16.7981	0.7250
30	17.9000	17.4750	0.237	15.5519	19.3981	0.4250
30	18.3000	17.4750	0.237	15.5519	19.3981	0.8250
30	18.6000	17.4750	0.237	15.5519	19.3981	1.1250
30	17.8000	17.4750	0.237	15.5519	19.3981	0.3250
40	19.2000	20.0750	0.362	18.0645	22.0855	-0.8750
40	19.6000	20.0750	0.362	18.0645	22.0855	-0.4750
40	18.9000	20.0750	0.362	18.0645	22.0855	-1.1750
40	20.0000	20.0750	0.362	18.0645	22.0855	-0.0750

Sum of Residuals	0
Sum of Squared Residuals	10.4700
Predicted Resid SS (Press)	14.0487

The 95% confidence and prediction bands for the seed yields are shown in Figure 10.10. ■

Exercises

10.13 Refer to the chemical influx data given in Example 10.3.

a Using a simple linear regression model for Y = chemical influx as a function of $X = \log d$, construct a 95% confidence interval for the expected chemical influx when the drug dose is 100 nM. Interpret the interval.

F I G U R E   **10.10**

95% confidence and prediction bands for seed yields

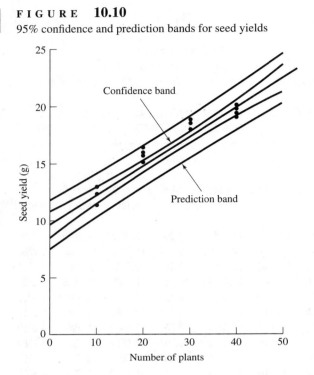

b   Use the data to verify the claim that quadrupling the drug dose $(d_2/d_1 = 4)$ will result in an increase in chemical influx of more than 5 nmol/mg protein. Answer this question using both hypothesis-testing and confidence-interval approaches.

c   Use a one-way ANOVA model (see Exercise 10.1) to construct the 95% confidence interval for the expected chemical influx when the drug dose is 100 nM. Compare this interval with the interval obtained in (a).

d   Using the simple linear regression model, construct a 90% confidence band for the expected chemical influx. On that basis, determine the 90% confidence intervals for the expected chemical influxes for drug doses $d = 100$ and $d = 150$. Interpret the meaning of these two intervals.

e   Use the simple linear regression model to construct a 99% lower bound to the mean chemical influx for the population of all rats receiving a drug dose 10 nM. Interpret the interval.

f   On the basis of a simple linear regression model, perform a statistical test to see if the data suggest that drug dose and chemical influx are associated.

g   Do the data support the claim that, on average, rats receiving a dose $d = 1$ will have a chemical influx of less than 25 nM?

10.14   Refer to the soil water content data in Exercise 10.2.

a   Using a simple linear regression model for Y = soil water content as a function of $X = \log d$, where d = soil depth, construct a 95% upper bound for the mean soil water content at soil depth 60 cm. On the basis of this upper bound, is it reasonable to conclude that, on average, the water concentration at depth 60 cm is more than $0.20 \text{ cm}^3/\text{cm}^3$? Explain.

b Repeat (a) using the one-way ANOVA model and compare the two bounds.

c Construct a 95% confidence band for the mean soil water content. Examine the confidence band to determine the depth at which the narrowest confidence interval will be obtained.

d Do these data support the hypothesis that the soil water content decreases by more than $0.003 \text{ cm}^3/\text{cm}^3$ for every 10-cm increase in soil depth?

e Calculate a 90% lower prediction bound for the mean measured soil water content of ten plots at a depth of 50 cm. Interpret the bound.

10.15 Refer to the weight gain data in Exercise 10.3.

a Using a simple linear regression model, calculate a 95% lower bound for the expected change in a mother's gain in weight for an increase of six months in her age. Interpret the bound.

b Use the data to construct a 95% confidence interval for the expected weight gain of a 35-year-old mother. Interpret this interval. Explain why there may be some questions about the validity of this interval.

c On the basis of the given data, what can you say—at the 99% confidence level—about the upper limit for the weight gain of a 24-year-old mother?

10.16 Refer to the disease index data in Exercise 10.7.

a Examine the data to determine the value of the crop interval for which the expected disease index can be determined with the smallest bound on the error of estimation.

b Construct a 95% confidence interval for the mean disease index when the crop index value is as determined in (a).

c Use the data to construct a graph showing the 95% confidence and prediction bands.

10.5
Simple linear regression in an ANOVA setting

As we saw in Section 10.3, a simple linear regression model and a one-way ANOVA model share several common characteristics. In both models, observed responses are expressed as the sum of an expected response and a random error. Standard techniques for making statistical inferences about these models assume that the errors have independent normal distributions with a common variance. The null hypothesis H_{01}: $\tau_1 = \cdots = \tau_t$ (no difference between treatment effects) in the one-way ANOVA model is equivalent to the null hypothesis H_{02}: $\beta_1 = 0$ (slope parameter is zero), because both null hypotheses imply that all levels of the experimental factor have the same effect on the response. We shall refer to H_{01} and H_{02} as the *hypothesis of no treatment effects* and the *hypothesis of no regression effect*, respectively. In Chapter 8, we saw that, under the ANOVA assumptions, an F-statistic of the form

$$F_{trt} = \frac{\text{MS[T]}}{\text{MS[E]}},$$

where MS[T] and MS[E] are, respectively, the mean squares for treatment and the mean squares for error in a one-way ANOVA, can be used to test the hypothesis of no treatment effects against the research hypothesis H_{11} that the treatment effects are not all equal. If H_{01} is true, both MS[T] and MS[E] estimate the error variance; if H_{01} is not true, however, MS[T] estimates a quantity larger than the quantity estimated by MS[E]. Thus, large values of F_{trt} indicate that we should reject the hypothesis of no treatment effects.

In this section, we'll see that the test of the hypothesis of no regression effect can also be set in an ANOVA format. Viewing regression analysis in the ANOVA setting helps us understand why the ANOVA and regression models can be considered as special cases of general linear models (Chapter 12).

From our discussions in Section 10.4, we see that the null hypothesis H_{02} can be tested against the research hypothesis H_{12}: $\beta_1 \neq 0$ using the test statistic

$$t_{\hat{\beta}_1} = \frac{\hat{\beta}_1}{\hat{\sigma}_{\hat{\beta}_1}},$$

where $\hat{\beta}_1$ is the least squares estimate of β_1 and

$$\hat{\sigma}_{\hat{\beta}_1} = \sqrt{\frac{\text{MS[E]}}{S_{xx}}} \tag{10.8}$$

is the estimated standard error of $\hat{\beta}_1$. This is a two-sided t-test with $\nu = n - 2$ degrees of freedom. A two-sided t-test is equivalent to a one-sided F-test, and so we can also perform an F-test of H_{02}, with $\nu_1 = 1$ degree of freedom for the numerator and $\nu_2 = n - 2$ degrees of freedom for the denominator. The test statistic for the F-test is the square of the test statistic for the t-test

$$F_{\hat{\beta}_1} = t_{\hat{\beta}_1}^2 = \frac{\hat{\beta}^2}{\hat{\sigma}_{\hat{\beta}_1}^2}.$$

Substituting for $\hat{\sigma}_{\hat{\beta}_1}$ from Equation (10.8), we get

$$F_{\hat{\beta}_1} = \frac{\hat{\beta}_1^2 S_{xx}}{\text{MS[E]}}.$$

As a means of establishing an analogy with the ANOVA terminology, let's refer to

$$\text{SS[R]} = \hat{\beta}_1^2 S_{xx} \tag{10.9}$$

as the *sum of squares for regression* and assign to it $\nu_1 = 1$ degree of freedom. Then the F-test for the regression effect can be considered as a ratio of two mean squares

$$F_{\hat{\beta}_1} = \frac{\text{MS[R]}}{\text{MS[E]}},$$

where

$$\text{MS[R]} = \frac{\text{SS[R]}}{\nu_1}, \qquad \nu_1 = 1,$$

$$\text{MS[E]} = \frac{\text{SS[E]}}{\nu_2}, \qquad \nu_2 = n - 2.$$

Thus, the *F*-statistic for testing the significance of the slope parameter in a simple linear regression model has the same form as the *F*-statistic for testing the significance of the difference between treatment effects in a one-way ANOVA model. Indeed, the analogy can be further extended, as follows.

1 In both models (see the expression for SS[E] in Box 10.3), the sum of squares for testing the hypothesis of no effects and the sum of squares for error add up to S_{yy}, the total sum of squares; that is

$$\text{SS[TOT]} = \text{SS[R]} + \text{SS[E]}$$

in the simple linear regression model and

$$\text{SS[TOT]} = \text{SS[T]} + \text{SS[E]}$$

in the one-way ANOVA model.

2 In both models, the degrees of freedom for the hypothesis sum of squares (SS[R] or SS[T]) and the error sum of squares add up to the degrees of freedom for the total sum of squares; that is

$$df\,(\text{SS[R]}) + df\,(\text{SS[E]}) = df\,(\text{SS[TOT]}),$$

$$1 + (n - 2) = n - 1$$

in the simple linear regression model and

$$df\,(\text{SS[T]}) + df\,(\text{SS[E]}) = df\,(\text{SS[TOT]}),$$

$$(t - 1) + (n - t) = n - 1$$

in the one-way ANOVA model.

3 As in the one-way ANOVA model, each sum of squares associated with the simple linear regression model can be interpreted as a measure of the variability in the observed responses due to the effect of a particular source of variation. To see this, let $\hat{y}_i$ and e_i denote, respectively, the predicted response and the residual at $X = x_i$. Also, let $\bar{\hat{y}}$ and $\bar{e}$ denote, respectively, the means of the $\hat{y}_i$ and the e_i. Then it can be shown that

$$\text{SS[R]} = \sum \left(\hat{y}_i - \bar{\hat{y}} \right)^2,$$

$$\text{SS[E]} = \sum \left(e_i - \bar{e} \right)^2.$$

The expressions for the regression and error sums of squares indicate that these two quantities are measures of the variability among the predicted values and among the residual values, respectively. The predicted values are estimates of the expected responses obtained using a simple linear regression model, and so more variability in the predicted values means that the model is more sensitive to changes in the values of the independent variable. Because SS[TOT] quantifies the variability in the observed responses, large values of SS[R] indicate that a large amount of variation in the observed responses can be explained by the simple linear regression model. The variability of the residuals is the variability of the observed responses that cannot be explained by the model. The regression and error sum of squares can be regarded as the components of the total observed variation that can be ascribed to the association of the independent and error variables with the observed responses.

The interpretation of SS[R], SS[E], and SS[TOT] in a simple regression analysis is illustrated in Figure 10.11, which shows a set of data points and the corresponding least squares regression line. Since the y-coordinates of the data points are the observed responses, SS[TOT] is a measure of the variability of these y-coordinates. Each circle on the estimated line locates the predicted response corresponding to a setting of the independent variable at which a response has been observed. The regression sum of squares SS[R] is a measure of the variability of the y-coordinates of the points enclosed in circles. From Figure 10.11, we see that the regression sum of squares will never exceed the total sum of squares. The difference between the total sum of squares and the regression sum of squares, which equals the error sum of squares, represents the amount of variability in the observed responses that cannot be explained by the fitted line. Note that, if the observed responses lie on a straight line, then SS[R] = SS[TOT] (SS[E] = 0), which implies that the observed variability is completely explained by the fitted least squares line.

F I G U R E   10.11

Partitioning the total sum of squares

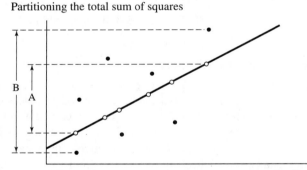

A: Extent of variability of predicted values (measured by SS[R]).
B: Extent of variability of observed values (measured by SS[TOT]).

4   Theorem B.2 (Appendix B) can be used to show that MS[E] and MS[R] are unbiased estimates of σ^2 and $\sigma^2 + \beta_1^2 S_{xx}$, respectively. Thus, analogously to the one-way ANOVA setting in Section 8.3, MS[E] is an unbiased estimate of σ^2 regardless of whether H_{02} is true, whereas MS[R] is an unbiased estimate of σ^2 only if H_{02} is true. If H_{02} is not true, then $\beta_1^2 S_{xx} > 0$, and MS[R] will estimate a quantity larger than σ^2. Thus, MS[R] and MS[E] play roles similar to those of MS[T] and MS[E] in a one-way ANOVA. As in the one-way ANOVA model, large values of $F_{\hat{\beta}_1}$ favor rejection of the null hypothesis that there is no regression effect.

5   The degrees of freedom associated with the sums of squares in the two models have similar interpretations. The $t - 1$ degrees of freedom for the treatment sum of squares in a one-way ANOVA account for the fact that this sum of squares can be expressed in terms of $t - 1$ mutually orthogonal linear combinations of the observed responses. It can be verified that the sum of squares for regression can be expressed in terms of a single linear combination as

$$SS[R] = \left(c_1 y_1 + \cdots + c_n y_n\right)^2,$$

where

$$c_i = \frac{x_i - \bar{x}}{\sqrt{S_{xx}}}, \qquad i = 1, \ldots, n.$$

Therefore, SS[R] has a single degree of freedom.

The various sums of squares in the simple linear regression analysis are often displayed as in Table 10.3. The next example illustrates the ANOVA table for regression analysis.

T A B L E   10.3
ANOVA table for testing H_0: $\beta_1 = 0$

Source	Degrees of freedom	Sum of squares	Mean squares
Regression	$\nu_1 = 1$	$SS[R] = \hat{\beta}_1^2 S_{xx}$	$MS[R] = \dfrac{SS[R]}{\nu_1}$
Error	$\nu_2 = n - 2$	$SS[E] = S_{yy} - \hat{\beta}_1^2 S_{xx}$	$MS[E] = \dfrac{SS[E]}{\nu_2}$
Total	$n - 1$	$SS[TOT] = S_{yy}$	

EXAMPLE   10.13   Suppose we want to fit a simple linear regression function to the seed yield data in Example 10.11. The portion of the output obtained by the REG procedure of SAS that contains the corresponding sums of squares is as follows.

```
--------------------------------------------------------------------------------

                          Analysis of Variance
                          Sum of        Mean
        Source      DF    Squares      Square     F Value    Prob>F

        Model        1   135.20000    135.20000   180.783    0.0001
        Error       14    10.47000      0.74786
        C Total     15   145.67000


--------------------------------------------------------------------------------
```

The sums of squares may be displayed as the following ANOVA table:

Source	df	SS	MS	F	p
Regression	1	135.2000	135.2000	180.783	0.0001
Error	14	10.4700	0.7479		
Total	15	145.6700			

Recall that there were $n = 16$ responses. Thus, the degrees of freedom for SS[TOT] and SS[R] are, respectively, $n - 1 = 15$ and $n - 2 = 14$. The table also shows the calculated value of the F-statistic for testing the hypothesis of no regression effect. In this case, the hypothesis of no regression effect is equivalent to the hypothesis that the number of plants per plot has no effect on the expected quantity of seeds produced. The calculated value of F is relatively large, which indicates that there is a regression effect. Indeed, when compared to the critical value of an F-distribution with $v_1 = 1$ and $v_2 = 14$ degrees of freedom, the calculated value favors rejection ($p = 0.0001$) of the null hypothesis.

Alternatively, the hypothesis of no regression effect can be tested with a two-sided t-test. Again, from the computer output in Example 10.12, we see that the t-statistic for testing H_0: $\beta_1 = 0$ is 13.446, which is equal to the square root of the F-statistic for testing the same null hypothesis. As expected, the two-sided t-test leads to the same conclusion as the F-test. ■

The coefficient of determination

The fraction of SS[TOT], the total variation in the observed responses, that can be explained by a simple linear regression model is called the *coefficient of determination*. Thus, the coefficient of determination for the simple linear regression model

$$Y = \beta_0 + \beta_1 x + E$$

is defined as

$$R_{Y.X}^2 = \frac{SS[R]}{SS[TOT]} = \frac{SS[R]}{SS[R] + SS[E]}. \tag{10.10}$$

It can be verified that $0 \leq R_{Y.X}^2 \leq 1$ and that $R_{Y.X}^2 = 1$ when the responses lie on a straight line (SS[E] = 0) and $R_{Y.X}^2 = 0$ when SS[R] = 0. We know that SS[R] = $\hat{\beta}^2 S_{xx}$, and so $R_{Y.X}^2 = 0$ when the best fitting line is parallel to the x-axis ($\hat{\beta}_1 = 0$)— that is, when all the values of the independent variable yield the same predicted value for the response. When there is no confusion about independent and dependent variables, we will drop the subscripts from $R_{Y.X}^2$ and denote the coefficient of determination by R^2. Figure 10.12 shows some patterns of responses corresponding to different values of R^2.

Let

$$s_{obs}^2 = \frac{\sum (y_i - \bar{y})^2}{n - 1}$$

FIGURE   **10.12**

The coefficient of determination as an index of straight line fit

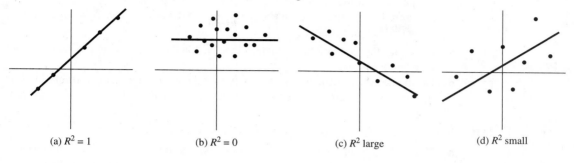

(a) $R^2 = 1$ (b) $R^2 = 0$ (c) R^2 large   (d) R^2 small

and

$$s_{pred}^2 = \frac{\sum (\hat{y}_i - \bar{y})^2}{n - 1}$$

denote, respectively, variances of the observed and predicted values. We see that

$$s_{obs}^2 = \frac{SS[TOT]}{n - 1} \quad \text{and} \quad s_{pred}^2 = \frac{SS[R]}{n - 1},$$

and so the sample coefficient of determination can be expressed as

$$R_{Y.X}^2 = \frac{s_{pred}^2}{s_{obs}^2}. \tag{10.11}$$

Thus, the sample coefficient of determination can be interpreted as the fraction of the sample variance that is accounted for by the variance of the predicted values. If $R_{Y.X}^2 = 1$, then $s_{pred}^2 = s_{obs}^2$, so that all of the observed variation in the responses can be explained by the linear regression model for predicting the responses. If $R_{Y.X}^2 = 0$, then $s_{pred}^2 = 0$, which implies that the linear regression model is not useful for explaining the variability in the observed responses.

EXAMPLE **10.14**   On the basis of the sums of squares for simple linear regression analysis in Example 10.13, the coefficient of determination for the seed yield data is

$$R^2 = \frac{135.20}{145.67} = 0.9281.$$

Therefore, 92.81% of the variation in the observed seed yield can be explained by the straight-line relationship between the expected seed yield and the number of plants per plot.   ■

Exercises

10.17   Refer to the chemical influx data in Example 10.3.

   a   Set up the ANOVA tables for simple linear regression analysis:

i   with Y = chemical influx as the dependent variable and $X = d$ as the independent variable;

ii   with Y = chemical influx as the dependent variable and $X = \log d$ as the independent variable.

b   Calculate the coefficients of determination for the two models. Which fits the data better? Why?

c   Do the necessary calculations to verify that the coefficients of determination for the two models are the ratios of the variances of the observed and predicted responses.

10.18   Refer to the soil water content data in Exercise 10.2.

a   Calculate the coefficients of determination for three simple linear regression models in which the dependent variable Y is the soil water content and the independent variable X is:

i   the soil depth;

ii   the logarithm (to base 10) of the soil depth;

iii   the square of the soil depth.

b   Which of the three models fits the data best? Why?

c   Do the necessary calculations to verify that the coefficients of determination for the three models are the ratios of the variances of the observed and predicted responses.

10.19   Refer to the weight gain data in Exercise 10.3. Set up the ANOVA table for a simple linear regression analysis of the mothers' weight gains as a function of their ages. Calculate R^2 and interpret the result.

10.6
Simultaneous inferences

The error rates or the confidence coefficients claimed for any of the inferential methods described thus far in this chapter are valid provided the method is applied only once in any application. As in the case of comparing population means (see Chapter 9), these methods need some modification in order to make several inferences simultaneously.

EXAMPLE **10.15**   When using a simple linear regression model, we might be interested in making inferences about several functions of the regression parameters or we might want to predict the responses at several levels of the independent variable. For instance, on the basis of the rat weight data in Example 10.7, we might choose to estimate or predict the average age of rats at three different body weights: $X = 100$ g, $X = 200$ g, and $X = 300$ g. If $\mu(x) = \beta_0 + \beta_1 x$ denotes the mean age of rats weighing x g, then the estimation problem would require simultaneous confidence intervals for the parameters $\mu(100)$, $\mu(200)$, and $\mu(300)$, while the prediction problem would require constructing a set of three prediction intervals at the three body weights.   ∎

Two methods of constructing simultaneous confidence intervals with prescribed confidence coefficients are described in Box 10.8. These methods can be used with the simple linear regression model to construct confidence intervals for several expected responses simultaneously.

BOX 10.8

Simultaneous confidence intervals for several expected responses

Let $\mu(x) = \beta_0 + \beta_1 x$ denote the expected response when $X = x$ and $\hat{\mu}(x) = \hat{\beta}_0 + \hat{\beta}_1 x$ be the least squares estimate of $\mu(x)$. Let

$$\hat{\sigma}_{(x)} = \sqrt{\left[\frac{1}{n} + \frac{(x - \bar{x})^2}{S_{xx}}\right] MS[E]}$$

denote the least squares estimate of the standard error of $\hat{\mu}(x)$. Then we may state the following properties.

1 A set of $100(1 - \alpha)\%$ Bonferroni simultaneous confidence intervals for $\mu(x_i)$, $i = 1, \ldots, k$, is given by

$$\hat{\mu}(x_i) \pm t(n - 2, \alpha, k)\, \hat{\sigma}(x_i), \quad i = 1, \ldots, k,$$

where $t(v, \alpha, k)$ is the critical value—at level $\alpha' = \alpha/2k$—of a t-distribution with v degrees of freedom.

2 Let a and b $(a < b)$ be two numbers in the range of values of X. A set of $100(1 - \alpha)\%$ Working-Hotelling simultaneous confidence intervals for $\mu(x)$, $a \leq x \leq b$, is given by

$$\hat{\mu}(x) \pm \sqrt{2F(2, n - 2, \alpha)}\, \hat{\sigma}(x), \quad a \leq x \leq b.$$

Compare the simultaneous confidence intervals in Box 10.8 with the simultaneous intervals for contrasts given in Box 9.9. In both instances, the Bonferroni intervals are designed for simultaneous coverage of a finite number k of parameters. The parameters are population contrasts in Box 9.9 and population mean responses in Box 10.8. The Bonferroni method guarantees that the k intervals will simultaneously cover their respective parameters with a probability at least as large as the associated confidence coefficient. The Working-Hotelling simultaneous intervals are designed to cover all (infinitely many) expected responses in a given range $a \leq x \leq b$ of the independent variable. As such, these intervals are similar to the Scheffe intervals for all contrasts.

The region enclosed by the endpoints of the Working-Hotelling intervals is called a *Working-Hotelling band*. If the interval (a, b) covers the entire range of the values of x, then the corresponding Working-Hotelling band can be interpreted as a confidence region for the true regression line; we have $100(1 - \alpha)\%$ confidence that the band will contain the true regression line. Figure 10.13 shows the 95%

FIGURE 10.13

The 95% Working-Hotelling band for the expected ages of rats

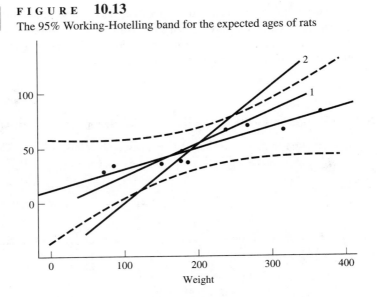

Working-Hotelling band constructed from the rat weight data in Example 10.7. Line 1 is within the band and hence is a plausible (at the 95% confidence level) line for representing the relationship between the age and weight of the rats. Line 2 is not a plausible regression line because it is not within the Working-Hotelling band. Increase in width of the band is accompanied by increase in the variety of plausible straight lines that can represent the relationship of the expected response to the independent variable.

The Working-Hotelling band can be used to construct simultaneous intervals for $\mu(x)$ at a finite number k of values of x. Unless k is large, the Working-Hotelling intervals will be wider (and therefore more conservative) than the corresponding Bonferroni intervals. How large should k be to make the Working-Hotelling intervals better than the Bonferroni intervals? The answer depends on factors such as the values of n, k, and α. For a given problem, it is best to try both methods (if applicable) and to use the one that yields narrower intervals.

EXAMPLE 10.16   For the study in Example 10.7, confidence intervals for the mean ages of the rats with body weights $x_1 = 100$ and $x_2 = 200$ g can be constructed in three ways:

a   by constructing confidence intervals for $\mu(100)$ and $\mu(200)$ one at a time using the confidence interval for a single expected response in Box 10.6;

b   by constructing confidence intervals for the two expected responses simultaneously using the Bonferroni method;

c   by constructing confidence intervals for the two expected responses simultaneously using the Working-Hotelling method.

Each of the three confidence intervals has the form $\hat{\mu}(x_i) \pm A\hat{\sigma}(x_i)$, where A is the critical value applicable to the particular method. For the rat weight data, the critical values for 90% confidence intervals are $A = t(8, 0.05) = 1.86$ for the one-at-a-time method, $A = t(8, 0.10, 2) = 2.306$ for the Bonferroni method, and $A = \sqrt{2F(2, 8, 0.10)} = 2.495$ for the Working-Hotelling method. It can be verified that the least squares estimates of $\mu(100)$ and $\mu(200)$ and their estimated standard errors are $\hat{\mu}(100) = 29.93$, $\hat{\sigma}(100) = 3.39$, $\hat{\mu}(200) = 48.97$, $\hat{\sigma}(200) = 2.18$. The three sets of 90% confidence intervals and their widths may be summarized as follows:

Method	Parameter $\mu(100)$		Parameter $\mu(200)$	
	CI	Width	CI	Width
One-at-a-time	(23.62, 36.24)	12.62	(44.92, 53.02)	8.10
Bonferroni	(22.11, 37.75)	15.64	(43.94, 54.00)	10.06
Working-Hotelling	(21.47, 38.39)	16.92	(43.53, 54.41)	10.88

The narrowest confidence interval corresponds to the one-at-a-time method and the widest to the Working-Hotelling method. The one-at-a-time method will not guarantee the probability of simultaneous coverage; the confidence level of 90% refers to each interval separately. The Bonferroni and Working-Hotelling methods guarantee 90% confidence that both intervals will contain their respective parameters. For instance, on the basis of the Bonferroni method, we have at least 90% confidence that both the following statements are true:

1 The average age of rats with a body weight of 100 g is between 22.11 and 38.39 days.

2 The average age of rats with a body weight of 200 g is between 43.94 and 54.00 days.

From the table, we see that the intervals for $x = 100$ are wider than the intervals for $x = 200$. The reason for this is that the estimated standard error $\hat{\sigma}(x)$ increases with increase in $(x - \bar{x})^2$. For the rat weight data, $\bar{x} = 207$, and so the value of $(x - \bar{x})^2$ is 49 when $x = 200$ and 11,449 when $x = 100$. ■

Obvious modifications of the Bonferroni and Working-Hotelling simultaneous confidence intervals lead to simultaneous prediction intervals (Box 10.9).

BOX **10.9**

Simultaneous prediction intervals in simple linear regression

Let $\mu(x) = \beta_0 + \beta_1 x$ denote the expected response when $X = x$ and $\hat{\mu}(x)$ be the least squares estimate (based on n responses) of $\mu(x)$. If $\bar{Y}_x$ denotes the mean of a random sample of m observed responses at $X = x$, then we can state the following properties:

1 A set of $100(1 - \alpha)\%$ Bonferroni simultaneous prediction intervals for $\overline{Y}_{x_i}$, for $i = 1, \ldots, k$, is given by

$$\hat{\mu}(x_i) - t(n - 2, \alpha, k)\hat{\sigma}_P(x_i) \leq \overline{Y}_{x_i} \leq \hat{\mu}(x_i) + t(n - 2, \alpha, k)\hat{\sigma}_P(x_i),$$

where $t(\nu, \alpha, k)$ is the critical value at the level $\alpha' = \alpha/2k$ of a t-distribution with ν degrees of freedom and $\hat{\sigma}_P(x)$ is the standard error in Box 10.7.

2 Let a and b $(a \leq b)$ be two values in the range of X. Then a set of $100(1 - \alpha)\%$ simultaneous prediction intervals for $\overline{Y}_x$ for $a \leq x \leq b$ is given by

$$\hat{\mu}(x) - \sqrt{2F(2, n - 2, \alpha)}\hat{\sigma}_P(x) \leq \overline{Y}_x \leq \hat{\mu}(x) + \sqrt{2F(2, n - 2, \alpha)}\hat{\sigma}_P(x).$$

The construction and interpretation of simultaneous prediction intervals and prediction bands are straightforward extensions of the corresponding methods for simultaneous confidence intervals and confidence bands.

Exercises

10.20 Refer to the chemical influx data in Example 10.3. Answer the following questions on the basis of a simple linear regression of Y = chemical influx on X = log (dose).

  a Construct a 95% Working-Hotelling confidence band for the expected chemical influx in the brains of rats.

  b Let $\hat{\mu}_{min}(30-120)$ and $\hat{\mu}_{max}(30-120)$ denote the minimum and maximum of the influxes that are contained in the portion of the confidence band in (a) covering the doses between $d = 30$ and $d = 120$. Determine and interpret the values of $\hat{\mu}_{min}(30-120)$ and $\hat{\mu}_{max}(30-120)$.

  c Construct a 95% Working-Hotelling prediction band for the expected chemical influx in the brains of rats.

  d Let $Y_{min}(30-120)$ and $Y_{max}(30-120)$ denote the minimum and maximum of the influxes that are contained in the portion of the prediction band in (c) covering the doses between $d = 30$ and $d = 120$. Determine and interpret the values of $Y_{min}(30-120)$ and $Y_{max}(30-120)$.

10.21 In Example 10.16, we constructed three sets of simultaneous confidence intervals for the expected ages of rats with body weights of 100 and 200 g. Construct the corresponding three sets of 90% prediction intervals for the same two body weights. Interpret the intervals.

10.22 Refer to the disease index data in Exercise 10.7.

  a Suppose we want to determine a set of 90% simultaneous confidence intervals for the expected disease indices for crop intervals $x = 0, 2, 4, 6, 8$, and 10 years. Would you use the Bonferroni method or the Working-Hotelling method? Why?

    b Which method would you use to determine a set of 90% simultaneous prediction intervals for these data? Why?

10.23 Refer to the species density data in Example 10.9.

    a Plot a 95% Working-Hotelling confidence band for the expected species density as a function of the altitude.

    b Use the confidence band in (a) to determine the maximum and minimum slopes of the straight lines that lie entirely within the band.

    c Interpret the slopes determined in (b).

10.7
Checking assumptions

Implicit in the simple linear regression analyses discussed thus far are the following assumptions:

1 The responses are independent and normally distributed.

2 The population variances are equal.

3 The regression function is a linear function of a quantitative independent variable.

The first two assumptions are the same as the ANOVA assumptions needed to justify procedures based on the one-way ANOVA model. The third assumption is needed to justify a simple linear regression analysis.

    Residual plots such as those described in Chapter 8 can be used to see if the ANOVA assumptions are reasonable for the errors in a simple linear regression model. A normal q-q plot can be used to check the assumption that the errors are independent and normally distributed; the shape of the q-q plot indicates the type of departure from normality, as we saw in Figure 8.2. The assumption that the errors have a common variance can be checked by plotting the residuals against the predicted values. Recall that, for one-way classified data, violation of the equal-variance assumption can be checked by comparing the residual variances at distinct levels of the experimental factor. This is not always possible in simple linear regression analysis because replicate observations are usually not available at several levels of the independent variable. However, the equal-variance assumption is usually violated because the variances of the responses increase (or decrease) with increase in the expected response. In that case, a violation of the equal variance assumption will give rise to a plot of the residual versus the predicted value—an $e \bigvee \hat{y}$ plot—in which the points cover a cone-shaped region symmetric about the line $e = 0$, as shown in Figure 10.14.

    The third assumption—that the regression function represents a straight-line relationship—can be checked by a scatter plot of the data. A scatter plot will also suggest alternative forms for the regression function. Instead of the scatter plot, a plot of the residuals against the value of the independent variable—an $e \bigvee x$ plot—can be used to determine possible alternatives to the straight-line form for the regression

FIGURE   10.14

Pattern of residuals

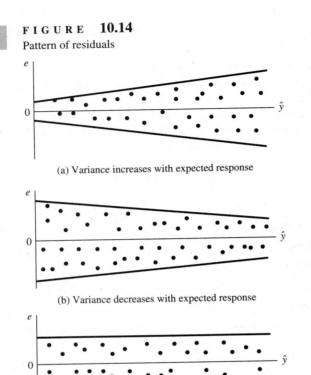

(a) Variance increases with expected response

(b) Variance decreases with expected response

(c) Variance does not change with expected response

function. As we'll see in Chapter 11, an $e \bigvee x$ plot is particularly useful for a multiple regression analysis involving more than one experimental factor.

To see how an $e \bigvee x$ plot can be used to check the appropriateness of a straight-line form for the regression function, let's consider one of the most frequently used alternatives to a simple linear regression function.

Quadratic regression functions

The simple quadratic regression function

$$\mu(x) = \beta_0 + \beta_1 x + \beta_2 x^2 \qquad \textbf{(10.12)}$$

has three regression parameters β_0, β_1, and β_2, in contrast to the two in a simple linear regression function. Plotting the simple quadratic regression function produces a bowl-shaped curve that increases (or decreases) initially and then decreases (or increases). As with the simple linear regression function, the parameters determine the location and the rates of increase and decrease of a simple quadratic regression function. Clearly, a simple quadratic regression function is a natural extension of the simple linear regression function, which can be obtained by setting $\beta_2 = 0$ in Equation (10.12).

We see in Figure 10.15 how the location and shape of the graph of a simple quadratic regression function changes with the values of the regression parameters. In each case, the y-intercept at $x = 0$ equals the value of β_0. In Figure 10.15a, both β_1 and β_2 are positive, so that the function increases as x increases from 0 to ∞. In Figure 10.15b, β_1 is negative and β_2 is positive; the function decreases initially and then increases. The regression function in Figure 10.15c first increases and then decreases, because β_1 is positive and β_2 is negative. Finally, both β_1 and β_2 are negative in Figure 10.15d, and so the regression function decreases as a function of x.

As an example, Figure 10.16a shows a scatter plot of a set of data that have been computer-simulated using the model

$$Y_i = 4 + 0.4x_i + 0.1x_i^2 + E_i,$$

where the E_i are independent and normal with a common variance $\sigma^2 = 1$. Notice how the points fit the quadratic function shown by the curve in Figure 10.16a. The straight line in the figure is the simple linear regression line estimated from the same data. Clearly, a quadratic function describes the data better than a straight-line function. We say that the scatter plot indicates a *lack of fit* of the straight line to the observed responses.

The residuals based on the best-fitting quadratic line and the simple linear regression line are plotted against x in Figures 10.16b and 10.16c, respectively. These two $e \setminus x$ plots differ noticeably. When a quadratic function is fitted to the data, the $e \setminus x$ plot shows a horizontal band of constant width that is symmetric about the line $e = 0$. This pattern seems consistent with what would be expected in the case of independent random errors with zero mean and constant variance. The $e \setminus x$ plot for the simple linear regression line shows a band of constant width whose location changes with x. This suggests that, while the variance appears constant (the band width is constant), the difference between the observed and predicted responses depends on x. The fact that the inclusion of an additional term involving x^2 to the simple linear regression model changed the residual plot into a band symmetric about the x-axis is a clear indication that a substantial improvement of the fit can be obtained by adding the term x^2 to the simple linear regression model.

FIGURE 10.15

Shapes of simple quadratic regression functions

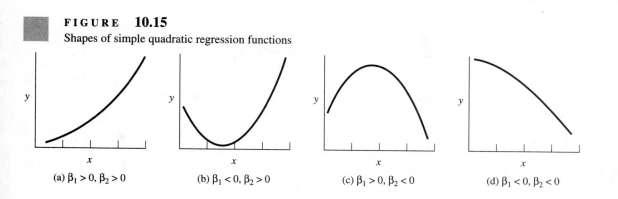

(a) $\beta_1 > 0,\ \beta_2 > 0$ (b) $\beta_1 < 0,\ \beta_2 > 0$ (c) $\beta_1 > 0,\ \beta_2 < 0$ (d) $\beta_1 < 0,\ \beta_2 < 0$

FIGURE 10.16

Checking for lack of fit

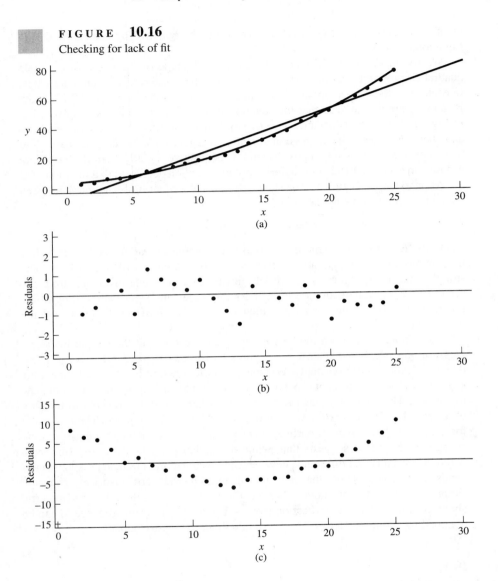

(a)

(b)

(c)

If a plot of e versus $\hat{y}$ yields the patterns of residuals such as those in Figure 10.17, we can expect that a quadratic term will improve the fit.

The next example illustrates how residual plots can be used to check for violation of assumptions in a simple linear regression analysis.

EXAMPLE 10.17   Let's consider three sets of computer-generated data, each containing $n = 25$ responses. Each data set is created by simulating one response at each of the $n = 25$ distinct values—$x_1 = 1, x_2 = 2, \ldots, x_{25} = 25$—for the independent variable X. The following models are used to simulate responses for the three data sets:

FIGURE   10.17

Typical $e \bigvee \hat{y}$ plots that indicate a need for a quadratic term

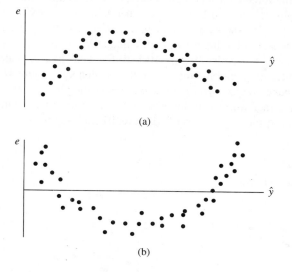

(a)

(b)

1   Model for data set 1

$$Y_i = 2 + 0.4x_i + E_i, \quad i = 1, \ldots, 25,$$

where the E_i are independent $N(0, 1)$.

2   Model for data set 2

$$Y_i = 2 + 0.4x_i + E_i, \quad i = 1, \ldots, 25,$$

where the E_i are independent and normally distributed with zero mean. The variances of the E_i were such that E_1–E_5 had variance 1, E_6–E_{10} had variance 2, E_{11}–E_{15} had variance 4, E_{16}–E_{20} had variance 8, and E_{21}–E_{25} had variance 10.

3   Model for data set 3

$$Y_i = 2 + 0.4x_i + 0.2x_i^2 + E_i, \quad i = 1, \ldots, 25,$$

where the E_i are independent $N(0, 1)$.

Data set 1 does not violate any of the four assumptions in a simple linear regression model. Note that, for this data set, $\beta_0 = 2$, $\beta_1 = 0.4$, and $\sigma^2 = 1$. Data set 2 satisfies all but the equal-variance assumption. In this case, the error variance increases with the expected response. Finally, data set 3 does not satisfy the assumption that the regression function is a linear function of the independent variable. The regression function appropriate for this data set has the form:

$$\mu(x) = 2 + 0.4x + 0.2x^2,$$

so that the relationship between the expected response and the independent variable is a quadratic function of x.

Figure 10.18 shows six residual plots obtained by fitting simple linear regression models to the three data sets. As expected, the normal q-q plot (Figure 10.18a) and the $e \bigvee \hat{y}$ plot (Figure 10.18b) for data set 1 indicate no violation of the assumptions. The q-q plot shows a straight-line pattern, and the $e \bigvee \hat{y}$ plot is a horizontal band of constant width that is symmetric about the x-axis. The normal q-q plot (Figure 10.18c) and $e \bigvee \hat{y}$ plot (Figure 10.18d) for data set 2 indicate a violation of the equal-variance assumption. Notice that the residuals in Figure 10.18d are contained in a cone-shaped region symmetric about $e = 0$; the plot gives a clear indication that variances are increasing with increasing expected responses. Finally, the pattern of the normal q-q plot for data set 3 (Figure 10.18e) suggests a systematic departure

FIGURE   10.18

Residual plots for checking assumptions

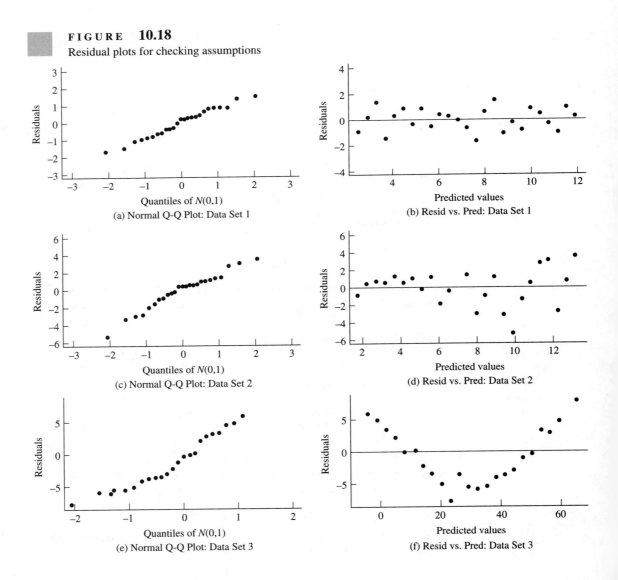

from a straight-line pattern, and the bowl-shaped pattern of the $e \bigvee \hat{y}$ plot in Figure 10.18f is an indication that a quadratic model will fit the data better than will a simple linear regression model.   ■

Exercises

10.24   Refer to the chemical influx data in Example 10.3.

a   Using appropriate residual plots, check for assumption violations for simple linear regression models in which:

i $Y =$ chemical influx is the dependent variable, and $X = d$ is the independent variable;

ii $Y =$ chemical influx is the dependent variable, and $X = \log d$ is the independent variable.

b   On the basis of analysis of the residual plots in (a), which model fits the data better? Why?

10.25   Assuming a simple linear regression model with $Y =$ disease index and $X =$ number of plants, check for assumption violations for the data in Exercise 10.7.

10.26   Refer to the soil water content data in Exercise 10.2.

a   Using residual plots, check for the adequacy of a simple linear regression model in which:

i $Y =$ soil water content is the response variable, and $X =$ soil depth is the independent variable;

ii $Y =$ soil water content is the response variable, and $X = \log_{10}$ (depth) is the independent variable;

iii $Y =$ soil water content is the response variable, and $X = (\text{depth})^2$ is the independent variable.

b   Which of the three models in (a) is the best for describing the soil water content data? Why?

10.27   Refer to the weight gain data in Exercise 10.3. Use residual plots to check for assumption violations when using a simple linear regression model with $Y =$ weight gain during pregnancy and $X =$ age.

10.8
The correlation coefficient

The simple linear regression model is a statistical model for predicting the value of a dependent variable Y given the value of an independent variable X. As noted in Section 10.2, strict validity of the applications of the simple linear regression model considered thus far requires that the data consist of observations of Y at fixed values of X. Under that assumption, the data can be treated as a collection of independent random samples from several populations. In the following situations, the data cannot be treated as a collection of independent random samples of observations at known fixed levels of the independent variable.

1 The response variable Y is measured at fixed values of X, but the actual value of X cannot be determined precisely, on account of nonnegligible measurement error. Consider a greenhouse study to determine how well the level of exposure to artificial sunlight (independent variable) predicts the height of a 12-week-old plant (dependent variable). In this case, the measurement error in the independent variable is often negligible, because the level of light exposure can be controlled precisely in a greenhouse. On the other hand, in a study to see how the weight (dependent variable) of a human subject is affected by his or her daily calorie intake (independent variable), the measured value of the independent variable is likely to be imprecise, because calorie intake is often estimated on the basis of diary entries maintained by the study subjects.

2 Both X and Y are measured responses; that is, the data constitute a random sample from a bivariate population. Independent measurements on a pair of response variables are useful for assessing how strongly the two variables are associated. Here are some typical examples of questions about a pair of response variables.

    a How closely are the height and weight of newborn babies associated?

    b Are the stem length and stem diameter of four-week-old plants positively associated? In other words, is there a tendency for stem length and stem diameter to increase or decrease together?

    c Is it reasonable to claim that body weight and the daily amount of physical exercise by adult males are negatively associated? In other words, is increase in the body weight of an adult male associated with a decrease in the amount of exercise?

The problem of measurement errors in independent variables will not be addressed in this book; for information on this topic, see Davies and Hutton (1975), Seber (1977), and Fuller (1987), among others. In this section, we confine our attention to the case where both X and Y are response variables and the degree to which they are associated is of interest.

EXAMPLE **10.18**   An experiment was conducted to see if a new method (N) can replace the standard method (S) of assaying for serum concentration of a chemical in pediatric patients. The standard method is reliable and widely accepted, but the new method is less invasive, less expensive, and more convenient to use. The serum concentrations (μg/ml) of the chemical measured by the two methods in a random sample of $n = 13$ patients are as follows:

Method							Patient						
	1	2	3	4	5	6	7	8	9	10	11	12	13
N	5.3	7.1	7.9	9.0	12.3	20.8	26.9	14.3	16.8	14.9	8.6	11.9	9.8
S	5.0	7.2	8.3	8.7	10.4	22.1	28.2	13.4	16.4	15.3	7.3	12.3	9.4

The primary objective of the study was to see how closely the results given by the two methods are associated. If a strong association was established, then the investigators wanted to explore the feasibility of replacing the standard method with the new method. Figure 10.19 shows a scatter plot of the serum concentration data.

F I G U R E   10.19

Scatter plot of serum concentration data

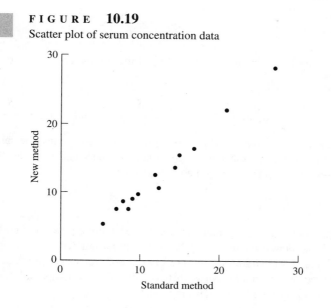

The tight clustering of the data around a straight line with a positive slope suggests that there is a strong linear association between the results from the two methods and that the linear association can be regarded as positive, in the sense that a large (small) value measured by one method tends to be associated with a large (small) value measured by the other method. In other words, the results from the two methods tend to increase or decrease together.

Because the presence of linear association will enable us to use the convenient simple linear regression model to describe one variable in terms of a simple linear function of the other variable, it is desirable to have a numerical index that measures the strength and direction of the linear association between a pair of interval variables.   ■

From a practical point of view, it is reasonable to expect that a useful index of linear association must have two characteristics:

1 Its magnitude should be a measure of how well a straight line will describe the observed association between the two variables.

2 Its sign should indicate whether the association is positive or negative.

As we will now see, the correlation coefficient is such an index.

The sample correlation coefficient

Let (x_1, y_1), (x_2, y_2), ..., (x_n, y_n) denote a random sample of n pairs of observed values of a pair of continuous variables (X, Y), and let S_{xx}, S_{xy}, and S_{yy} be defined as in Box 10.3. Then the sample correlation coefficient between X and Y is

$$r_{XY} = \frac{S_{xy}}{\sqrt{S_{xx}S_{yy}}}.$$

Let's look at some useful properties of the sample correlation coefficient.

1 Let

$$s_x^2 = \frac{S_{xx}}{n-1}, \qquad s_y^2 = \frac{S_{yy}}{n-1}, \qquad \text{and} \qquad s_{xy} = \frac{S_{xy}}{n-1} \qquad \textbf{(10.13)}$$

denote, respectively, the sample variance of the x values, the sample variance of the y values, and the sample covariance of the x and y values. Then the sample correlation coefficient can be expressed as

$$r_{XY} = \frac{s_{xy}}{s_x s_y}.$$

2 The correlation coefficient is a unit-free number, because both the numerator and the denominator in r_{XY} have the same unit (corresponding to the product of X and Y). Therefore, the correlation coefficient can be used to compare the degree of association between two pairs of continuous variables, regardless of the underlying units of measurement.

3 Given a bivariate random sample of the values of X and Y, we can compute two least squares regression lines: one by treating Y as the response (dependent) variable, and the other by treating X as the response variable. Let $\hat{\beta}_{Y.X}$ and $\hat{\beta}_{X.Y}$ denote, respectively, the slopes of the least squares lines fitted by regarding Y and X as dependent variables; that is, let

$$\hat{\beta}_{Y.X} = \frac{S_{xy}}{S_{xx}} \qquad \text{and} \qquad \hat{\beta}_{X.Y} = \frac{S_{xy}}{S_{yy}}.$$

Clearly, $\hat{\beta}_{Y.X}$ and $\hat{\beta}_{X.Y}$ are both positive or both negative depending on whether S_{xy} is positive or negative. Furthermore, straightforward algebraic manipulations show that

$$r_{XY} = \hat{\beta}_{Y.X}\sqrt{\frac{S_{xx}}{S_{yy}}} = \hat{\beta}_{X.Y}\sqrt{\frac{S_{yy}}{S_{xx}}} = \pm\sqrt{\hat{\beta}_{Y.X}\hat{\beta}_{X.Y}}, \qquad \textbf{(10.14)}$$

where the sign of the final expression is the same as the common sign of $\hat{\beta}_{Y.X}$ and $\hat{\beta}_{X.Y}$.

4 When r_{XY} is positive, both $\hat{\beta}_{Y.X}$ and $\hat{\beta}_{X.Y}$ are positive, so that the slopes of the least squares prediction lines are positive. Thus, a positive correlation coefficient is an indication of the tendency for both variables to increase or decrease in tandem. For example, a positive correlation would be expected for data on the

age and stem circumference of a random sample of a species of plant. A negative correlation coefficient indicates a tendency for the variables to change in opposite directions. For instance, a pair of negatively correlated variables is $X =$ the amount of daily exercise and $Y =$ the amount of body fat.

5 From Equation (10.14), we get

$$r_{XY}^2 = \frac{\hat{\beta}_{Y.X}^2 S_{xx}}{S_{yy}} = \frac{\hat{\beta}_{X.Y}^2 S_{yy}}{S_{xx}} = \hat{\beta}_{Y.X} \hat{\beta}_{X.Y}. \tag{10.15}$$

Let $SS_{Y.X}[R]$ and $SS_{Y.X}[TOT]$ denote, respectively, the regression and total sums of squares when a straight line is fitted with Y as the dependent variable and X as the independent variable. Let $SS_{X.Y}[R]$, $SS_{X.Y}[TOT]$ denote the corresponding quantities when X is the dependent variable and Y is the independent variable. Then, it follows from Equation (10.9) that

$$SS_{Y.X}[R] = \hat{\beta}_{Y.X}^2 S_{xx}, \qquad SS_{Y.X}[TOT] = S_{yy},$$

$$SS_{X.Y}[R] = \hat{\beta}_{X.Y}^2 S_{yy}, \qquad SS_{X.Y}[TOT] = S_{xx}.$$

Substituting these results into Equation (10.15), we find that

$$r_{XY}^2 = \frac{SS_{Y.X}[R]}{SS_{Y.X}[TOT]} = \frac{SS_{X.Y}[R]}{SS_{X.Y}[TOT]}. \tag{10.16}$$

Looking back at the definition in Equation (10.10), we conclude that the coefficients of determination in the simple linear regressions of Y on X and X on Y are both equal to the square of the correlation coefficient between X and Y. *Thus the square of the correlation coefficient is a measure of how well a straight line describes the association between X and Y.*

Yet another interpretation of the correlation coefficient can be obtained by substituting from Equation (10.11) into Equation (10.16). We have

$$1 - r_{XY}^2 = 1 - \frac{s_{pred}^2}{s_{obs}^2} = \frac{s_{obs}^2 - s_{pred}^2}{s_{obs}^2}.$$

Thus $1 - r_{XY}^2$ equals the proportional reduction in the variation of the observed responses resulting from the use of the independent variable as the predictor in a simple linear regression model.

6 The positive square root of the coefficient of determination equals the correlation coefficient between the observed responses and the predicted values based on the best-fitting least squares prediction equation. In other words, if y_i and $\hat{y}_i$ denote the observed and predicted values, then the correlation coefficient, $r_{Y\hat{Y}}$ between y_i and $\hat{y}_i$ is equal to $|r_{XY}|$, the positive square root of $R_{X.Y}^2$. That is,

$$r_{Y\hat{Y}} = |r_{XY}| = \sqrt{R_{Y.X}^2}.$$

7 Since the square of the sample correlation coefficient equals the coefficient of determination, the maximum value attainable by r_{XY}^2 is 1. Thus, the sample correlation coefficient itself is a value between -1 and $+1$

$$-1 \leq r_{XY} \leq +1.$$

The value $r_{XY}^2 = 1$ $(r_{XY} = \pm 1)$ occurs when the regression sum of squares equals the total sum of squares—that is, when the scatter plot exhibits a perfect straight-line relationship. In that case, X and Y are said to be *perfectly correlated* in the sample. The value $r_{XY} = 0$ results when the best-fitting straight line does not depend on the observed values of the independent variable. This is the case when the best-fitting line is parallel to the axis of the independent variable (that is, when its slope is zero), so that the predicted value of the response variable is the same for all values of the independent variable. Typical scatter plot patterns are shown in Figure 10.20, along with the associated values of r_{XY}.

FIGURE   10.20

Correlation coefficients and associated scatter plots

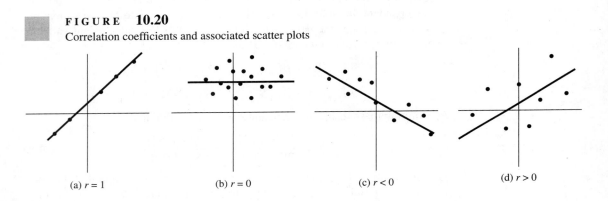

(a) $r = 1$ (b) $r = 0$ (c) $r < 0$ (d) $r > 0$

The correlation coefficient is a measure of the strength of *linear association* between two variables. Absence of correlation $(r_{XY} = 0)$ does not mean that the variables are not associated. It only means that the variables are not linearly associated. As an example, consider the data in Table 10.4, which consist of $n = 7$ pairs of values of X and Y. It may be verified by direct calculation that the data satisfy a quadratic relationship of the form

$$y = 3 + 2x^2$$

and, thus, there is a perfect nonlinear association between X and Y. Yet direct calculations show that the correlation coefficient is zero for this data set. A scatter plot of the data is shown in Figure 10.21, along with the least squares line fitted with Y as the dependent variable.

Several properties of the correlation coefficient r_{XY} are illustrated in the next example.

TABLE   10.4

Hypothetical data with $r_{XY} = 0$ and perfect nonlinear association

x	-3	-2	-1	0	1	2	3
y	21	11	5	3	5	11	21

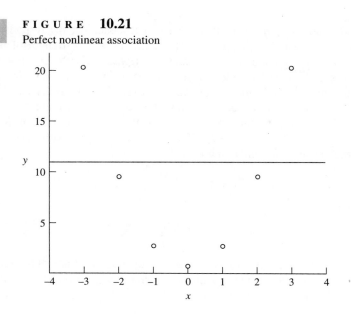

FIGURE   10.21
Perfect nonlinear association

EXAMPLE 10.19   To study the effects of air temperature on the growth characteristics of sour orange trees exposed to various concentrations of atmospheric carbon dioxide (CO_2), data were collected every other month, over a two-year period, on the dry weight per leaf and the mean air temperature of the preceding month, for trees exposed to an extra 300 μliter/liter of CO_2. The results obtained are as follows (Idso, Kimball, & Hendrix, 1983, Figure 4):

Year	1990	1990	1990	1990	1990	1990
Mean air temperature (°C)	12.0	18.4	23.2	26.4	33.5	35.0
Dry weight per leaf (g)	1.00	1.10	1.16	1.28	1.29	1.34

Year	1991	1991	1991	1991	1991	1991
Mean air temperature (°C)	13.6	18.4	23.2	22.4	33.5	33.6
Dry weight per leaf (g)	1.14	1.20	1.30	1.42	1.39	1.38

Let $X =$ the mean air temperature and $Y =$ dry weight per leaf. The following statistics can be calculated from the data

$$\bar{x} = 24.43, \quad \bar{y} = 1.25, \quad S_{xx} = 715.8867, \quad S_{xy} = 9.0440, \quad \text{and} \quad S_{yy} = 0.1862.$$

The sample correlation coefficient between X and Y is

$$r_{XY} = \frac{9.0440}{\sqrt{(715.8867)(0.1862)}} = 0.7833.$$

The data suggest a positive linear association between temperature and leaf dry weight. The positive sign for the correlation coefficient indicates that higher (lower) temperatures are associated with higher (lower) leaf dry weight. The presence of a linear association means that there is a tendency for the response pairs to cluster around a straight line. A numerical measure of the strength of linear association is given by the square of the sample correlation coefficient. For the present data, the square of the sample correlation coefficient is $r_{XY}^2 = (0.7833)^2 = 0.6136$, so that approximately 61% of the observed variation in the dry weight per leaf can be explained by its linear association with mean air temperature.

The regression sums of squares for the regression of Y on X and X on Y are $\mathrm{SS}_{Y.X}[R] = 0.1143$ and $\mathrm{SS}_{X.Y}[R] = 439.2800$, respectively. The corresponding coefficients of determination are

$$R_{Y.X}^2 = \frac{\mathrm{SS}_{Y.X}[R]}{\mathrm{SS}_{Y.X}[\mathrm{TOT}]} = \frac{0.1143}{0.1862} = 0.6139$$

and

$$R_{X.Y}^2 = \frac{\mathrm{SS}_{X.Y}[R]}{\mathrm{SS}_{X.Y}[\mathrm{TOT}]} = \frac{439.28}{715.8887} = 0.6136.$$

Thus, the coefficient of determination is the same for both regressions (the difference in the calculated values is due to rounding errors) and equals the square of the coefficient of correlation. ■

The population correlation coefficient

The population correlation coefficient between X and Y is defined analogously to the sample correlation coefficient. Let σ_X^2, σ_Y^2, and σ_{XY} denote, respectively, the variance of X, the variance of Y, and the covariance of X and Y in the bivariate population of values of (X, Y). Then the population correlation coefficient between X and Y is defined as

$$\rho_{XY} = \frac{\sigma_{XY}}{\sigma_X \sigma_Y}.$$

All the properties of r_{XY} are also properties of ρ_{XY}, with the proviso that the properties now refer to the population of values of (X, Y). For instance, ρ_{XY} is a measure of the tendency of the population of values of (X, Y) to cluster around a straight line called the *true regression line*. The closer these values are to the true regression line, the closer will be the value of ρ_{XY} to ± 1. The value $\rho_{XY} = 0$ implies that there is no linear association between X and Y. In that case, the true regression line will be parallel to the axis of the independent variable.

The population correlation coefficient is estimated by the sample correlation coefficient. Therefore, the sample correlation coefficient can be used to make inferences about the population correlation coefficient. In Box 10.10, a procedure is given for testing the null hypothesis that there is no linear association between X and Y. The test can be applied with any sample size, but requires some assumptions about the distribution of the bivariate population from which the sample was selected. In Box 10.10, the population distribution is assumed to be bivariate normal, but the results are valid under some conditions in which this assumption is relaxed further.

BOX 10.10

Exact sample test of H_0: $\rho_{XY} = 0$

Let $(x_1, y_1), \ldots, (x_n, y_n)$ denote a random sample from a bivariate normal distribution. Let r_{XY} and ρ_{XY} denote, respectively, the sample and population correlation coefficients. Then a test of the null hypothesis H_0: $\rho_{XY} = 0$ can be based on the test statistic

$$t_{r_{XY}} = \frac{r_{XY}\sqrt{n-2}}{\sqrt{1 - r_{XY}^2}}.$$

The rejection region for a test based on $t_{r_{XY}}$ is as follows:

1 Two-sided test: Reject H_0 in favor of H_1: $\rho_{XY} \neq 0$ if

$$t_{r_{XY}} \leq -t(n-2, \alpha/2) \quad \text{or} \quad t_{r_{XY}} \geq t(n-2, \alpha/2).$$

2 One-sided test: Reject H_0 in favor of H_1: $\rho_{XY} > 0$ if

$$t_{r_{XY}} \geq t(n-2, \alpha).$$

3 One-sided test: Reject H_0 in favor of H_1: $\rho_{XY} < 0$ if

$$t_{r_{XY}} \leq -t(n-2, \alpha).$$

It can be shown that the t-test in Box 10.10 is equivalent to the t-test of the null hypothesis that the slope is zero in the regression of Y on X (X on Y); that is, the test in Box 10.10 is the same as the t-test of the null hypothesis that the mean of the conditional distribution of Y (X) does not depend on the value of X (Y).

EXAMPLE 10.20

The value $r_{XY} = 0.7833$ calculated in Example 10.19 is an estimate of ρ_{XY}, the population correlation coefficient between X = mean air temperature and Y = leaf dry weight. The statistical significance of the observed correlation can be tested by testing the null hypothesis H_0: $\rho_{XY} = 0$ against the two-sided alternative hypothesis H_1: $\rho_{XY} \neq 0$. Under the assumption that the data are a random sample of size $n = 12$ from a bivariate normal distribution, the t-test in Box 10.10 can be used for this purpose. The calculated value of the test statistic is

$$t_{r_{XY}} = \frac{\sqrt{(12-2)}(0.7833)}{\sqrt{1 - (0.7833)^2}} = 3.985, \qquad df = 12 - 2 = 10.$$

It is simple to verify that, for testing H_0: $\rho_{XY} = 0$ against H_1: $\rho_{XY} \neq 0$, the p-value is quite small (< 0.01), so that the null hypothesis can be rejected. Thus, it is reasonable to conclude that there is a linear association between the leaf dry weight and temperature.

The exact test in Box 10.10 can be implemented using the CORR procedure in SAS. For the details of the CORR procedure, see Figure 10.14 in the companion text by Younger (1997), where it is shown that the exact *p*-value for testing H_0 against H_1 is $p = 0.0026$.   ∎

Unfortunately, the procedure in Box 10.10 cannot be used to test the null hypothesis that the value of ρ_{XY} is some specified nonzero value; nor can it be used to construct confidence intervals for ρ_{XY}. To avoid this difficulty, it is useful to consider a transformation of ρ_{XY}, the Fisher transformation.

The *Fisher transformations* of r_{XY} and ρ_{XY} are, respectively

$$Z_{XY} = \frac{1}{2} \log \left(\frac{1 + r_{XY}}{1 - r_{XY}} \right)$$

and

$$\psi_{XY} = \frac{1}{2} \log \left(\frac{1 + \rho_{XY}}{1 - \rho_{XY}} \right).$$

In Box 10.11, we describe a large-sample procedure for making inferences about ψ_{XY}. As we'll see, an inference about ψ_{XY} can readily be converted to an inference about ρ_{XY}.

BOX 10.11

> ### Large-sample inferences about the Fisher transformation of a correlation coefficient
>
> Let r_{XY} denote the sample correlation coefficient based on a random sample of size n from a continuous bivariate population. Let ρ_{XY} denote the population correlation coefficient between X and Y. If n is large, approximate tests of the null hypothesis H_0: $\psi_{XY} = \psi_{XY0}$ can be based on the test statistic
>
> $$Z_{r_{XY}} = \sqrt{n - 3} \left(Z_{XY} - \psi_{XY0} \right).$$
>
> The rejection region for a test based on $Z_{r_{XY}}$ is as follows:
>
> **1**   Two-sided test: Reject H_0 in favor of H_1: $\psi_{XY} \neq 0$ if
>
> $$Z_{r_{XY}} \leq -z(\alpha/2) \quad \text{or} \quad Z_{r_{XY}} \geq z(\alpha/2).$$
>
> **2**   One-sided test: Reject H_0 in favor of H_1: $\psi_{XY} > 0$ if
>
> $$Z_{r_{XY}} \geq z(\alpha).$$
>
> **3**   One-sided test: Reject H_0 in favor of H_1: $\psi_{XY} < 0$ if
>
> $$Z_{r_{XY}} \leq -z(\alpha).$$
>
> **4**   A $100(1 - \alpha)\%$ confidence interval for ψ_{XY} is given by
>
> $$Z_{XY} - z(\alpha/2)/\sqrt{n - 3} \leq \psi_{XY} \leq Z_{XY} + z(\alpha/2)/\sqrt{n - 3}.$$

Unlike the *t*-test in Box 10.10, the inferential procedures in Box 10.11 do not require any assumptions about the form of the bivariate population of values of (X, Y). However, the procedures are restricted to large samples.

As already noted, the conclusions derived from hypothesis tests and confidence intervals for ψ_{XY} lead to conclusions about ρ_{XY}. A connection between the two sets of conclusions can be established by using the inverse of the Fisher transformation to express r_{XY} and ρ_{XY} in terms of Z_{XY} and ψ_{XY}, respectively.

The *inverse Fisher transformations* of Z_{XY} and ψ_{XY} are defined as

$$r_{XY} = \frac{e^{2Z_{XY}} - 1}{e^{2Z_{XY}} + 1}, \qquad \rho_{XY} = \frac{e^{2\psi_{XY}} - 1}{e^{2\psi_{XY}} + 1}.$$

The next example demonstrates how the procedures in Box 10.11 can be combined with the inverse Fisher transformations to make inferences about a population correlation coefficient.

EXAMPLE **10.21**   In a study, a correlation coefficient of $r_{XY} = 0.623$ between hematocrit (percentage of red blood cells) and age was obtained in a sample of $n = 37$ Sprague-Dawley male rats (Brandt, Waters, Muron, & Block, 1982). Is this information sufficient for us to conclude that the population correlation coefficient between hematocrit and age is less than 0.7? In other words, is it reasonable to conclude that, in the population of Sprague-Dawley male rats, less than $100(0.7)^2 = 49\%$ of the variability in hematocrit can be explained by its linear association with age?

We can answer the question by testing the null hypothesis H_0: $\rho_{XY} \geq 0.7$ against the research hypothesis H_1: $\rho_{XY} < 0.7$. The Fisher transformation of 0.7 is

$$\psi_{XY} = \frac{1}{2} \log \left(\frac{1 + 0.7}{1 - 0.7} \right) = 0.8673,$$

and so we can proceed by testing the hypothesis H_0: $\psi_{XY} \geq 0.8673$ against H_1: $\psi_{XY} < 0.8673$. From the results in Box 10.11, the calculated value of the test statistic is $Z_{Y_{XY}} = (\sqrt{37 - 3})(0.7299 - 0.8673) = -0.8012$, where 0.7299 is the value of Z_{XY} corresponding to $r_{XY} = 0.623$. The (one-sided) *p*-value associated with -0.8012 is fairly large ($p = 0.212$), and so it is not reasonable to conclude in favor of H_1. There is not sufficient evidence to conclude that the population correlation coefficient between hematocrit and age is less than 0.7.

A $100(1 - \alpha)\%$ confidence interval for ρ_{XY} can be constructed as follows. First we construct a confidence interval for ψ_{XY} using the procedure in Box 10.11. Then we transform the confidence interval for ψ_{XY} to a confidence interval for ρ_{XY} by applying the inverse Fisher transformation to the endpoints.

A 95% confidence interval for ψ_{XY} is

$$0.7299 \pm (1.96)\frac{1}{\sqrt{37 - 3}} \qquad \Rightarrow \qquad 0.3938 \leq \psi_{XY} \leq 1.066.$$

A 95% confidence interval for ρ_{XY} is obtained by applying the inverse Fisher transformation to the endpoints 0.3938 and 1.066 of the confidence interval for ψ_{XY}. The required interval is

$$\frac{e^{2(0.3938)} - 1}{e^{2(0.3938)} + 1} \leq \rho_{XY} \leq \frac{e^{2(1.066)} - 1}{e^{2(1.066)} + 1} \qquad \Rightarrow \qquad 0.3746 \leq \rho_{XY} \leq 0.7879.$$

Hence, we can conclude with 95% confidence that, in the population of all male Sprague-Dawley rats, the correlation coefficient between hematocrit and age is somewhere between 0.3764 and 0.7879. Therefore, between 14% $(0.3746^2 = 0.1403)$ and 62% $(0.7879^2 = 0.6208)$ of the hematocrit variation in male Sprague-Dawley rats can be explained by its linear association with age. ∎

Correlation and causation

A question that often arises in practice is whether an observed strong association between two variables is sufficient to conclude that there is a cause–effect relationship between them. For example, is an observed positive correlation between anxiety level and blood pressure sufficient to conclude that anxiety causes high blood pressure? Similarly, does an observed negative correlation between the level of fungicide application and the level of leaf infestation imply that fungicide is effective in reducing leaf infestation?

Strong association in itself is not sufficient to claim that a causal relationship exists between two variables X and Y. Two additional requirements must be satisfied to support such claims: X and Y must be related by an appropriate time order, and causal relationship must be the *only* explanation for the strong association between X and Y. In other words, to justify the claim that X has a causal effect on Y, the following three conditions must be satisfied.

1   There is a strong association between X and Y.

2   The variable X precedes the variable Y in the sense that values of X refer to a time before the time for the corresponding values of Y.

3   The only possible explanation for association between X and Y is that X causes Y.

As an example of how a correlation coefficient can be used in a causal analysis, suppose we have data for several subjects showing a significant positive correlation between X and Y, where X is the change in the subject's anxiety level (measured on a numerical scale) and Y is the corresponding change in the subject's blood pressure level. What other information is needed to support the claim that increases in anxiety level cause increases in blood pressure levels?

The presence of significant correlation satisfies condition 1 listed above so that the claim that X causes Y can be justified if we can verify that X and Y satisfy conditions 2 and 3. For the claim that anxiety causes high blood pressure, conditions 2 and 3 require that the changes in blood pressure values of the study subjects occurred *after* the corresponding anxiety levels changed, and, in the particular data set, *only* a change in the anxiety level can cause a change in blood pressure. The conclusion that increase in anxiety level causes increase in blood pressure is not appropriate if there is a possibility that changes in both anxiety and blood pressure levels can be the result of some other cause such as an experimental condition under which the study subjects were observed.

Exercises

10.28   Verify that the relationships between the sample correlation coefficient and the estimated slopes are as in Equation (10.14).

10.29   Refer to the weight gain data in Exercise 10.3. Let Y = mother's gain in weight and X = mother's age.

   a   Determine the simple linear regression lines for predicting:

      i Y as a function of X;   **ii** X as a function of Y.

   b   Use the estimated regression lines in (a) to determine the correlation coefficient between mother's weight gain and mother's age.

   c   In the data, the mothers' ages are expressed in years, and the mothers' weight gains are in kilograms. Convert the data by expressing the mothers' ages in months and the mothers' weight gains in pounds. Use the results in (a) to determine:

      i   the slope of the estimated line for predicting a mother's gain in weight as a function of her age;

      ii   the correlation coefficient between mother's age and mother's weight gain.

   d   Interpret the values of r_{XY}^2 and $1 - r_{XY}^2$.

   e   Do the data support the conclusion that there is a linear association between a mother's age and her weight gain during pregnancy?

10.30   Refer to the disease index data in Exercise 10.7. Let Y = disease index and X = crop interval.

   a   Set up the ANOVA tables for regressions of Y on X and X on Y.

   b   Use the ANOVA tables in (a) to verify that the same value for the coefficient of determination is obtained regardless of whether Y is regressed on X or X is regressed on Y.

   c   Explain why the information in the ANOVA tables in (a) is not sufficient to determine the correlation coefficient between X and Y.

   d   By actual computation, show that the calculated value of the t-statistic is the same for testing the three null hypotheses

$$H_{01}\colon \rho_{XY} = 0, \qquad H_{02}\colon \beta_{Y.X} = 0, \qquad H_{03}\colon \beta_{X.Y} = 0.$$

   e   Write appropriate conclusions based on the calculated value in (d).

10.31   Show that the correlation coefficient is zero for the data in Table 10.4.

10.32   Data for $n = 62$ plants of a particular variety gave a correlation coefficient of $r_{XY} = 0.73$ between stem circumference and dry plant weight.

   a   Obtain a 95% confidence interval for the population correlation coefficient between stem circumference and dry weight.

   b   Interpret the interval obtained in (a).

   c   Obtain a 90% lower bound for the population correlation coefficient between stem circumference and dry weight.

   d   Interpret the lower bound calculated in (c).

   e   Suppose the stem circumference is used to predict plant dry weight. Estimate the proportional reduction in the variation of plant weight that can be obtained by using the stem circumference as the independent variable in a simple linear regression.

10.33 A random sample of $n = 76$ subjects gave a correlation coefficient of $r_{XY} = 0.393$ between $X =$ the subject's weight and $Y =$ the subject's systolic blood pressure.

   a Construct an appropriate bound to the population correlation coefficient that will enable you to determine whether the observed correlation coefficient supports the claim that more than 30% of the variation in the systolic blood pressures in the population can be explained by its linear association with the weights of the subjects in the population.

   b Interpret the bound calculated in (a).

   c Suppose we are interested in predicting the systolic blood pressure of a subject using a simple linear regression of Y on X. Construct a 95% confidence interval for the coefficient of determination for the prediction equation. Interpret the interval.

10.34 Serum cholesterol levels were measured for ten newly diagnosed cardiac patients. For each patient, a reading was taken soon after diagnosis (the pretreatment level), and another was taken after a four-month program of cardiac rehabilitation (the posttreatment level). The cholesterol levels obtained (mg/dliter) were as follows:

Patient	1	2	3	4	5	6	7	8	9	10
Pretreatment level	187	193	206	234	272	201	189	206	241	225
Posttreatment level	143	174	203	181	260	193	174	190	202	189

   a Using a simple linear regression analysis, develop a prediction equation for a patient's posttreatment cholesterol level as a function of his or her pretreatment cholesterol level.

   b Construct a 95% confidence interval for the slope of the regression equation and interpret the result.

   c Compute the correlation coefficient between the pretreatment and posttreatment levels and interpret the result.

10.35 Refer to the test statistic $t_{r_{XY}}$ in Box 10.10.

   a Verify the relationship

$$t_{r_{XY}}^2 = \frac{MS[R]}{MS[E]},$$

   where MS[R] and MS[E] are the mean squares for regression and error, respectively, and argue that testing H_0: $\rho_{XY} = 0$ is equivalent to testing H_0: $\beta_1 = 0$.

   b Calculate the value of $t_{r_{XY}}$ for the following combinations of r_{XY} and n:

   i $r_{XY} = 0.7$ and $n = 2, 4, 6, 10$, and 50;

   ii $n = 6$ and $r_{XY} = 0.2, 0.4, 0.6, 0.8$ and 0.9.

   c On the basis of the values calculated in (b), argue that:

   i a large correlation coefficient can be statistically insignificant if the sample size is small;

   ii a small correlation coefficient can be statistically significant if the sample size is large.

   What practical conclusions can be drawn from these observations?

10.9
Determining sample sizes

The problem of determining sample sizes in experiments with one, two, or t treatments was addressed in Sections 7.5 and 9.8. In this section we describe a method for selecting sample sizes in a *simple regression experiment*, an experiment in which a simple linear regression model is used to analyze the data. In a typical regression experiment, in contrast to a completely randomized experiment, sample size determination is a two-stage process, in which the levels of the independent variable are first selected and then the number of replications at each level is determined.

Some useful terminology

Experimental region

The range of the values of the independent variable over which the responses will be measured is called the *experimental region*. For instance, in Example 10.9, the responses (species densities) were measured at different altitudes in the range 0.64–1.72 (in units of 1000 m). Thus, the experimental region for this study is the interval of altitudes extending from 640 m to 1720 m.

Design points

The values of the independent variable at which the responses will be measured are called the *design points*. The design points will be denoted by $d_1, d_2, \ldots, d_t$. For instance, the study in Example 10.9 had $t = 6$ design points: $d_1 = 640$, $d_2 = 860$, $d_3 = 890$, $d_4 = 1220$, $d_5 = 1450$, and $d_6 = 1720$.

Replications of design points

The number of responses measured at a given design point is called the *number of replications* of that design point. The number of replications of d_i will be denoted by n_i. In the data in Example 10.9, there were two replications of each design point, so that $n_1 = \cdots = n_t = 2$. Note that the total number of responses is $N = n_1 + \cdots + n_t$. If each design point is replicated equally often—that is, if $n_1 = \cdots = n_t = n$—then $N = nt$. Also, it is important to distinguish between x_i, the level of X at which response i will be measured, and d_i, the i-th design point. In Example 10.9, there were $t = 6$ values of d_i and $N = 12$ values of x_i. The values of d_i are always distinct, but the values of x_i may not be distinct. In Example 10.9, we had $x_1 = x_2 = 640$, $x_3 = x_4 = 860$, and so on.

A completely randomized experiment can be visualized as a study in which the design points are the levels of the experimental factor (treatments) at which the responses will be measured. The treatments to be studied are determined by the experimental objectives, and so the choice of the design points is not a part of sample size determination in a completely randomized design. For instance, in a study to compare several experimental diets, the choice of diets that will be studied does not enter into the question of sample size determination. In contrast, the study objective in a typical simple regression experiment specifies the experimental region but not the design points at which the responses are to be measured. For instance, in Example 10.9, the investigators were interested in the relationship between plant density and altitude over a range of altitude values. The actual altitudes (design points) at which the responses were measured had to be determined as part of the study plan. Therefore, while sample size determination for a one-way ANOVA entails choosing the replication numbers for the treatments, the corresponding problem for a simple regression experiment involves the selection of the design points as well as their replication numbers.

A variety of objective methods can be used to select the location of the design points. One method is to select the design points such that $\sigma_{\hat{\beta}_1}^2 = \sigma^2 / S_{xx}$, the variance of the estimate of the slope parameter, is as small as possible. In this method, the design points are selected so as to maximize $S_{xx} = \sum (x_i - \bar{x})^2$. Another method is to select the design points so as to minimize

$$\sigma_{\hat{\mu}(x_0)}^2 = \left\{ \frac{1}{N} + \frac{(x_0 - \bar{x})^2}{S_{xx}} \right\} \sigma^2,$$

the variance of the estimated expected response at a specified value x_0 of the independent variable X. Often, in practice, no method based on a single criterion will satisfy all the objectives of a simple regression experiment. In that case, it is best to select design points equally spaced over the experimental region. Spreading the design points evenly over the region of interest enables us to study the response pattern uniformly over the entire range.

Suppose we decide to select t equally spaced design points. Let n be the number of replications at each of the t design points. Then the desired values of t and n can be determined by approaches similar to those in Sections 7.5 and 9.8. Either a hypothesis-testing approach or confidence interval approach can be used to determine t and n.

The hypothesis-testing approach

In the hypothesis-testing approach, we are interested in ensuring that the test of a null hypothesis about a selected parameter—such as the expected response at a given value of X or the slope of the regression line—is of sufficiently high power with respect to a specified alternative hypothesis. In Section 9.8, we saw that, in a completely randomized experiment with given sample sizes, meaningful alternatives to the null hypothesis $H_0 : \mu_1 = \cdots = \mu_t = 0$ can be specified in terms of the value of the noncentrality parameter λ defined in Equation (9.32). Once the noncentrality parameter has been specified, the power of a test of H_0 can be determined with respect to the selected alternative, and appropriate sample sizes can be chosen. A similar approach is possible in simple regression experiments. For example, suppose that the experimental region is confined to the interval (a, b) and that we plan to observe n responses at each of the t equally spaced values of X in the interval (a, b). Let

$$\theta = c_0\beta_0 + c_1\beta_1 \tag{10.17}$$

be a given linear combination of the regression parameters. Recall that two important linear combinations of the regression parameters are: (1) the slope parameter β_1, which corresponds to $c_0 = 0$ and $c_1 = 1$; and (2) the expected response $\mu(x)$, which corresponds to $c_0 = 1$ and $c_1 = x$. Then it can be shown that the power of the t-test of H_0: $\theta = \theta_0$ against H_1: $\theta = \theta_0 + \Delta$ depends on three quantities; α, the level of the test; Δ, the difference between the values of θ under H_0 and H_1; and the noncentrality parameter

$$\lambda = \left[\frac{(t+1)R^2}{(t+1)c_0^2 R^2 + 12(t-1)(c_1 - m_0 c_0)^2} \right] \left(\frac{nt\Delta^2}{2\sigma^2} \right), \tag{10.18}$$

where σ^2 is the error variance $m_0 = (a+b)/2$ and $R = b - a$ are, respectively, the midpoint and range of the experimental region. For given values of α and Δ, Equation (10.18) can be used to determine the power of the t-test of H_0 for several combinations of n and t, until a desired combination is found. It is best to determine the powers using statistical computing software; for details, see Section 7.5.

EXAMPLE **10.22**   The performance of lungs is often measured by a quantity called the forced expiratory volume (FEV). There is known to be an association between FEV and height in humans. Suppose an investigator wants to verify the research hypothesis that the expected FEV for individuals who are 175 cm tall is more than 3 liters.

Let X and Y denote, respectively, the height and the FEV values for a human subject. Assume that the simple linear regression model

$$Y = \beta_0 + \beta_1 x + E,$$

where E is $N(0, \sigma^2)$, may be used to describe the relationship between FEV and height in healthy humans whose heights range between $a = 150$ cm and $b = 200$ cm. Then, with $\theta = \beta_0 + 175\beta_1$, the primary objective of the study is to test the null hypothesis H_0: $\theta = 3$ against the research hypothesis H_1: $\theta > 3$.

Suppose that the investigators want to use a 0.01-level t-test ($\alpha = 0.01$) and that they want at least 80% power if $\theta = 3.3$ or larger ($\Delta = 0.3$). Suppose further that they expect the standard deviation σ of the error to be somewhere between 0.5 and 0.6. In Exercise 10.36, you will be asked to verify that the noncentrality parameter can be expressed as

$$\lambda = \frac{nt\Delta^2}{2\sigma^2}.$$

In the following, we show the SAS codes and the corresponding output for an SAS Program to compute the power of the one-sided t-test for combinations of $t = 4, 5$ and $n = 1, 2, 3, 4$.

```
-----------------------------------------------------------------------
              SAS codes for computing powers of one-sided t-test
data a;
input a b c0 c1 del sigma t n;
lines;
150 200 1 175 .3 .5 5 4
150 200 1 175 .3 .5 5 3
150 200 1 175 .3 .5 5 2
150 200 1 175 .3 .5 5 1

150 200 1 175 .3 .6 5 4
150 200 1 175 .3 .6 5 3
150 200 1 175 .3 .6 5 2
150 200 1 175 .3 .6 5 1

150 200 1 175 .3 .5 4 4
150 200 1 175 .3 .5 4 3
150 200 1 175 .3 .5 4 2
150 200 1 175 .3 .5 4 1


150 200 1 175 .3 .6 4 4
150 200 1 175 .3 .6 4 3
150 200 1 175 .3 .6 4 2
150 200 1 175 .3 .6 4 1
;
data a;
  set a;
  nu     = n*t-2;
  m      = (del**2)/(sigma**2);
  nr     = n*t*((b-a)**2)*(t+1);
  dr     = ((c0)**2)*((b-a)**2)*(t+1) + 3*(t-1)*(2*c1 - c0*(a+b))**2;
  lambda = (m*nr)/(2*dr);
  x      = tinv(.99, nu);
  power  = 1-probt(x, nu, 2*lambda);
  put x;
  put power;
proc print;
  var n t nu  sigma del lambda  power;
title 'Power of one-sided t-test';
run;
-----------------------------------------------------------------------
```

                    Power of one-sided t-test

OBS	N	T	NU	SIGMA	DEL	LAMBDA	POWER
1	4	5	18	0.5	0.3	3.600	0.99999
2	3	5	13	0.5	0.3	2.700	0.99330
3	2	5	8	0.5	0.3	1.800	0.74231

4	1	5	3	0.5	0.3	0.900	0.11522
5	4	5	18	0.6	0.3	2.500	0.98870
6	3	5	13	0.6	0.3	1.875	0.84673
7	2	5	8	0.6	0.3	1.250	0.40432
8	1	5	3	0.6	0.3	0.625	0.06370
9	4	4	14	0.5	0.3	2.880	0.99772
10	3	4	10	0.5	0.3	2.160	0.91646
11	2	4	6	0.5	0.3	1.440	0.46601
12	1	4	2	0.5	0.3	0.720	0.05888
13	4	4	14	0.6	0.3	2.000	0.89874
14	3	4	10	0.6	0.3	1.500	0.60451
15	2	4	6	0.6	0.3	1.000	0.22592
16	1	4	2	0.6	0.3	0.500	0.03769

From the output we see that if $\sigma = .5$, two replications at each of five equally spaced design points will yield 74.2% power, and three replications at each of four equally spaced points will yield 91.6% power. If the standard deviation is as large as $\sigma = .6$, then three replications at five equally spaced points will yield 84.6% power, whereas four replications at each of four equally spaced design points will result in 89.9% power. ■

The confidence interval approach

In the confidence interval approach, a bound for the error is specified for the estimate of a selected parameter, and we seek to ensure that the width of the confidence interval for that parameter is less than the specified bound. It can be seen from property 4 in Box 10.5 that the width of a $100(1 - \alpha)\%$ confidence interval for $\theta = c_0\beta_0 + c_1\beta_1$ is $W = 2t(\nu, \alpha/2)\sigma_{\hat{\theta}}$, where $\nu = n - 2$ denotes the degrees of freedom for the error and $\sigma_{\hat{\theta}}$ is the standard error of the least squares estimate of θ. Straightforward algebraic manipulations show that, if there are n replications at each of the t equally spaced design points over the interval (a, b), then

$$W = 2t(nt - 2, \alpha/2) \left[\frac{(t + 1)c_0^2 R^2 + 12(t - 1)(c_1 - m_0 c_0)^2}{nt(t + 1)R^2} \right]^{\frac{1}{2}} \sigma. \qquad \textbf{(10.19)}$$

For each selected combination of α, n, t, and σ, Equation (10.19) can be used to determine the halfwidth of the confidence interval. The following SAS output shows B, the half-width of 99% confidence interval for several combinations of n, t, and σ.

```
        Lengths of 99% confidence intervals
        OBS   EN   T   NU   SIGMA     B
         1     4    5   18    0.5    0.32182
         2     3    5   13    0.5    0.38888
         3     2    5    8    0.5    0.53053
```

4	1	5	3	0.5	1.30607
5	4	5	18	0.6	0.38618
6	3	5	13	0.6	0.46666
7	2	5	8	0.6	0.63664
8	1	5	3	0.6	1.56728
9	4	4	14	0.5	0.37211
10	3	4	10	0.5	0.45745
11	2	4	6	0.5	0.65539
12	1	4	2	0.5	2.48121
13	4	4	14	0.6	0.44653
14	3	4	10	0.6	0.54893
15	2	4	6	0.6	0.78646
16	1	4	2	0.6	2.97745

Exercises

10.36   Verify that, when $t = 5$ and $\Delta = 0.1$, the noncentrality parameter in Example 10.22 can be expressed as $\lambda = nt\Delta^2/2\sigma^2$.

10.37   Refer to Example 10.22.

a   Suppose we want to estimate the expected FEV for humans who are 175 cm tall, with an error of estimation less than 0.3 liters. If we decide to use five equally replicated design points $(t = 5)$, calculate the sample size needed to ensure that the error of estimation is within the specified bound with 90% confidence.

b   Suppose we are interested in simultaneously estimating the expected FEV for humans of height 160 and 170 cm. Determine the sample sizes corresponding to $t = 5$ design points and estimates that are within 0.3 liter of the true values with 90% confidence.

c   Determine the replication sizes for $t = 5$ design points if, in (a), we desire to predict the FEV of a single subject.

d   Determine the replication sizes for $t = 5$ design points if, in (b), we desire to predict the FEV values for two subjects, one of height 160 cm and the other of height 170 cm.

10.38   Use the chemical influx data in Example 10.3 to determine the sample sizes for a future study in each of the following cases. Assume that the investigator is interested in the region (1, 10,000) for the drug dose and that the data will be analyzed using the regression of $Y =$ chemical influx on $x = \log_{10}$ (dose).

a   Calculate the sample sizes needed in the case where the investigator:

i   wants to use five doses: 1, 10, 100, 1000, and 10,000 nM;

ii   wants to test the null hypothesis H_0 that the drug dose does not affect the expected chemical influx against the alternative hypothesis H_1 that increasing drug dose is associated with increasing expected chemical influx;

iii   would like to ensure at least an 80% chance that a 0.05-level test will reject the null hypothesis if increasing the drug dose by a factor of 10 (from d

to 10*d*) corresponds to an increase in the expected chemical influx of at least 5 nmol/mg protein.

b Calculate the sample sizes needed in the case where the investigator:

  i   wants to use five doses: 1, 10, 100, 1000, and 10,000 nM;
  ii wants to estimate the expected change in chemical influx when the drug dose is increased by a factor of 10;
  iii wants to ensure 95% probability of estimating the change with less than 3% error.

10.10
Regression when ANOVA assumptions are violated

Simple linear regression analysis requires two assumptions: (1) that the expected response is a linear function of the independent variable x—that is, that $\mu(x) = \beta_0 + \beta_1 x$; and (2) that the random errors have independent normal distributions with a common variance (the ANOVA assumption). In Chapter 11, we'll consider some options when the expected response is not a linear function of the independent variable. In this section we consider some alternatives to simple linear regression analysis that can be used when the validity of the ANOVA assumption is in doubt.

When the responses are measured on an interval scale, one approach to violations of the ANOVA assumption is to look for a transformation of the data, as described in Section 8.7. Often, a simple transformation such as a log transformation or a square-root transformation will solve the problem. When transforming the data is not a viable option, either for lack of a suitable transformation or because of difficulties in interpreting the parameters of the transformed model, it may be useful to employ weighted regression, rank regression, or logistic regression. Let's look at these methods.

Weighted regression

As in the case of one-way ANOVA analysis, simple linear regression analysis is particularly sensitive to violations of the key ANOVA assumption that the errors have equal variances. When such a violation is suspected, a possible remedy is to perform a weighted version of simple linear regression analysis.

In a weighted regression analysis, different responses are given different weights depending on their variances. The regression parameters are estimated by minimizing the weighted residual sum of squares

$$\mathrm{SS}_w[\mathrm{E}] = \sum_{i=1}^{n} w_i \left[y_i - (\hat{\beta}_0 + \hat{\beta}_1 x_i) \right]^2,$$

where the $w_i \geq 0$ are the weights given to the residuals. The resulting estimates are called *weighted least squares estimates*.

The w_i are selected to ensure that the contribution of a residual reflects the precision of the response that generated that residual. Intuitively, it makes sense to select the weights in such a way that smaller weights are associated with larger error variances. When the error variances are all equal, the same weight ($w_1 = w_2 = \ldots = w_n = w$) is given to all the residuals. In that case, estimating the regression parameters by minimizing the weighted residual sum of squares is the same as estimating them by minimizing the unweighted residual sum of squares

$$\text{SS[E]} = \sum_{i=1}^{n} \left[y_i - (\hat{\beta}_0 + \hat{\beta}_1 x_i) \right]^2 .$$

The resulting estimates $\hat{\beta}_0$ and $\hat{\beta}_1$ are the same as the least squares estimates in Box 10.3. These estimates are often called *ordinary least squares estimates.*

Let σ_i^2 denote the error variance for the i-th response. Then it can be shown that, when the errors are independent, choosing $w_i = 1/\sigma_i^2$ will yield regression parameter estimates that are unbiased and have minimum variances. Unfortunately, the practical use of weighted regression analysis is somewhat limited, because the error variances are usually unknown and it is difficult to obtain reliable values for the weights. In typical cases, the weights must be estimated from the observed data either by examining suitable residual plots or by exploiting special design structures that may be present in a particular data set. For instance, if multiple responses are available at every setting of the independent variable, then the variance of a response at a given setting can be estimated by the variance of the observed responses at that setting. The reliability of such estimates will depend on the number of observations available for their calculation. When reasonable estimates of error variances are available, statistical computing software such as SAS can be used for a weighted regression analysis of the data. For more details, see Draper and Smith (1981), Deaton, Reynolds, and Myers (1983), and Myers (1986).

Rank regression

Rank regression analysis can be used in circumstances where the assumption of normally distributed errors may be violated. Like the rank-based methods already discussed (for example, the Wilcoxon rank sum test and the Kruskal-Wallis test), rank regression analysis is particularly appropriate when there is a strong possibility that the responses are subject to gross errors.

The rank regression method in Box 10.12 can be used when the data consist of exactly one response at each distinct setting of the independent variable and the assumption that the errors are independently distributed with a common continuous distribution is reasonable. A modification of the method for the case where there is more than one response at some settings of the independent variable is described in Sen (1968). A rank regression method suitable for handling some general simple linear regression models is described in Rao and Thornby (1969). Useful information on rank regression analysis can also be found in Hollander and Wolfe (1973), Daniel (1990), and Sprent (1993).

BOX **10.12**

Rank regression analysis with a single independent variable

Let (x_i, y_i), $i = 1, 2, \ldots, n$, where $x_1 < x_2 < \cdots < x_n$, denote the measured values of a response variable Y at n values of an independent variable X. Suppose that, when $X = x_i$, the response Y_i satisfies the model

$$Y_i = \beta_0 + \beta_1 x_i + E_i,$$

where the E_i are independent and have a common continuous distribution with zero median. For $i < j$, let

$$\hat{\beta}_{ij} = \frac{y_j - y_i}{x_j - x_i}$$

denote the slope of the line joining the pair of points (x_i, y_i) and (x_j, y_j). Note that there are $N = n(n-1)/2$ slopes $\hat{\beta}_{ij}$.

Inferences about β_1 can be based on the test statistic of the form

$$S = \left(\text{number of } \hat{\beta}_{ij} > \beta_{1(0)}\right) - \left(\text{number of } \hat{\beta}_{ij} < \beta_{1(0)}\right),$$

where $\beta_{1(0)}$ is a specified value of β_1. Table C.13 in Appendix C contains the α-level critical values $S(n, \alpha)$ of S for testing the null hypothesis H_0: $\beta_1 = \beta_{1(0)}$. For values of n outside the range in Table C.13, approximate critical values can be calculated from the formula $S(n, \alpha) = z(\alpha)\sigma_S$, where

$$\sigma_S = \sqrt{\frac{n(n-1)(2n+5)}{18}}$$

is the standard error of S.

Hypothesis test

Let S_c denote the calculated value of S. Then the following are the rejection criteria for an α-level test:

1   Reject H_0 in favor of H_1: $\beta_1 > \beta_{1(0)}$ if $S_c > S(n, \alpha)$.

2   Reject H_0 in favor of H_1: $\beta_1 < \beta_{1(0)}$ if $S_c < -S(n, \alpha)$.

3   Reject H_0 in favor of H_1: $\beta_1 \neq \beta_{1(0)}$ if $S_c > S(n, \alpha/2)$ or $S_c < -S(n, \alpha/2)$.

Point estimation

Let $\hat{\beta}_{(1)} < \hat{\beta}_{(2)} < \cdots < \hat{\beta}_{(N)}$ denote the ordered values of the $\hat{\beta}_{ij}$. The median of the $\hat{\beta}_{ij}$ is a point estimate of β_1; that is, β_1 can be estimated by $\hat{\beta}_1$, where

$$\hat{\beta}_1 = \begin{cases} \hat{\beta}_{\left(\frac{N+1}{2}\right)} & \text{if } N \text{ is an odd number} \\[2ex] \frac{1}{2}\left(\hat{\beta}_{\left(\frac{N}{2}\right)} + \hat{\beta}_{\left(\frac{N}{2}+1\right)}\right) & \text{if } N \text{ is an even number.} \end{cases}$$

Interval estimation

A $100(1 - \alpha)\%$ confidence interval for β_1 is

$$\hat{\beta}_{(L)} \leq \beta_1 \leq \hat{\beta}_{(U)},$$

where

$$L = \frac{N - S\left(n, \dfrac{\alpha}{2}\right)}{2} + 1,$$

$$U = \frac{N + S\left(n, \dfrac{\alpha}{2}\right)}{2}.$$

The method in Box 10.12 can be justified intuitively as follows. Notice that, for $i < j$, $y_j - y_i$ is the observed change in the value of Y when the value of X is increased from x_i to x_j. Therefore, $\hat{\beta}_{ij}$ is an estimate of the change in the value of Y per unit increase in the value of X; that is, $\hat{\beta}_{ij}$ is an estimate of β_1 based on the pair of observations y_i and y_j. For this reason, we refer to $\hat{\beta}_{ij}$ as a *simple slope estimate*. The rank regression method described in Box 10.12 utilizes the $N = n(n - 1)/2$ simple slope estimates to make inferences about the true slope β_1. For example, the test statistic S compares the number of simple slope estimates that are less than $\beta_{1(0)}$ with the number that are larger than $\beta_{1(0)}$. Under the null hypothesis, we would expect approximately half of the simple slopes to be greater than $\beta_{1(0)}$, so that the value of S will be close to zero. A large positive value for S favors the hypothesis $H_1: \beta_1 > \beta_{1(0)}$ because such a value for S implies that substantially more than half of the simple estimates exceed $\beta_{1(0)}$. The point estimate $\hat{\beta}_1$, being the median of the simple slope estimates, is a natural way of estimating the unknown slope. Similarly, the confidence interval for β_1 is formed by discarding a selected number of simple slopes that have the highest and the lowest values.

EXAMPLE **10.23**   Suppose we are interested in the variation of the mineral content of soil over a large region. The observed values of Y, the mineral content (vol. %), for $n = 7$ sites located at selected distances (values of X, km) from a fixed base point in the region are as follows:

x	10	20	30	40	50	60	70
y	30	38	43	49	61	63	74

The values of the $N = 7(7 - 1)/2 = 21$ simple slopes are displayed in the following table, each entry in which is the simple slope of the line joining the pair of (x, y) values in the corresponding row and column; the number in parentheses is the rank of the slope.

x	10	20	30	40	50	60	70
y	30	38	43	49	61	63	74

x	y						
10	30						
20	38	0.800 (17)					
30	43	0.650 (7)	0.500 (2)				
40	49	0.633 (6)	0.550 (3)	0.600 (4)			
50	61	0.775 (15)	0.767 (14)	0.900 (19)	1.200 (21)		
60	63	0.660 (9)	0.625 (5)	0.667 (10)	0.700 (11)	0.200 (1)	
70	74	0.733 (13)	0.720 (12)	0.775 (16)	0.833 (18)	0.650 (8)	1.100 (20)

As an example of the calculation of the simple slopes, notice that the slope with rank 3 is calculated from the two pairs $(x, y) = (40, 49)$ and $(x, y) = (20, 38)$ as

$$\hat{\beta}_{ij} = \frac{49 - 38}{40 - 20} = 0.550.$$

Because N is odd, the median of the simple slopes is

$$\hat{\beta}_1 = \hat{\beta}_{\left(\frac{N+1}{2}\right)} = \hat{\beta}_{(11)} = 0.700.$$

Thus, we estimate that, for each 1-km increase in the distance from the base point, we can expect a 0.7% increase in the soil mineral content.

To obtain a 95% confidence interval for β_1, we establish from Table C.13 that $S(7, 0.025) = 15$. Therefore, $L = (21 - 15)/2 + 1 = 4$ and $U = (21 + 15)/2 = 18$, so that the required interval is $\hat{\beta}_{(4)} \leq \beta_1 \leq \hat{\beta}_{(18)}$. We can claim with 95% confidence that between 0.600% and 0.833% increase in soil mineral content may be expected for every 1-km increase in the distance from the base point. ∎

Logistic regression

Like simple linear regression analysis, the weighted and rank regression analyses assume that the responses are measured on an interval scale. Frequently, regression analysis must be performed when the response variable is measured on a nominal scale. In the following example, we are interested in relating a binary response variable to a quantitative independent variable.

EXAMPLE 10.24   Bone marrow transplantation (BMT) is a curative therapy for acute leukemias and other blood-related diseases. One problem associated with BMT is that the medications administered when the patient is being prepared for marrow transplantation can cause life-threatening toxicities. Let X be the age (in years) of the BMT patient, and Y be a binary variable defined as

$$Y = \begin{cases} 1 & \text{if the patient has acute pulmonary toxicity within 100 days after BMT} \\ 0 & \text{otherwise.} \end{cases}$$

In a study, the survival times and events associated with various drug toxicities for 73 BMT patients suffering from acute lymphocytic leukemia (ALL) were analyzed (Kumar et al., 1994). The following table shows the values of X and Y for 24 of the patients:

Patient number i	01	02	03	04	05	06	07	08
Age x_i	34	13	31	10	13	29	07	16
Toxicity y_i	0	1	1	1	0	0	1	1
Patient number i	09	10	11	12	13	14	15	16
Age x_i	25	26	14	55	04	46	24	15
Toxicity y_i	1	0	1	0	0	0	0	0
Patient number i	17	18	19	20	21	22	23	24
Age x_i	32	10	03	05	05	21	35	05
Toxicity y_i	1	1	1	1	1	0	1	0

The investigators wanted to study the association between pulmonary toxicity and the age of the BMT patient. In particular, they wondered whether pulmonary toxicity is more likely among younger patients than among older patients.

Thus, in this study, the association between a qualitative variable and a quantitative variable was of interest. ■

Logistic regression analysis can be used to study the association between a qualitative variable Y and a quantitative variable X, as in Example 10.24.

Let A be a given event (such as pulmonary toxicity) whose occurrence may be associated with the value of a quantitative independent variable X (such as age). Let $P(x)$ denote the probability of occurrence of A when $X = x$. Then a logistic regression analysis is based on the assumption that the probability of A when $X = x$ can be expressed as

$$P(x) = \frac{e^{g(x)}}{1 + e^{g(x)}}, \tag{10.20}$$

where $g(x)$ is a function of x. One of the most frequently used $g(x)$ is

$$g(x) = \beta_0 + \beta_1 x, \tag{10.21}$$

where β_0 and β_1 are unknown parameters. The resulting model for $P(x)$ takes the form

$$P(x) = \frac{e^{\beta_0 + \beta_1 x}}{1 + e^{\beta_0 + \beta_1 x}}. \tag{10.22}$$

In what follows, an analysis based on Equation (10.22) will be referred to as a *simple logistic regression analysis*. For example, a simple logistic regression analysis of the toxic death data in Example 10.24 is an analysis based on the assumption that $P(x)$, the probability that a BMT patient of age $X = x$ will experience pulmonary toxicity, can be expressed in terms of two parameters β_0 and β_1, as in

Equation (10.22). Once the two regression parameters have been estimated from the data, the probability of the event can be estimated for any given value of X.

Why did we use Equation (10.22) to express the relationship between $P(x)$ and x? We could simply have chosen the straight-line relationship

$$P(x) = \beta_0 + \beta_1 x. \tag{10.23}$$

Note that $P(x)$ is a probability and hence cannot take a value outside the interval $[0, 1]$. In Equation (10.23), $P(x)$ can take values that are outside the permissible range. Indeed, as you will be asked to show in Exercise 10.45, Equation (10.23) can give negative values for $P(x)$. On the other hand, Equation (10.22) will always give a value of $P(x)$ that is within the interval $[0, 1]$, because both the numerator and denominator are positive and the denominator is always at least as large as the numerator.

Interpretation of β_0 and β_1 in Equation (10.22)

Once again, β_0 acts as an intercept parameter, because knowledge of β_0 will permit the determination of the probability of A when $x = 0$. When $x = 0$, the probability of the event is

$$P(0) = \frac{e^{\beta_0}}{1 + e^{\beta_0}}.$$

The parameter β_1 indicates the strength of association between the probability of A and the value of the independent variable X. If $\beta_1 = 0$, the probability is not associated with the value of X. Figure 10.22 shows two simple logistic regression functions, one with $\beta_1 > 0$ and the other with $\beta_1 < 0$. When $\beta_1 > 0$, the probability

FIGURE   10.22
Simple logistic regression functions

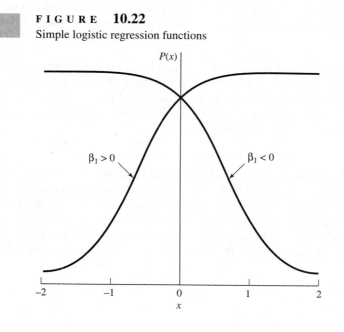

increases with increasing values of X. When $\beta_1 < 0$, the probability decreases with increasing values of X. When both β_0 and β_1 are zero, the probability of occurrence is 0.5; that is, the event is as likely to occur as not.

Logistic regression analysis

Typical data sets requiring logistic regression analysis fall into one of two categories: (1) data sets in which only one response is available at each level of the independent variable, such as the toxic death data in Example 10.24; (2) data sets—usually obtained from designed experiments—in which multiple responses are available at every distinct level of the independent variable.

EXAMPLE **10.25**   In a study, a suspected carcinogen, aflatoxin B_1, was fed to test animals. The following data for liver tumors were obtained (Gaylor, 1987):

Dose (ppb)	0	1	5	15	50	100
Number of test animals N	18	22	22	21	25	28
Number of animals with tumors R	0	2	1	4	20	28

Let the response variable Y be defined as $Y = 1$ if the test animal developed a liver tumor and $Y = 0$ otherwise. Then the tumor incidence data consist of $n = 18 + 22 + 22 + 21 + 25 + 28 = 136$ observations of the qualitative response variable Y at $t = 6$ settings of the quantitative independent variable $X =$ dose. There are multiple responses at each setting of the independent variable. For instance, at $X = 15$, we have 21 responses (four 1s and 17 0s). A logistic regression analysis can be used to investigate the dose-response relationship between the probability of developing a liver tumor and the dose of the carcinogen.   ■

Regardless of the type of data available, a logistic regression analysis is best performed using statistical computing software. In the next example, the use of the LOGISTIC procedure of SAS is demonstrated for the data in Example 10.25. For details of the SAS codes needed to implement this procedure, see Section 10.5 in the companion text by Younger (1997).

EXAMPLE **10.26**   The following is a portion of the output resulting from the use of the LOGISTIC procedure of SAS to analyze the tumor incidence data in Example 10.25. The first part of the output shows the estimates of the regression parameters, their standard errors, and the associated two-sided p-values (in the column headed Pr > Chi-Square). The second part of the output shows the estimate of the probability of a tumor and the corresponding 95% confidence bounds at each dose tested in the study.

```
------------------------------------------------------------------------
              Logistic Regression Analysis of Tumor Incidence Data
Response Variable (Events): R
Response Variable (Trials): N


              Parameter   Standard    Pr >
   Variable  DF  Estimate    Error    Chi-Square
   INTERCPT   1   -3.0360    0.4823    0.0001
   DOSE       1    0.0901    0.0146    0.0001
------------------------------------------------------------------------

                                        95%        95%
    OBS    DOSE    R    N    PROB.EST  LOWER.CB   UPPER.CB

     1       0    0   18   0.04582   0.01832    0.10999
     2       1    2   22   0.04993   0.02043    0.11691
     3       5    1   22   0.07007   0.03141    0.14900
     4      15    4   21   0.15647   0.08559    0.26880
     5      50   20   25   0.81281   0.62719    0.91809
     6     100   28   28   0.99746   0.97686    0.99973
------------------------------------------------------------------------
```

The estimates $\hat{\beta}_0 = -3.0360$ and $\hat{\beta}_1 = 0.0901$ are both significantly ($p = 0.0001$) different from zero. In fact, given that $\hat{\beta}_1$ is positive, the p-value for testing the null hypothesis H_0: $\beta_1 = 0$ against the one-sided alternative hypothesis H_1: $\beta_1 > 0$ is $0.0001/2 = 0.00005$; this indicates a significant positive association between the level of exposure and the probability of a tumor. Thus, it is reasonable to conclude that increasing exposure to aflatoxin increases the probability of developing a tumor.

The estimated tumor probability can be expressed as a function of the aflatoxin dose x by substituting the estimates for β_0 and β_1 into Equation (10.22)

$$\hat{P}(x) = \frac{e^{-3.0360+0.0901x}}{1 + e^{-3.0360+0.0901x}}. \tag{10.24}$$

A plot of the estimated probability is shown in Figure 10.23.

The estimated probability of a tumor for an unexposed animal (zero dose) is obtained by setting $x = 0$ in Equation (10.24)

$$\hat{P}(0) = \frac{e^{-3.0360}}{1 + e^{-3.0360}} = 0.046.$$

Thus, approximately 4.6% of the unexposed animals develop a tumor. From the second part of the output, we see that a 95% confidence interval for $P(0)$ is (0.018, 0.110) so that, with 95% confidence, we can claim that between 1.8% and 11% of the unexposed animals will develop a tumor. ∎

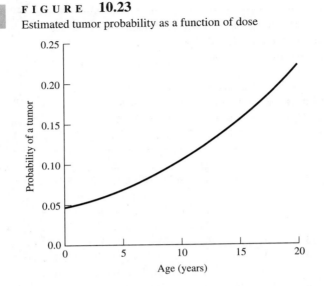

FIGURE 10.23
Estimated tumor probability as a function of dose

Exercises

10.39 For the chemical influx data in Example 10.3, what graphical method can be used to see if a weighted regression analysis is more appropriate than an unweighted analysis? Suggest reasonable weights that can be used in a weighted regression analysis.

10.40 Repeat Exercise 10.39 for the soil water content data in Exercise 10.2.

10.41 Explain why there is no obvious way of using the data in Exercise 10.7 to estimate the weights needed for a weighted regression analysis of the disease indices.

10.42 Refer to the weight gain data in Exercise 10.3.

    a State the assumptions needed to justify the use of a rank regression analysis of a mother's weight gain with her age as the independent variable.

    b Use the data to construct an approximate 95% confidence interval for the slope parameter in the regression of a mother's weight gain on her age.

    c Compare the confidence interval constructed in (b) with the corresponding interval constructed under the assumption that the ANOVA assumptions are satisfied. Describe the conditions under which the rank regression interval will be preferable over the simple linear regression interval.

10.43 Refer to the serum cholesterol data in Exercise 10.34.

    a Use rank regression analysis to estimate the slope parameter in the regression of the posttreatment cholesterol level on the corresponding pretreatment cholesterol level.

    b Clearly state the assumptions needed to justify the estimation method used in (a).

    c Use rank regression analysis to test the null hypothesis that there is a positive association between the pretreatment and posttreatment cholesterol levels.

    d Compare the results of your test in (c) with those of simple linear regression analysis. Describe the conditions under which rank regression will be preferable to simple linear regression.

10.44 Refer to the mineral data in Example 10.23.

a Use simple linear regression analysis to construct a 95% confidence interval for the slope parameter in the regression of the mineral content on the distance from the base point.

b Compare the interval in (a) with that constructed in Example 10.23. Describe conditions under which the interval in (a) is preferable.

c Perform appropriate statistical tests to see if the data support the conclusion that the soil mineral content can be expected to increase by 1% for every 1-km increase in the distance from the base point. Perform the test using both rank regression and simple linear regression. Comment on the results.

10.45 By selecting suitable values for β_0 and β_1, show that Equation (10.23) can give a negative value for a probability.

10.46 Suppose that, in Example 10.26, the estimate of β_1 was -0.0901, with a two-sided p-value of 0.0001. Determine the p-value for testing the null hypothesis $H_0: \beta_1 = 0$ against the alternative hypothesis $H_1: \beta_1 < 0$.

10.47 The following is a portion of the SAS output obtained when a logistic regression is fitted to the data for the 73 BMT patients in Example 10.24.

```
--------------------------------------------------------------------
      Logistic Regression Analysis of Tumor Incidence Data
Response Variable Y = Pulmonary Toxicity
                      Parameter  Standard      Pr >
        Variable  DF  Estimate    Error     Chi-Square

        INTERCPT  1    -1.6033    0.4783      0.0008
        AGE       1     0.0901    0.0146      0.0226
--------------------------------------------------------------------

                                        95%      95%
        OBS    AGE    Y    PROB.EST  LOWER.CB UPPER.CB
         1     34     0     0.5560    0.3295   0.7614
         2     13     1     0.2882    0.1906   0.4104
         3     31     1     0.5159    0.3189   0.7081
         4     10     1     0.2563    0.1585   0.3866
         5     13     0     0.2882    0.1906   0.4104
--------------------------------------------------------------------
```

a Write a simple logistic regression model relating the probability of pulmonary toxicity (PT) to the age of the patient.

b Interpret the parameter estimates given in the output.

c Use the computer output to obtain a formula for the estimated probability of PT as a function of age.

d Use the formula in (c) to estimate the probability of PT for a 15-year-old patient.

e Interpret the p-values in the output (in the column headed Pr > Chi-Square).

f Use the output to obtain a 95% confidence interval for the probability of PT for a 13-year-old patient. Interpret the confidence interval.

10.11
Overview

Like the one-way ANOVA model, the simple linear regression model is very frequently used in statistics. It is based on the assumption that the relationship between the expected value of a quantitative response variable and the levels of a corresponding quantitative independent variable can be expressed as a linear combination of two parameters β_0 and β_1, in which the level of the independent variable acts as the coefficient of β_1. The model has a simple structure, is easy to interpret, and gives at least approximate results over a wide range of applications. Under assumptions similar to the ANOVA assumptions (Chapter 8), the simple linear regression model can be used to make inferences about expected and predicted responses similar to those derived using the one-way ANOVA models.

Conceptually, there is a close connection between the one-way ANOVA model and the simple linear regression model. In this chapter, we have seen how the simple linear regression model can be viewed as a specialization of the one-way ANOVA model for the case when the independent variable is quantitative. In Chapter 12, we'll see that the two models can be regarded as special cases of a general linear model.

The concept of correlation coefficients is closely related to the notion of a simple linear regression model. While simple linear regression models can be used to model a straight-line relationship between the expected response and the level of an independent variable, correlation coefficients can be used as a quantitative measure of the tendency of a pair of quantitative variables to take values that cluster close to a straight line. Several important relationships between correlation coefficients and the corresponding regression parameters are described in this chapter.

The assumptions underlying a simple linear regression model can be violated in several ways. Either or both of the independent and dependent variables may not be quantitative, the errors may not have independent normal distributions, or the regression function may not be a straight line. In such cases, weighted regression, logistic regression, and multiple regression are some useful alternatives to simple linear regression.

11

Multiple Linear Regression

11.1 Introduction

As we saw in Chapter 10, a simple linear regression model can be used to explore the linear relationship between the expected value of a response variable and the value of a quantitative explanatory variable. A natural extension of the simple linear regression model is the multiple linear regression model, in which the expected response is a linear function of the values of several quantitative explanatory variables. In this chapter, we'll see how multiple linear regression models can be used to examine possible relationships between observed responses and the values of possible explanatory variables.

As might be expected, many of the procedures related to multiple linear regression analysis (such as estimation based on least squares, inferences about linear combinations of parameters, and prediction and confidence bands) are straightforward extensions of the procedures in simple linear regression analysis. For this reason, many of our discussions in this chapter will draw on the corresponding portions of the preceding chapter. You should review the chapter on simple linear regression models before proceeding further.

11.2
The multiple linear regression model

In a multiple linear regression model, the linear relationship of the expected responses to the independent variables is specified by a multiple linear regression function (Box 11.1).

BOX **11.1**   ***The multiple linear regression function***

Let $\mu(x_1, \ldots, x_p) = \mathcal{E}(Y|x_1, \ldots, x_p)$ denote the expected value of a response variable Y given the values $x_1, \ldots, x_p$ of p quantitative independent variables $X_1, \ldots, X_p$. Then $\mu(x_1, \ldots, x_p)$ is called a multiple linear regression function if it can be expressed as a linear combination

$$\mu(x_1, \ldots, x_p) = \beta_0 + \beta_1 x_1 + \cdots + \beta_p x_p$$

of $p + 1$ parameters $\beta_0, \ldots, \beta_p$. The parameters β_i, $i = 0, 1, \ldots, p$, are called the *regression parameters*.

The next two examples illustrate how multiple linear regression functions can be used to model the expected responses.

EXAMPLE **11.1**   Enzymes are organic substances produced by living cells. Because enzymes are capable of inducing chemical changes in other substances without themselves being changed in the process, biologists are very interested in the capacity of enzymes to adhere to substances found in the human body.

In an experiment on the adsorption (adhesion to a solid) of a particular enzyme on ultrafine particles of silica suspended in a liquid medium, two variables were measured: C, the concentration of the enzyme (g/dm^3) in the medium; and Y, the amount of enzyme adsorbed (expressed as a percentage of the amount of silica in the medium). The following results were obtained:

c	0.05	0.10	0.30	0.50	0.60	0.80	1.0
y	2.0	5.0	14.5	18.6	20.1	22.0	23.5

A primary objective of the study was to see how the amount of enzyme adsorbed varied with its concentration in the medium. Figure 11.1 shows a scatter plot of the data.

A comparison of the plots of quadratic functions in Figure 10.15 with the plot in Figure 11.1 suggests that the relationship between the mean percentage enzyme adsorbed and the concentration of the enzyme is better described by the quadratic regression function

$$\mathcal{E}(Y|c) = \mu(c) = \beta_0 + \beta_1 c + \beta_2 c^2 \qquad \textbf{(11.1)}$$

than by a simple linear regression function. In Equation (11.1), the expected response is expressed as a quadratic function of a single explanatory variable C. Alternatively, the relationship in Equation (11.1) can be expressed in terms of a multiple linear regression function of $p = 2$ independent variables $X_1 = C$ and $X_2 = C^2$

$$\mathcal{E}(Y|X_1 = x_1, X_2 = x_2) = \mu(x_1, x_2) = \beta_0 + \beta_1 x_1 + \beta_2 x_2. \qquad \textbf{(11.2)}$$

Equations (11.1) and (11.2) are two different ways of formulating the same relationship between the expected adsorption and the enzyme concentration. From a user's viewpoint, Equation (11.1) seems preferable, because it expresses the expected responses directly in terms of the explanatory variable C. From the viewpoint of statistical analysis of the relationship between the adsorption and the enzyme

FIGURE 11.1

Scatter plot of enzyme adsorption data

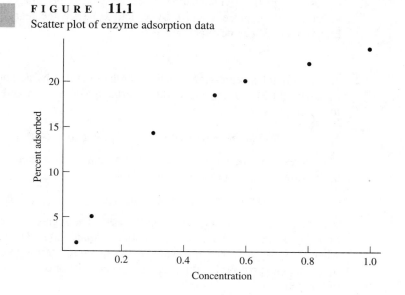

concentration, however, Equation (11.2) has many advantages, because the theory and applications of linear models are highly developed in statistics.   ∎

In Example 11.1, the two independent variables in the multiple linear regression function were generated by a single explanatory variable—C, the enzyme concentration. The next example introduces multiple linear regression functions in which the independent variables depend on more than one explanatory variable.

EXAMPLE **11.2**   In an experiment to study the total oxygen demand in dairy wastes, the data consisted of values of the response variable $Y = \log$ (oxygen demand) along with the values of five explanatory variables: A, the biological oxygen demand; B, the total Kjeldahl nitrogen; C, the total solids; D, the total volatile solids; and F, the chemical oxygen demand (Moore, 1975). The six variables were measured over a 220-day period on a sample kept in suspension in a laboratory. Also, see Weisberg (1981).

A simple linear regression function can be used to study the linear association between Y and any one of the five explanatory variables, but a more complex function will be needed to investigate the association between Y and all five of the variables.

A wide variety of relationships may be postulated between the expected value of Y and the values of A–F. For instance, here are three possible multiple linear regression functions to describe the expected response as a function of A and B.

1   Let $\mathcal{E}(Y|a, b)$ denote the expected response when $A = a$ and $B = b$. Then one of the simplest regression functions involving a and b is

$$\mathcal{E}(Y|a, b) = \mu(a, b) = \beta_0 + \beta_1 a + \beta_2 b. \qquad \textbf{(11.3)}$$

To emphasize that a and b are the values of the independent variables in the regression function (which happen to be the same as the values of the explanatory variables), we replace a and b by x_1 and x_2. Then the regression function in Equation (11.3) can be expressed as

$$\mu(x_1, x_2) = \beta_0 + \beta_1 x_1 + \beta_2 x_2.$$

2   A more complex regression function can be obtained by adding more terms to Equation (11.3). For instance, adding terms quadratic in a and b will yield the function

$$\mathcal{E}(Y|a, b) = \mu(a, b) = \beta_0 + \beta_1 a + \beta_2 b + \beta_3 a^2 + \beta_4 b^2. \qquad \textbf{(11.4)}$$

The regression function in Equation (11.4) can also be expressed as a function of the values of $p = 4$ independent variables

$$\mu(x_1, x_2, x_3, x_4) = \beta_0 + \beta_1 x_1 + \beta_2 x_2 + \beta_3 x_3 + \beta_4 x_4,$$

where $x_1 = a$, $x_2 = b$, $x_3 = a^2$, and $x_4 = b^2$. The regression function in Equation (11.4) is an extension of the function in Equation (11.3), to which it reduces when $\beta_3 = \beta_4 = 0$. The inclusion of the quadratic terms makes the function more complex, but the resulting function is capable of describing a wider variety of relationships.

3   Another extension of Equation (11.3) can be obtained by adding the crossproduct term ab

$$\mathcal{E}(Y|a, b) = \mu(a, b) = \beta_0 + \beta_1 a + \beta_2 b + \beta_3 ab. \tag{11.5}$$

This regression function can be expressed in terms of the values of $p = 3$ independent variables

$$\mu(x_1, x_2, x_3) = \beta_0 + \beta_1 x_1 + \beta_2 x_2 + \beta_3 x_3,$$

where $x_1 = a$, $x_2 = b$, and $x_3 = ab$.

Regression functions involving more than two explanatory variables can be expressed in a similar manner. For example, the regression function with three explanatory variables

$$\mathcal{E}(Y|a, b, c) = \beta_0 + \beta_1 a + \beta_2 b + \beta_3 c + \beta_4 a^2 + \beta_5 b^2 + \beta_6 c^2$$

can be expressed in terms of six independent variables as

$$\mu(x_1, x_2, x_3, x_4, x_5, x_6) = \beta_0 + \beta_1 x_1 + \beta_2 x_2 + \beta_3 x_3 + \beta_4 x_4 + \beta_5 x_5 + \beta_6 x_6. \quad \blacksquare$$

The three regression functions in Example 11.2 are just a few of the many possible multiple linear regression functions that can be used to relate the expected response to the values of two independent variables. Later in the chapter, we'll look at how to select an appropriate multiple linear regression function for a particular situation.

Successful practical use of multiple linear regression functions requires that the user be able to interpret them correctly.

Interpretation of regression parameters

As in simple linear regression, the regression coefficient β_0 is the expected response when all the independent variables are zero and is known as the *intercept parameter*.

In simple linear regression, the slope parameter is the expected change in the response corresponding to unit increase in x. In multiple linear regression, by contrast, the parameter β_i $(i = 1, \ldots, p)$ is the expected change in the response when the value of X_i is increased by one unit without changing the values of the other independent variables. In symbols

$$\beta_i = \mu(x_1, \ldots, x_{i-1}, x_i + 1, x_{i+1}, \ldots, x_p) - \mu(x_1, \ldots, x_{i-1}, x_i, x_{i+1}, \ldots, x_p)$$

$$= \left\{ \beta_0 + \beta_1 x_1 + \cdots + \beta_{i-1} x_{i-1} + \beta_i(x_i + 1) + \beta_{i+1} x_{i+1} + \cdots + \beta_p x_p \right\}$$

$$- \left\{ \beta_0 + \beta_1 x_1 + \cdots + \beta_{i-1} x_{i-1} + \beta_i x_i + \beta_{i+1} x_{i+1} + \cdots + \beta_p x_p \right\}.$$

Because β_i equals the expected change in the response when X_i is the only variable whose value has increased by one unit, the parameter β_i is called the *partial slope* with respect to X_i.

As we saw in Example 11.1, a multiple linear regression function can be thought of in two ways: as a function of the values of a set of p independent variables; or as a function of the values of a set of k explanatory variables. Representation in terms of the values of independent variables is useful for data analysis, because it permits the use of powerful tools of the theory of least squares for multiple linear regression analysis. However, when interpreting the results of regression analysis, we should think of the regression function in terms of the values of the explanatory variables.

Care should be exercised when interpreting the partial slopes. Interpreting a partial slope—say, β_i—as the expected change makes sense only if it is possible to change the value of X_i without changing the values of any of the other independent variables. For instance, in Example 11.1, the independent variables were $X_1 = C$ and $X_2 = C^2$, where C is the concentration of the enzyme being studied. Clearly, it does not make sense to talk about increasing X_1 by 1 unit without changing the value of X_2 at the same time. In the regression function in Equation (11.3), on the other hand, the interpretation of β_2 as the expected change in log (oxygen demand) corresponding simply to an increase of 1 g/liter in the level of B, the total Kjeldahl nitrogen, is practically meaningful, because the level of B could be increased in practice without changing the level of A, the biological oxygen demand.

Interpretation of expected responses

When considering the implications of a particular regression function or interpreting the results of a linear regression analysis (simple or multiple), we can think of a regression function as representing a surface in a $(k + 1)$-dimensional space, where k is the number of explanatory variables. This surface—called an *expected response surface*—is the region in the $(k + 1)$-dimensional space over which the expected responses can be varied by varying the levels of the explanatory variables. The shape, location, and orientation of an expected response surface help us to understand how the mean responses will vary with the levels of the explanatory variables.

For a simple linear regression function, the expected response surface is easy to visualize. There is only one ($k = 1$) explanatory variable, and so the surface is a one-dimensional region (a straight line) in a two-dimensional space. The location and orientation of the surface are determined by the intercept and slope parameters (β_0 and β_1) in the regression function.

Figure 11.2 shows some two-dimensional expected response surfaces that can be represented by multiple linear regression functions involving two explanatory variables; the a and b axes represent the levels of the explanatory variables, while the z axis represents the expected response $\mathcal{E}(Y|a, b)$.

Figure 11.2a shows an expected response surface when the regression function takes the form

$$\mathcal{E}(Y|a, b) = \beta_0 + \beta_1 a + \beta_2 b,$$

where β_0, β_1, and β_2 are all positive. Notice that the surface is a flat plane sloping upwards in the positive directions of the a and b axes. The actual magnitudes and

FIGURE 11.2

Two-dimensional expected response surfaces

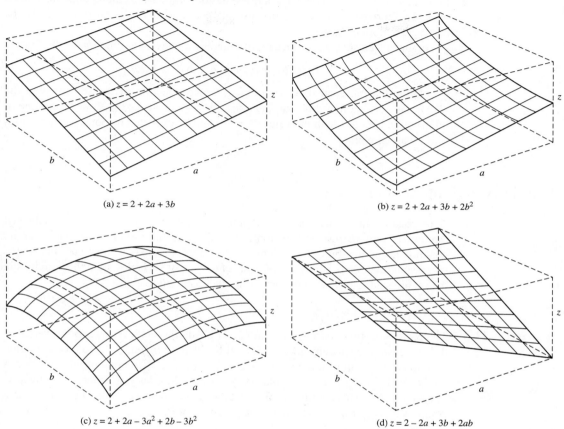

(a) $z = 2 + 2a + 3b$

(b) $z = 2 + 2a + 3b + 2b^2$

(c) $z = 2 + 2a - 3a^2 + 2b - 3b^2$

(d) $z = 2 - 2a + 3b + 2ab$

signs of the slopes are determined by the magnitudes and signs of β_1 and β_2. The magnitude and sign of β_0 determines the height of the surface at $a = b = 0$.

Figure 11.2b shows an expected response surface when the regression function is of the form

$$\mathcal{E}(Y|a, b) = \beta_0 + \beta_1 a + \beta_2 b + \beta_3 b^2,$$

where β_0, β_1, β_2, and β_3 are all positive. The addition of the quadratic term b^2 to the regression function has caused the surface to bend upwards along an axis perpendicular to the (a, z) plane. The form of the bend, as seen in the curve traced by the intersection of the expected response surface and the (b, z) plane, corresponds to a bowl-shaped quadratic curve, because regression coefficient β_3 is positive.

Figure 11.2c shows an expected response surface represented by a regression function of the form

$$\mathcal{E}(Y|a, b) = \beta_0 + \beta_1 a + \beta_2 a^2 + \beta_3 b + \beta_4 b^2,$$

where β_2 and β_4 are negative. The presence of the quadratic terms in a and b is the reason that the expected response surface has quadratic bends along both the a and b axes. The surface is dome-like because the coefficients of the quadratic terms (β_2 and β_4) are both negative.

Figure 11.2d shows a surface represented by a regression function of the form

$$\mathcal{E}(Y|a, b) = \beta_0 + \beta_1 a + \beta_2 b + \beta_3 ab.$$

Notice that, at each fixed value of a, the intersection of the surface with the (b, z) plane is a straight line but, in contrast to Figure 11.2a, the slope of the line of intersection changes with the value of a, so that the expected response surface looks like a warped plane.

Additivity and interaction

A regression function is said to be *additive* with respect to a set of variables if the terms in the regression function can be divided into groups such that no group contains terms involving the values of more than one explanatory variable. If the regression function is additive with respect to a set of variables, then the variables in the set are said to have *additive effects* on the expected response.

EXAMPLE **11.3**   The regression function in Equation (11.3) is additive with respect to the variables A, biological oxygen demand, and B, total Kjeldahl nitrogen, because this regression function is the sum of three components β_0, $\beta_1 a$, and $\beta_2 b$, none of which depends on the level of more than one explanatory variable. Similarly, the regression function in Equation (11.4) is additive with respect to A and B, because it is the sum of three components: β_0, $\beta_1 a + \beta_3 a^2$, and $\beta_2 b + \beta_4 b^2$.   ∎

When a regression function is additive with respect to a set of variables, the change in the expected response corresponding to a change in the level of only one of the variables remains the same at all levels of the other variables. For example, consider the additive regression function

$$\mathcal{E}(Y|a, b) = 1 + 3a + 2b \tag{11.6}$$

obtained by setting $\beta_0 = 1$, $\beta_1 = 3$, and $\beta_2 = 2$ in Equation (11.3). If the level of A is changed by one unit, the expected change in the response is

$$\mathcal{E}(Y|a + 1, b) - \mathcal{E}(Y|a, b) = [1 + 3(a + 1) + 2b] - (1 + 3a + 2b)$$

$$= 3,$$

which remains the same regardless of the value of b. Similarly, the result of increasing the level of A by one unit in the additive regression function

$$\mathcal{E}(Y|a, b) = 2 + a + 3a^2 + 2b^2 \tag{11.7}$$

is

$$\mathcal{E}(Y|a+1, b) - \mathcal{E}(Y|a, b)$$
$$= [2 + (a+1) + 3(a+1)^2 + 2b^2] - [2 + a + 3a^2 + 2b^2]$$
$$= 4 + 6a,$$

which depends on the level of *A* but is independent of the level of *B*.

Additivity in these two regression models is illustrated in Figure 11.3. In Figure 11.3a, the regression function in Equation (11.6) is used to plot the expected response against *b* when *a* = 1, 2, and 3. Notice that the three lines are parallel, with a vertical distance of 3 units between the lines. This indicates that the difference between the expected responses remains constant across values of *b*. Figure 11.3b shows a similar plot for the regression function in Equation (11.7). Again the lines are parallel, which indicates additivity. However, the change resulting from an increase of one unit in the value of *a* will be different when *a* = 1, 2, and 3. For instance, increasing the value of *a* from 1 to 2 will cause a change of 4 + 6(1) = 10 units in the expected response, whereas increasing *a* from 2 to 3 will produce a change of 4 + 6(2) = 16.

A regression function is said to be *interactive* with respect to a set of factors if the result of changing the level of one factor will depend on the levels of the other factors. An example of an interactive regression function is seen in Equation (11.5). The term $\beta_3 ab$ cannot be expressed as a sum of components that depend on the level of (at most) a single factor, and so the regression function is nonadditive. This regression function is shown in Figure 11.4 for *a* = 1, 2, and 3. Notice that the regression lines are not parallel; this indicates that the difference between the expected responses at any two values of *A* will depend on the value of *B*. A set of variables with respect to which a regression function is interactive is said to have an *interaction effect* on the expected response.

The next example illustrates an important practical implication of the additivity of a regression function.

EXAMPLE 11.4 An investigator was interested in modeling the expected yield *Y* of a particular crop as a function of *A*, the mean daily temperature, and *B*, the mean daily rainfall during the growing season. Two regression functions were considered as possible candidates for describing this relationship: function 1

$$\mathcal{E}(Y|a, b) = \mu(a, b) = \beta_0 + \beta_1 a + \beta_2 b$$

and function 2

$$\mathcal{E}(Y|a, b) = \mu(a, b) = \beta_0 + \beta_1 a + \beta_2 b + \beta_3 ab.$$

As we have already seen, function 1 is additive. Its additivity can be interpreted as follows. First, notice that the effect on the expected response of increasing *A* alone by one unit is

$$\mu(\Delta A) = \mu(a+1, b) - \mu(a, b)$$
$$= [\beta_0 + \beta_1(a+1) + \beta_2 b] - [\beta_0 + \beta_1 a + \beta_2 b] = \beta_1.$$

FIGURE   11.3

Plots of additive regression functions

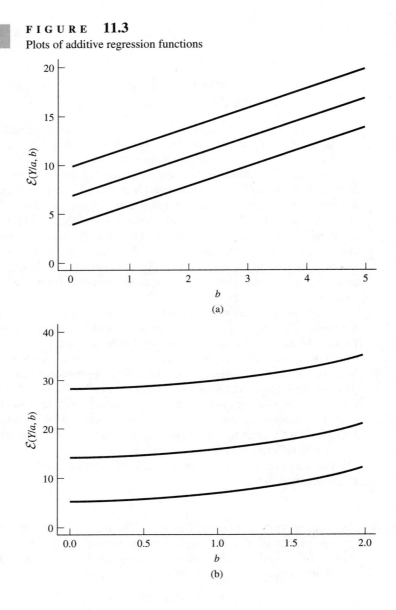

Similarly, $\mu(\Delta B)$, the corresponding effect of increasing B alone by one unit, is β_2. Finally, the effect of simultaneously increasing both A and B by one unit each—that is, changing (A, B) from (a, b) to $(a + 1, b + 1)$—is

$$
\begin{aligned}
\mu(\Delta A, \Delta B) &= \mu(a + 1, b + 1) - \mu(a, b) \\
&= \left[\beta_0 + \beta_1 (a + 1) + \beta_2 (b + 1) \right] - \left[\beta_0 + \beta_1 a + \beta_2 b \right] \\
&= \beta_1 + \beta_2 \\
&= \mu(\Delta A) + \mu(\Delta B).
\end{aligned}
$$

FIGURE 11.4

Plot of the nonadditive regression function $E(Y|a, b) = 1 + 3a + 2b - 2ab$

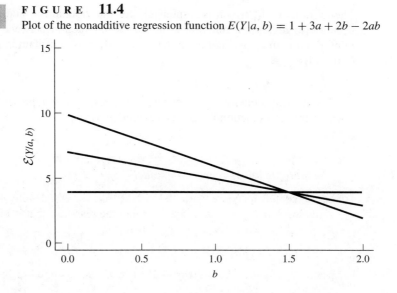

Thus, for regression function 1, the effect on the expected crop yield of simultaneously increasing the daily mean temperature and the daily mean rainfall by one unit each is equal to the sum of the effects of increasing by one unit the daily mean temperature and the daily mean rainfall individually. As an example, let $\beta_1 = 2$ and $\beta_2 = 3$. Then we can state the following conclusions:

1 If only the daily mean temperature is increased by one unit, then the expected yield will increase by two units.

2 If only the daily rainfall is increased by one unit, then the expected yield will increase by three units.

3 If the daily mean temperature and the daily mean rainfall are increased simultaneously by one unit each, then the expected crop yield will increase by $2 + 3 = 5$ units.

For regression function 2, the effect of increasing A alone by one unit is

$$\mu(\Delta A) = \beta_1 + \beta_3 b;$$

the effect of increasing B alone by one unit is

$$\mu(\Delta B) = \beta_2 + \beta_3 a;$$

and the effect of increasing A and B simultaneously by one unit is

$$\mu(\Delta A, \Delta B) = (\beta_1 + \beta_3 b) + (\beta_2 + \beta_3 a) + \beta_3$$
$$= \mu(\Delta A) + \mu(\Delta B) + \beta_3.$$

Thus, the effect of simultaneously increasing the values of A and B is not the same as the sum of the effects of increasing A and B individually. In symbols

$$\mu(\Delta A, \Delta B) \neq \mu(\Delta A) + \mu(\Delta B).$$

The extra term β_3, which is responsible for the violation of additivity, is the contribution to regression function 2 due to the interaction between temperature and rainfall. Of course, depending on the sign of β_3, this contribution could be positive or negative.   ■

A statistical model in which the expected response is represented by a multiple linear regression function is called a *multiple linear regression model* (Box 11.2).

BOX **11.2**

The multiple linear regression model

Let Y be the response variable and $X_1, X_2, \ldots, X_p$ denote p independent variables. For $i = 1, \ldots, n$, let Y_i denote the response that is observed when $X_1 = x_{i1}, X_2 = x_{i2}, \ldots, X_p = x_{ip}$. Then

$$Y_i = \beta_0 + \beta_1 x_{i1} + \beta_2 x_{i2} + \cdots + \beta_p x_{ip} + E_i, \qquad i = 1, \ldots, n,$$

where E_i is the random error in Y_i. The E_i are assumed to be independent $N(0, \sigma^2)$.

The assumptions in the multiple linear regression model in Box 11.2 are similar to those in one-way ANOVA and simple linear regression models. In each case, there are two sets of assumptions: a set of assumptions about the expected responses; and a set of assumptions about the errors. The assumptions about the errors, which we call the ANOVA assumptions, are the same for all three models. The differences between the three models arise from the assumptions about the expected responses.

The data needed in multiple linear regression analysis consist of n observations of the dependent (response) variable Y at n settings of the values of p independent variables $X_1, \ldots, X_p$. As in simple linear regression, the settings of the independent variables will be called the *design points*. Let y_i denote the observed value of Y when $X_1 = x_{i1}, \ldots, X_p = x_{ip}$. Then the observed data can be displayed as in Table 11.1.

TABLE   11.1

Data for fitting a multiple linear regression model

Observation	Observed response	Settings of the independent variables (design points)				
1	y_1	x_{11}	$\cdots$	x_{1j}	$\cdots$	x_{1p}
$\vdots$	$\vdots$	$\vdots$		$\vdots$		$\vdots$
i	y_i	x_{i1}	$\cdots$	x_{ij}		x_{ip}
$\vdots$	$\vdots$	$\vdots$		$\vdots$		$\vdots$
n	y_n	x_{n1}	$\cdots$	x_{nj}	$\cdots$	x_{np}

EXAMPLE **11.5**   On the basis of the quadratic regression function in Equation (11.2), a multiple linear regression model for the enzyme adsorption data in Example 11.1 can be written as

$$Y_i = \beta_0 + \beta_1 x_{i1} + \beta_2 x_{i2} + E_i, \qquad i = 1, 2, \ldots, 7,$$

where Y_i is the enzyme adsorption (%); x_{i1} is the concentration of the enzyme in the i-th medium; $x_{i2} = x_{i1}^2$; and E_i is the random error in the i-th measured response.

The data needed for a multiple linear regression analysis of enzyme adsorption on the basis of a quadratic regression function of enzyme concentration are as follows:

Observation number i	y_i	x_{i1}	x_{i2}
1	2.0	0.05	0.0025
2	5.0	0.10	0.0100
3	14.5	0.30	0.0900
4	18.6	0.50	0.2500
5	20.1	0.60	0.3600
6	22.0	0.80	0.6400
7	23.5	1.00	1.0000

■

There are $p + 2$ parameters ($p + 1$ regression parameters $\beta_0, \beta, \ldots, \beta_p$ and the error variance σ^2) in a multiple linear regression model. The methods for making inferences about these parameters are similar to those for a simple linear regression model, but the formulas are more complicated. The expressions for the multiple linear regression formulas may be considerably simplified by the use of matrix algebra. Although most of our discussion of multiple linear regression models in this book can be understood without a knowledge of matrix notation, we strongly urge you to learn some of the basic concepts of this powerful mathematical tool outlined in the next section. Familiarity with matrix notation will enable you to fully appreciate the power of regression analysis as applied to general linear models.

Exercises

11.1   Refer to the enzyme concentration data in Example 11.1.

a   Consider the *cubic* (so called because of the term c^3) regression function

$$\mathcal{E}(Y|C = c) = \mu(c) = \beta_0 + \beta_1 c + \beta_2 c^2 + \beta_3 c^3$$

as a possible model of the relationship between the expected adsorption and the enzyme concentration. Examine the shape of the regression function by plotting $\mathcal{E}(Y|C = c)$ against c for several selected values of $\beta_0, \beta_1, \beta_2$, and β_3. Comment on the appropriateness of the cubic regression function as compared to the linear and quadratic regression functions.

b   Show that the cubic regression function in (a) can be represented as a multiple linear regression function with $p = 3$ independent variables.

11.2   Show that each of the following regression functions can be expressed as a multiple linear regression function. In each case, indicate clearly how the independent variables will be defined.

a $\mathcal{E}(Y|a, b, c, d, f) = \beta_0 + \beta_1 a + \beta_2 b + \beta_3 c + \beta_4 d + \beta_5 f$;

b $\mathcal{E}(Y|a, b, c) = \beta_0 + \beta_1 a + \beta_2 b + \beta_3 c + \beta_4 ab + \beta_5 ac + \beta_6 bc$;

c $\mathcal{E}(Y|a, b, c) = \beta_0 + \beta_1 a + \beta_2 b + \beta_3 c + \beta_4 ab + \beta_5 ac + \beta_6 bc + \beta_7 a^2 + \beta_8 b^2 + \beta_9 c^2$;

d $\mathcal{E}(Y|a, b, c, d, f) = \beta_0 + \beta_1 a + \beta_2 b + \beta_3 c + \beta_4 d + \beta_5 f + \beta_6 ab + \beta_7 ac + \beta_8 ad + \beta_9 af + \beta_{10} bc + \beta_{11} bd + \beta_{12} bf + \beta_{13} cd + \beta_{14} cf + \beta_{15} df$.

11.3   Consider the four different regression functions in Exercise 11.2 as possible candidates for expressing the expected log (oxygen demand) as a function of five explanatory variables in Example 11.2.

a   Classify each regression function as additive or interactive. Give reasons.

b   Interpret the parameters β_1 and β_2 in Exercise 11.2a.

c   Interpret the quantity $\beta_1 + \beta_2 + \beta_3$ in Exercise 11.2a.

d   Explain why, in Exercise 11.2b, the parameter β_4 can be interpreted as a measure of the effect of the interaction between A and B.

11.4   Refer to regression function 2 in Example 11.4. For the purpose of this problem, assume that $\beta_0 = 2$, $\beta_1 = 3$, $\beta_2 = 4$, and $\beta_3 = -1$.

a   For $b = 0, 1, 2, 3, 4, 5$, calculate the change in the expected crop yield resulting from changing the value of (A, B) from (a, b) to $(a + 1, b)$. Plot the six calculated values against the values of B. Notice that the change in the expected yield is not the same at all values of B.

b   For $a = 0, 1, 2, 3, 4, 5$, calculate the change in the expected crop yield resulting from changing the value of (A, B) from (a, b) to $(a, b + 1)$. Plot the six calculated values against the values of A. Notice that the change in the expected yield is not the same at all values of A.

c   For $(a, b) = (0, 0), (1, 1), \ldots, (5, 5)$, calculate the change in the expected yield when the value of (A, B) is changed from (a, b) to $(a + 1, b + 1)$.

d   On the basis of the values calculated in (a)–(c), argue that the regression function is not additive.

11.5   Give the multiple linear regression models that correspond to the four regression functions in Exercise 11.2.

11.3
Matrix algebra

Matrix algebra is very useful when working simultaneously with several linear relationships. In this section, we look briefly at some aspects of matrix algebra that prove convenient in the statistical theory of linear models.

Definition and notation

A *matrix* is a rectangular array of objects called *elements* arranged in rows and columns.

EXAMPLE 11.6   Here are some examples of matrices

$$A = \begin{bmatrix} 2 & 4 & 6 \end{bmatrix}; \quad B = \begin{bmatrix} 3 \\ 2 \\ 5 \end{bmatrix}; \quad C = \begin{bmatrix} 3 & 2 \\ 6 & 4 \end{bmatrix}; \quad D = \begin{bmatrix} 3 & 8 \\ 8 & 0 \end{bmatrix};$$

$$E = \begin{bmatrix} 3 & 4 & 1 & 0 \\ 2 & 0 & 6 & 2 \end{bmatrix}; \quad F = \begin{bmatrix} 0 & 2 & 6 \\ 1 & 4 & 0 \\ 3 & 0 & 1 \end{bmatrix}; \quad G = \begin{bmatrix} 2 & 6 & 3 \\ 6 & 4 & 1 \\ 3 & 1 & 7 \end{bmatrix}; \quad H = \begin{bmatrix} 4 & 0 & 0 \\ 0 & 2 & 0 \\ 0 & 0 & 5 \end{bmatrix};$$

$$I = \begin{bmatrix} 1 & 0 & 0 \\ 0 & 1 & 0 \\ 0 & 0 & 1 \end{bmatrix}; \quad O = \begin{bmatrix} 0 & 0 & 0 & 0 \\ 0 & 0 & 0 & 0 \end{bmatrix}.$$

As we see, a matrix can have any number of rows and columns. For instance, A is a matrix with one row and three columns, whereas E is a matrix with two rows and four columns.   ■

A matrix with r rows and c columns is said to be of *dimension* $r \times c$. For instance, in Example 11.6, the dimensions of A, B, and E are 1×3, 3×1, and 2×4, respectively.

A matrix that has the same number of rows and columns is called a *square matrix*. A square matrix with r rows (and r columns) is said to be of dimension r. In Example 11.6, C, D, F, G, I, and H are square matrices of dimensions 2, 2, 3, 3, 3, and 3, respectively.

A matrix with a single row of c elements is called a *row vector* of dimension c. In Example 11.6, A is a row vector of dimension 3. A matrix with a single column of r elements is called a *column vector* of dimension r. In Example 11.6, B is a column vector of dimension 3.

The *transpose* of a matrix M, denoted by M', is defined as a matrix whose columns are the rows of M. Therefore, if M has dimension $r \times c$, then M' has dimension $c \times r$. Also, the transpose of a row vector is a column vector, and vice versa. The transposes of A, B, C, and E in Example 11.6 are

$$A' = \begin{bmatrix} 2 \\ 4 \\ 6 \end{bmatrix}; \quad B' = \begin{bmatrix} 3 & 2 & 5 \end{bmatrix}; \quad C' = \begin{bmatrix} 3 & 6 \\ 2 & 4 \end{bmatrix}; \quad E' = \begin{bmatrix} 3 & 2 \\ 4 & 0 \\ 1 & 6 \\ 0 & 2 \end{bmatrix}.$$

The element in the i-th row and the j-th column of a matrix is called the *ij*-element and is usually denoted by an appropriate symbol with two subscripts, the first for the row number and the second for the column number. For instance, the *ij*-element of a matrix M might be denoted by the symbol m_{ij}. Thus, a typical element of matrix E in Example 11.6 can be denoted as e_{ij}, where i takes the values 1 and 2 and j takes the values 1, 2, 3, and 4. The element common to the first row and the third column is 1, so that $e_{13} = 1$. Similarly, $e_{22} = 0$, $e_{24} = 2$, and so on. The elements of E can be displayed as follows

$$\begin{bmatrix} 3 & 4 & 1 & 0 \\ 2 & 0 & 6 & 2 \end{bmatrix} = \begin{bmatrix} e_{11} & e_{12} & e_{13} & e_{14} \\ e_{21} & e_{22} & e_{23} & e_{24} \end{bmatrix}.$$

The *ii*-element of a square matrix is called the *i*-th *diagonal element*. In Example 11.6, the first and second diagonal elements of C are 3 and 4, respectively. A *diagonal matrix* is a square matrix whose nondiagonal elements are all 0. In Example 11.6, H and I are both diagonal matrices.

An *identity matrix* is a diagonal matrix in which all the diagonal elements are 1. We denote an identity matrix of dimension r as I or $I(r)$. In Example 11.6, I is an identity matrix of dimension 3 and may also be denoted as $I(3)$.

A matrix in which all the elements are 0 is called a *null matrix*. A null matrix of dimension $(r \times c)$ will be denoted by either O or $O(r \times c)$. In Example 11.6, O is a 2×4 null matrix and may also be denoted as $O(2 \times 4)$. In Exercise 11.7, you will be asked to verify that the identity and null matrices play roles similar to those played by the numbers 1 and 0.

A square matrix is said to be *symmetric* if its *ij*-element is the same as its *ji*-element; that is, the element in the *i*-th row and *j*-th column of a symmetric matrix is the same as the element in its *j*-th row and *i*-th column. Therefore, if (m_{ij}) is a symmetric matrix, then $m_{ij} = m_{ji}$ for all i and j. Also, a symmetric matrix is equal to its transpose. Thus, if M is symmetric, then $M = M'$. The reader should verify that G in Example 11.6 is a symmetric matrix of dimension 3.

Matrix operations

Let $M = (m_{ij})$ and $N = (n_{ij})$ be two matrices of dimension $r \times c$.

Multiplication by a constant

If k is a constant, multiplying M by k is the same as multiplying every element of M by k. Thus $kM = P$, where the *ij*-element of P is $p_{ij} = km_{ij}$. For example, the result of multiplying E in Example 11.6 by $k = 4$ is

$$4E = \begin{bmatrix} 4(3) & 4(4) & 4(1) & 4(0) \\ 4(2) & 4(0) & 4(6) & 4(2) \end{bmatrix} = \begin{bmatrix} 12 & 16 & 4 & 0 \\ 8 & 0 & 24 & 8 \end{bmatrix}.$$

Addition of matrices

Two matrices of identical dimension are added by adding their corresponding elements. Thus, the *ij*-element of $P = M + N$ is $p_{ij} = m_{ij} + n_{ij}$. For instance, the sum of C and D in Example 11.6 is

$$C + D = \begin{bmatrix} 3 & 2 \\ 6 & 4 \end{bmatrix} + \begin{bmatrix} 3 & 8 \\ 8 & 0 \end{bmatrix}$$

$$= \begin{bmatrix} 3+3 & 2+8 \\ 6+8 & 4+0 \end{bmatrix}$$

$$= \begin{bmatrix} 6 & 10 \\ 14 & 4 \end{bmatrix}.$$

Subtraction of matrices

The difference between two matrices of the same dimension is obtained by taking the difference between the corresponding elements. For instance, the difference between C and D in Example 11.6 is

$$C - D = \begin{bmatrix} 3 & 2 \\ 6 & 4 \end{bmatrix} - \begin{bmatrix} 3 & 8 \\ 8 & 0 \end{bmatrix}$$

$$= \begin{bmatrix} 3-3 & 2-8 \\ 6-8 & 4-0 \end{bmatrix}$$

$$= \begin{bmatrix} 0 & -6 \\ -2 & 4 \end{bmatrix}.$$

Multiplication of matrices

As its name implies, *row-column multiplication* is an operation in which a row vector is multiplied by a column vector. Let

$$R = [a_1, a_2, \ldots, a_r], \qquad C = \begin{bmatrix} b_1 \\ b_2 \\ \vdots \\ b_r \end{bmatrix}$$

be a row vector and a column vector, both of the same dimension r. The product of R and C, denoted by RC, is the sum of the products of the corresponding elements of R and C.

$$RC = a_1 b_1 + a_2 b_2 + \cdots + a_r b_r. \tag{11.8}$$

For instance, the row vector A and the column vector B in Example 11.6 both have dimension $r = 3$, so that the row-column product AB can be computed

$$AB = (2)(3) + (4)(2) + (6)(5) = 44.$$

Note that a row-column product is an *ordered operation*, in which the row vector precedes the column vector to produce the product. In general, the column-row product CR, in which the column precedes the row, is not the same as the row-column product RC.

Let M and N be two matrices such that the number of columns in M is equal to the number of rows in N. Assume that the dimensions of M and N are $r \times k$ and $k \times c$, respectively. Let $R_1, R_2, \ldots, R_r$ denote the rows of M and $C_1, C_2, \ldots, C_c$ denote the columns of N. Then the product of M by N, taken in that order, is denoted by MN. The product MN is an $r \times c$ matrix whose ij-element is the row-column product of the i-th row of M and the j-th column of N; that is, $MN = (p_{ij})$, where $p_{ij} = R_i C_j$.

As an example of matrix multiplication, consider the matrices C (of dimension 2×2) and E (of dimension 2×4) in Example 11.6. Since the number of columns ($k = 2$) in C equals the number of rows in E, the product CE can be calculated. The product is a matrix of dimension 2×4. Let

$$R_1 = \begin{bmatrix} 3 & 2 \end{bmatrix}; \quad R_2 = \begin{bmatrix} 6 & 4 \end{bmatrix}$$

denote the rows of C and

$$C_1 = \begin{bmatrix} 3 \\ 2 \end{bmatrix}; \quad C_2 = \begin{bmatrix} 4 \\ 0 \end{bmatrix}; \quad C_3 = \begin{bmatrix} 1 \\ 6 \end{bmatrix}; \quad C_4 = \begin{bmatrix} 0 \\ 2 \end{bmatrix}$$

denote the columns of E.

Then

$$CE = \begin{bmatrix} R_1 C_1 & R_1 C_2 & R_1 C_3 & R_1 C_4 \\ R_2 C_1 & R_2 C_2 & R_2 C_3 & R_2 C_4 \end{bmatrix},$$

where the elements are calculated using the row-column multiplication rule. For instance, the element in the first row and second column of CE is

$$R_1 C_2 = (3)(4) + (2)(0) = 12.$$

Straightforward calculations show that

$$CE = \begin{bmatrix} 13 & 12 & 15 & 4 \\ 26 & 24 & 30 & 8 \end{bmatrix}.$$

As a second example, consider the column-row product BA. Since B and A are 3×1 ($r = 3$, $k = 1$) and 1×3 ($k = 1$, $c = 3$) matrices, the product is a 3×3 matrix. Actual multiplication can be carried out by multiplying the one-dimensional row vectors of B by the one-dimensional column vectors of A. The result is

$$BA = \begin{bmatrix} 3 \\ 2 \\ 5 \end{bmatrix} \begin{bmatrix} 2 & 4 & 6 \end{bmatrix} = \begin{bmatrix} (3)(2) & (3)(4) & (3)(6) \\ (2)(2) & (2)(4) & (2)(6) \\ (5)(2) & (5)(4) & (5)(6) \end{bmatrix} = \begin{bmatrix} 6 & 12 & 18 \\ 4 & 8 & 12 \\ 10 & 20 & 30 \end{bmatrix}.$$

As we have seen, a row-column product requires the row and column to have the same dimension. By contrast, a column-row product requires no such restrictions. Indeed, whereas a row-column product is a single number, the *column-row product* of an r-dimensional column and a c-dimensional row is a matrix of dimension $r \times c$ (see Exercise 11.8).

In the next example, we see how matrix algebra can be used in the case of simple linear regression. Of course, as already noted, there is no need to use matrix notation in that case; its benefits become apparent when working with multiple linear regression models.

EXAMPLE 11.7   For $i = 1, \ldots, n$, let Y_i denote the response that will be observed at the value x_i of the independent variable. Then the simple linear regression model in Box 10.2 can be expressed as a set of n equations

$$Y_1 = \beta_0 + \beta_1 x_1 + E_1,$$

$$Y_2 = \beta_0 + \beta_1 x_2 + E_2,$$

$$\vdots \quad \vdots \qquad \vdots \quad \vdots$$

$$Y_i = \beta_0 + \beta_1 x_i + E_i,$$

$$\vdots \quad \vdots \qquad \vdots \quad \vdots$$

$$Y_n = \beta_0 + \beta_1 x_n + E_n.$$

Recall that β_0 and β_1 are the regression parameters and the E_i are the random errors.

Note that the set of n responses on the left-hand side of the equations can be expressed as a column vector

$$Y = \begin{bmatrix} Y_1 \\ Y_2 \\ \vdots \\ Y_n \end{bmatrix},$$

whereas the right-hand side can be expressed as the sum of two column vectors:

$$\begin{bmatrix} \beta_0 + \beta_1 x_1 \\ \beta_0 + \beta_1 x_2 \\ \vdots \\ \beta_0 + \beta_1 x_n \end{bmatrix} + \begin{bmatrix} E_1 \\ E_2 \\ \vdots \\ E_n \end{bmatrix}. \tag{11.9}$$

Now let

$$X = \begin{bmatrix} 1 & x_1 \\ 1 & x_2 \\ \vdots & \vdots \\ 1 & x_n \end{bmatrix} \qquad \text{and} \qquad \beta = \begin{bmatrix} \beta_0 \\ \beta_1 \end{bmatrix}.$$

Then the first column vector in Equation (11.9) is the product $X\beta$. Therefore, if we let

$$E = \begin{bmatrix} E_1 \\ E_2 \\ \vdots \\ E_n \end{bmatrix}$$

be a vector of errors that are independent $N(0, \sigma^2)$, the simple linear regression model can be expressed as

$$Y = X\beta + E.$$

As we saw in Equation (10.5), the normal equations for estimation of the regression parameters take the form

$$\sum y_i = n\hat{\beta}_0 + \left(\sum x_i\right)\hat{\beta}_1,$$
$$\sum x_i y_i = \left(\sum x_i\right)\hat{\beta}_0 + \left(\sum x_i^2\right)\hat{\beta}_1.$$

Let

$$y = \begin{bmatrix} y_1 \\ y_2 \\ \vdots \\ y_n \end{bmatrix}$$

denote the column vector of observed responses. Then the left-hand side of the normal equations can be expressed in the form

$$X'y = \begin{bmatrix} 1 & 1 & \cdots & 1 \\ x_1 & x_2 & \cdots & x_n \end{bmatrix} \begin{bmatrix} y_1 \\ y_2 \\ \vdots \\ y_n \end{bmatrix}.$$

Let

$$\hat{\beta} = \begin{bmatrix} \hat{\beta}_1 \\ \hat{\beta}_2 \end{bmatrix}$$

denote the column vector of the estimates of regression parameters. Then, noting that

$$X'X = \begin{bmatrix} 1 & 1 & \cdots & 1 \\ x_1 & x_2 & \cdots & x_n \end{bmatrix} \begin{bmatrix} 1 & x_1 \\ 1 & x_2 \\ \vdots & \vdots \\ 1 & x_n \end{bmatrix} = \begin{bmatrix} n & \sum x_i \\ \sum x_i & \sum x_i^2 \end{bmatrix},$$

the right-hand side of the normal equations can be expressed as the product

$$X'X\hat{\beta} = \begin{bmatrix} n & \sum x_i \\ \sum x_i & \sum x_i^2 \end{bmatrix} \begin{bmatrix} \hat{\beta}_0 \\ \hat{\beta}_1 \end{bmatrix}.$$

Therefore, the normal equations take the matrix form

$$X'y = X'X\hat{\beta}. \quad \blacksquare$$

A similar matrix formulation can be obtained for the multiple linear regression model (Box 11.3).

BOX **11.3**   *Matrix formulation of multiple linear regression model*

Let Y_i denote the value that will be observed for the dependent variable Y

when the independent variables have the values $X_1 = x_{i1}, \ldots,$ and $X_p = x_{ip}$, and define

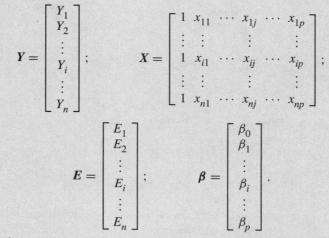

$$Y = \begin{bmatrix} Y_1 \\ Y_2 \\ \vdots \\ Y_i \\ \vdots \\ Y_n \end{bmatrix}; \qquad X = \begin{bmatrix} 1 & x_{11} & \cdots & x_{1j} & \cdots & x_{1p} \\ \vdots & \vdots & & \vdots & & \vdots \\ 1 & x_{i1} & \cdots & x_{ij} & \cdots & x_{ip} \\ \vdots & \vdots & & \vdots & & \vdots \\ 1 & x_{n1} & \cdots & x_{nj} & \cdots & x_{np} \end{bmatrix};$$

$$E = \begin{bmatrix} E_1 \\ E_2 \\ \vdots \\ E_i \\ \vdots \\ E_n \end{bmatrix}; \qquad \beta = \begin{bmatrix} \beta_0 \\ \beta_1 \\ \vdots \\ \beta_i \\ \vdots \\ \beta_p \end{bmatrix}.$$

Then the multiple linear regression model with regression parameters $\beta_0, \beta_1, \ldots, \beta_p$ and errors $E_1, \ldots, E_n$ can be represented as

$$Y = X\beta + E,$$

where E is a column vector of errors that are independent $N(0, \sigma^2)$.

For obvious reasons, the column vectors Y, β, and E in Box 11.3 are called the *observation vector*, the *parameter vector*, and the *error vector*, respectively. The matrix X is called the *design matrix*, because its rows identify the *design points*, the settings of the independent variables at which the responses are to be measured.

EXAMPLE 11.8 The $n = 8$ measured values of a response variable Y at the $n = 8$ settings of two independent variables X_1 and X_2 are as follows:

i	1	2	3	4	5	6	7	8
x_{1i}	4.2	3.6	4.7	5.2	4.6	5.0	5.4	7.0
x_{2i}	6.9	2.3	7.6	5.8	8.1	8.6	1.7	6.9
y_i	1.7	1.9	2.1	2.8	2.2	2.9	3.4	4.1

Let Y_i denote the response that will be observed at the settings $X_1 = x_{i1}$ and $X_2 = x_{i2}$. A multiple linear regression model for the Y_i can be expressed as

$$Y_i = \beta_0 + \beta_1 x_{i1} + \beta_2 x_{i2} + E_i, \qquad i = 1, \ldots, 8,$$

or as

$$Y = X\beta + E,$$

where

$$X = \begin{bmatrix} 1 & 4.2 & 6.9 \\ 1 & 3.6 & 2.3 \\ 1 & 4.7 & 7.6 \\ 1 & 5.2 & 5.8 \\ 1 & 4.6 & 8.1 \\ 1 & 5.0 & 8.6 \\ 1 & 5.4 & 1.7 \\ 1 & 7.0 & 6.9 \end{bmatrix}$$

is the design matrix and

$$Y = \begin{bmatrix} Y_1 \\ Y_2 \\ \vdots \\ Y_8 \end{bmatrix}, \qquad \beta = \begin{bmatrix} \beta_0 \\ \beta_1 \\ \beta_2 \end{bmatrix}, \qquad \text{and} \qquad E = \begin{bmatrix} E_1 \\ E_2 \\ \vdots \\ E_8 \end{bmatrix}$$

are, respectively, the response vector (of dimension 8), the parameter vector (of dimension 3), and the error vector (of dimension 8). The vector of observed responses is a column vector of dimension 8

$$y = \begin{bmatrix} 1.7 \\ 1.9 \\ 2.1 \\ 2.8 \\ 2.2 \\ 2.9 \\ 3.4 \\ 4.1 \end{bmatrix}. \quad \blacksquare$$

Exercises

11.6 Determine the result of performing the following operations on the matrices in Example 11.6, if appropriate. If not, so indicate.

a $2C + 4D$;   b $4F - G$;   c $2G$;   d $E + F$;
e $A' + 2B$;   f $A + 2B'$;   g IG;   h IE;
i EE' and $E'E$;   j AA' and $A'A$.

11.7 The results of matrix operations with identity and null matrices are similar to the results of algebraic operations involving the numbers 1 and 0. Verify the following results for the matrices in Example 11.6.

a $E + O = E$ (adding a null matrix to any matrix will not alter that matrix);

b $OE' = O$ and $O'E = O'$ (multiplying any matrix by a null matrix will reduce that matrix to a null matrix);

c $FI = F$ and $IG = G$ (multiplying any matrix by an identity matrix will not alter that matrix).

11.8   Let

$$
C = \begin{bmatrix} c_1 \\ \vdots \\ c_r \end{bmatrix}, \quad R = \begin{bmatrix} r_1 & \cdots & r_c \end{bmatrix}
$$

denote a column vector of dimension r and a row vector of dimension c, respectively. Use the definition of a matrix product to exhibit the elements of CR. What is the dimension of CR?

11.9   For the design matrix in Example 11.8, display the matrices $X\beta$ and $X'y$.

11.10   Display the design matrix associated with the multiple linear regression model in Example 11.5.

11.4
Estimating parameters
Estimation of regression parameters

The method of least squares can be used to estimate the parameters in a multiple linear regression model. As in the case of a simple linear regression model, $\hat{\beta}_0, \ldots, \hat{\beta}_p$, the least squares estimates of the regression coefficients, are determined by minimizing the sum of squares of the residuals

$$
\mathrm{SS[E]} = \sum_{i=1}^{n} \left(y_i - \hat{y}_i \right)^2
$$

$$
= \sum_{i=1}^{n} \left[y_i - \left(\hat{\beta}_0 + \hat{\beta}_1 x_{i1} + \cdots + \hat{\beta}_p x_{ip} \right) \right]^2. \tag{11.10}
$$

It can be shown that the least squares estimates can be obtained by solving the $p + 1$ equations

$$
\sum_{k=1}^{n} y_k = n\hat{\beta}_0 + \left(\sum_{k=1}^{n} x_{k1} \right) \hat{\beta}_1 + \cdots + \left(\sum_{k=1}^{n} x_{kj} \right) \hat{\beta}_j + \cdots + \left(\sum_{k=1}^{n} x_{kp} \right) \hat{\beta}_p,
$$

$$
\sum_{k=1}^{n} x_{k1} y_k = \left(\sum_{k=1}^{n} x_{k1} \right) \hat{\beta}_0 + \left(\sum_{k=1}^{n} x_{k1}^2 \right) \hat{\beta}_1 + \cdots + \left(\sum_{k=1}^{n} x_{k1} x_{kj} \right) \hat{\beta}_j + \cdots + \left(\sum_{k=1}^{n} x_{k1} x_{kp} \right) \hat{\beta}_p,
$$

$$
\vdots \qquad \vdots \qquad \vdots \qquad \vdots \qquad \vdots
$$

$$
\sum_{k=1}^{n} x_{kj} y_k = \left(\sum_{k=1}^{n} x_{kj} \right) \hat{\beta}_0 + \left(\sum_{k=1}^{n} x_{kj} x_{k1} \right) \hat{\beta}_1 + \cdots + \left(\sum_{k=1}^{n} x_{kj}^2 \right) \hat{\beta}_j + \cdots + \left(\sum_{k=1}^{n} x_{kj} x_{kp} \right) \hat{\beta}_p,
$$

$$
\vdots \qquad \vdots \qquad \vdots \qquad \vdots \qquad \vdots \tag{11.11}
$$

$$
\sum_{k=1}^{n} x_{kp} y_k = \left(\sum_{k=1}^{n} x_{kp} \right) \hat{\beta}_0 + \left(\sum_{k=1}^{n} x_{kp} x_{k1} \right) \hat{\beta}_1 + \cdots + \left(\sum_{k=1}^{n} x_{kp} x_{kj} \right) \hat{\beta}_j + \cdots + \left(\sum_{k=1}^{n} x_{kp}^2 \right) \hat{\beta}_p.
$$

These are called the *normal equations*. When $p = 1$, they reduce to the normal equations for the simple linear regression model. Solving the normal equations for a multiple linear regression model can be a tedious computational task. Whereas simple formulas for the least squares estimates are available for the simple linear regression model (Box 10.3), the corresponding calculations for the multiple linear regression model are best performed using an appropriate statistical software package. The next example illustrates the calculation of the least square estimates for the regression parameters by the SAS software; details of the corresponding SAS codes are given in Chapter 11 of Younger (1997).

EXAMPLE **11.9**   As noted in Example 11.5, the quadratic model for the enzyme adsorption data in Example 11.1 can be written as

$$Y_i = \beta_0 + \beta_1 x_{i1} + \beta_2 x_{i2} + E_i, \qquad i = 1, 2, \ldots, 7,$$

where Y_i is the enzyme adsorption (%); x_{i1} is the concentration of the enzyme in the i-th medium; $x_{i2} = x_{i1}^2$; and E_i is the random error in the i-th measured response.

The normal equations for fitting the quadratic model can be obtained by setting $n = 7$ and $p = 2$ in Equation (11.11)

$$\sum_{i=1}^{7} y_i = 7\hat{\beta}_0 + \left(\sum_{i=1}^{7} x_{i1}\right)\hat{\beta}_1 + \left(\sum_{i=1}^{7} x_{i2}\right)\hat{\beta}_2,$$

$$\sum_{i=1}^{7} x_{i1}y_i = \left(\sum_{i=1}^{7} x_{i1}\right)\hat{\beta}_0 + \left(\sum_{i=1}^{7} x_{i1}^2\right)\hat{\beta}_1 + \left(\sum_{i=1}^{7} x_{i1}x_{i2}\right)\hat{\beta}_2,$$

$$\sum_{i=1}^{7} x_{i2}y_i = \left(\sum_{i=1}^{7} x_{i2}\right)\hat{\beta}_0 + \left(\sum_{i=1}^{7} x_{i2}x_{i1}\right)\hat{\beta}_1 + \left(\sum_{i=1}^{7} x_{i2}^2\right)\hat{\beta}_2.$$

Straightforward calculations using the data in Example 11.5 (see Exercise 11.13) show that the required normal equations are

$$105.700 = 7\hat{\beta}_0 + 3.35\hat{\beta}_1 + 2.3525\hat{\beta}_2,$$

$$67.410 = 3.35\hat{\beta}_0 + 2.3525\hat{\beta}_1 + 1.8811\hat{\beta}_2,$$

$$50.826 = 2.3525\hat{\beta}_0 + 1.8811\hat{\beta}_1 + 1.6099\hat{\beta}_2.$$

Thus, to estimate the regression coefficients, we need to solve three simultaneous equations. The following is a portion of the output when the SAS REG procedure is used to analyze these data.

```
                      Part I
                Parameter Estimates
                                 Parameter
          Variable  DF           Estimate

          INTERCEP   1           0.172006
          X1         1          51.261041
```

```
        X2        1     -28.577485

                  Part II
              Dep Var   Predict
       Obs      Y        Value  Residual
        1     2.0000     2.6636  -0.6636
        2     5.0000     5.0123  -0.0123
        3    14.5000    12.9783   1.5217
        4    18.6000    18.6582  -0.0582
        5    20.1000    20.6407  -0.5407
        6    22.0000    22.8912  -0.8912
        7    23.5000    22.8556   0.6444
```

```
Sum of Residuals                   0
Sum of Squared Residuals         4.2614
```

The least squares estimates of β_0, β_1, and β_2 are seen to be

$$\hat{\beta}_0 = 0.17, \quad \hat{\beta}_1 = 51.26, \quad \text{and} \quad \hat{\beta}_2 = -28.58.$$

The estimated least squares prediction equation is

$$\hat{y} = 0.17 + 51.26x_1 - 28.58x_2,$$

where x_1 and x_2 are, respectively, the enzyme concentration x and its square x^2. The least squares equation can be written in terms of x alone as

$$\hat{y} = 0.17 + 51.26x - 28.58x^2.$$

Figure 11.5 shows a plot of the estimated quadratic regression function, along with the observed adsorption data. The quadratic function seems to fit the data very well.

As we saw in Chapter 10, how well the estimated line fits the data can be evaluated by examining the residuals. The SAS printout gives the observed value y_i, the predicted value $\hat{y}_i$, and the residual value $e_i = y_i - \hat{y}_i$ for $i = 1, \ldots, 7$. For example, when $i = 3$, we have $y_3 = 14.5$, $\hat{y}_3 = 0.17 + 51.26(0.30) - 28.58(0.30)^2 = 12.9783$, and $e_3 = 14.5 - 12.9783 = 1.5217$. The sum of squares of the residuals is $SS[E] = \sum e_i^2 = 4.2614$. ■

Estimation of error variance

In addition to the estimates of the regression parameters, an estimate of the error variance σ^2 is needed in a regression analysis. Theorem B.6d in Appendix B can be used to show that, if $n \geq p + 2$ and the normal equations have a unique solution, then an unbiased estimate of σ^2 is

$$\hat{\sigma}^2 = MS[E] = \frac{SS[E]}{n - (p + 1)}, \tag{11.12}$$

FIGURE 11.5

Least squares quadratic regression function

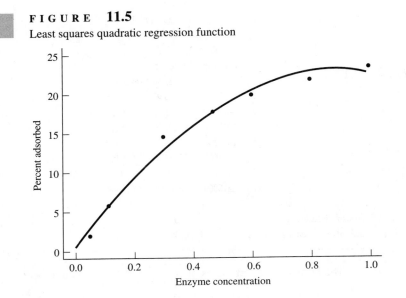

where n and p are, respectively, the number of observations and the number of independent variables used to fit the multiple linear regression model. Notice that $p + 1$ is the number of regression parameters in the model and, if $p = 1$, the estimate in Equation (11.12) reduces to the estimate of the error variance in a simple linear regression model.

To ensure that $\hat{\sigma}^2$ is an unbiased estimate of the error variance, two conditions must be satisfied. The first condition—that the number of observations n must be at least $p + 2$—ensures that the data contain enough information to estimate the $p + 2$ parameters in the multiple linear regression model. The second condition—that the normal equations have a unique solution—rules out the possibility that more than one set of $\hat{\beta}$ values may satisfy the normal equations. It can be shown that a set of normal equations can always be solved, but it may have more than one solution in some instances. Specifically, it will have more than one solution when the $p + 1$ normal equations are linearly dependent—that is, when one or more of the equations can be obtained as linear combinations of the others. For example, suppose we want to fit the model

$$Y = \beta_0 + \beta_1 x_1 + \beta_2 x_2 + E$$

to the following data:

i	x_{1i}	x_{2i}	y_i
1	1	2	2
2	2	3	2
3	3	4	3
4	4	5	3

The corresponding normal equations are

$$10 = 4\hat{\beta}_0 + 10\hat{\beta}_1 + 14\hat{\beta}_2,$$
$$27 = 10\hat{\beta}_0 + 30\hat{\beta}_1 + 40\hat{\beta}_2,$$
$$37 = 14\hat{\beta}_0 + 40\hat{\beta}_1 + 54\hat{\beta}_2.$$

The normal equations are linearly dependent, because the third equation is the sum of the first two equations. The solution to these equations is not unique, because $(\hat{\beta}_0 = 3/2, \hat{\beta}_1 = 2/5, \hat{\beta}_2 = 0)$ and $(\hat{\beta}_0 = 1/2, \hat{\beta}_1 = -3/5, \hat{\beta}_2 = 1)$ both satisfy the equations.

If the solutions to the normal equations are nonunique, it means that the available data are not sufficient for unique estimation of the regression parameters. In this situation, the multiple linear regression model is said to be of less than full rank. In such a case, we should either redefine the regression function with fewer parameters or collect more data so as to ensure unique estimates of the regression parameters. Throughout this chapter, it will be assumed that the multiple linear regression models have full ranks. Models of less than full rank will be considered in the next chapter.

EXAMPLE 11.10   For the enzyme adsorption data, as we saw in Example 11.9, $n = 7$, $p = 2$, and SS[E] = 4.2614. Therefore, the estimated error variance is

$$\hat{\sigma}^2 = \text{MS[E]} = \frac{4.2614}{7 - (2 + 1)} = 1.0654. \quad \blacksquare$$

Multiple linear regression in an ANOVA setting

In Section 10.5, we saw that, in a simple linear regression setting, the total sum of squares SS[TOT] can be partitioned into two components: the regression sum of squares SS[R]; and the error sum of squares SS[E]. These two sums of squares can be interpreted as the portions of the variation in the observed responses that can be explained by the regression model and the random error, respectively. Similar partitioning is possible for a multiple linear regression model. Table 11.2 is the

TABLE   11.2

ANOVA table for testing H_0: $\beta_i = 0$, $i = 1, \ldots, p$

Source	Degrees of freedom	Sum of squares	Mean squares
Regression	p	SS[R]	$\text{MS[R]} = \dfrac{\text{SS[R]}}{p}$
Error	$n - p - 1$	SS[E]	$\text{MS[E]} = \dfrac{\text{SS[E]}}{n - p - 1}$
Total	$n - 1$	SS[TOT]	

ANOVA table for partitioning the total sum of squares into the regression sum of squares and the error sum of squares.

In the case of simple linear regression analysis, the ratio of the mean squares for the regression and the error can be used to test the null hypothesis H_0: $\beta_1 = 0$. It can be shown using Theorem B.6g in Appendix B that the ratio

$$F_{reg} = \frac{MS[R]}{MS[E]} \tag{11.13}$$

can be used to test the null hypothesis H_0: $\beta_1 = \beta_2 = \cdots = \beta_p = 0$. The value of F_{reg} should be compared to the critical value of an F-distribution with p degrees of freedom for the numerator and $n - (p + 1)$ degrees of freedom for the denominator. Rejection of the null hypothesis implies that at least one of the p regression parameters $\beta_1, \ldots, \beta_p$ is not zero and hence that at least one of the p independent variables in the model is useful for predicting Y. Notice that, when $p = 1$, the F-test based on F_{reg} in Equation (11.13) is the same as the F-test based on the ANOVA table in Section 10.5.

EXAMPLE **11.11**   In Example 11.9, we fitted a multiple linear regression model to $n = 7$ observations of enzyme adsorption. Two ($p = 2$) independent variables X_1 = enzyme concentration and $X_2 = X_1^2$ were used in the study. When the REG procedure in SAS is used to fit the model

$$Y = \beta_0 + \beta_1 x_1 + \beta_2 x_2 + E$$

to the observed data, the output includes the following ANOVA table:

```
-------------------------------------------------------------------
                        Analysis of Variance
                        Sum of          Mean
      Source      DF    Squares         Square     F Value    Prob>F

      Model        2    425.13863      212.56931   199.531    0.0001
      Error        4      4.26137        1.06534
      C Total      6    429.40000
-------------------------------------------------------------------
```

The model sum of squares in the ANOVA table is the same as SS[R], the regression sum of squares. Notice that $p = 2$ degrees of freedom are associated with SS[R]. The F-statistic for testing H_0: $\beta_1 = \beta_2 = 0$ has the value

$$F_{reg} = \frac{212.56931}{1.06534} = 199.531.$$

The calculated value should be compared with the critical values of an F-distribution with two degrees of freedom for the numerator and $n - (p + 1) = 7 - (2 + 1) = 4$ degrees of freedom for the denominator. From the output, we see that the associated p-value is 0.0001. It appears that at least one of the two terms x_1 and x_1^2 is useful for predicting enzyme adsorption. To determine whether we need both terms and, if not, which is preferable, we must resort to further analysis of the data. ∎

The *S*-matrix

The formulas in multiple linear regression analyses can be simplified by means of the *S*-matrix. Its use reveals the common features of inferences based on a large class of statistical models called general linear models, which will be discussed in the next chapter.

Suppose we are fitting a multiple linear regression model to a set of data consisting of n observed responses, where the i-th response y_i was observed at the settings $X_1 = x_{i1}, X_2 = x_{i2}, \ldots, X_j = x_{ij}, \ldots, X_p = x_{ip}$ of p independent variables. The design matrix associated with the data is the $n \times (p+1)$ matrix X in Box 11.3. Each row of X corresponds to one observation in the data and contains the coefficients of the regression parameters in the linear combination that represents the expected value of that response. In the i-th row, the first element is the coefficient of β_0, the intercept parameter; the remaining p elements in that row are the coefficients of the partial slopes (the β_j).

EXAMPLE 11.12 The design matrix for the enzyme adsorption data in Example 11.1, with $n = 7$ rows and $p + 1 = 2 + 1 = 3$ columns, takes the form

$$\mathbf{X} = \begin{bmatrix} 1 & 0.05 & 0.0025 \\ 1 & 0.10 & 0.0100 \\ 1 & 0.30 & 0.0900 \\ 1 & 0.50 & 0.2500 \\ 1 & 0.60 & 0.3600 \\ 1 & 0.80 & 0.6400 \\ 1 & 1.00 & 1.0000 \end{bmatrix}.$$

Each row of X contains the coefficients of the linear combination of the regression parameters representing the expected response for an observation. For example, the third observation y_3 was made at $X_1 = 0.30$ and $X_2 = 0.09$, so that the expected value for that response is the linear combination

$$\mathcal{E}(Y|X_1 = 0.3, X_2 = 0.09) = \beta_0 + 0.30\beta_1 + 0.09\beta_2,$$

with coefficients 1, 0.30, and 0.09. Note that these coefficients are the elements in the third row of X. ∎

A $(p+1) \times (p+1)$ symmetric matrix—an *S*-matrix—is associated with each design matrix. The *ij*-element of the *S*-matrix is denoted by s_{ij} (symmetry implies that $s_{ij} = s_{ji}$), and so the matrix is displayed as

$$S = \begin{bmatrix} s_{00} & s_{01} & \cdots & s_{0j} & \cdots & s_{0p} \\ s_{10} & s_{11} & \cdots & s_{1j} & \cdots & s_{1p} \\ \vdots & \vdots & & \vdots & & \vdots \\ s_{i0} & s_{i1} & \cdots & s_{ij} & \cdots & s_{ip} \\ \vdots & \vdots & & \vdots & & \vdots \\ s_{p0} & s_{p1} & \cdots & s_{pj} & \cdots & s_{pp} \end{bmatrix}.$$

Notice the special numbering convention used in S: The row and column numbers start with 0 instead of 1. Thus, the element in the first row and first column is s_{00}, rather than s_{11}. Similarly, the element in the $(p + 1)$-th row and the $(p + 1)$-th column is s_{pp}, rather than $s_{(p+1)(p+1)}$. The advantage of this numbering convention is that the range for the row and column numbers of S is the same as the range of the subscript in β_i.

The least squares estimates of the regression parameters bear a simple relationship to the elements of the S-matrix. For $i = 0, \ldots, p$, the least squares estimate of β_i is the row-column product of the i-th row of S and the column $X'y$

$$\hat{\beta}_i = s_{i0} \sum y_k + s_{i1} \sum x_{k1} y_k + \cdots + s_{ip} \sum x_{kp} y_p, \qquad i = 0, 1, 2, \ldots, p. \quad \textbf{(11.14)}$$

Notice that in Equation (11.14), $\hat{\beta}_i$ is a linear combination of the quantities on the left-hand side of the normal equations—Equation (11.11). The coefficients in the linear combination for estimating β_i are the elements of the i-th row of S. The relationships in Equation (11.14) between the least squares estimates and the S-matrix can be expressed in matrix notation as

$$\hat{\beta} = SX'y. \qquad \textbf{(11.15)}$$

Calculating an S-matrix can be laborious, because it involves a computationally intensive procedure called matrix inversion. Indeed, in the theory of linear equations S is called an inverse of $X'X$. The degree of difficulty in computing S increases with p, the number of independent variables in the regression model. Fortunately, a user of multiple linear regression does not need to know how to compute the S-matrix, because this task can be performed by means of appropriate software—for instance, the REG and GLM procedures in SAS.

EXAMPLE 11.13   In Example 11.9, on the basis of the enzyme adsorption data in Example 11.5, we used the REG procedure in SAS to estimate the regression parameters. The REG procedure can also be used to determine the S-matrix (called $X'X$ Inverse in the output). For these data, $p = 2$, so that S is a $(2 + 1) \times (2 + 1) = 3 \times 3$ matrix:

$$S = \begin{bmatrix} 0.817941 & -3.183435 & 2.524516 \\ -3.183435 & 18.863957 & -17.390099 \\ 2.524516 & -17.390099 & 17.251952 \end{bmatrix}.$$

In order to see how the elements of S are related to the least squares estimators, recall from Example 11.9 that the left-hand side of the normal equations ($X'y$ in matrix notation) is the column vector

$$X'y = \begin{bmatrix} 105.700 \\ 67.410 \\ 50.826 \end{bmatrix}.$$

Thus, the estimate of β_i is a linear combination of the quantities 105.700, 67.410, and 50.826 with coefficients given by the elements in the i-th row of S. For example, to estimate β_1, we use the row $i = 1$ of S (the middle row)

$$\hat{\beta}_1 = (-3.183435)(105.700) + (18.863957)(67.410) + (-17.390099)(50.826)$$

$$= 51.26109.$$

As an exercise, verify that, except for some rounding errors, this value of $\hat{\beta}_1$ is the same as the least squares estimate of β_1 in the REG output in Example 11.9. ∎

To illustrate the role of the S-matrix in regression analysis, let's now look at the special case of fitting a simple linear regression model.

EXAMPLE **11.14** The design matrix for fitting a simple linear regression model to n responses will have n rows and $p + 1 = 2$ columns. Let $x_1, x_2, \ldots, x_n$ be the levels of the independent variable X, at which observations of Y will be made. (Because there is only one independent variable, one subscript is sufficient to identify its levels.)

The design matrix is

$$\mathbf{X} = \begin{bmatrix} 1 & x_1 \\ 1 & x_2 \\ \vdots & \vdots \\ 1 & x_i \\ \vdots & \vdots \\ 1 & x_n \end{bmatrix}.$$

The corresponding S-matrix has two rows and two columns

$$S = \begin{bmatrix} s_{00} & s_{01} \\ s_{10} & s_{11} \end{bmatrix}.$$

Note that, on account of the symmetry of S, element s_{10} will equal element s_{01}. For a simple linear regression model, in contrast to a multiple linear regression model, the elements of S can be expressed in a form suitable for easy hand computation. Indeed, we have already encountered expressions for these elements in Box 10.5. The quantities s_{00}, s_{01}, and s_{11} in the formula for the standard error of $\hat{\theta} = c_0 \hat{\beta}_0 + c_1 \hat{\beta}_1$ in Box 10.5 are the same as the elements in the S-matrix. Thus, in the simple linear regression case, the S-matrix can be expressed as

$$S = \begin{bmatrix} \dfrac{1}{n} + \dfrac{\bar{x}^2}{S_{xx}} & -\dfrac{\bar{x}}{S_{xx}} \\[2ex] -\dfrac{\bar{x}}{S_{xx}} & \dfrac{1}{S_{xx}} \end{bmatrix}.$$

It is evident from Box 10.5 that the elements of the S-matrix influence the standard errors associated with the hypothesis tests, confidence intervals, and prediction intervals. Because the S-matrix is completely determined by the design matrix X, the choice of the design matrix plays a critical role in controlling the magnitudes of the standard errors of interest. ∎

As we'll see in Chapter 12, our conclusions in Example 11.14 on the influence of the S-matrix on inferences can be extended to general linear models. Thus, to evaluate a particular study design, careful assessment of the associated S-matrix is required.

Exercises

11.11   In Example 11.8, we considered a study in which responses were measured at $n = 8$ settings of two independent variables X_1 and X_2. Suppose that we are considering two models for the study data: model 1

$$Y_i = \beta_0 + \beta_1 x_{i1} + \beta_2 x_{i2} + E_i;$$

and model 2

$$Y_i = \beta_0 + \beta_1 x_{i1} + \beta_2 x_{i2} + \beta_3 x_{i1} x_{i2} + E_i.$$

Portions of the SAS output from fitting the two models to the observed data are as follows.

```
-----------------------------------------------------------------
                           MODEL 1
                     Analysis of Variance
                     Sum of        Mean
   Source    DF     Squares      Square     F Value    Prob>F
   Model     2      4.40039      2.20020     34.555     0.0012
   Error     5      0.31836      0.06367
   C Total   7      4.71875


                     Parameter Estimates
                 Parameter      Standard     T for H0:
   Variable  DF   Estimate        Error     Parameter=0    Prob > |T|
   INTERCEP  1    -0.861140     0.50011530    -1.722        0.1457
   X1        1     0.800067     0.09657595     8.284        0.0004
   X2        1    -0.078780     0.03714975    -2.121        0.0874
```

```
                 Dep Var   Predict
           Obs     Y       Value    Residual
            1    1.7000    1.9556    -0.2556
            2    1.9000    1.8379     0.0621
            3    2.1000    2.3004    -0.2004
            4    2.8000    2.8423    -0.0423
            5    2.2000    2.1811     0.0189
            6    2.9000    2.4617     0.4383
            7    3.4000    3.3253     0.0747
            8    4.1000    4.1958    -0.0958
```

MODEL 2

Analysis of Variance

Source	DF	Sum of Squares	Mean Square	F Value	Prob>F
Model	3	4.40325	1.46775	18.609	0.0082
Error	4	0.31550	0.07887		
C Total	7	4.71875			

Parameter Estimates

Variable	DF	Parameter Estimate	Standard Error	T for H0: Parameter=0	Prob > |T|
INTERCEP	1	-0.597752	1.49077841	-0.401	0.7089
X1	1	0.743553	0.31560506	2.356	0.0780
X2	1	-0.126096	0.25185920	-0.501	0.6429
X1X2	1	0.009976	0.05237927	0.190	0.8582

----

a   Argue that, while the parameter β_0 has the same meaning in both models, the meanings of the parameters β_1 and β_2 are different in the two models.

b   Determine the estimated least squares prediction equation using model 1.

c   Explain how the error sum of squares for model 1 is related to the residuals resulting from fitting this model.

d   Use the given ANOVA table to test the null hypothesis H_0: $\beta_1 = \beta_2 = 0$ in model 1. Write your conclusion.

e   Determine the estimated least squares prediction equation based on model 2.

f   Use the estimated prediction equation to calculate the residuals for model 2.

g   On a single graph, plot the residuals from the two models against the observed values. Compare the two plots, and comment on the relative adequacy of the two models.

h   Obtain estimates of the error variance based on the two models. Do the two estimates lead you to prefer one model over the other?

11.12   In analyzing X_1, patient's age in years, and X_2, perceived uncertainty about outcome of hospitalization (measured by the Mishel Uncertainty in Illness Scale; Mishel, 1981), as predictors of Y, stress in hospitalized patients (measured by the Hospital Stress Rating Scale; Volicer & Bohannon, 1975), the three variables were measured on a sample of $n = 25$ hospitalized patients.

A portion of the computer output obtained for model 1

$$Y = \beta_0 + \beta_1 x_1 + \beta_2 x_2 + E$$

is as follows.

```
-------------------------------------------------------------------
                        Analysis of Variance
                    Sum of          Mean
    Source    DF    Squares        Square      F Value      Prob>F
    Model      2 181369.07387   90684.53693     4.665       0.0205
    Error     22 427625.96613   19437.54392
    C Total   24 608995.04000
                        Parameter Estimates
                    Parameter     Standard    T for H0:
    Variable  DF    Estimate       Error     Parameter=0   Prob > |T|

    INTERCEP   1  426.033719  252.81248037     1.685         0.1061
    X1         1   -6.341394    2.61207977    -2.428         0.0238
    X2         1    3.375532    1.92991857     1.749         0.0942
```

A portion of the corresponding output obtained for model 2

$$Y = \beta_0 + \beta_1 x_1 + \beta_2 x_2 + \beta_3 x_1 x_2 + E$$

is as follows.

```
                        Analysis of Variance
                    Sum of          Mean
    Source    DF    Squares        Square      F Value      Prob>F

    Model      3 261163.93806   87054.64602     5.256       0.0073
    Error     21 347831.10194   16563.38581
    C Total   24 608995.04000
                        Parameter Estimates
                    Parameter     Standard    T for H0:
    Variable  DF    Estimate       Error     Parameter=0   Prob > |T|

    INTERCEP   1 -2441.150690  1326.9820750   -1.840         0.0800
    X1         1    41.528133    21.94241282    1.893         0.0723
    X2         1    30.875288    12.65501245    2.440         0.0237
    X1X2       1    -0.459044     0.20914211   -2.195         0.0396
-------------------------------------------------------------------
```

a   Argue that models 1 and 2 are both multiple linear regression models. Identify
the independent variables in each.

b   Determine the estimated prediction equation for each of the two models.

c   Interpret the estimated values of β_1 and β_2 in model 1. What is the significance
of the negative value for $\hat{\beta}_1$ in model 1?

d   Interpret the estimated value of β_3 in model 2.

  e In Exercise 11.11, it was argued that the parameter β_0 has the same interpretation in both models. How would you explain the big difference in the estimated values of β_0 in the two models in the present case? (Note the standard errors of the estimates.)

  f Test the null hypothesis H_0: $\beta_1 = \beta_2 = 0$ in model 1. Interpret the results.

  g Test the null hypothesis H_0: $\beta_1 = \beta_2 = \beta_3 = 0$ in model 2. Interpret the results.

11.13 Verify that the normal equations for estimating the parameters in the quadratic model for the enzyme adsorption are as in Example 11.9.

11.14 In Example 11.7, we used matrix notation to express the normal equations for fitting a simple linear regression model. Using a similar logic, show that the normal equations for fitting a multiple linear regression can be expressed as

$$X'y = X'X\hat{\beta},$$

where X is the design matrix (of dimension $n \times p$), y is the observation vector (of dimension n), and $\hat{\beta}$ is the vector of the estimated regression parameters (of dimension $p + 1$).

11.15 Display the matrices X, $X'X$, and $X'y$ for the two models in Exercise 11.11.

11.5
Inferences about expected and future responses

As in the case of a simple linear regression model, inferences about expected responses based on a multiple linear regression model can be viewed as inferences about linear combinations of the regression parameters. In typical cases, the desired inference is an inference about the parameter

$$\theta = c_0\beta_0 + c_1\beta_1 + \cdots + c_p\beta_p,$$

where the coefficients $c_0, c_1, \ldots, c_p$ are constants selected so that θ is the expected value of interest. The results needed to make inferences about θ are summarized in Box 11.4.

BOX **11.4**

Properties of the least squares estimate of a linear combination of regression parameters

For $i = 1, \ldots, n$, let Y_i denote the response that will be observed when the independent variable X_j has the value $x_{ij}, j = 1, \ldots, p$. Assume that the responses satisfy the multiple linear regression model

$$Y_i = \beta_0 + \beta_1 x_{i1} + \cdots \beta_p x_{ip} + E_i, \qquad i = 1, \ldots, n,$$

where the E_i are a random sample from a population with zero mean and variance σ^2. (Note that a normal distribution for the errors is not assumed here.) For given constants $c_0, c_1, \ldots, c_p$, let

$$\theta = c_0\beta_0 + c_1\beta_1 + \cdots + c_p\beta_p$$

denote a linear combination of the regression parameters. Then we can state the following properties.

1 An unbiased estimate of θ is

$$\hat{\theta} = c_0\hat{\beta}_0 + c_1\hat{\beta}_1 + \cdots + c_p\hat{\beta}_p.$$

2 The standard error of $\hat{\theta}$ is $\mathrm{SE}(\hat{\theta}) = \sigma_{\hat{\theta}}$, where

$$\sigma_{\hat{\theta}}^2 = \sigma^2 \left(\sum_{i=0}^{p} \sum_{j=0}^{p} c_i c_j s_{ij} \right),$$

where the s_{ij} are the elements of the S-matrix. The values of the s_{ij} depend on the design points—that is, the settings of the independent variables at which the responses are observed.

3 If the E_i are independent $N(0, \sigma^2)$, then $\hat{\theta}$ has a normal distribution with mean θ and variance $\sigma_{\hat{\theta}}^2$; that is, $\hat{\theta}$ is $N(\theta, \sigma_{\hat{\theta}}^2)$.

4 Let $\hat{\sigma}^2$ denote the least squares estimate of the error variance—that is, $\hat{\sigma}^2 = \mathrm{MS[E]}$—and let

$$\hat{\sigma}_{\hat{\theta}}^2 = \hat{\sigma}^2 \left(\sum_{i=0}^{p} \sum_{j=0}^{p} c_i c_j s_{ij} \right)$$

denote the estimated variance of $\hat{\theta}$ obtained by replacing σ^2 by $\hat{\sigma}^2$. Then the null hypothesis H_0: $\theta = \theta_0$ can be tested using the test statistic

$$t_{\hat{\theta}} = \frac{\hat{\theta} - \theta_0}{\hat{\sigma}_{\hat{\theta}}}.$$

The value of the test statistic should be compared to an appropriate critical value of the $t(n - p - 1)$-distribution.

5 A $100(1 - \alpha)\%$ confidence interval for θ is given by

$$\hat{\theta} - t(n - p - 1, \alpha/2)\hat{\sigma}_{\hat{\theta}} \leq \theta \leq \hat{\theta} + t(n - p - 1, \alpha/2)\hat{\sigma}_{\hat{\theta}}.$$

Inferences about regression parameters

Methods for making inferences about a single regression parameter are easily obtained from the results in Box 11.4. If we set $c_i = 1$ and all other coefficients equal to zero, we can deduce the following results concerning inferences about the regression parameter β_i (see Exercise 11.16).

1 Regardless of whether the errors are normally distributed, we know that:
   a the least squares estimate $\hat{\beta}_i$ is an unbiased estimate of β_i;

b the estimated standard error of $\hat{\beta}_i$ is

$$\hat{\sigma}_{\hat{\beta}_i} = \hat{\sigma}\sqrt{s_{ii}}.$$

2 If the errors are independent $N(0, \sigma^2)$, then:

a the least squares estimate $\hat{\beta}_i$ has a normal distribution with mean β_i and variance $\sigma^2_{\hat{\beta}_i} = \sigma^2 s_{ii}$;

b tests of the null hypothesis H_0: $\beta_i = \beta_{i(0)}$ can be based on the test statistic

$$t_{\hat{\beta}_i} = \frac{\hat{\beta}_i - \hat{\beta}_{i(0)}}{\hat{\sigma}_{\hat{\beta}_i}},$$

where the observed value of the test statistic should be compared to the critical value of the $t(n - p - 1)$-distribution;

c a $100(1 - \alpha)\%$ confidence interval for β_i is

$$\hat{\beta}_i - t(n - p - 1, \alpha/2)\hat{\sigma}_{\hat{\beta}_i} \leq \beta \leq \hat{\beta}_i + t(n - p - 1, \alpha/2)\hat{\sigma}_{\hat{\beta}_i}.$$

The next example illustrates how the results in Box 11.4 can be used to make inferences about linear combinations of regression parameters.

EXAMPLE 11.15 In Example 11.2, we described a study on oxygen demand in dairy wastes. In this study, the response variable is $Y = \log(\text{oxygen demand})$ (the oxygen demand is measured in mg/liter). The five explanatory variables are: X_1, the biological oxygen demand (g/liter); X_2, the total Kjeldahl nitrogen (g/liter); X_3, the total solids (g/liter); X_4, the total volatile solids (g/liter); and X_5, the chemical oxygen demand (g/liter). The data for these six variables are given in Table 11.3.

For purposes of illustration, we'll use a multiple linear regression analysis to model the biological oxygen demand as a function of X_1 and X_2 only. The following is part of the output obtained when the SAS REG procedure is used to fit the multiple linear regression model

$$Y_i = \beta_0 + \beta_1 x_{1i} + \beta_2 x_{2i} + E_i, \qquad i = 1, \dots, 20,$$

to the data.

PART I

Parameter Estimates

Variable	DF	Parameter Estimate	Standard Error	T for H0: Parameter=0	Prob > |T|
INTERCEP	1	-1.420604	0.36669193	-3.874	0.0012
X1	1	1.422058	0.23361056	6.087	0.0001
X2	1	2.907873	1.33480559	2.178	0.0437

TABLE   11.3

Data on oxygen demand in dairy wastes

i	x_{1i}	x_{2i}	x_{3i}	x_{4i}	x_{5i}	y_i
0	1.125	0.232	7.160	0.0859	8.905	1.5563
7	0.920	0.268	8.804	0.0865	7.388	0.8976
15	0.835	0.271	8.108	0.0852	5.348	0.7482
22	1.000	0.237	6.370	0.0838	8.056	0.7160
29	1.150	0.192	6.441	0.0821	6.960	0.3130
37	0.990	0.202	5.154	0.0792	5.690	0.3617
44	0.840	0.184	5.896	0.0812	6.932	0.1139
58	0.650	0.200	5.336	0.0806	5.400	0.1139
65	0.640	0.180	5.041	0.0784	3.177	−0.2218
72	0.583	0.165	5.012	0.0793	4.461	−0.1549
80	0.570	0.151	4.825	0.0787	3.901	0.0000
86	0.570	0.171	4.391	0.0780	5.002	0.0000
93	0.510	0.243	4.320	0.0723	4.665	−0.0969
100	0.555	0.147	3.709	0.0749	4.642	−0.2218
107	0.460	0.286	3.969	0.0744	4.840	−0.3979
122	0.275	0.198	3.558	0.0725	4.479	−0.1549
129	0.510	0.196	4.361	0.0577	4.200	−0.2218
151	0.165	0.210	3.301	0.0718	3.410	−0.3979
171	0.244	0.327	2.964	0.0725	3.360	−0.5229
220	0.079	0.334	2.777	0.0719	2.599	−0.0458

PART II

Obs	Dep Var Y	Predict Value	Residual
1	1.5563	0.8538	0.7025
2	0.8976	0.6670	0.2306
3	0.7482	0.5548	0.1934
4	0.7160	0.6906	0.0254
5	0.3130	0.7731	−0.4601
6	0.3617	0.5746	−0.2129
7	0.1139	0.3090	−0.1951
8	0.1139	0.0853	0.0286
9	−0.2218	0.0129	−0.2347
10	−0.1549	−0.1117	−0.0432
11	0	−0.1709	0.1709
12	0	−0.1128	0.1128
13	−0.0969	0.0113	−0.1082
14	−0.2218	−0.2039	−0.0179
15	−0.3979	0.0652	−0.4631
16	−0.1549	−0.4538	0.2989
17	−0.2218	−0.1254	−0.0964

```
18    -0.3979   -0.5753    0.1774
19    -0.5229   -0.1227   -0.4002
20    -0.0458   -0.3370    0.2912
```

Sum of Residuals   0
Sum of Squared Residuals   1.5809

PART III

X´X Inverse

	INTERCEP	X1	X2
INTERCEP	1.4459	-0.5577	-4.7453
X1	-0.5577	0.5868	0.8463
X2	-4.7453	0.8463	19.1588

The output consists of three parts. The first shows the least squares estimate of β_i, its estimated standard error $(\hat{\sigma}_{\hat{\beta}_i})$, and the p-value for two-sided tests of the null hypothesis H_{0i}: $\beta_i = 0$, for $i = 0, 1, 2$. From this portion of the output, we see that the estimated regression function is

$$\hat{y} = -1.4206 + 1.4221 x_1 + 2.9079 x_2.$$

Since the p-value for each β_i is small, it is reasonable to conclude that each of the three null hypotheses H_{0i}: $\beta_i = 0$, $i = 0, 1, 2$, can be rejected. Thus, it appears that each of the two independent variables, X_1 and X_2 (with partial slopes β_1 and β_2) has a demonstrable effect on oxygen demand.

The estimated standard errors can be used to construct confidence intervals for the regression coefficients. The confidence intervals are constructed using critical values of t with $\nu = n - p - 1 = 20 - 2 - 1 = 17$ degrees of freedom. For example, a 99% confidence interval for β_1 is

$$\hat{\beta}_1 \pm t(17, 0.005)\hat{\sigma}_{\hat{\beta}_1} \implies 1.4221 \pm (2.898)(0.2336) \implies 0.7451 \le \beta_1 \le 2.0991.$$

Thus, with 99% confidence, we can conclude that every increase of 1 g/liter in biological oxygen demand is associated with an increase of between 0.7451 and 2.0991 in log (oxygen demand) (where oxygen demand is expressed in mg/liter).

The second part of the output shows the observed response y_i, the predicted response $\hat{y}_i$, and the residual $e_i = y_i - \hat{y}_i$ for $i = 1, \ldots, 20$. The sum of squares of the residuals SS[E] equals 1.5809, and so an unbiased estimate of the error variance is

$$\hat{\sigma}^2 = \text{MS[E]} = \frac{\text{SS[E]}}{n - (p + 1)} = \frac{1.5809}{20 - (2 + 1)} = \frac{1.5809}{17} = 0.0930.$$

The third part of the output shows an array of three rows and three columns, called $X'X$ inverse. This is the S-matrix; it contains the values of the s_{ij} needed to calculate the standard errors. There are $p + 1 = 3$ regression parameters in the model, and so the S-matrix is an array of three rows and three columns. Notice that the matrix is symmetric $(s_{ij} = s_{ji})$; that is, the element common to row i and column j is the same as the element common to column i and row j. For example, $s_{01} = s_{10} = -0.5577$ and $s_{12} = s_{21} = 0.8463$. Using the S-matrix and MS[E], we can obtain the estimate of the standard error of the least squares estimate of a given linear combination $\theta = c_0\beta_0 + c_1\beta_1 + c_2\beta_2$. The standard error can then be used to make inferences about θ. For instance, a 90% lower bound for the expected change in log (oxygen demand) when X_1 is increased by 0.6 g/liter and X_2 is increased by 0.2 g/liter can be constructed as follows.

First, note that the linear combination of interest is the differential expected response

$$\theta = \mu(x_1 + 0.6, \ x_2 + 0.2) - \mu(x_1, \ x_2) = 0.6\beta_1 + 0.2\beta_2,$$

and so the parameter of interest is a linear combination with $c_0 = 0$, $c_1 = 0.6$, and $c_2 = 0.2$. The estimated mean change in log (oxygen demand) is the corresponding linear combination of the least squares estimates. Thus

$$\hat{\theta} = 0.6\hat{\beta}_1 + 0.2\hat{\beta}_1 = 0.6(1.4221) + 0.2(2.9079) = 1.4348$$

is the least squares estimate of the expected change. The standard error of the estimate is

$$
\begin{aligned}
\hat{\sigma}_{\hat{\theta}} &= \hat{\sigma}\left(\sqrt{\sum_{i=0}^{2}\sum_{j=0}^{2} c_i c_j s_{ij}}\right) \\
&= \hat{\sigma}\left(\sqrt{c_0^2 s_{00} + 2c_0 c_1 s_{01} + 2c_0 c_2 s_{02} + c_1^2 s_{11} + 2c_1 c_2 s_{12} + c_2^2 s_{22}}\right) \\
&= \sqrt{0.0930}\Big[(0)^2(1.4459) + 2(0)(0.6)(-0.5577) + 2(0)(0.2)(-4.7453) \\
&\qquad + (0.6)^2(0.5868) + 2(0.6)(0.2)(0.8463) + (0.2)^2(19.1588)\Big]^{\frac{1}{2}} \\
&= (0.3050)(1.0866) = 0.3314.
\end{aligned}
$$

A 90% lower bound is

$$\hat{\theta}_L = \hat{\theta} - t(17, 0.10)(0.3314) = 1.4348 - (1.333)(0.3314) = 0.9930. \quad \blacksquare$$

Inferences about mean responses

The procedures for inferences about a linear combination of regression parameters can be used to make inferences about an expected (population mean) response. Let $\mu(x_0)$ denote the expected response when the independent variables have the values $X_1 = x_{01}, X_2 = x_{02}, \dots, X_p = x_{0p}$. Then

$$\mu(x_0) = \beta_0 + \beta_1 x_{01} + \cdots + \beta_p x_{0p},$$

and so a formula for constructing confidence intervals for this expected response can be obtained by setting $c_0 = 1, c_1 = x_{01}, \dots, c_p = x_{0p}$ in the formula for the confidence intervals in Box 11.4. The corresponding expression is presented in Box 11.5.

Inferences about the means of future responses can be made in exactly the same way as in simple linear regression. The prediction interval for the mean of m future responses at $X_1 = x_{01}, X_2 = x_{02}, \dots, X_p = x_{0p}$ is an interval for $\mu(x_0) + \overline{E}_{x_0}$, where $\overline{E}_{x_0}$ is the mean of the error components in the m responses. The corresponding formula is also given in Box 11.5; the prediction interval for a single response is obtained by setting $m = 1$.

Confidence intervals for the expected response and prediction intervals for the mean of m future responses

1 Let $\mu(x_0)$ denote the expected response at $X_1 = x_{01}, \ldots, X_p = x_{0p}$. Then the $100(1 - \alpha)\%$ confidence interval for $\mu(x_0)$ is

$$\hat{\mu}(x_0) - t(n - p - 1, \alpha/2)\hat{\sigma}_{(x_0)} \leq \mu(x_0)$$

$$\leq \hat{\mu}(x_0) + t(n - p - 1, \alpha/2)\hat{\sigma}_{(x_0)},$$

where $\hat{\mu}(x_0) = \hat{\beta}_0 + \hat{\beta}_1 x_{01} + \cdots + \hat{\beta}_p x_{0p}$ is the predicted value of Y when $X_j = x_{0j}$, $j = 1, \ldots, p$, and

$$\hat{\sigma}^2_{(x_0)} = \left[\frac{1}{n} + \sum_{i=1}^{p} \sum_{j=1}^{p} (x_{0i} - \overline{x}_i)(x_{0j} - \overline{x}_j) s_{ij} \right] \hat{\sigma}^2;$$

$\overline{x}_j$ is the mean of the n settings of X_j at which the responses are measured, and $\hat{\sigma}^2$ is the least squares estimate of error variance σ^2.

2 Let $\overline{Y}_{x_0}$ denote the mean of a random sample of m responses observed at $X_j = x_{0j}$, $j = 1, \ldots, p$. Then a $100(1 - \alpha)\%$ prediction interval for $\overline{Y}_{x_0}$ is given by

$$\hat{\mu}(x_0) - t(n - p - 1, \alpha/2)\hat{\sigma}_P(x_0) \leq \overline{Y}_{x_0}$$

$$\leq \hat{\mu}(x_0) + t(n - p - 1, \alpha/2)\hat{\sigma}_P(x_0),$$

where

$$\hat{\sigma}^2_P(x_0) = \left[\frac{1}{m} + \frac{1}{n} + \sum_{i=1}^{p} \sum_{j=1}^{p} (x_{0i} - \overline{x}_i)(x_{0j} - \overline{x}_j) s_{ij} \right] \hat{\sigma}^2.$$

Compare the expressions for the confidence and prediction intervals in Box 11.5 with those in Chapter 10 for the simple linear regression model (Boxes 10.6 and 10.7). The structures of the formulas are similar in the two settings, but the computations for the multiple linear regression model are more complex. In Exercise 11.20, you will be asked to verify that, when $p = 1$, the expressions in Box 11.5 reduce to the corresponding expressions for a simple linear regression model.

The interpretation of the confidence and prediction intervals in multiple regression is the same as their interpretation in simple linear regression. Statistical software packages such as SAS have procedures for quick and easy computation of these intervals.

Matrix formulation

The formulas for θ, $\hat{\theta}$, and the standard error of $\hat{\theta}$ in Box 11.4 can be expressed in matrix notation. Let c denote a $(p+1)$-dimensional column vector whose elements are $c_0, c_1, \ldots$, and c_p. Then the following expressions can be derived.

1   The linear combination θ and its least squares estimate can be expressed as

$$\theta = c'\beta, \qquad \hat{\theta} = c'\hat{\beta},$$

where β and $\hat{\beta}$ are, respectively, the column vectors of the regression constants and their least squares estimates.

2   The standard error of $\hat{\theta}$ can be expressed as

$$\sigma_{\hat{\theta}} = \sigma\sqrt{c'Sc}.$$

The expressions for $\hat{\sigma}^2_{(x_0)}$ and $\hat{\sigma}^2_P(x_0)$ in Box 11.5 can be further simplified by introducing matrix notation. If S is the S-matrix and x_0 is the column vector

$$x_0 = \begin{bmatrix} 1 \\ x_{01} \\ x_{02} \\ \vdots \\ x_{0p} \end{bmatrix},$$

then

$$\hat{\sigma}^2_{(x_0)} = (x_0'Sx_0)\hat{\sigma}^2 \tag{11.16}$$

and

$$\hat{\sigma}^2_P(x_0) = \left(\frac{1}{m} + x_0'Sx_0\right)\hat{\sigma}^2. \tag{11.17}$$

In Exercise 11.21, you will be asked to use the matrix formulations of the confidence and prediction intervals to compute the lower bound in Example 11.15.

Exercises

11.16   **a**   Using the results in Box 11.4, verify that, if the errors have independent normal distributions, then the sampling distribution of $\hat{\beta}_i$ is a normal distribution with mean β_i and variance $\sigma^2_{\hat{\beta}_i} = \sigma^2 s_{ii}$.

b   Compare the result in (a) to the corresponding results for the least squares estimates of the regression parameters in a simple linear regression model.

c   Use the sum of squares of the residuals and the $X'X$ inverse matrix in the SAS output in Example 11.15 to estimate the standard errors of the regression parameters in the regression of $Y = \log$ (oxygen demand) on X_1 = biological oxygen demand and X_2 = total Kjeldahl nitrogen. Compare the result with the standard errors given in the first part of the SAS output.

11.17   The following $X'X$ inverses correspond to the data in Exercise 11.11:

X´X Inverse for Model 1

	INTERCEP	X1	X2
INTERCEP	3.9282004789	-0.668364892	-0.081242539
X1	-0.668364892	0.146484386	-0.009781023
X2	-0.081242539	-0.009781023	0.021675301

X´X Inverse for Model 2

	INTERCEP	X1	X2	X1X2
INTERCEP	28.176689801	-5.87126557	-4.437356806	0.9184038348
X1	-5.87126557	1.2628498267	0.9248928684	-0.197058211
X2	-4.437356806	0.9248928684	0.804228443	-0.164986445
X1X2	0.918403834	-0.197058211	-0.164986445	0.0347842537

a   Is there evidence to suggest that, in model 1, the independent variable X_1 is useful for predicting Y? In other words, is there evidence to suggest that $\beta_1 \neq 0$?

b   Construct a 95% confidence interval for β_2 in model 1 and interpret the result.

c   Construct a 99% confidence interval for the parameter $\theta = \beta_1 - \beta_2$ in model 1 and interpret the result.

d   Use model 1 to construct a 95% lower confidence bound for the expected response at $X_1 = 5$ and $X_2 = 3$. (Note that the design points at which the responses were measured are given in Example 11.8.) Interpret the bound.

e   Use model 1 to construct a 95% interval for an observed response when $X_1 = 5$ and $X_2 = 3$. Interpret the interval.

f   Will the interval in (e) be wider or narrower than the corresponding 95% confidence interval for the expected response at $X_1 = 5$ and $X_2 = 3$. Explain.

g   In model 2, test the null hypothesis that the change in the expected response resulting from a 1-unit increase in X_1 is the same at all values of X_2.

11.18   Refer to the hospital stress data in Exercise 11.12.

a   If the ranges of the independent variables in the stress data are 35–74 for age and 79–135 for uncertainty score, explain why you would expect a large standard error for the estimate of β_0 in the two models. [Hint: The point $(X_1 = 0, X_2 = 0)$ is far outside the experimental region.]

b   How would you explain that the partial slope of X_1 is negative in model 1 and positive in model 2? [Hint: The interpretations of β_1 in the two models are not the same.]

11.19   In Example 11.9, we fitted a quadratic model to $n = 7$ measurements of enzyme adsorption. The $X'X$ inverse for the quadratic model is as follows:

X´X Inverse

	INTERCEP	X1	X2
INTERCEP	0.8179405091	-3.183435412	2.5245165003
X1	-3.183435412	18.863957296	-17.39009949
X2	2.5245165003	-17.39009949	17.251952306

a Compute the standard errors of the least squares estimates of β_0, β_1, and β_2.

b Use the standard errors in (a) to test the two null hypotheses H_{01}: $\beta_1 = 0$ and H_{02}: $\beta_2 = 0$, each at the level $\alpha = 0.05$. What conclusions can be drawn from your tests?

c Let $\mu(c)$ denote the expected enzyme adsorption when the enzyme concentration is $X_1 = c$; that is, let

$$\mu(c) = \beta_0 + \beta_1 c + \beta_2 c^2.$$

Then the rate at which the expected adsorption changes with increasing enzyme concentration is the mathematical derivative of $\mu(c)$. It can be shown that this derivative is

$$\mu'(c) = \beta_1 + 2\beta_2 c.$$

Construct 95% confidence intervals for $\mu'(c)$ when $c = 0.05, 0.15, \ldots, 0.95$. Plot these intervals against the values of c, and connect the endpoints of the intervals with straight lines. Propose an interpretation for the region enclosed by the two lines.

11.20 Verify that, when $p = 1$, the expressions for the intervals in Box 11.5 are the same as the corresponding intervals for a simple linear regression model.

11.21 Show that, with

$$c = \begin{bmatrix} 0 \\ 0.6 \\ 0.2 \end{bmatrix},$$

the matrix representation in Equations (11.16) and (11.17) yields the value for the lower bound given in Example 11.15.

11.6
Testing simultaneous hypotheses

The t-test described in Section 11.5 is useful for a hypothesis about a single regression coefficient. In multiple regression analysis, however, we often want to test several hypotheses simultaneously. Here's an example.

EXAMPLE **11.16** In Example 11.15, we analyzed some data collected in a study on oxygen demand in dairy wastes. The data consisted of $n = 20$ measurements of the response variable $Y = \log(\text{oxygen demand})$ and $p = 5$ independent variables: X_1, biological oxygen demand; X_2, total Kjeldahl nitrogen; X_3, total solids; X_4, total volatile solids; and X_5, chemical oxygen demand. A first step in analyzing these data is to see if any of the five independent variables is useful as a predictor of the response.

Let's assume, for the purpose of this example, that the multiple linear regression model

$$Y = \beta_0 + \beta_1 x_1 + \cdots + \beta_5 x_5 + E$$

is reasonable for these data. Then, to determine whether at least one of the five independent variables is useful for predicting the response, we can test the null hypothesis H_0 that none of the five variables is useful for predicting Y against the alternative hypothesis H_1 that at least one of the five variables is useful for predicting Y. In terms of the regression parameters, the null hypothesis of interest is the hypothesis of no regression effect—H_0: $\beta_1 = \cdots = \beta_5 = 0$. Clearly, testing H_0 is the same as simultaneously testing a collection of five null hypotheses—H_{i0}: $\beta_i = 0, i = 1, \cdots, 5$. ■

A useful and widely applicable method of testing simultaneous hypotheses about regression parameters involves a class of models called nested models.

Nested statistical models

Consider two statistical models A and B. If model B can be derived from model A by imposing additional restrictions on model A, then model B is said to be *nested* in model A. We refer to models A and B as the *full* and *reduced* models, respectively.

EXAMPLE 11.17 Consider the following linear regression models connecting a dependent variable Y with three independent variables X_1, X_2, and X_3

1. $\quad Y = \beta_0 + \beta_1 x_1 + \beta_2 x_2 + \beta_3 x_3 + E;$
2. $\quad Y = \beta_0 + \beta_1 x_1 + \beta_2 x_2 + E;$
3. $\quad Y = \beta_0 + \beta_1 x_1 + \beta_3 x_3 + E;$
4. $\quad Y = \beta_0 + \beta_1 x_1 + E;$
5. $\quad Y = \beta_0 + \beta_1 x_1 + \beta_2 (x_2 + x_3) + E;$
6. $\quad Y = \beta_0 + \beta_1 x_1 + \beta_2 x_2^2 + E.$

We can identify several nested pairs. For example, model 2 is nested in model 1, because it can be obtained by imposing the restriction $\beta_3 = 0$ on model 1. Thus, model 1 is the full model, and model 2 is the reduced model. Similarly, both model 3 and model 4 are nested in model 1, because the restrictions $\beta_2 = 0$ and $\beta_2 = \beta_3 = 0$ reduce model 1 to model 3 and model 4, respectively; model 3 and model 4 are reduced versions of model 1. Model 5 is also nested in model 1; the corresponding restriction on model 1 is $\beta_2 = \beta_3$.

By contrast, model 1 and model 6 are nonnested models. No restrictions on model 1 will reduce it to model 6, which, unlike model 1, contains the independent variable X_2^2. Similarly, model 2 and model 3 are also a pair of nonnested models. ■

Various interesting hypotheses about regression parameters can be stated in terms of nested models. With the assumed regression model as the full model, the null hypothesis states that it satisfies a specific set of restrictions, whereas the alternative hypothesis states that the restrictions are not satisfied. For example, consider the model

$$Y = \beta_0 + \beta_1 x_1 + \beta_2 x_2 + \beta_3 x_3 + E, \tag{11.18}$$

where Y is the birth-weight of a newborn child, and x_1, x_2, and x_3 are, respectively, the mother's age, the mother's weekly coffee consumption, and the mother's weekly cigarette consumption. If we are interested in seeing whether at least one of two factors, mother's coffee consumption and mother's cigarette consumption, is useful for predicting child's birth-weight, then we would want to test the null hypothesis

$$H_0: \beta_2 = \beta_3 = 0 \tag{11.19}$$

against the alternative hypothesis H_1 that at least one of β_2, and β_3 is not zero. The restrictions specified by H_0 in Equation (11.19) imply that, if the null hypothesis is true, then the reduced model

$$Y = \beta_0 + \beta_1 x_1 + E \tag{11.20}$$

is the appropriate model. Thus, the problem of testing the hypothesis in Equation (11.19) can be regarded as a problem of comparing the reduced model in Equation (11.20) with the full model in Equation (11.18).

The effect of the restrictions in Equation (11.19) is to reduce the full model by deleting two independent variables.

The simple deletion of a subset of independent variables is a frequently encountered restriction. The null hypothesis

$$H_0: \beta_1 = \beta_2 = \beta_3 \tag{11.21}$$

places more complex constraints on the regression parameters. It claims that the change in the expected response per unit change in a given independent variable (when all others are fixed) is the same for all three independent variables. For example, suppose that the model in Equation (11.18) relates Y, the yield of a particular crop, to x_1, x_2, x_3, the amount (kg per plot) of applied nitrogen, phosphate, and potash, respectively. Then the hypothesis $H_0: \beta_1 = \beta_2 = \beta_3$ implies that the expected change in yield associated with increasing the amount of nitrogen by 1 kg per plot is the same as that expected when either of the other two factors—phosphate and potash—is increased by 1 kg per plot.

If the null hypothesis in Equation (11.21) is true, the three partial slopes in Equation (11.18) can be replaced by a single regression parameter—say, β_1—leading to the reduced model

$$\begin{aligned} Y &= \beta_0 + \beta_1 x_1 + \beta_1 x_2 + \beta_1 x_3 + E \\ &= \beta_0 + \beta_1 (x_1 + x_2 + x_3) + E \\ &= \beta_0 + \beta_1 x + E, \end{aligned} \tag{11.22}$$

where $x = x_1 + x_2 + x_3$ is the value of $X = X_1 + X_2 + X_3$. Thus, the reduced model in Equation (11.22) is a simple linear regression model with independent variable $X = X_1 + X_2 + X_3$.

A hypothesis-testing problem in which the null hypothesis places a specific set of restrictions on the regression parameters can be regarded as a problem of comparing two nested models. Imposing restrictions on a model can only reduce its ability to predict responses, and so the hypotheses for comparing two nested models are, essentially, H_0 that the reduced model is as good as the full model and H_1 that the full model is better than the reduced model.

Nested models obtained by linear restrictions

The restrictions on the regression parameters imposed by the null hypotheses in Equations (11.19) and (11.21) can be expressed by equations of the general form

$$c_0\beta_0 + c_1\beta_1 + \cdots + c_p\beta_p = 0, \tag{11.23}$$

where $c_0, c_1, \ldots, c_p$ are suitably chosen constants. For example, the null hypothesis in Equation (11.19) imposes the restrictions

$$(0)\beta_1 + (1)\beta_2 + (0)\beta_3 = 0 \qquad (c_1 = 0, \ c_2 = 1, \ c_3 = 0),$$

$$(0)\beta_1 + (0)\beta_2 + (1)\beta_3 = 0 \qquad (c_1 = c_2 = 0, \ c_3 = 1),$$

whereas the restrictions in Equation (11.21) can be expressed as (see Exercise 11.23)

$$(1)\beta_1 + (0)\beta_2 + (-1)\beta_3 = 0 \qquad (c_1 = 1, \ c_2 = 0, \ c_3 = -1),$$

$$(0)\beta_1 + (1)\beta_2 + (-1)\beta_3 = 0 \qquad (c_1 = 0, \ c_2 = 1, \ c_3 = -1).$$

A restriction of the form in Equation (11.23) is called a *linear restriction* or a *linear constraint*. A hypothesis that specifies linear constraints on a set of parameters is called a *linear hypothesis*.

Linear null hypotheses are encountered frequently in multiple linear regression analysis. We have already seen two examples in Equations (11.19) and (11.21). The linear null hypothesis that a specified subset of $p - q$ regression parameters are all equal to zero can be tested by the method in Box 11.6.

BOX 11.6

Testing a subset of regression parameters

Suppose we have n independent responses satisfying the multiple linear regression model with $p \ (> q)$ independent variables

$$Y = \beta_0 + \beta_1 x_1 + \cdots + \beta_q x_q + \beta_{q+1} x_{q+1} + \cdots + \beta_p x_p + E;$$

this is the full model. We want to test the null hypothesis

$$H_0: \beta_{q+1} = \cdots = \beta_p = 0$$

against the research hypothesis H_1 that at least one of the $p - q$ regression parameters $\beta_{q+1}, \ldots, \beta_p$ is not zero. Then the reduced model under H_0 is

$$Y = \beta_0 + \beta_1 x_1 + \cdots + \beta_q x_q + E.$$

Let $SS[E]_f$ and $SS[E]_r$ denote the error sum of squares for the full and reduced models, respectively. Also, let

$$SS[H_0] = SS[E]_r - SS[E]_f$$

denote the extra error sum of squares resulting from fitting the reduced model instead of the full model. Then a test of H_0 against H_1 can be based on the test statistic

$$F_{extra} = \frac{MS[H_0]}{MS[E]_f},$$

where

$$MS[E]_f = \frac{SS[E]_f}{n - p - 1}$$

is the error mean square for the full model and

$$MS[H_0] = \frac{SS[E]_r - SS[E]_f}{p - q} = \frac{SS[H_0]}{p - q}.$$

If F_{extra} exceeds the α-level critical value of an $F(p - q, n - p - 1)$-distribution, then an α-level test will reject H_0 in favor of H_1.

According to Box 11.6, a large value of the test statistic F_{extra} favors rejection of H_0. This can be understood intuitively by writing the full model in Box 11.6 as

$$Y = \beta_0 + \beta_1 x_1 + \cdots + \beta_q x_q + E^*,$$

where

$$E^* = \beta_{q+1} x_{q+1} + \cdots + \beta_p x_p + E$$

is the error term. Thus, fitting the reduced model is the same as fitting the full model in which the term $\beta_{q+1} x_{q+1} + \cdots + \beta_p x_p$ is pooled with the error term, so that the extra error sum of squares $SS[H_0]$ is the portion of the variability in the observed responses $SS[TOT]$ that can be ascribed to the part of the full model that contains the $p - q$ independent variables specified in H_0. If H_0 is true—that is, if $X_{q+1}, \ldots, X_p$ do not have any effect on the observed responses—then $SS[H_0]$ will be affected only by the random errors, so that $MS[H_0]$ and $MS[E]$ are both estimates of σ^2. Thus, when H_0 is true, values of F_{extra} that are close to 1 would be expected. When H_0 is not true, $SS[H_0]$ will be influenced by the random errors plus those independent variables specified in H_0 whose partial slopes are nonzero. In that case, $MS[H_0]$ will estimate a quantity larger than σ^2, so that the observed value of F_{extra} will tend to be larger than 1.

As we will see later, the method in Box 11.6 can be adapted easily to the case of null hypotheses specified by more complex linear restrictions on the regression parameters.

Extra regression sum of squares

It is useful to interpret $SS[H_0]$ as the extra variation that can be explained by the term $\beta_{q+1}x_{q+1} + \cdots + \beta_p x_p$ in addition to, or in excess of, that explained by the term $\beta_0 + \beta_1 x_1 + \cdots + \beta_q x_q$. Thus, $SS[H_0]$ is often referred to as the *extra regression sum of squares* obtained by adding the parameters $\beta_{q+1}, \ldots, \beta_p$ to a model containing $\beta_0, \beta_1, \ldots, \beta_q$. We will use the notation

$$SS[H_0] = R(\beta_{q+1}, \ldots, \beta_p | \beta_0, \beta_1, \ldots, \beta_q) \tag{11.24}$$

when we wish to emphasize the interpretation of $SS[H_0]$ as the extra regression sum of squares.

For the model

$$Y = \beta_0 + \beta_1 x_1 + \cdots + \beta_p x_p + E, \tag{11.25}$$

the null hypothesis $H_0: \beta_1 = \cdots = \beta_p = 0$ can be interpreted as the null hypothesis that none of the p independent variables is useful for predicting Y. The sum of squares for testing this null hypothesis is the extra error sum of squares resulting from fitting the reduced model

$$Y = \beta_0 + E. \tag{11.26}$$

In Exercise 11.24, you are asked to verify that the error sum of squares from fitting the model in Equation (11.26) is the same as the total sum of squares calculated from the observed responses; that is, $SS[E]_r = SS[TOT]$. Therefore, the sum of squares for testing $H_0: \beta_1 = \cdots = \beta_p = 0$ is

$$SS[H_0] = SS[E]_r - SS[E]_f = SS[TOT] - SS[E], \tag{11.27}$$

where $SS[E]$ denotes the error sum of squares for the full model in Equation (11.25). Notice that $SS[H_0]$ in Equation (11.27) is the same as the regression sum of squares $SS[R]$ in Table 11.2. Thus, a multiple linear regression analysis of a given set of responses can be viewed as an analysis of variance of the data, in which the primary step is to divide $SS[TOT]$ into two components

$$SS[TOT] = SS[R] + SS[E],$$

where $SS[R]$ is the component accounting for the variation due to the independent variables in the model and $SS[E]$ is the component accounting for the variation due to random errors.

EXAMPLE **11.18**   Table 11.4 presents data on the percentages of nitrogen (X_1), chlorine (X_2), potassium (X_3), and the log of leafburn time in seconds (Y) for a sample of $n = 30$ tobacco leaves (Steel & Torrie, 1980). Assuming the multiple linear regression model

$$Y = \beta_0 + \beta_1 x_1 + \beta_2 x_2 + \beta_3 x_3 + E, \tag{11.28}$$

we want to test the null hypothesis that addition of chlorine and potassium will not increase the predictive power of a model that contains only nitrogen as the

TABLE 11.4

Leafburn data for tobacco leaves

i	x_{1i}	x_{2i}	x_{3i}	y_i
1	3.05	1.45	5.67	0.34
2	4.22	1.35	4.86	0.11
3	3.34	0.26	4.19	0.38
4	3.77	0.23	4.42	0.68
5	3.52	1.10	3.17	0.18
6	3.54	0.76	2.76	0.00
7	3.74	1.59	3.81	0.08
8	3.78	0.39	3.23	0.11
9	2.92	0.39	5.44	1.53
10	3.10	0.64	6.16	0.77
11	2.86	0.82	5.48	1.17
12	2.78	0.64	4.62	1.01
13	2.22	0.85	4.49	0.89
14	2.67	0.90	5.59	1.40
15	3.12	0.92	5.86	1.05
16	3.03	0.97	6.60	1.15
17	2.45	0.18	4.51	1.49
18	4.12	0.62	5.31	0.51
19	4.61	0.51	5.16	0.18
20	3.94	0.45	4.45	0.34
21	4.12	1.79	6.17	0.36
22	2.93	0.25	3.38	0.89
23	2.66	0.31	3.51	0.91
24	3.17	0.20	3.08	0.92
25	2.79	0.24	3.98	1.35
26	2.61	0.20	3.64	1.33
27	3.74	2.27	6.50	0.23
28	3.13	1.48	4.28	0.26
29	3.49	0.25	4.71	0.73
30	2.94	2.22	4.58	0.23

independent variable—in symbols, the hypothesis H_0: $\beta_2 = \beta_3 = 0$. The reduced model under H_0 is

$$Y = \beta_0 + \beta_1 x_1 + E.$$

The following output is obtained when the REG procedure of SAS is used to fit the reduced and full models to the leafburn data.

```
PART I
Full Model


Analysis of Variance
                        Sum of       Mean
Source          DF      Squares      Square     F Value     Prob>F
```

Model	3	5.50473	1.83491	40.267	0.0001
Error	26	1.18479	0.04557		
C Total	29	6.68952			

Parameter Estimates

Variable	DF	Parameter Estimate	Standard Error	T for H0: Parameter=0	Prob > |T|
INTERCEP	1	1.811043	0.27951935	6.479	0.0001
X1	1	-0.531455	0.06957678	-7.638	0.0001
X2	1	-0.439636	0.07303727	-6.019	0.0001
X3	1	0.208975	0.04064022	5.142	0.0001

PART II

Reduced Model

Analysis of Variance

Source	DF	Sum of Squares	Mean Square	F Value	Prob>F
Model	1	3.44601	3.44601	29.748	0.0001
Error	28	3.24351	0.11584		
C Total	29	6.68952			

Parameter Estimates

Variable	DF	Parameter Estimate	Standard Error	T for H0: Parameter=0	Prob > |T|
INTERCEP	1	2.625704	0.36102426	7.273	0.0001
X1	1	-0.591614	0.10846980	-5.454	0.0001

The first part of the output shows the ANOVA table and the parameter estimates for the full model, and the second part gives the same information for the reduced model.

Suppose that we start with Equation (11.28) as the model for predicting $Y =$ log (leafburn time) as a function of the three independent variables: X_1, the percentage of nitrogen; X_2, the percentage of chlorine; and X_3, the percentage of potassium. We want to test the null hypothesis H_0: $\beta_2 = \beta_3 = 0$. This hypothesis implies that, when X_1 is used as an independent variable, including either X_2 or X_3 in the model provides no additional advantage in terms of explaining the observed variability in log (leafburn time). Note that the null hypothesis H_0 does not imply that log (leafburn time) is not associated with X_2 or X_3. Rejection of H_0 leads to the conclusion that the model can be improved by adding at least one of X_2 and X_3 to X_1.

The sum of squares for testing H_0 is the difference between the error sums of squares for the full and reduced models. The required sums of squares and their degrees of freedom are found in the ANOVA tables in the REG output. We have $p - q = 3 - 1 = 2$ and

$$SS[H_0] = SS[E]_r - SS[E]_f$$

$$= 3.24351 - 1.18479 = 2.05872.$$

The mean square for testing H_0 is

$$MS[H_0] = \frac{SS[H_0]}{p-q} = \frac{2.05872}{2} = 1.02936,$$

and the calculated value of the test statistic is

$$F_{extra} = \frac{MS[H_0]}{MS[E]_f} = \frac{1.02936}{0.04557} = 22.589.$$

Comparison with the critical values for an $F(2, 26)$-distribution shows that the null hypothesis can be rejected ($p < 0.0001$). Thus, it is reasonable to conclude that, in addition to the nitrogen content, either the chlorine content or the potassium content (or both) will be useful for predicting leafburn time. ■

Simultaneous and one-at-a-time tests

When $p - q > 1$, the F-test in Box 11.6 is a simultaneous test of $p - q$ separate null hypotheses. For instance, in Example 11.18, testing the null hypothesis H_0: $\beta_2 = \beta_3 = 0$ is the same as simultaneously testing two separate null hypotheses H_{02}: $\beta_2 = 0$ and H_{03}: $\beta_3 = 0$. The next example illustrates the importance of distinguishing between simultaneous and one-at-a-time tests of regression parameters.

EXAMPLE **11.19**   For the leafburn data in Example 11.18, we want to consider an equation that is quadratic in the nitrogen content

$$Y = \beta_0 + \beta_1 x_1 + \beta_2 x_1^2 + E$$

as the model for predicting log (leafburn times). The ANOVA table and parameter estimates for the quadratic model obtained by the SAS REG procedure are as follows.

Analysis of Variance

Source	DF	Sum of Squares	Mean Square	F Value	Prob>F
Model	2	3.58632	1.79316	15.602	0.0001
Error	27	3.10320	0.11493		
C Total	29	6.68952			

Parameter Estimates

Variable	DF	Parameter Estimate	Standard Error	T for H0: Parameter=0	Prob > |T|
INTERCEP	1	4.698461	1.91009316	2.460	0.0206
X1	1	-1.852492	1.14625409	-1.616	0.1177
X1SQ	1	0.186069	0.16840072	1.105	0.2789

The p-value of 0.0001 in the ANOVA table indicates that the null hypothesis H_0: $\beta_1 = \beta_2 = 0$ can be rejected in favor of the conclusion that at least one of the two independent variables X_1 and $X_2 = X_1^2$ is useful for predicting Y. Yet the p-values of 0.1177 and 0.2789 for the null hypotheses H_{01}: $\beta_1 = 0$ and H_{02}: $\beta_2 = 0$ indicate that the data do not favor the rejection of either of these hypotheses. At first sight, this might look like a contradiction of the result of the F-test of H_0. The simultaneous test indicates that at least one of the two variables X_1 and X_1^2 should be retained in the model, whereas the individual tests seem to suggest that neither is needed.

However, the result of the simultaneous test does not contradict the results from the one-at-a-time tests. The simultaneous test examines the null hypothesis that both of the independent variables can be simultaneously dropped from the model. The one-at-a-time tests, on the other hand, examine two separate null hypotheses H_{01} and H_{02}, where H_{0i} claims that the variable X_i can be dropped from the model provided the other independent variable is retained in the model. Considering the result of the one-at-a-time tests in conjunction with the result of the simultaneous test, we conclude that only one of the two variables is needed in the model.

To determine which of the two variables is a better predictor of Y, we can fit each variable separately to the data. The ANOVA table in Part I of the output in Example 11.18 shows that, when only X_1 is used in the model, the p-value for testing H_0: $\beta_1 = 0$ is 0.0001. Thus, X_1 by itself is a useful predictor of Y. When $X_2 = X_1^2$ alone is fitted to the data—that is, when the model

$$Y = \beta_0 + \beta_1 x_1^2 + E$$

is used to predict Y—we get the following ANOVA table.

Analysis of Variance

Source	DF	Sum of Squares	Mean Square	F Value	Prob>F
Model	1	3.28613	3.28613	27.035	0.0001
Error	28	3.40339	0.12155		
C Total	29	6.68952			

Thus, X_1^2 alone is also a useful predictor of Y. Which model is better—the one with X_1 alone or the one with X_1^2 alone? How to judge the adequacy of a predictor variable will be addressed in Section 11.10. One approach—not necessarily the best—is to examine the coefficient of determination for the two models; the model yielding the higher coefficient is better. In this instance, the coefficients of determination for X_1 and X_1^2 are 52% and 49%, respectively. Thus, judging from their fit to the observed data, the model with X_1 alone seems slightly better than the model with X_1^2 alone.   ■

Exercises

11.22   For each of the six models in Example 11.17 in turn, determine whether the other five models can be classified as

a   the full model;   **b**   the reduced model;   **c**   neither.

11.23   Show that the null hypothesis in Equation (11.21) is equivalent to the linear restrictions

$$\beta_1 - \beta_3 = 0 \text{ and } \beta_2 - \beta_3 = 0, \qquad \text{(11.29)}$$

by noting that Equation (11.21) implies, and is implied by, Equation (11.29). Display another set of linear restrictions of the form in Equation (11.29) that is equivalent to the null hypothesis in Equation (11.21).

11.24   By setting $p = 0$ in Equation (11.11), show that the least squares estimate of β_0 in the model $Y = \beta_0 + E$ is $\hat{\beta}_0 = \bar{y}$. Hence, argue that the residual sum of squares for the model $Y = \beta_0 + E$ is the same as the total sum of squares for the observed responses; that is,

$$SS[E] = \sum (y_i - \hat{y}_i)^2 = \sum (y_i - \bar{y})^2 = SS[TOT].$$

11.25   When the quadratic model

$$Y = \beta_0 + \beta_1 x_1 + \beta_2 x_2 + \beta_3 x_1 x_2 + \beta_4 x_1^2 + \beta_5 x_2^2 + E$$

is fitted to the hospital stress data in Exercise 11.12, a residual sum of squares $SS[E] = 339{,}971.7162$ is obtained.

  a Given that the total sum of squares does not depend on the assumed model, calculate the regression sum of squares for the quadratic model.

  b Set up the ANOVA table showing the degrees of freedom, the sum of squares, and the mean squares associated with the quadratic model.

  c Test the null hypothesis H_0: $\beta_1 = \beta_2 = \beta_3 = \beta_4 = \beta_5 = 0$ and write your conclusions.

  d Calculate the extra regression sum of squares $R(\beta_3, \beta_4, \beta_5 | \beta_0, \beta_1, \beta_2)$ and interpret this number.

  e Test the null hypothesis H_0: $\beta_3 = \beta_4 = \beta_5 = 0$ and write your conclusions.

  f Calculate the extra regression sum of squares $R(\beta_4, \beta_5 | \beta_0, \beta_1, \beta_2, \beta_3)$ and interpret this number.

  g Test the null hypothesis H_0: $\beta_4 = \beta_5 = 0$ and write your conclusions.

11.7
Multiple and partial correlation coefficients

The simple correlation coefficient is useful to measure linear association between a quantitative response variable Y and a quantitative independent variable X. Analogously, the multiple and partial correlation coefficients are useful for assessing the linear association between a quantitative response variable Y and a set of p quantitative independent variables $\{X_1, \ldots, X_p\}$.

The multiple correlation coefficient

The multiple correlation coefficient is best described in terms of the coefficient of determination associated with a multiple linear regression model. We have already encountered the coefficient of determination for a simple linear regression model; an obvious extension leads to the corresponding definition for a multiple linear regression model. The *sample coefficient of determination* for the multiple linear regression model

$$Y = \beta_0 + \beta_1 x_1 + \cdots + \beta_p x_p + E \tag{11.30}$$

is defined as

$$R^2_{Y.X_1,X_2,\ldots,X_p} = \frac{SS[R]}{SS[TOT]},$$

where SS[R] and SS[TOT] are the regression and total sums of squares obtained by fitting the model to a given data set. The positive square root of the coefficient of determination

$$R_{Y.X_1,X_2,\ldots,X_p} = \sqrt{\frac{SS[R]}{SS[TOT]}}$$

is called the *sample multiple correlation coefficient* between the dependent variable Y and the set of independent variables $\{X_1, X_2, \ldots, X_p\}$. Several useful properties of the coefficient of determination and multiple correlation coefficient can be deduced by direct analogy with simple linear regression.

Upper and lower bounds

The coefficient of determination and the multiple correlation coefficient can take only values between 0 and 1; that is

$$0 \le R^2_{Y.X_1,X_2,\ldots,X_p} \le 1 \qquad \text{and} \qquad 0 \le R_{Y.X_1,X_2,\ldots,X_p} \le 1.$$

The strength of linear association

As in simple linear regression, the coefficient of determination $R^2_{Y.X_1,\ldots,X_p}$ is the fraction of SS[TOT], the variability in the observed responses that can be explained by the variability in the values predicted by the best-fitting multiple linear regression model in Equation (11.30). As such, the coefficient of determination $R^2_{Y.X_1,X_2,\ldots,X_p}$ can be regarded as an overall measure of the strength of linear association between the response variable Y and the set of independent variables $\{X_1, X_2, \ldots, X_p\}$. The coefficient of determination measures linear association because the assumption of a multiple linear regression model implies that the expected responses are linear functions of the independent variables.

Ratio of variances

The sample coefficient of determination can be expressed as the ratio of the variance of the predicted values to the variance of the observed values. In other words, if y_i and $\hat{y}_i$, $i = 1, \ldots, n$, denote the observed and predicted values, then the coefficient of determination is the ratio of the variance of the $\hat{y}_i$ to the variance of the y_i. Recall that a similar interpretation of the coefficient of determination was presented for a simple linear regression model.

Correlation between observed and predicted values

In simple linear regression, the positive square root of the coefficient of determination equals the sample correlation coefficient between the observed responses and the values predicted on the basis of the best-fitting simple linear regression model. A similar property holds in the multiple linear regression model. The positive square root of the coefficient of determination—that is, the multiple correlation coefficient—equals the correlation coefficient between the observed responses and the predicted responses. Thus, the multiple correlation coefficient is the correlation between the observed values and the values predicted by the best-fitting (as judged by least squares) multiple linear regression model.

Population multiple correlation coefficient

The population multiple correlation coefficient between Y and $\{X_1, \ldots, X_p\}$, denoted by $\rho_{Y.X_1, \ldots, X_p}$, is defined as the correlation coefficient between the population of values of Y and the population of values of its best (as judged by least squares) linear predictor of the form $\beta_0 + \beta_1 X_1 + \cdots + \beta_p X_p$. In other words, the population multiple correlation coefficient can be conceptualized as the correlation coefficient between the observed and predicted values when the prediction equation is obtained by fitting the multiple linear regression model to the population of all values of $(Y, X_1, \ldots, X_p)$. It can be shown that, if the assumptions of the multiple linear regression analysis are valid, then testing the null hypothesis H_0: $\rho_{Y.X_1, \ldots, X_p} = 0$ is equivalent to testing H_0: $\beta_1 = \beta_2 = \cdots = \beta_p = 0$. Thus, the null hypothesis that a multiple correlation coefficient is zero can be tested using the test statistic F_{reg} in Equation (11.13).

EXAMPLE **11.20**　In Example 11.18, data were given on $n = 30$ observed values of the dependent variable $Y = \log$ (leafburn time) and three independent variables: X_1, the percentage of nitrogen; X_2, the percentage of chlorine; and X_3, the percentage of potassium. From the first ANOVA table in the REG output in Example 11.18, we can calculate the coefficient of determination for the model

$$Y = \beta_0 + \beta_1 x_1 + \beta_2 x_2 + \beta_3 x_3 + E. \tag{11.31}$$

We find that

$$R^2_{Y.X_1, X_2, X_3} = \frac{5.50473}{6.68952} = 0.8229.$$

Thus, approximately 82% of the variation in the observed log (leafburn time) can be explained by their linear association with the nitrogen, chlorine, and potassium contents of the leaves. The sample multiple correlation coefficient between Y and $\{X_1, X_2, X_3\}$ is $\sqrt{0.8229} = 0.9071$. Thus, the correlation coefficient between the observed log (leafburn time) and the values predicted by the best model of the form in Equation (11.31) is 0.9071. In other words, 0.9071 is the correlation coefficient between the observed log (leafburn time) and the corresponding predicted values based on the prediction equation

$$\hat{y} = 1.811043 - 0.531455x_1 - 0.439636x_2 + 0.208975x_3.$$

If the responses are predicted using any other prediction equation of the form in Equation (11.31), the resulting correlation between the observed and predicted values will be less than 0.9071. In Exercise 11.26, you are asked to demonstrate this by computing the correlation coefficient between the observed values and the values predicted by a different version of the equation.

The null hypothesis H_0: $\rho_{Y.X_1,X_2,X_3} = 0$ can be rejected on the basis of the p-value of 0.0001 for testing H_0: $\beta_1 = \beta_2 = \beta_3 = 0$ in the first ANOVA table of the REG output in Example 11.18. ∎

The partial correlation coefficient

The partial correlation coefficient may be used to assess the advantages of adding a new independent variable X to a multiple linear regression model that already contains p independent variables $\{X_1, \ldots, X_p\}$. For $i = 1, \ldots, n$, let $e_{Yi.X_1,\ldots,X_p}$ and $e_{Xi.X_1,\ldots,X_p}$ denote, respectively, the i-th residual from fitting the models

$$Y = \beta_0 + \beta_1 x_1 + \ldots + \beta_p x_p + E \tag{11.32}$$

and

$$X = \beta_0 + \beta_1 x_1 + \ldots + \beta_p x_p + E \tag{11.33}$$

to a set of n observations of the variables $\{Y, X, X_1, \ldots, X_p\}$. Let $R_{YX.X_1,\ldots,X_p}$ denote the correlation coefficient between the n pairs of residuals $(e_{Yi.X_1,\ldots,X_p}, e_{Xi.X_1,\ldots,X_p})$, $i = 1, \ldots, n$. Then $R_{YX.X_1,\ldots,X_p}$, is called the *sample partial correlation coefficient* between Y and X adjusted for $X_1, X_2, \ldots, X_p$.

A practical interpretation of a sample partial correlation coefficient is as follows. The residuals $e_{Yi.X_1,\ldots,X_p}$ and $e_{Xi.X_1,\ldots,X_p}$ are, respectively, the portions of the i-th observed values of Y and X that are not explained by the independent variables $X_1, \ldots, X_p$. A large absolute value of the correlation coefficient between the two sets of residuals implies that there is a strong linear relationship between the unexplained portions of Y and X, and so a large value of the sample partial correlation coefficient is an indication that additional information about Y can be obtained by adding X to a model containing $X_1, \ldots, X_p$. The next example illustrates the calculation and interpretation of the partial correlation coefficient.

EXAMPLE 11.21 Let's calculate the sample partial correlation coefficient between Y and X adjusted for X_1 and X_2 on the basis of the following $n = 8$ observations:

Y	2.5	1.1	3.8	4.8	2.0	0.8	3.2	1.3
X_1	3.0	4.0	3.3	3.8	3.5	3.7	3.8	3.5
X_2	1.5	1.6	0.3	0.2	0.8	1.6	0.4	1.1
X	25.0	5.2	49.0	82.0	14.0	3.2	38.0	10.0

The first step in the calculation is to fit the two models

$$Y = \beta_0 + \beta_1 x_1 + \beta_2 x_2 + E$$

and

$$X = \beta_0 + \beta_1 x_1 + \beta_2 x_2 + E$$

to the given data. Any statistical software can be used to carry out this step. The least squares equations for predicting Y and X in terms of X_1 and X_2 are

$$\hat{y} = 6.942163 - 0.716979 x_1 - 2.070895 x_2$$

and

$$\hat{x} = 77.321858 - 3.863996 x_1 - 37.555276 x_2.$$

The residuals for the two models are the differences between the observed and predicted values. Straightforward calculations show that the eight pairs of residuals are as follows:

i	1	2	3	4	5	6	7	8
$e_{Yi.X_1,X_2}$	0.8151	0.3392	−0.1549	0.9965	−0.7760	−0.1759	−0.1893	−0.8548
$e_{Xi.X_1,X_2}$	15.6030	3.4226	−4.3041	26.8724	−19.7537	0.2634	−9.6166	−12.4871

The sample partial correlation between Y and X adjusted for X_1 and X_2 is the same as the correlation coefficient between the eight pairs of residuals. Figure 11.6 shows a plot of these pairs of residuals.

It can be verified that this correlation coefficient is 0.9561, so that

$$R_{YX.X_1,X_2} = 0.9561.$$

FIGURE 11.6

Plot of the residuals of Y and X after adjusting for X_1 and X_2

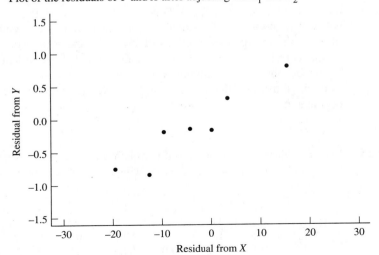

There is a strong linear association between the portions of Y and X left unexplained by X_1 and X_2. It appears that additional useful information about Y can be gained by adding X to a model containing X_1 and X_2.   ∎

Let's look at some properties of partial correlation coefficients.

Partial correlation and partial slope

The slope of the estimated regression line in the simple linear regression of $e_{Y.X_1,\ldots,X_p}$ on $e_{X.X_1,\ldots,X_p}$ is the same as the least squares estimate of the partial slope of X in the regression of Y on $\{X, X_1, \ldots, X_p\}$.

EXAMPLE **11.22**   When fitted to the data on Y, X, X_1, and X_2 in Example 11.21, the model

$$Y = \beta_0 + \beta_1 x_1 + \beta_2 x_2 + \beta x + E$$

yields the prediction equation

$$\hat{y} = 3.6732 - 0.5536 x_1 - 0.4832 x_2 + 0.0423 x,$$

so that the estimated partial slope of X is $\hat{\beta} = 0.0423$. In Exercise 11.26, you will be asked to verify that, if the simple linear regression model

$$e_{Y.X_1,X_2} = \beta_0 + \beta_1 e_{X.X_1,X_2} + E$$

is fitted to the eight pairs of residuals in Example 11.21, the slope of the estimated line will also be $\hat{\beta}_1 = 0.0423$.   ∎

Strength of linear association

As we have seen, the square of the multiple correlation coefficient (the coefficient of determination) is a measure of the linear association between a quantitative variable Y and a set of quantitative variables $\{X_1, \ldots, X_p\}$. Similarly, the square of a partial correlation coefficient is a measure of the linear association between those portions of Y and X that are not explained by their linear association with $\{X_1, \ldots, X_p\}$. Thus, $R^2_{YX.X_1,\ldots,X_p}$ is a measure of the strength of linear association between Y and X after eliminating (adjusting for) their linear association with $\{X_1, \ldots, X_p\}$. We refer to $R^2_{YX.X_1,\ldots,X_p}$ as a *partial coefficient of determination*.

Partial correlation and extra regression sum of squares

Suppose that we fit the model

$$Y = \beta_0 + \beta_1 x_1 + \cdots + \beta_p x_p + \beta x + E$$

to the given data on the variables $\{Y, X_1, \ldots, X_p, X\}$. It can be shown that

$$R^2_{YX.X_1,\ldots,X_p} = \frac{R(\beta|\beta_0, \beta_1, \ldots, \beta_p)}{\text{SS[TOT]} - R(\beta_1, \ldots, \beta_p|\beta_0)},\qquad (11.34)$$

where $R(\beta|\beta_0, \beta_1, \ldots, \beta_p)$ and $R(\beta_1, \ldots, \beta_p|\beta_0)$ are extra regression sums of squares defined as in Equation (11.24). For instance, $R(\beta|\beta_0, \beta_1, \ldots, \beta_p)$ is the extra regression sum of squares obtained by adding β to a model that already contains $\beta_0, \beta_1, \ldots, \beta_p$, and $R(\beta_1, \ldots, \beta_p|\beta_0)$ is the regression sum of squares obtained when fitting the model

$$Y = \beta_0 + \beta_1 x_1 + \cdots + \beta_p x_p + E.\qquad (11.35)$$

The denominator in Equation (11.34) is the unexplained variation (error sum of squares) when the model in Equation (11.35) is fitted to the data, and so the square of the sample partial correlation coefficient can be interpreted as the proportion of the variation in Y not explained by $\{X_1, \ldots, X_p\}$, that can be explained by X. This interpretation of the partial coefficient of determination in Equation (11.34) is illustrated in Figure 11.7.

F I G U R E   11.7

Interpretation of the partial coefficient of determination

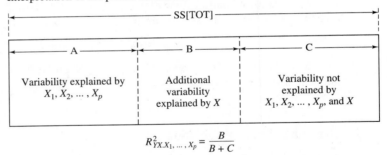

$$R^2_{YX.X_1,\ldots,X_p} = \frac{B}{B + C}$$

Population partial correlation coefficient

The population partial correlation coefficient between Y and X adjusted for $\{X_1, \ldots, X_p\}$ is denoted by $\rho_{YX.X_1,\ldots,X_p}$ and defined as the correlation coefficient between the residuals when the multiple linear regression models in Equations (11.32) and (11.33) are fitted to the population of all values of $\{Y, X, X_1, \ldots, X_p\}$.

If $\beta_{YX.X_1 X_2,\ldots,X_p}$ is the partial slope associated with X in the multiple linear regression model relating Y to $X_1, \ldots, X_p$, and X, then it can be shown that, under the assumptions for a multiple linear regression model, testing the null hypothesis H_0: $\rho_{YX.X_1,\ldots,X_p} = 0$ is the same as testing the null hypothesis H_0: $\beta_{YX.X_1 X_2,\ldots,X_p} = 0$.

EXAMPLE 11.23   Suppose that we want to fit the model

$$Y = \beta_0 + \beta_1 x_1 + \beta_2 x_2 + \beta_3 x_3 + E \qquad \text{(11.36)}$$

to the data in Example 11.18 on $Y = \log$ (leafburn time), X_1 = percentage of nitrogen, X_2 = percentage of chlorine, and X_3 = percentage of potassium. The corresponding output obtained using the GLM procedure in SAS is as follows.

PART I

Dependent Variable: Y

Source	DF	Sum of Squares	Mean Square	F Value	Pr > F
Model	3	5.5047341	1.8349114	40.27	0.0001
Error	26	1.1847859	0.0455687		
Corrected Total	29	6.6895200			

PART II

Source	DF	Type I SS	Mean Square	F Value	Pr > F
X1	1	3.4460076	3.4460076	75.62	0.0001
X2	1	0.8538448	0.8538448	18.74	0.0002
X3	1	1.2048816	1.2048816	26.44	0.0001

Source	DF	Type III SS	Mean Square	F Value	Pr > F
X1	1	2.6587124	2.6587124	58.35	0.0001
X2	1	1.6510626	1.6510626	36.23	0.0001
X3	1	1.2048816	1.2048816	26.44	0.0001

The output is in two parts. Part I shows the ANOVA table for fitting the model, which is the same as the ANOVA table in Part I of the output in Example 11.18.

Part II of the output shows two types of extra regression sums of squares—Type I and Type III—that can be calculated using the GLM procedure of SAS. The Type I SS column contains the extra regression sums of squares obtained by sequentially adding to the model the three variables $X_1, X_2,$ and X_3, in that order. For instance, the Type I SS for X_1 is the extra regression sum of squares resulting from adding X_1 to a model containing no independent variables ($Y = \beta_0 + E$); that is, the Type I SS for X_1 is the same as $R(\beta_1 | \beta_0)$. Similarly, the Type I SS for X_2 is the extra regression sum of squares obtained by adding X_2 to a model that contains X_1 ($Y = \beta_0 + \beta_1 x_1 + E$), so that the Type I SS for X_2 is the same as $R(\beta_2 | \beta_0, \beta_1)$. Because of the way it is calculated, a Type I SS is usually called a *sequential sums of squares.*

The Type III SS for a variable is the extra regression sum of squares resulting from adding that particular variable to a model containing the other two variables. For example, the Type III SS for X_1 is the extra regression sum of squares resulting from the addition of X_1 to a model containing X_2 and X_3. The Type III SS for X_1 is the same as $R(\beta_1 | \beta_0, \beta_2, \beta_3)$. A Type III sum of squares is usually called a *partial sum of squares.* The various Type I and Type III sums of squares in the GLM output are as follows:

Variable	Sequential SS	Partial SS
X_1	$R(\beta_1\|\beta_0) = 3.4460$	$R(\beta_1\|\beta_0, \beta_2, \beta_3) = 2.6587$
X_2	$R(\beta_2\|\beta_0, \beta_1) = 0.8538$	$R(\beta_2\|\beta_0, \beta_1, \beta_3) = 1.6511$
X_3	$R(\beta_3\|\beta_0, \beta_1, \beta_2) = 1.2049$	$R(\beta_3\|\beta_0, \beta_1, \beta_2) = 1.2049$

Recalling that $R(\beta_1, \beta_2, \ldots, \beta_p\|\beta_0)$ is the regression sum of squares for the model in Equation (11.35), we can summarize some useful algebraic relationships for Type I SS as follows:

Variable	Type I SS symbol	Expression
X_1	$R(\beta_1\|\beta_0)$	Regression sum of squares for $Y = \beta_0 + \beta_1 X_1$
X_2	$R(\beta_2\|\beta_0, \beta_1)$	$R(\beta_1, \beta_2\|\beta_0) - R(\beta_1\|\beta_0)$
X_3	$R(\beta_3\|\beta_0, \beta_1, \beta_2)$	$R(\beta_1, \beta_2, \beta_3\|\beta_0) - R(\beta_1, \beta_2\|\beta_0)$

Analagous expressions for Type III SS are as follows:

Variable	Type III SS symbol	Expression
X_1	$R(\beta_1\|\beta_0, \beta_2, \beta_3)$	$R(\beta_1, \beta_2, \beta_3\|\beta_0) - R(\beta_2, \beta_3\|\beta_0)$
X_2	$R(\beta_2\|\beta_0, \beta_1, \beta_3)$	$R(\beta_1, \beta_2, \beta_3\|\beta_0) - R(\beta_1, \beta_3\|\beta_0)$
X_3	$R(\beta_3\|\beta_0, \beta_1, \beta_2)$	$R(\beta_1, \beta_2, \beta_3\|\beta_0) - R(\beta_1, \beta_2\|\beta_0)$

To calculate the square of the partial correlation coefficient (the partial coefficient of determination) between Y and X_3 adjusted for X_1 and X_2, we can use the formula

$$R^2_{YX_3 \cdot X_1, X_2} = \frac{R(\beta_3\|\beta_0, \beta_1, \beta_2)}{SS[TOT] - R(\beta_1, \beta_2\|\beta_0)}.$$

The quantity in the numerator—the partial sum of squares for X_3—is seen from the SAS output to be 1.2049. The total sum of squares SS[TOT] needed to calculate the denominator is found in the ANOVA table in Part I. The quantity $R(\beta_1, \beta_2\|\beta_0)$, which is the extra regression sum of squares due to the addition of X_1 and X_2 to a model containing no independent variables, can be obtained by adding the Type I SS for X_1 and X_2

$$R(\beta_1, \beta_2\|\beta_0) = R(\beta_1\|\beta_0) + R(\beta_2\|\beta_0, \beta_1)$$

$$= 3.4460 + 0.8538 = 4.2998.$$

Therefore, the amount of the observed variation in log (leafburn time) not explained by its linear association with the nitrogen and chlorine contents of the leaf is $SS[TOT] - R(\beta_1, \beta_2\|\beta_0) = 6.6895 - 4.2998 = 2.3897$. The partial coefficient of determination

$$R^2_{YX_3 \cdot X_1, X_2} = \frac{1.2049}{2.3897} = 0.5042$$

is the proportion of unexplained variation in log (leafburn time) that can be explained by adding $X_3 =$ percentage of potassium to a model that already contains the percentage of nitrogen and the percentage of chlorine as the independent variables.

The SAS output in Example 11.18 shows that the least squares estimate of β_3 in the model in Equation (11.36) is $\hat{\beta}_3 = 0.208975$. Since the estimated partial slope for X_3 is positive, the sample partial correlation coefficient $R_{YX_3.X_1,X_2}$ is also positive. Therefore, we have

$$R_{YX_3.X_1,X_2} = +\sqrt{0.5042} = 0.7101. \quad \blacksquare$$

Regression analysis has a wealth of applications; for more information about this important topic, consult a specialized text such as Seber (1977), Draper and Smith (1981), or Myers (1986). Some further aspects of multiple regression are briefly described in the next section.

Exercises

11.26 The SAS output obtained by fitting the model

$$Y = \beta_0 + \beta_1 x_1 + \beta_2 x_2 + \beta x + E$$

to the data in Example 11.21 includes the following information:

Source	DF	Squares	Square	F Value	Pr > F
Model	3	13.704038	4.568013	66.51	0.0007
Error	4	0.274712	0.068678		
Total	7	13.978750			

Source	DF	Type I SS	Mean Square	F Value	Pr > F
X1	1	0.183785	0.183785	2.68	0.1772
X2	1	10.598855	10.598855	154.33	0.0002
X	1	2.921398	2.921398	42.54	0.0029

Source	DF	Type III SS	Mean Square	F Value	Pr > F
X1	1	0.2170814	0.2170814	3.16	0.1501
X2	1	0.1841850	0.1841850	2.68	0.1768
X	1	2.9213975	2.9213975	42.54	0.0029

Parameter	Estimate	T for H0: Parameter=0	Pr > |T|	Std Error of Estimate
INTERCEPT	3.673237345	2.97	0.0413	1.23870108
X1	-0.553621186	-1.78	0.1501	0.31139424
X2	-0.483175489	-1.64	0.1768	0.29504412
X	0.042276870	6.52	0.0029	0.00648211

Observation	Observed Value	Predicted Value	Residual
1	2.50000000	2.34453230	0.15546770
2	1.10000000	0.90551154	0.19448846
3	3.80000000	3.77290141	0.02709859
4	4.80000000	4.93954507	-0.13954507
5	2.00000000	1.94089898	0.05910102
6	0.80000000	0.98704416	-0.18704416
7	3.20000000	2.98272770	0.21727230
8	1.30000000	1.62683885	-0.32683885

a Calculate the variances s_{obs}^2 and s_{pred}^2 of the observed and predicted values, respectively.

b Verify that the sample coefficient of determination calculated from the ANOVA table is the same as the ratio of the variances calculated in (a); that is, verify that

$$R_{Y.X_1,X_2,X}^2 = \frac{s_{pred}^2}{s_{obs}^2}.$$

c Verify that the correlation coefficient between the observed and the predicted values is the same as the multiple correlation coefficient between Y and $\{X_1, X_2, X\}$.

d Suppose we predict the values of Y using the equation

$$\hat{y}^* = 3 - 0.6x_1 - 0.4x_2 + 0.1x.$$

Will the correlation coefficient between the observed values and the values based on this prediction equation be less than, equal to, or more than the correlation coefficient in (c)? Explain the reasoning that supports your answer.

e Do the necessary calculations to verify your answer in (d).

f Calculate the partial correlation coefficient between Y and X after adjusting for the linear association of Y with X_1 and X_2.

g Interpret the number you calculated in (f).

h Verify that the slope estimate in Example 11.22 equals $\hat{\beta}_1 = 0.0423$.

i Explain, giving reasons, why the sum of three Type I SS equals the regression sum of squares, but the same is not true of the Type III SS.

j Use the data to test the null hypothesis that both X_2 and X can be dropped from the model; that is, perform a test to see if the data support the conclusion that at least one of the variables X_2 and X is a useful addition to the model

$$Y = \beta_0 + \beta_1 x_1 + E.$$

11.27 In Example 11.20, we found that the least squares equation for predicting log (leafburn time) on the basis of the model

$$Y = \beta_0 + \beta_1 x_1 + \beta_2 x_2 + \beta_3 x_3 + E$$

for the leafburn data in Example 11.18 is

$$\hat{y} = 1.811043 - 0.531455x_1 - 0.439636x_2 + 0.208975x_3.$$

The observed and least squares predicted values of log (leafburn time) are as follows:

OBS NO.	OBSERVED	PREDICTED	RESIDUAL
1	0.34000000	0.73752196	-0.39752196
2	0.11000000	-0.00958716	0.11958716
3	0.38000000	0.79728306	-0.41728306
4	0.68000000	0.63001067	0.04998933
5	0.18000000	0.11917223	0.06082777
6	0.00000000	0.17233942	-0.17233942
7	0.08000000	-0.07942528	0.15942528
8	0.11000000	0.30567378	-0.19567378
9	1.53000000	1.22456077	0.30543923
10	0.77000000	1.16945208	-0.39945208
11	1.17000000	1.07576371	0.09423629
12	1.01000000	1.01769581	-0.00769581
13	0.89000000	1.19582047	-0.30582047
14	1.40000000	1.16455663	0.23544337
15	1.05000000	0.97303237	0.07696763
16	1.15000000	1.15352328	-0.00352328
17	1.49000000	1.37232124	0.11767876
18	0.51000000	0.45853138	0.05146862
19	0.18000000	0.21513193	-0.03513193
20	0.34000000	0.44921266	-0.10921266
21	0.36000000	0.12387627	0.23612373
22	0.89000000	0.85030609	0.03969391
23	0.91000000	0.99458766	-0.08458766
24	0.92000000	0.68204602	0.23795398
25	1.35000000	1.05449138	0.29550862
26	1.33000000	1.09668716	0.23331284
27	0.23000000	0.18376596	0.04623404
28	0.26000000	0.39134078	-0.13134078
29	0.73000000	0.83062828	-0.10062828
30	0.23000000	0.22967940	0.00032060

a Calculate the variances s_{obs}^2 and s_{pred}^2 of the observed and predicted values, respectively.

b Calculate the sample coefficient of determination as the ratio of these two variances.

c Verify that the simple correlation coefficient between the observed and the least squares predicted values is the same as the multiple correlation coefficient between Y and $\{X_1, X_2, X_3\}$.

d Will the correlation coefficient between the observed values and the values based on the prediction equation

$$\hat{y} = 3.0 - 0.3x_1 - 0.6x_2 + 0.5x_3$$

be less than, equal to, or more than the correlation coefficient in (c)? Explain.

e   Calculate the partial correlation coefficient between log (leafburn time) and the percentage of nitrogen after adjusting for the percentage of chlorine and the percentage of potassium.

f   Interpret the number you calculated in (e).

11.8
Residual plots in multiple linear regression

The role of a residual plot as a diagnostic tool in multiple linear regression is very similar to its role in simple linear regression. As in a simple linear regression model, the assumption that the errors are independent and normal can be checked with a normal q-q plot of the residuals. The equal variance assumption can be checked by plotting the residuals against the predicted values.

The residual plot needed to check the assumption that the regression function is a linear function of the independent variables is slightly different in multiple linear regression analysis than in simple linear regression analysis. When the regression function involves only one independent variable, as in a simple linear regression function, a plot of the residuals against the levels of the independent variable can be used as a visual aid in deciding whether the regression function can be represented by a straight line. Because a multiple linear regression function involves more than one independent variable, the plots of the residuals against the independent variables will be in a space of three or more dimensions. Such plots are not only difficult to construct but also difficult to interpret. An alternative to plotting residuals against multiple independent variables is to construct residual plots known as partial regression leverage plots. As we will see, partial regression leverage plots are two-dimensional residual plots that are helpful in evaluating the linearity of a multiple linear regression function.

For $i = 1, \ldots, n$, let the residuals $e_{Yi.X_1,\ldots,X_p}$ and $e_{Xi.X_1,\ldots,X_p}$ be defined as in Section 11.7. The plot of $e_{Yi.X_1,\ldots,X_p}$ against $e_{Xi.X_1,\ldots,X_p}$ is called a *partial regression leverage plot* of Y against X adjusted for $\{X_1,\ldots,X_p\}$. As noted in Section 11.7, the pattern of association between these two sets of residuals helps us to visualize whether the variable X will contribute additional information about Y in a model that already contains the variables $X_1,\ldots,X_p$. The observed pattern of points in the partial regression leverage plot for a candidate independent variable suggests the form in which that variable should enter the regression function. For instance, a straight-line pattern for X_1 will suggest linear inclusion of the corresponding term in the regression function; that is, the regression function should contain the term $\beta_1 x_1$. Similarly, a quadratic pattern suggests the inclusion of the term $\beta_1 x_1^2$ in the regression function.

Figure 11.8 shows partial regression leverage plots for the independent variables X_1 and X_2 on the basis of a multiple linear regression model with a dependent variable Y. The plots indicate a straight-line pattern for X_1 and quadratic pattern for X_2. Correspondingly, a model with only a linear term for X_1 and linear and quadratic terms for X_2 is suggested in this case

$$Y = \beta_0 + \beta_1 x_1 + \beta_2 x_2 + \beta_3 x_2^2 + E.$$

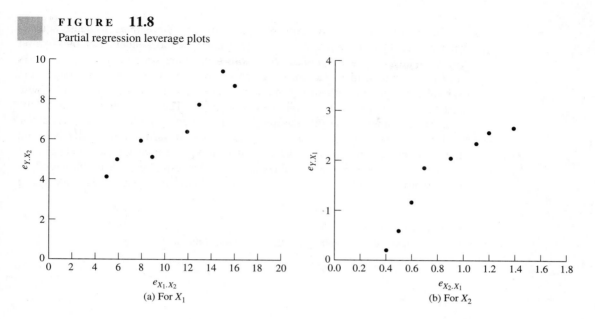

FIGURE 11.8

Partial regression leverage plots

(a) For X_1

(b) For X_2

The residuals can also be used to identify data points that are not consistent with the observed pattern in the rest of the data set. Such points, often referred to as *outliers*, must be examined in detail, and causes for their occurrence must be identified. The presence of outliers may be an indication of the violation of such model assumptions as: (1) the failure of the random error associated with the outliers to have zero mean; or (2) the failure of the variance of the random error associated with the outliers to be the same as the variance of the rest of the observations. It is also possible that the outliers have arisen purely by chance. Whether an outlying data point should be deleted from regression analysis requires careful evaluation. For more information about residual plotting, see Cook and Weisberg (1982) and Belsley, Kuhn, and Welsch (1980).

11.9
Multicollinearity

In multiple linear regression problems, we need an alternative to the least squares analysis if the underlying assumptions of the analysis are violated. In that case, possible remedies are transformation of the variables, weighted regression analysis, rank-based regression analysis, and logistic regression analysis. The basic principles of these techniques in multiple linear regression are the same as for simple linear models (Chapter 10), but the implementation of the methods and the interpretation of their results are much more complicated. The use of rank-based methods in multiple linear regression is further limited because no suitable software is available. For applications of weighted and logistic multiple regression models, see Draper and Smith (1981) and Myers (1986).

In multiple regression problems, furthermore, alternatives to least squares analysis are often recommended when the data on the independent variables exhibit a pattern known as *multicollinearity*, in which there are strong linear dependencies among the independent variables. As an example, a set of data on independent variables X_1 and X_2 with strong multicollinearity is compared in Figure 11.9 with a data set that does not exhibit multicollinearity. The crosshatched area on the x_1, x_2 plane shows the region covered by the settings of the independent variables. In Figure 11.9a, the x_1, x_2 values are confined to a narrow rectangular band on the x_1, x_2 plane. This implies that there is a strong linear relationship between the settings of the independent variables at which the responses will be measured; in other words, the settings of the independent variables in Figure 11.9a show strong

F I G U R E   11.9

Independent variables exhibiting marked and minimal multicollinearity

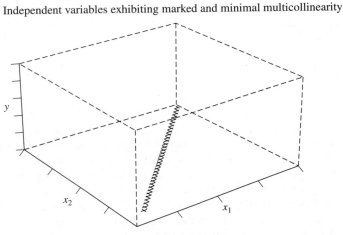

(a) Strong indication of multicollinearity

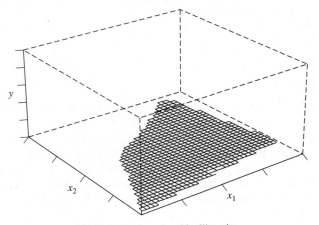

(b) No indication of multicollinearity

multicollinearity. For the design points in Figure 11.9b, multicollinearity is not likely to be a problem, because the x_1, x_2 values are spread out over the region and do not indicate a strong tendency for linear association.

Intuitively, we see that estimation of the expected response surface over a wide region on the basis of responses covering a narrow region is likely to be less reliable than estimation based on responses over the entire region of interest. In practice, we find that the presence of strong multicollinearity tends to reduce the reliability of the least squares prediction equations; the estimated regression parameters tend to have large variances.

Detection of multicollinearity is not difficult in the case of two independent variables; it is enough to plot the x_1, x_2 values or look at the correlation coefficient between them. In typical cases, unfortunately, multicollinearity arises when we are dealing with a large number of independent variables and we have no control over the settings at which the responses will be measured.

There are many diagnostic tools for detecting multicollinearity. In the presence of multicollinearity, the data may be subjected to *ridge regression analysis*, *principal-component regression analysis*, or other techniques. For more information, see Myers (1986) and Draper and Smith (1981).

11.10
Selecting independent variables

The selection of the independent variables for a multiple linear regression model can be conducted in two steps: (1) selecting a set of candidate independent variables; and (2) selecting a subset of the candidate variables for inclusion in the final model.

Selecting a candidate pool of independent variables is not easy. One of the primary considerations in establishing such a pool is the strength of association between the independent and dependent variables. Usually, we would want to include in the candidate pool all variables that have strong association with the dependent variable. For instance, if a variable X is believed to have a strong linear association with the dependent variable Y, then X is a natural candidate for inclusion in the pool. If a quadratic relationship is suspected between X and Y, on the other hand, then both X and X^2 should be included in the set. However, the strength of association is not the only consideration when establishing a candidate pool. The objective of a particular study may preclude the inclusion of a variable even if it is known to have a high association with the response of interest. For instance, when developing a regression model to study the systolic blood pressure Y as a function of various suspected risk factors—such as X_1, age; X_2, body weight; X_3, amount of daily exercise; and X_4, amount of daily calorie intake—the independent variables that will actually be used in the model will depend on the purpose for which the model is being developed. If the objective is to see how well the blood pressure can be predicted using only body weight and age, the only possible candidates for inclusion in the model are X_1 and X_2 and functions related to them, such as X_1^2, X_2^2, and $X_1 X_2$. If the study objective is to identify all risk factors for blood pressure, the candidate list may include many more variables.

Once the set of candidate independent variables has been identified, the next step is to select a subset for inclusion in the final model. A multiple linear regression function containing a subset of the candidate variables is called a *subset regression function*.

The best criterion to use for selecting a subset regression function has received much attention recently. Clearly, the criterion should depend on the intended use of the regression model. Will the regression function be used to describe the observed pattern in the data, or will it be used to identify the nature of the relationship—for example, linear or quadratic—between the dependent and independent variables? Will the model be used to estimate and predict future mean responses? Will the prediction be confined to the region of values of the independent variables in the current data set, or will there be extrapolation outside this region? Questions such as these must be answered before we can decide on a criterion for selecting a subset regression function. As noted by Hocking (1976), various criteria can be used to select subset regression functions; no single criterion is the best under all circumstances. We focus here on a subset selection criterion based on the coefficient of determination R^2 and a statistic known as Mallows C_p.

A subset regression selection problem can be formulated in terms of three types of models: a true model; a model that is underspecified; and a model that is overspecified. A model in which the regression function is specified correctly is called a *true model*. If the regression function contains only a subset of the independent variables in the true model, then the model is said to be *underspecified*. A model in which the regression function contains independent variables in addition to those contained in the true model is said to be *overspecified*. For example, if the model with p variables

$$Y = \beta_0 + \beta_1 x_1 + \cdots + \beta_p x_p + E$$

is the true model, then the model

$$Y = \beta_0 + \beta_1 x_1 + \cdots + \beta_{p-1} x_{p-1} + E$$

is underspecified because it contains one less independent variable. On the other hand, the model

$$Y = \beta_0 + \beta_1 x_1 + \cdots + \beta_p x_p + \beta_{p+1} x_{p+1} + E$$

is overspecified because it contains the additional term $\beta_{p+1} x_{p+1}$. As noted by Hocking (1976) and Myers (1986), an underspecified model introduces bias into the estimates of regression parameters and expected responses, whereas an overspecified model increases the variances of these estimates. Thus, we want a subset selection criterion that yields the best possible approximation to the true model. As a rule of thumb, it is better to select an underspecified model than an overspecified model because, in many instances, the disadvantage of small biases is outweighed by the advantage of reduced variance.

The coefficient of determination R^2 is a natural candidate for a subset selection criterion. Unfortunately, although R^2 can be interpreted as the proportion of the total observed variation due to the variation in the predicted responses, it has some

TABLE 11.5
Small but significant R^2

Source	df	SS	MS	F	R^2	p
Model	2	10	5	10	0.10	0.0001
Error	180	90	0.5			
Total	182	100				

TABLE 11.6
Large but nonsignificant R^2

Source	df	SS	MS	F	R^2	p
Model	5	30	6	3	0.79	0.1547
Error	4	8	2			
Total	9	38				

serious drawbacks as a subset selection criterion. Crocker (1972) noted that neither the magnitude of R^2 nor its statistical significance is completely reliable for the evaluation of a particular regression model. As we see in Tables 11.5 and 11.6, it is possible to obtain a small value of R^2 that is statistically significant or a large value of R^2 that is not statistically significant. Clearly, this is undesirable, because it implies that the magnitude of R^2 alone is not sufficient to determine the adequacy of a subset regression function.

Another drawback of R^2 is that, depending on the steepness of the expected response surface, the same value of the residual sum of squares can give different values of R^2 (Barrett, 1974). Scatter plots and the best-fitting least squares straight lines for two sets of data on an independent variable Y and a dependent variable X are shown in Figure 11.10. Since the vertical distances of the data points from the least squares straight line—that is, the residuals—are the same in both plots, the residual sum of squares SS[E] associated with a simple linear regression model is the same in both sets of data. However, the data with the steeper fitted straight line (Figure 11.10a) has the larger SS[TOT] because the spread of the values of Y is larger in this case. Using the formula

$$R^2 = 1 - \frac{SS[E]}{SS[TOT]},$$

we see that, for any fixed value of SS[E], the data set with the larger SS[TOT] will have a larger R^2. Thus, R^2 is larger for the data with a steeper fitted regression line (Figure 11.10a). This is a drawback of R^2 as a measure of the adequacy of a regression function, because the steepness of a regression function should have no influence on deciding how well the regression model fits the data set.

FIGURE   11.10

Two regression lines with the same SS[E] but different R^2

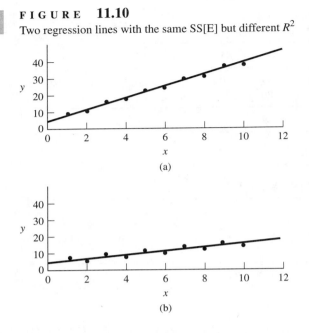

(a)

(b)

The C_p criterion for subset selection

At present, a widely accepted method of subset selection is to use R^2 in conjunction with the C_p criterion suggested by Mallows (1973). Using R^2 with C_p offers a reasonable compromise between minimizing bias and minimizing standard errors.

Suppose that there are m variables in the candidate pool, and let MS[E: p] denote the estimated error variance (error mean squares) obtained by fitting a subset regression model containing p variables. For a data set containing n responses, Mallows suggested

$$C_p = p + 1 + \frac{(\text{MS[E: } p] - \text{MS[E: } m])(n - p - 1)}{\text{MS[E: } m]}$$

as a criterion for evaluating a subset regression containing p variables ($p + 1$ parameters). Models with $C_p \leq p + 1$ are preferable over those for which $C_p > p + 1$. Among the models that are acceptable on the basis of the C_p criterion, a model that has a small number (p) of independent variables and a large value of R^2 is preferable.

EXAMPLE 11.24   For the leafburn data in Example 11.18, we want to select a set of independent variables to model $Y = \log(\text{leafburn time})$ as a function of three variables: X_1, the percentage of nitrogen; X_2, the percentage of chlorine; and X_3, the percentage of potassium. For the purpose of this example, let's assume that $\{X_1, X_2, X_3, X_1^2, X_2^2, X_3^2, X_1X_2, X_1X_3, X_2X_3\}$ is the candidate set of independent variables. Our objective is to use the R^2 and C_p criteria to select a subset of the nine candidate variables.

With nine variables in the candidate set, there are $2^9 - 1 = 511$ possible subset regressions (9 containing a single variable, 36 containing two variables, 84 containing three variables, and so on). The SAS RSQUARE procedure can be used for easy and quick computation of the R^2 and C_p values for all of the 511 subset regressions. The values of C_p are plotted against p in Figure 11.11, which also shows the line at which $C_p = p + 1$. Points below line correspond to subsets with $C_p < p + 1$, whereas points above the line correspond to subsets with $C_p > p + 1$. Thus, the subsets that are desirable on the basis of the C_p criterion correspond to points close to and below the line $C_p = p + 1$. Among all such subsets, the one with the smallest number of variables (smallest p) and the largest value of R^2 is the desirable subset.

FIGURE   11.11

Plot of C_p vs. p+1 for leafburn data

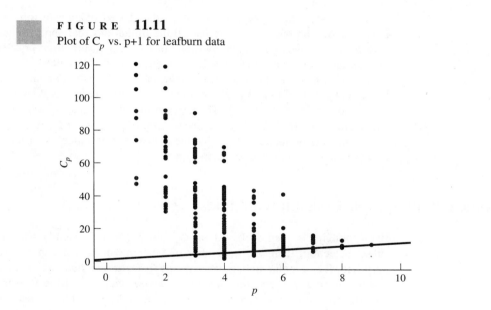

From Figure 11.11, it can be seen that the smallest value of p that gives a C_p value close to but less than $p + 1$ is $p = 3$. Therefore, according to the C_p criterion, a subset containing three variables seems to be adequate for explaining Y. Among these three-variable subsets (there are 84 of them), we want to select the one for which $C_p \leq p + 1 = 3 + 1$ and R^2 is as high as possible. The following is a portion of the SAS output showing the R^2 and C_p values for some three-variable subsets that have the highest values for R^2.

Number in Model	R-square	Cp	Variables in Model
3	0.83194018	3.54831	X1, X2, X1X3
3	0.82288925	4.92423	X1, X2, X3
3	0.82159696	5.12068	X2, X1X3, X1SQ
3	0.82007581	5.35192	X1, X2, X3SQ
3	0.81992440	5.37494	X1, X1X3, X2X3

3	0.81504323	6.11697	X2, X3, X1SQ
3	0.80994057	6.89267	X1X3, X2X3, X1SQ
3	0.80975233	6.92129	X2, X1SQ, X3SQ
3	0.80154080	8.16960	X1, X2X3, X3SQ

Notice that the subset containing the variables X_1, X_2, and X_1X_3 has the highest possible R^2 among the three-variable models and is such that the associated C_p is close to and below the cutoff value of 4. Thus, the best subset for explaining Y contains the variables X_1, X_2, and X_1X_3, with the associated multiple linear regression model

$$Y = \beta_0 + \beta_1 x_1 + \beta_2 x_2 + \beta_3 x_1 x_3 + E.$$

Of course, the C_p criterion is only a data-based diagnostic tool, and any diagnosis based on one set of data has all the limitations that appear when a sample statistic is regarded as a population value. In verifying assumptions or selecting models, nothing can replace the considered judgments of experienced investigators and statisticians. For example, the investigator may have reason to believe that a model based on the subset $\{X_1, X_2, X_3\}$ has more scientific credibility than a model based on the subset $\{X_1, X_2, X_1X_3\}$. Then, given that the C_p and R^2 values for the two subsets are close to each other, the investigator might select $\{X_1, X_2, X_3\}$. That subset has the additional advantage that the corresponding model is easier to interpret, because there is no interaction term.   ■

Some common subset selection methods

Even though the method based on C_p and R^2 is a frequently used method for subset selection, there are a number of other methods that can be used for this purpose. More popular among these are the *forward selection* method, the *backward elimination* method, and the *stepwise selection* method. In the forward selection method, we start with a model containing no independent variables and add new independent variables one at a time, making sure that the added variable significantly (as judged by an appropriate F test) improves the predictive ability of the model. In the backward elimination method, we start with a model containing all independent variables in the candidate pool and eliminate, one at a time, all those variables that do not make significant contribution to the predictive ability of the model. The stepwise method is similar to the forward selection method in that the variables are added one by one to the model, but differs from the forward selection method in that the variables already in the model need not necessarily stay there. As in the forward selection method, the variable to be added to the model must be significant at a specified level. Once a new variable is added to the model, the stepwise method looks at all those variables already in the model and deletes all those variables that are not significant at a specified level.

EXAMPLE **11.25**   In Example 11.24 we saw how the C_p and R^2 criteria can be used in combination to select a set of independent variables to model log (leafburn time). The following shows the codes and a portion of the output when the stepwise method in the REG procedure of SAS is used for the same purpose.

```
------------------------------------------------------------------------
    PART 1  Codes for STEPWISE option in REG procedure of SAS

Data leafbrn1;
infile 'leafbrn.dat';
input obsn x1 x2 x3 y;
x1sq=x1*x1; x2sq=x2*x2; x3sq=x3*x3;
x1x2=x1*x2; x1x3=x1*x3; x2x3=x2*x3;
run;
proc reg;
model y=x1 x2 x3 x1x2 x1x3 x2x3 x1sq x2sq x3sq
          / method=stepwise sle=.10 sls=.05;
run;
------------------------------------------------------------------------
                    PART 2 Output


       Stepwise Procedure for Dependent Variable Y

Step 1   Variable X1 Entered
                 R-square = 0.51513526   Cp = 47.70872455


              Parameter    Standard
Variable      Estimate     Error          F       Prob>F

INTERCEP     2.62570402    0.36102426     52.90   0.0001
X1          -0.59161367    0.10846980     29.75   0.0001


          ----------------------------------------------


Step 2   Variable X2 Entered
                 R-square = 0.64277444   Cp = 30.30512594


              Parameter    Standard
Variable      Estimate     Error          F       Prob>F

INTERCEP     2.65313172    0.31569264     70.63   0.0001
X1          -0.52854935    0.09696245     29.71   0.0001
X2          -0.28996442    0.09335597      9.65   0.0044


          ----------------------------------------------


Step 3   Variable X1X3 Entered
                 R-square = 0.83194018   Cp = 3.54831144
```

Variable	Parameter Estimate	Standard Error	F	Prob>F
INTERCEP	2.81116779	0.22258374	159.51	0.0001
X1	-0.83761159	0.08864063	89.29	0.0001
X2	-0.44384158	0.07118282	38.88	0.0001
X1X3	0.06396689	0.01182441	29.27	0.0001

Step 4   Variable X1X2 Entered
                 R-square = 0.85602137   Cp = 1.88750833

Variable	Parameter Estimate	Standard Error	F	Prob>F
INTERCEP	3.42752155	0.36741719	87.02	0.0001
X1	-1.01577141	0.12079559	70.71	0.0001
X2	-1.28629282	0.41743124	9.50	0.0050
X1X2	0.25065706	0.12258008	4.18	0.0515
X1X3	0.06178217	0.01121227	30.36	0.0001

Step 5   Variable X1X2 Removed
                 R-square = 0.83194018   Cp = 3.54831144

Variable	Parameter Estimate	Standard Error	F	Prob>F
INTERCEP	2.81116779	0.22258374	159.51	0.0001
X1	-0.83761159	0.08864063	89.29	0.0001
X2	-0.44384158	0.07118282	38.88	0.0001
X1X3	0.06396689	0.01182441	29.27	0.0001

All variables left in the model are significant at the 0.0500 level.
The stepwise method terminated because the next variable to be entered
was just removed.

The SAS codes sle $= .10$ and sls $= .05$ set the level for entering a new variable at
$\alpha = 0.10$ and the level for deleting an existing variable at $\alpha = 0.05$. In other words,
the codes imply that in the stepwise selection procedure, a new variable will be
entered into the model if it is significant at $\alpha = 0.10$ and an existing variable will be
deleted from the model if it is *not* significant at $\alpha = 0.05$ level. As can be seen from
the SAS output, the variables X_1, X_2, X_1X_3, and X_1X_2 were entered into the model in

the first four steps. However, in the fifth step, the variable $X_1 X_2$ was removed from the model because it was not significant at the 0.05 level (p-value $= 0.0515$).   ■

Exercises

11.28   Refer to the data in Exercise 11.26.

   a   Construct the necessary residual plots to check for violations of the assumption that the errors are independent $N(0, \ \sigma^2)$.

   b   Calculate the R^2 and C_p values for the subsets $\{X_1\}$, $\{X_1, X_2\}$, and $\{X_1, X_2, X\}$. Comment on the adequacies of these three subsets as predictors of the dependent variable.

   c   Suppose you want to use the data to select the best three-variable subset from the candidate set $\{X_1, X_2, X_3, X_1 X_2, X_2 X_3, X_2 X_3\}$. Explain how you would do so.

   d   Using appropriate computer software, perform the subset selection in (c).

11.29   Refer to the computer output in Exercise 11.27.

   a   Construct the necessary residual plots to see if it is reasonable to assume that the errors are independent $N(0, \ \sigma^2)$.

   b   Using the computer output in Example 11.23, calculate the R^2 and C_p values associated with the subsets $\{X_1\}$, $\{X_1, X_2\}$, and $\{X_1, X_2, X_3\}$. On the basis of these values, comment on the adequacy of these three subsets as explanatory variables for log (leafburn time).

11.11
Overview

The multiple linear regression model is frequently used to study relationships between a quantitative response variable and a set of quantitative independent variables. The model is based on the assumption that the expected response at a given setting of the independent variable is a known (in the sense that the coefficients are known) linear function of a set of regression parameters. The coefficients in the linear function are determined by the settings of the independent variables at which the response will be observed.

Because multiple linear regression models are generalizations of simple linear regression models, all of the concepts developed for simple linear models have their counterparts for multiple linear models. In particular, concepts such as the regression sum of squares, the coefficient of determination, and confidence and prediction intervals have similar interpretations in both settings.

A key difference is that, in inferences based on multiple linear regression models, we need to study the association of the response to multiple independent variables, as opposed to a single independent variable in simple linear regression. Consequently, when using multiple linear regression models, we employ concepts specifically designed to distinguish between simultaneous and one-at-a-time inferences—for example, the notions of the extra sum of squares, the partial correlation coefficient, and multicollinearity.

12

The General Linear Model

12.1
Introduction

In Chapter 8, we saw that the one-way ANOVA model can be used to describe the relationship between a quantitative response variable and a qualitative independent variable. In Chapters 10 and 11, we looked at the simple and multiple linear regression models, which describe relationships between a quantitative response variable and one or more quantitative independent variables. The ANOVA and regression models have one thing in common; both are models in which the response variable is quantitative. The difference between the two types of models is that the independent variables in an ANOVA model are qualitative, whereas those in a regression model are quantitative. The purpose of this chapter is to introduce a unified class of models—called general linear models—in which the independent variables can be qualitative or quantitative, in any combination. As we would expect, the ANOVA and regression models are special cases of general linear models.

12.2

A regression format for the ANOVA model

Let's begin by expressing a one-way ANOVA model in the form of a multiple linear regression model.

EXAMPLE 12.1   In an experiment to compare the effects of three therapies (treatments) for improving mental capacity, 30 subjects were randomly divided into three groups of ten; the three treatments were randomly assigned to the three groups. Each subject was given a test (pretest) requiring mental addition prior to his or her assigned treatment and another test (posttest) after the treatment was complete. The pretest (Z) and posttest (Y) scores for the 30 experimental subjects are as follows:

Treatment 1	Z	24	28	38	42	24	39	45	19	19	22
	Y	45	50	59	60	47	66	76	50	39	36
Treatment 2	Z	23	33	31	34	18	24	41	34	30	39
	Y	28	39	36	39	22	28	49	39	33	43
Treatment 3	Z	27	27	44	38	32	26	24	13	36	52
	Y	34	31	55	43	44	28	33	13	39	58

As always, the first step in the analysis of these data is to select an appropriate statistical model. Two identifiable factors may affect the response (posttest score): X, the type of treatment received by the subject (a qualitative factor); and Z, the subject's pretest score (a quantitative factor). Neither the ANOVA models nor the multiple linear regression models are equipped to handle qualitative and quantitative independent variables simultaneously, and so they are not suitable for analyzing the posttest scores. What we need is a model that can handle qualitative and quantitative explanatory variables at the same time. The general linear model is such a model, and its use to analyze the posttest scores will be discussed in detail in the next section. In this example, we will ignore the effects of the pretest scores and illustrate how the ANOVA model for relating the posttest scores to the treatment types can be expressed as a general linear model.

Recall that the effect of a treatment on the posttest score can be modeled using a one-way ANOVA model involving $t = 3$ treatments. If Y_{ij} denotes the posttest score for the j-th subject in the i-th treatment group ($i = 1, 2, 3$; $j = 1, \ldots, 10$), then a one-way ANOVA model for Y_{ij} takes the form

$$Y_{ij} = \mu + \tau_i + E_{ij}, \qquad i = 1, 2, 3; j = 1, \ldots 10. \tag{12.1}$$

Here μ is the overall mean posttest score; τ_i is the effect of the i-th treatment; and E_{ij} is the random error in Y_{ij}. In the usual ANOVA setting, the τ_i add to zero ($\tau_1 + \tau_2 + \tau_3 = 0$), and the E_{ij} are assumed to be independent $N(0, \sigma^2)$.

The model in Equation (12.1) can be expressed in a form that looks like a regression model with two independent variables. Define the independent variables X_1

and X_2 as

$$X_1 = \begin{cases} 1 & \text{for a subject in treatment group 1} \\ 0 & \text{otherwise;} \end{cases}$$

$$X_2 = \begin{cases} 1 & \text{for a subject in treatment group 2} \\ 0 & \text{otherwise.} \end{cases}$$

Number the subjects from 1 to 30. Let Y_i denote the posttest score for the i-th subject and E_i be the associated random error. If x_{1i} and x_{2i} are, respectively, the values of X_1 and X_2 for the i-th subject, then the one-way ANOVA model in Equation (12.1) is equivalent to the model

$$Y_i = \beta_0 + \beta_1 x_{1i} + \beta_2 x_{2i} + E_i, \qquad i = 1, \ldots, 30. \tag{12.2}$$

The only difference between the models in Equations (12.2) and (12.1) is in the interpretation of the parameters. To see this, compare the expected responses as expressed by the two models, as follows:

Treatment	(x_{1i}, x_{2i})	Expected response for ANOVA form	Expected response for regression form
1	$(1, 0)$	$\mu_1 = \mu + \tau_1$	$\beta_0 + \beta_1$
2	$(0, 1)$	$\mu_2 = \mu + \tau_2$	$\beta_0 + \beta_2$
3	$(0, 0)$	$\mu_3 = \mu + \tau_3$	β_0

A comparison of the expected responses for treatment 3 yields

$$\mu_3 = \mu + \tau_3 = \beta_0, \tag{12.3}$$

so that β_0 in Equation (12.2) is the same as $\mu + \tau_3 = \mu_3$ in Equation (12.1). Similarly, by comparing the expected responses for treatment 2, we get

$$\mu_2 = \mu + \tau_2 = \beta_0 + \beta_2. \tag{12.4}$$

Therefore, subtracting Equation (12.3) from Equation (12.4), we get

$$\beta_2 = \mu_2 - \mu_3 = (\mu + \tau_2) - (\mu + \tau_3) = \tau_2 - \tau_3,$$

so that β_2 is the difference between the expected responses for treatment 2 and treatment 3 or, equivalently, is the difference between the effects of treatment 2 and treatment 3. Similarly, $\beta_1 = \mu_1 - \mu_3 = \tau_1 - \tau_3$. Notice that, in Equation (12.2), the treatment effects are measured by regarding treatment 3 as the reference treatment.

The model in Equation (12.2) is a multiple linear regression model with $p = 2$ independent variables, except that the independent variables are not quantitative. The values 0 and 1 taken by X_1 and X_2, though numerical, have no quantitative interpretation. The numerical values of X_1 and X_2 simply convey information about the assignment of the treatments to the experimental subjects. Indeed, the main difference between Equations (12.1) and (12.2) is in the way a subject's treatment group is identified. In the ANOVA model, the treatment assignments are indicated by the first subscript i in Y_{ij}, whereas in Equation (12.2) the treatment groups are identified by the values of X_1 and X_2. The values $(X_1, X_2) = (1, 0)$, $(0, 1)$, and $(0, 0)$ correspond to treatment groups 1, 2, and 3, respectively.   ■

The model in Equation (12.2), which has the general form

$$Y = \beta_0 + \beta_1 x_1 + \beta_2 x_2 + E,$$

is a *general linear model*. The class of general linear models is an extension of the class of multiple linear regression models to cases where the independent variables are not necessarily quantitative. Like the ANOVA and multiple linear regression models, a general linear model is best regarded as the sum of two components: a *general linear function* indicating the assumed form of the expected response; and a *random error*. A formal description of a general linear function and the associated general linear model is given in Box 12.1.

BOX **12.1**   *The general linear model*

Let $\mu(x_1, \ldots, x_p) = \mathcal{E}(Y | x_1, \ldots, x_p)$ denote the expected value of a response variable Y, given the values $x_1, \ldots, x_p$ of p numerically valued (but not necessarily quantitative) independent variables $X_1, X_2, \ldots, X_p$. Then $\mu(x_1, \ldots, x_p)$ is called a general linear regression function if it can be expressed as a linear combination

$$\mu(x_1, \ldots, x_p) = \beta_0 + \beta_1 x_1 + \cdots + \beta_p x_p$$

of $p + 1$ parameters $\beta_0, \beta_1, \ldots, \beta_p$. As in a multiple linear regression function, β_0 is called the *intercept parameter*, and $\beta_1, \ldots, \beta_p$ are called the *slope parameters*.

A general linear model for Y takes the form

$$Y = \beta_0 + \beta_1 x_1 + \cdots + \beta_p x_p + E,$$

where E is the random error in Y. In particular, for $i = 1, \ldots, n$, let Y_i denote the value of Y that will be observed at the value $X_1 = x_{i1}$, $X_2 = x_{i2}$, $\ldots$, $X_p = x_{ip}$. Then a general linear model for Y_i takes the form

$$Y_i = \beta_0 + \beta_1 x_{i1} + \beta_2 x_{i2} + \cdots + \beta_p x_{ip} + E_i, \qquad i = 1, \ldots, n,$$

where E_i is the random error in Y_i. The E_i are assumed to be independent $N(0, \sigma^2)$.

Notice that the structure of the general linear model in Box 12.1 is exactly the same as the structure of the multiple linear regression model in Box 11.2. The only difference is that a multiple linear regression model contains only quantitative independent variables, whereas any numerically valued independent variable is permitted in a general linear model.

The statistical theory of least squares for general linear models (Graybill, 1976; Seber, 1977) can be used to show that the methods of multiple linear regression analysis are applicable to data under general linear models. However, because a general linear model can have independent variables that are not quantitative, careful attention must be paid to the differences in the interpretation of the regression parameters in the two types of models. Let's look at some specific cases.

The general linear form of a one-way ANOVA model

Consider the one-way ANOVA model with t treatments

$$Y_{ij} = \mu + \tau_i + E_{ij}, \qquad j = 1, \ldots, n_i; \; i = 1, \ldots, t. \tag{12.5}$$

This model can be written as a general linear model with $p = t - 1$ qualitative independent variables, as follows. For $k = 1, 2, \ldots, p$, define X_k as

$$X_k = \begin{cases} 1 & \text{if the response is for treatment } k \\ 0 & \text{otherwise.} \end{cases} \tag{12.6}$$

Let $n = \sum_{i=1}^{t} n_i$ be the total number of responses. For $i = 1, 2, \ldots, n$, let Y_i and $x_{i1}, x_{i2}, \ldots, x_{ip}$ denote, respectively, the i-th response and the associated values of $X_1, X_2, \ldots, X_p$. Then, as the following discussion will show, the general linear model

$$Y_i = \beta_0 + \beta_1 x_{i1} + \cdots + \beta_p x_{ip} + E_i, \qquad i = 1, \ldots, n, \tag{12.7}$$

where E_i is the random error associated with Y_i, is equivalent to the one-way ANOVA model in Equation (12.5).

Dummy independent variables

The relationships between the parameters in a one-way ANOVA model and the parameters in the corresponding general linear model can be obtained as in Example 12.1. Let $\mu_i = \mu + \tau_i$ be the expected response for treatment i. Then

$$
\begin{aligned}
\beta_0 &= \mu + \tau_t &= \mu_t \\
\beta_1 &= \tau_1 - \tau_t &= \mu_1 - \mu_t, \\
&\;\;\vdots &\vdots \\
\beta_i &= \tau_i - \tau_t &= \mu_i - \mu_t, \\
&\;\;\vdots &\vdots \\
\beta_{t-1} &= \tau_{t-1} - \tau_t &= \mu_{t-1} - \mu_t.
\end{aligned} \tag{12.8}
$$

Thus, in the general linear model in Equation (12.7), the slope parameters measure the difference between treatment effects when treatment t is the reference treatment. Equation (12.8) implies that, in a general linear model, the interpretation of a regression parameter can be quite different from its interpretation in a multiple linear regression model. For instance, the parameters $\beta_1, \ldots, \beta_p$ in Equation (12.7), though called slope parameters, do not correspond to the usual interpretation of slopes; that is, β_i does not measure the change in the expected response per unit increase in the value of X_i. Rather, β_i equals the differential expected response $\mu_i - \mu_t$. For this reason, the independent variable X_k in Equation (12.6) is often called a *dummy independent variable*.

Notice that only $t - 1$ dummy variables are needed to identify the t levels of the experimental factor. For instance, in Example 12.1, two dummy variables were used to identify three treatments. From Equation (12.8), we see that using only $t - 1$ dummy variables causes a certain asymmetry in the interpretation of the β_i. The level that is left out in the definition of the dummy variables acts as the reference level in the interpretation of the β_i. For example, when the dummy variables are defined in terms of the first $t - 1$ levels, the slope parameters in the corresponding general linear model become the differential expected responses at the first $t - 1$ levels; the difference is measured with respect to the expected response at the t-th level. Two questions arise here:

1   How do we select the $t - 1$ levels for defining the dummy variables?

2   Why not write the model in terms of t (instead of $p = t - 1$) dummy independent variables as

$$Y = \beta_0 + \beta_1 x_1 + \cdots + \beta_t x_t + E, \qquad \textbf{(12.9)}$$

where x_k is the value of the dummy variable X_k defined by Equation (12.6)? By adopting this formulation of the model, we can avoid choosing which level will be left out when selecting the dummy variables.

The answer to the first question may seem obvious. It doesn't matter which level is left out in the definition of the dummy variables, as long as the interpretation of the model parameters is consistent with the definition. The answer to the second question has to do with the theory of least squares, which forms the basis of regression analysis. Many of the methods in Chapter 11 required that the normal equations for determining the estimates of the regression parameters have a unique solution. The normal equations for estimating the parameters in Equation (12.9) do not have a unique solution, whereas the parameters in Equation (12.7) can be estimated uniquely. Consequently, if we want to use the results in Chapter 11, it is best to express the one-way ANOVA model as a general linear model with $p = t - 1$ dummy variables.

Note that Equation (12.6) is only one of many ways of defining dummy variables to represent the qualitative variables in regression models. In Exercise 12.9, you will be asked to interpret the parameters when the dummy variables are defined in another way. The choice of a particular form of dummy variables will depend on the user's preferences. Regardless of how they are defined, only $t - 1$ dummy variables are needed to guarantee uniqueness of the solution of the normal equations. In this book, we use Equation (12.6), because this is also the definition adopted in SAS, one of the most frequently used statistical software packages.

The regression and ANOVA forms

From the viewpoint of statistical interpretations, either the ANOVA form in Equation (12.5) or the regression form in Equation (12.7) can be used to express the assumptions implied by a one-way ANOVA model. The ANOVA form is intuitively appealing, because it expresses an observed response directly in terms of the expected response and the random error; we can write the model without creating dummy variables. The regression form, on the other hand, has all the advantages resulting from the use of a single method—multiple linear regression analysis—to handle both qualitative and quantitative explanatory variables.

When performing a one-way ANOVA using a general linear model, it is often necessary to reformulate statements about the parameters of the ANOVA model in terms of the parameters of the corresponding general linear model. In such cases, we can use Equation (12.8) to transform statements about the parameters in one model into statements about the parameters of the other. For example, in Exercise 12.1, you will be asked to use Equation (12.8) to verify that the three null hypotheses

$$H_{0\beta}: \beta_1 = \beta_2 = \cdots = \beta_{t-1} = 0,$$
$$H_{0\tau}: \tau_1 = \tau_2 = \cdots = \tau_t = 0, \qquad \textbf{(12.10)}$$
$$H_{0\mu}: \mu_1 = \mu_2 = \cdots = \mu_t$$

are equivalent. The equivalence of these null hypotheses implies, among other things, that testing the null hypothesis of equal treatment effects in a one-way ANOVA model is the same as testing the null hypothesis of no regression effects in the corresponding general linear model. Also, it is easily verified that Equation (12.8) can be expressed as

$$\begin{aligned}
\mu_t &= \beta_0, \\
\mu_1 &= \beta_1 + \beta_0, \\
&\vdots \\
\mu_{t-1} &= \beta_{t-1} + \beta_0,
\end{aligned} \qquad \textbf{(12.11)}$$

so that the expected responses for individual treatments in a one-way ANOVA can be estimated by estimating the regression parameters in the corresponding general linear model.

Various statistical software packages, including the GLM procedure in SAS, may be used to perform a general linear model analysis. In the next example, we describe output from the GLM procedure in the one-way ANOVA context.

EXAMPLE **12.2**   Suppose we want to fit the general linear model

$$Y = \beta_0 + \beta_1 x_1 + \beta_2 x_2 + E$$

to the posttest scores in Example 12.1 using the SAS GLM procedure; a detailed description of the corresponding computer codes can be found in Section 12.2 of Younger (1997). Portions of the output obtained are as follows.

```
PART 1
Dependent Variable: POSTSCR
                              Sum of         Mean
Source              DF        Squares        Square   F Value    Pr > F
Model                2       1752.2667      876.1333    6.71      0.0043
Error               27       3527.6000      130.6519
Corrected Total     29       5279.8667
```

```
PART 2
                              T for H0:      Pr > |T|    Std Error of
Parameter    Estimate      Parameter=0                    Estimate
INTERCEPT  37.80000000 B      10.46          0.0001      3.61457953
TRT      1  15.00000000 B       2.93          0.0067      5.11178739
         2  -2.20000000 B      -0.43          0.6703      5.11178739
         3   0.00000000 B        .             .             .
```

```
PART 3
                    X´X Generalized Inverse (g2)
                  INTERCEPT      TRT 1      TRT 2       TRT 3
INTERCEPT           0.1          -0.1       -0.1          0
TRT 1              -0.1           0.2        0.1          0
TRT 2              -0.1           0.1        0.2          0
TRT 3               0             0          0            0
```

The output is divided into three parts. Part 1 shows the ANOVA table for testing the null hypothesis $H_{0\beta}$: $\beta_1 = \beta_2 = 0$. This null hypothesis is equivalent to the null hypothesis $H_{0\tau}$ that there is no difference between the expected responses for the three therapies, and so the ANOVA table in Part 1 is the same as the ANOVA table obtained by direct calculation of the various sums of squares. It can be verified (see Exercise 12.3) that the treatment and error sums of squares for the one-way ANOVA of the posttest scores are the same as the model and error sums of squares in the ANOVA table in Part 1. The p-value of 0.0043 for testing $H_{0\beta}$ provides strong evidence that there is a difference between the expected posttest scores for the three treatments.

Part 2 of the output shows some key statistics needed to make inferences about individual regression parameters. For each β_i, the output contains the estimate $\hat{\beta}_i$; the value of the t-statistic for testing the null hypothesis $H_{0\beta_i}$: $\beta_i = 0$; the two-sided p-value for testing $H_{0\beta_i}$; and the estimated standard error of $\hat{\beta}_i$. The intercept parameter β_0 is the same as the expected posttest score for treatment 3, and so the INTERCEPT row contains statistics needed for inferences about the expected posttest score for the third treatment. The next two rows in Part 2 contain statistics needed for inferences about the two slope parameters β_1 and β_2. We know that β_i equals the difference between the expected responses for treatment i and treatment 3, and so the statistics in these rows can be used to make inferences about these differences. For example, on the basis of the p-values of 0.0067 and 0.6703, we can conclude that there is a statistically significant difference between the mean scores for treatments 1 and 3 but no demonstrable difference between the mean scores for treatments 2 and 3. The estimates of the standard errors of $\hat{\beta}_1$ and $\hat{\beta}_2$ can be used to construct confidence intervals for the corresponding expected differences. For instance, a 95% confidence interval for the expected difference $\beta_1 = \mu_1 - \mu_3$ between the posttest scores for treatments 1 and 3 is

$$\hat{\beta}_1 \pm t(27, 0.025)\hat{\sigma}_{\hat{\beta}_1} = 15.00 \pm (2.052)(5.11) \iff 4.51 \leq \mu_1 - \mu_3 \leq 25.49.$$

Accordingly, we can conclude with 95% confidence that a difference of at least 4.51 points and at most 25.49 points can be expected between the posttest scores for subjects receiving these two treatments.

The fourth row (TRT 3) shows the parameter estimate as zero and has missing values (indicated by periods) for other entries. This row is the SAS system's way of reminding the user that the dummy variables corresponding to the parameters are defined with reference to treatment 3. Just as the rows TRT 1 and TRT 2 contain estimates of $\beta_1 = \tau_1 - \tau_3$ and $\beta_2 = \tau_2 - \tau_3$, respectively, the row for TRT 3 shows the estimate of the dummy parameter $\beta_3 = \tau_3 - \tau_3$. By definition, the dummy parameter is zero, and so its estimate is always zero, regardless of the data. Thus, the TRT 3 row doesn't contain any information provided by the data and can be ignored for all practical purposes.

Part 3 of the SAS output shows the S-matrix. Once again, the S-matrix in the output has four rows and four columns, rather than the three rows and three columns that we would expect for a model with three regression parameters. Like the TRT 3 row in Part 2, the last column and last row in the S-matrix can be ignored in all calculations. Ignoring these two columns, we see that the S-matrix is

$$S = \begin{bmatrix} 0.1 & -0.1 & -0.1 \\ -0.1 & 0.2 & 0.1 \\ -0.1 & 0.1 & 0.2 \end{bmatrix}. \tag{12.12}$$

As we saw in Box 11.4, the S-matrix can be used to make inferences about linear combinations of the regression parameters. For example, suppose that we are interested in constructing a 95% lower confidence bound for $\theta = \mu_1 - \mu_2$, the difference between the expected responses for treatments 1 and 2. From Equation (12.11), $\mu_1 - \mu_2 = (\beta_1 + \beta_0) - (\beta_2 + \beta_0) = \beta_1 - \beta_2$, and so we are looking for a lower bound for the linear combination $\theta = c_0\beta_0 + c_1\beta_1 + c_2\beta_2$, where $c_0 = 0$, $c_1 = 1$, and $c_2 = -1$. This bound can be calculated as $\hat{\theta} - t(27, 0.05)\hat{\sigma}_{\hat{\theta}}$, where $\hat{\theta} = \hat{\beta}_1 - \hat{\beta}_2 = 15 - (-2.2) = 17.2$ is the estimate of θ and $\hat{\sigma}_{\hat{\theta}}$ is the standard error of $\hat{\theta}$. From Box 11.4, we have

$$\hat{\sigma}_{\hat{\theta}} = \sqrt{\hat{\sigma}^2 \sum_{i=0}^{3} \sum_{i=0}^{3} c_i c_j s_{ij}},$$

where the c_i are just defined; $\hat{\sigma}^2$ is the mean square for error ($\hat{\sigma}^2 = 130.6519$); and the s_{ij} are the elements of the S-matrix in Equation (12.12). Direct calculations show that $\hat{\sigma}_{\hat{\theta}} = 5.11$, so that the required lower bound is $17.2 - (1.703)(5.11) = 8.5$. With 95% confidence, we can conclude that the expected posttest score for treatment 1 is at least 8.5 points higher than that for treatment 2. Exercise 12.4 contains several questions concerning inferences about linear combinations of β_0, β_1, and β_2. ∎

Exercises

12.1   Verify that the null hypotheses $H_{0\beta}$, $H_{0\tau}$, and $H_{0\mu}$ in Equation (12.10) are equivalent.

12.2   Use the S-matrix in Equation (12.12) to obtain the standard errors of the estimated regression parameters given in the SAS output in Example 12.2.

12.3   Refer to the mental capacity data in Example 12.1.

   a Construct the one-way ANOVA table for comparing the three treatments when the pretest scores are ignored.

   b Compare the ANOVA table obtained in (a) with that in Part 1 of the SAS output in Example 12.2. Comment on your findings.

12.4   Refer to the results in the SAS output in Example 12.2.

   a Construct a set of 95% simultaneous confidence intervals for the expected posttest scores for the three treatments. Interpret the results.

   b At the 95% confidence level, construct an interval that will contain the posttest score of a subject who has received treatment 3. Interpret the interval. Explain how it differs from a 95% confidence interval for the expected posttest score for a subject receiving treatment 3.

   c Following the usual notation, let μ_1, μ_2, and μ_3 denote the expected posttest scores for treatments 1, 2, and 3, respectively. Obtain a 99% lower bound for the contrast $\theta = \mu_3 - \frac{1}{2}(\mu_1 + \mu_2)$. Interpret the bound.

12.5   Refer to the plant height data in Example 8.2.

   a Write a general linear model using dummy variables, as in Equation (12.7).

   b Interpret the parameters of the model in (a) in terms of the expected response for the five light conditions.

   c For each regression parameter in the model in (a), use the results in Table 8.1 to determine the parameter estimate and the value of the t-statistic for testing the null hypothesis that the parameter is zero.

   d State the conclusions that can be based on the results in (c).

12.6   The following is a portion of the output obtained when the GLM procedure of SAS is used to analyze the plant height data in Example 8.2.

Part 1

Source	DF	Sum of Squares	Mean Square	F Value	Pr > F
Model	4	41.080770	10.270192	9.41	0.0005
Error	15	16.376550	1.091770		
Corrected Total	19	57.457320			

Part 2

Parameter		Estimate	T for H0: Parameter=0	Pr > |T|	Std Error of Estimate
INTERCEPT		34.29250000 B	65.64	0.0001	0.52243899
SFLT	1	-0.27250000 B	-0.37	0.7174	0.73884031
	2	-2.39250000 B	-3.24	0.0055	0.73884031
	3	-3.45250000 B	-4.67	0.0003	0.73884031

| 4 | 0.01500000 B | 0.02 | 0.9841 | 0.73884031 |
| 5 | 0.00000000 B | . | . | . |

Part 3

X'X Generalized Inverse (g2)

	INTERCEPT	SFLT 1	SFLT 2	SFLT 3	SFLT 4	SFLT 5
INTERCEPT	0.25	-0.25	-0.25	-0.25	-0.25	0
SFLIGHT 1	-0.25	0.5	0.25	0.25	0.25	0
SFLIGHT 2	-0.25	0.25	0.5	0.25	0.25	0
SFLIGHT 3	-0.25	0.25	0.25	0.5	0.25	0
SFLIGHT 4	-0.25	0.25	0.25	0.25	0.5	0
SFLIGHT 5	0	0	0	0	0	0

---

a Using this output, construct a 90% lower prediction bound for the four-week height of a plant grown under safelight D. What conclusions can be drawn from the result?

b Using the output, construct a 99% confidence interval for:

i a contrast that can be used to compare the mean four-week height of plants exposed to safelight A with the mean four-week height of plants exposed to safelight B;

ii a contrast that can be used to determine whether the difference between the expected responses for the two light types (A and B) is the same at the two light intensities (H and L).

c State the conclusions that can be drawn from the intervals constructed in (b).

12.7 Refer to the crisis frequency data in Exercise 8.16.

a Write a general linear model suitable for comparing the frequencies of painful crises in high and normal phosphate sickle cell anemia patients.

b Calculate the estimate of the slope parameter in your model and construct a 90% lower confidence bound for this parameter.

c Interpret the bound calculated in (b).

12.8 Using the matrix formulation of the formula for the estimated standard error of the estimate of a linear combination of regression parameters in Section 11.5, calculate $\sigma_{\hat{\theta}}$ in Example 12.2.

12.9 Refer to the one-way ANOVA model in Equation (12.1). Define independent variables X_1 and X_2 as

$$X_1 = \begin{cases} 0 & \text{if treatment 1} \\ 1 & \text{if treatment 2} \\ -1 & \text{if treatment 3,} \end{cases}$$

$$X_2 = \begin{cases} 1 & \text{if treatment 1} \\ 0 & \text{if treatment 2} \\ -1 & \text{if treatment 3.} \end{cases}$$

a   Use these independent variables to express the model in Equation (12.1) as a general linear model.

b   Relate the parameters of the model you write in (a) to the parameters of the models in Equations (12.1) and (12.2).

c   On the basis of the relationship between the dummy variables used to generate the SAS output in Example 12.2 and the dummy variables defined here, determine how the entries in part 2 of the SAS output will change when the GLM procedure of SAS is used to fit the model in (a) to the mental capacity data in Example 12.1.

12.3
Models with nominal and interval variables

Even though a one-way ANOVA can be handled easily in the setting of a general linear model, there is no compelling reason to do so. Neither the calculation of the needed statistics nor the interpretation of the results is easier in a general linear model setting than in an ANOVA setting. The real advantage of the general linear model approach is seen when the underlying model contains both nominal (qualitative) and interval (quantitative) independent variables.

EXAMPLE 12.3   In Example 12.2, we analyzed the posttest scores by considering treatment assignment as the sole identifiable explanatory factor. Let's now look at how we can perform the analysis when the model includes a second explanatory factor—the pretest score.

The posttest scores are plotted against the pretest scores for each of the three treatments in Figure 12.1, which also shows, for each treatment, the straight line that best approximates the relationship between pretest and posttest scores. It appears from Figure 12.1 that, within each treatment, the relationship of the posttest scores to the pretest scores is well approximated by a straight line and that the slopes of the three lines are nearly equal. Thus, in analyzing the posttest scores, it seems reasonable to assume that the expected posttest score for a given treatment has a straight-line relationship with the subject's pretest score and that the straight lines corresponding to the three treatments have a common slope. Under that assumption, a model for the observed posttest scores can be developed as follows.

Let Y_{ij} denote the posttest score that will be observed for the j-th subject who receives treatment i, and let z_{ij} denote the pretest score for this subject. Then, in symbols, our basic assumption takes the form

$$Y_{1j} = \beta_0^1 + \beta z_{1j} + E_{1j}, \qquad j = 1, \ldots, 10;$$

$$Y_{2j} = \beta_0^2 + \beta z_{2j} + E_{2j}, \qquad j = 1, \ldots, 10; \qquad \textbf{(12.13)}$$

$$Y_{3j} = \beta_0^3 + \beta z_{3j} + E_{3j}, \qquad j = 1, \ldots, 10,$$

where E_{ij} is the random error in Y_{ij}. The E_{ij} are independent $N(0, \sigma^2)$.

F I G U R E 12.1

Test scores of $n = 30$ subjects

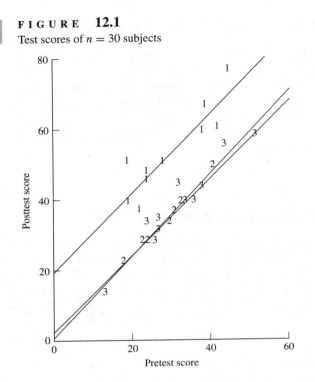

The model in Equation (12.13) can be used to analyze the observed posttest scores. It consists of three simple linear regression models, which must have a common slope β and a common error variance σ^2. The intercepts (β_0^1, β_0^2, and β_0^3) of the three regression lines can be different.

How can we make inferences about the parameters in Equation (12.13)? At first sight, we might be tempted to say that three separate straight lines should be fitted to the three sets of data for the three treatments, and the resulting parameter estimates should be used to make the required inferences. Unfortunately, that approach does not take into consideration the implicit requirement of the model that the straight lines have a common slope and a common error variance. If estimates of the slopes and error variances are obtained for each of the three lines separately, then they should be combined to form what might be called pooled estimates of β and σ^2. In this connection, recall our discussion of the pooled estimate of the error variance in a one-way ANOVA model (Chapter 8). The general linear model approach provides a unified method in which the familiar multiple linear regression setting is used to obtain the appropriate pooled estimates.

A general linear model that is equivalent to Equation (12.13) can be written in terms of two dummy variables X_1 and X_2 defined as in Example 12.2. If we number the experimental subjects from 1 to 30, and let Y_j, x_{1j}, x_{2j}, and z_j denote, respectively, the values of Y, X_1, X_2, and Z for the j-th subject, then the model in Equation (12.13) is the same as the model

$$Y_j = \beta_0 + \beta_1 x_{1j} + \beta_2 x_{2j} + \beta z_j + E_j, \qquad j = 1, \ldots, 30, \qquad \textbf{(12.14)}$$

where E_j is the random error in Y_j; the E_j are independent $N(0, \sigma^2)$. By comparing the expected responses for the two forms of the model, we can establish the relationship between the parameters in Equations (12.13) and (12.14). In the following table, the symbol $\mu(x_1, x_2, z)$ stands for the expected posttest score when the independent variables X_1, X_2, and Z have the values x_1, x_2, and z, respectively.

Treatment	(x_1, x_2)	$\mu(x_1, x_2, z)$ in Equation (12.13)	$\mu(x_1, x_2, z)$ in Equation (12.14)
1	(1, 0)	$\beta_0^1 + \beta z$	$\beta_0 + \beta_1 + \beta z$
2	(0, 1)	$\beta_0^2 + \beta z$	$\beta_0 + \beta_2 + \beta z$
3	(0, 0)	$\beta_0^3 + \beta z$	$\beta_0 + \beta z$

The functional forms of the expected posttest scores imply that the relationships between the parameters in the two forms of the model are quite similar to those between the parameters in the ANOVA and general linear models in Example 12.1. From the expected responses under treatment 3, we see that $\beta_0 = \beta_0^3$, so that β_0 is the intercept parameter in the simple linear regression model for treatment 3. Similarly, since $\beta_0^1 = \beta_0 + \beta_1$ and $\beta_0^2 = \beta_0 + \beta_2$, the parameters β_1 and β_2 are, respectively, (treatment 1 intercept) − (treatment 3 intercept) and (treatment 2 intercept) − (treatment 3 intercept). Thus, the slope parameters corresponding to the dummy variables in Equation (12.14) measure the treatment effects with reference to treatment 3. The hypothesis $H_{0\tau}: \beta_1 = \beta_2 = 0$ is the same as the hypothesis that there are no treatment differences, because the regression functions for the three treatments are identical under $H_{0\tau}$. Similarly, the hypothesis $H_{0\beta}: \beta = 0$ implies that the posttest scores are not associated with the pretest scores.

We can use the GLM procedure in SAS to fit the model in Equation (12.14) to the posttest scores; the details of the corresponding computer code are given in Younger (1997). Portions of the SAS output obtained are as follows.

```
-----------------------------------------------------------------
PART 1
Dependent Variable: POSTSCR

                         Sum of        Mean
Source          DF      Squares       Square   F Value  Pr > F
Model            3    4899.2881    1633.0960    111.57  0.0001
Error           26     380.5786      14.6376
Corrected Total 29    5279.8667
-----------------------------------------------------------------

PART 2
Source          DF    Type I SS  Mean Square  F Value  Pr > F
TRT              2    1752.2667     876.1333    59.85  0.0001
```

PRESCR	1	3147.0214	3147.0214	215.00	0.0001

Source	DF	Type III SS	Mean Square	F Value	Pr > F
TRT	2	2052.2270	1026.1135	70.10	0.0001
PRESCR	1	3147.0214	3147.0214	215.00	0.0001

PART 3

Parameter		Estimate	T for HO: Parameter=0	Pr > |T|	Std Error of Estimate
INTERCEPT		1.79986240 B	0.66	0.5166	2.73712476
TRT	1	17.14420882 B	9.98	0.0001	1.71723972
	2	-0.84576285 B	-0.49	0.6257	1.71349283
	3	0.00000000 B	.	.	.
PRESCR		1.12853096	14.66	0.0001	0.07696604

PART 4

X'X Generalized Inverse (g2)

	INTERCEPT	TRT 1	TRT 2	TRT 3	PRESCR
INTERCEPT	0.511821125	-0.124528531	-0.11549170	0	-0.012909753
TRT 1	-0.124528531	0.201460947	0.10092270	0	0.000768919
TRT 2	-0.115491704	0.100922703	0.20058276	0	0.000485633
TRT 3	0	0	0	0	0
PRESCR	-0.012909753	0.000768919	0.00048563	0	0.000404694

The output is divided into four parts. Part 1 contains the relevant ANOVA table. Notice that there are three degrees of freedom for the model sum of squares, one for each of the $p = 3$ slope parameters β_1, β_2, and β. The p-value of 0.0001 for the model sum of squares applies to tests of the null hypothesis H_0: $\beta_1 = \beta_2 = \beta = 0$ (no regression effects). This p-value lends high credibility to the research hypothesis that at least one of the three slope parameters is not zero. It also lends credence to the conclusion that at least one of the following null hypotheses is not true:

- $H_{0\tau}$: The three treatments have the same effects ($\beta_1 = \beta_2 = 0$).
- $H_{0\beta}$: The posttest score is not associated with the pretest score ($\beta = 0$).

Part 2 in the SAS output shows the results needed to test the two null hypotheses $H_{0\tau}$ and $H_{0\beta}$. As explained in Section 11.7, these tests can be based on either Type I (sequential) or Type III (partial) sums of squares, which can be computed using the GLM procedure in SAS. As we saw, tests based on Type I sums of squares are useful for examining the effects of new independent variables, when they are brought into the model in a predetermined sequence. Type III sums of squares, on the other hand, are useful for testing whether deletion of a set of variables has a significant effect on the predictive ability of the model. In the following table, the notation in Equation (11.24) is used to show how the Type I and Type III sums of squares in Part 2 can be interpreted.

Source	df	Type I SS	Type III SS
TRT	2	$R(\beta_1, \beta_2|\beta_0)$	$R(\beta_1, \beta_2|\beta_0, \beta)$
PRESCR	1	$R(\beta|\beta_0, \beta_1, \beta_2)$	$R(\beta|\beta_0, \beta_1, \beta_2)$

Comparison shows that the Type I sum of squares for TRT in Part 2 of the SAS output here is the same as the model sum of squares in Part 1 of the SAS output in Example 12.2. Both of these sums of squares are equal to $R(\beta_1, \beta_2|\beta_0)$, the extra sum of squares obtained by adding the parameters β_1 and β_2 to a model containing β_0. Thus, the conclusion based on the p-value for the Type I sum of squares for TRT is the same as the conclusion from the ANOVA table in Example 12.2: that the information about treatment assignments (contained in the values of X_1 and X_2) by itself is a useful predictor of the posttest scores. The Type III sum of squares for TRT is the extra regression sum of squares obtained by adding β_1 and β_2 to a model containing β_0 and β. The p-value of 0.0001 for this sum of squares implies that better predictions of the posttest scores can be made by using the information about treatment assignments in addition to the information in the pretest scores.

Exercises 12.10 and 12.11 provide opportunities to explore the information in the SAS output in greater detail. For the present, notice that the p-values for testing $H_{0\tau}$ and $H_{0\beta}$ on the basis of Type III sums of squares are both equal to 0.0001; this provides strong support for rejection of the two null hypotheses. We conclude not only that the treatments have different effects on the posttest scores, but also that the pretest score has a linear association with the posttest score.

Part 3 of the SAS output contains information useful for inferences about the estimates of the regression parameters. The estimated prediction equation for the posttest score when $X_1 = x_1$, $X_2 = x_2$, and $Z = z$ takes the form

$$\hat{y} = 1.7999 + 17.1442x_1 - 0.8458x_2 + 1.1285z. \tag{12.15}$$

By setting $x_1 = 1$, $x_2 = 0$ in Equation (12.15), we get the prediction equation for a subject whose pretest score is z and who received treatment 1

$$\hat{y} = 1.7999 + 17.1442 + 1.1285z$$

$$= 18.9441 + 1.1285z.$$

Similarly, the prediction equations for subjects receiving treatments 2 and 3 are obtained by setting $(x_1, x_2) = (0, 1)$ and $(x_1, x_2) = (0, 0)$, respectively. We obtain

$$\hat{y} = 0.9541 + 1.1285z$$

for treatment 2, and

$$\hat{y} = 1.7999 + 1.1285z$$

for treatment 3.

The information in Part 3 can also be used to draw conclusions about individual regression parameters. For instance, the p-value of 0.6257 for treatment 2 suggests that the data do not provide sufficient evidence to reject the null hypothesis $H_{0\beta_2}$: $\beta_2 = 0$. In other words, the data do not demonstrate a significant difference between the expected posttest scores for treatments 2 and 3. The estimated standard error of 0.07697 for β can be used to construct a confidence interval for the common slope of the three regression lines.

Part 4 of the SAS output contains the S-matrix needed to make inferences about specified linear combinations of the regression parameters. Once again, the row and column containing all 0s should be deleted from this matrix before it is used. Exercise 12.11 illustrates the use of the information in Part 4 to make inferences. ∎

Exercises

12.10 In Example 12.3, we considered the posttest scores of children who received one of three therapies. In the general linear model in Equation (12.14), we now delete the subject-specific subscript j

$$Y = \beta_0 + \beta_1 x_1 + \beta_2 x_2 + \beta_3 z + E. \tag{12.16}$$

a Argue that testing the null hypothesis $H_{0\tau}$: $\beta_1 = \beta_2 = 0$ is the same as testing the null hypothesis that the three therapies have identical effects on the posttest scores.

b Write the reduced model under $H_{0\tau}$ in (a).

c Using the information in the SAS output in Example 12.3, determine the model and error sums of squares under the reduced model in (b).

d In what way does the null hypothesis H_0: $\beta_1 = \beta_2$ differ from $H_{0\tau}$ in (a)?

e Write the reduced model under H_0 in (d).

f Given that the model sum of squares for the reduced model in (e) is 3282.6955, test the null hypothesis H_0 and write your conclusions.

g Use the $X'X$ generalized inverse matrix in the SAS output in Example 12.3 to test H_0 in (d). Explain why this test is equivalent to the F-test in (f).

h The model in Equation (12.16) was selected on the basis of the pattern of parallel straight lines observed in the posttest scores. Suppose the investigators want to see if a model with the quadratic term z^2

$$Y = \beta_0 + \beta_1 x_1 + \beta_2 x_2 + \beta_3 z + \beta_4 z^2 + E \tag{12.17}$$

significantly improves the fit. Using plots of the expected posttest scores under the two models, describe how the model in Equation (12.17) changes the assumptions of the model in Equation (12.16).

i Use the GLM procedure in SAS or any other suitable regression software to fit the model in Equation (12.17). Carry out an appropriate test to see if the addition of the quadratic term improves the model in Equation (12.16) significantly.

12.11 Refer to the SAS output from the analysis of posttest scores in Example 12.3.

a Explain why the F-value (215.00) in the PRESCR row for the Type I sum of squares in Part 2 approximately equals the square of the value of the t-statistic (14.66) in the PRESCR row in Part 3.

b   If you used the results in this SAS output to work Exercise 12.4, would you be surprised if you obtained different answers? Explain.

c   Work Exercise 12.4 using the model in Equation (12.16).

d   Use the model in Equation (12.16) to construct a 95% confidence interval for the expected posttest score of a child whose pretest score is 30 and who received treatment 1. Interpret the interval.

e   Use the model in Equation (12.16) to construct a set of 90% simultaneous lower prediction bounds for the posttest scores of the children with pretest score 30 in each of the three treatment groups. Interpret the bounds.

12.12   Augustin and Clarke (1991) investigated the effects on Y = calcium ion (Ca^{2+}) activities (mM) of three explanatory factors:

- TRT, the preheat treatment applied during the manufacture of milk powder, at six levels (none, low heat, medium heat, high heat, indirect UHT, and direct UHT);

- TYP, the type of milk, at three levels (9% total solids, 19.6% total solids, and 26% total solids);

- X, the pH value of the milk.

The measured values of X and Y for five samples of 9% total solid milk subjected to the six preheating treatments are summarized in Table 12.1. Suppose we want to fit the data to a general linear model with seven independent variables: five dummy variables for the six preheat treatments; and two quantitative variables x and x^2 representing the influence of the pH values. Portions of the resulting SAS output are as follows.

```
----------------------------------------------------------------
Dependent Variable: Y

                       Sum of        Mean
Source          DF     Squares      Square   F Value  Pr > F
Model            7   9.4288464   1.3469781   1506.26  0.0001
Error           22   0.0196736   0.0008943
Corrected Total 29   9.4485200


Source          DF   Type I SS  Mean Square  F Value  Pr > F
TRT              5   0.0063600   0.0012720      1.42  0.2553
X                1   8.9505438   8.9505438  10008.93  0.0001
XSQR             1   0.4719426   0.4719426    527.75  0.0001


Source          DF  Type III SS  Mean Square  F Value  Pr > F
TRT              5   0.0202401   0.0040480      4.53  0.0055
X                1   0.5633601   0.5633601    629.98  0.0001
XSQR             1   0.4719426   0.4719426    527.75  0.0001


                       T for HO:   Pr > |T|   Std Error of
Parameter     Estimate  Parameter=0             Estimate
INTERCEPT   77.50180082 B    27.66    0.0001    2.80146202
TRT       1 -0.00001944 B    -0.00    0.9992    0.01892421
```

TABLE **12.1**
Ca^{2+} activity (mM) and pH values

Treatment	Sample	pH (z)	Ca^{2+} (y)
None (TRT = 1)	1	6.07	2.21
	2	6.35	1.39
	3	6.52	1.05
	4	6.71	0.78
	5	6.92	0.61
Low heat (TRT = 2)	1	6.08	2.19
	2	6.36	1.39
	3	6.53	1.09
	4	6.72	0.89
	5	6.93	0.60
Medium heat (TRT = 3)	1	6.08	2.20
	2	6.37	1.43
	3	6.54	1.07
	4	6.73	0.78
	5	6.94	0.60
High heat (TRT = 4)	1	6.06	2.17
	2	6.35	1.40
	3	6.53	1.05
	4	6.72	0.77
	5	6.93	0.57
Indirect UHT (TRT = 5)	1	6.06	2.15
	2	6.34	1.39
	3	6.52	1.06
	4	6.70	0.78
	5	6.92	0.58
Direct UHT (TRT = 6)	1	6.08	2.20
	2	6.35	1.40
	3	6.52	1.04
	4	6.71	0.78
	5	6.98	0.56

	2	0.04189145 B	2.21	0.0375	0.01892195
	3	0.03727011 B	1.97	0.0615	0.01891808
	4	0.01532813 B	-0.81	0.4264	0.01891641
	5	-0.02946003 B	-1.56	0.1338	0.01892302
	6	0.00000000 B	.	.	.
X		-21.65821430	-25.10	0.0001	0.86289987
XSQR		1.52378627	22.97	0.0001	0.06633004

a   Write the model that was fitted to the data. Explain the meaning of all the terms in the model. Call this model A.

b   For each of the six treatment groups, determine the prediction equation for the calcium activity.

c   The tests for treatment differences based on Type I and Type III sums of squares yield different results ($p = 0.2553$ and $p = 0.0055$). How would you interpret that?

d   The p-values for TRT = 2 ($p = 0.0375$) and TRT = 3 ($p = 0.0615$) are the only two treatment differences (relative to TRT = 6) with p-values small enough to suggest statistical significance. As a result, the investigators decided to fit two more models by regrouping the data as follows:

- Model B: Data are divided into three groups; group 1 contains data for treatments 1, 4, 5, and 6, and the data for treatments 2 and 3 make up the other two groups.

- Model C: Data are divided into two groups; group 1 is the same as in model B, and the rest of the data make up the second group.

The model sum of squares for the two models was 9.4258428 for model B and 9.4257892 for model C. Using this information, perform appropriate statistical tests to see if there are significant differences between the predictive abilities of models A, B, and C. On the basis of your tests, which model would you use to predict calcium activities?

e   The following output was obtained by employing the GLM procedure of SAS to fit model C to the data.

```
--------------------------------------------------------------------
Dependent Variable: Y
                         Sum of        Mean
Source           DF     Squares       Square  F Value   Pr > F
Model             3   9.4257892    3.1419297  3593.81   0.0001
Error            26   0.0227308    0.0008743
Corrected Total  29   9.4485200
--------------------------------------------------------------------
Source           DF  Type III SS  Mean Square  F Value  Pr > F
T                 1    0.0171829    0.0171829    19.65   0.0001
X                 1    0.5648195    0.5648195   646.05   0.0001
XSQR              1    0.4732614    0.4732614   541.33   0.0001
--------------------------------------------------------------------
                          T for H0:      Pr > |T|   Std Error of
Parameter    Estimate   Parameter=0                  Estimate
INTERCEPT   77.58934262 B   28.02        0.0001      2.76885299
T        1  -0.05078067 B   -4.43        0.0001      0.01145436
         2   0.00000000 B     .            .             .
X          -21.67343697     -25.42       0.0001      0.85269457
XSQR         1.52499028      23.27       0.0001      0.06554463
--------------------------------------------------------------------
```

X´X Generalized Inverse (g2)

	INTERCEPT	T 1	T2	X	XSQR
INTERCEPT	8769.173764	-0.610484288	0	-2699.90798	207.39308982
T 1	-0.610484288	0.150072414	0	0.15311266	-0.01145250
T 2	0	0	0	0	0
X	-2699.907977	0.153112659	0	831.6601129	-63.91323421
XSQR	207.393089	-0.011452503	0	-63.9132342	4.91397665

Obtain the estimated prediction equation for each of the two groups.

f Use the X′X inverse matrix in (e) to construct 95% confidence bands for the expected responses in the two treatment groups.

g Show that, under model C, the change in the expected response when x is increased from x to $x + 1$ is the same for the three groups and is

$$\theta(x) = \beta_2 + (2x + 1)\beta_3,$$

where β_2 and β_3 are the regression coefficients of x and x^2, respectively.

h Plot the estimates and the endpoints of the 95% confidence interval for $\theta(x)$ against the values of x.

i Explain the type of information that can be obtained from the plot in (h).

12.4
Modeling interactions

We discussed additive and interactive effects in Section 11.2 in the context of quantitative factors. The notion of interaction can easily be extended to any set of factors regardless of whether the set consists of factors that are all quantitative, all qualitative or some qualitative and some quantitative. In this section, we look briefly at how the general linear model enables us to extend the methods for handling interactions in multiple linear regression analysis to interactions between qualitative and quantitative factors.

A review of the definitions of additive and interactive effects in Section 11.2 suggests that nothing requires us to limit them to quantitative independent variables. Consequently, in the remainder of this book, we adopt the definitions of interaction and additivity in Chapter 11, but without the requirement that the explanatory factors are quantitative.

Interaction effect

Two explanatory factors (both qualitative, both quantitative, or one qualitative and one quantitative) will be considered to have an interaction effect on the response if the change in the expected response that results from changing the level of one of the factors depends on the level of the other factor.

The next example illustrates how general linear models handle interactions between qualitative and quantitative explanatory factors.

EXAMPLE **12.4**   In an experiment to compare the uptake of a chemical from two dietary supplements, 14 animals were randomly divided into two groups of seven. The following data show the initial body weight (kg) and the chemical uptake (g/day) of the experimental animals.

Diet 1	Initial weight X	12.1	10.2	13.6	14.8	16.1	14.3	12.0
	Uptake Y	12.7	10.3	19.8	17.4	18.9	16.1	12.7
Diet 2	Initial weight X	11.3	12.4	13.6	14.1	15.0	14.9	10.9
	Uptake Y	41.4	43.3	49.7	51.6	51.5	50.5	38.4

The measured values of the chemical uptake are plotted against the initial weights in Figure 12.2. An examination of Figure 12.2 suggests that the relationship between the chemical uptake and initial weight can be approximated by two straight lines that are not parallel (of different slopes). The different slopes of the lines suggest that the difference between the expected responses for the two diets is not the same at different initial weights. Indeed, it appears that the differential response to the two diets increases with the initial weight. Thus, Figure 12.2 indicates an interaction between diet (a qualitative explanatory factor) and initial weight (a quantitative explanatory factor), because the difference between the expected responses that results from changing the type of diet appears to depend on the initial weight.

The following general linear model represents the expected responses for the two diets by means of two simple linear regression functions with possibly unequal

FIGURE 12.2
Chemical uptake for two diets

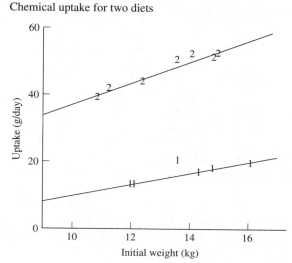

slopes. Define the dummy variable X_1 as

$$X_1 = \begin{cases} 1 & \text{if diet 1} \\ 0 & \text{if diet 2.} \end{cases}$$

Number the experimental animals from 1 to 14, and let Y_j, x_{1j}, and z_j denote the chemical uptake, the type of diet, and the initial weight for the j-th animal. Then

$$Y_j = \beta_0 + \beta_1 x_{1j} + \beta_2 z_j + \beta_3 x_{1j} z_j + E_j, \qquad \text{(12.18)}$$

where the E_j are independent $N(0, \sigma^2)$, is a model that combines two simple linear regression functions for predicting the chemical uptake on the basis of diet and body weight. In this model, either the intercepts or the slopes of the two regression lines may differ. To see this, we can substitute the appropriate values for the dummy variable ($x_{1j} = 0$ or 1) in Equation (12.18) and write the specific models for diet 1

$$Y_j = (\beta_0 + \beta_1) + (\beta_2 + \beta_3)z_j + E_j$$

and for diet 2

$$Y_j = \beta_0 + \beta_2 z_j + E_j.$$

Thus, $\beta_0 + \beta_1$ and β_0 are, respectively, the intercepts of the two simple linear regression functions corresponding to the two diets, and $\beta_2 + \beta_3$ and β_2 are the corresponding slopes. The parameters β_1 and β_3 are, respectively, the difference between the intercepts and the difference between the slopes of the two regression lines. As we have seen, unequal slopes of the regression lines imply an interaction between diet and initial weight, and so the hypothesis $H_{0\beta_3}$: $\beta_3 = 0$ that the two regression lines are parallel is the same as the hypothesis that the diet and the initial weight do not have an interaction effect on chemical uptake. The inclusion of the product term $\beta_3 x_{1j} z_j$ in the model enables us to test for an interaction between diet and initial weight. The value of β_3 provides a measure of the intensity of the interaction between the two factors. The relevant parts of the SAS output obtained on fitting the model in Equation (12.18) to the chemical uptake data are as follows.

```
------------------------------------------------------------------
PART 1
Dependent Variable: UPTKE
                             Sum of        Mean
Source           DF         Squares      Square  F Value   Pr > F
Model             3      3634.7500    1211.5833   389.66   0.0001
Error            10        31.0936       3.1094
Corrected Total  13      3665.8436
------------------------------------------------------------------
PART 2
Source           DF     Type I SS  Mean Square  F Value   Pr > F
DIET              1     3410.1607    3410.1607  1096.74   0.0001
WT                1      199.5358     199.5358    64.17   0.0001
WT*DIET           1       25.0535      25.0535     8.06   0.0176
```

Source	DF	Type III SS	Mean Square	F Value	Pr > F
DIET	1	5.93516	5.93516	1.91	0.1972
WT	1	218.89542	218.89542	70.40	0.0001
WT*DIET	1	25.05349	25.05349	8.06	0.0176

PART 3

Parameter		Estimate	T for HO: Parameter=0	Pr > |T|	Std Error of Estimate
INTERCEPT		4.99785297 B	0.87	0.4038	5.73350839
DIET	1	-10.36759736 B	-1.38	0.1972	7.50408117
	2	0.00000000 B	.	.	.
WT		3.16068361 B	7.31	0.0001	0.43234794
DIET*WT	1	-1.59797458 B	-2.84	0.0176	0.56295176
	2	0.00000000 B	.	.	.

PART 4

X´X Generalized Inverse (g2)

	INTERCEPT	DIET 1	DIET 2	WT	WT*DIET 1	WT*DIET 2
INTERCEPT	10.5723	-10.5723	0	-0.7918	0.7918	0
DIET 1	-10.5723	18.1102	0	0.7918	-1.3478	0
DIET 2	0	0	0	0	0	0
WT	-0.7918	0.7918	0	0.0601	-0.0601	0
DIET*WT 1	0.7918	-1.3478	0	-0.0601	0.1019	0
DIET*WT 2	0	0	0	0	0	0

The interpretation of this output is similar to that for the output in Example 12.3. Notice that, because there are three slope parameters (β_1, β_2, and β_3) in the general linear model, the model sum of squares in Part 1 has three degrees of freedom. The p-value of 0.0001 for model sum of squares provides strong support for the conclusion that either the type of diet or the body weight is a useful predictor of chemical uptake.

Part 2 of the output shows the Type I and Type III sums of squares for testing the significance of the null hypotheses concerning the difference between diets (H_0: $\beta_1 = 0$), the effect of body weight (H_0: $\beta_2 = 0$), and the effect of interaction between diet and body weight (H_0: $\beta_3 = 0$). The p-value of 0.0001 for the Type I sum of squares for diet suggests that diet, by itself, is a useful predictor of chemical uptake. However, the p-value of 0.1972 for the Type III sum of squares for diet implies that adding a term to represent the effect of diet will not significantly increase the predictive ability of a model that contains terms representing the effect of weight and the effect of interaction between weight and diet.

In Part 3, the rows with missing values (indicated by periods) correspond to parameters that are equal to zero by definition, and the p-values in the rows DIET, WT, and DIET*WT correspond to testing the null hypotheses $H_{0\beta_i}$: $\beta_i = 0$ for $i = 1$,

2, and 3, respectively. On the basis of the *p*-value of 0.0176 for testing $H_{0\beta_3}$: $\beta_3 = 0$, we conclude that there is strong evidence of interaction between the effects of diet and weight—that is, that changes in the expected uptake per unit increase in the initial weight (the slopes of the regression lines) are not the same for the two diets. In other words, the difference between the expected uptake will be different for different initial weights.

The standard error of $\hat{\beta}_3$ can be used to construct a confidence interval for β_3, the differential rate of change.

From Part 3 of the display, the estimated regression equation is

$$\hat{y}_j = 4.9979 - 10.3676x_{1j} + 3.1607z_j - 1.5980x_{1j}z_j,$$

so that, for a subject on diet 1 ($x_{1j} = 1$), the prediction equation is

$$\hat{y}_j = (4.9979 - 10.3676) + (3.1607 - 1.5980)z_j = -5.3697 + 1.5627z_j.$$

The corresponding prediction equation for a subject on diet 2 is

$$\hat{y}_j = 4.9979 + 3.1607z_j.$$

Finally, the *S*-matrix in Part 4 can be used to make inferences about linear combinations of the regression parameters in the general linear model. In Exercise 12.13, you will be asked to make such inferences. ■

So far, we have seen examples of general linear models that are one-way ANOVA models (Example 12.1) and models obtained by combining several simple linear regression models (Examples 12.3 and 12.4). In the next example, we examine how a general linear model can be obtained by combining several multiple linear regression models.

EXAMPLE 12.5 In an experiment to evaluate two minerals—sodium selenite and calcium selenite—as dietary sources of selenium (Se) for broiler chickens. Two diets—diet A and diet B—were formed by supplementing the standard basal diet with known amounts of sodium and calcium selenite, respectively. Among other things, the investigators were interested in studying how Y, the selenium concentration ($\mu g/g$) in the liver, was affected by four experimental factors:

- X_1, the type of diet: 1 for A and 0 for B;
- X_2, the gender of the animal: 1 for female (F) and 0 for male (M);
- X_3, the Se concentration ($\mu g/g$) in the diet;
- X_4, the feeding period (weeks)—that is, the time period over which the diet was fed to the animal.

For a detailed description of the study design and the resulting conclusions, see Tarla, Henry, Ammerman, Rao, and Miles (1991).

The data collected in this study have been modified for illustration purposes—to ensure that certain interactions are significant. The resulting hypothetical data set is shown in Table 12.2.

TABLE 12.2

Hypothetical data on dietary selenium in broiler chickens

X_1	X_2	X_3	X_4	Y	X_1	X_2	X_3	X_4	Y	X_1	X_2	X_3	X_4	Y
1	1	0.1	1	1.21	1	1	0.2	1	1.59	1	1	0.3	1	1.65
1	1	0.1	1	1.18	1	1	0.2	1	1.56	1	1	0.3	1	1.59
1	0	0.1	1	1.13	1	0	0.2	1	1.21	1	0	0.3	1	1.81
1	0	0.1	1	1.16	1	0	0.2	1	1.29	1	0	0.3	1	1.70
1	1	0.1	3	1.69	1	1	0.2	3	2.00	1	1	0.3	3	1.97
1	1	0.1	3	1.70	1	1	0.2	3	1.93	1	1	0.3	3	2.03
1	0	0.1	3	1.62	1	0	0.2	3	1.73	1	0	0.3	3	1.89
1	0	0.1	3	1.58	1	0	0.2	3	1.82	1	0	0.3	3	1.93
1	1	0.1	5	2.33	1	1	0.2	5	2.11	1	1	0.3	5	2.85
1	1	0.1	5	2.40	1	1	0.2	5	2.08	1	1	0.3	5	2.76
1	0	0.1	5	2.54	1	0	0.2	5	2.56	1	0	0.3	5	3.95
1	0	0.1	5	2.63	1	0	0.2	5	2.67	1	0	0.3	5	3.84
0	1	0.1	1	1.29	0	1	0.2	1	1.25	0	1	0.3	1	1.71
0	1	0.1	1	1.36	0	1	0.2	1	1.46	0	1	0.3	1	1.61
0	0	0.1	1	1.15	0	0	0.2	1	1.06	0	0	0.3	1	1.68
0	0	0.1	1	1.03	0	0	0.2	1	1.14	0	0	0.3	1	1.60
0	1	0.1	3	1.69	0	1	0.2	3	1.83	0	1	0.3	3	1.70
0	1	0.1	3	1.53	0	1	0.2	3	1.90	0	1	0.3	3	1.82
0	0	0.1	3	1.78	0	0	0.2	3	1.77	0	0	0.3	3	2.53
0	0	0.1	3	1.60	0	0	0.2	3	1.61	0	0	0.3	3	2.65
0	1	0.1	5	1.94	0	1	0.2	5	2.87	0	1	0.3	5	3.13
0	1	0.1	5	2.03	0	1	0.2	5	3.01	0	1	0.3	5	3.46
0	0	0.1	5	2.83	0	0	0.2	5	2.98	0	0	0.3	5	4.35
0	0	0.1	5	2.94	0	0	0.2	5	3.23	0	0	0.3	5	4.25

Some questions that were of interest to the investigators are as follows:

1 Is there a difference between the liver Se concentration for the two diets?

2 Does the effect of diet on liver Se depend on the animal's gender? In other words, is there an interaction between diet and gender?

3 Does the effect of diet on liver Se depend on the feeding period? In other words, is there an interaction between diet and feeding period?

4 Is there an effect of the Se concentration in the diet? Is there an interaction between feeding period and dietary Se concentration?

The first step in the analysis of the liver concentration data is to formulate a statistical model for Y, the measured liver Se concentration, in terms of the four explanatory variables X_1, X_2, X_3, and X_4. Since two of the explanatory variables (X_1 and X_2) are dummy variables, it is best to conceptualize the model as a combination of several models, one for each group formed by combining the possible values of the dummy variables. There are four groups corresponding to the four possible combinations of values of X_1 and X_2— (1, 1), (1, 0), (0, 1), and (0, 0)—and so the required general linear model can be obtained by combining four models representing the relationships of Y to X_3 and X_4 in the four groups. Within each group, the model can be expressed as a multiple linear regression model relating the observed value of Y to the values of the quantitative independent variables X_3 and X_4. Obviously, a wide variety of models can be selected as within-group models; the appropriateness of any model should be judged by the assumptions needed to justify its use. The following are some examples of general linear models obtained by combining four within-group models.

Adopting the usual notation, let $\mu(x_1, x_2, x_3, x_4)$ denote the expected response when $X_1 = x_1, X_2 = x_2, X_3 = x_3$, and $X_4 = x_4$. Then one of the simplest general linear forms for $\mu(x_1, x_2, x_3, x_4)$ is

$$\mu(x_1, x_2, x_3, x_4) = \beta_0 + \beta_1 x_1 + \beta_2 x_2 + \beta_3 x_3 + \beta_4 x_4. \tag{12.19}$$

The assumptions implied by Equation (12.19) are most easily seen by examining its form at specific combinations of values of the dummy variables. For example, setting $x_1 = x_2 = 0$, we find that the expected response for a male receiving diet B is

$$\mu(0, 0, x_3, x_4) = \beta_0 + \beta_3 x_3 + \beta_4 x_4.$$

Thus, according to Equation (12.19), the expected response for a male chicken receiving diet B can be expressed as a multiple linear regression function with two independent variables X_3 and X_4. For a chicken in this group, β_0 is the intercept parameter of the regression function. The parameters β_3 and β_4 have the usual interpretation as the partial slopes of X_3 and X_4, respectively. For instance, β_3 is the expected change in the liver Se concentration per unit increase in the dietary Se concentration. The form of $\mu(x_1, x_2, x_3, x_4)$ for each of the four groups defined by X_1 and X_2 may be summarized as follows.

Group Diet	Group Gender	Expected response when $X_3 = x_3, X_4 = x_4$ Symbol	Expression
A	F	$\mu(1, 1, x_3, x_4)$	$(\beta_0 + \beta_1 + \beta_2) + \beta_3 x_3 + \beta_4 x_4$
A	M	$\mu(1, 0, x_3, x_4)$	$(\beta_0 + \beta_1) + \beta_3 x_3 + \beta_4 x_4$
B	F	$\mu(0, 1, x_3, x_4)$	$(\beta_0 + \beta_2) + \beta_3 x_3 + \beta_4 x_4$
B	M	$\mu(0, 0, x_3, x_4)$	$\beta_0 + \beta_3 x_3 + \beta_4 x_4$

From the expressions for the within-group expected responses, we see that the regression functions for the four groups differ only in the intercepts: $\beta_0 + \beta_1 + \beta_2$ for diet A, female; $\beta_0 + \beta_1$ for diet A, male; $\beta_0 + \beta_2$ for diet B, female; and β_0 for diet B, male. Thus, in this model, the expected response surface for each combination of diet and gender is a two-dimensional plane represented by an equation of the form

$$y = \text{intercept} + \beta_3 x_3 + \beta_4 x_4,$$

in which the intercept depends on the particular diet–gender combination. Because the four response surfaces differ only in their intercepts, they represent parallel planes in two dimensions. Two such surfaces are shown in Figure 12.3 for a set of selected values of the β_i.

At any given point on the x_3–x_4 plane, the differences between the heights of the expected response surfaces represent the expected differences between the responses for the four combinations of diet and gender. Because the surfaces are parallel, we conclude that the expected differences do not depend on x_3 and x_4 and hence that the treatments do not interact with the dietary Se concentration X_3 and the feeding period X_4.

If $\beta_1 = \beta_2 = 0$, the expected response surfaces for the four groups coincide, and so the null hypothesis H_0: $\beta_1 = \beta_2 = 0$ is equivalent to the null hypothesis that there is no difference in the expected responses for the four groups. Furthermore, a comparison of the expressions for the within-group expected responses shows that β_1 is the difference between the expected liver Se concentrations for two groups that differ only in the type of diet, while β_2 is the corresponding difference between two groups that differ only in the gender of the chicken. In symbols

$$\beta_1 = \mu(1, 1, x_3, x_4) - \mu(0, 1, x_3, x_4) = \mu(1, 0, x_3, x_4) - \mu(0, 0, x_3, x_4),$$

$$\beta_2 = \mu(1, 1, x_3, x_4) - \mu(1, 0, x_3, x_4) = \mu(0, 1, x_3, x_4) - \mu(0, 0, x_3, x_4).$$

(12.20)

The relationships in Equation (12.20) also imply that the changes in the expected response that result from changing the level of X_1 (diet) from 1 to 0 do not depend on the level of X_2 (gender) and that the changes in the expected response that result from changing the level of X_2 (gender) from 1 to 0 do not depend on the level of X_1 (diet). In other words, Equation (12.20) implies that there is no interaction between diet and gender. Thus, the model in Equation (12.19) not only implies no interaction between dietary Se and feeding period, but also implies no interaction between diet and gender.

FIGURE 12.3

Expected response surfaces for the model in Equation (12.19)

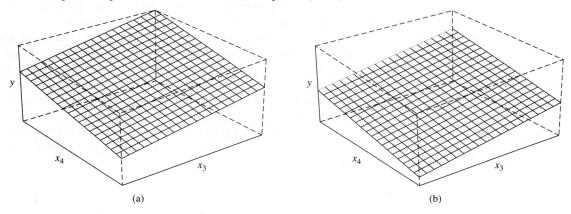

(a)   (b)

If needed, a model that allows for interaction between diet and gender can be obtained by introducing the product term $\beta_{12}x_1x_2$ in Equation (12.19). The resulting regression function takes the form

$$\mu(x_1, x_2, x_3, x_4) = \beta_0 + \beta_1 x_1 + \beta_2 x_2 + \beta_{12}x_1 x_2 + \beta_3 x_3 + \beta_4 x_4. \qquad \text{(12.21)}$$

The corresponding expected responses are as follows:

Group		Expected response when $X_3 = x_3, X_4 = x_4$	
Diet	Gender	Symbol	Expression
A	F	$\mu(1, 1, x_3, x_4)$	$(\beta_0 + \beta_1 + \beta_2 + \beta_{12}) + \beta_3 x_3 + \beta_4 x_4$
A	M	$\mu(1, 0, x_3, x_4)$	$(\beta_0 + \beta_1) + \beta_3 x_3 + \beta_4 x_4$
B	F	$\mu(0, 1, x_3, x_4)$	$(\beta_0 + \beta_2) + \beta_3 x_3 + \beta_4 x_4$
B	M	$\mu(0, 0, x_3, x_4)$	$\beta_0 + \beta_3 x_3 + \beta_4 x_4$

As in Equation (12.19), the four groups differ only in the intercepts of the regression functions; in contrast to Equation (12.19), however, Equation (12.21) allows for the possibility of an interaction between diet and gender. For diet A ($X_1 = 1$), the difference between the expected liver Se concentrations for female and male chickens (changing X_2 from 1 to 0) is

$$\mu(1, 1, x_3, x_4) - \mu(1, 0, x_3, x_4) = \beta_2 + \beta_{12}. \qquad \text{(12.22)}$$

The corresponding difference for diet B ($X_1 = 0$) is

$$\mu(0, 1, x_3, x_4) - \mu(0, 0, x_3, x_4) = \beta_2. \qquad \text{(12.23)}$$

Subtracting Equation (12.23) from Equation (12.22), we get

$$\beta_{12} = \left\{ \mu(1, 1, x_3, x_4) - \mu(1, 0, x_3, x_4) \right\} \\ - \left\{ \mu(0, 1, x_3, x_4) - \mu(0, 0, x_3, x_4) \right\}, \qquad \text{(12.24)}$$

so that β_{12} is the amount by which the difference between expected liver concentrations for female and male chickens changes when the diet is changed from A to B. In other words, $\beta_{12} \neq 0$ implies that there is an interaction between diet and gender. Consequently, the presence of an interaction between diet and gender can be tested by testing the null hypothesis $H_{0\beta_{12}}: \beta_{12} = 0$. The parameter β_{12} provides a measure of the intensity of interaction between diet and sex.

Equation (12.21) allows for the possibility of interaction between the qualitative variables X_1 and X_2. This regression function can be extended further to include interactions between other variables—qualitative, quantitative, or both. We simply add appropriate terms that are products of the relevant explanatory variables. For instance, the regression function

$$\mu(x_1, x_2, x_3, x_4) = \beta_0 + \beta_1 x_1 + \beta_2 x_2 + \beta_{12}x_1 x_2 \\ + \beta_3 x_3 + \beta_{13}x_1 x_3 + \beta_{23}x_2 x_3 + \beta_4 x_4 \qquad \text{(12.25)}$$

is obtained by adding the terms $\beta_{13}x_1x_3$ and $\beta_{23}x_2x_3$ to Equation (12.21). The resulting within-group regression functions are as follows:

Group			Expected response when $X_3 = x_3, X_4 = x_4$	
Diet	Gender	Symbol		Expression
A	F	$\mu(1, 1, x_3, x_4)$		$(\beta_0 + \beta_1 + \beta_2 + \beta_{12}) + (\beta_3 + \beta_{13} + \beta_{23})x_3 + \beta_4x_4$
A	M	$\mu(1, 0, x_3, x_4)$		$(\beta_0 + \beta_1) + (\beta_3 + \beta_{13})x_3 + \beta_4x_4$
B	F	$\mu(0, 1, x_3, x_4)$		$(\beta_0 + \beta_2) + (\beta_3 + \beta_{23})x_3 + \beta_4x_4$
B	M	$\mu(0, 0, x_3, x_4)$		$\beta_0 + \beta_3x_3 + \beta_4x_4$

As in the previous two models, the within-group regression functions in Equation (12.25) are multiple linear regression functions with independent variables X_3 and X_4. Again, the parameters of the model can be interpreted by examining the within-group multiple linear regression functions. In Exercise 12.15, you will be asked to argue that β_{13} and β_{23} measure the effects of the interaction between diet and dietary Se concentration and between gender and dietary Se concentration, respectively.

The three models described thus far are only a few of the general linear models available for analyzing the Se concentration data. For instance, the model based on the regression function

$$\mu(x_1, x_2, x_3, x_4) = \beta_0 + \beta_1x_1 + \beta_2x_2 + \beta_{12}x_1x_2 + \beta_3x_3 + \beta_{13}x_1x_3 + \beta_{23}x_2x_3$$
$$+ \beta_4x_4 + \beta_{14}x_1x_4 + \beta_{24}x_2x_4, \qquad \textbf{(12.26)}$$

which is obtained by adding two terms to the model in Equation (12.25), accounts for interaction between diet and the feeding period and between gender and feeding period. Finally, the model

$$\mu(x_1, x_2, x_3, x_4) = \beta_0 + \beta_1x_1 + \beta_2x_2 + \beta_{12}x_1x_2 + \beta_3x_3 + \beta_4x_4 + \beta_{13}x_1x_3$$
$$+ \beta_{23}x_2x_3 + \beta_{14}x_1x_4 + \beta_{24}x_2x_4 + \beta_{34}x_3x_4, \qquad \textbf{(12.27)}$$

extends Equation (12.26) by adding a term to account for possible interaction between the Se concentration in the diet and the feeding period. ∎

In Example 12.5, several possible general linear models that allow for interaction between pairs of explanatory factors were described. If no interaction term in a model contains a quantitative factor—as in Equations (12.19) and (12.21)—then the expected response surfaces corresponding to the groups formed by combining the levels of the qualitative factors will be parallel. The distances between the surfaces will equal the differences between the expected responses for the corresponding groups. Figure 12.3 shows two such surfaces for models involving two quantitative variables. If an interaction term contains one or more quantitative factors—as in Equations (12.25), (12.26), and (12.27)—then the corresponding expected response surfaces will not be parallel. Figure 12.4 displays two examples of such expected response surfaces.

FIGURE 12.4

Expected response surfaces in the presence of interactions involving quantitative variables

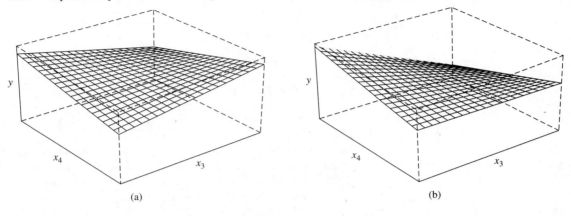

(a)   (b)

The next example illustrates the use of general linear models in statistical data analysis.

EXAMPLE **12.6**   Let's analyze the liver Se concentration data in Example 12.5. The first step is to select a model describing how the response variable Y = liver Se concentration is related to the four experimental factors: X_1, diet; X_2, gender; X_3, dietary Se concentration; and X_4, feeding period. In Section 11.10, we looked at methods of selecting a suitable subset from a set of candidate independent variables for inclusion in a multiple linear regression model. The same methods can be used to select the independent variables for a general linear model. For illustration purposes, let's assume that the general linear model with $p = 10$ independent variables

$$Y = \beta_0 + \beta_1 x_1 + \beta_2 x_2 + \beta_3 x_3 + \beta_4 x_4 + \beta_{12} x_1 x_2 + \beta_{13} x_1 x_3$$
$$+ \beta_{14} x_1 x_4 + \beta_{23} x_2 x_3 + \beta_{24} x_2 x_4 + \beta_{34} x_3 x_4 + E$$

is selected for the analysis of the liver Se concentration data. When the GLM procedure of SAS is used to fit the model to the data, the output obtained includes the following portions.

Part 1

Source	DF	Sum of Squares	Mean Square	F Value	Pr > F
Model	10	37.533670	3.753367	47.73	0.0001
Error	61	4.797058	0.078640		
Corrected Total	71	42.330728			

Part 2

Source	DF	Type III SS	Mean Square	F Value	Pr > F
X1	1	0.1956967	0.1956967	2.49	0.1199
X2	1	0.7344025	0.7344025	9.34	0.0033

X3	1	0.4248922	0.4248922	5.40	0.0234
X4	1	2.8893725	2.8893725	36.74	0.0001
X1*X2	1	0.0648000	0.0648000	0.82	0.3676
X1*X3	1	0.1323000	0.1323000	1.68	0.1995
X1*X4	1	0.5292000	0.5292000	6.73	0.0119
X2*X3	1	0.3780750	0.3780750	4.81	0.0322
X2*X4	1	1.8018750	1.8018750	22.91	0.0001
X3*X4	1	0.8160031	0.8160031	10.38	0.0020

Part 3

Parameter	Estimate	T for H0: Parameter=0	Pr > |T|	Std Error of Estimate
INTERCEPT	0.173993056	0.73	0.4697	0.23917641
X1	0.351666667	1.58	0.1199	0.22292666
X2	0.681250000	3.06	0.0033	0.22292666
X3	2.375520833	2.32	0.0234	1.02197917
X4	0.368020833	6.06	0.0001	0.06071461
X1*X2	0.120000000	0.91	0.3676	0.13219539
X1*X3	-1.050000000	-1.30	0.1995	0.80952813
X1*X4	-0.105000000	-2.59	0.0119	0.04047641
X2*X3	-1.775000000	-2.19	0.0322	0.80952813
X2*X4	-0.193750000	-4.79	0.0001	0.04047641
X3*X4	0.798437500	3.22	0.0020	0.24786636

In the ANOVA table in Part 1, the model sum of squares has ten degrees of freedom, one for each of the ten independent variables in the model. The F-test based on the model sum of squares yields a p-value of 0.0001, thereby providing strong evidence in support of the conclusion that at least one of the ten independent variables has an effect on the liver Se concentration. The p-values for the tests for the significance of the individual independent variables are given in Part 2. Seven out of the ten independent variables show significant association at the 5% level.

Part 3 shows the estimates of the regression parameters. The estimated equation for predicting the liver Se concentration is

$$\hat{y} = 0.174 + 0.352x_1 + 0.681x_2 + 2.376x_3 + 0.368x_4 + 0.120x_1x_2 - 1.050x_1x_3$$
$$- 0.105x_1x_4 - 1.775x_2x_3 - 0.194x_2x_4 + 0.798x_3x_4.$$

By assigning specific values to the dummy variables, estimated prediction equations for the four individual groups can be obtained. For example, the equation for predicting the liver Se for female chickens ($X_2 = 1$) on diet A ($X_1 = 1$) is

$$\hat{y} = 0.174 + 0.352 + 0.681 + 2.376x_3 + 0.368x_4 + 0.120 - 1.050x_3$$
$$- 0.105x_4 - 1.775x_3 - 0.194x_4 + 0.798x_3x_4$$
$$= (0.174 + 0.352 + 0.681 + 0.120) + (2.376 - 1.050 - 1.775)x_3$$
$$+ (0.368 - 0.105 - 0.194)x_4 + 0.798x_3x_4$$
$$= 1.327 - 0.449x_3 + 0.069x_4 + 0.798x_3x_4.$$

As an exercise, verify that the prediction equations for the four groups are as follows:

| Group | | Prediction equation |
Diet	Gender	when $X_3 = x_3$ and $X_4 = x_4$
A	F	$\hat{y} = 1.327 - 0.449x_3 + 0.069x_4 + 0.798x_3x_4$
A	M	$\hat{y} = 0.526 + 1.326x_3 + 0.263x_4 + 0.798x_3x_4$
B	F	$\hat{y} = 0.855 + 0.601x_3 + 0.174x_4 + 0.798x_3x_4$
B	M	$\hat{y} = 0.174 + 2.376x_3 + 0.368x_4 + 0.798x_3x_4$

∎

Exercises

12.13   Refer to the SAS output in Example 12.4 for the analysis of chemical uptake data.

   a   Plot the two prediction lines as functions of the initial weight. Notice that the estimated lines are not parallel. What does this say about the absence or presence of interaction between weight and diet? Use the plot to argue that, under the assumed general linear model, the null hypothesis H_{01} that the regression lines are parallel is the same as the null hypothesis H_{02} that there is no interaction between weight and diet.

   b   Write the expressions for the reduced and full models appropriate for computing the Type I and Type III sums of squares for diet. Use these expressions to explain the difference between the p-values associated with these two sums of squares for diet.

   c   Use the model in Equation (12.18) to show that, for an animal of initial weight $Z = x$, the expected difference between uptake from the two diets (diet 1 − diet 2) is

$$\theta(x) = \beta_1 + \beta_3 x.$$

   d   Plot $\theta(x)$ as a function of x and comment on the way that the expected differential uptake changes with the initial weight.

   e   Obtain an expression for the least squares estimate $\hat{\theta}(x)$ of $\theta(x)$.

   f   Calculate the value of $\hat{\theta}(x)$ when the initial weights are $x = 10$, $x = 13$, and $x = 16$.

   g   Use the given S-matrix to construct a set of 90% simultaneous lower bounds for $\theta(z)$ when $x = 10$, $x = 13$, and $x = 16$. What conclusions can be drawn from these bounds?

12.14   Argue, using the definition of interaction, that the model in Equation (12.19) implies no interaction between diet (X_1) and the dietary Se concentration (X_3).

12.15   Argue that, in the model in Equation (12.25), the parameters β_{13} and β_{23} measure the effects of the interaction between diet and dietary Se concentration and between gender and dietary Se concentration, respectively.

12.16   Refer to the SAS output in Example 12.6 for the analysis of liver Se concentration data.

   a   How will you interpret the negative sign of the estimate of the regression coefficient for X_2X_4?

b   Notice that the p-values for X_1, $X_1 X_2$, and $X_1 X_3$ exceed 0.05. Does this mean that the reduced model obtained by dropping these three variables will predict liver Se concentrations as well as the full model (at the 5% significance level)? Explain.

c   Given the following Type I sums of squares, test the null hypothesis that, under the assumed model, the four explanatory variables—diet, gender, dietary Se, and feeding period—do not interact.

Source	DF	Type I SS	Mean Square	F Value	Pr > F
X1	1	0.231200	0.231200	2.94	0.0915
X2	1	0.684450	0.684450	8.70	0.0045
X3	1	5.413633	5.413633	68.84	0.0001
X4	1	27.482133	27.482133	349.47	0.0001
X1*X2	1	0.064800	0.064800	0.82	0.3676
X1*X3	1	0.132300	0.132300	1.68	0.1995
X1*X4	1	0.529200	0.529200	6.73	0.0119
X2*X3	1	0.378075	0.378075	4.81	0.0322
X2*X4	1	1.801875	1.801875	22.91	0.0001
X3*X4	1	0.816003	0.816003	10.38	0.0020

d   Using the GLM procedure in SAS or otherwise, fit the reduced model obtained by deleting the three variables in (b) and test whether the full model is as good as the reduced model. Write your conclusions.

12.17   In a study, eight insulin-dependent diabetic children received a new treatment (N) and another eight received the standard treatment (S). For each patient, the values of gycosolated hemoglobin (HbAlc), an index of average blood glucose level over 3 months, were measured before and after the treatment. Table 12.3 gives the pretreatment and posttreatment HbAlc for each patient, along with data on gender, age, and treatment assignment.

a   Let

$$X_1 = \begin{cases} 1 & \text{if female} \\ 0 & \text{if male,} \end{cases}$$

$$X_2 = \begin{cases} 1 & \text{if new treatment} \\ 0 & \text{if standard treatment,} \end{cases}$$

$X_3 = $ age, $X_4 = $ pretreatment HbAlc, and $Y = $ posttreatment HbAlc. Consider the model

$$Y = \beta_0 + \beta_1 x_1 + \beta_2 x_2 + \beta_3 x_3 + \beta_4 x_4 + \beta_{34} x_3 x_4 + E. \qquad \textbf{(12.28)}$$

Use the definition of interaction to demonstrate that the model in Equation (12.28) takes no account of interaction between gender and treatment but does take account of interaction between age and pretreatment HbAlc.

b   Explain the meanings of the parameters in Equation (12.28).

c   Using the model in Equation (12.28) and the following SAS output, draw conclusions about the effects of X_1, X_2, X_3, and X_4 on Y.

TABLE   12.3

Study data for insulin-dependent children

Subject	Age (years)	Gender	Treatment	Pretreatment HbAlc	Posttreatment HbAlc
1	11	F	S	16.90	18.10
2	14	F	S	10.87	10.57
3	18	M	S	10.70	10.20
4	12	M	N	12.87	9.15
5	16	F	N	14.60	13.30
6	10	M	N	13.25	9.90
7	14	F	N	11.35	9.60
8	17	F	S	9.83	10.70
9	12	M	S	9.90	9.10
10	11	F	N	11.20	8.90
11	7	M	S	10.58	7.90
12	7	M	S	9.14	7.95
13	15	M	N	14.30	10.10
14	18	F	N	14.50	11.50
15	16	M	N	14.40	12.35
16	11	M	S	11.25	12.30

```
-----------------------------------------------------------------------
Part 1
                              Sum of         Mean
Source             DF         Squares        Square     F Value   Pr > F
Model              5          83.0273752     16.6054750  15.84    0.0002
Error              10         10.4859998     1.0486000
Corrected Total    15         93.5133750
-----------------------------------------------------------------------
Part 2
Source             DF    Type III SS    Mean Square  F Value   Pr > F
GENDER             1     3.5456384      3.5456384    3.38      0.0958
TREAT              1     18.8733820     18.8733820   18.00     0.0017
AGE                1     0.9635804      0.9635804    0.92      0.3604
PRESCR             1     9.2784847      9.2784847    8.85      0.0139
AGE*PRESCR         1     0.7056604      0.7056604    0.67      0.4311
-----------------------------------------------------------------------
Part 3
                             T for H0:    Pr > |T|   Std Error of
Parameter        Estimate    Parameter=0             Estimate
INTERCEPT      -7.259643918 B   -1.24     0.2417     5.83314860
GENDER    F     1.035270659 B    1.84     0.0958     0.56300416
          M     0.000000000 B    .        .          .
TREAT     N    -2.723247790 B   -4.24     0.0017     0.64189977
          S     0.000000000 B    .        .          .
AGE             0.440035621      0.96     0.3604     0.45903815
```

| PRESCR | 1.499960256 | 2.97 | 0.0139 | 0.50425028 |
| AGE*PRESCR | -0.032125570 | -0.82 | 0.4311 | 0.03916137 |

---

d Using the GLM procedure in SAS or other appropriate software, derive the information in the output in (c).

12.5
Analysis of covariance

In Chapter 7 we saw that, at the planning stage, the principle of blocking can be used, under certain conditions, to ensure that the observed differences between mean responses are not influenced by the effects of selected confounding factors. Unfortunately, blocking on the basis of every potential confounding factor is not always a viable option in practice. In that case, the effects of confounding factors on the observed mean differences need to be statistically adjusted at the time of data analysis. In this section, we examine a statistical technique called *analysis of covariance (ACOVA)*, which can often be used to compare responses to treatments when a set of quantitative confounding factors—called *covariates*—are held fixed (adjusted) at given values.

EXAMPLE **12.7** In Example 12.3, the primary objective was to compare the responses (posttest scores) to three treatments designed to enhance children's mental capacities. Suspecting that the mental capacities of the children before they received the treatment may influence their posttest scores, the investigators wanted to compare the treatment effects on the basis of the posttest scores of children with similar pretreatment mental capacities. Since the pretest score is an indicator of pretreatment mental capacity, the investigators decided to compare the treatments after adjusting the posttest scores to account for the differences between the pretest scores. Under certain assumptions (described later) the analysis of covariance can be used to compare the responses to the three therapies after adjusting the values of the covariate, the pretest score. ∎

As we will see, the general linear model approach enables us to make simple statistical adjustments for the effects of covariates, provided the experimental factors do not interact with the covariates. For Example 12.3, this assumption implies that the pretest scores do not interact with the effects of the therapies—that is, that the differences between the effects of the therapies do not depend on the child's pretest score.

Analysis of covariance model

A general linear model in which the experimental and confounding factors do not interact is called an *analysis of covariance model*. The mathematical structure of analysis of covariance models plays a key role in their practical interpretations.

Let Y denote the response variable. Suppose that the independent variables in the general linear model for Y can be divided into two groups: a group containing r variables $X_1, X_2, \ldots, X_r$ that represent the experimental factors; and another group containing s variables $Z_1, Z_2, \ldots, Z_s$ that represent the confounding factors. Then the model for Y is an analysis of covariance model if, for given values $X_i = x_i$ $(i = 1, \ldots, r)$ and $Z_j = z_j$ $(j = 1, \ldots, s)$, it can be expressed as

$$Y = \mu_e(x_1, \ldots, x_r) + \mu_c(z_1, \ldots z_s) + E, \tag{12.29}$$

where E is the random error, which is assumed to be $N(0, \sigma^2)$; $\mu_e(x_1, \ldots, x_r)$ is a function of $x_1, \ldots, x_r$ alone; and $\mu_c(z_1, \ldots, z_s)$ is a function of $z_1, \ldots, z_s$ alone. That Equation (12.29) implies no interaction between the experimental and confounding factors can be seen as follows. At any given values of the confounding factors—say, $Z_1 = z_1, \ldots, Z_s = z_s$—the differential expected response when the levels of $X_1, \ldots, X_r$ are changed from $X_1 = x_1, \ldots, X_r = x_r$ to $X_1 = x_1', \ldots, X_r = x_r'$ is

$$\mu(x_1, \ldots, x_r, z_1, \ldots, z_s) - \mu(x_1', \ldots, x_r', z_1, \ldots, z_s)$$
$$= \{\mu_e(x_1, \ldots, x_r) + \mu_c(z_1, \ldots, z_s)\} - \{\mu_e(x_1', \ldots, x_r') + \mu_c(z_1, \ldots, z_s)\}$$
$$= \mu_e(x_1, \ldots, x_r) - \mu_e(x_1', \ldots, x_r'), \tag{12.30}$$

which does not depend on the $z_1, \ldots, z_s$. Thus, in Equation (12.29), the change in the expected response that results from changing the levels of the experimental factors does not depend on the levels of the confounding factors.

Estimating adjusted mean responses

An important practical implication of Equation (12.29) is that the expected response

$$\mu(x_1, \ldots, x_r, z_1, \ldots, z_s)$$

is the sum of two quantities: $\mu_e(x_1, \ldots, x_r)$, the effect of the experimental factors; and $\mu_c(z_1, \ldots, z_s)$, the effect of the confounding factors. When it is important to emphasize that the expected responses are to be compared at different levels of the experimental factors, the quantity $\mu(x_1, \ldots, x_r, z_1, \ldots, z_s)$ is often referred to as the *adjusted population mean response* at $(x_1, \ldots, x_r)$ when the covariates are fixed at $(z_1, \ldots, z_s)$. The quantity $\mu_c(z_1, \ldots, z_s)$ is the amount by which the effect of the experimental factors will be adjusted at the covariate values $(z_1, \ldots, z_s)$.

We know that the differential expected response defined in Equation (12.30) does not depend on the covariate values, and so the expected responses may be compared by fixing the covariate values at a convenient set of reference values. The most frequently used reference values are $(\bar{z}_1, \ldots, \bar{z}_s)$, the averages of the observed values of the covariates in the data being analyzed. The quantity $\mu(x_1, \ldots, x_r, \bar{z}_1, \ldots, \bar{z}_s)$ can be regarded as a typical expected response at $(x_1, \ldots, x_r)$, in the sense that it is the expected response when the covariate values are equal to their mean values. We refer to $\mu(x_1, \ldots, x_r, \bar{z}_1, \ldots, \bar{z}_s)$ as the adjusted population mean response at $(x_1, \ldots, x_r)$. Any estimate of the adjusted population mean response will be referred to as an *adjusted mean response*.

EXAMPLE **12.8**   In a study to compare four drugs for treating hypertensive patients, 40 patients are divided into four groups of ten. The four drugs are assigned at random to the four groups, and each patient is treated with the drug assigned to his or her group. There is a possibility that the response—Y, the reduction in systolic blood pressure over two weeks of treatment—may be influenced by two confounding factors: Z_1, the patient's age; and Z_2, the patient's systolic blood pressure before treatment initiation. Accordingly, the investigators want to ensure that the comparisons of the effects of the drugs are not influenced by the possible confounding effects of Z_1 and Z_2. On the basis of an extensive literature review and their own past experience, the investigators are willing to make the assumption that the covariates Z_1 and Z_2 do not interact with the drugs. Thus, an analysis of covariance model can be used to compare the effects of the drugs on a patient's posttreatment systolic blood pressure after eliminating the effects of the covariates, the patient's age and the pretreatment systolic blood pressure.

Let X_k ($k = 1, 2, 3$) be dummy variables defined as in Equation (12.6). Then X_1, X_2, and X_3 are the independent variables representing the experimental factor (type of drug), and the assumption that the drugs do not interact with age and pretreatment blood pressure is equivalent to the assumption that the experimental factors X_1, X_2, and X_3 do not interact with the confounding factors Z_1, Z_2. Thus, an analysis of covariance model for Y in terms of the explanatory variables X_1, X_2, X_3, Z_1, and Z_2 can be expressed as in Equation (12.29), with $r = 3$ and $s = 2$

$$Y = \mu_e(x_1, x_2, x_3) + \mu_c(z_1, z_2) + E.$$

For example, the model

$$Y = \beta_0 + \beta_1 x_1 + \beta_2 x_2 + \beta_3 x_3 + \beta_4 z_1 + \beta_5 z_2 + E,$$

with $\mu_e(x_1, x_2, x_3) = \beta_0 + \beta_1 x_1 + \beta_2 x_2 + \beta_3 x_3$ and $\mu_c(z_1, z_2) = \beta_4 z_1 + \beta_5 z_2$, and the model

$$Y = \beta_0 + \beta_1 x_1 + \beta_2 x_2 + \beta_3 x_3 + \beta_4 z_1 + \beta_5 z_2 + \beta_{45} z_1 z_2 + E,$$

with $\mu_e(x_1, x_2, x_3) = \beta_0 + \beta_1 x_1 + \beta_2 x_2 + \beta_3 x_3$ and $\mu_c(z_1, z_2) = \beta_4 z_1 + \beta_5 z_2 + \beta_{45} z_1 z_2$, are both analysis of covariance models. On the other hand, the model

$$Y = \beta_0 + \beta_1 x_1 + \beta_2 x_2 + \beta_3 x_3 + \beta_4 z_1 + \beta_5 z_2 + \beta_{14} x_1 z_1 + E$$

is not an analysis of covariance model, because the term $\beta_{14} x_1 z_1$ means that the expected response cannot be expressed as a sum of $\mu_e(x_1, x_2, x_3)$ and $\mu_c(z_1, z_2)$. ∎

In the next example, we reexamine Example 12.3 from the viewpoint of analysis of covariance.

EXAMPLE **12.9**   In Example 12.3, we described a study to evaluate the effects of three treatments on mental capacity. The analysis of posttest scores was based on the assumption that the relationship between the expected response and the level of the confounding factor can be represented by straight lines with a common slope. As we know, the assumption of parallel lines implies that there is no interaction between the treatments and the pretest scores, and so the general linear model in Example 12.3 is also an analysis of covariance model.

Let $\mu(x_1, x_2, z)$ denote the expected posttest score for children with a pretest score of z who receive treatment (x_1, x_2). Then $\mu(x_1, x_2, z)$, the adjusted population mean response at (x_1, x_2) when $Z = z$, may be regarded as the population mean response for treatment (x_1, x_2) when the pretest score is set equal to (or adjusted to be) z. The assumption that there is no interaction between the drugs and the pretest mental capacity implies that the difference between the two adjusted means does not depend on the value of Z at which the means are calculated. For example, the model in Equation (12.14) implies that the population mean adjusted posttest score for treatment (x_1, x_2) is

$$\mu(x_1, x_2, z) = \mu_e(x_1, x_2) + \mu_c(z),$$

where $\mu_e(x_1, x_2) = \beta_0 + \beta_1 x_1 + \beta_2 x_2$ and $\mu_c(z) = \beta z$. This adjusted population mean can be estimated by

$$\hat{\mu}(x_1, x_2, z) = \hat{\beta}_0 + \hat{\beta}_1 x_1 + \hat{\beta}_2 x_2 + \hat{\beta} z,$$

where $\hat{\beta}_i$ $(i = 0, 1, 2)$ and $\hat{\beta}$ are the least squares estimates of the regression parameters (given in the SAS output in Example 12.3). The estimated adjusted expected posttest score for treatment (x_1, x_2) when $Z = z$ is

$$\hat{\mu}(x_1, x_2, z) = 1.7999 + 17.1442 x_1 - 0.8458 x_2 + 1.1285 z,$$

from which the estimates of the adjusted population mean posttest scores for the three treatment groups can be obtained by substitution of the appropriate values for x_1 and x_2, as follows.

Treatment	Symbol	Estimated adjusted means when $Z = z$ Expression
1	$\hat{\mu}(1, 0, z)$	$18.9441 + 1.1285z$
2	$\hat{\mu}(0, 1, z)$	$0.9541 + 1.1285z$
3	$\hat{\mu}(0, 0, z)$	$1.7999 + 1.1285z$

Notice that the term containing z is the same in each of the three adjusted means, so that the difference between two adjusted means does not depend on the value of Z at which they are calculated. For example

$$\hat{\mu}(1, 0, z) - \hat{\mu}(0, 1, z) = (18.9441 + 1.1285z) - (0.9541 + 1.1285z)$$

$$= 18.9441 - 0.9541 = 17.99,$$

and so the differential expected mean posttest score for treatments 1 and 2 can be estimated as 17.99. This estimated difference is the same regardless of the subject's pretest score.

The estimated adjusted expected posttest score $\hat{\mu}(x_1, x_2, z)$ is the estimate of the mean posttest score of the population of all children with a pretest score of z who received treatment (x_1, x_2). If $\bar{z}$ denotes the mean of the pretest scores of all 30 children enrolled in the study, $\hat{\mu}(x_1, x_2, \bar{z})$ is the adjusted mean for treatment (x_1, x_2).

These adjusted means can be computed by letting $z = \bar{z} = 30.87$ in the expressions for the adjusted means at $Z = z$ in the preceding table. These adjusted means for the three treatment groups and the corresponding unadjusted means—obtained by calculating the mean posttest scores for each of the three groups—are as follows:

Treatment	Adjusted mean	Unadjusted mean
1	53.8	52.8
2	35.8	35.6
3	36.6	37.8

Adjusted means can be interpreted as the estimated mean posttest scores for a conceptual population of children who have an average mental capacity, in the sense that their pretest score equals the average score in the study sample. For example, the adjusted mean of 53.8 for treatment 1 is an estimate of the expected posttest score for an average child—a child with a pretest score of 30.87—who receives treatment 1. The corresponding estimate for a similar child receiving treatment 2 is 35.8. The difference between these two adjusted means is an estimate of the expected difference in posttest scores of two children with the same pretest score, one of whom receives treatment 1, whereas the other receives treatment 2. The unadjusted means, on the other hand, are estimates of the expected posttest scores for the treatments administered to children with different pretest scores. If the pretest score is indeed a confounding factor, the observed difference between the unadjusted means will be affected by the pretest scores of the children in the three groups. A comparison of the adjusted and unadjusted means shows that the adjustments to account for the differences in the children's pretest scores resulted in an upward adjustment of the mean for treatment 1 (from 52.8 to 53.8) and treatment 2 (from 35.6 to 35.8) and a downward adjustment of the mean for treatment 3 (from 37.8 to 36.6, respectively). To understand this intuitively, notice, first, that the common slope ($\hat{\beta} = 1.1285$) of the regression lines is positive, and so higher pretest scores are associated with higher posttest scores. Next, the mean pretest scores of $\bar{z}_1 = 30$ and $\bar{z}_2 = 30.7$ are both smaller than $\bar{z} = 30.87$, and so the observed means for treatment 1 and treatment 2 are both adjusted upwards. Similarly, because $\bar{z}_3$ is larger than $\bar{z} = 30.87$, the observed mean for treatment 3 is adjusted downward. ∎

Selecting covariates

Note that care is necessary in the selection of covariates. Only variables whose values reflect the characteristics of the experimental units are appropriate candidates. Variables whose values are affected by the treatments should not be used as covariates, because arbitrarily fixing their values amounts to imposing arbitrary constraints on the treatment effects. For example, the pretest score is an appropriate covariate, because its value is an indicator of a characteristic (mental capacity) of the experimental units and the treatment in no way affects the value of this covariate. But consider an investigator who wants to compare the yields for four fertilizer treatments of plants grown in field plots. The plots were homogeneous at the start of the experiment, but the investigator noticed at the end of the experiment that different experimental units had different quantities of weeds growing in them. Aware that the quantity of weeds in a plot will most likely affect the yield, the investigator wondered whether to compare the fertilizer effects by treating the quantity of weeds as a covariate. The answer depends on whether or not the fertilizers (treatments) affected the quantity of weeds. If the quantity of weeds is affected by the fertilizers, comparison of fertilizer effects by holding fixed the quantity of weeds in the plots does not make sense, because arbitrarily fixing the quantity of weeds at values other than those observed will imply that the treatments compared are not the treatments tested.

Exercises

12.18 In a study, four bags of ten oysters each were randomly placed at each of five locations in the cooling water canal of a power plant (Output 4.1 in Freund, Littell, & Spector, 1981). Table 12.4 gives the initial and final (one month later) weights of the oyster bags in the five environments.

a Write an analysis of covariance model for predicting the final weight as a function of the treatment and initial weight. Clearly describe the assumptions you are making and explain how the parameters in the model can be interpreted.

TABLE   12.4

Data on oyster weights in five environments

Location	Bag	Initial weight	Final weight
1	1	27.2	32.6
1	2	32.0	36.6
1	3	33.0	37.7
1	4	26.8	31.0
2	1	28.6	33.8
2	2	26.8	31.7
2	3	26.5	30.7
2	4	26.8	30.4
3	1	28.6	35.2
3	2	22.4	29.1
3	3	23.2	28.9
3	4	24.4	30.2
4	1	29.3	35.0
4	2	21.8	27.0
4	3	30.3	36.4
4	4	24.3	30.5
5	1	20.4	24.6
5	2	19.6	23.4
5	3	25.1	30.3
5	4	18.1	21.8

b   The SAS output obtained when the GLM procedure is used to perform an analysis of covariance of the oyster weight data is as follows:

```
-----------------------------------------------------------------
Dependent Variable: FIN_WT
                        Sum of        Mean
Source            DF    Squares      Square   F Value   Pr > F
Model              5   354.44718   70.88944   235.05    0.0001
Error             14     4.22232    0.30159
Corrected Total   19   358.66950

-----------------------------------------------------------------
Source            DF   Type I SS  Mean Square  F Value   Pr > F
LOC                4   198.40700    49.60175   164.47    0.0001
INI_WT             1   156.04018   156.04018   517.38    0.0001


Source            DF  Type III SS  Mean Square  F Value   Pr > F
LOC                4    12.08936     3.02234    10.02     0.0005
INI_WT             1   156.04018   156.04018   517.38    0.0001
-----------------------------------------------------------------

                              T for H0:  Pr > |T|   Std Error of
Parameter      Estimate     Parameter=0              Estimate
INTERCEPT    2.494859769 B      2.43     0.0293     1.02786287
LOC      1  -0.244459378 B     -0.42     0.6780     0.57658196
         2  -0.280271345 B     -0.57     0.5786     0.49290825
         3   1.654757698 B      3.85     0.0018     0.42943036
         4   1.107113519 B      2.35     0.0342     0.47175112
         5   0.000000000 B        .        .            .
INI_WT       1.083179819       22.75     0.0001     0.04762051
-----------------------------------------------------------------
                   X'X Generalized Inverse (g2)


              INTERCEPT       LOC 1         LOC 2         LOC 3
INTERCEPT   3.5030546261  1.1497518704  0.7470299635  0.3521278995
LOC 1       1.1497518704  1.1022970788  0.6790104891  0.5090886875
LOC 2       0.7470299635  0.6790104891  0.805580097   0.4345464115
LOC 3       0.3521278995  0.5090886875  0.4345464115  0.6114515583
LOC 4       0.6297323208  0.6285386669  0.5196294973  0.412835069
LOC 5                 0             0             0             0
INI_WT     -0.156396857  -0.067295763  -0.047934133  -0.028948457
FIN_WT      2.4948597692 -0.244459378  -0.280271345   1.6547576977


                LOC 4      LOC 5     INI_WT         FIN_WT
INTERCEPT   0.6297323208     0   -0.156396857   2.4948597692
LOC 1       0.6285386669     0   -0.067295763  -0.244459378
LOC 2       0.5196294973     0   -0.047934133  -0.280271345
LOC 3       0.412835069      0   -0.028948457   1.6547576977
LOC 4       0.73790838       0   -0.042294823   1.1071135193
```

```
LOC 5          0           0         0          0
INI_WT   -0.042294823      0    0.0075190797  1.0831798188
FIN_WT    1.1071135193     0    1.0831798188  4.2223232546
```

For each of the five locations, determine the equation for predicting the final weight as a function of the initial weight.

c   Calculate the adjusted means for comparing the effects of locations on the final weights of the oysters. Compare the adjusted means with the corresponding unadjusted means and give an intuitive explanation for the pattern of adjustments.

d   Construct a 90% confidence interval for the expected difference between the final weights of oysters who have the same initial weight but are grown in two different locations—location 1 and location 5. What conclusions can be drawn from this interval?

e   Describe how the *p*-values associated with the Type I and Type III sums of squares for the variable LOC can be interpreted.

f   Perform a statistical test to see if differences in the initial weight are associated with differences in the final weight.

g   Construct a 95% confidence interval for a parameter that measures the association in (f). On the basis of this interval, what can you say about the initial weight as a predictor of the final weight?

h   Construct a 95% lower prediction bound for the final weight of a bag of oysters with an initial weight of 25 in location 3. Interpret this lower bound.

12.19   Refer to the oyster data in Exercise 12.18. The five locations differed as follows in terms of the water temperature and the height from the bottom of the canal:

Location	Height	Water temperature
1	Low	Low
2	High	Low
3	Low	High
4	High	High
5	Medium	Medium

a   Let Y and Z denote, respectively, the pretreatment and posttreatment weights of the oyster bags and let

$$X_1 = \begin{cases} 1 & \text{if low height} \\ 0 & \text{otherwise,} \end{cases}$$

$$X_2 = \begin{cases} 1 & \text{if low temperature} \\ 0 & \text{otherwise,} \end{cases}$$

$$X_3 = \begin{cases} 1 & \text{if medium height and medium temperature} \\ 0 & \text{otherwise.} \end{cases}$$

Explain the meanings of the parameters in the model

$$Y = \beta_0 + \beta_1 x_1 + \beta_2 x_2 + \beta_{12} x_1 x_2 + \beta_3 x_3 + \beta z + E. \qquad \textbf{(12.31)}$$

b Use the output in Exercise 12.18 to obtain the estimates and standard errors of the estimates of the parameters in Equation (12.31). [Hint: By comparing the expected responses, express the parameters in Equation (12.31) as linear combinations of the parameters of the model that generated the output in Exercise 12.18.]

c Using the estimates and standard errors in (b), obtain 95% confidence intervals for the parameters in Equation (12.31). State the conclusions that can be drawn from these confidence intervals.

12.20 Using the GLM procedure in SAS or otherwise, fit the model in Equation (12.31) to the oyster weight data. In what essential way does this output differ from that in Exercise 12.18?

12.6
Overview

In this chapter, general linear models (GLM) are introduced as a class of models that includes the ANOVA and multiple linear regression models.

The benefit of the GLM framework is that it allows us to regard ANOVA and regression models as having a common form, expressed in terms of: a response variable Y; given values $x_1, x_2, \ldots, x_p$ of a set of independent variables $X_1, X_2, \ldots, X_p$; and a set of unknown regression parameters $\beta_0, \beta_1, \ldots, \beta_p$. Despite this common GLM form, however, the interpretation of the regression parameters in the two types of models will depend on whether the independent variables are qualitative, as in ANOVA models, or quantitative, as in linear regression models.

The GLM also provides a convenient means of modeling expected responses by means of independent variables, some of which are qualitative, while others are quantitative. An important example of models involving both qualitative and quantitative independent variables is the analysis of covariance (ACOVA) model. Under the assumption that the experimental and confounding factors do not interact, these models can be used to compare the effects of qualitative experimental factors after adjusting for the effects of (quantitative or qualitative) confounding factors.

13

Completely Randomized Factorial Experiments

13.1
Introduction

In Chapter 8, we introduced analysis of variance (ANOVA) as a method for testing the null hypothesis that there are no differences among the expected responses at t levels of an experimental factor. In Chapter 9, we described how the ANOVA method can be used to compare the expected responses at different factor levels. In this chapter, we consider the use of the ANOVA to study how experimental factors combine to affect the measured values.

EXAMPLE **13.1**   In Example 9.14 the weight gain of fish was measured in four aquarium conditions: SC (still cold water), SW (still warm water), FC (flowing cold water), and FW (flowing warm water). The data were analyzed using a one-way ANOVA with four treatments: SC, SW, FC, and FW. Analyzing the data on the basis of a single factor is appropriate provided the objective is to compare the expected weight gains (responses) in the four aquarium conditions.

In addition to the four aquarium conditions, we may also want to study the interaction between water movement and water temperature. In that case, we regard water movement and water temperature as two separate experimental factors and study how one factor—say, water temperature—modifies the effect of the other factor—water movement—on weight gain of the fish. For example, we might ask the following questions:

1   How does the effect of a given level of water temperature (T) change when the level of water movement (M) is changed?

2   Does the differential mean weight gain of fish grown in cold and warm water (levels of T) change with the level of water movement (M)?

Suppose we believe that flowing water is beneficial when used with warm water but detrimental when used with cold water. In that case, it would be of interest to see whether the data confirm the research hypothesis

$$H_1: \ \mu_{FW} - \mu_{SW} > 0 \quad \text{and} \quad \mu_{FC} - \mu_{SC} < 0,$$

where $\mu_{FW}, \mu_{SW}, \mu_{FC}, \mu_{SC}$ denote the expected weight gains for the four combinations of levels of the two factors.

Hypothesis H_1 can be checked by testing suitably defined comparisons among the mean responses. For instance, if $\theta_1 = \mu_{FW} - \mu_{SW}$ and $\theta_2 = \mu_{FC} - \mu_{SC}$, then H_1 can be verified by simultaneously testing

$$H_{0\theta_1}: \ \theta_1 \leq 0 \text{ against } H_{1\theta_1}: \ \theta_1 > 0$$

and

$$H_{0\theta_2}: \ \theta_2 \geq 0 \text{ against } H_{1\theta_2}: \ \theta_2 < 0.$$

Data analyses in which the focus is on verifying hypotheses such as H_1 are called factorial analyses. In Exercise 13.1, you will be asked to perform a factorial analysis of the weight gain data by testing suitably defined contrasts in a one-way ANOVA setting.   ∎

In *factorial analysis*, data are regarded as responses at combinations of levels of one or more experimental factors. The main focus in such an analysis is on the way in which differences between the effects of the levels of individual factors depend upon the levels of other factors.

As can be seen from Example 13.1, factorial analysis of data gathered in completely randomized experiments with only four treatment combinations can easily be performed by the methods described in Chapter 9 for comparing means in a one-way ANOVA setting. However, in a study design that involves combinations of the levels of several factors, where each factor occurs at several levels, factorial analysis in a one-way ANOVA setting can be quite complicated. In such cases, conceptual and theoretical advantages result from analyzing the data in a setting in which each measured response is classified into categories that depend on the levels of the factors affecting that response. As we'll see, such an analysis leads to a so-called k-way ($k \geq 2$) ANOVA of the data.

Exercises

13.1 Refer to Example 13.1.

a Explain how the Bonferroni method of simultaneous hypothesis tests can be performed to verify research hypothesis H_1.

b Use the data in Example 9.14 to perform the test described in (a). Write your conclusions.

c Construct a set of 95% simultaneous confidence intervals for θ_1 and θ_2. Interpret the confidence intervals.

d Suppose that the investigator conjectures that the differential weight gains resulting from still and flowing water are not the same at cold and warm temperatures. Set up an appropriate contrast to verify this conjecture.

e Use the contrast defined in (d) to see if the conjecture is supported by the data in Example 9.14.

13.2 In Example 9.12, sheep were fed four types of hay, and the following data were given on the organic matter content of the sheep feces:

Type of hay	JP	SP	JM	SM
Sample size	6	6	6	6
Sample mean	60.3	63.8	67.2	67.1

SS[E] = 63.95

The data can be regarded as resulting from a study with two experimental factors: time of harvest A, at two levels—a_1 = June and a_2 = September; and type of hay B, at two levels—b_1 = Pensacola and b_2 = Mott.

a Construct two contrasts θ_1 and θ_2 such that θ_1 measures the effect of harvest time on the organic matter content for Pensacola—that is, compares JP with SP—and θ_2 measures the effect of harvest time on the organic matter content for Mott—that is, compares JM with SM. Show that $\theta = \theta_1 - \theta_2$ is a contrast. What does this contrast measure?

b Test the significance of the contrast θ defined in (a) and write your conclusions.

c Plot the measured mean responses as four points in a graph in which the horizontal axis shows the levels of B and the vertical axis shows the observed mean response. Use straight lines to connect, the pairs of points that correspond to the same level of A. Describe how the degree of interaction between A and B can be

visually assessed on the basis of the degree to which the two lines corresponding
to levels a_1 and a_2 are parallel.

d   Construct a 95% confidence interval for the contrast θ defined in (a) and interpret
the interval.

13.2
Factorial designs: Notation and definitions

A study design in which responses are measured at different combinations of levels
of one or more experimental factors is called a *factorial design*. We use capital letters
to denote the factors, and the corresponding lowercase letters to denote their levels.
Thus, the levels of A will be denoted as $a_1, a_2, \ldots$, and so on.

In the factorial notation, the treatment representing a particular combination of
factor levels is denoted by the same combination of the lowercase letters that represent the corresponding levels. Thus, with three factors A, B, and C, the treatment
corresponding to the combination of levels a_1 of A, b_3 of B, and c_2 of C is denoted
by $a_1 b_3 c_2$.

An experiment in which responses are measured at all combinations of the levels
of the factors is called a *complete factorial experiment*. A complete factorial experiment in which there are a levels of factor A, b levels of factor B, and so on is called
an $a \times b \times \cdots$ *factorial experiment*. Clearly, the total number of treatments in an
$a \times b \times c \cdots$ complete factorial experiment is $t = abc \cdots$.

EXAMPLE 13.2   The design of the digestibility trial in Example 9.12 is a 2×2 factorial design. There
are $k = 2$ experimental factors: A, the time of planting; and B, the type of hay. Each
factor occurs at two levels, with $a = 2$ levels for A and $b = 2$ levels for B. There are
$t = ab = (2)(2) = 4$ treatments in this study. The actual levels of A are $a_1 =$ June
and $a_2 =$ September, while the levels of B are $b_1 =$ Pensacola and $b_2 =$ Mott. In
the factorial notation, the four treatments JP, SP, JM, and SM are $a_1 b_1$, $a_2 b_1$, $a_1 b_2$,
and $a_2 b_2$, respectively.   ∎

EXAMPLE 13.3   Nakabayashi and Wada (1991) conducted an experiment to study the effects of
iron compounds on the Pg (a green colored fraction of humic acid) content in a
Japanese dark brown forest soil. The experiment used a factorial design, in which
20-g portions of soil samples placed in Petri dishes were incubated at 25°C under
95% humidity conditions. The Pg contents of the soil samples were measured at the
end of the incubation period. The experimental factors were:

- A, the added Pg, at two levels ($a_1 = 0$, $a_2 = 0.46$ mg/g moist soil);

- B, glucose, at three levels ($b_1 = 0$, $b_2 = 2.9$, and $b_3 = 5.8$ mg/g moist soil);

- C, ferric iron, at four levels ($c_1 = 0.00$, $c_2 = 0.03$, $c_3 = 0.06$, $c_4 = 0.09$ mg/g
moist soil);

- D, the incubation period, at three levels ($d_1 = 1$, $d_2 = 5$, and $d_3 = 25$ days).

Thus, the study is based on a $2 \times 3 \times 4 \times 3$ complete factorial design with $t = (2)(3)(4)(3) = 72$ treatments. The treatment $a_2 b_3 c_1 d_2$ is the result of adding 0.46 mg/g of Pg, 5.8 mg/g of glucose, and no ferric iron to the soil sample and incubating it for five days.

Factorial analysis of the data from this study may be used to answer such questions as the following:

1 Do the differences between the measured Pg contents at a given incubation period depend on the amount of added ferric iron?

2 Does the effect of a given combination of levels of added glucose and ferric iron on the measured Pg content change with the incubation period? ■

Exercises

13.3 To investigate how a subject's ability to recall words differs as a function of age and the level of processing (degree of mental processing) at the initial exposure, 50 young and 50 old subjects were chosen for a study. The subjects in each age group were randomly assigned to five processing levels (called counting, rhyming, adjective, imagery, and intentional) in such a way that ten subjects were assigned to each level. Each subject was asked to read a list of 27 words three times and to process the word each time at the assigned level (Eysenck, 1974).

a Identify the factors and their levels.

b Explain the notation appropriate for describing the study.

c Suggest two research hypotheses that can be verified using a factorial analysis of the data.

13.4 To determine the reasons for the apparent disappearance of certain *Drosophila* species of flies in winter in some temperate regions, Izquierdo (1991) conducted a study of the survival pattern of *D. melanogaster* in temperate regions under unfavorable winter conditions.

Samples of 250 male and 250 female flies were selected from three population cages: one kept at 24°C (cage I); one at 17°C (cage II); and one at the site of fly collection (cage III), where the flies were exposed to natural outdoor temperatures. Flies selected from each cage were placed in bottles, and the bottles were kept at 4°C under natural light conditions. Each bottle contained 50 flies; 15 bottles (5 per cage) contained male flies, and another 15 contained female flies. Table 13.1 shows the number of survivors at the end of 45 days.

a Identify the factors and their levels in this study.

b Explain the notation appropriate for describing this study.

c Plot the mean responses in a graph in which the cages are marked on the horizontal axis. Join the three mean responses for each gender with a straight line. Examine the graph to see if the difference between the mean survival rates of males and females varies with the type of cage.

d Suggest two research hypotheses that can be verified using a factorial analysis of these data.

TABLE   13.1

Data on fly survival

Type of cage	Bottle	Gender Female	Male	Total
	1	26	22	48
	2	22	11	33
I (24°C)	3	27	8	35
	4	24	14	38
	5	27	14	41
	Total	126	69	195
	1	25	11	36
	2	35	16	51
II (17°C)	3	27	14	41
	4	32	16	48
	5	32	18	50
	Total	151	75	226
	1	33	6	39
	2	32	14	46
III (environmental)	3	33	12	45
	4	25	5	30
	5	28	16	44
	Total	151	53	204

13.5   In a greenhouse study, the drought resistance of four varieties of a field crop was compared. After a pretreatment to stimulate root growth, each cutting of a given variety was exposed to a drought condition. There were three pretreatments, and each pretreatment was applied to three cuttings. Table 13.2 gives the average root lengths after four months of growth in the greenhouse.

a   Identify the factors and their levels.

b   Suggest a set of notation to describe this study.

c   Suggest two research hypotheses that can be evaluated using a factorial analysis of these data.

13.3
Factorial effects in 2 × 2 experiments

Factorial analysis of data is best described in terms of three types of factorial effects: simple effects, interaction effects, and main effects. In this section, we consider the three factorial effects in the context of a 2 × 2 factorial experiment.

T A B L E   13.2

Root length data for cuttings

Pretreatment	Cutting	Variety 1	2	3	4	Total
	1	11	26	17	08	62
1	2	05	13	30	05	53
	3	07	15	21	05	48
	Total					163
	1	15	20	15	15	65
2	2	17	21	29	12	79
	3	04	20	28	10	62
	Total					206
	1	03	05	06	10	24
3	2	01	04	03	10	18
	3	06	04	04	05	19
	Total					61
Total		69	128	153	80	430

Simple effects

Consider a 2×2 factorial with two factors: A at levels a_1, a_2; and B at levels b_1, b_2. Let μ_{ij} denote the expected response for treatment $a_i b_j$; that is, μ_{11}, μ_{12}, μ_{21}, and μ_{22} denote, respectively, the expected responses for the treatments $a_1 b_1$, $a_1 b_2$, $a_2 b_1$, and $a_2 b_2$. The *simple effect* of A at level b_1 of B is defined as

$$\mu[AB_1] = \mu_{21} - \mu_{11}.$$

In words, the simple effect of A at level b_1 of B is the amount of change in the expected response when the level of A is changed from a_2 to a_1 and the level of B is held fixed at b_1. Thus, the simple effect of A at $b = b_1$ measures the effect of changing the level of A while holding the level of B fixed at b_1. The simple effect of A at level b_2 of B is defined in a similar manner

$$\mu[AB_2] = \mu_{22} - \mu_{12}.$$

Table 13.3 shows the four simple effects that can be defined in a 2×2 study.

Interaction effects

If the simple effect of A changes with the level of B, then we can say that the level of B has an impact on the simple effect of A. In that case, the factors A and B are said to *interact* with each other. Let

$$\mu[AB] = \frac{1}{2} \left(\mu[AB_2] - \mu[AB_1] \right). \tag{13.1}$$

TABLE **13.3**

Population means and simple effects in a 2 × 2 factorial

| Level of A | Level of B | | Simple effect of B |
	b_1	b_2	$\mu[A_iB]$
a_1	μ_{11}	μ_{12}	$\mu[A_1B] = \mu_{12} - \mu_{11}$
a_2	μ_{21}	μ_{22}	$\mu[A_2B] = \mu_{22} - \mu_{21}$
Simple effect of A $\mu[AB_j]$	$\mu[AB_1] =$ $\mu_{21} - \mu_{11}$	$\mu[AB_2] =$ $\mu_{22} - \mu_{12}$	

Clearly, if there is no interaction between A and B, then $\mu[AB] = 0$. The existence of interaction between these two factors is indicated by a nonzero value of $\mu[AB]$. The magnitude of $\mu[AB]$ measures the degree (intensity) of the interaction between A and B. Larger $\mu[AB]$ indicates greater interaction between A and B.

For obvious reasons, $\mu[AB]$ is called the effect of interaction between A and B, or simply the *AB interaction effect*. In Exercise 13.6, you will be asked to show that the interaction effect can also be defined as the difference between the simple effects of B; that is

$$\mu[AB] = \frac{1}{2} \left(\mu[A_2B] - \mu[A_1B] \right). \tag{13.2}$$

Thus, as we would expect, the definition of interaction between two factors does not depend on which factor is used to compare the simple effects.

Let's now look at the concept of interaction using some hypothetical numbers for the expected responses (population means). Table 13.4 illustrates three distinct cases: no interaction and two types of interaction. It is easy to verify the values of the simple and interaction effects in Table 13.4. For instance, in Table 13.4a, the simple effect of B at level a_1 of A is $\mu[A_1B] = 30 - 10 = 20$. Similarly, the simple effect of A at level b_1 of B is $\mu[AB_1] = 16 - 10 = 6$. Finally, the interaction effect of A and B is given by Equation (13.1): $\mu[AB] = \frac{1}{2}(\mu[A_2B] - \mu[A_1B]) = \frac{1}{2}(20 - 20) = 0$. Of course, the interaction effect can also be calculated using Equation (13.2): $\mu[AB] = \frac{1}{2}(\mu[AB_2] - \mu[AB_1]) = \frac{1}{2}(6 - 6) = 0$.

In Table 13.4a, the simple effect of A is the same at both levels of B, leading to the value $\mu[AB] = 0$. Thus, Table 13.4a depicts a situation where there is no interaction between A and B. In Table 13.4b, the simple effect of A changes with the level of B, thereby showing interaction between A and B. The interaction effect is $\mu[AB] = -7$. Table 13.4c also shows AB interaction, but notice the difference between the type of interaction here and that in Table 13.4b. In Table 13.4b, the simple effects of each factor are both positive or both negative. For instance, the simple effects of B are both positive; hence, at both levels of A, the expected response increases when the level of B is changed from b_1 to b_2. The presence of interaction is indicated by the fact that the magnitudes of the increases are not the same. In Table 13.4c, by contrast, the expected response decreases at a_1 (the simple effect of B is negative) and increases at a_2 (the simple effect of B is positive). Thus, the interaction is due to the difference between the signs of the simple effects. Often, the interaction

T A B L E   13.4

Simple and interaction effects associated with hypothetical population means (expected responses)

(a) Identical simple effects

	b_1	b_2	$\mu[A_iB]$	
a_1	10	30	20	$\mu[AB] = 0$
a_2	16	36	20	
$\mu[AB_j]$	6	6		

(b) Unequal simple effects with the same signs

	b_1	b_2	$\mu[A_iB]$	
a_1	12	32	20	$\mu[AB] = -7$
a_2	4	10	6	
$\mu[AB_j]$	-8	-22		

(c) Unequal simple effects with opposite signs

	b_1	b_2	$\mu[A_iB]$	
a_1	20	2	-18	$\mu[AB] = 15$
a_2	3	15	12	
$\mu[AB_j]$	-17	13		

depicted in Table 13.4b is called a *quantitative interaction*, because changing the levels of a given factor results in a change in the magnitude (not the direction) of the simple effect of the other factor. The interaction depicted in Table 13.4c is called a *qualitative interaction*, because changing the level of one factor results in a change in the direction (sign) of the simple effect of the other factor.

Plotting interactions

A convenient graphical representation of interaction can be obtained by plotting the expected responses against the levels of one of the factors. In such a plot, often called an *interaction plot*, the levels of one of the factors—say, factor A—are marked on the horizontal axis. The expected responses (population means) for the four treatments are plotted against the levels of A, and the points corresponding to the same level of the other factor—factor B—are joined by straight lines. Figure 13.1 shows interaction plots of the hypothetical expected responses in Table 13.4.

Figure 13.1a shows parallel lines, because the difference between the expected responses at a_1 is the same as the difference between the expected responses at a_2.

Interaction plots of the population means in Table 13.4

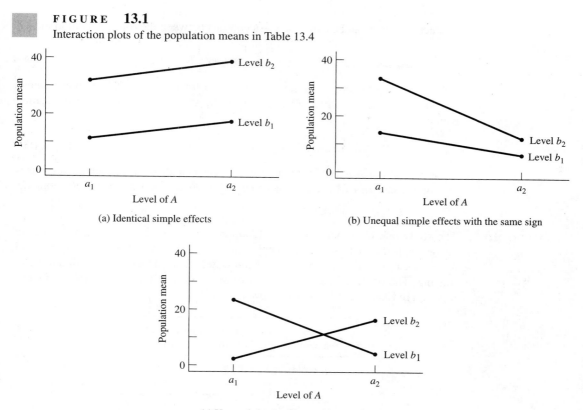

(a) Identical simple effects

(b) Unequal simple effects with the same sign

(c) Unequal simple effects with opposite signs

Thus, the absence of interaction is indicated by parallel lines. The lines in Figures 13.1b and 13.1c are not parallel, because the two factors interact with each other. In Figure 13.1b, both lines have downward slopes because, at both levels of B, the expected responses decrease as the level of A is changed from a_1 to a_2. The lines are not parallel because the magnitudes of the decrease are different at the two levels of B. Recall that the interaction in Table 13.4b is called a quantitative interaction. Therefore, a quantitative interaction is indicated by a pattern in which the lines are not parallel but the direction of their slopes is the same. Finally, in Figure 13.1c, the line corresponding to b_1 has a downward slope but the line corresponding to b_2 has an upward slope. Qualitative interaction is indicated by a pattern of nonparallel lines in which the direction of the slopes of the lines is different.

Testing for interaction

The presence of interaction can be tested by testing the null hypothesis H_0: $\mu[AB] = 0$ against the research hypothesis H_1: $\mu[AB] \neq 0$. By substituting the expressions for $\mu[AB_1]$ and $\mu[AB_2]$ into Equation (13.1) or Equation (13.2), we see that the interaction effect can be expressed as a contrast between the four population means

$$\mu[AB] = \frac{1}{2}\mu_{22} - \frac{1}{2}\mu_{12} - \frac{1}{2}\mu_{21} + \frac{1}{2}\mu_{11}. \tag{13.3}$$

Let $\overline{Y}_{ij+}$ denote the observed mean response for treatment $a_i b_j$. Then an estimate of the contrast in Equation (13.3) is given by the same contrast between the sample means

$$\hat{\mu}[AB] = \frac{1}{2}\overline{Y}_{22+} - \frac{1}{2}\overline{Y}_{12+} - \frac{1}{2}\overline{Y}_{21+} + \frac{1}{2}\overline{Y}_{11+}. \tag{13.4}$$

Provided the ANOVA assumptions are reasonable for the observed responses, the significance of $\hat{\mu}[AB]$ can be tested using an F-test, as described in Section 9.2.

EXAMPLE **13.4**   The study of weight gain in fish in Example 9.14 corresponds to a 2×2 factorial design with the factors: A, the temperature of the water, at two levels $a_1 = $ cold, $a_2 = $ warm; and B, the movement of the water, at two levels $b_1 = $ still and $b_2 = $ flowing. The observed mean responses (based on $n = 3$ replications per treatment) reported in Example 9.14 are as follows:

Level of temperature	Level of movement	
	b_1 (still)	b_2 (flowing)
a_1 (cold)	1.55	1.08
a_2 (warm)	1.59	2.01

The corresponding one-way ANOVA table from Example 9.14 is as follows:

Source	df	SS	MS
Aquarium condition	3	1.302	0.434
Error	8	0.347	0.043

The effect of interaction between temperature and movement is estimated by the contrast

$$\hat{\mu}[AB] = \frac{1}{2}\overline{Y}_{22+} - \frac{1}{2}\overline{Y}_{12+} - \frac{1}{2}\overline{Y}_{21+} + \frac{1}{2}\overline{Y}_{11+} = 0.445.$$

The sum of squares for testing the significance of this contrast can be calculated using $c_1 = +\frac{1}{2}, c_2 = -\frac{1}{2}, c_3 = -\frac{1}{2}, c_4 = +\frac{1}{2}$, and $n_1 = n_2 = n_3 = n_4 = 3$ in Equation (9.9). The required sum of squares is

$$SS[AB] = 0.594,$$

so that the test statistic for testing the significance of AB interaction has the calculated value

$$F_c = \frac{MS[AB]}{MS[E]} = \frac{0.594}{0.043} = 13.81.$$

This calculated value is significant at the level $\alpha = 0.01$ ($p = 0.006$), and so we infer that there is a demonstrable interaction between water movement and water temperature on the weight gain of fish grown in aquariums. We can conclude, at the level $\alpha = 0.01$, that the effect of changing the level of water movement (temperature) is not the same at the two levels of water temperature (movement).   ∎

Main effects

The main effect of A is denoted by $\mu[A]$ and is defined as the average of the two simple effects $\mu[AB_1]$ and $\mu[AB_2]$

$$\mu[A] = \frac{1}{2}(\mu[AB_2] + \mu[AB_1])$$

$$= \frac{1}{2}(\mu_{22} - \mu_{12} + \mu_{21} - \mu_{11}). \tag{13.5}$$

The main effect of A is the average change in the expected response (the population mean response) when the level of A is changed from a_2 to a_1. From Equation (13.5), the main effect of A is a contrast among the population means. This contrast can be estimated by the corresponding contrast among the sample means

$$\hat{\mu}[A] = \frac{1}{2}(\overline{Y}_{22+} - \overline{Y}_{12+} + \overline{Y}_{21+} - \overline{Y}_{11+}). \tag{13.6}$$

Similarly, the main effect of B is defined as the average of the two simple effects of B

$$\mu[B] = \frac{1}{2}(\mu[A_2B] + \mu[A_1B])$$

$$= \frac{1}{2}(\mu_{22} - \mu_{21} + \mu_{12} - \mu_{11}), \tag{13.7}$$

with the corresponding estimate

$$\hat{\mu}[B] = \frac{1}{2}(\overline{Y}_{22+} - \overline{Y}_{21+} + \overline{Y}_{12+} - \overline{Y}_{11+}). \tag{13.8}$$

If the ANOVA assumptions are reasonable for the data, inferences about $\mu[A]$ and $\mu[B]$ can be made by the methods in Chapter 9.

From Equation (13.5), we see that, if A and B do not interact—that is, if $\mu[AB_1] = \mu[AB_2]$—then the main effect of A is equal to the common value of simple effects of A at the two levels of B. Thus, in the absence of AB interaction, $\mu[A]$ measures the expected change in response resulting from changing the level of A. For instance, in Table 13.4a, $\mu[A] = \frac{1}{2}(6 + 6) = 6$ is the expected change in response at either of the two levels of B. If there is AB interaction, practical interpretation of $\mu[A]$ is difficult, because in that case the main effect is an average of two unequal simple effects. Indeed, it is possible that a large positive simple effect at b_1 and a large negative simple effect at b_2 can result in a very small value for the main effect of A. For instance, in Table 13.4c, $\mu[A] = \frac{1}{2}(-17 + 13) = -2$, a value much smaller than the simple effects of A. Thus, in the presence of interaction, a small value for the main effect of A does not imply that factor A has a minor effect on the expected response.

In Figure 13.2, the simple effects of A are plotted for the three hypothetical populations whose means are given in Table 13.4. In Figure 13.2a, the simple effect of A is 6 at both levels of B, so that the main effect of A is also $\mu[A] = 6$. In other words, when we change the level of A from a_1 to a_2, we can expect an increase of response equal to 6 units at each level of B. Thus, when there is no interaction between A and B, the main effect of A can be interpreted as the increase (decrease, if negative) in the expected response resulting from changing the level of A.

FIGURE 13.2

Simple effects of A for the three populations in Table 13.4

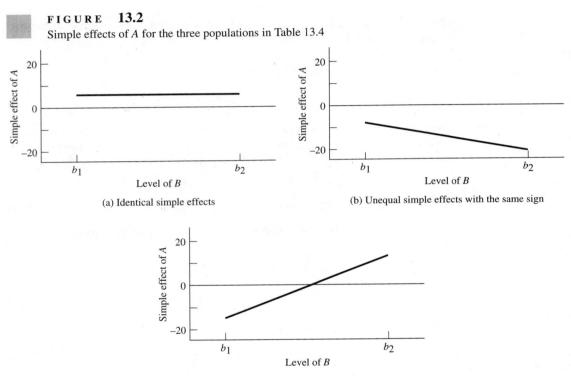

(a) Identical simple effects

(b) Unequal simple effects with the same sign

(c) Unequal simple effects with opposite signs

AB interaction is present in Figure 13.2b. At each level of B, changing the level of A from a_1 to a_2 results in a decrease in the expected response, but the magnitudes of the decrease are not the same. We can expect a decrease of 22 units at b_2 but only 8 units at b_1. The main effect of A is the average of these two expected increases: $\mu[A] = \frac{1}{2}(22 + 8) = 15$. Thus, in contrast to the no-interaction case, the main effect of A does not have a practically meaningful interpretation, because 15 is not close to either 22 or 8.

In Figure 13.2c, not only are the expected changes in response different at the two levels of B, but also the direction of change is reversed. The effect of changing a_1 to a_2 is to decrease (by 17 units) the expected response at b_1 and to increase (by 13 units) the expected response at b_2. Thus, the value $\mu[A] = \frac{1}{2}(-17 + 13) = -2$

for the main effect of A is not of much help in describing the effect of changing the level of A on the expected response.

Strategy for 2 × 2 factorial analysis

On the basis of the above considerations, the following strategy is usually adopted in analyzing the results of 2 × 2 factorial experiments. First, a test is performed to see if there is an interaction between the two factors. If a statistically significant interaction is indicated, the simple effects of each of the two factors are examined separately. If there is no demonstrable interaction, then inferences are made about each of the two main effects. The next two examples illustrate how to make inferences about simple and main effects in a 2 × 2 factorial analysis.

EXAMPLE **13.5**   In Example 13.4, a test of interaction indicated that the (simple) effects of water movement (factor B) at the two water temperatures are significantly different ($p = 0.005$). Consequently, we must investigate the simple effect of movement at each level of temperature. Let's first look at $\mu[A_1 B]$, the simple effect of B in cold water. This effect can be estimated by the sample contrast

$$\hat{\mu}[A_1 B] = \overline{Y}_{12+} - \overline{Y}_{11+} = 1.08 - 1.55 = -0.47$$

and tested for significance using an F-test based on the test statistic in Equation (9.11).

The calculated value of the test statistic F_c is 7.71 ($df = 1, 8$), which is significant at the level $\alpha = 0.05$ ($p = 0.024$). We conclude that still and flowing water have different effects on the weight gains of fish grown in aquariums with cold water. The formula for a confidence interval for a contrast in Box 9.2 can be used to verify that $(-0.86, -0.08)$ is a 95% confidence interval for $\mu[A_1 B]$, the simple effect of B at a_1. Thus, with 95% confidence, we can assert that, if the water is cold, movement of water will decrease the weight gain of fish by an amount between 0.08 g and 0.86 g, on average.

Inferences about the effect of water movement in warm water can be made in a similar manner. The estimated value of the simple effect of B at a_2 is

$$\hat{\mu}[A_2 B] = \overline{Y}_{22+} - \overline{Y}_{21+} = 2.01 - 1.59 = 0.42$$

and hence we obtain a calculated value $F_c = 6.15$ ($df = 1, 8$) for the test statistic. This value is significant at $\alpha = 0.05$ ($p = 0.038$). Thus, still and flowing water also produce significantly ($p = 0.038$) different weight gains at a high water temperature.

The observed effect of the interaction between water movement and water temperature can be graphically displayed by plotting the observed mean responses against the levels of one of the factors. Figure 13.3 shows a plot of the observed mean weight gains against the levels of water movement. We see in Figure 13.3 that, in warm water, water movement increased the mean weight gain of fish. In cold water, by contrast, the water movement decreased the mean weight gain.

FIGURE   13.3

Observed interaction between movement and temperature

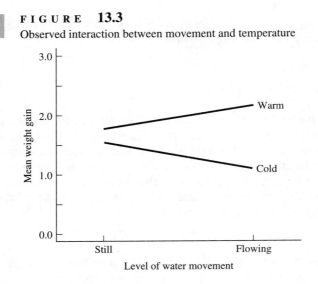

An alternative method of exhibiting the observed interaction is to plot the observed simple effects against the level of one of the factors, as in Figure 13.4.   ■

FIGURE   13.4

Observed simple effects of water movement

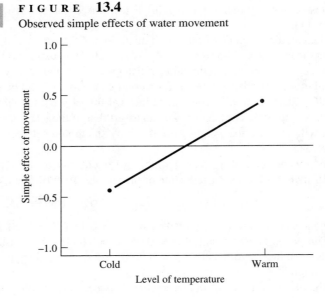

EXAMPLE 13.6   In a study, the percentage water content in the tissues of snails grown under six different experimental conditions was measured. The six conditions were obtained by combining three levels of humidity (factor B) with two levels of temperature (factor A). There were $n = 4$ replications of each treatment combination. The

results obtained for the percentage water content are as follows (Underwood, 1981, Table XVI):

Temperature (°C)	Humidity (%)					
	45		75		100	
20	76	64	72	82	100	96
	79	71	86	86	92	100
30	72	72	72	75	100	94
	64	70	82	84	98	99

The tissue water content data can be regarded as the results from a 3 × 2 factorial experiment. The treatment means, are as follows:

Temperature	Humidity (%)		
	$b_1 = 45$	$b_2 = 75$	$b_3 = 100$
$a_1 = 20°C$	72.50	81.50	97.00
$a_2 = 30°C$	69.50	78.25	97.75

For the time being, let's disregard the data for 100% humidity and analyze the remaining data as the result of a completely randomized experiment with $t = 4$ treatments. The corresponding ANOVA table is as follows:

Source	df	MS	F	p-value
Treatments	3	118.06	3.54	0.048
Error	12	33.31		

In Exercise 13.13, you will be asked to verify that the interaction between humidity and temperature is not statistically significant ($p > 0.10$). Knowing that humidity and temperature do not interact, we proceed to examine the main effects of these factors. First, consider $\mu[B]$, the main effect of humidity, which is estimated by the contrast

$$\hat{\mu}[B] = \frac{1}{2}(\overline{Y}_{22+} - \overline{Y}_{21+} + \overline{Y}_{12+} - \overline{Y}_{11+})$$

$$= 8.875.$$

The sum of squares for testing this contrast is $SS[B] = 315.06$, and so the calculated value of the F-statistic is $F_c = 9.46$ with 1 and 12 degrees of freedom, which is highly significant ($p < 0.01$). Calculations for the main effect of temperature yield $\hat{\mu}[A] = -3.13$, $SS[A] = 39.06$, $F_c = 1.17$, and $p = 0.30$.

It appears that the differential mean tissue water contents at 30 and 20°C are not demonstrably different at the two humidities; that is, the differential mean tissue water content at the two temperatures is not affected by humidity. The expected tissue water content is significantly ($p < 0.01$) higher at 75% humidity than at 40%

humidity. However, the expected tissue water content at 30°C is not significantly ($p = 0.30$) different from that expected at 20°C. ■

ANOVA for 2×2 factorial experiment

In a 2×2 factorial analysis, the main focus is on inferences about three factorial effects that measure the interaction and main effects of the two factors under consideration. As we have seen, these effects are estimated by the contrasts in Equation (13.4), (13.6), and (13.8). The significance of these contrasts can be established by the F-tests described in Section 9.2. The required sums of squares are usually displayed as in Table 13.5, where N is the total number of responses in the corresponding completely randomized experiment.

The calculation of the quantities needed to construct Table 13.5 is straightforward. All standard statistical computing packages are able to produce the ANOVA tables associated with factorial analyses; the companion text by Younger (1997) shows how SAS can be used to accomplish this task. However, it is instructive to see how closely the sums of squares in Table 13.5 are tied to the quantities we encountered when discussing one-way ANOVA in Chapters 8 and 9.

First, SS[TOT] and SS[E] are the total and error sum of squares in a one-way ANOVA in which there are $t = 4$ levels for the experimental factor. The sums of squares SS[A], SS[B], and SS[AB] are, respectively, the sums of squares for the contrasts $\hat{\mu}[A]$, $\hat{\mu}[B]$, and $\hat{\mu}[AB]$ defined in Equations (13.6), (13.8), and (13.4) respectively. Let n_{ij} denote the number of replications of the treatment combination $a_i b_j$, and suppose that none of the n_{ij} is zero. Then the general formula in Equation (9.9) for the sum of squares of a contrast can be used to show that in the case of a 2×2 experiment

$$SS[AB] = \frac{4(\hat{\mu}[AB])^2}{\left(\dfrac{1}{n_{11}} + \dfrac{1}{n_{12}} + \dfrac{1}{n_{21}} + \dfrac{1}{n_{22}} \right)},$$

$$SS[A] = \frac{4(\hat{\mu}[A])^2}{\left(\dfrac{1}{n_{11}} + \dfrac{1}{n_{12}} + \dfrac{1}{n_{21}} + \dfrac{1}{n_{22}} \right)}, \tag{13.9}$$

$$SS[B] = \frac{4(\hat{\mu}[B])^2}{\left(\dfrac{1}{n_{11}} + \dfrac{1}{n_{12}} + \dfrac{1}{n_{21}} + \dfrac{1}{n_{22}} \right)}.$$

If every treatment is replicated the same number of times—that is, if $n_{ij} = n$—then we can use Equation (9.10) to further simplify Equation (13.9)

$$SS[AB] = n \left(\hat{\mu}[AB] \right)^2,$$

$$SS[A] = n \left(\hat{\mu}[A] \right)^2, \tag{13.10}$$

$$SS[B] = n \left(\hat{\mu}[B] \right)^2.$$

TABLE 13.5

ANOVA table for a 2 × 2 factorial experiment

Source	df	SS	MS	F_c
A	1	SS[A]	MS[A]	F_A
B	1	SS[B]	MS[B]	F_B
AB	1	SS[AB]	MS[AB]	F_{AB}
Error	$N-4$	SS[E]	MS[E]	
Total	$N-1$	SS[TOT]		

Missing subclasses

It is important to note that Equations (13.9) and (13.10) give meaningful quantities only when the study design is a *complete factorial*—that is, only when there is at least one replication of every treatment combination ($n_{ij} > 0$ for all i, j). If $n_{ij} = 0$ for some treatment combinations, then the experiment will have some *missing subclasses*. To see the consequence of missing subclasses, let's consider an experiment in which no responses are measured for a_1b_1 ($n_{11} = 0$). If there are no responses for treatment a_1b_1, it is not possible to estimate μ_{11}, the expected response for this treatment. Consequently, neither $\mu[AB_1] = \mu_{21} - \mu_{11}$ (the simple effect of A at b_1) nor $\mu[A_1B] = \mu_{12} - \mu_{11}$ (the simple effect of B at a_1) can be estimated from the available data. Therefore, if $n_{11} = 0$, the factorial effects

$$\mu[AB] = \frac{1}{2}(\mu[A_2B] - \mu[A_1B]),$$

$$\mu[A] = \frac{1}{2}(\mu[A_2B] + \mu[A_1B]),$$

$$\mu[B] = \frac{1}{2}(\mu[AB_2] + \mu[AB_1]),$$

cannot be estimated using the study results.

EXAMPLE 13.7 The ANOVA table for a factorial analysis of the $N = 12$ measurements of weight gain in fish in Example 9.14 is as follows:

Source	df	SS	MS	F_c
Temperature (A)	1	0.706	0.706	23.53
Movement (B)	1	0.002	0.002	< 1
Interaction (AB)	1	0.594	0.594	13.81
Error	8	0.240	0.030	
Total	11			

■

Experiments with equal subclass numbers

In Exercise 13.15, you will be asked to verify that, if $n_{ij} = n$ for all i, j, then the contrasts in Equations (13.4), (13.6), and (13.8) for estimating the three factorial effects are mutually orthogonal. Consequently, the result in Box 9.3 implies that the sum of squares for treatment (SS[T]) can be expressed in terms of the sums of squares for the factorial effects as

$$SS[T] = SS[A] + SS[B] + SS[AB]. \tag{13.11}$$

Thus, when the treatments are equally replicated, the process of performing a 2×2 factorial analysis involves subdividing the treatment sum of squares into component sums of squares corresponding to the three mutually orthogonal contrasts—$\hat{\mu}[A]$, $\hat{\mu}[B]$, and $\hat{\mu}[AB]$—between the four treatment means. However, when the replications are unequal, the three contrasts for estimating the factorial effects may not be orthogonal (see Exercise 13.15b), and the treatment sum of squares may not be equal to the sum of SS[A], SS[B] and SS[AB].

Another consequence of equal subclass numbers is that the sums of squares for testing the factorial effects can be expressed in a form similar to that of the sums of squares in a one-way ANOVA. To demonstrate this, let's adopt the following notation, which is a direct extension of that in Section 8.2:

- Y_{ijk} = the k-th response for the treatment combination $a_i b_j$;
- Y_{ij+} = the total of the n responses for the treatment combination $a_i b_j$;
- Y_{i++} = the total of the $2n$ responses observed at the i-th level of A;
- Y_{+j+} = the total of the $2n$ responses observed at the j-th level of B;
- Y_{+++} = the total of all $N = 4n$ responses.

These totals may be organized in a 2×2 array as follows:

	b_1	b_2	Total
a_1	Y_{11+}	Y_{12+}	Y_{1++}
a_2	Y_{21+}	Y_{22+}	Y_{2++}
Total	Y_{+1+}	Y_{+2+}	Y_{+++}

The sums of squares can be computed using the formulas

$$SS[T] = \frac{Y_{11+}^2}{n} + \frac{Y_{12+}^2}{n} + \frac{Y_{21+}^2}{n} + \frac{Y_{22+}^2}{n} - \frac{Y_{+++}^2}{4n},$$

$$SS[A] = \frac{Y_{1++}^2}{2n} + \frac{Y_{2++}^2}{2n} - \frac{Y_{+++}^2}{4n},$$

$$SS[B] = \frac{Y_{+1+}^2}{2n} + \frac{Y_{+2+}^2}{2n} - \frac{Y_{+++}^2}{4n}, \tag{13.12}$$

$$SS[AB] = SS[T] - SS[A] - SS[B].$$

EXAMPLE 13.8   The following are the treatment totals for the tissue water content data (with the data for 100% humidity deleted) in Example 13.3:

	$b_1 = 45°C$	$b_2 = 75°C$	
$a_1 = 20°C$	290	326	616
$a_2 = 30°C$	278	313	591
	568	639	1207

Since each treatment is replicated $n = 4$ times, there are four responses in each of the four subclasses. From Equation (13.12), the various uncorrected sums of squares are

$$\text{SS}[T] = \frac{(290)^2}{4} + \frac{(326)^2}{4} + \frac{(278)^2}{4} + \frac{(313)^2}{4} - \frac{(1207)^2}{4 \times 4} = 354.19,$$

$$\text{SS}[A] = \frac{(616)^2}{2 \times 4} + \frac{(591)^2}{2 \times 4} - \frac{(1207)^2}{4 \times 4} = 39.06,$$

$$\text{SS}[B] = \frac{(568)^2}{2 \times 4} + \frac{(639)^2}{2 \times 4} - \frac{(1207)^2}{4 \times 4} = 315.06,$$

$$\text{SS}[AB] = 354.19 - 39.06 - 315.06 = 0.07. \quad \blacksquare$$

Exercises

13.6   Show that the definition of interaction between two factors in Equation (13.1) does not depend on which factor is used to compare the simple effects; that is, verify that $\mu[AB_2] - \mu[AB_1] = \mu[A_2B] - \mu[A_1B]$.

13.7   For the data in Table 13.4, construct interaction plots with levels of B plotted on the horizontal axis. Compare your plots with those in Figure 13.1. Do the two plots lead to different conclusions? Explain.

13.8   The following are tables of hypothetical population means in four 2×2 experiments with factors A and B. For each table, check whether AB interaction is indicated. If interaction is present, identify its type (qualitative or quantitative).

a

	b_1	b_2
a_1	16	23
a_2	34	23

b

	b_1	b_2
a_1	33	63
a_2	22	52

c

	b_1	b_2
a_1	16	23
a_2	34	42

d

	b_1	b_2
a_1	33	13
a_2	22	42

13.9   The effect of a new cholesterol-reducing diet was studied in four groups of subjects: group 1, young males; group 2, young females; group 3, old males; group 4, old females. The response measured was the percentage reduction in cholesterol level after six months on the experimental diet. The results obtained are as follows:

Group	1	2	3	4
Sample size	8	10	9	10
Sample mean	26	21	12	14
Sample S.D.	3.2	3.5	2.9	2.8

a   Perform an ANOVA to see if there is evidence suggesting that the four groups differ with respect to the mean reduction in their cholesterol levels.

b   Set up a contrast of sample means that is suitable for estimating the effect of interaction between age and gender.

c   Test the significance of the contrast in (b) and write a conclusion about the effect of interaction between age and gender.

d   Construct a 95% confidence interval for the interaction effect and interpret it.

13.10   In a greenhouse study to determine the effects of two types of environmental conditions—amount of light (10 hr and 12 hr) and daily temperature (70°F and 80°F)—on the growth of a certain species of plant, three plants were grown under each of the four combinations of light and temperature conditions. The measured dry weights of the plant material after ten weeks of growth are as follows:

Treatment	Environmental condition	Dry weight (g)
1	10 hr, 70°F	10.6 9.8 10.1 10.7
2	10 hr, 80°F	18.3 20.1 19.6 20.2
3	12 hr, 70°F	11.3 10.9 10.5 10.2
4	12 hr, 80°F	34.2 33.8 34.6 33.5

a   Perform an analysis of variance to test the null hypothesis that there is no difference between the effects of the four environmental conditions.

b   Estimate the effect of interaction between the amount of light and daily temperature.

c   Test the significance of the contrast you estimated in (b) and write your conclusions.

13.11   Refer to Exercise 13.9.

a   Set up contrasts of sample means to estimate the main effects of age and gender.

b   Test the significance of the contrasts in (a) and interpret the results. Remember to pay attention to the significance or otherwise of the interaction.

c   Construct a set of 95% simultaneous confidence intervals for the interaction and the main effects of age and gender. Interpret the intervals.

13.12   Refer to Exercise 13.10.

    a Estimate the main effects of the amount of light and the daily temperature.

    b Test the significance of the main effects of light and temperature and interpret the results.

13.13 Refer to Example 13.6

    a Test for the significance of the interaction between temperature and humidity.

    b What would be your conclusion about the interaction and the main effects of temperature and humidity if you wanted to make sure that the probability of at least one Type I error is less than 1%. [Hint: use the Bonferroni method to simultaneously test the three null hypotheses.]

13.14 Explain why it is not meaningful to test the significance of the main effects in Example 13.5.

13.15 **a** Verify that, in a 2 × 2 factorial with equal subclass numbers, the contrasts that estimate the three factorial effects are mutually orthogonal.

    b Consider a 2 × 2 factorial with $n_{11} = 2$, $n_{12} = 3$, $n_{21} = 3$, $n_{22} = 3$. Show that the three contrasts that measure the three factorial effects are not mutually orthogonal.

13.16 In Example 13.6, suppose that there was some problem in the temperature settings under which the two measurements of 86% water content were made, so that the validity of these two measurements is questionable. Delete these two measurements and answer the following questions for the remaining 14 measurements.

    a Construct the two-way ANOVA table and show that the sum of SS[A], SS[B], and SS[AB] does not equal the treatment sum of squares.

    b Test the significance of AB interaction and write your conclusions.

    c Construct appropriate simultaneous confidence intervals for contrasts that measure the effects suggested by your test in (b).

13.17 Refer to Exercise 13.9.

    a Use Equation (13.12) to calculate the sums of squares for interaction and main effects in the cholesterol reduction data. Compare your results to the corresponding sums of squares calculated in Exercises 13.9c and 13.11b. Explain why these two sets of values are (or are not) different.

    b Set up the ANOVA table for a 2 × 2 analysis of the cholesterol reduction data.

    c Write a summary of the conclusions that can be drawn from the ANOVA table in (b).

13.18 Refer to Exercise 13.10.

    a Use Equation (13.12) to calculate the sums of squares for the interaction and the main effects in the dry weight data. Compare your results to the corresponding sums of squares calculated in Exercises 13.10c and 13.12b. Explain why these two sets of values are (or are not) different.

    b Set up the ANOVA table for a 2 × 2 analysis of the dry weight data.

    c Write a summary of conclusions that can be drawn from the ANOVA table in (b).

    d Construct appropriate plots of treatment means to illustrate the presence or absence of simple effects and interaction.

13.4
Analyzing $a \times b$ factorial experiments

The method of analyzing a 2×2 experiment can be extended to an $a \times b$ experiment in a straightforward manner. As in the 2×2 case, the definitions of the factorial effects can be given in terms of the expected responses (population means). Let's look at the basic ideas in the context of 3×3 experiments.

Simple effects

Let μ_{ij} denote the population mean response for the treatment combination $a_i b_j$. As in the case of a 2×2 experiment, a simple effect of a given factor—say, factor A—is defined as the differential expected response resulting from changing the level of A while holding the level of B fixed. There are six simple effects corresponding to the six ways in which the level of A can be changed from one value to another (from 1 to 2, 2 to 1, 1 to 3, 3 to 1, 2 to 3 and 3 to 2). However, the six simple effects resulting from these six ways of changing levels can be expressed in terms of two appropriately chosen simple effects. So as to be specific, let's consider the two simple effects resulting from changing level 3 of A to the other two levels:

1 $\mu_1[AB_j] = \mu_{3j} - \mu_{1j}$, which measures the simple effect of changing the level of A from a_3 to a_1, with the level of B at b_j.

2 $\mu_2[AB_j] = \mu_{3j} - \mu_{2j}$, which measures the simple effect of changing the level of A from a_3 to a_2, with the level of B at b_j.

We refer to $\mu_1[AB_j]$ and $\mu_2[AB_j]$ as the *components of the simple effect* of A at the j-th level of B.

Any simple effect of A at a given level of B can be expressed in terms of its corresponding two components. For instance, at level b_2 of B, the simple effect of A when a_2 changes to a_1 is $\mu_{22} - \mu_{12}$, which can be expressed as the difference $(\mu_{32} - \mu_{12}) - (\mu_{32} - \mu_{22})$. Thus, at the second level of B, the simple effect of changing the level of A from a_2 to a_1 is $\mu_1[AB_2] - \mu_2[AB_2]$. In Table 13.6, the components of the simple effects of A are expressed in terms of population means.

Interaction effects

As in the case of a 2×2 factorial, we say that there is no interaction between A and B if the differential expected response resulting from changing the level of A does not depend on the level of B. In other words, if there is no interaction between A and B, then the simple effects in any row of Table 13.6 should be equal. The presence of interaction in a 3×3 experiment can be tested by testing the null hypothesis

$$H_{0AB}: \mu_1[AB_1] = \mu_1[AB_2] = \mu_1[AB_3] \text{ and } \mu_2[AB_1] = \mu_2[AB_2] = \mu_2[AB_3].$$

TABLE   13.6

Components of simple effects of A

Component	$j = 1$	$j = 2$	$j = 3$
$\mu_1[AB_j]$	$\mu_{31} - \mu_{11}$	$\mu_{32} - \mu_{12}$	$\mu_{33} - \mu_{13}$
$\mu_2[AB_j]$	$\mu_{31} - \mu_{21}$	$\mu_{32} - \mu_{22}$	$\mu_{33} - \mu_{23}$

Let's compare H_{0AB} with the null hypothesis for testing the interaction in a 2×2 experiment. The test of interaction in a 2×2 experiment is based on one contrast among the means: $\hat{\mu}[AB] = \frac{1}{2}\overline{Y}_{22+} - \frac{1}{2}\overline{Y}_{21+} - \frac{1}{2}\overline{Y}_{12+} + \frac{1}{2}\overline{Y}_{11+}$. The corresponding null hypothesis in a 3×3 experiment involves testing four contrasts:

1 $\overline{Y}_{33+} - \overline{Y}_{13+} - \overline{Y}_{31+} + \overline{Y}_{11+}$ for testing the null hypothesis $\mu_1[AB_3] = \mu_1[AB_1]$.

2 $\overline{Y}_{33+} - \overline{Y}_{13+} - \overline{Y}_{32+} + \overline{Y}_{12+}$ for testing the null hypothesis $\mu_1[AB_3] = \mu_1[AB_2]$.

3 $\overline{Y}_{33+} - \overline{Y}_{23+} - \overline{Y}_{31+} + \overline{Y}_{21+}$ for testing the null hypothesis $\mu_2[AB_3] = \mu_2[AB_1]$.

4 $\overline{Y}_{33+} - \overline{Y}_{23+} - \overline{Y}_{32+} + \overline{Y}_{22+}$ for testing the null hypothesis $\mu_2[AB_3] = \mu_1[AB_1]$.

The concept of interaction in a 3×3 experiment is best described with some hypothetical examples. Table 13.7 shows three sets of hypothetical population means

TABLE   13.7

Hypothetical population means and components of the simple effects of A

(a) No interaction	$j = 1$	$j = 2$	$j = 3$
μ_{1j}	30	40	20
μ_{2j}	20	30	10
μ_{3j}	60	70	50
$\mu_1[AB_j]$	30	30	30
$\mu_2[AB_j]$	40	40	40

(b) Quantitative interaction	$j = 1$	$j = 2$	$j = 3$
μ_{1j}	30	40	20
μ_{2j}	20	40	10
μ_{3j}	60	90	50
$\mu_1[AB_j]$	30	50	30
$\mu_2[AB_j]$	40	50	40

(c) Qualitative interaction	$j = 1$	$j = 2$	$j = 3$
μ_{1j}	30	40	20
μ_{2j}	20	90	10
μ_{3j}	60	20	50
$\mu_1[AB_j]$	30	-20	30
$\mu_2[AB_j]$	40	-70	40

and the corresponding components of the simple effect of *A*. In Table 13.7a, the components of simple effects are the same within rows. Hence, the effect of changing the level of *A* does not depend on the level of *B*. In Table 13.7b, the signs of the components of the simple effect are the same within rows, but their magnitudes differ. This is a quantitative interaction because, even though the differential expected responses change with the changing level of *B*, the direction of change stays the same within the levels of *B*. Finally, Table 13.7c illustrates a qualitative interaction between *A* and *B*, because the simple effects within rows are different and have different signs.

Interaction plots for the three hypothetical sets of means in Table 13.7 are shown in Figure 13.5. In each case, the population mean responses are plotted against the levels of *A*, and the points corresponding to the same level of *B* are joined by straight lines. As in the 2 × 2 case, parallel lines indicate no interaction. If the lines are not parallel, but the direction of the slopes of the corresponding segments (the lines joining two adjacent points) is the same for all levels of *B*, then a quantitative interaction is indicated. If the lines are not parallel and the direction of the slopes of the corresponding segments joining two adjacent points is not the same for all levels of *B*, then the graph indicates a qualitative interaction.

FIGURE 13.5

Interaction plots of the population means in Table 13.7

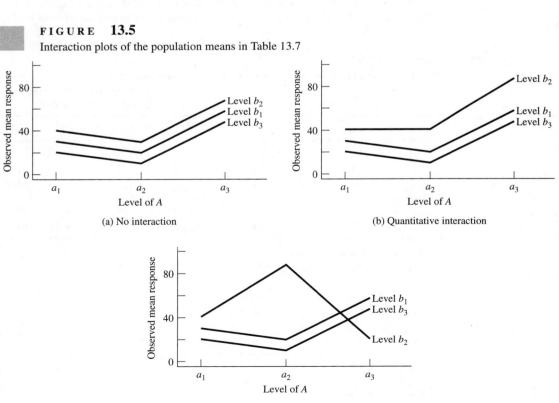

(a) No interaction

(b) Quantitative interaction

(c) Qualitative interaction

Main effects

The definition of the main effects in a 2×2 factorial can be extended to cover a 3×3 factorial in a straightforward manner. For example, the main effect of A is defined in terms of two components that are the averages (over the levels of B) of the corresponding components of the simple effects of A. In the notation of Table 13.6, the two components of the main effect of A are

$$\mu_1[A] = \frac{1}{3}(\mu_1[AB_1] + \mu_1[AB_2] + \mu_1[AB_3])$$

$$= \frac{1}{3}\left[(\mu_{31} - \mu_{11}) + (\mu_{32} - \mu_{12}) + (\mu_{33} - \mu_{13})\right]$$

$$= \frac{1}{3}(\mu_{31} + \mu_{32} + \mu_{33}) - \frac{1}{3}(\mu_{11} + \mu_{12} + \mu_{13}),$$

$$\mu_2[A] = \frac{1}{3}(\mu_2[AB_1] + \mu_2[AB_2] + \mu_2[AB_3])$$

$$= \frac{1}{3}\left[(\mu_{31} - \mu_{21}) + (\mu_{32} - \mu_{22}) + (\mu_{33} - \mu_{23})\right]$$

$$= \frac{1}{3}(\mu_{31} + \mu_{32} + \mu_{33}) - \frac{1}{3}(\mu_{21} + \mu_{22} + \mu_{23}).$$

We see that $\mu_1[A]$ is the difference between the average expected responses at the third and first levels of A, while $\mu_2[A]$ is the difference between the average expected responses at the third and second levels of A. The components of the main effect of B are defined in a similar manner.

If there is no AB interaction, the components of the main effect of A will be equal to the common value (across levels of B) of the corresponding components of the simple effects of A

$$\mu_1[A] = \mu_1[AB_1] = \mu_1[AB_2] = \mu_1[AB_3],$$

$$\mu_2[A] = \mu_2[AB_1] = \mu_2[AB_2] = \mu_2[AB_3].$$

Then $\mu_1[A]$ and $\mu_2[A]$ are equal to the expected differential mean responses resulting from changing the level of A from a_3 to a_1 and from a_3 to a_2, respectively. Testing for the presence of the main effect of A is equivalent to testing the null hypothesis that the two components $\mu_1[A]$ and $\mu_2[A]$ are both zero. Therefore, instead of testing a single contrast, as in a 2×2 factorial experiment, the test for a main effect in a 3×3 factorial involves testing two contrasts.

Even though the definition of the factorial effects in a 2×2 experiment is easily extended to an $a \times b$ experiment, the computing formulas similar to Equation (13.9) do not exist in the case of a general $a \times b$ factorial. This is because, unlike in a 2×2 experiment, testing factorial effects in $a \times b$ experiments involves more than one contrast of treatment means. It is best to use statistical computing software such as SAS for this purpose. For details of how to analyze data from factorial experiments using SAS, see Section 13.3 in the companion text by Younger (1997). A partial SAS output for factorial analysis of the snail tissue content data is shown following Example 13.9.

If the subclass numbers are equal—that is, if the treatments have equal replications—simple formulas similar to Equation (13.12) can be used to calculate the sums of squares needed for testing the factorial effects. Suppose we have n replications per treatment, and let Y_{ijk} denote the k-th response for the treatment combination $a_i b_j$. Then, using a notation similar to that in Equation (13.12), the sums of squares in an $a \times b$ factorial experiment can be expressed as follows

$$SS[T] = \frac{Y_{11+}^2}{n} + \frac{Y_{12+}^2}{n} + \cdots + \frac{Y_{ab+}^2}{n} - \frac{Y_{+++}^2}{abn},$$

$$SS[E] = SS[TOT] - SS[T],$$

$$SS[A] = \frac{Y_{1++}^2}{bn} + \frac{Y_{2++}^2}{bn} + \cdots + \frac{Y_{a++}^2}{bn} - \frac{Y_{+++}^2}{abn}, \qquad \textbf{(13.13)}$$

$$SS[B] = \frac{Y_{+1+}^2}{an} + \frac{Y_{+2+}^2}{an} + \cdots + \frac{Y_{+b+}^2}{an} - \frac{Y_{+++}^2}{abn},$$

$$SS[AB] = SS[T] - SS[A] - SS[B].$$

Two features of Equation (13.13) are worth noting. First, if we set $a = 2$ and $b = 2$, the formulas will match the formulas for a 2×2 factorial in Equation (13.12). Second, in each of these formulas, every square of a total is divided by the number of measurements that are summed to get that total. For instance, the divisor for Y_{11+}^2 is n, because this quantity is the square of a total of n responses. Similarly, Y_{+1+}^2 has the divisor an since this quantity is the square of a total of an responses (n responses at each of the a levels of A).

In a 2×2 experiment, the sum of squares for testing each factorial effect had one degree of freedom because each sum of squares is used to test the significance of one contrast among the sample means. The corresponding degrees of freedom in an $a \times b$ experiment are $a - 1$ for SS[A], $b - 1$ for SS[B], and $(a - 1)(b - 1)$ for SS[AB]. An informal justification for these degrees of freedom can be given as follows.

Degrees of freedom for factorial sums of squares

Consider the main effect of A. Recall that, if A has three levels ($a = 3$), the test of the main effect of A is a test of the null hypothesis that the two components $\mu_1[A]$ and $\mu_2[A]$ are both zero. In general, the test of the main effect of A is a test of the null hypothesis that each of the $a - 1$ components of the main effect of A is zero. Therefore, of the $ab - 1$ degrees of freedom for the treatment sum of squares, $a - 1$ degrees of freedom are associated with the sums of squares for testing the main effect of A. Similarly, the degrees of freedom for the sum of squares for testing the main effect of B is $b - 1$. The remaining $(ab - 1) - (a - 1) - (b - 1) = (a - 1)(b - 1)$ degrees of freedom are associated with the sum of squares for testing the interaction effect. Note that, when $a = b = 2$, these formulas yield one degree of freedom for testing each of the three factorial effects.

The ANOVA table for analyzing an $a \times b$ factorial experiment is shown in Table 13.8.

TABLE 13.8

ANOVA table for an $a \times b$ factorial experiment

Source	df	SS	MS	F_c
A	$a - 1$	SS[A]	MS[A]	F_A
B	$b - 1$	SS[B]	MS[B]	F_B
AB	$(a - 1)(b - 1)$	SS[AB]	MS[AB]	F_{AB}
Error	$ab(n - 1)$	SS[E]	MS[E]	
Total	$abn - 1$	SS[TOT]		

Strategy for $a \times b$ factorial analysis

The strategy for analyzing an $a \times b$ experiment is exactly the same as the strategy for 2×2 experiments. The first step is to test for interaction between the two factors. If interaction is significant, then we proceed to analyze the simple effects of each of the two factors. If interaction is not significant, then an analysis of the main effects of the individual factors would be appropriate. The next example illustrates the factorial analysis of a 2×3 experiment.

EXAMPLE 13.9 In Example 13.6, we described a factorial experiment with two factors: A, the temperature, at $a = 2$ levels ($20°C$ and $30°C$); and B, the humidity, at $b = 3$ levels (45%, 75%, and 100%). There were $n = 4$ replications per treatment, resulting in $N = abn = 24$ responses. The objective of the experiment was to study the effects of temperature and humidity on the tissue water content of snails. In Exercise 13.20, you will be asked to verify that the total sum of squares for these data is SS[TOT] = 3386.5. The totals of the $n = 4$ responses for the $(a)(b) = (2)(3) = 6$ treatments are as follows:

Level of A	Level of B			Total
	$b_1 = 45\%$	$b_2 = 75\%$	$b_3 = 100\%$	
$a_1 = 20°C$	290	326	388	1004
$a_2 = 30°C$	278	313	391	982
Total	568	639	779	1986

The corresponding sums of squares are

$$\text{SS[T]} = \frac{(290)^2}{4} + \cdots + \frac{(391)^2}{4} - \frac{(1986)^2}{2 \times 3 \times 4} = 2922.0,$$

$$df_T = ab - 1 = (2)(3) - 1 = 6 - 1 = 5,$$

$$\text{SS[E]} = \text{SS[TOT]} - \text{SS[T]} = 3386.5 - 2922.0 = 464.5,$$

$$df_E = ab(n-1) = (4-1)(2)(3) = 18,$$

$$\text{SS}[A] = \frac{(1004)^2}{3 \times 4} + \frac{(982)^2}{3 \times 4} - \frac{(1986)^2}{2 \times 3 \times 4} = 20.17,$$

$$df_A = a - 1 = 2 - 1 = 1,$$

$$\text{SS}[B] = \frac{(568)^2}{2 \times 4} + \frac{(639)^2}{2 \times 4} + \frac{(779)^2}{2 \times 4} - \frac{(1986)^2}{2 \times 3 \times 4} = 2881.75,$$

$$df_B = b - 1 = 3 - 1 = 2,$$

$$\text{SS}[AB] = 2922.0 - 20.17 - 2881.75 = 20.08,$$

$$df_{AB} = (a-1)(b-1) = (2-1)(3-1) = 2.$$

The resulting ANOVA table for the snail tissue water content data is as follows:

Source	df	SS	MS	F_c
Temperature (A)	1	20.17	20.17	0.78
Humidity (B)	2	2881.75	1440.88	55.85
Interaction (AB)	2	20.08	10.04	0.39
Error	18	464.5	25.81	
Total	23	3386.5		

From the ANOVA table, we see that the interaction between temperature and humidity is not significant, because the calculated value of the F-statistic is 0.39, which is less than 1. We conclude that the data do not support the hypothesis of an interaction between temperature and humidity.

Since AB interaction is not significant, we proceed to analyze the main effects of temperature and humidity. This can be done by analyzing two types of differences:

a   the differences between the expected tissue water content at the two levels of temperature;

b   the differences between the expected tissue water content at the three levels of humidity.

These analyses will involve tests of main effects, the determination of confidence intervals for selected differences between population means, and multiple pairwise comparisons of selected sample means.

Let's first consider testing the significance of the main effects. The F-test for the main effect of A is not significant ($F_c < 1$), and so there is insufficient evidence to conclude that the expected tissue water contents are different at the two temperatures. However, the F-test for the main effect of humidity is highly significant ($p < 0.0001$), which implies that the data support the conclusion that the expected tissue water contents are not the same at the three humidity levels.

A plot of the observed treatment means (an interaction plot) provides a convenient method of graphically displaying the results of a factorial experiment. The observed treatment mean tissue water content are shown in the following table. In Figure 13.6, these means are plotted against the levels of humidity.

		Level of B	
Level of A	$b_1 = 45\%$	$b_2 = 75\%$	$b_3 = 100\%$
$a_1 = 20°C$	72.5	81.5	97.0
$a_2 = 30°C$	69.5	78.25	97.75

The means corresponding to the same level of temperature are joined by solid lines; the dotted line joins the average observed responses at the three levels of humidity.

F I G U R E   13.6

Plot of the means based on snail tissue water content data

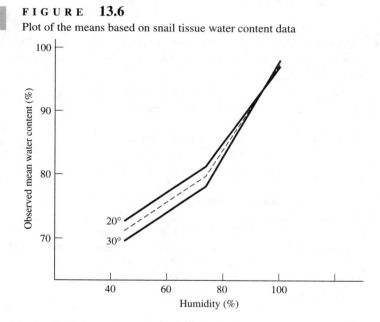

The solid lines in the interaction plot are practically parallel; this supports the result of the F-test for interaction between temperature and humidity. Furthermore, the proximity of the two solid lines indicates that the difference between the observed mean responses at the two temperatures is likely to be nonsignificant; that conclusion is supported by the F-test for the main effect of temperature. A graphical check for the presence of the main effect of humidity is provided by the orientation of the dotted line. If such a main effect is not present, then the dotted line should be nearly parallel to the horizontal axis. Figure 13.6 clearly shows that this is not the case. The F-test for the main effect of B supports this conclusion.

Since the expected responses at the three humidity levels are unequal, we might want to examine in detail the differences between the observed mean responses at the three humidity levels. The methods described in Chapter 9 can be used for this purpose. For instance, if the subclass numbers are equal, a multiple pairwise comparison of the three means can be performed using any of the methods described in Section 9.5. Note that each humidity level is replicated $n = 8$ times and MS[E] is equal to 25.81, with 18 degrees of freedom. In Exercise 13.20, you will be asked to perform further analyses of the humidity means.   ■

As already noted, the calculations needed for factorial analysis can be performed using any of the existing statistical software packages. The SAS GLM procedure can be used to perform a wide variety of analyses, including factorial analyses. When the subclass numbers are equal, the SAS ANOVA procedure may be preferable, because of its increased efficiency.

EXAMPLE **13.10**   When the ANOVA procedure is applied to the snail tissue water content data, the SAS output obtained include the following ANOVA information.

```
-------------------------------------------------------------------
                    Analysis of Variance Procedure
                       Class Level Information


                   Class    Levels   Values
                   TEMP         2     20 30
                   HUMI         3     45 75 100


                 Number of observations in data set = 24
Dependent Variable: WATER
                              Sum of         Mean  F Value   Pr > F
       Source         DF     Squares       Square    22.65   0.0001
       Model           5    2922.0000     584.4000
       Error          18     464.5000      25.8056
       Corrected Total 23    3386.5000


       Source         DF    ANOVA SS   Mean Square  F Value   Pr > F
       TEMP            1      20.1667      20.1667     0.78    0.3883
       HUMI            2    2881.7500    1440.8750    55.87    0.0001
       TEMP*HUMI       2      20.0833      10.0417     0.39    0.6832
-------------------------------------------------------------------
```
■

Example 13.9 illustrated the essentials of an $a \times b$ factorial analysis when the interaction effect is not significant. In the next example, we examine an analysis in which the *AB* interaction is significant.

EXAMPLE **13.11**   In Example 9.14, we described a 2×2 factorial experiment to study the effects of water temperature and water movement on the weight gain of fish. The treatments were combinations of two levels of water temperature (cold and warm) and two levels of water movement (still and flowing). The data for these four treatments were analyzed for factorial effects in Example 13.4 and 13.5. In Example 13.7, we looked at the ANOVA table for these data.

The weight gain data in Example 9.14 did not include data for a third level of temperature investigated in the original study. The actual study was a 3×2 factorial, in which there were three levels of water temperature (cold, lukewarm, warm) and two levels of water movement (still, flowing). The treatment totals for the 3×2 experiment are as follows:

Level of temperature i	Level of movement j 1 (still)	2 (flowing)	Total
1 (cold)	4.65	3.24	7.89
2 (lukewarm)	5.61	5.28	10.89
3 (warm)	4.77	6.03	10.80
Total	15.03	14.55	29.58

The following is the corresponding ANOVA table, all entries of which, except for SS[TOT], can be calculated from the treatment totals:

Source	df	SS	MS	F_c
Temperature (A)	2	0.971	0.486	19.44
Movement (B)	1	0.013	0.013	< 1
Interaction (AB)	2	0.601	0.301	12.04
Error	12	0.301	0.025	
Total	17	1.886		

As in Examples 13.4 and 13.5, we begin with a test of the interaction between temperature and movement. Since the calculated value $F_c = 12.04$ is highly significant ($p = 0.0014$), we conclude that there is interaction between water temperature and water movement; in other words, the effect of changing temperature depends on the level of water movement, and vice versa. Therefore, instead of testing the main effects, we proceed to examine simple effects of the two factors.

The interaction between temperature and movement implies that the differences between the expected responses for still and flowing water (levels of B) are not the same at the three temperatures (levels of A). Thus, we might be interested in examining the difference between the expected responses at each level of water temperature. At each level of A, we have $2n$ measurements consisting of $n = 3$ replications for each of the two treatments. These data can be used to construct confidence intervals (see Exercise 13.21) or perform significance tests.

The sum of squares for testing the significance of the observed mean responses is computed using Equation (13.13). For instance, for cold water ($i = 1$), the sum of squares for testing the difference between still ($j = 1$) and flowing ($j = 2$) water can

be calculated as

$$SS[A_1B] = \frac{Y_{11+}^2}{n} + \frac{Y_{12+}^2}{n} - \frac{(Y_{11+}^2 + Y_{12+}^2)}{2n}$$

$$= \frac{(4.65)^2}{3} + \frac{(3.24)^2}{3} - \frac{(7.89)^2}{6}$$

$$= 0.331.$$

Two mean responses are compared at each level of A, and so the degrees of freedom for $SS[A_1B]$ is $df_{A_1B} = 2 - 1 = 1$. The corresponding test statistic is

$$F_c(A_1B) = \frac{MS[A_1B]}{MS[E]},$$

where $MS[E]$ is the mean square for error in the ANOVA table for all data. Therefore

$$F_c(A_1B) = \frac{0.331}{0.025} = 13.24$$

with $df = 1, 12$. This is significant at $\alpha = 0.01$ ($p = 0.0034$), as can be seen by comparing 13.24 to the critical value of F with 1 and 12 degrees of freedom.

Similar calculations using data for the lukewarm ($i = 2$) and warm ($i = 3$) temperatures give

$$SS[A_2B] = 0.018; \quad F_c(AB_1) = 0.72(<1); \quad p = 0.4127;$$

$$SS[A_3B] = 0.265; \quad F_c(AB_2) = 10.60; \quad p = 0.0069.$$

On the basis of these calculations, we can conclude that the weight gains for still and flowing water are significantly different if cold or warm water is used but are not significantly different if lukewarm water is used. The variation in observed mean weight gain within temperature levels is shown in Figure 13.7.

F I G U R E   13.7

Observed mean weight gain of fish plotted against the levels of water movement

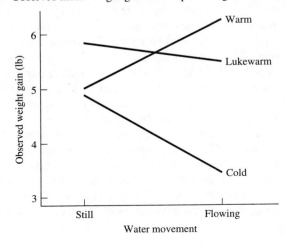

When the level of water movement is changed from still to moving, the weight gain decreases in cold water, shows a nonsignificant change in lukewarm water, and increases in warm water. Thus, the pattern reverses as the water temperature increases from cold to warm: Water movement reduces the weight gain in cold water but increases it in warm water. Further analysis of the differential effects of water movement at each temperature might include the construction of confidence intervals for the differential mean weight gains at the three levels of water temperature (see Exercise 13.21).

Significant interaction between temperature and movement also implies that the pattern of differences between the expected responses at the three temperatures (levels of A) is not the same in still and flowing water (at thelevels of B). Thus, we might wish to compare the expected responses at each level of water movement. Such an analysis can be performed by noting that, at each level of B, we have nine measurements, consisting of data for $n = 3$ replications of each of the three treatments. Suppose we want to test the null hypothesis that, for still water ($j = 1$), there is no difference between the expected weight gains at the three temperatures. The required sum of squares, the associated degrees of freedom, and the mean squares are

$$SS[AB_1] = \frac{Y_{11+}^2}{3} + \frac{Y_{21+}^2}{3} + \frac{Y_{31+}^2}{3} - \frac{(Y_{11+} + Y_{21+} + Y_{31+})^2}{9}$$

$$= \frac{(4.65)^2}{3} + \frac{(5.61)^2}{3} + \frac{(4.77)^2}{3} - \frac{(15.03)^2}{9} = 0.183,$$

$$df_{AB_1} = 3 - 1 = 2,$$

$$MS[AB_1] = \frac{SS[AB_1]}{df_{AB_1}} = \frac{0.183}{2} = 0.092.$$

The calculated test statistic is

$$F_c(AB_1) = \frac{MS[AB_1]}{MS[E]} = \frac{0.092}{0.025} = 3.68$$

with $df = 2, 12$. This is not significant ($p = 0.0567$) at $\alpha = 0.05$. Thus, the evidence is not sufficient to suggest that, for still water, the expected responses are different at the three temperatures. Similar calculations for flowing water give

$$F_c = 27.797, \qquad p < 0.0001,$$

and so there is reason to believe that, for flowing water, the expected responses are different at the three temperatures. The variation in the observed mean weight gain within the levels of water movement is plotted in Figure 13.8.

In Exercise 13.21c, you will be asked to perform multiple comparisons of the means at each level of water movement. ∎

Exercises

13.19 Refer to Table 13.7.

a Construct tables showing the components of the simple effect of B in each of the three hypothetical cases.

FIGURE **13.8**

Observed mean weight gain of fish plotted against the levels of water temperature

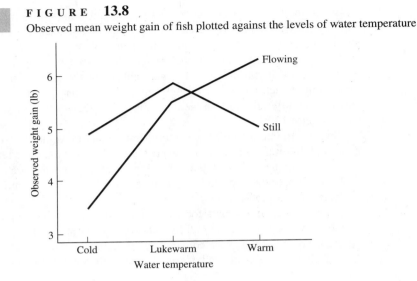

b   Recall that, in a 2 × 2 experiment, the interaction effect can be defined in terms of the simple effects of *A* or the simple effects of *B*; the meaning of interaction does not depend on which simple effects are used in its definition. Using the simple effects of *A* in Table 13.7 and those in the tables constructed in (a), argue that the meaning of interaction in a 3 × 3 experiment does not depend on which set of simple effects is used for its definition.

13.20   Refer to Example 13.9.

a   Verify that the total sum of squares equals 3386.5.

b   Using the 0.05-level Bonferroni method, perform a multiple pairwise comparison of the observed mean tissue water contents of snails exposed to the three humidity levels. Interpret the results.

c   Construct a set of 95% simultaneous confidence intervals for $\mu_1[B]$ and $\mu_2[B]$, the two components of the main effect of humidity. Interpret these intervals.

13.21   Refer to Example 13.11.

a   Construct a set of 90% simultaneous confidence intervals for the three simple effects of water movement (*B*).

b   Compare the information about the significance of the three simple effects that can be derived from the intervals in (a) to the conclusions derived from the corresponding significance tests in Example 13.11.

c   Use the Tukey multiple comparison method to compare the three mean weight gains at each water movement level.

13.22   Refer to the survival data for *Drosophila melanogaster* in Exercise 13.4.

a   Perform a one-way analysis of the data to see if there is a difference between the true mean survival times of the six groups of flies resulting from combining three cage types (levels of *A*) with two genders (levels of *B*).

b Construct an appropriate plot of the treatment means. On the basis of your plot, comment on the possible existence of various types of factorial effects (such as the interaction effect and the main effect).

c Construct the ANOVA table appropriate for testing the significance of the factorial effects.

d Perform all appropriate hypothesis tests using the ANOVA table constructed in (c). Write your conclusions.

e Construct a 95% confidence interval (or a set of simultaneous intervals) for the difference between survival rates (the ratio of the number surviving to the number exposed) for males and females. Note that the type of confidence interval (single or simultaneous set) that is appropriate will depend on whether there is interaction between gender and type of cage.

f Use 95% simultaneous confidence intervals to compare the survival rate in environmental conditions with the survival rates at 24°C and 17°C. (Be sure to take into consideration the presence or absence of interaction between gender and type of cage.)

13.23 Refer to the root length data in Exercise 13.5.

a Perform a one-way analysis of the data to see if there is a difference between the expected root lengths for the 12 treatment combinations resulting from combining three types of pretreatment (levels of A) with four varieties (levels of B).

b Construct an appropriate plot of the treatment means. On the basis of your plot, comment on the possible existence of various types of factorial effects (such as the interaction effect and the main effect).

c Construct the ANOVA table needed to test the significance of the factorial effects.

d Perform an appropriate factorial analysis of the data and write your conclusions.

13.5
The two-way ANOVA model

In an $a \times b$ factorial experiment, the ANOVA F-test for treatment effects is based on the assumption that the responses satisfy the one-way ANOVA model in Box 8.2. Since an $a \times b$ factorial experiment has $t = ab$ treatments, the model can be expressed as

$$Y_{ij} = \mu + \tau_i + E_{ij}, \qquad i = 1, 2, \ldots, ab; \quad j = 1, 2, \ldots, n_i. \qquad \textbf{(13.14)}$$

Here Y_{ij} is the j-th response for the i-th treatment; n_i is the number of replications of the i-th treatment; and μ, τ_i, and E_{ij} are the overall mean, the effect of the i-th treatment, and the random error, respectively. The E_{ij} are assumed to be independent $N(0, \sigma^2)$. As noted in Chapter 8, the model is called a one-way classification (or a one-way ANOVA) model because it assigns the responses to the ab treatment groups on the basis of one classification criterion: the treatment that generated the response.

Factorial analysis of an $a \times b$ experiment with a completely randomized design is based on a two-way classification model. In such a model, the responses are assigned to treatment groups on the basis of two classification criteria: the levels of the two factors that are used in forming the treatment combinations. A two-way classification model that is often used to analyze $a \times b$ completely randomized factorial experiments is given in Box 13.1.

BOX **13.1**

The two-way ANOVA model with fixed treatment effects

Let Y_{ijk} denote the k-th $(k = 1, 2, \ldots, n_{ij})$ response for the treatment combination $a_i b_j$ in an $a \times b$ completely randomized factorial experiment, with n_{ij} replications for the treatment combination $a_i b_j$ $(i = 1, \ldots, a; j = 1, \ldots, b)$. Then

$$Y_{ijk} = \mu + \alpha_i + \beta_j + (\alpha\beta)_{ij} + E_{ijk},$$

where the following notation is employed:

- μ is the overall mean expected response;
- α_i is the effect of the i-th level of A;
- β_j is the effect of the j-th level of B;
- $(\alpha\beta)_{ij}$ is the joint effect of the i-th level of A and the j-th level of B;
- E_{ijk} is the random error in Y_{ijk}.

The E_{ijk} are independent $N(0, \sigma^2)$, and the parameters $\alpha_i, \beta_j, (\alpha\beta)_{ij}$ satisfy the conditions

$$\sum_{i=1}^{a} n_{i+}\alpha_i = \sum_{j=1}^{b} n_{+j}\beta_j = \sum_{i=1}^{a} n_{ij}(\alpha\beta)_{ij} = \sum_{j=1}^{b} n_{ij}(\alpha\beta)_{ij} = 0,$$

where n_{i+}, n_{+j} denote the total number of responses at the i-th level of A and the j-th level of B, respectively.

The following comments may be made about this two-way ANOVA model.

1 A comparison of the one-way ANOVA model in Equation (13.14) with the two-way model in Box 13.1 shows that there is only one essential difference between the two models. The expected response $\mu_i = \mu + \tau_i$ in the one-way model is replaced by $\mu_{ij} = \mu + \alpha_i + \beta_j + (\alpha\beta)_{ij}$ in the two-way model. In the two-way model, the effect of a particular treatment combination is expressed as the sum of three components representing the individual and joint effects of the corresponding factor levels.

2 The mathematical constraints

$$\sum_{i=1}^{a} n_{i+}\alpha_i = \sum_{j=1}^{b} n_{+j}\beta_j = \sum_{i=1}^{a} n_{ij}(\alpha\beta)_{ij} = \sum_{j=1}^{b} n_{ij}(\alpha\beta)_{ij} = 0 \tag{13.15}$$

are the analogs of the condition $\sum_{i=1}^{t} n_i \tau_i = 0$ in the one-way ANOVA model. If the treatments are equally replicated—that is, if $n_{ij} = n$—these constraints can be written in the form

$$\sum_{i=1}^{a} \alpha_i = 0; \quad \sum_{j=1}^{b} \beta_j = 0; \quad \sum_{i=1}^{a} (\alpha\beta)_{ij} = \sum_{j=1}^{b} (\alpha\beta)_{ij} = 0. \quad \text{(13.16)}$$

3 It can be shown that the null hypotheses for testing the factorial effects in an $a \times b$ experiment can be stated in terms of the parameters in the two-way model. For example, the null hypothesis that there is no AB interaction is equivalent to the null hypothesis

$$H_{0AB}: (\alpha\beta)_{ij} = 0, \quad i = 1, \ldots, a; \quad j = 1, \ldots, b.$$

Similarly, the null hypothesis that there is no main effect of A is the same as the null hypothesis

$$H_{0A}: \alpha_i = 0, i = 1, \ldots, a.$$

EXAMPLE **13.12** Let's construct a two-way ANOVA model for the weight gain data in Example 9.14. If Y_{ijk} denotes the weight gain of the k-th ($k = 1, 2, 3$) fish grown in an aquarium in which the temperature (A) and water movement (B) conditions are at the i-th ($i = 1$ for cold and 2 for warm) and j-th ($j = 1$ for still and $j = 2$ for flowing) levels, respectively, then

$$Y_{ijk} = \mu + \alpha_i + \beta_j + (\alpha\beta)_{ij} + E_{ijk}, \quad i = 1, 2; j = 1, 2; k = 1, 2, 3.$$

The notation employed here is as follows:

- μ is overall mean weight gain;
- α_i is the effect of the i-th level of water temperature;
- β_j is the effect of the j-th level of water movement;
- $(\alpha\beta)_{ij}$ is the joint effect of the i-th level of water temperature and the j-th level of water movement;
- E_{ijk} is the random error.

The E_{ijk} are independent $N(0, \sigma^2)$. The subclass numbers are equal, and so the parameters in the model should satisfy the constraints in Equation (13.16):

$$\alpha_1 + \alpha_2 = 0;$$
$$\beta_1 + \beta_2 = 0;$$
$$(\alpha\beta)_{11} + (\alpha\beta)_{12} = 0;$$
$$(\alpha\beta)_{21} + (\alpha\beta)_{22} = 0;$$
$$(\alpha\beta)_{11} + (\alpha\beta)_{21} = 0;$$
$$(\alpha\beta)_{12} + (\alpha\beta)_{22} = 0 \quad \blacksquare$$

Exercises

13.24   Show that, if each treatment combination is replicated n times, then the constraints in Equation (13.15) are equivalent to those in Equation (13.16).

13.25   Write an appropriate two-way ANOVA model for the cholesterol reduction data in Exercise 13.9.

13.26   Write an appropriate two-way ANOVA model for the dry weight data in Exercise 13.10.

13.27   Write an appropriate two-way ANOVA model for the water content data in Example 13.9.

13.6
Factorial experiments with more than two factors

The principles underlying the analysis of an $a \times b$ factorial experiment can be extended directly to the analysis of more complex factorial experiments. As an example, consider a factorial experiment with three factors, A, B, and C. The factorial effects can be divided into three types:

1   *Second-order interaction effect.* This is the effect of interaction between the three factors and is denoted by ABC.

2   *First-order interaction effects.* These are the effects of interactions between two factors. There are three first-order interactions: AB, AC, and BC.

3   *Main effects.* There are three main effects: A, B, and C.

The definitions of the main effect and first-order interactions are similar to those in Section 13.4. A second-order interaction can be thought of as an interaction between a main effect and a first-order interaction. In the presence of a second-order interaction, the nature of the first-order interaction between any two factors will depend on the level of the third factor; that is, each factor has an impact on the interaction between the other two factors.

EXAMPLE **13.13**   Consider a three-factor experiment in which the factor C has two levels. Let ABC_k ($k = 1, 2$) denote the first order interaction between A and B when only treatment combinations that contain the k-th level of C are considered. For instance, in a $2 \times 2 \times 2$ experiment, ABC_1 denotes the interaction between A and B when the four treatments $a_1b_1c_1$, $a_1b_2c_1$, $a_2b_1c_1$, and $a_2b_2c_1$ are used to define interaction. If the components of ABC_k are the same for all k, we say that there is no second-order interaction between A, B, and C.

In Figure 13.9, each pair of graphs shows the plots of ABC_1 and ABC_2 for a hypothetical set of population means in $2 \times 3 \times 2$ experiments. The presence or absence of second order interaction can be graphically checked by comparing the

FIGURE 13.9

Types of second order interactions

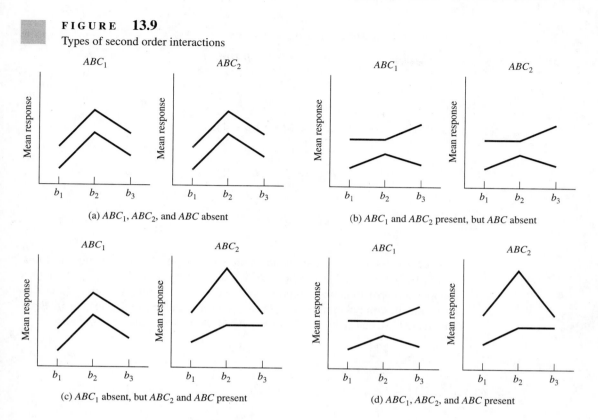

(a) ABC_1, ABC_2, and ABC absent

(b) ABC_1 and ABC_2 present, but ABC absent

(c) ABC_1 absent, but ABC_2 and ABC present

(d) ABC_1, ABC_2, and ABC present

plots of ABC_1 and ABC_2. In Figure 13.9a, AB interaction is absent at both levels of C; the lines are parallel. In Figure 13.9b, by contrast, AB interaction is present at both levels of C; the lines are not parallel. However, the pattern of interaction (as depicted by the pattern of lines) is identical. Indeed, in Figures 13.9a and 13.9b, the plots of the first-order interactions are identical at the two levels of C, and hence the way in which A and B interact does not depend on the level of C. Thus, Figures 13.9a and 13.9b depict patterns showing no second order interaction.

In Figures 13.9c and 13.9d, ABC interaction is present. In Figure 13.9c, AB interaction is absent at c_1 (the lines are parallel) but present at c_2. In Figure 13.9d, AB interaction is present at both levels of C but the way in which A and B interact changes with the level of C. ■

Factorial analysis of an experiment with k factors is based on a k-way ANOVA model; which can be written analogously to the two-way model in Box 13.1. The degrees of freedom associated with the sums of squares in a k-way ANOVA are obtained in exactly the same way as in a two-way ANOVA. For instance, in an $a \times b \times c$ experiment with n replications per treatment combination, the degrees of freedom for the factorial sums of squares are

$$df_A = a - 1,$$
$$df_B = b - 1,$$
$$df_C = c - 1,$$
$$df_{AB} = (a - 1)(b - 1),$$
$$df_{AC} = (a - 1)(c - 1),$$
$$df_{BC} = (b - 1)(c - 1),$$
$$df_{ABC} = (a - 1)(b - 1)(c - 1),$$
$$df_E = abc(n - 1),$$
$$df_{TOT} = nabc - 1.$$

(13.17)

If the subclass numbers are equal, simple computing formulas similar to Equation (13.13) are available for the ANOVA sums of squares. Suppose that we have n replications per treatment combination, and let Y_{ijkl} denote the l-th response for the treatment combination $a_i b_j c_k$. Then, using a notation similar to that in Equation (13.13), we can write the three-way ANOVA sums of squares in the form

$$CF = \frac{Y_{++++}^2}{abcn},$$

$$SS[TOT] = Y_{1111}^2 + \cdots + Y_{abcn}^2 - CF,$$

$$SS[T] = \frac{Y_{111+}^2}{n} + \frac{Y_{112+}^2}{n} + \cdots + \frac{Y_{abc+}^2}{n} - CF,$$

$$SS[A] = \frac{Y_{1+++}^2}{bcn} + \frac{Y_{2+++}^2}{bcn} + \cdots + \frac{Y_{a+++}^2}{bcn} - CF,$$

$$SS[B] = \frac{Y_{+1++}^2}{acn} + \frac{Y_{+2++}^2}{acn} + \cdots + \frac{Y_{+b+}^2}{acn} - CF$$

$$SS[C] = \frac{Y_{++1+}^2}{abn} + \frac{Y_{++2+}^2}{abn} + \cdots + \frac{Y_{++c+}^2}{abn} - CF,$$

(13.18)

$$SS[AB] = \frac{Y_{11++}^2}{cn} + \frac{Y_{12++}^2}{cn} + \cdots + \frac{Y_{ab++}^2}{cn} - CF - SS[A] - SS[B],$$

$$SS[AC] = \frac{Y_{1+1+}^2}{bn} + \frac{Y_{1+2+}^2}{bn} + \cdots + \frac{Y_{a+c+}^2}{bn} - CF - SS[A] - SS[C],$$

$$SS[BC] = \frac{Y_{+11+}^2}{an} + \frac{Y_{+12+}^2}{an} + \cdots + \frac{Y_{+bc+}^2}{bn} - CF - SS[B] - SS[C],$$

$$SS[ABC] = SS[T] - SS[A] - SS[B] - SS[C] - SS[AB] - SS[AC] - SS[BC].$$

These expressions are similar to Equation (13.13). In particular, note that, in each formula, the square of a total is divided by the number of terms added to obtain that total.

In practice, computations using these formulas tend to be tedious and error-prone, and so it is best to employ statistical computing software; for more details, see Chapter 13 in the companion text by Younger (1997).

EXAMPLE **13.14**   In Example 13.11, we presented the weight gain data for fish grown under six experimental conditions obtained by combining three levels of water temperature (cold, lukewarm, warm) and two levels of water flow (still, flowing). The data were analyzed as the results of a 3×2 factorial experiment.

For the data in Example 13.11, the aquarium light condition was maintained at a low level (16 hr/day). The following summary of treatment totals includes the data in Example 13.11 along with data for six new experimental conditions in which the aquarium light condition was maintained at a high level (24 hr/day):

Light condition (k)	Temperature (i)	Movement (j)		Total
		1 (still)	2 (flowing)	
1 (low)	1 (cold)	4.65	3.24	7.89
	2 (lukewarm)	5.61	5.28	10.89
	3 (warm)	4.77	6.03	10.80
	Total	15.03	14.55	29.58
2 (high)	1 (cold)	5.55	4.08	9.63
	2 (lukewarm)	6.09	5.01	11.10
	3 (warm)	6.42	5.55	11.97
	Total	18.06	14.64	32.70

Let's analyze these data by regarding them as the result of a $3 \times 2 \times 2$ experiment with three factors: A, the water temperature at $a = 3$ levels; B, the water movement at $b = 2$ levels; and C, the light condition at $c = 2$ levels. The experimental design has equal subclass numbers with $n = 3$ replications per treatment. A three-way ANOVA model for the data can be written using four subscripts: i to identify the levels of A; j to identify the levels of B; k to identify the levels of C; and m to identify the responses for a given treatment combination. Let Y_{ijkm} denote the m-th ($m = 1, 2, 3$) response for the treatment combination $a_i b_j c_k$ ($i = 1, 2, 3; j = 1, 2; k = 1, 2$). Then the ANOVA model is

$$Y_{ijkm} = \mu + \alpha_i + \beta_j + \gamma_k + (\alpha\beta)_{ij} + (\alpha\gamma)_{ik} + (\beta\gamma)_{jk} + (\alpha\beta\gamma)_{ijk} + E_{ijkm},$$

where the notation is as follows:

- μ is the overall mean expected response;
- α_i is the effect of the i-th level of A;
- β_j is the effect of the j-th level of B;
- γ_k is the effect of the k-th level of C;
- $(\alpha\beta)_{ij}$ is the joint effect of the i-th level of A and the j-th level of B;
- $(\alpha\gamma)_{ik}$ is the joint effect of the i-th level of A and the k-th level of C;

- $(\beta\gamma)_{jk}$ is the joint effect of the j-th level of B and the k-th level of C;
- $(\alpha\beta\gamma)_{ijk}$ is the joint effect of the i-th level of A, the j-th level of B, and the k-th level of C;
- E_{ijkm} is the random error.

The E_{ijkm} are independent $N(0, \sigma^2)$. The parameters in the model (α_i, β_j, and so on) satisfy constraints similar to those in Equation (3.16).

The ANOVA table associated with this model is as follows:

Source	df	SS	MS	F_c	p
Treatment	11	3.002	0.273	11.870	0.0001
Temperature (A)	2	1.338	0.669		
Movement (B)	1	0.423	0.423		
Light Condition (C)	1	0.270	0.270		
Interaction AB	2	0.446	0.223		
Interaction AC	2	0.099	0.050		
Interaction BC	1	0.240	0.240		
Interaction ABC	2	0.186	0.093	4.04	0.0307
Error	24	0.552	0.023		
Total	35	3.554			

Note that the computing formulas in Equation (13.18) can be used to calculate all of the sums of squares in this ANOVA table, except for SS[TOT]. Individual responses (not given here) are needed to compute SS[TOT]. Equation (13.17) can be used to determine the degrees of freedom.

The first step in the analysis is to examine the second-order interaction. The observed mean responses are as follows. In Figure 13.10, these means are plotted against the level of water movement (factor B).

Light condition (k)	Temperature (i)	Movement (j)	
		1 (still)	2 (flowing)
1 (low)	1 (cold)	1.55	1.08
	2 (lukewarm)	1.87	1.76
	3 (warm)	1.59	2.01
2 (high)	1 (cold)	1.85	1.36
	2 (lukewarm)	2.03	1.67
	3 (warm)	2.14	1.85

From Figure 13.10, it appears that there is second order interaction between the three factors, because the two first order interaction plots indicate that AB is present at c_1 (ABC_1 is present) and absent at c_2 (ABC_2 is absent). The p-value for ABC interaction in the ANOVA table confirms our conjecture based on the first-order interaction plots. We can conclude at $\alpha = 0.05$ (indeed at any level greater than 0.0307) that the

FIGURE   13.10

Observed mean gains in weight of fish

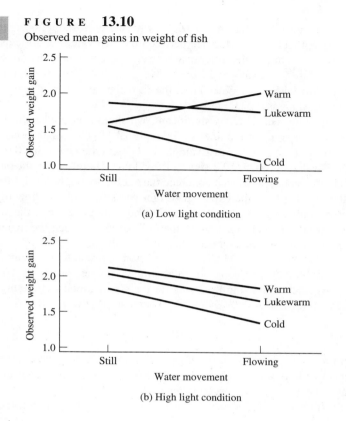

(a) Low light condition

(b) High light condition

data support the hypothesis that there is a second-order interaction between water temperature, water movement, and light condition. More specifically, we can state the following conclusions:

1   The interaction between temperature (*A*) and movement (*B*) depends on the light condition (*C*).

2   The interaction between temperature (*A*) and light condition (*C*) depends on water movement (*B*).

3   The interaction between movement (*B*) and light condition (*C*) depends on water temperature (*A*).

In view of the significance of the second-order interaction, the next step is to analyze the first order interactions at each level of the third factor. First, consider the interaction between *A* and *B*. In Example 13.11, we saw that the sum of squares for *AB* at the low level of *C* is 0.601; that is, $SS[ABC_1] = 0.601$ with two degrees of freedom. The test statistic for ABC_1 is

$$F_c(ABC_1) = \frac{MS[ABC_1]}{MS[E]} = \frac{0.301}{0.023} = 13.086$$

with $df = 2, 24$. This is highly significant ($p = 0.0001$).

Compare this value of $F_c(ABC_1)$ with the corresponding F_c value of 12.04 in Example 13.11. The difference between the two values and the associated degrees of freedom is due to the difference between the corresponding error mean squares in the two cases. In Example 13.11 the MS[E] of 0.025 was based on the 18 responses for the low level of C. When all of the 36 responses are used to compute MS[E], the resulting value is 0.023.

Knowing that ABC_1 is significant, we should proceed with the analysis at the low level of C along the lines in Example 13.11. The conclusions remain unaltered if we repeat the analysis of Example 13.11 with the new value 0.023 in place of the MS[E] value of 0.025 (12 degrees of freedom). Therefore, on the basis of an analysis of all 36 responses, we can conclude that, at the low light level, there is a significant difference between the expected weight gains for still and flowing water when the water temperature is low or warm, but no difference when the water temperature is lukewarm. Conclusions about the differences between the mean weight gains at different water temperatures can be derived similarly.

Next let's analyze ABC_2, the interaction between temperature and movement at the high level of light condition. The sum of squares for testing ABC_2 is the same as SS[AB] calculated using the totals for the six treatments at the high level of C. We find that SS[ABC_2] = 0.031 ($df = 2$), so that MS[BC_2] = 0.016 and

$$F_c(ABC_2) = \frac{0.016}{0.023} = 0.696$$

with $df = 2, 24$. Thus, in the high light condition, the water temperature does not appear to have any effect on the differential mean weight gain in still and flowing water. Therefore, we proceed to analyze AC_2 and BC_2, the main effects of A and B at the high level of C. Calculations using the data for the high level of C yield

$$SS[AC_2] = 0.466, \quad MS[AC_2] = 0.233, \quad F_c(AC_2) = 10.130, \quad df = 2, 24;$$

$$SS[BC_2] = 0.650, \quad MS[BC_2] = 0.650, \quad F_c(BC_2) = 28.261, \quad df = 1, 24.$$

The calculated values of the test statistics are highly significant ($p < 0.01$). Hence, at the high level of the light condition, there is a significant difference between the expected weight gains at the three temperatures and also at the two levels of water movement. ■

In Example 13.14, we found a significant second-order interaction. Here's an example in which the second-order interaction is not significant.

EXAMPLE **13.15**   In a $4 \times 2 \times 3$ factorial experiment, with $n = 2$ replications per treatment, the growth of a variety of fungus over four time periods (A, with levels a_1, a_2, a_3, a_4), in two environments (B, with levels b_1, b_2), using three growth media (C, with levels c_1, c_2, and c_3) was measured. The treatment totals obtained are as follows:

	b_1			b_2		
	c_1	c_2	c_3	c_1	c_2	c_3
a_1	5.31	6.32	8.51	6.28	11.01	5.61
a_2	5.01	6.53	10.48	6.01	10.02	6.36
a_3	6.33	7.92	10.50	6.51	11.31	7.48
a_3	7.01	7.53	10.92	6.53	18.30	8.00

The corresponding ANOVA table is as follows:

Source	df	Mean square	F_c	p
Time (A)	3	4.00	1.942	0.1498
Environment (B)	1	2.54	1.233	0.2778
Medium (C)	2	14.33	6.956	0.0041
AB	3	0.88	0.427	0.7355
AC	6	0.93	0.451	0.8371
BC	2	19.66	9.830	0.0008
ABC	6	1.20	0.583	0.7403
Error	24	2.06		

The ANOVA table shows that the second-order interaction between the three factors is not significant ($p = 0.7403$). Knowing that there is no demonstrable second-order interaction between the three factors, we proceed to examine the first-order interactions between each of the three pairs of factors.

Among the first-order interactions, only BC is significant ($p = 0.0008$), meaning that neither the environment (B) nor the medium (C) appears to interact with time (A). Consequently, the main effect of time can be examined without regard to the effects of the other two factors. Since the main effect of time is not significant ($p = 0.1498$), we conclude that the data do not provide sufficient evidence in support of the hypothesis that the expected growth of the fungus is different for the four time periods. Because the interaction between the environment (B) and the medium (C) is significant, the effects of the different environments (mediums) must be compared separately at each level of the medium (environment). Such an analysis can be performed along the lines of the analysis of ABC_1 and ABC_2 in Example 13.13. In Exercise 13.29, you will be asked to carry out such an analysis.   ■

Analysis when the subclass numbers are unequal

As already noted, the analysis of factorial experiments with unequal but nonzero subclass numbers—that is, when n_{ij} are unequal and no n_{ij} is zero—is more complicated than in the case of equal subclass numbers. In a 2×2 factorial, as we saw in Exercise 13.15, unequal subclass numbers cause the contrasts measuring the factorial effects to be nonorthogonal. In such cases, Equation (13.13) may not be appropriate for computing the ANOVA sums of squares. In the presence of unequal subclass numbers, the problem of nonorthogonality can arise in a multifactorial experiment, and the required calculations are best done using statistical computing software; any of the existing software packages can be used. In the next example, the GLM procedure of SAS is used to analyze a 3×2 experiment with unequal subclass numbers.

EXAMPLE 13.16 To study the effects of two methods of fertilization (a_1, a_2) and three methods of irrigation (b_1, b_2, b_3) on the yield (bushel per acre) of a variety of corn, four plots were randomized to receive each of the six treatment combinations. However, the plants in two plots receiving $a_1 b_3$ and one plot receiving $a_2 b_3$ were destroyed before the harvest time. The results obtained are as follows:

Fertilization method		Irrigation method					
	b_1		b_2		b_3		
a_1	17.0	18.1	16.8	15.3	22.1	23.4	
	17.5	16.9	14.6	16.1	—	—	
a_2	18.3	17.8	14.9	15.8	24.2	26.2	
	16.8	18.9	14.2	14.8	25.1	—	

The computer codes and the data input for analyzing this data set using the GLM procedure of SAS are as follows:

```
data corn;
input fert irrig yield @@;
cards;
1 1 17.0 1 1 18.1 1 1 17.5 1 1 16.9 1 2 16.8 1 2 15.3 1 2 14.6
1 2 16.1 1 3 22.1 1 3 23.4 2 1 18.3 2 1 17.8 2 1 16.8 2 1 18.9
2 2 14.9 2 2 15.8 2 2 14.2 2 2 14.8 2 3 24.2 2 3 26.2 2 3 25.1
;
proc glm;
class fert irrig;
model yield=fert irrig fert*irrig;
run;
```

The corresponding output of the SAS GLM procedure is as follows:

```
Dependent Variable:  YIELD
                           Sum of      Mean
Source           DF      Squares     Square   F Value   Pr > F

Model             5    257.56143   51.51229    75.93    0.0001
Error            15     10.17667    0.67844
Corrected Total  20    267.73810


Source    DF   Type III SS   Mean Square   F Value   Pr > F

FERT       1       2.68015       2.68015      3.95    0.0654
```

IRRIG	2	228.17006	114.08503	168.16	0.0001
FERT*IRRIG	2	7.66891	3.83446	5.65	0.0148

The computer codes for implementing the GLM procedure are very similar to those for implementing the ANOVA procedure. As we saw in Chapters 11 and 12, the GLM procedure produces two types of sums of squares (Type I and Type III). In Chapter 11, the difference between these two types of sums of squares was explained in the multiple linear regression context. When analyzing ANOVA models, the Type III sums of squares are used to test the significance of the factorial effects. For any particular factorial effect—for instance, the main effect of A—the Type III sum of squares is the difference between the regression sums of squares for the full model and the model appropriate under the null hypothesis being tested. Thus, the sum of squares for testing the main effect of the fertilizer is 2.68015, whereas the sum of squares for testing the interaction between the fertilizer and the irrigation method is 7.66891.   ■

Note that the case of missing subclasses—when some $n_{ij} = 0$—must be handled differently from the case when the numbers of subclasses are unequal. In the case of missing subclasses, as we saw in Section 13.3, it is impossible to estimate certain components of the simple effects, and so at least some of the usual measures of factorial effects cannot be estimated.

Exercises

13.28   Refer to Example 13.14.

  a   Use Equation (13.18) or appropriate software to compute the sums of squares in the ANOVA table.

  b   Construct a set of 95% simultaneous confidence intervals for $\mu[AC_2]$ and $\mu[BC_2]$. Interpret the result.

  c   Analyze the data to see how the interaction between water temperature and light condition affects the weight gain of the fish.

  d   Analyze the data to see how the interaction between water movement and light condition affects the weight gain of fish.

13.29   Refer to Example 13.15.

  a   Construct interaction plots that are useful for visually evaluating second- and first-order interaction between the experimental factors. What conclusions can you draw from these graphs?

  b   Write a model for the growth data.

  c   Perform appropriate tests to see if there are differences between the two environments for each of the three growth mediums. For each growth medium, construct a 95% confidence interval for the difference between the effects of the two environments. Interpret the intervals.

  d   Perform appropriate tests to see if there are differences between the three mediums for each of the two environments. For each environment, perform an appropriate multiple comparison of the three mean responses for the three growth mediums.

13.30 An integrated pest management study involved a nematocide (factor A at two levels: present, absent), a herbicide (factor B at two levels: present, absent), and an insecticide (factor C at two levels: present, absent). The investigators assigned 32 field plots at random to the eight treatment combinations, in such a way that each combination received four plots. A particular variety of corn was grown in each of these plots. The treatment means calculated from the measured yields are as follows:

	c_1		c_2	
	b_1	b_2	b_1	b_2
a_1	20.3	22.0	24.9	28.0
a_2	21.4	22.1	25.2	26.3

The corresponding ANOVA table takes the following form:

Source	Sum of squares
Nematocide (A)	0.02
Herbicide (B)	21.78
Insecticide (C)	172.98
AB	4.50
AC	3.38
BC	1.62
ABC	0.50
Error	46.32

a Plot graphs that may be used to visually evaluate whether there are first- and second-order interactions between the experimental factors. What conclusions can you draw from your graphs?

b Test the significance of the various interaction and main effects. Write your conclusions.

c Carry out all relevant follow-up analyses suggested by your findings in (b).

13.31 An experiment to study the effect of sleep deprivation on the ability to perform mental tasks was conducted with 36 subjects, all of whom had similar mental capacities as measured by their IQ scores. Of the 36 subjects, 18 were males and 18 were females. In each gender group, nine subjects were between 20 and 30 years of age, while the other nine were between 30 and 40 years of age. The nine subjects in each of the four age-gender groups were divided into three groups of three subjects each, and the three groups were assigned to three treatments (sleep deprivation levels) at random. Over a period of 48 hours, every subject was observed in the act of performing a set of tasks—reading, writing, exercising, and resting. The resting activities differed from group to group. Subjects in treatment group 1 were allowed 8 hours of sleep at the end of a 24-hour period whereas subjects in treatment groups 2 and 3 were allowed 6 and 4 hours of sleep, respectively. At the end of 48 hours, the subjects were asked to take a test that measured their mental capacity. The mean test scores obtained are as follows:

	Male			Female		
	4 hrs	6 hrs	8 hrs	4 hrs	6 hrs	8 hrs
Young	35	72	80	45	78	90
Old	34	28	60	10	32	80

The corresponding ANOVA table may be written in the following form:

Source	Mean square
Age	6084
Gender	169
Treatment	6499
Age×Gender	169
Age×treatment	819
Gender×treatment	364
Age×gender×treatment	388
Error	143

a   Draw the appropriate interaction plots to display the nature of second- and first-order interactions between the three factors.

b   Use the plots in (a) to draw conclusions about the interactions between age, gender, and sleep deprivation.

c   Perform the appropriate F-tests to analyze the factorial effects at the 5% significance level.

d   Does your analysis in (c) confirm the conclusions in (b)?

e   Use a multiple comparison method to compare the mean test scores at the three sleep deprivation levels at each of the two age groups. Comment on the results.

f   Is the multiple comparison in (e) justified by the analysis in (c)? Explain.

13.32   Refer to Example 13.16.

a   Use the computer output to fill in the following ANOVA table.

Source	df	Sum of source	Mean square	F_c
Fertilizer				
Irrigation				
Fertilizer×Irrigation				
Error				
Total				

b   Use the ANOVA table in (a) to draw appropriate conclusions about the factorial effects of fertilizer and irrigation.

    c Explain why, in contrast to the analysis of variance of an $a \times b$ factorial with equal subclass numbers, the sums of squares for the factorial and error effects do not add to the total sums of squares.

13.33 Refer to Example 13.16.

    a Write the appropriate model for the yield data.

    b Write the reduced models associated with the following null hypotheses:

       i H_{0A}: the main effect of the fertilizer types is not significant;

      ii H_{0B}: the main effect of the irrigation methods is not significant;

    iii H_{0AB}: the interaction between fertilizer types and irrigation methods is not significant.

    c Use the reduced models in (b) to identify each Type III sum of squares in the computer printout as the difference between the regression sums of squares associated with the full model and an appropriately chosen reduced model.

13.7
Experiments with nested factors

Thus far, we have considered complete factorial experiments in which responses were measured at every possible combination of levels of the experimental factors. In this section, our objective is to describe a type of experiment—called a *nested experiment*—in which responses are measured at a subset of the combinations of factor levels. To begin, let's look at two important ways in which factors can be configured in the design of a factorial study.

Crossed and nested factors

The levels of factor A are said to be *crossed* with the levels of factor B if every level of A occurs in combination with every level of B.

EXAMPLE **13.17** In the weight gain study in Example 13.4, the levels of the factors are crossed. Recall that there were two factors: A, the water temperature, with two levels $a_1 = $ cold, $a_2 = $ warm; and B, the water movement, with two levels $b_1 = $ still, $b_2 = $ flowing. The experiment used four treatment combinations: $a_1 b_1$, $a_1 b_2$, $a_2 b_1$, $a_2 b_2$. An examination of the composition of the four treatment combinations reveals that A is crossed with B, because every level of A occurs with each of the two levels of B.

    Similarly, in the $4 \times 2 \times 3$ factorial experiment in Example 13.15, each of the three factors—time (factor A), environment (factor B), and growth medium (factor C)—is crossed with every other factor. To verify this, we simply need to observe that the 24 treatments in this experiment correspond to the 24 possible ways of combining all of the levels of the three factors. ■

The levels of factor B are said to be *nested* within the levels of factor A if the levels of B can be divided into subsets (nests) such that every level in any given subset occurs with exactly one level of A.

EXAMPLE 13.18 A study was conducted to investigate the effects of a new chemotherapy (factor B) for a particular type of childhood cancer on the survival time of two types of patient populations (factor A). The two populations (levels of A) of interest were: population 1 (level a_1), consisting of subjects classified as low-risk patients at the time of diagnosis; and population 2 (level a_2), consisting of high-risk patients. Four doses (levels of B) of chemotherapy were compared in the study; two lower doses (b_1 and b_2) were used to treat low-risk patients, and two higher doses (b_3 and b_4) were used to treat high-risk patients. Thus, the study involved four treatments: a_1b_1, a_1b_2, a_2b_3, and a_2b_4. Consequently, the study is not a complete factorial experiment, because responses were measured for only four out of the eight possible treatment combinations. For instance, responses were not measured for the treatment combination a_1b_3.

In this study, the levels of factor B are nested in the levels of factor A. The levels of B can be divided into two subsets $\{b_1, b_2\}$ and $\{b_3, b_4\}$. The levels in $\{b_1, b_2\}$ occur only with a_1, whereas the levels in $\{b_3, b_4\}$ occur only with a_2.

The tree diagrams in Figure 13.11 display the differences between crossed and nested factors. The notation introduced in Figure 13.11 may conveniently be used to distinguish between the two types of factor configurations. Design I shows the treatment configuration in a 2×2 experiment with treatments a_1b_1, a_1b_2, a_2b_1, and a_2b_2. Note that, in the treatment combinations a_1b_1 and a_1b_2, a_1 occurs with b_1 and b_2, whereas, in the treatment combinations a_2b_1 and a_2b_2, a_2 occurs with b_1 and b_2. Therefore, each of the two levels of A occurs with each of the two levels of B; in other words, the levels of A are crossed with the levels of B. Design II in

FIGURE 13.11

Crossed and nested factor configurations

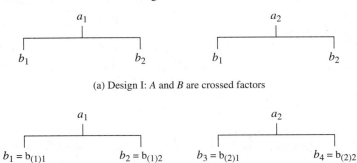

(a) Design I: A and B are crossed factors

(a) Design II: B is nested in A

Figure 13.11 shows an experiment with two factors: A with two levels; and B with four levels. The experiment has four treatments a_1b_1, a_1b_2, a_2b_3, and a_2b_4, such that the first two levels of B (b_1 and b_2) are nested within the first level of A and the second two levels of B (b_3 and b_4) are nested within the second level of A.   ■

The notion of interaction, which is fundamental to the analysis of crossed factors, is not useful for anaylzing nested factors. Consider the two designs in Figure 13.11. In both designs, there are two factors and four treatment combinations. Design I is a 2×2 experiment and, as we see from Equation (13.2), the effect of interaction AB is equal to half of the difference between the simple effects of B at a_1 and a_2. Recall that the interaction effect can be interpreted as a measure of the impact of A on the effect of changing the level of B from b_2 to b_1.

In design II, by contrast, the simple effects of B at the two levels of A correspond to different subsets of the levels of B. At a_1, the simple effect of B measures the effect of changing the level of B from b_2 to b_1; at a_2, however, it measures the effect of changing the level of B from b_4 to b_3. Consequently, the two simple effects of B measure the effect of changing the levels of B within the two nests defined by the levels of A. Thus, when one of the factors is nested within the levels of the other factor, there is no way of measuring interaction as defined in Equations (13.1) and (13.2).

Notation for crossed and nested factors

To distinguish between nested and crossed factors, it is convenient to have a notation that clearly indicates whether or not two factors are nested within each other. In Figure 13.11, the symbol $b_{(1)2}$ in design II denotes the second level of B in the nest formed by the first level of A. Thus, $b_{(1)2}$ is the same as b_2. Similarly, $b_{(2)1}$ is the first level of B in the nest formed by the second level of A; $b_{(2)1}$ is the same as b_3. In general, the symbol $b_{(i)j}$ is the j-th level of B nested within the i-th level of A. The treatment obtained by combining a_i with $b_{(i)j}$ will be denoted by $a_ib_{(i)j}$. Thus, in design II, the symbols a_2b_4 and $a_2b_{(2)2}$ denote the same treatment.

ANOVA for nested experiments

A two-way ANOVA may be used to analyze the results of a two-factor nested experiment on the basis of the model in Box 13.2.

BOX **13.2**   *Two-way ANOVA fixed-effects model with one nested factor*

Consider a two-factor completely randomized experiment in which factor B is nested within factor A. Suppose that b levels of B are nested within each of the a levels of A. Let a_i ($i = 1, \ldots, a$) denote the i-th level of A and $b_{(i)j}$ ($j = 1, \ldots, b$) denote the j-th level of B nested in the i-th level of A. Thus, the experiment involves ab treatment combinations.

Let Y_{ijk} denote the k-th ($k = 1, \ldots, n_{ij}$) response for the treatment combination $a_i b_{(i)j}$. Then a two-way ANOVA model for Y_{ijk} is

$$Y_{ijk} = \mu + \alpha_i + \beta_{(i)j} + E_{ijk}, \quad i = 1, \ldots, a; \quad j = 1, \ldots, b; \quad k = 1, \ldots, n_{ij};$$

where μ is the overall mean; α_i is the effect of the i-th level of A; $\beta_{(i)j}$ is the effect of the j-th level of B nested in the i-th level of A; and E_{ijk} is the random error. The E_{ijk} are independent $N(0, \sigma^2)$, and the parameters α_i and $\beta_{(i)j}$ satisfy the constraints

$$\sum_{i=1}^{a} n_{i+}\alpha_i = 0 \quad \text{and} \quad \sum_{j=1}^{b} n_{ij}\beta_{(i)j} = 0, \quad i = 1, \ldots, a.$$

Comments on the nested-effects model in Box 13.2

1 In both the two-way nested-effects model in Box 13.2 and the two-way crossed-effects model in Box 13.1, the expected response for a treatment is represented as the sum of a set of interpretable factorial effects. The nature of that sum is the fundamental difference between the two models. In the crossed-effects model, the expected response for treatment $a_i b_j$ is the sum of the overall mean μ and three effects α_i, β_j, and $(\alpha\beta)_{ij}$; in the nested-effects model, by contrast, the expected response for treatment $a_i b_{(i)j}$ is the sum of the overall mean μ and two effects α_i and $\beta_{(i)j}$.

2 The usual null hypotheses that are tested with the nested model are

$$H_{0B(A)}: \beta_{(i)j} = 0, \qquad i = 1, \ldots, a; \quad j = 1, \ldots, b; \tag{13.19}$$

and

$$H_{0A}: \alpha_i = 0, \qquad i = 1, \ldots, a. \tag{13.20}$$

The null hypothesis $H_{0B(A)}$ is equivalent to the hypothesis that there is no difference between the effects of the levels of B nested within levels of A—that is, that the expected responses are the same for treatment combinations containing the same level of A. Rejection of $H_{0B(A)}$ implies that, within at least one level of A, different levels of B do not have same effects.

In practice, rejection of $H_{0B(A)}$ leads to specific comparisons of the levels of B within the levels of A. If $H_{0B(A)}$ is not rejected—that is, if there is no demonstrable difference between the effects of the levels of B within the levels of A—then a test of the main effect of A is performed. This can be done by testing H_{0A}, which states that there is no difference between the average (over the levels of B) expected responses at different levels of A.

3   The sums of squares for testing H_{0A} and $H_{0B(A)}$ can be derived using the general linear model approach of the theory of least squares. The sums of squares and their associated degrees of freedom in the analysis of variance corresponding to the nested model in Box 13.2 may be summarized as follows:

- SS[A] is the sum of squares for A. This is a measure of the variability of the observed mean responses at different levels of A. The degrees of freedom for this sum of squares are

$$df_A = a - 1.$$

- $SS[B(A)]$ is the sum of squares for $B(A)$. This is a measure of the variability of the mean responses at the levels of B within the levels of A. There are b levels of B nested within each level of A, and so the sum of squares within each nest will have $b - 1$ degrees of freedom. There are a nests (levels of A), and so the degrees of freedom associated with SS[B(A)] are

$$df_{B(A)} = a(b - 1).$$

- SS[E] is the sum of squares for the error. This is a measure of the variability of the responses within treatment combinations. The n_{ij} replications for treatment combination $a_i b_{(i)j}$ result in $n_{ij} - 1$ degrees of freedom for the estimate of the variance within this particular treatment group. Therefore, the degrees of freedom for the pooled estimate are

$$df_E = \sum (n_{ij} - 1) = N - t,$$

where $N = \sum n_{ij}$ and $t = ab$ are the total number of responses and the total number of treatments, respectively. If treatments have equal replications, then $N = abn$ and $N - ab = ab(n - 1)$.

The ANOVA table associated with the nested model in Box 13.2 is shown in Table 13.9.

T A B L E   13.9

ANOVA table for two-factor nested experiments

Source	df	SS	MS	F_c
A	$a - 1$	SS[A]	MS[A]	MS[A]/MS[E]
B(A)	$a(b - 1)$	SS[B(A)]	MS[B(A)]	MS[B(A)]/MS[E]
Error	$N - ab$	SS[E]	MS[E]	
Total	$N - 1$	SS[TOT]		

4   As in the analysis of a complete factorial experiment, computation of the sums of squares in Table 13.9 can be tedious if the subclass numbers n_{ij} are unequal. In that case, we recommend the use of statistical computing software such as SAS. In the special case of equal subclass numbers—that is, when $n_{ij} = n$ and $N = nab$—these sums of squares can be expressed as

$$SS[TOT] = \sum_{i=1}^{a} \sum_{j=1}^{b} \sum_{k=1}^{n} Y_{ijk}^2 - \frac{Y_{+++}^2}{nab}, \quad df_{TOT} = nab - 1,$$

$$SS[T] = \sum_{i=1}^{a} \sum_{j=1}^{b} \frac{Y_{ij+}^2}{n} - \frac{Y_{+++}^2}{nab}, \quad df_T = ab - 1,$$

$$SS[A] = \sum_{i=1}^{a} \frac{Y_{i++}^2}{nb} - \frac{Y_{+++}^2}{nab}, \quad df_A = a - 1, \tag{13.21}$$

$$SS[B(A)] = SS[T] - SS[A], \quad df_{B(A)} = (ab - 1) - (a - 1) = a(b - 1),$$

$$SS[E] = SS[TOT] - SS[T], \quad df_E = (abn - 1) - (ab - 1) = ab(n - 1).$$

Notice from Equation (13.21) that

$$SS[T] = SS[A] + SS[B(A)],$$

so that, the two-factor nested ANOVA subdivides the observed variation between treatment means (measured by the treatment sum of squares) into two components: a component that measures the variation between means at different levels of A; and a component that measures the variation between means at levels of B within levels of A. This subdivision of the treatment sum of squares is possible because, when the subclass numbers are equal, the contrasts that estimate the components of the main effects of A are orthogonal to the contrasts that estimate the main effects of B within any level of A.

EXAMPLE **13.19**   The amount of readily soluble phosphorus in a large number of soil samples was to be determined in a laboratory that employed six technicians. Three technicians worked in the morning shift, and the other three worked in the evening shift. In a study to determine whether the measured values of phosphorus (lb/acre) were affected by (1) the time of day (A.M. and P.M.), and (2) the technician making the measurement, 24 identical specimen samples were assigned to the six technicians at random, in such a way that each received four samples. Each technician was asked to independently analyze the four assigned samples. The results obtained are as follows:

Time (i)		1 (A.M.)			2 (P.M.)		
Technician		1	2	3	1	2	3
	1	42	43	47	50	49	47
	2	44	44	46	49	48	51
Specimen (k)	3	43	45	47	52	49	46
	4	44	42	43	50	47	48
Total (Y_{ij+})		173	174	183	201	193	192
Total (Y_{i++})		530			586		
Total (Y_{+++})			1116				

Let's assume a two-way nested model, as in Box 13.2

$$Y_{ijk} = \mu + \alpha_i + \beta_{(i)j} + E_{ijk}, \quad i = 1, 2; \ j = 1, 2, 3; \ k = 1, 2, 3, 4.$$

The notation here is as follows:

- Y_{ijk} is the measured soluble phosphorus in the k-th specimen analyzed by the j-th technician working in the i-th shift;
- μ is the overall mean;
- α_i is the effect of the i-th shift, and $\alpha_1 + \alpha_2 = 0$;
- $\beta_{(i)j}$ is the effect of the j-th technician in the i-th shift, and

$$\beta_{(1)1} + \beta_{(1)2} = 0; \quad \beta_{(2)1} + \beta_{(2)2} = 0;$$

- E_{ijk} is the random error.

The E_{ijk} are independent $N(0, \sigma^2)$.

The necessary calculations can be performed using Equation (13.21). The following computer output contains the sums of squares obtained when these data are analyzed by the ANOVA procedure in SAS (for details on the computer codes required to obtain this output, see Section 13.7 in Younger, 1997):

```
Nested ANOVA:  Soluble Phosphorus Data


                  Analysis of Variance Procedure


Dependent Variable:  PHOS
                       Sum of        Mean
Source           DF   Squares       Square    F Value   Pr > F
Model             5   158.00000    31.60000    14.22    0.0001
Error            18    40.00000     2.22222
Corrected Total  23   198.00000


Source          DF   ANOVA SS   Mean Square   F Value   Pr > F
SHIFT            1   130.66667   130.66667     58.80    0.0001
TECH (SHIFT)     4    27.33333     6.83333      3.07    0.0429
```

On the basis of this output, we can construct the following ANOVA table:

Source	df	SS	MS	F_c	p
Between shifts A	1	130.67	130.67	58.80	0.0001
Between Technicians $B(A)$	4	27.33	6.83	3.07	0.0429
Error	18	40.00	2.22		
Total	23	198.00			

The first step in the analysis is to test the significance of the differences between technicians within shifts. The p-value of 0.0429 for $B(A)$ suggests that the null hypothesis $H_{0B(A)}$ that there is no difference between technicians within shifts, can be rejected at any level $\alpha \geq 0.0429$. Thus, it is reasonable to conclude ($p = 0.0429$) that the true mean values for technicians within shifts differ among themselves. Having concluded that the technicians within shifts appear to produce different results, we go on to compare the mean responses for the three technicians within each shift. Such comparisons can be conducted by the procedures in Chapter 9 if we note that each mean is based on four replications and $MS[E] = 2.22$ with $df = 18$. In Exercise 13.34, you are asked to complete the analysis. ■

Factors may also be nested in studies that involve more than two factors. Analysis of such studies follows the pattern adopted here for nested experiments with two factors; for details, see Searle, Casella, and McCulloch (1992), or Graybill (1976). In Chapter 14, the analysis of factorial experiments with nested and crossed factors will be discussed in a general context.

Exercises

13.34 Refer to Example 13.9.

    a The lab manager suspects that the technicians in the evening shift differ with respect to the mean values they measure. Perform a test to check whether the manager's suspicion is justified.

    b Construct a set of 95% simultaneous confidence intervals for the pairwise differences between the means for the three technicians in the morning shift. Interpret the intervals you obtain.

    c The F-test for the main effect of shift gave a highly significant ($p = 0.0001$) result. How would you interpret this result?

13.35 A study to evaluate the effects of the type of drug (factor A) and the method of administration (factor B) on the fasting blood sugar of diabetic patients was based on a nested factorial design, with three levels of A—a_1, brand I tablet; a_2, brand II tablet; and a_3, insulin injection—and six levels of B: $b_{(1)1} = 30$ mg once a day; $b_{(1)2} = 15$ mg twice a day; $b_{(2)1} = 20$ mg once a day; $b_{(2)2} = 10$ mg twice a day; $b_{(3)1} =$ injection before breakfast; and $b_{(3)2} =$ injection before supper. The change in the fasting blood sugar (mg/dliter) was measured for 18 patients (three per treatment combination) who were on their assigned therapy for six months, with the following results:

Level of A		1		2		3
Level of B	1	2	1	2	1	2
Replications 1	18	23	20	17	30	33
Replications 2	16	19	19	15	28	36
Replications 3	13	17	21	20	26	30
Totals	47	59	60	52	84	99
Factor-A Totals		106		112		183

a   Draw a diagram to display the study design.

b   Write a two-way ANOVA model for the data.

c   Construct the ANOVA table and test for the significance of the nested effects. State the conclusion that can be drawn from your test.

d   What further analyses are suggested by your conclusion in (c)?

e   Construct a set of 90% simultaneous confidence intervals for the simple effects of the method of drug administration. Interpret these intervals.

f   Using the 0.05-level Tukey multiple comparison procedure, compare the mean change in the blood sugar levels resulting from the three drugs. What practically meaningful conclusions (if any) can be drawn from the multiple comparison test?

13.8
Overview

Methods for investigating the way in which the levels of several factors combine to produce an effect on a response are discussed in this chapter. Such data analyses, called factorial analyses, are based on ANOVA models in which the effects of treatments are represented as sums of component effects—main effects and interaction effects. An interaction effect measures the impact of changing the levels of a given subset of factors on the expected response at a fixed combination of levels of the remaining factors. A main effect, on the other hand, can be interpreted as the average change in the expected response that results from changing the levels of a given factor; the average is taken over all combinations of levels of the other factors.

In a completely randomized experiment with more than one factor, factorial analysis of the data implies the use of a multiway ANOVA rather than a one-way ANOVA. However, once the overall ANOVA is complete, the follow-up analysis that is appropriate will depend on the study objectives and on whether the factors are crossed or nested. The possible types of follow-up analysis for crossed and nested factors are discussed in this chapter. In most cases, the methods in Chapters 4 and 9 can be used for the follow-up analysis.

14

Random- and
Mixed-Effects ANOVA

14.1
Introduction

In Chapters 8 and 9, we discussed methods for the design and analysis of studies to compare the effects of a set of t treatments. In such studies, as we have seen, responses can be regarded as measurements at t levels of an experimental factor. The investigator selects (fixes), in advance, the actual levels at which measurements are to be made. For instance, Example 8.2 concerned a photomorphogenetic study in which the objective was to compare the effects of five treatments (types of light sources). In that study, the investigators had selected (fixed) $t = 5$ levels of one experimental factor—the light source.

Recall that the one-way ANOVA model in Box 8.2 is called a one-way fixed-effects model because it is used to describe data classified into groups on the basis of the levels of one experimental factor. Similarly, the two-way classification models in Boxes 13.1 and 13.2 are called two-way fixed-effects models with crossed and nested effects, respectively.

If investigators are interested in the effects of many levels of an experimental factor, measuring responses at every level may be difficult, impossible, or prohibitively expensive. In such cases, it is often convenient to regard the set of all levels (treatments) under consideration as a statistical population in its own right and to draw conclusions about this population on the basis of the observed responses to a random sample of levels selected from this population. Let's look at an example.

EXAMPLE **14.1**   To assess the batch-to-batch variation in the protein content of 250-ml cartons of pasteurized buttermilk produced by a dairy farm, four batches of cartons were randomly selected from the production line. From each batch, the protein contents were determined for three randomly selected cartons. The resulting data (percentage protein content) are as follows:

Batch i	percentage protein content			Total
1	3.42	3.41	3.57	10.40
2	3.05	3.14	3.23	9.42
3	3.23	3.48	3.37	10.08
4	3.46	3.59	3.23	10.28
Total				40.18

In this study, the experimental factor is the batch, because the objective was to see how the protein contents of buttermilk cartons vary from one batch to another. It was impossible to make protein content measurements for samples selected from every batch of production, and so the investigators decided to confine the study to four randomly selected batches from the population of all batches. In other words, the investigators decided to restrict their study to $t = 4$ treatments (batches) selected at random from a population consisting of a large (infinite) number of treatments.

Even though only four treatments were included for observation, the objective of the study extended beyond the effects of these treatments. The investigators wanted to study the carton-to-carton variability of protein content over all batches of cartons produced by the dairy farm.   ■

In this chapter, we examine methods for the design and analysis of studies that measure responses at a random sample of levels of one or more experimental factors. In Section 14.2, we consider one-way classification models with random treatment effects, while in Section 14.3 we examine the ANOVA method based on such models. The problem of sample size determination is addressed in Section 14.4. In Section 14.5 we describe the analyses of variance based on two-way classification models in which one factor has a random effect. Designs in which more than one response is measured for each experimental unit are described in Section 14.6. Finally, in Section 14.7, a set of rules is given for determining the so-called EMS values when the subclass numbers are equal; as we'll see, determining EMS values is a necessary step in the analysis of models containing random effects.

14.2
The one-way random-effects model

A one-way random-effects model can be used to describe the responses in a completely randomized experiment in which the levels of the experimental factor constitute a random sample from a population of levels. The responses observed in such experiments can be regarded as random samples selected in two stages. In the first stage, a random sample of a particular size—say, t—is selected from a population of levels of the experimental factor. In the second stage, a random sample of n_i responses is observed for the i-th treatment selected in the first stage. Then $N = n_1 + n_2 + \cdots + n_t$ is the total number of responses in the experiment. In the important special case of a balanced experiment in which each treatment is replicated the same number of times n, the total number of responses is $N = nt$.

In Example 14.1, the first stage sample consisted of $t = 4$ levels of the experimental factor, the batch. In the second stage, a random sample of $n = 3$ responses was observed for each of the four batches, so that in this example $n_1 = n_2 = n_3 = n_4 = 3$ and $N = 12$.

The structure of the one-way random-effects model is very similar to the structure of the one-way fixed-effects model in Equation (8.10). Let μ be the overall mean, and M_i be the mean of the population of all responses for the i-th randomly selected treatment. Keep in mind that the population mean (expected response) M_i is a random quantity, because it is associated with the randomly selected treatment; the purpose of switching from the Greek letter μ_i to the English letter M_i is to emphasize that M_i is a random quantity. Then $T_i = M_i - \mu$, where μ is the overall mean, is the effect of the i-th treatment. Again, unlike the treatment effect τ_i in the fixed-effects model in Equation (8.10), the treatment effect T_i is a random quantity.

Let Y_{ij} denote the j-th observed response for the i-th treatment. Then the one-way random effects model can be expressed as

$$Y_{ij} = \mu + T_i + E_{ij}, \qquad\qquad \textbf{(14.1)}$$

where μ is the overall mean; T_i is the random effect of the i-th treatment; and E_{ij} is the random error.

In Equation (14.1), $\mu + T_i$ is the true mean response (or the expected response) for the i-th randomly selected treatment; Y_{ij}, the observed response, is the sum of E_{ij}, the random error associated with the response, and the corresponding expected response. It can be shown that, under suitable assumptions about the T_i and E_{ij} (normal distributions, equal variances, and so on), the one-way ANOVA approach of Chapter 8 can be used to make inferences about the treatment effects in Equation (14.1). The corresponding assumptions are listed in Box 14.1.

BOX **14.1**   *The one-way ANOVA model with random effects*

Let Y_{ij} denote the j-th response for treatment i ($j = 1, 2, \ldots, n_i$; $i = 1, 2, \ldots, t$). Then

$$Y_{ij} = \mu + T_i + E_{ij},$$

where μ is the overall mean expected response; T_i is the effect of the i-th randomly selected treatment; and E_{ij} is the random error. The T_i and the E_{ij} are, respectively, independent $N(0, \sigma_T^2)$ and independent $N(0, \sigma^2)$. Furthermore, T_i and E_{ij} are mutually independent.

EXAMPLE **14.2**   A one-way random-effects model (model A) for the protein content data in Example 14.1 may be written in the form

$$Y_{ij} = \mu + T_i + E_{ij}, \qquad j = 1, 2, 3; \quad i = 1, 2, 3, 4.$$

Here Y_{ij} is the measured protein content of the j-th carton selected from the i-th batch; μ is the overall mean protein content of buttermilk produced at the dairy farm; T_i is the effect of the i-th randomly selected batch; and E_{ij} is the random error. The T_i are independent $N(0, \sigma_T^2)$, and the E_{ij} are independent $N(0, \sigma^2)$. Also, T_i and E_{ij} are mutually independent. The corresponding fixed effects model (model B) takes the form

$$Y_{ij} = \mu + \tau_i + E_{ij}, \quad j = 1, 2, 3; \quad i = 1, 2, 3, 4.$$

Here Y_{ij} is the measured protein content of the j-th carton selected from the i-th batch; μ is the overall mean protein content of buttermilk produced in the four batches; τ_i is the fixed effect of the i-th batch; and E_{ij} random error. The E_{ij} are independent $N(0, \sigma^2)$, and $\sum_{i=1}^{4} \tau_i = 0$.

In model A, μ is the average protein content of the buttermilk produced at the dairy farm, and making inferences about μ is a problem of practical interest. In contrast, inferences about μ in model B are usually of no interest because, as we see from Equation (8.6), the parameter does not have a practically meaningful interpretation.

In model A, the effect of the i-th batch T_i is the difference between the mean protein content of the population of cartons produced in the i-th randomly selected batch and the mean protein content of the population of cartons produced at the dairy farm.

The key difference between models A and B is that, under model A, we are interested in comparing the mean protein contents of all batches produced at the farm, whereas, under model B, we compare the mean protein contents of the cartons produced in the four batches selected for observation.

In a fixed-effects analysis based on model B, inferences about the differences between the mean protein contents of the four selected batches can be made by comparing the corresponding treatment effects. Thus, in model B, inferences about the differences between the τ_i are of primary concern. Such inferences can be made by testing the null hypothesis H_0: $\tau_1 = \tau_2 = \tau_3 = \tau_4 = 0$ or by constructing confidence intervals for specific contrasts among the τ_i.

In model A, the parameter σ_T^2, which is the variance of the true mean protein contents of all batches produced at the farm, is a measure of the batch-to-batch variability of the protein contents. The value $\sigma_T^2 = 0$ implies that the protein contents do not differ from one batch to another. As σ_T^2 increases, the batch-to-batch variability of the protein content increases. Accordingly, inferences about σ_T^2 are of interest when using model A.

The error variance σ^2 has the same interpretation in the two models. It is the variance of the measured protein contents of cartons within any single batch.   ■

Differences between fixed- and random-effects models

The following differences between fixed- and random-effects models can be illustrated by comparing the experimental design that generated the protein content data of Example 14.1 with the experimental design that generated the plant height data of Example 8.2.

1   If we repeat an experiment in which a fixed-effects model is being used, the same set of treatments will be used in each repetition. In a repetition of an experiment with a random-effects model, a different set of treatments can—and very likely will—enter in different repetitions.

2   A one-way fixed effects model is specified in terms of the parameters $\mu, \tau_1, \tau_2, \ldots, \tau_t$, and σ^2. The corresponding one-way random-effects model is specified in terms of the parameters μ, σ_T^2, and σ^2.

3   The null hypothesis H_0 that the treatment effects are equal can be stated as

$$H_0: \tau_1 = \tau_2 = \cdots = \tau_t = 0$$

in a fixed-effects model, and as

$$H_0: \sigma_T^2 = 0$$

in a random-effects model.

Variance components

In Exercise 14.4, you will be asked to use the results in Box 9.1 to show that for the one-way random-effects model, the variance of an observed response is equal to the sum of the variances of T_i and E_{ij}

$$\mathrm{Var}(Y_{ij}) = \sigma_T^2 + \sigma^2. \tag{14.2}$$

Equation (14.2) implies that the variability of an observed response can be expressed as the sum of two components σ_T^2 and σ^2. Of these, σ_T^2 accounts for the variability among treatment effects; this component contributes to the variance of the observed response because the observed response depends on the effect of a randomly selected treatment. The second component σ^2 accounts for the variability of the replicate measurements for the same treatment. The terms σ_T^2 and σ^2 are called the *components of variance* (or variance components) in a one-way random effects model.

In contrast to a one-way random effects model, the random variation of a measurement in a fixed-effects model is affected by a single component σ^2.

Exercises

14.1   To determine the variability of the protein content of seeds produced in the F_3 generation of a soybean cross, a study was conducted in 30 experimental plots with similar soil and environmental characteristics. Seeds from each of a random sample of ten F_2 generation plants were planted in three randomly selected plots. Suppose that the percentage protein contents of the seeds produced by the F_3 plants are as follows:

Plant	Plot 1	Plot 2	Plot 3	Total
1	42.4	41.0	39.6	123.0
2	28.6	36.3	42.2	107.1
3	43.2	42.1	40.2	125.5
4	40.8	41.0	38.9	120.7
5	41.0	38.3	41.1	120.4
6	39.4	39.5	37.2	116.1
7	39.6	40.4	38.9	118.9
8	38.1	38.3	37.9	114.3
9	35.9	36.1	35.6	107.6
10	39.6	39.9	39.7	119.2
Total				1172.8

a   Write a one-way ANOVA model for the data.

b   Using the parameter(s) of the model, describe the null hypothesis H_0 that the protein content of seeds is constant over the F_2 generation plants.

c   Using the parameters of the model, describe the null hypothesis H_0 that the average protein content of seeds from F_3 generation plants is 40%.

d   Which parameter in the model measures the amount of variability in the measured protein contents of F_3 plants grown from the seeds of a single F_2 plant?

14.2   In a preliminary study of the serum cholesterol levels of the population in a particular region, a random sample of eight normal subjects was selected. A one-sixth aliquot serum sample from each subject was independently tested twice for serum cholesterol level. The measured values (mg/dliter) are as follows:

Subject	Sample 1	Sample 2	Total
1	191	193	384
2	185	189	374
3	175	172	347
4	200	198	398
5	168	171	339
6	177	176	353
7	182	178	360
8	173	175	348
Total			2903

a   Let Y_{ij} denote the cholesterol level measured in the j-th test for the i-th subject. Consider the model

$$Y_{ij} = \mu + T_i + E_{ij},$$

where μ is the overall mean, and T_i and E_{ij} denote, respectively, the effects of subject and error on the measured value. Interpret the meanings of μ, T_i, and E_{ij}.

b   State a set of assumptions that will justify an ANOVA approach for the given data.

c   Let σ_T^2 and σ^2 denote, respectively, the variances of T_i and E_{ij}. What do the quantities σ_T^2 and σ^2 measure? In the context of this problem, what is the implication if σ^2 is twice as large as σ_T^2?

14.3   In a study of the cation-exchange capacity of the surface soil in a region containing a particular type of soil, each of a random sample of five cores was thoroughly mixed and divided into four parts. The cation-exchange capacity (m eq./100 g) was measured for each part, with the following results:

Core	Sample 1	Sample 2	Sample 3	Sample 4	Total
1	19.0	18.6	20.1	19.6	77.3
2	17.8	18.5	21.0	17.9	75.2
3	19.3	17.9	19.5	*	56.7
4	20.3	22.1	*	*	42.4
5	18.8	17.9	20.1	18.6	75.4
Total					327.0

An asterisk denotes data that are missing on account of problems encountered during cation-exchange analysis.

a   Write a model for the data and interpret the parameters in your model.

b   In terms of the parameters of the model, formulate the null hypothesis that the mean cation-exchange capacity of the soil in the region is more than 18.00 m eq./100 g.

14.4   Assume that Y_{ij} satisfies the one-way random-effects model in Box 14.1. Use the results in Box 9.1 to show that the mean (expected value) and variance of Y_{ij} are, respectively, μ and $\sigma_T^2 + \sigma^2$. [Hint: T_i and E_{ij} can be regarded as independent random samples of size $n = 1$ from populations with zero means and variances σ_T^2 and σ^2.]

14.3
ANOVA for one-way random-effects models

It turns out that the calculations needed to test the null hypothesis H_0: $\sigma_T^2 = 0$ in a one-way random-effects model are exactly the same as the calculations to test H_0: $\tau_1 = \tau_2 = \cdots = \tau_t$ in a one-way fixed-effects model.

EXAMPLE 14.3   Suppose that the protein content data in Example 14.1 satisfy the assumptions for a one-way random effects model in Box 14.1. Do these data indicate a significant batch-to-batch variation in the protein content of buttermilk cartons?

Let σ_T^2 denote the true between-batch variance of the percentage protein contents of cartons produced in the dairy farm. We want to test the null hypothesis H_0: $\sigma_T^2 = 0$ against the research hypothesis H_1: $\sigma_T^2 > 0$. As we'll see, H_0 can be tested by means of the one-way ANOVA F-test that is applied to the hypothesis of no treatment differences in a one-way fixed-effects model. When the ANOVA procedure of SAS is used to analyze the data, the portion of the SAS output that contains the ANOVA table is as follows.

Dependent Variable: PROTEIN

Source	DF	Sum of Squares	Mean Square	F Value	Pr > 4
Model	3	0.1910333	0.0636778	3.91	0.0545
Error	8	0.1301333	0.0162667		
Corrected Total	11	0.3211667			

Source	DF	Anova SS	Mean Square	F Value	Pr > 4
BATCH	3	0.1910333	0.0636778	3.91	0.0545

As we see from the SAS output, the p-value associated with $F_c = 3.91$ is 0.0545. Thus, the null hypothesis $\sigma_T^2 = 0$ can be rejected in favor of the research hypothesis $\sigma_T^2 > 0$ at a level close to 0.05. Therefore, on the basis of the observed data, we conclude ($p \approx 0.05$) that there is a demonstrable batch-to-batch variability in the protein content of the buttermilk cartons produced at the dairy farm.   ∎

It may seem puzzling that exactly the same procedure can be used to test the null hypothesis of no treatment effects in both random-effects and fixed-effects models. It turns out that this is only true under some special circumstances. One such case is when we are dealing with one-way ANOVA models. Even in this case, the procedures for data analysis—beyond testing for differences between the treatment effects—are different for the two models. This is because, in general, the typical research questions associated with a random-effects model are not the same as those that arise in the case of fixed-effects models.

Expected mean squares for one-way ANOVA models

Recall that the mean of the sampling distribution of a statistic is called its *expected value*. Thus, according to the definition of an unbiased estimate in Section 3.8, a statistic is always an unbiased estimator of its expected value. In particular, MS[T] and MS[E] in a one-way ANOVA table are the unbiased estimators of their expected values. Because the sampling distributions of MS[T] and MS[E] will depend on the assumptions about the data used to calculate these quantities, their expected values have to be determined on the basis of assumptions about the underlying model. In Box 14.2, we give the *expected mean squares* (EMS) for one-way random- and fixed-effects models; for theoretical derivations of these EMS values, see Example B.4 (Appendix B).

BOX 14.2

Expected mean squares in one-way ANOVA models

Fixed effects

Let Y_{ij} denote the j-th response for the i-th treatment and assume the model

$$Y_{ij} = \mu + \tau_i + E_{ij}, \qquad j = 1, 2 \ldots, n_i; i = 1, 2, \ldots t.$$

Here μ, τ_i, and E_{ij} are as in Box 8.2. Then the expected values of MS[E] and MS[T] are

$$\text{EMS[E]} = \sigma^2,$$

$$\text{EMS[T]} = \sigma^2 + n_0 \psi_T^2,$$

where

$$n_0 = \frac{1}{t-1}\left(N - \frac{\sum n_i^2}{N}\right),$$

$$\psi_T^2 = \frac{1}{n_0(t-1)}\sum_{i=1}^{t} n_i(\mu_i - \mu)^2 = \frac{1}{n_0(t-1)}\sum_{i=1}^{t} n_i \tau_i^2.$$

Random effects

Assume the model

$$Y_{ij} = \mu + T_i + E_{ij} \qquad j = 1, 2, \ldots, n_i; i = 1, 2, \ldots t.$$

Here, μ, T_i, and E_{ij} are as in Box 14.1. Then the expected values of MS[E] and MS[T] are

$$\text{EMS[E]} = \sigma^2,$$

$$\text{EMS[T]} = \sigma^2 + n_0 \sigma_T^2.$$

Comments on the EMS values in Box 14.2

1   If each treatment is replicated the same number of times—say, n—then $N = nt$ and $n_0 = n$. For example, if $t = 4$ and $n_1 = n_2 = n_3 = n_4 = 3$ (as in Example 14.3), then $N = 3 \times 4 = 12$ and

$$n_0 = \frac{1}{4 - 1} \left(12 - \frac{3^2 + 3^2 + 3^2 + 3^2}{12} \right) = 3.$$

2   If all the n_i are equal, then

$$\psi_T^2 = \frac{1}{t - 1} \sum \tau_i^2 = \frac{1}{df_T} \sum \tau_i^2, \tag{14.3}$$

where df_T denotes the degrees of freedom for MS[T]. The noncentrality parameter λ in Equation (9.32) can be expressed in terms of ψ_T^2 as

$$\lambda = \frac{n(t - 1)\psi_T^2}{2\sigma^2}.$$

3   Regardless of whether the n_i are equal or not, ψ_T^2 will be zero if and only if every τ_i is zero; that is, ψ_T^2 will be zero if and only if the null hypothesis H_0: $\tau_1 = \tau_2 = \cdots = \tau_t = 0$ is true. Otherwise, ψ_T^2 will be greater than zero. Thus, in a fixed effects model, testing the null hypothesis that there is no difference between the treatment effects against the research hypothesis that the treatment effects are different is same as testing H_0: $\psi_T^2 = 0$ against H_1: $\psi_T^2 > 0$. Compare this with the problem of testing H_0: $\sigma_T^2 = 0$ against H_1: $\sigma_T^2 > 0$ in the random-effects model.

4   The expressions for the EMS values are similar for the two models. The EMS values for the random-effects model can be obtained by replacing ψ_T^2 with σ_T^2 in the expression for EMS[T] for the fixed-effects model. Indeed, the interpretation of ψ_T^2 in a one-way fixed-effects model is similar to the interpretation of σ_T^2 in the one-way random-effects model. Whereas ψ_T^2 is a measure of the variability among the effects of the t treatments for which responses will be measured in the study, σ_T^2 is a measure of the variability among the effects of all treatments, including those for which responses will not be measured in the study.

5   The error mean square MS[E] is an unbiased estimate of the error variance σ^2 in both models.

6   If the null hypothesis H_0 that there is no difference between the treatment effects is true, then the treatment mean square MS[T] is an unbiased estimate of σ^2

under both models. If H_0 is not true, then MS[T] will be an unbiased estimate of different quantities under the two models. However, both of these quantities are larger than σ^2. Therefore, in both models, increase in the difference between MS[T] and MS[E] increases the likelihood that the treatment effects are different. Thus, in both models, large values of the ANOVA F-statistic support the rejection of H_0. In Section 8.3, we provided an intuitive justification of this conclusion for the fixed-effects model.

EMS values and estimation of variance components

The EMS values in Box 14.2 may be used to estimate the variance components σ^2 and σ_T^2.

Let $\hat{\sigma}_T^2$ and $\hat{\sigma}^2$ denote, respectively, the estimates of σ_T^2 and σ^2. The results in Box 14.2 imply that MS[T] and MS[E] are unbiased estimates of $\sigma_T^2 + n_0\sigma^2$ and σ^2, respectively, and so it is reasonable to require that the estimates of the variance components satisfy the estimating equations

$$\hat{\sigma}^2 + n_0\hat{\sigma}_T^2 = \text{MS[T]},$$

$$\hat{\sigma}^2 = \text{MS[E]}. \qquad \textbf{(14.4)}$$

Solving Equation (14.4), we get the estimates of variance components in a one-way random-effects model as

$$\hat{\sigma}^2 = \text{MS[E]} \quad \text{and} \quad \hat{\sigma}_T^2 = \frac{\text{MS[T]} - \text{MS[E]}}{n_0}. \qquad \textbf{(14.5)}$$

Occasionally, it might happen that MS[T] is less than MS[E], and so a negative value is obtained for $\hat{\sigma}_T^2$. In that case, we let $\hat{\sigma}_T^2 = 0$.

EXAMPLE 14.4 In a crop survey, 20 farms were randomly selected from a large number of farms under the crop. Three 1-acre plots were randomly selected from each of the 20 farms and yields (bushels per acre) were measured for each farm. The following statistics were calculated from the crop yield data

$$\bar{y}_{++} = 689.23,$$

$$\text{MS[T]} = 4297.77,$$

$$\text{MS[E]} = 153.33.$$

The crop yield data can be modeled using a one-way random-effects model with $t = 20$ randomly selected treatments (farms) and $n = 3$ replications per treatment. Let

$$Y_{ij} = \mu + T_i + E_{ij}, \quad j = 1, 2, 3; \quad i = 1, 2, \ldots, 20.$$

Here Y_{ij} is the yield from the j-th plot in the i-th farm; μ is the overall mean yield (the true mean yield for the population of all farms); T_i is the effect of the i-th farm;

and E_{ij} is the random quantity that accounts for the difference between yields from different plots within a single farm. The E_{ij} are independent $N(0, \sigma^2)$. In addition, the T_i and E_{ij} are independent.

In this model, the variance components σ_T^2 and σ^2 measure, respectively, the farm-to-farm and plot-to-plot variation in yields. The values of σ^2 and σ_T^2 can be estimated using Equation (14.4) or Equation (14.5). The ANOVA table for the crop yield data is as follows:

Source	df	MS	EMS
Farms (treatments)	19	4297.77	$\sigma^2 + 3\sigma_T^2$
Plots in farms (error)	40	153.33	σ^2
Total	59		

For these data, each treatment has $n = 3$ replications, and so we have $n_0 = 3$ in Box 14.2.

Each mean square in the ANOVA table is an unbiased estimate of the corresponding EMS value. Therefore, an unbiased estimate of σ^2 is $\hat{\sigma}^2 = 153.33$. Similarly, the mean square for farms MS[T] is an unbiased estimate of $\sigma^2 + 3\sigma_T^2$, so that an estimate for σ_T^2 can be obtained by solving Equation (14.4)

$$\hat{\sigma}_T^2 = \frac{4297.77 - 153.33}{3} = 1381.48. \quad \blacksquare$$

When interpreting the variance components σ^2 and σ_T^2, we need to remember that they are population variances. Thus, smaller values of σ_T^2 and σ^2 indicate less variability in the measurements in the corresponding populations. In Example 14.4, a small value of σ^2 indicates that the plot-to-plot variability within farms is small. Similarly, a small value of σ_T^2 implies that the variability of the mean yields of the farms is small. Of course, the criterion for a small value will vary from one problem to another. In this case, it appears that farm-to-farm variation is the major cause of the variation between yields.

The coefficient of variation

Often, variances are most meaningful when expressed in relative terms. We have already encountered the coefficient of variation (CV), which is the value of the standard deviation expressed in units of the mean. For a one-way random-effects model, the coefficient of variation of the observed responses is

$$\text{CV} = \frac{(\text{standard deviation of } Y_{ij})}{|\mu|} = \frac{\sqrt{\sigma_T^2 + \sigma^2}}{|\mu|}.$$

An estimate of CV can be calculated as

$$\widehat{CV} = \frac{\sqrt{\hat{\sigma}_T^2 + \hat{\sigma}^2}}{|\hat{\mu}|},$$

where $\hat{\sigma}_T^2$ and $\hat{\sigma}^2$ are estimated variance components and $\hat{\mu}$ is an estimate of the overall mean. For the crop yield data, $\hat{\sigma}^2 + \hat{\sigma}_T^2 = 153.33 + 1381.48 = 1534.81$ is an estimate of the variance of the yields in the region. As we'll see, $\overline{Y}_{++}$, the mean yield from the sampled plots, is an unbiased estimate of μ. Thus, the estimated μ for the crop yield data is $\bar{y}_{++} = 689.23$ bushels per acre. An estimate of the coefficient of variation is given by

$$\widehat{CV} = \frac{\sqrt{1531.81}}{689.23} = 0.057.$$

It follows that the standard deviation of the crop yields is about 5.7% of the true mean yield. We can conclude that the variability of yields in these farms is fairly small. Note that our value of 5.7% for CV is an estimate based on a random sample of farms and plots. To assess the reliability of $\widehat{CV}$, we need a standard error for this estimate. Under suitable assumptions (large sample sizes, normal distributions, and so on), it is possible to derive a formula for computing an approximate standard error for $\widehat{CV}$. The details fall outside the scope of the present work.

The intraclass correlation coefficient

In a one-way random-effects model, the intraclass correlation between observations within treatments is defined as

$$\rho_I = \frac{\sigma_T^2}{\sigma^2 + \sigma_T^2}.$$

The parameter ρ_I is the proportion of the variability of the observed responses that can be ascribed to the variability of the treatment effects. The value of ρ_I never exceeds 1 and is never less than 0; that is, $0 \le \rho_I \le 1$. A large value of ρ_I implies that σ_T^2 is large compared to σ^2, which, in turn, implies that responses for the same level of the experimental factor are less variable than the responses at different levels. Thus, ρ_I is a measure of the association (correlation) between responses for the same treatment (the same level of the experimental factor).

The variance component estimates calculated from the crop yield data can be used to compute an estimate of the intraclass correlation coefficient between yields from a single farm

$$\hat{\rho}_I = \frac{\hat{\sigma}_T^2}{\hat{\sigma}_T^2 + \hat{\sigma}^2} = \frac{1381.48}{1381.48 + 153.33} = 0.90.$$

Thus, we estimate that 90% of the variation in crop yields is due to farm-to-farm variability.

A procedure for computing confidence intervals for ρ_I in the important special case where each treatment has the same number of replications is given in Box 14.3.

BOX **14.3**

Confidence interval for ρ_I

Let Y_{ij} denote the j-th response for the i-th treatment ($j = 1, 2, \ldots, n$; $i = 1, 2, \ldots, t$). Assume that the Y_{ij} satisfy the one-way random-effects model in Box 14.1. Let

$$F_c = \frac{MS[T]}{MS[E]} \quad \text{and} \quad F_\alpha = F(t-1, t(n-1), \alpha).$$

Then a $100(1-\alpha)\%$ confidence interval for ρ_I is

$$\frac{F_c - F_{\alpha/2}}{F_c + (n-1)F_{\alpha/2}} \leq \rho_I \leq \frac{F_c - F_{1-\alpha/2}}{F_c + (n-1)F_{1-\alpha/2}}.$$

The next example demonstrates the computation of a confidence interval for the intraclass correlation coefficient.

EXAMPLE **14.5**   Let's use the crop yield data in Example 14.4 to construct a 95% confidence interval for ρ_I, the coefficient of intraclass correlation between crop yields from the same farm.

In Example 14.4, we computed the estimates $\hat{\sigma}_T^2 = 1381.48$, $\hat{\sigma}^2 = 153.33$. Also, from the ANOVA table in Example 14.4, we see that

$$F_c = \frac{4297.27}{153.33} = 28.03.$$

For the crop yield data, $t = 20$ and $n = 3$. Therefore, $t - 1 = 19$, and $t(n-1) = 40$. Also, on the basis of Table C.4 and Equation (3.8), we have for $\alpha = 0.05$

$$F_{\alpha/2} = F_{0.025} = 2.07 \quad \text{and} \quad F_{1-\alpha/2} = F_{0.975} = 0.43.$$

Therefore, the desired confidence interval for ρ_I can be constructed as in Box 14.3

$$\frac{28.03 - 2.07}{28.03 + 2(2.07)} \leq \rho_I \leq \frac{28.03 - 0.43}{28.03 + 2(0.43)},$$

which yields the interval

$$0.81 \leq \rho_I \leq 0.96.$$

Thus, we can conclude, with 95% confidence, that between 81% and 96% of the variation in crop yields is due to the variability of the mean yields for the individual farms. ■

Estimation of the overall mean

We have already seen why making inferences about the overall mean in a one-way random-effects model might be of interest. In Box 14.4, a procedure is given for making inferences about the overall mean μ in a one-way random effects model using data with equal replications per treatment.

BOX 14.4

Inferences about μ in a one-way random-effects model

Assume that Y_{ij} $(j = 1, 2, \ldots, n;\ i = 1, 2, \ldots, t)$ satisfies the one-way random-effects model in Box 14.1.

Confidence interval

A $100(1 - \alpha)\%$ confidence interval for the overall mean μ is given by

$$\overline{Y}_{++} - t(t - 1, \alpha/2)\sqrt{\frac{MS[T]}{nt}} \leq \mu \leq \overline{Y}_{++} + t(t - 1, \alpha/2)\sqrt{\frac{MS[T]}{nt}},$$

where $\overline{Y}_{++}$ and MS[T] are the overall sample mean and the treatment mean square, respectively.

Hypothesis test

The null hypothesis H_0: $\mu = \mu_0$ can be tested using the test statistic

$$t_c = \frac{\overline{Y}_{++} - \mu_0}{\sqrt{MS[T]/(nt)}}.$$

The calculated value t_c should be compared to the appropriate critical value of a t-distribution with $t - 1$ degrees of freedom.

The next example illustrates the construction and interpretation of the confidence interval described in Box 14.4.

EXAMPLE 14.6   Let's use the crop yield data in Example 14.4 to construct a 90% confidence interval for μ, the overall mean crop yield.

From Example 14.4, we have $t = 20$, $n = 3$, $\bar{y}_{++} = 689.23$, and MS[T] $= 4297.77$. Also, from Table C.2, $t(19, 0.05) = 1.729$. Therefore, a 90% confidence interval for μ is

$$689.23 - (1.729)\sqrt{\frac{4297.77}{(3)(20)}} \leq \mu \leq 689.23 + (1.729)\sqrt{\frac{4297.77}{(3)(20)}},$$

so that

$$674.60 \leq \mu \leq 703.86.$$

Thus, we have 90% confidence that the true mean yield for all farms under the crop is somewhere between 674.60 and 703.86 bushels per acre.   ∎

Note that the confidence interval in Box 14.4 may be written in the general form

$$\hat{\theta} - t(\nu, \alpha/2)\hat{\sigma}_{\hat{\theta}} \leq \theta \leq \hat{\theta} + t(\nu, \alpha/2)\hat{\sigma}_{\hat{\theta}}, \qquad \textbf{(14.6)}$$

where $\nu = t - 1$, $\hat{\theta} = \overline{Y}_{++}$, and

$$\hat{\sigma}_{\hat{\theta}} = \sqrt{\frac{\text{MS[T]}}{nt}}.$$

The structure of the interval in Equation (14.6) is identical to the structure of the confidence interval in Box 9.2 for linear functions of the means of normally distributed populations. Of course, the assumption of independent means and a common population variance in Box 9.2 implies a one-way fixed-effects model, whereas the procedure in Box 14.4 is applicable under the assumption of a one-way random-effects model. Using techniques beyond the scope of this book, it can be shown that, under some rather general conditions, the inferential procedures in Box 9.2 can be used for models other than the one-way fixed-effects model. The procedures in Box 14.5 can be used for inferences about population means under a set of conditions that are less restrictive than those in Box 9.2. As we'll see, the procedures in Box 14.4 can be derived from those in Box 14.5. Furthermore, the procedures in Box 14.5 will be useful in subsequent chapters, where we encounter situations in which inferences about population means under more complex random-effects models are needed.

BOX **14.5**

Inferences about linear functions of population means: unequal population variances

Let $\overline{Y}_1, \overline{Y}_2, \ldots, \overline{Y}_t$ denote the means of independent random samples from normal populations with means $\mu_1, \mu_2, \ldots, \mu_t$ and variances $\sigma_1^2, \sigma_2^2, \ldots, \sigma_t^2$. Let $c_0, c_1, \ldots, c_t$ be known constants, and consider the linear function

$$\hat{\theta} = c_0 + c_1\overline{Y}_1 + \cdots + c_t\overline{Y}_t.$$

Then $\hat{\theta}$ is an unbiased estimate of

$$\theta = c_0 + c_1\mu_1 + \cdots + c_t\mu_t.$$

The variance of $\hat{\theta}$ is

$$\sigma_{\hat{\theta}}^2 = \frac{c_1^2\sigma_1^2}{n_1} + \frac{c_2^2\sigma_2^2}{n_2} + \cdots + \frac{c_t^2\sigma_t^2}{n_t}.$$

If an unbiased estimate of $\sigma_{\hat{\theta}}^2$ can be expressed in terms of a single mean square in an appropriate ANOVA table, the following procedures can be used for inferences about θ.

Confidence interval

A $100(1 - \alpha)\%$ confidence interval for θ is given by

$$\hat{\theta} - t(\nu, \alpha/2)\hat{\sigma}_{\hat{\theta}} \leq \theta \leq \hat{\theta} + t(\nu, \alpha/2)\hat{\sigma}_{\hat{\theta}},$$

where $\hat{\sigma}_{\hat{\theta}}$ is the estimated standard error of $\hat{\theta}$, and ν denotes the degrees of freedom associated with the mean square used to estimate the standard error of $\hat{\theta}$.

Hypothesis test

The null hypothesis H_0: $\theta = \theta_0$ can be tested using the test statistic

$$t_c = \frac{\hat{\theta} - \theta_0}{\hat{\sigma}_{\hat{\theta}}}.$$

The calculated value of the test statistic should be compared to the appropriate critical value of a t-distribution with ν degrees of freedom.

EXAMPLE 14.7 Let's examine how the inferential procedures of Box 14.4 can be derived from the results in Box 14.5.

Under the one-way random-effects model with n replications per treatment, the observed response has the representation

$$Y_{ij} = \mu + T_i + E_{ij}, \qquad i = 1, \ldots, t, j = 1, \ldots, n.$$

By equating the average of nt values on the two sides of this set of nt equations, the overall mean response $\overline{Y}_{++}$ can be expressed as

$$\overline{Y}_{++} = \mu + \overline{T}_{+} + \overline{E}_{++},$$

where $\overline{T}_{+}$ and $\overline{E}_{++}$ denote, respectively, the mean of a random sample of t treatment effects and the mean of a random sample of nt error effects. Under the assumptions of a one-way random-effects model, the T_i and E_{ij} are independent and normally distributed, and so the quantity $\mu + \overline{T}_{+} + \overline{E}_{++}$ is a linear function of the means of $k = 2$ independent random samples from normally distributed populations. The coefficients in this linear function are $c_0 = \mu$, $c_1 = 1$, and $c_2 = 1$.

The populations of the T_i and E_{ij} have zero means, and so we conclude from Box 14.5 that $\overline{Y}_{++}$ is an unbiased estimate of $\theta = \mu + (1)(0) + (1)(0) = \mu$.

Now $\overline{T}_{+}$ and $\overline{E}_{++}$ are the means of independent samples—of respective sizes $n_1 = t$ and $n_2 = nt$—from populations with respective variances $\sigma_1^2 = \sigma_T^2$ and $\sigma_2^2 = \sigma^2$, and so the results in Box 14.5 imply that

$$\sigma_{\overline{Y}_{++}}^2 = \frac{(1)^2\sigma_T^2}{t} + \frac{(1)^2\sigma^2}{nt} = \frac{\sigma^2 + n\sigma_T^2}{nt}. \tag{14.7}$$

Finally, an examination of the EMS values in the one-way random-effects model shows that the numerator on the right-hand side of this expression is the same as the expected mean square for treatments. In other words, MS[T] is an unbiased estimate of $\sigma^2 + n\sigma_T^2$, so that the standard error of $\overline{Y}_{++}$ can be estimated by

$$\hat{\sigma}_{\overline{Y}_{++}} = \sqrt{\frac{MS[T]}{nt}}.$$

The degrees of freedom associated with the critical value are $\nu = t - 1$; that is, they are equal to the degrees of freedom associated with the mean square used to estimate the standard error of $\overline{Y}_{++}$. ∎

Exercises

14.5 Assume that the one-way random-effects model in Box 14.1 is appropriate for the protein content data in Exercise 14.1.

a Construct the ANOVA table showing the mean squares and their expected values.

b On the basis of the ANOVA table constructed in (a), would it be reasonable to conclude that there is a significant plant-to-plant variation in the protein content of the seeds produced by the F_3 generation plants?

c Estimate the components of variance, and comment on their relative magnitudes.

d Calculate an estimate of the coefficient of variation of the protein content measurements, and comment on its value.

e Construct a 95% confidence interval for the proportion of the variance of the measured protein content values that can be ascribed to plant-to-plant variation. Interpret this interval.

f On the basis of the observed data, would it be reasonable to conclude that the average protein content of seeds produced in the F_3 generation is more than 40%?

14.6 Refer to the serum cholesterol study in Exercise 14.2.

a Assuming a one-way random-effects model, estimate the variance components σ_S^2 and σ^2. Comment on the relative magnitudes of σ_S^2 and σ^2.

b Construct a 95% lower bound for the coefficient of intraclass correlation between measured cholesterol values for the same subject. On the basis of that bound, what can you say about the subject-to-subject variation in the serum cholesterol levels?

c Obtain a 99% confidence interval for the true mean serum cholesterol level for all subjects in the region. Interpret the interval.

14.7 Assume that a one-way random-effects model is appropriate for the cation-exchange capacity data in Exercise 14.3.

a Perform an analysis of variance, and test the null hypothesis that cation-exchange capacity does not vary over the region.

b Estimate the components of variance due to differences between the cores and differences between parts of the same core. Compare the relative magnitudes of the variance components and comment.

c Estimate the mean cation-exchange capacity for the region.

d Determine the form of an unbiased estimate of the standard error of $\overline{Y}_{++}$, the mean measured cation-exchange capacity. Estimate this standard error using the estimates of the variance components.

e Construct a 95% upper bound for the mean cation-exchange capacity for the region. Interpret the bound you calculate.

14.8 To assess the average energy value of oranges in a grove, a random sample of t trees was selected from the grove, and a random sample of n oranges was selected from each tree. Suppose that the following ANOVA table resulted from the energy value (kcal) measurements on the sampled oranges.

Source	df	SS	MS	EMS
Trees	19	1282		
Oranges	60	625		
Total	79	1907		

a Fill in the MS and EMS columns in the above ANOVA table.

b Do these data indicate that the energy values of the oranges vary from tree to tree?

c Define a parameter that assesses the variation of the energy values of oranges within trees relative to the variation of energy values between trees.

d Use the data to estimate the value of the parameter defined in (c).

e Construct a 95% confidence interval for the parameter estimated in (d).

f Suppose that the average energy value of all sampled oranges was 26.8 kcal. On the basis of this information and the ANOVA table, would it be reasonable to conclude that the average energy value for oranges in the region does not exceed 27.5 kcal?

14.4
Determining sample sizes

There are important differences between determining sample sizes in one-way random-effects models and determining sample sizes in one-way fixed-effects models. In the one-way fixed-effects model, only one sample size—the number of replications per treatment—is determined. In the one-way random-effects model, there are two sample sizes: the number of treatments and the number of replications per treatment. In fixed-effects models, sample sizes are usually determined in order to control the reliability of inferences about population means. In random-effects models, sample sizes are determined with a view to making inferences either about the overall mean μ or about a function of the variance components σ^2 and σ_T^2. In this section, we examine three methods of determining sample sizes for one-way random-effects models:

1 the hypothesis-testing approach for inferences about variance components;

2 the confidence-interval approach for inferences about μ;

3 the cost-function approach for inferences about μ.

In each case, we assume that the objective is to plan a study with t treatments and n replications per treatment in which a one-way random-effects model will be used.

Hypothesis-testing approach

Suppose we want to determine n and t such that the ANOVA F-test of H_0: $\sigma_T^2 = 0$ against H_1: $\sigma_T^2 > 0$ has a prescribed power if the actual value of σ_T^2 is substantially higher than zero. One way in which such a value could be specified is to select a constant δ such that any value of σ_T^2 that exceeds $\delta\sigma^2$ is considered substantially higher than zero. How do we specify δ in practice? Consider Example 14.4 where yields of a crop were measured on a random sample of $n = 3$ plots selected from each of a random sample of $t = 4$ farms. If σ_T^2 and σ^2 denote, respectively, the variance of the mean yields of the farms and variance of yields of plots within farms, then we might think that it's reasonable to choose the criterion $\sigma_T^2 > 3\sigma^2$, which means that the farm-to-farm variability is at least three times as high as the variability within any single farm.

The required sample sizes should guarantee that, if $\sigma_T^2 > \delta\sigma^2$, then the power of the α-level test of H_0 is at least $1 - \beta$. The sample sizes satisfying this requirement can be determined by a trial-and-error method in which we use the result in Box 14.6 to calculate the power of the one-way ANOVA F-test for various combinations of t and n, until we arrive at a combination for which the power is at least $1 - \beta$.

BOX **14.6**

> ### The power of the ANOVA F-test for one-way random-effects models
>
> Consider a study using a one-way random-effects model with t treatments and n replications per treatment and suppose that we wish to test the null hypothesis H_0: $\sigma_T^2 = 0$, using the ANOVA F-test with the test statistic
>
> $$F_c = \frac{MS[T]}{MS[E]}.$$
>
> Let $\delta \geq 0$ be a given number. If $\sigma_T^2 = \delta\sigma^2$, then the probability that the α-level test based on F_c will reject H_0 will equal the area to the right of $(1 + n\delta)^{-1}F(t - 1, t(n - 1), \alpha)$ under an F-distribution with $t - 1$ degrees of freedom for the numerator and $t(n - 1)$ degrees of freedom for the denominator.

The result in Box 14.6 states that the power of the test of H_0: $\sigma_T^2 = 0$ equals an appropriate right-tail area under an F-distribution. The power is shown as an area under a F-distribution in Figure 14.1. For given values of n, t, α, and δ, the value of $1 - \beta$ can be determined by determining the shaded area in Figure 14.1. Table C.4 (Appendix C) gives tail areas for a number of F-distributions, but unfortunately it

FIGURE   14.1

The power of an α-level F-test of H_0: $\sigma_T^2 = 0$

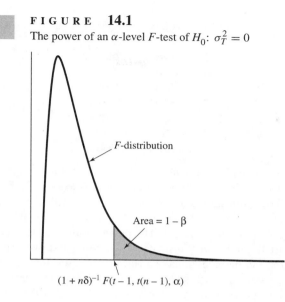

F-distribution

Area = $1 - \beta$

$(1 + n\delta)^{-1} F(t - 1, t(n - 1), \alpha)$

is generally not of much help in determining $1 - \beta$. For exact calculation of $1 - \beta$, we need one of the many computer packages capable of determining areas under F-distributions. Two functions FINV and PROBF in the SAS software can be used to determine areas under F-distributions; Section 2.3 in the companion text by Younger (1997) gives details on how to use these functions.

EXAMPLE   14.8   The carbon content of soil is a measure of the amount of organic matter present in the soil. Suppose we want to assess the variability of carbon content in a particular type of soil over a large region. We plan to select a random sample of t sites in the region and then determine the carbon contents (ppm) of n randomly selected core samples in each region. Let σ_T^2 and σ^2 denote, respectively, the variance of the true mean carbon contents of the sites in the region and the variance of the carbon contents of the core samples within a site; that is, σ_T^2 and σ^2 are, respectively, the between-site and within-site variances of the carbon contents in the region. If $\sigma_T^2 = 0$, we conclude that the organic matter is evenly distributed across the region. With increase in σ_T^2, the variability of the organic matter content across sites increases. For instance, if $\sigma_T^2 = 3\sigma^2$, then

$$\rho_I = \frac{\sigma_T^2}{\sigma^2 + \sigma_T^2} = \frac{3\sigma^2}{\sigma^2 + 3\sigma^2} = \frac{3\sigma^2}{4\sigma^2} = 0.75, \qquad \textbf{(14.8)}$$

so that 75% of the variation in soil carbon content is due to site-to-site variability. Let's determine t and n such that, if $\sigma_T^2 = 3\sigma^2$, then a 0.05-level ANOVA F-test will have at least 80% chance of rejecting H_0: $\sigma_T^2 = 0$.

We have $\alpha = 0.05$, and $\delta = 3$. We want to find t and n such that the power is at least 0.80. Using the FINV and PROBF functions in SAS, we obtain the following values for the power of the F-test with various combinations of t and n.

T	N	NDF	DDF	FALPHA	FDELTA	POWER
2	2	1	2	18.5128	2.64469	0.24541
2	3	1	4	7.7086	0.77086	0.42952
3	2	2	3	9.5521	1.36458	0.37892
3	3	2	6	5.1433	0.51433	0.62207
3	4	2	9	4.2565	0.32742	0.72902
4	3	3	8	4.0662	0.40662	0.75253
4	4	3	12	3.4903	0.26848	0.84688

For example, when $t = 2$ and $n = 2$, the numerator degrees of freedom (NDF) are $t - 1 = 1$; the denominator degrees of freedom (DDF) are $t(n - 1) = 2(2 - 1) = 2$. The corresponding critical value (FALPHA) is $F(1, 2, 0.05) = 18.5128$, so that $(1 + n\delta)^{-1} F(t - 1, t(n - 1), \alpha) = 2.64469$; this quantity is denoted by FDELTA in the SAS output. The power of the test is equal to the area to the right of FDELTA under an F-distribution with one and two degrees of freedom for the numerator and the denominator respectively. This area is equal to 0.24541, so that the sample sizes of $t = 2$ and $n = 2$ are inadequate for our purpose.

On the basis of these power calculations, it follows that the sample sizes $t = 4$ and $n = 4$ will be adequate for our purpose. Of course, other combinations of t and n that meet the power requirement might be more convenient from a practical standpoint. For instance, we might want to keep the number of sites t as small as possible. In that case, we should try values such as $(t, n) = (2, 5)$, $(2, 6)$, and so on. ■

Confidence interval approach

The procedure in Box 14.7 allows us to determine sample sizes such that the expected width of a $100(1 - \alpha)\%$ confidence interval for μ is less than a specified value $W = 2B$. The justification for this procedure resembles that for the sample size procedure in Box 7.4.

BOX **14.7**

Sample sizes needed to attain a specified bound on the width of the confidence interval for μ

Suppose we use the procedure in Box 14.4 to construct a $100(1 - \alpha)\%$ confidence interval for μ, the overall mean in a one-way random-effects model. Then, for any $0 < \alpha < 1$ and $L > 0$, the values of t and n that satisfy the inequality

$$t \geq F(1, t - 1, \alpha) \left(\frac{\sigma_T^2}{B^2} + \frac{1}{n} \frac{\sigma^2}{B^2} \right)$$

are sample sizes such that the expected width of the confidence interval will be less than $2B$. In practice, σ^2 and σ_T^2 are usually unknown and are estimated on the basis of past experience or other available information.

A trial-and-error strategy can be used to implement the procedure in Box 14.7, as in the next example.

EXAMPLE **14.9**   In Example 14.8, we used the hypothesis testing approach to determine the sample sizes needed to study the carbon content of soil in a region. Let's see how the estimation approach can be used to determine sample sizes for the same study.

Assume that, on the basis of a pilot study or by some other means, we have estimated the between-sites and within-site variances of the soil carbon contents as approximately $\sigma_T^2 = 456$ (ppm)^2 and $\sigma^2 = 200$ (ppm)^2. Suppose also that we want to use the study results to estimate the overall mean carbon content within 25 ppm with 95% confidence; that is, we would like to be 95% confident that the mean carbon content can be estimated with a margin of error less than $B = 25$ ppm.

We can use the procedure in Box 14.7, with $\alpha = 0.05$, $B = 25$, $\sigma_T^2 = 456$, and $\sigma^2 = 200$. We have,

$$\frac{\sigma_T^2}{B^2} = \frac{456}{(25)^2} = 0.73,$$

$$\frac{\sigma^2}{B^2} = \frac{200}{(25)^2} = 0.32,$$

so that the sample sizes t and n should satisfy the inequality

$$t \geq F(1, t-1, 0.05)\left(0.73 + \frac{0.32}{n}\right). \tag{14.9}$$

When $t = 2$ and $n = 2$, the right-hand side of Equation (14.9) equals $(161.44)(0.89) = 143.69$. Since 2 is less than 143.69, the sample sizes $(t, n) = (2, 2)$ do not satisfy Equation (14.9). Therefore, we must try other values for t and n. The following table shows the right-hand side of Equation (14.9) for some other combinations of t and n:

t	n			
	2	4	6	8
2	143.69	130.77	126.46	124.31
3	16.48	14.99	14.50	14.25
4	9.01	8.20	7.93	7.80
5	6.86	6.24	6.04	5.94
6	5.88	5.35	5.18	5.09

An examination of this table shows that any of the combinations $(t, n) = (6, 2)$, $(t, n) = (6, 4)$, or $(t, n) = (6, 6)$, among others, will meet the requirements for the sample size. Of course, there are other acceptable combinations of t and n but, as we can see, those combinations will require a large number of core samples per site. Our choice of sample size will depend on considerations such as cost and convenience.   ■

Cost-function approach

In the cost-function approach, the sample sizes are calculated by first determining pairs (n, t) that satisfy given cost constraints and then selecting a pair on the basis of requirements on the standard error of the estimate of μ. To illustrate such a procedure, we'll assume a simple cost constraint that is frequently used.

Let C_0 and C_1 denote, respectively, the cost per treatment (for instance, the cost of sampling a farm) and the cost per replication of a treatment (for instance, the cost of measuring the yield of a single plot). Then the total cost of a study that includes t treatments and n replications per treatment is

$$C = C_0 t + C_1 n t. \tag{14.10}$$

We want to determine t and n such that the cost of the study is held fixed at a given value C, and the standard error of $\overline{Y}_{++}$, which is the estimate of the overall mean, is as small as possible. Using some straightforward calculus, we can show that the values

$$n = \sqrt{\left(\frac{C_0}{C_1}\right)\left(\frac{\sigma^2}{\sigma_T^2}\right)},$$

$$t = \frac{C}{C_0 + nC_1} \tag{14.11}$$

satisfy Equation (14.10) and minimize the standard error of $\overline{Y}_{++}$. In Equation (14.11), n depends on the ratio $\gamma = \sigma^2/\sigma_T^2$, which is typically unknown. In practice, estimates of γ are usually based on preliminary studies or a reasonable guess. The next example illustrates the calculations involved in determining the sample size on the basis of Equation (14.11).

EXAMPLE 14.10 Let's use Equation (14.11) to determine the sample size for the crop yield data in Example 14.4. Suppose that the overhead cost of sampling a farm is $100, and the cost of measuring the yield of a plot in a farm is $5. If a total of $5000 is available for the study, determine the number of farms and the number of plots within a farm that should be sampled to minimize the standard error for the estimate of the overall mean yield for the region.

We have $C_0 = 100$, $C_1 = 5$, and $C = 5000$. From the data in Example 14.4, we have the estimates $\hat{\sigma}_T^2 = 1381.48$ and $\hat{\sigma}^2 = 153.33$. Therefore, we take

$$\frac{\sigma^2}{\sigma_T^2} \approx \frac{\hat{\sigma}^2}{\hat{\sigma}_T^2} = \frac{153.33}{1381.48} = 0.11$$

and determine the sample sizes as

$$n = \sqrt{\left(\frac{100}{5}\right)(0.11)} = 1.48 \approx 2,$$

$$t = \frac{5000}{100 + 5(2)} = 45.45 \approx 46.$$

Thus, we should select 46 farms in the region and 2 plots in each farm. The total cost of conducting such a study is $5060. Note that, due to rounding of the sample sizes, the total cost exceeds the available budget by $100. If this is not satisfactory, we can modify the sample sizes to ensure that the actual cost is less than $5000.

On the basis of the results in Example 14.4, a preliminary estimate of the standard error for $\overline{Y}_{++}$, the estimator of μ, in the new study can be calculated using Equation (14.7)

$$\hat{\sigma}_{\overline{Y}_{++}} = \sqrt{\frac{\hat{\sigma}^2 + n\hat{\sigma}_T^2}{nt}} = \sqrt{\frac{153.33 + (2)(1381.48)}{(2)(46)}} = 5.63. \quad \blacksquare$$

Exercises

14.9   Refer to the protein content data in Exercise 14.1.

a   Determine the values for t and n such that a 0.10-level test of the null hypothesis H_0 that the true mean protein content of the seeds is the same for all F_2-generation soybean plants will be rejected with 80% probability if the variability of protein contents between F_2-generations is at least twice as large as the variability within the F_2-generation.

b   Determine t and n such that, with 90% confidence, the average protein content of the seeds produced by F_3 generation soybean plants can be estimated with a margin of error less than 5%.

c   Suppose that it costs $50 and $150, respectively, to include one F_2 generation and one F_3 generation plant in the study. Determine n and t such that the true average protein content of the seeds from F_3-generation plants can be estimated with the highest possible precision—that is, the smallest possible standard error—at a total cost of no more than $2000.

d   Let μ denote the true average protein content of the seeds of F_3 generation soybean plants. Using the data in Exercise 14.1, estimate the standard error of $\overline{Y}_{++}$, the unbiased estimator for μ, in a study that will use the values of n and t determined in (c). What information does this standard error give about the accuracy of $\overline{Y}_{++}$ that can be attained in the study?

14.10   Refer to Exercise 14.2. Suppose we are interested in developing a new sampling design in which n independent determinations of cholesterol levels will be made on each of a random sample of t subjects.

a   Using SAS or otherwise, determine n and t such that, if the between-subjects variance of the measured cholesterol levels is at least 60% of the overall variance, then there is at least 80% probability that a 0.05-level F-test will support the hypothesis of significant variation in the cholesterol levels of the subjects in the study population.

b   Determine n and t such that the halfwidth of a 95% confidence interval for the true average cholesterol level for the population does not exceed 5 mg/dliter.

c   Suppose we conduct a new study using the values of n and t calculated in (b). Using the data from the present study, determine an approximate value for the standard error of $\overline{Y}_{++}$ in the new study.

d   Assume that it costs \$100 to recruit a subject and \$5 to perform one serum analysis. Given that the total budget allocated for the study is \$2000, determine n and t for a new study that will permit the estimation of the average cholesterol level with the minimum possible standard error.

e   For a study that uses the sample sizes determined in (d), what can you say about the error of estimation of the true mean cholesterol level for the population?

14.11   Using the data in Exercise 14.3, calculate the sample sizes for a new cation-exchange study to estimate the average cation-exchange capacity with a standard error of 1 m eq./100 g or less.

14.12   Refer to Exercise 14.8. Let σ_T^2 and σ^2 denote, respectively, the between-trees and between-oranges within-tree variances of the energy values. Suppose it is suspected that there is considerable variation in the energy values of the oranges from one tree to another. The investigators want the study to reject H_0: $\sigma_T^2 = 0$ with 80% power at the 0.05 level if the between-trees variation (σ_T^2) is at least 70% of the overall variation ($\sigma_T^2 + \sigma^2$). Determine the number of trees and the number of oranges per tree that must be sampled to achieve this objective.

14.5
ANOVA for two-way random- and mixed-effects models

Thus far, we have considered one-way ANOVA using models in which the effect of an experimental factor is regarded as random. In this section, we examine how one-way random-effects ANOVA can be extended to two-way ANOVA based on models containing two factors, of which at least one has random effects.

Two-way models

There are six types of two-way ANOVA models.

1   *Fixed-effects models with crossed factors.* In these models, both factors have fixed effects, and the levels of the factors are crossed (Sections 13.4, 13.5, and 13.6).

2   *Fixed-effects models with nested factors.* In these models, both factors have fixed effects, and the levels of one factor are nested in the levels of the other factor (Section 13.7).

3   *Random-effects models with crossed factors.* In these models, both factors have random effects, and their levels are crossed.

4   *Mixed-effects models with crossed factors.* These models contain the effects of two factors, of which one is random and one is fixed. The levels of the factors are crossed.

5   *Random-effects models with nested factors.* These are the same as type 3, except that the levels of one factor are nested in the levels of the other factor.

6   *Mixed-effects models with nested factors.* These are the same as type 4, except that the levels of one factor are nested in the levels of the other factor.

Let's look in more detail at models of types 3–6.

Two-way random-effects model with crossed factors

A two-way random-effects model with crossed factors is of the form

$$Y_{ijk} = \mu + A_i + B_j + (AB)_{ij} + E_{ijk}, \tag{14.12}$$

$$i = 1, \ldots, a; \ j = 1, \ldots, b; \ k = 1, \ldots, n_{ij}.$$

The notation here is as follows:

- Y_{ijk} is the k-th response for the treatment combination $a_i b_j$;
- μ is the overall mean;
- A_i is the random effect of the i-th level of A;
- B_j is the random effect of the j-th level of B;
- $(AB)_{ij}$ is the joint random effect of the i-th level of A and the j-th level of B;
- E_{ijk} is the random error.

The A_i are independent $N(0, \sigma_A^2)$; the B_j are independent $N(0, \sigma_B^2)$; the $(AB)_{ij}$ are independent $N(0, \sigma_{AB}^2)$; and the E_{ijk} are independent $N(0, \sigma^2)$. Also, the A_i, B_j, $(AB)_{ij}$, and E_{ijk} are mutually independent.

The model Equation (14.12) is appropriate for an $a \times b$ experiment provided the experimental levels of the two factors can be regarded as random samples from populations of levels. How does this model compare with the two-way fixed-effects model? Recall that a two-way fixed-effects model can be obtained from a one-way fixed-effects model by representing the treatment effect as

$$\tau_i = \alpha_i + \beta_j + (\alpha\beta)_{ij}.$$

Similarly, the two-way random-effects model in Equation (14.12) can be obtained by setting

$$T_i = A_i + B_j + (AB)_{ij}$$

in the one-way random-effects model in Box 14.1 .

The interpretations of the parameters μ, σ_A^2, σ_B^2, σ_{AB}^2, and σ^2 are similar to the interpretations of μ, σ_T^2, and σ^2 in the one-way random-effects model. For instance, since A_i, B_j, and $(AB)_{ij}$ are assumed to be independent with zero means, the results in Box 9.1 can be used to verify that the mean and variance of an observed response are

$$E(Y_{ijk}) = \mu,$$

$$\text{Var}(Y_{ijk}) = \sigma_A^2 + \sigma_B^2 + \sigma_{AB}^2 + \sigma^2.$$

Thus, under a two-way random-effects model, μ is the expected response, and σ_A^2, σ_B^2, σ_{AB}^2, and σ^2 are the components of variance of an observed response. The null hypothesis H_{0AB}: $\sigma_{AB}^2 = 0$ can be interpreted as the null hypothesis that there is no interaction between the two experimental factors.

EXAMPLE **14.11**   A study was conducted to evaluate the effects of baking temperature (factor A) and baking time (factor B) on the tenderness of cakes baked using a particular recipe. The primary objective of the study was to see how sensitive the tenderness of cake is to incorrect settings of baking temperature and baking time. The recommended baking temperature and baking time are 350°C and 120 min, respectively. To ensure that the conclusions can be applied to temperatures in the range 325°–375°C, and times in the range 100–140 min, the study was conducted using a 4 × 4 factorial experiment in which the levels of A and B were random samples of size $n = 4$ from the corresponding population of levels. Two cakes were baked at each combination of the selected levels of temperature and time. Each cake was slowly tilted from the horizontal to a vertical position, and the angle at which the first signs of breaking appeared was recorded.

Let Y_{ijk} denote the breaking angle for the cake baked using the i-th randomly selected temperature and the j-th randomly selected time ($i = 1, \ldots, 4$; $j = 1, \ldots, 4$; $k = 1, 2$). Then, under suitable assumptions, Equation (14.12) is a reasonable model for describing Y_{ijk}. ■

Two-way mixed-effects model with crossed factors

An obvious modification of the two-way random effects model in Equation (14.12) leads to a model for two crossed factors in which one factor is fixed and one factor is random.

$$Y_{ijk} = \mu + \alpha_i + B_j + (\alpha B)_{ij} + E_{ijk}, \tag{14.13}$$

$$i = 1, \ldots, a; \ j = 1, \ldots, b; \ k = 1, \ldots, n_{ij}.$$

The notation used here is as follows:

- Y_{ijk} is the k-th response for the treatment combination $a_i b_j$;
- μ is the overall mean;
- α_i is the fixed effect of the i-level of A, and $\Sigma \alpha_i = 0$;
- B_j is the random effect of the j-th level of B;
- $(\alpha B)_{ij}$ is the joint random effect of the i-level of A and the j-th level of B;
- E_{ijk} is the random error.

The B_j are independent $N(0, \sigma_B^2)$; the $(\alpha B)_{ij}$ are independent $N(0, \sigma_{\alpha B}^2)$; and the E_{ijk} are independent $N(0, \sigma^2)$. In addition, the B_j, $(\alpha B)_{ij}$, and E_{ijk} are mutually independent.

Recall from Chapter 13 that, in a two-way fixed-effects model, the presence of interaction implies that components of the simple effects of a factor will vary with the level of the other factor. Consequently, the main effect—defined as the average of the simple effects—does not have a simple, practically useful interpretation. In a mixed-effects model, by contrast, the main effect of the fixed factor makes perfect sense, regardless of whether the interaction component $\sigma_{\alpha B}^2$ is zero. This is because the main effects in a mixed-effects model are the averages of the simple effects over the population of all possible levels of the random factor (not just the levels included in the study). The average of all values in a population of random effects (interaction as well as main effects) is assumed to be zero, and so a main effect can be interpreted as the effect of changing the level of the fixed factor when the level of the other factor is selected at random.

Another distinction between two-way fixed-effects and mixed-effects models is that, in the fixed-effects model in Box 13.1, the interaction effects satisfy the restrictions

$$\sum_{i=1}^{a} n_{ij}(\alpha\beta)_{ij} = 0, \quad j = 1, \ldots, b, \text{ and } \sum_{j=1}^{b} n_{ij}(\alpha\beta)_{ij} = 0, \quad i = 1, \ldots, a,$$

but no such restrictions are placed on the interaction effects in Equation (14.13). Nevertheless, some researchers advocate similar restrictions on the interaction effects in two-way mixed-effects models. They argue that, under certain circumstances, it is logical to work with mixed models in which the interaction effects satisfy the restrictions

$$\sum_{i=1}^{a} n_{ij}(\alpha B)_{ij} = 0, \quad j = 1, \ldots, b.$$

For a good discussion of restricted mixed-effects models, see Searle, Casella, and McCulloch (1992).

EXAMPLE **14.12**   A 3×4 factorial study was designed to investigate the effects of the fertilization method (factor A, with three levels) and seeding rate (factor B, with levels varying in the range 10–20 kg per acre) on the yield of sugar beets. Because the investigators were interested in evaluating the variation in yield that can result from varying the seeding rate, they decided to conduct the study by selecting a random sample of four seeding rates from the range of interest. The study was conducted in 36 quarter-acre plots. Each treatment combination was replicated in three plots.

Let Y_{ijk} denote the yield from the k-th plot treated at the i-th method of fertilization with the j-th seeding rate. Then, under suitable assumptions, a reasonable model for Y_{ijk} is obtained by taking $a = 3$, $b = 4$, and $n_{ij} = 3$ in the mixed-effects model in Equation (14.13).   ■

Two-way random-effects model with nested factors

A two-way random-effects model with nested factors can be expressed as

$$Y_{ijk} = \mu + A_i + B_{(i)j} + E_{ijk}, \tag{14.14}$$

$$i = 1, \ldots, a; \ j = 1, \ldots, b; \ k = 1, \ldots, n_{ij}.$$

The notation here is as follows:

- Y_{ijk} is the k-th response for the treatment combination $a_i b_{(i)j}$;
- μ is the overall mean;
- A_i is the random effect of the i-level of A.
- $B_{(i)j}$ is the random effect of the j-th level of B nested within the i-th level of A;
- E_{ijk} is the random error.

The A_i are independent $N(0, \sigma_A^2)$; the $B_{(i)j}$ are independent $N(0, \sigma_{B(A)}^2)$; and the E_{ijk} are independent $N(0, \sigma^2)$. Also, the A_i, $B_{(i)j}$, and E_{ijk} are mutually independent.

EXAMPLE **14.13**   In Exercise 14.1, the investigators were interested in evaluating the variability of the protein contents of seeds produced in the F_3 generation of a soybean cross. Comparison of the protein content variabilities of seeds in successive generations of plants is of interest in plant breeding programs where the goal is to breed plants that consistently (with little variation) produce seeds with high protein content. The following study design is suitable for comparing the variability of protein contents in F_2 generation plants with the corresponding variability in the F_3 generation.

First, select a random sample of 15 plants of F_2 generation (factor A). Next, select a random sample of ten plants of the F_3 generation (factor B) from the progeny of each selected F_2 generation plant. Randomly divide the seeds from each selected F_3 generation plant into three parts, and subject each part to protein analysis. This is a nested design, with factor B nested in factor A. Both factors are random.

Let Y_{ijk} denote the measured protein content of the k-th analysis of the seeds from the j-th F_3 generation plant produced by the i-th F_2 generation plant. Then, under suitable assumptions, Equation (14.14) with $a = 15$, $b = 10$, and $n_{ij} = 3$ is a possible model for the Y_{ijk}.   ■

Two-way mixed-effects model with nested effects

A minor modification of the two-way random-effects model with nested factors leads to a two-way mixed-effects model with nested factors

$$Y_{ijk} = \mu + \alpha_i + B_{(i)j} + E_{ijk}, \tag{14.15}$$

$$i = 1, \ldots, a; \ j = 1, \ldots, b; \ k = 1, \ldots, n_{ij}.$$

The notation here is as follows:

- Y_{ijk} is the k-th response for the treatment combination $a_i b_{(i)j}$;
- μ is the overall mean;
- α_i the fixed effect of the i-level of A, and $\Sigma_{i=1}^a \alpha_i = 0$;
- $B_{(i)j}$ is the random effect of the j-th level of B nested within the i-th level of A;
- E_{ijk} is the random error.

The $B_{(i)j}$ are independent $N(0, \sigma_{B(A)}^2)$, and the E_{ijk} are independent $N(0, \sigma^2)$. Also, the $B_{(i)j}$ and E_{ijk} are mutually independent.

EXAMPLE **14.14**   A study was designed to evaluate how technicians using two measurement techniques affect the measured amount of radon gas at the earth-air interface; radon gas is a source of natural radiation that emanates from the earth. The study used 24 experimental plots (known to be more or less homogeneous with respect to the amount of radon gas present) and six technicians selected randomly from a large pool of technicians. The experimental plots were randomly assigned to the two methods such that each method received 12 plots. Within each method, the 12 plots were randomly assigned to three technicians in such a way that each technician was allocated four plots. In each of the 24 plots, the assigned technician made radon measurements using the assigned method.

This is a study with two factors A and B, in which A (method) is fixed and B (technician) is random. Also, the levels of B are nested in the levels of A. Let Y_{ijk} denote the radon level in the k-th plot measured by the j-th technician assigned to the i-th method. Then Equation (14.15) with $a = 2$, $b = 3$, and $n_{ij} = 4$ is a possible model for analyzing the Y_{ijk}. ∎

Two-way ANOVA with random- and mixed-effects models

Data analysis based on the two-way models in Equations (14.12)–(14.15) is analogous to analysis based on two-way models with fixed effects. The ANOVA tables for the random- and mixed-effects models are exactly the same as their counterparts in the fixed-effects case. The computing formulas in Equations (13.13) and (13.21) are appropriate for calculating the needed sums of squares, provided the subclass numbers are equal ($n_{ij} = n$). If the subclass numbers are unequal (but nonzero), simple computing formulas do not exist for non-nested models, but the required ANOVA tables can easily be constructed by means of appropriate software (such as SAS, Splus, or SPSS); Chapter 14 in the companion text by Younger (1997) contains details for SAS.

As with one-way models, the procedures for making inferences about $\mu, \alpha_i, \beta_j, \sigma_A^2, \sigma_B^2, \sigma_{AB}^2, \sigma_{B(A)}^2$, and σ^2 in two-way models depend on the expected mean squares (EMS) in the appropriate ANOVA table. Theoretical derivation of these expected values is analogous to the derivation of the EMS for one-way models in Appendix B (Example B.4). If the subclass numbers are equal, a simple set of rules may be used to determine the EMS for a general multiway model containing crossed and/or nested factors and fixed and/or random effects (Section 14.7). When the subclass numbers are unequal, the EMS may be determined using statistical software such as SAS.

Table 14.1 shows the EMS for the six types of two-way ANOVA models under the assumption that the subclass numbers are equal ($n_{ij} = n$). In Table 14.1, a is the number of levels of A; b is the number of levels of B; n is the number of replications per treatment; and

$$\psi_A^2 = \frac{1}{a-1} \Sigma \alpha_i^2,$$

$$\psi_B^2 = \frac{1}{b-1} \Sigma \beta_j^2,$$

TABLE   14.1

EMS for two-way ANOVA models

1. When A and B are crossed

Source	df	(a) When A and B are fixed	(b) When A and B are random	(c) When A is fixed and B is random
A	$a - 1$	$nb\psi_A^2 + \sigma^2$	$nb\sigma_A^2 + n\sigma_{AB}^2 + \sigma^2$	$nb\psi_A^2 + n\sigma_{\alpha B}^2 + \sigma^2$
B	$b - 1$	$na\psi_B^2 + \sigma^2$	$na\sigma_B^2 + n\sigma_{AB}^2 + \sigma^2$	$na\sigma_B^2 + n\sigma_{\alpha B}^2 + \sigma^2$
AB	$(a-1)(b-1)$	$n\psi_{AB}^2 + \sigma^2$	$n\sigma_{AB}^2 + \sigma^2$	$n\sigma_{\alpha B}^2 + \sigma^2$
Error	$ab(n-1)$	σ^2	σ^2	σ^2

2. When B is nested in A

Source	df	(a) When A and B are fixed	(b) When A and B are random	(c) When A is fixed and B is random
A	$a - 1$	$nb\psi_A^2 + \sigma^2$	$nb\sigma_A^2 + n\sigma_{B(A)}^2 + \sigma^2$	$nb\psi_A^2 + n\sigma_{B(A)}^2 + \sigma^2$
$B(A)$	$a(b-1)$	$n\psi_{B(A)}^2 + \sigma^2$	$n\sigma_{B(A)}^2 + \sigma^2$	$n\sigma_{B(A)}^2 + \sigma^2$
Error	$ab(n-1)$	σ^2	σ^2	σ^2

$$\psi_{AB}^2 = \frac{1}{(a-1)(b-1)} \Sigma(\alpha\beta)_{ij}^2,$$

$$\psi_{B(A)}^2 = \frac{1}{a(b-1)} \Sigma\beta_{(i)j}^2.$$

A comparison of the definition of the parameters ψ_A^2, ψ_B^2, ψ_{AB}^2, and $\psi_{B(A)}^2$ for the two-way models with the definition of ψ_T^2 in Equation (14.3) for the one-way models shows that they are analogous.

Possible uses of Table 14.1 are as follows.

1 *Hypothesis testing.* The four null hypotheses: H_{0A}: $\psi_A^2 = 0$, H_{0B}: $\psi_B^2 = 0$, H_{0AB}: $\psi_{AB}^2 = 0$, and $H_{0B(A)}$: $\psi_{B(A)}^2 = 0$ are equivalent to the null hypotheses H_{0A}: $\alpha_1 = \cdots = \alpha_a$ (no main effect of A), H_{0B}: $\beta_1 = \cdots = \beta_b$ (no main effect of B), H_{0AB}: $(\alpha\beta)_{11} = \cdots = (\alpha\beta)_{ab}$ (no interaction between A and B), and $H_{0B(A)}$: $\beta_{(1)1} = \cdots = \beta_{(a)b}$ (no difference between the effects of A within levels of B), respectively.

The EMS values in Table 14.1 can be used to select the appropriate test statistics for tests about the parameters in two-way ANOVA models. For instance, in a two-way mixed-effects model with crossed factors, the F-test for the null hypothesis H_{0A}: $\psi_A^2 = 0$ (no main effect of A) can be determined as follows.

From the EMS values in column 1c of Table 14.1, we see that, if H_{0A} is true—that is, if $\psi_A^2 = 0$—then the expected values of MS[A] and MS[AB] will both be $n\sigma_{\alpha B}^2 + \sigma^2$. Otherwise, the expected value of MS[A] will be larger (by the amount $nb\psi_A^2$) than the expected value of MS[AB]. Consequently, a test of the null hypothesis that there is no main effect of A can be based on a comparison of MS[A] and MS[AB]. Such a comparison is provided by an F-test based on the test statistic

$$F_c = \frac{MS[A]}{MS[AB]}.$$

The degrees of freedom for the F-test correspond to the degrees of freedom associated with the mean squares used to construct the test statistic. We reject H_{0A} at level α if F_c is larger than the α-level critical value of an F-distribution with $a - 1$ degrees of freedom for the numerator and $(a - 1)(b - 1)$ degrees of freedom for the denominator.

Similarly, an examination of the EMS values for the two-way fixed-effects model (columns 1a and 2a) reveals that the test statistic for testing H_{0A} is

$$F_c = \frac{MS[A]}{MS[E]}, \qquad df = a - 1, \ ab(n - 1).$$

Indeed, we see from Table 14.1 that, if both A and B are fixed factors, then the denominator of the F-statistic for testing any of the two-way ANOVA null hypotheses (such as H_{0A} and $H_{0B(A)}$) is MS[E], regardless of whether the factors are crossed or nested. That's why all the F-tests in Chapter 13 utilized MS[E] as the denominator for the test statistics.

Finally, in two-way nested models where B is random, regardless of whether the effects of A are fixed or random, an F-test of the null hypothesis H_0 that there is no difference between the effects of A—that is, that $\psi_A^2 = 0$ if A is fixed and $\sigma_A^2 = 0$ if A is random—can be based on the test statistic

$$F_c = \frac{MS[A]}{MS[B(A)]}.$$

As always, the degrees of freedom for the F-test correspond to the degrees of freedom associated with the mean squares used to construct the test statistic.

2   *Estimating variance components.* As in the one-way random-effects model, the EMS in Table 14.1 can be used to set up estimating equations for the variance components. Like Equation (14.4), these equations are obtained by equating each mean square in the ANOVA table to its expected value. Solution of the estimating equations yield estimates of the variance components. For instance, in a random-effects model with nested factors, the estimating equations are

$$\hat{\sigma}^2 + nb\hat{\sigma}_A^2 + n\hat{\sigma}_{B(A)}^2 + \hat{\sigma}^2 = MS[A],$$

$$\hat{\sigma}^2 + n\hat{\sigma}_{B(A)}^2 + \hat{\sigma}^2 = MS[B(A)],$$

$$\hat{\sigma}^2 = MS[E].$$

Solving these equations, we get

$$\hat{\sigma}^2 = MS[E],$$

$$\hat{\sigma}_{B(A)}^2 = \frac{MS[B(A)] - MS[E]}{n},$$

$$\hat{\sigma}_A^2 = \frac{MS[A] - MS[B(A)]}{nb}.$$

Of course, because negative values are inadmissible for variances, any negative solution is usually replaced by zero.

3 *Estimating fixed effects.* The expected mean squares are also useful for estimating the standard errors and the associated degrees of freedom needed for inferences about linear combinations of fixed effects (such as μ, $\mu + \alpha_1$, and $\alpha_1 - \alpha_2$) in the assumed model. In Chapters 9 and 13, we have seen many examples of inferences that can be made using linear combinations of fixed effects in models in which all effects except the error are considered fixed. The confidence interval for the overall mean μ in a one-way random-effects model (Section 14.3) is an example of an inference about a fixed parameter in a random-effects model.

For an example of an inference about linear combinations of fixed effects in mixed models, recall Example 14.12, where the effects of fertilizers (fixed) and seeding rate (random) on the yield of sugar beets were compared. A natural parameter that we might want to estimate is the expected yield when fertilization method 1 is used with a randomly selected seeding rate. Since each random effect that affects Y_{1jk}—that is, B_j, $(\alpha B)_{1j}$, and E_{1jk}—is assumed to have zero expected value, the expected value of Y_{1jk} is

$$\mathcal{E}(Y_{1jk}) = \mu + \alpha_1 + \mathcal{E}(B_j) + \mathcal{E}(\alpha B_{1j}) + \mathcal{E}(E_{1jk})$$
$$= \mu + \alpha_1 + 0 + 0 + 0$$
$$= \mu + \alpha_1.$$

Thus, estimation of $\mathcal{E}(Y_{1jk})$ is an inference about a linear combination of fixed parameters μ and α_1. Similarly, it can be shown that making an inference about the difference between the expected responses for fertilization methods 1 and 2 is the same as making inferences about the linear combination $\alpha_1 - \alpha_2$. A general method for making inferences about linear combinations of the fixed effects in multiway models will be described later in this section.

EXAMPLE 14.15   In Example 14.12, we considered a 3×4 factorial experiment on the yield of sugar beets. The experimental design utilized $a = 3$ levels of the fixed factor, fertilization method (factor A), and $b = 4$ levels of the random factor, seeding rate (factor B). There were $n = 3$ replications per treatment. The following yields (tons per acre) were obtained in the experiment.

Level of A (i)	Plot number (k)	1	2	3	4
			Level of $B(j)$		
	1	15.71	16.21	17.32	17.54
1	2	16.02	16.36	17.03	17.82
	3	15.90	16.33	17.22	17.62
	1	17.83	17.68	17.95	18.08
2	2	17.45	17.70	18.01	18.56
	3	16.96	17.52	18.41	18.90
	1	14.78	15.80	16.21	16.99
3	2	15.03	15.62	16.44	16.39
	3	14.63	15.77	16.32	17.02

Let Y_{ijk} denote the observed yield for the k-th plot receiving the i-th fertilization method and the j-th seeding rate. The following two-way mixed model with crossed effects is possible for the Y_{ijk}

$$Y_{ijk} = \mu + \alpha_i + B_j + (\alpha B)_{ij} + E_{ijk},$$

$$i = 1, \ldots, 3; \; j = 1, \ldots, 4; \; k = 1, 2, 3.$$

The notation here is as follows:

- μ is the overall mean;
- α_i is the effect of the i-th fertilization method ($\alpha_1 + \alpha_2 + \alpha_3 = 0$);
- B_j is the random effect of the j-th seeding rate;
- $(\alpha B)_{ij}$ is the random interaction effect;
- E_{ijk} is the random error.

The $(\alpha B)_{ij}$ are independent $N(0, \sigma_{\alpha B}^2)$, and the E_{ijk} are independent $N(0, \sigma^2)$. Also, B_j, $(\alpha B)_{ij}$ and E_{ijk} are mutually independent.

The ANOVA table can be constructed using Equation (13.13) or statistical computing software. The ANOVA table containing the mean squares and their expected values for the sugar beet data is as follows:

Source	df	SS	MS	EMS
Fertilizer (A)	2	24.3103	12.1552	$\sigma^2 + 3\sigma_{\alpha B}^2 + 12\psi_A^2$
Seeding Rate (B)	3	13.8179	4.6060	$\sigma^2 + 3\sigma_{\alpha B}^2 + 9\sigma_B^2$
Interaction (AB)	6	0.9326	0.1554	$\sigma^2 + 3\sigma_{\alpha B}^2$
Error	24	1.3903	0.0579	σ^2
Total	35	40.4512		

Note that the EMS values are obtained by setting $a = 3$, $b = 4$, and $n = 3$ in column 1c of Table 14.1. On the basis of the EMS values, we see that the following significance tests are possible:

1   *Test for variance component due to interaction:* $H_0 \colon \sigma_{\alpha B}^2 = 0$ versus $H_1 \colon \sigma_{\alpha B}^2 > 0$

$$F_{AB} = \frac{MS[AB]}{MS[E]} = \frac{0.1554}{0.0579} = 2.684, \quad df = 6, 24; \quad p = 0.0383.$$

2   *Test for main effect of A:* $H_0 \colon \psi_A^2 = 0$ versus $H_1 \colon \psi_A^2 > 0$

$$F_A = \frac{MS[A]}{MS[AB]} = \frac{12.1552}{0.1554} = 78.219, \quad df = 2, 6; \quad p < 0.0001.$$

3   *Test of variance component due to B:* $H_0 \colon \sigma_B^2 = 0$ versus $H_1 \colon \sigma_B^2 > 0$

$$F_B = \frac{MS[B]}{MS[AB]} = \frac{4.6060}{0.1554} = 29.640, \quad df = 3, 6; \quad p < 0.0001.$$

Thus, it appears that there is significant ($p = 0.0388$) variation due to interaction between the type of fertilizer and the seeding rate and there is also significant

($p < 0.0001$) variation due to variability of the rate. Also, there are significant differences between the expected yields for the four fertilizer types ($p < 0.0001$).

The estimates of the variance components can be obtained by solving the estimating equations

$$MS[E] = \hat{\sigma}^2,$$

$$MS[AB] = \hat{\sigma}^2 + 3\hat{\sigma}_{\alpha B}^2,$$

$$MS[B] = \hat{\sigma}^2 + 3\hat{\sigma}_{\alpha B}^2 + 9\hat{\sigma}_B^2,$$

so that

$$\hat{\sigma}^2 = 0.0579,$$

$$\hat{\sigma}_{\alpha B}^2 = \frac{MS[AB] - MS[E]}{3} = \frac{0.1554 - 0.0579}{3} = 0.0325,$$

$$\hat{\sigma}_B^2 = \frac{MS[B] - MS[AB]}{9} = \frac{4.6060 - 0.1554}{9} = 0.4945.$$

Therefore, the estimated total variation due to random effects of the seeding rate and error is

$$\hat{\sigma}_B^2 + \hat{\sigma}_{\alpha B}^2 + \hat{\sigma}^2 = 0.4945 + 0.0325 + 0.0579 = 0.5849,$$

of which 84.54% ($0.4944/0.5849 = 0.8454$) is due to variability in the seeding rates; 5.56% ($0.0325/0.5849$) is due to variability introduced by the interaction between the fertilization method and the seeding rate; and 9.90% ($0.0579/0.5849$) is due to variability attributable to experimental error. ∎

The next example illustrates data analysis based on a two-way random-effects model with nested effects.

EXAMPLE **14.16**   In a study to develop and evaluate a technique of measuring radon flux at the earth-air interface using charcoal canisters, the experimental area—a 44-acre tract of reclaimed phosphate land in Lakeland, Florida—was surveyed for gamma radiation, and a contour map showing contours of equal gamma radiation was prepared. Three contours were selected at random. Within each selected contour, two sites were selected at random. At each site, the radon flux ($pCi/m^2 \cdot hr$) was sampled by charcoal canisters over a 30-hr period. The log (radon flux) data obtained are as follows (Hagi, 1978):

Contour	Site 1			Site 2		
1	0.6130	−0.4050	−0.6445	0.8303	−0.1154	1.2507
2	0.9450	0.3500	0.6060	−1.3205	−0.7875	−1.3548
3	0.6136	0.0266	0.2776	−0.4636	−0.9782	−0.1335

Experience suggests that a normal distribution is a reasonable model for describing log (radon flux) measurements, and so we might analyze these data using a two-way random-effects model with nested factors. We have $a = 3$, $b = 2$, and $n = 3$. The model can be expressed as

$$Y_{ijk} = \mu + A_i + B_{(i)j} + E_{ijk}, \qquad i = 1, 2, 3; \; j = 1, 2; \; k = 1, 2, 3.$$

The notation here is as follows:

- Y_{ijk} is the log (radon flux) measured by the k-th canister in the i-th site in the j-th contour;
- μ is the overall mean log (radon flux) for the region;
- A_i is the random effect of the i-th randomly selected contour;
- $B_{(i)j}$ is the effect of the j-th randomly selected site nested within the i-th contour;
- E_{ijk} is the random error.

The A_i are independent $N(0, \sigma_A^2)$; the $B_{(i)j}$ are independent $N(0, \sigma_{B(A)}^2)$; and the E_{ijk} are independent $N(0, \sigma^2)$. Also, the A_i, $B_{(i)j}$, and E_{ijk} are mutually independent.

The ANOVA table can be constructed on the basis of Equation (13.21) or the NESTED procedure in SAS. The resulting ANOVA table and the associated EMS values are as follows:

Source	df	MS	F	p	EMS
A	2	0.4209	0.1859	0.8392	$6\sigma_A^2 + 3\sigma_{B(A)}^2 + \sigma^2$
B[A]	3	2.2642	9.7469	0.0015	$3\sigma_{B(A)}^2 + \sigma^2$
Error	12	0.2323			σ^2
Total	17				

These values of the F-statistics are consistent with the F-tests suggested by the EMS values. For instance, the F-statistic for testing H_0: $\sigma_A^2 = 0$ is

$$F_A = \frac{MS[A]}{MS[B(A)]} = \frac{0.4209}{2.2642} = 0.1859, \qquad df = 2, 3.$$

On the basis of the F-values in the ANOVA table, we can conclude that there is no demonstrable evidence against the null hypothesis H_{0A}: $\sigma_A^2 = 0$, but $H_{0B(A)}$: $\sigma_{B(A)}^2 = 0$ can be rejected at $\alpha = 0.01$ ($p = 0.0015$). It appears that the variability in gamma radiation contours is not a significant factor in the variability in the region of the measured log (radon flux). However, there is significant site-to-site variability within gamma radiation contours. The components of variance of log (radon flux) can be estimated by equating the mean squares to their expected values (Note: negative estimates are replaced by 0).

$$\hat{\sigma}_A^2 = \frac{0.4209 - 2.2642}{6} = -0.3702 \approx 0,$$

$$\hat{\sigma}_{B(A)}^2 = \frac{2.2642 - 0.2323}{3} = 0.6773,$$

$$\hat{\sigma}^2 = 0.2323.$$

An estimate of the overall variance of log (radon flux) is

$$\hat{\sigma}^2 = \hat{\sigma}_A^2 + \hat{\sigma}_{B(A)}^2 + \hat{\sigma}^2 = 0 + 0.6773 + 0.2323 = 0.9096.$$

Thus, approximately 75% (0.6773/0.9096) of the overall variation is due to variability within sites and the remaining 25% is caused by canister-to-canister differences.

As in a one-way random-effects model, we might be interested in inferences about μ, the overall mean log (radon flux) for the region. An unbiased estimate of μ is $\overline{Y}_{+++}$, the overall sample mean. For the radon flux data, $\overline{Y}_{+++} = -0.03834$. Therefore, the estimated mean log (radon flux) for the region is -0.0383. In Example 14.7, we expressed the overall mean response from a one-way random-effects model with nested factors as a sum of the means of the effects in the model. The corresponding expression for a two-way random-effects model with nested factors is

$$\overline{Y}_{+++} = \mu + \overline{A}_+ + \overline{B}_{(+)+} + \overline{E}_{+++},$$

where $\overline{A}_+$, $\overline{B}_{(+)+}$, and $\overline{E}_{+++}$ are, respectively, the mean of a random sample of size a from the population of effects of the contours, the mean of a random sample of size ab from a population of effects of the sites within contours, and the mean of a random sample of size nab from a population of the canisters within sites. Thus, following a reasoning similar to that in Example 14.7, the variance of $\overline{Y}_{+++}$ can be shown to be

$$\text{Var}(\overline{Y}_{+++}) = \frac{\sigma_A^2}{a} + \frac{\sigma_{B(A)}^2}{ab} + \frac{\sigma^2}{nab} = \frac{nb\sigma_A^2 + n\sigma_{B(A)}^2 + \sigma^2}{nab}.$$

A reference to the EMS values in column 2b of Table 14.1 shows that the variance of $\overline{Y}_{+++}$ can be estimated by

$$\widehat{\text{Var}(\overline{Y}_{+++})} = \frac{\text{MS}[A]}{nab} = \frac{0.4209}{3 \times 3 \times 2} = 0.0234.$$

Since MS[A] has two degrees of freedom, a $100(1 - \alpha)\%$ confidence interval for μ is $\overline{Y}_{+++} \pm t(2, \alpha/2) \, \text{SE}(\widehat{\overline{Y}_{+++}})$. A 95% confidence interval for the expected log (radon flux) is $-0.0383 \pm t(2, 0.025)(0.1529)$, so that the desired confidence interval is

$$-0.6962 < \mu < 0.6196. \quad \blacksquare$$

Inferences about linear combinations of population means

As in fixed-effects models, investigators are often interested in making inferences about linear combinations of fixed components in models containing random effects. If θ is a linear combination of fixed components, it can be expressed as a linear combination of population means, so that the corresponding linear combination of sample means $\hat{\theta}$ will be an unbiased estimator of θ. For instance, in the two-way mixed-effects model in Example 14.15, $\theta = \mu + \alpha_1$ is a linear combination of the fixed components μ and α_1. We know that θ is the mean of the population of all responses for fertilization method 1, and so a suitable unbiased estimator is $\hat{\theta} = \overline{Y}_{1++}$, the mean of all observed responses for method 1. Similarly, the linear combination of fixed effects $\theta = \alpha_1 - \alpha_2$ can be expressed as the linear combination of the population means $\theta = (\mu + \alpha_1) - (\mu + \alpha_2)$. An unbiased estimator of θ is $\hat{\theta} = \overline{Y}_{1++} - \overline{Y}_{2++}$.

In the two-way mixed-effects models in Equations (14.13) and (14.15), linear combinations of population mean responses may be written in the general form

$$\theta = c_1(\mu + \alpha_1) + c_2(\mu + \alpha_2) + \cdots + c_a(\mu + \alpha_a),$$

where $c_1, c_2, \ldots, c_a$ are appropriately chosen constants. The corresponding linear combination of the sample means

$$\hat{\theta} = c_1\overline{Y}_{1++} + c_2\overline{Y}_{2++} + \cdots + c_a\overline{Y}_{a++}$$

is an unbiased estimator of θ.

As in Section 14.3, hypothesis tests and confidence intervals for θ can be constructed provided the conditions in Box 14.5 are satisfied.

In Example 14.7, the standard error of $\hat{\theta} = \overline{Y}_{++}$, was determined by utilizing the fact that $\overline{Y}_{++}$ can be expressed as a linear function of the means of independent random samples. Such a representation enabled us to use the results in Box 14.5 to determine the required standard error. The same approach can be used to find the standard errors of linear combinations of the sample means in two-way mixed model analysis. In Box 14.8, we present the standard errors of the linear combinations of mean responses in two-way random- and mixed-effects models.

Standard errors of linear combinations of mean responses in two-way random- and mixed-effects models

For the two-way models in Equations (14.13) and (14.15), assume that the subclass numbers are equal ($n_{ij} = n$). Let

$$\theta = c_1(\mu + \alpha_1) + \cdots + c_a(\mu + \alpha_a)$$

be a given linear combination of population means. Then the variance of the unbiased estimator

$$\hat{\theta} = c_1\overline{Y}_{1++} + \cdots + c_a\overline{Y}_{a++}$$

is

$$\sigma_{\hat{\theta}}^2 = \left(\frac{n\sigma_{\alpha B}^2 + \sigma^2}{nb}\right)\left(\sum_{i=1}^{a} c_i^2\right) + \left(\frac{\sigma_B^2}{b}\right)\left(\sum_{i=1}^{a} c_i\right)^2$$

for crossed factors and

$$\sigma_{\hat{\theta}}^2 = \left(\frac{n\sigma_{B(A)}^2 + \sigma^2}{nb}\right)\left(\sum_{i=1}^{a} c_i^2\right)$$

for nested factors.

In practice, the variance components σ^2, σ_B^2, $\sigma_{B(A)}^2$, and $\sigma_{\alpha B}^2$ have to be estimated from the observed data, and the inferences about linear combinations in Box 14.8 are made using the estimated standard error $\hat{\sigma}_{\hat{\theta}}$, which is obtained from $\sigma_{\hat{\theta}}$ in

Box 14.8 by replacing the variance components with their estimates. The estimated standard error will typically involve more than one mean square in the ANOVA table, and so we also need an estimate of the degrees of freedom appropriate for inferential procedures based on *t*-tests. The most common method of obtaining such an estimate is based on the method in Box 14.9, which was suggested by Satterthwaite (1946) and Welch (1951).

BOX **14.9**

> ### *Estimate of the degrees of freedom for a linear combination of mean squares in two-way ANOVA*
>
> Assume a two-way ANOVA model, and let θ and $\hat{\theta}$ denote, respectively, a linear combination of population means and the corresponding linear combination of the sample means. Suppose that an estimate of the standard error of $\hat{\theta}$ is given by
>
> $$\hat{\sigma}_{\hat{\theta}} = \sqrt{p_1 MS_1 + p_2 MS_2 + \cdots + p_k MS_k,}$$
>
> where $p_1, p_2, \ldots, p_k$ are nonnegative constants, and $MS_1, MS_2, \ldots, MS_k$ are the mean squares in the ANOVA table. Then the degrees of freedom associated with $\hat{\sigma}_{\hat{\theta}}$ are approximately equal to ν, where ν is the largest integer less than or equal to
>
> $$\nu^* = \frac{\left(p_1 MS_1 + p_2 MS_2 + \cdots + p_k MS_k\right)^2}{p_1^2 \dfrac{MS_1^2}{\nu_1} + p_2^2 \dfrac{MS_2^2}{\nu_2} + \cdots + p_k^2 \dfrac{MS_k^2}{\nu_k}};$$
>
> ν_i denotes the degrees of freedom for MS_i.

Comments on inferences based on Box 14.9

1   If only one mean square—say, MS_1—is needed to estimate $\sigma_{\hat{\theta}}$, then ν in Box 14.9 will equal ν_1, the degrees of freedom for MS_1. In that case, the inferences will be exact, in the sense that, if the assumptions implied by the underlying model are satisfied, the confidence levels and *p*-values associated with the conclusions will be exact.

2   If more than one mean square is needed to estimate the standard error of θ, then inferences based on the results in Box 14.9 are approximate. Generally speaking, the best results may be expected when the degrees of freedom for the MS_i are large and close to each other.

3   The approximation in Box 14.9 requires the coefficients p_i to be positive. However, in some applications, negative coefficients p_i are encountered. In that case, we can either set them equal to zero (thereby increasing the standard error of the estimate) or use the given approximation with negative coefficients. Gaylor and Hopper (1969) investigate the appropriateness of the Satterthwaite formula under some settings.

EXAMPLE 14.17   In Example 14.15, we analyzed a 3×4 factorial experiment using a two-way mixed-effects model. The experiment involved $a = 3$ levels of the fixed factor A, fertilization method, and $b = 4$ levels of the random factor B, seeding rate.

First, suppose we are interested in estimating the average yield for fertilization methods 1 and 2. Then the parameter of interest is $\theta = 0.5(\mu + \alpha_1 + \mu + \alpha_2) = \mu + 0.5\alpha_1 + 0.5\alpha_2$, which can be expressed as a linear combination of $\mu + \alpha_1$, $\mu + \alpha_2$, and $\mu + \alpha_3$ with coefficients $c_1 = 1$, $c_2 = 1$, and $c_3 = 0$. By letting $a = 3$, $b = 4$, $n = 3$, $c_1 = 1$, $c_2 = 1$, and $c_3 = 0$ in Box 14.8, we get $\sum c_i = 2$ and $\sum c_i^2 = 3/2$. Thus

$$\sigma_{\hat{\theta}}^2 = \frac{(3\sigma_{\alpha B}^2 + \sigma^2)}{3 \times 4}\left(\frac{3}{2}\right) + \frac{\sigma_B^2}{4}(2)^2.$$

Now, from the EMS values for the two-way mixed-effects ANOVA, the variance component estimates are

$$\hat{\sigma}_B^2 = \frac{MS[B] - MS[AB]}{9},$$

$$\hat{\sigma}_{\alpha B}^2 = \frac{MS[AB] - MS[E]}{3},$$

$$\hat{\sigma}^2 = MS[E],$$

so that the estimated variance of $\hat{\theta}$ has the form

$$\hat{\sigma}_{\hat{\theta}}^2 = \frac{1}{8}\left(\frac{3(MS[AB] - MS[E])}{3} + MS[E]\right) + \frac{MS[B] - MS[AB]}{3 \times 3}$$

$$= \frac{1}{72}MS[AB] + \frac{1}{9}MS[B].$$

From the data in Example 14.15, we see that

$$\bar{y}_{1++} = 16.76, \quad \bar{y}_{2++} = 17.92, \quad \bar{y}_{3++} = 15.92, \quad MS[B] = 4.6060,$$

and

$$MS[AB] = 0.1554.$$

The estimated standard error of $\hat{\theta}$ is

$$\hat{\sigma}_{\hat{\theta}} = \sqrt{\frac{1}{72}MS[AB] + \frac{1}{9}MS[B]} = \sqrt{\frac{0.1554}{72} + \frac{4.6060}{9}} = 0.7169$$

We know that $MS_1 = MS[AB]$ and $MS_2 = MS[B]$ have $\nu_1 = 6$ and $\nu_2 = 3$ degrees of freedom, respectively, and so the degrees of freedom associated with the estimated standard error can be approximated by the largest integer less than or equal to

$$\nu^* = \frac{\left(\dfrac{0.1554}{72} + \dfrac{4.6060}{9}\right)^2}{\left(\dfrac{1}{72}\right)^2 \dfrac{(0.1554)^2}{6} + \left(\dfrac{1}{9}\right)^2 \dfrac{(4.6060)^2}{3}} = 3.03.$$

Thus, $\nu = 3$.

On the basis of the results in Box 14.8, an unbiased estimate of $\theta = \mu + 0.5\alpha_1 + 0.5\alpha_2$ is

$$\hat{\theta} = 0.5(\bar{y}_{1++} + \bar{y}_{2++}) = 0.5(16.76 + 17.92) = 17.34.$$

A 95% confidence interval for θ is

$$17.34 \pm t(3, 0.025)(0.7169) = 17.34 \pm 2.28.$$

Thus, at the 95% confidence level, the average expected response for fertilization methods 1 and 2 is somewhere between 15.06 and 19.62 tons/acre.   ■

Multiway ANOVA models with random effects

The notion of random effects is easily extended to ANOVA models with more than two factors. In multiway random-effects models, the levels of every factor constitute a random sample from a corresponding population of levels. In multiway mixed-effects models, the factorial effects will be a mixture of random and fixed effects; that is, levels of some factor are random, whereas other factors are fixed. Without going into details, we may simply note that the key to analysis based on these models is to figure out the ANOVA table and the corresponding EMS values associated with the models. Some examples of multiway mixed- and random-effects models will be encountered in the next section and in Chapters 15 and 16.

Exercises

14.13   Assuming a two-way mixed-effects model with crossed factors, show that the difference between the expected responses at levels 1 and 2 of the fixed factor is $\alpha_1 - \alpha_2$.

14.14   Let $\bar{Y}_{1++}$ denote the observed mean response for level 1 of factor A in a two-way mixed-effects model with crossed factors and n replications per treatment combination.

a   Following the usual notation, describe the meaning of the terms in the expression

$$\bar{Y}_{1++} = \mu + \alpha_1 + \bar{B}_+ + \overline{(\alpha B)}_{1+} + \bar{E}_{1++}.$$

b   Following the technique described in Section 14.3, use the expression in (a) to show that

$$\mathrm{Var}(\bar{Y}_{1++}) = \frac{(n\sigma_{\alpha B}^2 + \sigma^2)}{nb} + \frac{\sigma_B^2}{b}.$$

c   Use the variance formula in (b) to obtain an expression for the estimate of the standard error of $\bar{Y}_{1++}$.

d   Argue that the expression derived in (c) is the same as that resulting from the formula in Box 14.8.

14.15   Refer to Example 14.17.

a   Construct a 95% confidence interval for the contrast $\theta = \frac{1}{2}(\alpha_1 + \alpha_2) - \alpha_3$, and interpret the results.

b Construct a set of 95% simultaneous confidence intervals for all possible pair-wise differences between the true mean responses for the three fertilizers. Interpret the results.

14.16 To see how the effect of diet supplements on the weights of wethers varied with environmental conditions, a study was conducted in four randomly selected locations; each location represented a randomly selected environment. The experimenters randomly assigned 24 crossbred wethers to the four locations in such a way that each location had six wethers. Within each location, the animals were randomized to receive three diets, with two animals on each diet. The four-week weight gains of the wethers are as follows:

Diet	Location			
	1	2	3	4
1	2.10	2.02	2.16	1.98
	2.32	2.04	2.18	1.86
2	2.24	2.30	2.22	1.64
	2.22	2.12	2.18	1.73
3	2.28	2.14	2.26	1.83
	2.24	2.17	2.21	1.89

a Write a suitable ANOVA model for the data. Explain all the terms in your model.
b Obtain estimates of the components of the variance due to location, interaction between location and diet, and error. Interpret the results.
c Construct the appropriate ANOVA table for the data and perform all ANOVA F-tests. Interpret the results.
d Construct a 90% confidence interval for the true mean weight gain of animals fed diet 1 in a randomly selected location. (Use Exercise 14.14 and the results in Box 14.8.)
e Construct a set of Bonferroni 90% simultaneous confidence intervals for all possible difference between diet means. Interpret the intervals.

14.17 The experiment in Example 14.14 involved two methods of measuring radon flux and six technicians. The radon flux measurements obtained ($pCi/m^2 \cdot hr$) are as follows:

Method	Technician					
	1	2	3	4	5	6
1	42.4	41.6	39.3	43.2	40.1	41.0
	41.8	40.9	38.2	44.6	41.5	42.3
2	39.7	40.2	38.6	42.0	39.1	40.1
	40.3	41.1	37.6	42.9	41.2	40.6

a Write a model for the data. Explain all the terms in your model.
b Construct the appropriate ANOVA table. What conclusions can be drawn from this table?

c Obtain estimates of the components of variance due to variability between technicians within methods $(\hat{\sigma}^2_{B(A)})$ and variability between measurements within technicians $(\hat{\sigma}^2)$. Compare these estimates and draw appropriate conclusions about the variability between technicians relative to the variability between measurements by the same technician.

d Let $\overline{Y}_{1++}$ denote the observed mean radon flux measurement resulting from method 1. What does this quantity estimate?

e Let θ denote the quantity estimated by $\hat{\theta} = \overline{Y}_{1++}$. Using the technique described in Section 14.3, show that an estimate of the standard error of $\hat{\theta}$ is

$$\hat{\sigma}_{\hat{\theta}} = \sqrt{\frac{MS[B(A)]}{nb}}$$

(Note that, under the assumed model, $\overline{Y}_{1++} = \mu + \alpha_1 + \overline{B}_{(1)+} + \overline{E}_{1++}$, where $\overline{B}_{(1)+}$ and $\overline{E}_{1++}$ are the means of independent random samples.) Verify that this expression for $\hat{\sigma}_{\hat{\theta}}$ matches that obtained from Box 14.8.

f Using (d) and the results in Box 14.8 construct a 99% confidence interval for θ. Interpret the interval.

g Using the results in Box 14.8, construct a 95% upper confidence interval for the difference between the true mean radon fluxes measured by the two methods. Interpret this interval.

14.18 To estimate the average cost of hospital stay, a random sample of four hospitals (factor A) was selected from the population of all hospitals in a large region. For each hospital, patient admission records for three randomly selected days (factor B) were examined. Two patients were selected at random from all the patients admitted on each day to each hospital. On the basis of the average daily hospital bill (in dollars) for each selected patient, the following ANOVA table was constructed:

Source	Mean square
A (hospitals)	2240.05
$B(A)$ (days in hospitals)	312.32
Error (patients)	122.02

The average of all 24 patient bills was $380.24.

a Estimate the components of variability in the daily hospital cost due to differences between hospitals, due to differences between days, and due to difference between patients. Calculate the proportion of total variability in daily costs due to each of these sources. Comment on the results.

b Construct a 95% confidence interval for the expected daily cost for a patient admitted to one of the hospitals in the region.

c Construct a 95% lower confidence interval for the expected daily cost to a patient admitted to one of the hospitals in the region. Interpret the result.

14.19 In an experiment to compare the serum amylase values determined by four laboratories, 160 serum specimens with a known amylase value of 4.2 were used. The specimens were randomized to 16 technicians (four from each lab) in such a way that each technician was assigned ten specimens. The technicians made independent measurements of the amylase values of the specimens assigned to them. On the basis of the results, the following ANOVA table was constructed:

Source	Mean square
A (labs)	9.07
B(A) (technicians)	1.63
Error (Specimens)	1.02

The totals for each lab may be summarized as follows:

Lab	1	2	3	4	Total
Total	152	144	184	176	656

a Write a model for the data on the assumption that the four technicians in each lab constitute a random sample from the population of technicians working in that lab. Explain all the terms in the model.

b Perform a pairwise multiple comparison of the mean values for the four labs. Interpret the result.

c Let $\sigma^2_{B(A)}$ and σ^2 denote, respectively, the components of variance due to differences between technicians and due to differences between specimens. Compute an estimate of the quantity

$$\rho = \frac{\sigma^2_{B(A)}}{\sigma^2_{B(A)} + \sigma^2}.$$

What conclusions can be drawn from the estimated value of ρ?

d Let μ_i denote the true mean amylase values for lab i ($i = 1, \ldots, 4$). Construct a set of 90% simultaneous confidence intervals for the contrasts: $\theta_1 = \mu_1 - \mu_2$; $\theta_2 = \frac{1}{2}(\mu_1 + \mu_2) - \frac{1}{2}(\mu_3 + \mu_4)$; and $\theta_3 = \mu_3 - \mu_4$. Interpret the intervals.

14.6
ANOVA models with subsampling
Advantages of subsampling

The main difference between the fixed- and random-effects models considered thus far is the way in which the data are collected. In the case of random- and mixed-effects models, the data are collected by sampling at more than one stage, whereas only one sampling stage is involved when data are collected under fixed-effects models. For example, consider a study that uses a one-way ANOVA model. In the case of a random-effects model, data collection involves two stages. In the first stage, a random sample of size t is selected from the population of all treatments. In the second stage, a random sample of responses is selected for each of the t treatments selected in the first stage. In the case of a fixed-effects model, by contrast, sampling is done at only one stage—the stage at which the responses are measured for each of the t treatments. Similarly, in two-way mixed-effects models, there are two sampling stages. The first stage consists of selecting a random sample of the levels of B (the random factor). In the second stage, a random sample of responses is obtained at each combination of the levels of A and B—$a_i b_j$ for crossed factors and $a_i b_{(i)j}$ for nested factors. Sampling plans that lead to one-way fixed and random models with $t = 3$ and $n_1 = n_2 = n_3 = 3$ are shown in Figure 14.2.

FIGURE 14.2

Sampling plans for one-way random- and fixed-effects models

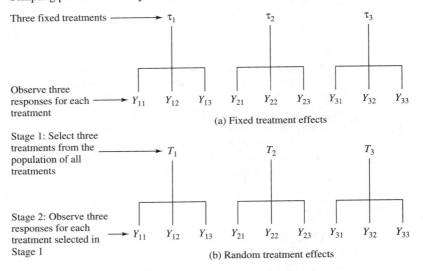

(a) Fixed treatment effects

(b) Random treatment effects

For all the models considered thus far, only one measurement is made on each experimental unit. Sometimes it is useful to observe a sample of more than one response for each experimental unit. For example, recall Example 8.2, where five sources of light—D, AL, AH, BL, and BH—were compared. Four randomly selected seedlings were assigned to each light source. In this experiment, each of the 20 seedlings is an experimental unit, because single seedlings were exposed to individual treatments. As already noted, a one-way fixed-effects model with one sampling stage is appropriate for this study design.

It may be more convenient (and less expensive) to conduct the same study in ten experimental units with two seedlings per unit. Each experimental unit will be treated with a single light source. Such a study could use 20 identical seedlings and ten pots. Two randomly selected pots (first-stage samples) could be assigned to each light source and two randomly selected seedlings (second-stage samples) could be grown in the pot. The four-week height will be measured for each of the 20 seedlings. The treatments (light sources) are applied to the pots, and so the pots are the experimental units. However, the heights are measured for each single seedling, and so the seedlings are the sampling (observational) units. The sampling design of such a study is shown in Figure 14.3.

F I G U R E   14.3

Sampling plan in a study with fixed treatments and two stages of sampling

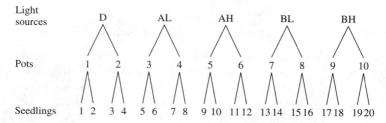

As an example of a design with random treatment effects and multiple measurements per experimental unit, consider the study of buttermilk protein content in Example 14.1. Even though it is easy to sample 250-ml cartons of buttermilk, the protein analysis of the whole sample may be cumbersome and expensive. The investigators might prefer to do protein analyses on 10-ml samples. In that case, it is desirable to include an additional stage of sampling in the study design. We can select, say, two 10 ml subsamples from each selected carton of buttermilk. The sampling plan of the resulting study design is shown in Figure 14.4.

F I G U R E   14.4

The sampling plan for a study with random treatments and two stages of sampling

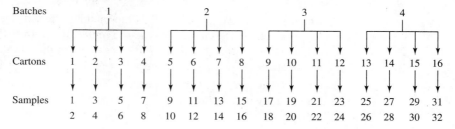

In general, *subsampling*—making multiple measurements per experimental unit—is needed when it is impractical or undesirable to measure the response from the whole experimental unit. In studies involving measuring yields from treated field plots, it may not be practical to make yield determination by harvesting whole plots. For example, experiments with forage crops usually require large plots (to permit grazing, irrigation, fertilization control, and so on), and harvesting the whole plot is inexpedient. The usual practice is to measure the yields in a random sample of subplots in each experimental unit. Subsampling is also necessary, for example, when counting nematodes in treated plots, determining the chemical composition of a treated sample, and measuring the serum concentration levels of drugs.

Models with subsampling

The ANOVA models described thus far are easily extended to include subsampling. A one-way fixed-effects model with subsampling is described in Box 14.10.

BOX **14.10**

One-way ANOVA models with subsampling

Consider a study involving t treatments in s_1 experimental units. Suppose that s_2 responses are measured for each treatment in each experimental unit. Let Y_{ijk} denote the observed response for the k-th sample in the j-th replication of the i-th treatment ($i = 1, 2, \ldots, t$; $j = 1, 2, \ldots, s_1$; $k = 1, 2, \ldots, s_2$). The treatment effects can be compared by ANOVA under the following two models for Y_{ijk}.

Fixed treatment effects

$$Y_{ijk} = \mu + \tau_i + R_{(i)j} + S_{ijk},$$

where μ is the overall mean; τ_i is the effect of the i-th treatment, and $\Sigma \tau_i = 0$; $R_{(i)j}$ is the error associated with the j-th replication of the i-th treatment; and S_{ijk} is the error associated with the k-th subsample in the j-th replication of the i-th treatment. The $R_{(i)j}$ are independent $N(0, \sigma_R^2)$, and the S_{ijk} are independent $N(0, \sigma_S^2)$. Also, the $R_{(i)j}$ and S_{ijk} are independent.

Random treatment effects

$$Y_{ijk} = \mu + T_i + R_{(i)j} + S_{ijk},$$

where μ is the overall mean; T_i is the random effect of the i-th treatment; $R_{(i)j}$ is the error associated with the j-th replication of the i-th treatment; S_{ijk} is the error associated with the k-th subsample in the j-th replication of the i-th treatment. The $T_{(i)}$ are independent $N(0, \sigma_T^2)$; the $R_{(i)j}$ are independent $N(0, \sigma_R^2)$, and the S_{ijk} are independent $N(0, \sigma_S^2)$. Also, the T_i, $R_{(i)j}$ and S_{ijk} are mutually independent.

Comments on the models in Box 14.10

1 The assumption that the S_{ijk} are independent implies that, within an experimental unit, the responses can be regarded as a random sample from a population of responses. For instance, in the context of the plant height example, independence of S_{ijk} means that the errors associated with the measured heights of two seedlings within a given pot are statistically independent. In other words, responses within experimental units can be regarded as subsamples of responses.

2 Comparison with Boxes 8.2 and 14.1 shows that the assumptions underlying the models in Box 14.10 are straightforward generalizations of the assumptions in the one-way random- and fixed-effects models. The assumptions in these models can be summarized as follows:

a Each response is the sum of a number of components (for instance, μ, $R_{(i)j}$, E_{ijk}), some of which are fixed, while others are random.

b Each random component in the model has a normal distribution, with zero mean and an unknown variance; for instance, the $R_{i(j)}$ are $N(0, \sigma_R^2)$.

c The random components are mutually independent; for instance, T_i, $R_{i(j)}$, and S_{ijk} are mutually independent.

3 Comparison with the two-way nested-effects models in Section 14.5 shows that the two models in Box 14.10 are the same as the two-way mixed- and random-effects models in Equations (14.15) and (14.14), respectively. The only difference is that, in the two-way nested effects models, there are—in addition to the overall mean—two effects of experimental factors (*A* and *B*) and one effect of the sampling factor (error). In the one-way model with subsampling, by contrast, we have one effect of the experimental factor (*T*) and two effects of sampling factors (*R* and *S*). Thus, analysis of the data under the models in Box 14.10 can be performed in exactly the same manner as analysis under two-way mixed-effects models. When we interpret results from such an analysis, however, we need to bear in mind the proper interpretation of the various effects.

EXAMPLE **14.18** In Example 8.2, the effects of five sources of light on the four-week heights of a variety of plant were compared. In this completely randomized experiment, the investigators exposed four randomly selected seedlings to each of five light sources. The study utilized the 20 seedlings as the experimental units. In Example 8.6, we used the following one-way fixed model to analyze these data

$$Y_{ij} = \mu + \tau_i + E_{ij}, \qquad i = 1, \ldots, 5; j = 1, \ldots, 4. \qquad \textbf{(14.16)}$$

Here Y_{ij} is the four-week height of the *j*-th seedling exposed to the *i*-th light source; μ is the overall mean four-week height; τ_i is the effect of the *i*-th light source, and $\Sigma \tau_i = 0$; and E_{ij} is the random error associated with the measurement on the *j*-th seedling receiving the *i*-th treatment. The E_{ij} are independent $N(0, \sigma^2)$.

Now, let's suppose that the study in Example 8.2 was actually conducted with ten pots as the experimental units and two seedlings per pot. The plant height data in Table 8.2, are shown in Table 14.2 with the appropriate pot identifications.

We have $t = 5$, $s_1 = 2$, and $s_2 = 2$. The total number of responses is $N = ts_1s_2 = 5 \times 2 \times 2 = 20$; the number of responses per treatment is $s_1s_2 = 2 \times 2 = 4$; the number of replications per treatment is the same as the number of experimental units (pots) per treatment—that is, $s_1 = 2$. The model is

$$Y_{ijk} = \mu + \tau_i + R_{(i)j} + S_{ijk}, \quad i = 1, \ldots, 5; \quad j = 1, 2; \quad k = 1, 2. \qquad \textbf{(14.17)}$$

The notation here is as follows:

■ Y_{ijk} is the four-week height of the *k*-th seedling in the *j*-th pot receiving the *i*-th treatment;

■ μ is the overall mean;

TABLE 14.2

Four-Week heights of seedlings, grouped in subsamples

Treatment (i)	Pot (j)	Four-week height (Y_{ijk}) Seedling ($k = 1$)	Seedling ($k = 2$)	Pot total (Y_{ij+})	Treatment total (Y_{i++})
1 (D)	1	32.94	35.98	68.92	136.08
	2	34.76	32.40	67.16	
2 (AL)	1	30.55	32.64	63.19	127.60
	2	32.37	32.04	64.41	
3 (AH)	1	31.23	31.09	62.32	123.36
	2	30.62	30.42	61.04	
4 (BL)	1	34.41	34.88	69.29	137.23
	2	34.07	33.87	67.94	
5 (BH)	1	35.61	35.00	70.61	137.17
	2	33.65	32.91	66.56	
Total (Y_{+++})					661.44

- τ_i is the effect of the i-th treatment, and $\Sigma \tau_i = 0$;
- $R_{(i)j}$ is the effect of the j-th replication (pot) receiving the i-th treatment;
- S_{ijk} is the effect of the k-th seedling in the j-th pot receiving the i-th treatment.

The $R_{(i)j}$ are independent $N(0, \sigma_R^2)$, and the S_{ijk} are independent $N(0, \sigma_S^2)$. Also the T_{ij} and S_{ijk} are independent.

As we see from a comparison of the models In Equations (14.16) and (14.17), the main difference between the two models is that the error effect E_{ij} in Equation (14.16) is replaced in Equation (14.17) by a sum of two independent effects: $R_{(i)j}$, the effect of the j-th pot (replication) receiving the i-th treatment; and S_{ijk}, the effect of the k-th seedling in the j-th pot receiving the i-th treatment. We assume that $E_{ij} = R_{(i)j} + S_{ijk}$, S_{ijk} and R_{ij} are independent, and so the variance of E_{ij} in Equation (14.16) is equal to the sum of the variances of $R_{(i)j}$ and S_{ijk} in Equation (14.17). Thus

$$\sigma^2 = \sigma_R^2 + \sigma_S^2. \tag{14.18}$$

This means that the variation due to the error component in Equation (14.16) is equal to the sum of the variance component due to $R_{(i)j}$ and the variance component due to S_{ijk}.

The computations for ANOVA based on the model in Equation (14.17) follow directly from the computing formulas for two-way nested-effects models. The relevant portions of output obtained when the NESTED procedure of SAS is used to analyze these data are as follows:

```
-----------------------------------------------------------------
              Coefficients of Expected Mean Squares
             Source        TRT        POT        ERROR
              TRT           4          2          1
              POT           0          2          1
              ERROR         0          0          1
        ----------------------------------------------------
Nested Random Effects Analysis of Variance for Variable HEIGHT


            Degrees
Variance      of         Sum of                            Error
Source      Freedom       Squares    F Value    Pr > F     Term

TOTAL          19        57.457320
TRT             4        41.080770    8.401      0.0192     POT
POT             5         6.112350    1.191      0.3793     ERROR
ERROR          10        10.264200

        ----------------------------------------------------
Variance                             Variance           Percent
Source      Mean Square              Component          of Total
TOTAL         3.024069               3.386376           100.0000
TRT          10.270193               2.261931            66.7950
POT           1.222470               0.098025             2.8947
ERROR         1.026420               1.026420            30.3103
            Mean                                 33.07200000
            Standard error of mean                0.71659486
-----------------------------------------------------------------
```

On the basis of this output, the ANOVA mean squares and their expected values (EMS) may be summarized as in Table 14.3.

To test the null hypothesis H_0: $\psi_T^2 = 0$ (no differences between treatment effects) against the research hypothesis H_1: $\psi_T^2 > 0$, we use the test statistic

$$F_T = \frac{\text{MS}[T]}{\text{MS}[R(T)]} = \frac{10.2701}{1.2225} = 8.401, \qquad df = 4, 5.$$

TABLE 14.3

Mean squares and EMS for the plant height data

Source	df	MS	EMS
T treatments (light sources)	4	10.2701	$\sigma_S^2 + 2\sigma_R^2 + 4\psi_T^2$
R(T) pots within treatments	5	1.2225	$\sigma_S^2 + 2\sigma_R^2$
S(R) seedlings within Pots	10	1.0264	σ_S^2
Total	19		

We can reject H_0 at $\alpha = 0.05$ and conclude that the five light sources have different effects on plant height. Indeed, as we see from the output, the p-value for the test is $p = 0.0192$, so that the null hypothesis can be rejected at a level as small as 0.0192.

The null hypothesis $H_0 : \sigma_R^2 = 0$ can be tested using the test statistic

$$F_R = \frac{MS[R(T)]}{MS[S(R)]} = \frac{1.2225}{1.0264} = 1.1911, \quad df = 5, 10.$$

The calculated value is not significant, even at a level as high as $\alpha = 0.25$. Indeed, the exact p-value corresponding to $F_R = 1.1911$ is 0.3793. We conclude that there is insufficient evidence to reject H_0. It appears that pot-to-pot variation within treatments is not a significant factor in the observed variability of plant heights.

Table 14.3 can also be used to obtain point estimators of σ_R^2 and σ_S^2. Let $\hat{\sigma}_R^2$ and $\hat{\sigma}_S^2$ denote the estimates of σ_R^2 and σ_S^2, respectively. Then

$$\hat{\sigma}_S^2 = MS[S(R)] = 1.0264,$$

$$\hat{\sigma}_R^2 = \frac{MS[R(T)] - MS[S(R)]}{2} = \frac{1.2225 - 1.0264}{2} = 0.0981.$$

Thus, an estimate of the variance of heights of seedlings exposed to the same light source is

$$\hat{\sigma}_R^2 + \hat{\sigma}_S^2 = 0.0981 + 1.0264 = 1.1245,$$

of which 0.0981 can be ascribed to the variability in heights of seedlings grown in different pots. Thus, about 8.7% of the variation between seedlings within treatments $(0.0981/1.1245 = 0.0872)$ can be ascribed to differences between the heights of seedlings grown in the same pot.

The method used to construct confidence intervals for the expected response at a given level of the fixed effect in a two-way mixed-effects model (see Example 14.17) can also be used for one-way models with fixed treatment effects and subsampling. For instance, a 95% confidence interval for μ_1, the true mean four-week height of seedlings grown in darkness, is given by

$$\overline{Y}_{1+} - t(t(s_1 - 1), 0.025)\sqrt{\frac{MS[R(T)]}{s_1 s_2}} \leq \mu_1$$

$$\leq \overline{Y}_{1+} + t(t(s_1 - 1), 0.025)\sqrt{\frac{MS[R(T)]}{s_1 s_2}}$$

Thus,

$$32.60 \leq \mu_1 \leq 35.44. \quad \blacksquare$$

Exercises

14.20   The objective of a greenhouse experiment was to compare the effects of four fertilizer treatments—1, none; 2, straw; 3, straw + PO_4; 4, straw + PO_4 + lime—on the yield (g) of a variety of wheat. The experimenters used 12 pots, each containing three wheat plants. Each treatment was assigned to three pots at random (Cornell, 1971).

Hypothetical wheat yields for this experiment are shown in Table 14.4. Assume the model

$$Y_{ijk} = \mu + \tau_i + R_{(i)j} + S_{ijk},$$

where μ, τ_i, $R_{(i)j}$ and S_{ijk} denote, respectively, the overall mean; the effect of the i-th treatment; the effect of the j-th pot in the i-th treatment; and the error associated with the measurement on the k-th plant in the j-th pot receiving the i-th treatment.

a   Explain the meanings of the terms in the model. State assumptions that will justify the application of ANOVA F- and t-test procedures to the wheat yield data.

b   Construct the ANOVA table, including a column showing the EMS.

c   Do these data indicate a difference between the effects of the four fertilizer treatments?

d   Use the Tukey multiple comparison method to simultaneously test, at the 0.05 level, all possible pairwise differences between the treatment means. Write a summary conclusion.

e   Using the Bonferroni method, construct a set of 95% simultaneous confidence intervals for the contrasts

$$\theta_1 = \tau_1 - \tau_2,$$

$$\theta_2 = \tau_1 - \frac{1}{3}(\tau_2 + \tau_3 + \tau_4).$$

Interpret the resulting intervals.

f   Obtain an estimate of the proportion of variance between yields within treatments that can be ascribed to differences between pots. Comment on the estimate.

TABLE   **14.4**

Wheat yields (g) in a greenhouse experiment

Treatment number	Treatment	Pot number	Plant number		
			1	2	3
1	None	1	20.6	22.3	19.8
		2	23.4	21.9	22.8
		3	21.8	20.6	21.3
2	Straw	1	13.6	13.9	14.2
		2	13.7	14.5	13.8
		3	12.9	13.1	13.4
3	Straw + PO$_4$	1	14.8	14.6	14.9
		2	14.3	13.9	13.5
		3	14.4	13.8	14.1
4	Straw + PO$_4$ + lime	1	14.1	13.8	14.3
		2	14.0	13.9	14.2
		3	14.4	14.1	13.6

14.21   Consider the one-factor random-effects model with subsampling

$$Y_{ijk} = \mu + T_i + R_{(i)j} + S_{ijk}.$$

a   Show that the variance of $\overline{Y}_{+++}$, the overall sample mean, is

$$\text{Var}(\overline{Y}_{+++}) = \frac{\sigma_S^2 + s_1 \sigma_R^2 + s_1 s_2 \sigma_T^2}{t s_1 s_2}.$$

[Hint: Note that $\overline{Y}_{+++} = \mu + \overline{R}_{++} + \overline{S}_{+++}$.]

b   Use the variance expression in (a) to suggest an estimator for the standard error of $\overline{Y}_{+++}$. [Hint: Examine the EMS values.]

14.7
Determining expected mean squares

We have seen many examples of the use of expected mean squares in the analysis of factorial experiments. In addition to their use in determining the appropriate denominators for ANOVA F-test statistics, the expected mean squares provide a convenient tool for estimating the standard errors needed to construct confidence intervals for linear functions of fixed effects. In this section, we consider a set of rules for determining the EMS values associated with a given model in the special—but important—case where the subclass numbers are equal. For more details about these rules, see Khuri (1982) and Searle, Casella, and McCulloch (1992).

The rules for finding EMS values require the underlying statistical model to be written using a combination of subscripts—one for each source of variation that affects the response. Thus, there should be one subscript corresponding to each experimental factor, each identifiable confounding variable, and each source that affects the error (such as sampling and experimental errors).

Every effect in a model has a number of subscripts and each subscript has a fixed range of possible values. For example, consider the two-way mixed effects model

$$Y_{ijk} = \mu + \alpha_i + B_j + (\alpha B)_{ij} + E_{ijk}. \tag{14.19}$$

Each of the four effects has its own subscripts—i for α, i and j for (αB), and so on—and each subscript has its own range: a for i; b for j; and n for k.

Every effect is associated with a parameter—called a *variance component* if the effect is random and an *effect size* if the effect is fixed. For example, in Equation (14.19), the random effect $(\alpha B)_{ij}$ is associated with the variance component $\sigma_{\alpha B}^2$, whereas the fixed effect α_i is associated with the effect size ψ_A^2.

Let MS[X] be the mean square associated with the effect X in the model. Then the rules for determining EMS[X] are as follows.

1   If X random, EMS[X] is a linear combination of those variance components corresponding to effects whose subscripts include the subscripts of X. For example, in Equation (14.19), the subscript of B is j, so that EMS[B] is a linear combination of the variance components associated with $(\alpha B)_{ij}$ and E_{ijk}; that is, EMS[B] is a linear combination of $\sigma_{\alpha B}^2$ and σ^2.

The coefficient of a particular variance component in a linear combination equals 1 if the variance component is the error variance σ^2. Otherwise, it is equal to the product of the ranges of the subscripts that do not appear in the effect associated with the variance component. As an example, consider the coefficient of $\sigma_{\alpha B}^2$. The subscripts of the effect corresponding to this variance component are i and j, so that the coefficient is the product of the ranges of the remaining subscripts—namely, k. Thus, the coefficient of $\sigma_{\alpha B}^2$ is n. Similarly, the coefficient of σ_B^2 is the product of the ranges of i and k, which is $a \times n = an$.

2 If X is fixed, we first determine EMS[X] by treating X as random and then replace the variance component for X by its effect size. To determine EMS[A] in Equation (14.19), we first notice that it is a linear combination of the variance components σ_A^2, $\sigma_{\alpha B}^2$, and σ^2, because the subscript i is contained in the subscripts of α_i, $(\alpha B)_{ij}$, and E_{ijk}. The corresponding coefficients are bn, n, and 1, respectively. Thus, the expected value when A is treated as random is

$$\text{EMS}[A] = bn\sigma_A^2 + n\sigma_{\alpha B}^2 + \sigma^2.$$

Now, if we replace the variance component of A with its effect size, we get

$$\text{EMS}[A] = bn\psi_A^2 + n\sigma_{\alpha B}^2 + \sigma^2,$$

which is the correct expected mean squares.

As, already noted, these rules for determining EMS are not applicable in the case of unequal subclass numbers. In that case, statistical software such as SAS can be used to determine EMS.

EXAMPLE 14.19 Consider the radon flux data in Example 14.16. Suppose that one value in contour 1, site 1 (-0.4050) and two values in contour 3, site 2 ($-0.9782, -0.1335$) are missing. The resulting data will have unequal subclass numbers, and so the EMS values cannot be determined using the rules in this section. Though we could determine the EMS values from first principles, it is best to use statistical software such as SAS. The control codes required to apply the NESTED procedure of SAS to the modified radon flux data are as follows.

```
Data height;
Input cont site cannist logradon @@;
cards;
1 1 1   0.6130 1 1 3 -0.6445 1 2 1 0.8303 1 2 2 -0.1154 1 2 3   1.2507
2 1 1   0.9450 2 1 2   0.3500 2 1 3 0.6060 2 2 1 -1.3205 2 2 2 -0.7875
2 2 3 -1.3548 3 1 1   0.6136 3 1 2 0.0266 3 1 3   0.2776 3 2 1 -0.4636
;
Proc sort;
by cont site;
proc nested;
class cont site;
var logradon;
run;
```

For more details on how to implement this procedure, see Section 14.4 in Younger (1997). The output from the NESTED procedure includes the following table.

Coefficients of Expected Mean Squares

Source	CONT	SITE	ERROR
CONT	4.933333333	2.683333333	1.000000000
SITE	0.000000000	2.300000000	1.000000000
ERROR	0.000000000	0.000000000	1.000000000

From the output, the expected mean squares for contours, sites in contours, and canisters within sites (E) are

$$\text{EMS}[A] = 4.93\sigma_A^2 + 2.68\sigma_{B(A)}^2 + \sigma^2,$$

$$\text{EMS}[B(A)] = 2.3\sigma_{B(A)}^2 + \sigma^2$$

$$\text{EMS}[E] = \sigma^2 \quad \blacksquare$$

Exercises

14.22   Assuming equal subclass numbers, use the EMS rules to verify the EMS values for the one-way models in Box 14.2.

14.23   Use the EMS rules to derive the expected mean squares for two-way models in Table 14.1.

14.24   Consider the three-way mixed-effects model with crossed factors

$$Y_{ijk} = \mu + \alpha_i + B_j + \gamma_k + (\alpha B)_{ij} + (\alpha\gamma)_{ik} + (B\gamma)_{ik} + (\alpha B\gamma)_{ijk} + E_{ijkl},$$

in which A ($a = 3$ levels) and C ($c = 2$ levels) are fixed, and B ($b = 2$ levels) is random.

a   Assuming that there are $n = 4$ replications per treatment combination, derive the EMS values.

b   Determine the forms of the test statistic—that is, determine the numerator and the denominator of the F-statistic—for each factorial effect included in the model.

14.8
Overview

In this chapter, we focus on the design and analysis of experiments based on one- and two-way models with random and mixed effects. Models with random effects are needed in situations where measuring responses at every level of a factor is either impracticable or very expensive. Random effects can occur in a model whether the factors are crossed or nested.

In the next two chapters, we turn our attention to models that are more complex and more general. In order to understand the methods appropriate for handling such complex models, we must understand the similarities and differences between the various models for studying one or two experimental factors. While the similarities help us to appreciate the general logic underlying ANOVA models , the differences highlight the need to know precisely what quantities can be estimated by the various mean squares in an ANOVA table. The expected values of the mean squares (EMS) can be used to answer this question. The EMS values also provide useful clues for constructing intuitively reasonable and theoretically sound procedures for making inferences with ANOVA models.

The main difference between fixed- and random-effects models can be illustrated for the case of one-way models. In a one-way fixed-effects model, the treatment means ($\mu_i = \mu + \tau_i$) and the error variance (σ^2) are the parameters of primary interest; in a one-way random-effects model, by contrast, the overall mean (μ) and the variance components (σ_T^2 , σ^2) are the key parameters. Consequently, the procedures that are used to make inferences for fixed-effects models are generally different from those used for random-effects models. In identifying the procedures that are appropriate, we look at the EMS values, which play a critical role in untangling the relationships of ANOVA mean squares to the parameters in the underlying model.

Section 14.6 is devoted to the effect of subsampling on the analysis of one-way models. In subsampling, a random sample of the responses observed on each experimental unit is considered. The presence of subsampling does not complicate the statistical analysis of the data; single-factor models with subsampling can be handled similarly to the corresponding models without subsampling. The main difference is that subsampling permits the identification of an additional source of variation in the measured values. Consequently, the error sum of squares can be subdivided into two parts: the component due to differences between replications (experimental units) within treatments; and a component due to differences between samples (responses) within replications. The net effect of subsampling is to require appropriate modifications in the procedures for handling the models. Again, the EMS values play a key role in determining the type of alterations necessary.

15

ANOVA Models with Block Effects

15.1
Introduction

In Section 7.3, we saw how researchers can use blocking to eliminate the effects of confounding factors in the studies that they design. Blocks are groups of experimental units affected by a common level of a confounding factor. In Display 7.2, we described a block design in which the confounding factor was light condition and the experimental factor was soil moisture content.

This chapter focuses on two study designs: the randomized complete block design and the Latin square design. These designs have simple structures and are based on models that are reasonable under a variety of experimental conditions.

15.2
Single-factor studies in randomized complete blocks

Let's begin with an example.

EXAMPLE 15.1 A feed trial to compare three dietary supplements was conducted using 12 animals of approximately the same body weight. The 12 animals consisted of four litters, each containing three animals. Within a given litter, the three animals were randomly assigned to the three dietary supplements. The animals were housed in 12 identical pens and fed their assigned diets under identical conditions. The measured weight gains (g) after 12 weeks on the experimental diets are as follows:

Dietary supplement	Litter 1	2	3	4	Total
1	28.7	29.3	28.2	28.6	114.8
2	30.7	34.9	32.6	34.4	132.6
3	31.9	34.2	34.9	35.3	136.3
Total	91.3	98.4	95.7	98.3	383.7

The experimental design for the feed trial is an example of a randomized complete block design with three treatments and four blocks. The treatments are the levels of the experimental factor, the dietary supplement; the blocks are the levels of the confounding factor, the litter. Because animals in different litters may respond differently to a given dietary supplement, the litter is regarded as a confounding factor. The 12 experimental units (animals) are grouped into four blocks in such a way that, within each group, the three units are affected by the same level of the confounding factor. Because of the shared inherited characteristics of animals within a litter (block), their responses are likely to be relatively similar, whereas the responses of animals belonging to different litters will vary more widely; that is, the experimental units are more homogeneous within blocks than between blocks. ∎

Randomized complete block design

A *randomized complete block* (RCB) design with t treatments and r blocks utilizes $N = rt$ experimental units that are divided into r blocks of t units each. The units within a block are more homogeneous (similar in their ability to respond to the treatments) than units in different blocks. The t treatments are assigned at random to the t units within each block.

The RCB design in Example 15.1 required $N = 12$ experimental units divided into $r = 4$ blocks of $t = 3$ units each. The levels of the experimental factor were fixed in advance by the investigator, and so the study was a randomized complete block experiment with a fixed experimental factor. In the next example, we consider a RCB experiment with a random experimental factor.

EXAMPLE 15.2   Consider the soybean protein study in Exercise 14.1, which was conducted using a completely randomized design; 30 seeds (three seeds from each of a random sample of ten F_2 generation plants) were randomly assigned to 30 homogeneous experimental plots. Suppose that we decide to conduct the study by growing the plants in pots in a greenhouse. Suppose further that at most ten pots can be accommodated on a single bench and that, as is often the case, the bench effect is a potential confounding effect when evaluating the plant-to-plant variability of the soybean protein content. Provided the observed responses satisfy certain assumptions—described in Box 15.1—a RCB design in which the benches are the blocks may be used to eliminate the bench effects when making inferences about the plant-to-plant variability of soybean protein contents. Such a design will have $r = 3$ blocks and $t = 10$ treatments. Ten seeds, one from each plant, will be randomly assigned to the ten pots on a bench (block). Clearly, such a study will have a random experimental factor in a randomized complete block experiment.   ■

A model for a single-factor RCB design can be written by noting that the structure of the data generated by such designs is similar to the structure of the data from an $r \times t$ factorial experiment in a completely randomized design (CR design) in which each treatment combination is replicated $n = 1$ times. In both designs there are two factors. In the $r \times t$ factorial experiment in a CR design, both factors are experimental factors, whereas a single-factor RCB design includes one experimental factor with t levels (treatments) and one confounding factor with r levels (blocks). Thus, both designs generate two-way classified data that can be classified into rt classes according to the levels of two factors.

A model for a RCB design can be written along the lines of the two-way ANOVA models in Sections 13.5 and 14.5. There is $n = 1$ observation at each of the rt combinations of the levels of the two factors (treatment and block), and so the required models can be obtained by setting $n_{ij} = n = 1$ in the models in Sections 13.5 and 14.5. For instance, if both blocks and treatments are fixed, a model for a RCB design can be obtained from the model in Box 13.1

$$Y_{ijk} = \mu + \rho_i + \tau_j + (\rho\tau)_{ij} + E_{ijk}, \qquad i = 1, \ldots, r; \; j = 1, \ldots, t; \; k = 1. \quad \text{(15.1)}$$

The notation here is as follows:

- Y_{ijk} is the observed response for the j-th treatment in the i-th block;
- μ is the overall mean;
- ρ_i is the effect of the i-th block, and $\Sigma\rho_i = 0$;
- τ_j is the effect of the j-th treatment, and $\Sigma\tau_j = 0$;
- $(\rho\tau)_{ij}$ is the interaction effect; $\Sigma_i(\rho\tau)_{ij} = 0$, and $\Sigma_j(\rho\tau)_{ij} = 0$;
- E_{ijk} is the effect of random error.

The E_{ijk} are independent $N(0, \sigma^2)$.

The third subscript in Equation (15.1) k takes only one value, because there is only one observation for each treatment within a block. Thus, $k = 1$ in all cases, and this subscript is not necessary to distinguish responses from one another; it serves no useful purpose in the model. Accordingly, in what follows, models for RCB designs will be written with only two subscripts. For instance, Equation (15.1) will be written as

$$Y_{ij} = \mu + \rho_i + \tau_j + (\rho\tau)_{ij} + E_{ij}, \qquad i = 1, \ldots, r;\ j = 1, \ldots, t.$$

The terms Y_{ij}, μ, and so on have the same meanings as the corresponding terms in Equation (15.1).

Even though a CR design with two crossed factors and a RCB design with one experimental factor produce two-way classified data with classes determined by the combinations of levels of two factors, the methods of randomizing treatments to experimental units are different in the two designs. In a CR design, the treatments are assigned to experimental units completely at random. In contrast, randomization in RCB design is within blocks; this is sometimes called *restricted randomization*. Treatments are assigned to experimental units under the restriction that every block should contain exactly one unit assigned to each treatment. Consequently, some statisticians (Anderson, & McLean, 1974; Box, & Hunter, 1978) argue that the two-way ANOVA approach for analyzing the results of a two-factor CR design is not entirely appropriate for analyzing the results of a RCB design. They point out that, while the two-way ANOVA F-test for the differences between treatments (levels of the experimental factor) is valid in a RCB design, the corresponding F-test for differences between the block effects is, at best, approximate. However, in most applications of block designs, the primary focus is on inferences about treatment differences, and so a two-way ANOVA approach for analyzing a RCB design is adequate from a practical viewpoint. Consequently, in this book, we apply appropriate ANOVA approaches to RCB designs and other block designs involving restricted randomization on the understanding that inferences concerning block differences are only approximate.

Block-treatment interaction in RCB designs

The model in Equation (15.1) is based on the assumption that there is only one observation at each block-treatment combination—that is, that $n_{ij} = n = 1$. However, if $n = 1$, it is not possible to make meaningful inferences about the possible interaction between blocks and treatments. To illustrate this for the case of ANOVA based on the two-way models in Sections 13.5 and 14.5, let's examine the associated EMS values. These EMS values, which can be obtained by setting $a = r$, $b = t$, and $n = 1$ in part 1 of Table 14.1, are shown in Table 15.1. Notice the zero degrees of freedom for SS[E] in Table 15.1. That implies that the data provide no direct estimate of the error variance σ^2.

The mean square for error is the only quantity with an associated EMS value of σ^2, and so it is not possible to calculate a direct estimate of σ^2 using data from an RCB design. Intuitively, this makes sense because, as pointed out in Chapter 7, σ^2 is the variance of the population of measurements made under identical experimental conditions. In a RCB design, responses measured under identical experimental conditions are responses corresponding to a common block-treatment combination. Consequently, σ^2 is the variance of the conceptual population of responses that can be observed for a given treatment within a given block. As in a CR design—see

Equation (8.2)—the MS[E] in a RCB ANOVA is the pooled estimate of σ^2 obtained by combining the variances of the sample of measurements made under identical experimental conditions. In symbols, if n_{ij} and S_{ij}^2 denote, respectively, the sample size and the sample variance for the sample of responses for treatment j in block i, then

$$MS[E] = \frac{SS[E]}{df_E},$$

where

$$df = \sum_{i=1}^{r} \sum_{j=1}^{t} (n_{ij} - 1)$$

and

$$SS[E] = \sum_{i=1}^{r} \sum_{j=1}^{t} (n_{ij} - 1) S_{ij}^2.$$

In a RCB design, the response to each block-treatment combination constitutes a sample of size $n = 1$ ($n_{ij} = n = 1$), so that the variance of each sample will be zero—that is, $S_{ij}^2 = 0$. Consequently, both df_E and SS[E] will be zero. This implies that MS[E] is undefined and that, in a RCB design, we do not have a direct estimate of σ^2.

The EMS in Table 15.1 reveal that, if we are unable to obtain a direct estimate of σ^2, we cannot separately estimate the components of variance due to interaction (ψ_{RT}^2, σ_{RT}^2 and $\sigma_{R\tau}^2$) and due to error (σ^2). Some consequences of the fact that $n = 1$ in the analysis of RCB designs are as follows:

1   The mean square for interaction MS[RT] is an unbiased estimator of $\psi_{RT}^2 + \sigma^2$ if treatments and blocks are both fixed, of $\sigma_{RT}^2 + \sigma^2$ if they are both random, and of $\sigma_{R\tau}^2 + \sigma^2$ if treatments are fixed and blocks are random.

2   If both blocks and treatments are fixed, the hypotheses H_{0T}: $\psi_T^2 = 0$, H_{0R}: $\psi_R^2 = 0$, and H_{0RT}: $\psi_{RT}^2 = 0$ cannot be tested using ratios of two mean squares. However, if the assumption that there is no block-treatment interaction—that is, that $\psi_{RT}^2 = 0$—is reasonable, then the F-ratios

$$F_T = \frac{MS[T]}{MS[RT]} \quad \text{and} \quad F_R = \frac{MS[R]}{MS[RT]} \tag{15.2}$$

can be used to test H_{0T} and H_{0R}, respectively.

TABLE   15.1

EMS for two-way ANOVA of RCB experiments with interaction

Source	df	R, T fixed	R, T random	R random, T fixed
			EMS	
R (blocks)	$r - 1$	$t\psi_R^2 + \sigma^2$	$t\sigma_R^2 + \sigma_{RT}^2 + \sigma^2$	$t\sigma_R^2 + \sigma_{R\tau}^2 + \sigma^2$
T (treatments)	$t - 1$	$r\psi_T^2 + \sigma^2$	$r\sigma_T^2 + \sigma_{RT}^2 + \sigma^2$	$r\psi_T^2 + \sigma_{R\tau}^2 + \sigma^2$
RT (treatment $\times$ block)	$(r-1)(t-1)$	$\psi_{RT}^2 + \sigma^2$	$\sigma_{RT}^2 + \sigma^2$	$\sigma_{R\tau}^2 + \sigma^2$
E (error)	0	σ^2	σ^2	σ^2

3 If both blocks and treatments are random, then the hypothesis H_{0RT}: $\sigma_{RT}^2 = 0$ cannot be tested using a ratio of two mean squares. However, the F-ratios in Equation (15.2) can be used to test the hypotheses H_{0T}: $\sigma_T^2 = 0$ and H_{0R}: $\sigma_R^2 = 0$, respectively.

4 If the blocks are random and the treatments are fixed, then the hypothesis H_{0RT}: $\sigma_{R\tau}^2 = 0$ cannot be tested using a ratio of two mean squares. However, the F-ratios in Equation (15.2) can be used to test the hypotheses H_{0T}: $\psi_T^2 = 0$ and H_{0R}: $\sigma_R^2 = 0$, respectively.

These observations are summarized in Table 15.2.

Thus, if both blocks and treatments are fixed, RCB designs do not permit inferences based on ANOVA F-tests about any of the three factorial effects. Even when one of the factors (block or treatment) is random, the RCB design cannot be used to make inferences about possible interaction between treatments and blocks. Consequently, RCB designs are most useful when it is reasonable to assume that there is no interaction between blocks and treatments.

An obvious way to overcome the problem of block-treatment interaction in a RCB design is to make sure that, within each block, at least two responses are observed for each treatment. For instance, by setting $a = r$, $b = t$, and $n = 2$ in part 1 of Table 14.1, we can verify that, if two responses are observed for each treatment within each block, then the EMS associated with a two-way ANOVA with block-treatment interaction are as in Table 15.3.

As we see from Table 15.3, the problem of zero degrees of freedom for experimental error can be overcome by introducing two replications per treatment within each block. Clearly, data from RCB designs with several replications of

T A B L E   15.2

Hypotheses about RCB models with interaction that can be tested using F-ratios

	Null hypothesis		
Assumption	H_{0RT}	H_{0R}	H_{0T}
R and T fixed	No	No	No
R and T random	No	Yes	Yes
R random, T fixed	No	Yes	Yes

T A B L E   15.3

EMS for RCB designs with two replications per treatment within each block

		EMS		
Source	df	R and T fixed	R and T random	R random, T fixed
R (blocks)	$r - 1$	$t\psi_R^2 + \sigma^2$	$t\sigma_R^2 + 2\sigma_{RT}^2 + \sigma^2$	$t\sigma_R^2 + 2\sigma_{R\tau}^2 + \sigma^2$
T (treatments)	$t - 1$	$r\psi_T^2 + \sigma^2$	$r\sigma_T^2 + 2\sigma_{RT}^2 + \sigma^2$	$r\psi_T^2 + 2\sigma_{R\tau}^2 + \sigma^2$
RT (treatment × block)	$(r-1)(t-1)$	$2\psi_{RT}^2 + \sigma^2$	$2\sigma_{RT}^2 + \sigma^2$	$2\sigma_{R\tau}^2 + \sigma^2$
E (error)	rt	σ^2	σ^2	σ^2

each treatment within each block can be analyzed using a two-way ANOVA, as described in Chapters 13 and 14. Unfortunately, multiple replication of treatments within blocks implies an increase in the number of experimental units per block. This is often a practical problem. For instance, in the feed trial in Example 15.1, assigning each diet to two animals in each litter means that we need a litter size of 6. This can be a problem if the normal litter size for the animals being studied is less than 6. Clearly, as the number of treatments increases, so does the difficulty involved in ensuring that the experimental units within blocks are homogeneous. In this book, we confine our attention to RCB designs in which each treatment is represented only once in each block.

Models for RCB designs with no block-treatment interaction

Two-way ANOVA models for randomized complete block designs are summarized in Box 15.1 under the assumption that there is no block-treatment interaction.

BOX **15.1**   *Two-Way ANOVA models for randomized complete block designs*

Let Y_{ij} denote the response for treatment j in block i ($i = 1, \ldots, r; j = 1, \ldots, t$). Then the following are three possible models for Y_{ij}.

Fixed-effects model (treatments and blocks fixed)

$$Y_{ij} = \mu + \rho_i + \tau_j + E_{ij},$$

where μ the overall mean response; ρ_i is the fixed effect of block i, and $\Sigma \rho_i = 0$; τ_j is the fixed effect of treatment j, and $\Sigma \tau_j = 0$; and E_{ij} is the random error in Y_{ij}. The E_{ij} are independent $N(0, \sigma^2)$.

Mixed-effects model (treatments fixed, blocks random)

$$Y_{ij} = \mu + R_i + \tau_j + E_{ij},$$

where μ, τ_j, and E_{ij} are as in the preceding case; R_i is the random effect of block i. The R_i are independent $N(0, \sigma_R^2)$ and are independent of the E_{ij}.

Random-effects model (treatments and blocks random)

$$Y_{ij} = \mu + R_i + T_j + E_{ij},$$

where μ, R_i, and E_{ij} are as in the preceding case; T_j is the random effect of treatment j. The T_j are independent $N(0, \sigma_T^2)$ and are independent of the R_i and E_{ij}.

The EMS associated with the RCB models in Box 15.1 can be obtained by deleting the interaction components from the EMS in Table 15.1, as shown in Table 15.4.

TABLE 15.4

EMS for the RCB models in Box 15.1

Source	MS	df	R, T fixed	R, T random	R random, T fixed
				EMS	
Blocks	MS[R]	$r - 1$	$t\psi_R^2 + \sigma^2$	$t\sigma_R^2 + \sigma^2$	$t\sigma_R^2 + \sigma^2$
Treatments	MS[T]	$t - 1$	$r\psi_T^2 + \sigma^2$	$r\sigma_T^2 + \sigma^2$	$r\psi_T^2 + \sigma^2$
Error	MS[RT]	$(r - 1)(t - 1)$	σ^2	σ^2	σ^2

From Table 15.4, we see that, under the models in Box 15.1, MS[RT] is an unbiased estimator of σ^2. Therefore, the interaction sum of squares in a RCB design is often called the error sum of squares.

The ANOVA table for the analysis of a RCB design is given in Table 15.5. The sums of squares in Table 15.5 can be calculated using statistical software or Equation (13.13) with $n = 1$.

The next example illustrates the analysis of a RCB design with fixed treatment effects.

EXAMPLE 15.3 Let's analyze the weight gain data of Example 15.1 to answer the following questions.

1 Do the data indicate a difference between the three dietary supplements?

2 Do the data indicate that there is a litter effect?

3 On the basis of the study results, would you recommend that litters be used as blocks in the future?

4 What is a 95% lower confidence bound for the difference between the average of the expected weight gains for diet 2 and diet 3 and the expected weight gain under diet 1?

Assume that there is no interaction between diets and litters; that is, assume that the differential expected responses to the diets are the same across litters. The study objective is to compare the three diets, and so the treatments must be regarded as fixed. Also, the blocks should be regarded as random, because inferences about diet

TABLE 15.5

ANOVA table for RCB designs

Source	df	SS	MS
Blocks	$r - 1$	SS[R]	MS[R]
Treatments	$t - 1$	SS[T]	MS[T]
Error	$(r - 1)(t - 1)$	SS[E]	MS[E]
Total	$rt - 1$	SS[TOT]	

differences should refer to the population of all litters. The litters included in the experiment should be regarded as a random sample from the population of all litters. Therefore, it is reasonable to analyze these data using the model

$$Y_{ij} = \mu + R_i + \tau_j + E_{ij}, \qquad i = 1, \ldots, 4; \; j = 1, 2, 3.$$

Here μ is the overall mean; R_i is the random effect of the i-th litter; τ_j is the fixed effect of the j-th diet, and $\Sigma \tau_j = 0$; and E_{ij} is the random error. The R_i are independent $N(0, \sigma_R^2)$, and the E_{ij} are independent $N(0, \sigma^2)$.

The calculations for the ANOVA can be performed using Equation (13.13) with $a = t = 3$, $b = r = 4$, and $n = 1$. When performing these calculations, keep in mind that the interaction sum of squares acts as the error sum of squares in a RCB design; that is, SS[E] = SS[RT]. Thus

$$\text{SS[TOT]} = 12{,}353.35 - 12{,}268.81 = 84.54, \quad df_{TOT} = rt - 1 = 11;$$

$$\text{SS[R]} = 12{,}279.88 - 12{,}268.81 = 11.07, \quad df_R = r - 1 = 3;$$

$$\text{SS[T]} = 12{,}334.87 - 12{,}268.81 = 66.06, \quad df_T = t - 1 = 2;$$

$$\text{SS[RT]} = 84.54 - 66.06 - 11.07 = 7.41, \quad df_{RT} = (rt - 1) - (r - 1) - (t - 1)$$
$$= (r - 1)(t - 1) = 6.$$

The ANOVA table for the weight gain data is as follows:

Source	df	SS	MS	EMS	F_{cal}	p
Blocks (litters)	3	11.07	3.69	$3\sigma_R^2 + \sigma^2$	2.98	0.118
Treatments (diets)	2	66.06	33.03	$4\psi_T^2 + \sigma^2$	26.64	0.001
Error	6	7.41	1.24	σ^2		
Total	11	84.54				

On the basis of the EMS values, the null hypothesis, H_{0T}: $\psi_T^2 = 0$—in words, that there is no difference between the expected responses for the three diets—can be tested using the F-ratio

$$F_T = \frac{\text{MS[T]}}{\text{MS[E]}} = \frac{33.03}{1.24} = 26.64, \qquad df_T = 2, \; df_E = 6.$$

The calculated value is significant at $\alpha = 0.001$ ($p = 0.001$), and so we conclude that there is a difference between the effects of the three diets. Conclusions about the differences between litter effects can be based on the calculated F-value for blocks ($F_R = 2.98$; $df = 3, 6$; $p = 0.118$). There does not appear to be a significant litter-to-litter variation in weight gains. Of course, as already noted, the ANOVA F-test for blocks is only an approximate test even when all the assumptions are satisfied.

Should we use litters as blocks in similar studies in the future? The answer depends on whether the null hypothesis H_{0R}: $\sigma_R^2 = 0$ is true. If H_{0R} is true, then clearly there is no need to block on the basis of litters.

Although statistical hypothesis tests are not designed to prove null hypotheses, many practicing statisticians would recommend accepting H_0 if the associated *p*-value is sufficiently high. A rule of thumb that is often used is to accept H_0 if $p \geq 0.25$. For our data, the F-ratio for litters has a *p*-value of 0.118. Therefore, even though there is insufficient evidence to reject H_{0R}: $\sigma_R^2 = 0$, it is not a good idea to ignore the litter effects in future studies.

Inferences about linear combinations of population means can be drawn using the methods in Boxes 14.8 and 14.9. For instance, to construct the 95% lower bound for the difference between the average of the expected weight gains for diet 2 and diet 3 and the expected weight gain under diet 1 (question 4), we consider the linear combination

$$\theta = (-1)(\mu + \tau_1) + \frac{1}{2}\left[(\mu + \tau_2) + (\mu + \tau_3)\right] = (-1)\tau_1 + \frac{1}{2}(\tau_2 + \tau_3).$$

As we saw in Box 14.8, an unbiased estimate of θ is the corresponding linear combination of sample means

$$\hat{\theta} = (-1)\overline{Y}_{+1} + \frac{1}{2}\left(\overline{Y}_{+2} + \overline{Y}_{+3}\right) = (-1)\frac{114.8}{4} + \frac{1}{2}\left(\frac{132.6}{4} + \frac{136.3}{4}\right)$$

$$= 4.91.$$

Note that, in this expression for $\hat{\theta}$, we have used only two subscripts (as compared to three in Box 14.8) to denote the sample means. This is because, as already noted, only two subscripts are used in the model for a RCB design.

Under the assumption that there is no block-treatment interaction—that is, that $\sigma_{R\tau}^2 = 0$—Var($\hat{\theta}$) can be obtained by setting $\sum c_i = (-1) + (0.5) + (0.5) = 0$, $\sum c_i^2 = (-1)^2 + (0.5)^2 + (0.5)^2 = 1.5$, $a = 3$, $b = 4$, $n = 1$, $\sigma_{\alpha B}^2 = 0$ in the formula for $\sigma_{\hat{\theta}}$ for the crossed-factor two-way model given in Box 14.8. We get

$$\sigma_{\hat{\theta}}^2 = \frac{(1.5)\sigma^2}{4},$$

so that the estimated standard error of $\hat{\theta}$ is

$$\hat{\sigma}_{\hat{\theta}} = \sqrt{\left(\frac{1.5}{4}\right)\text{MS[E]}} = \sqrt{\left(\frac{1.5}{4}\right)1.24}$$

$$= 0.68.$$

Only one mean square is used to estimate of the standard error, and so the degrees of freedom associated with the estimated standard error of $\hat{\theta}$ are the same as the degrees of freedom for MS[E]. Thus, a 95% lower bound for θ can be calculated as

$$\hat{\theta} - t(6, 0.05)\hat{\sigma}_{\hat{\theta}} = 4.91 - (1.943)(0.68) = 3.59.$$

We conclude with 95% confidence that the average expected weight gain for diets 2 and 3 exceeds that for diet 1 by at least 3.59 g.   ■

In Example 15.3, an inference was desired about the linear combination of fixed effects

$$\theta = (-1)\tau_1 + \frac{1}{2}\tau_2 + \frac{1}{2}\tau_3.$$

An unbiased estimate of θ was given by

$$\hat{\theta} = (-1)\overline{Y}_{+1} + \frac{1}{2}\overline{Y}_{+2} + \frac{1}{2}\overline{Y}_{+3}.$$

Application of the results in Box 14.8 showed that only one mean square—MS[E]—was needed to estimate the standard error of $\hat{\theta}$. At first sight, it may seem surprising that, even though two sources of random variation (effect of the block and effect of error) affect the observed responses, only one source of random variation—the variation due to error—affects the estimate of θ.

To see why the standard error of $\hat{\theta}$ depends only on σ^2 (and not on σ_R^2), let's express $\hat{\theta}$ in terms of the effects in the underlying model. First of all, reasoning similar to the derivation of the standard error in Equation (14.7) shows that, under the model for a RCB design, the mean response for the j-th treatment can be expressed as

$$\overline{Y}_{+j} = \mu + \overline{R}_+ + \tau_j + \overline{E}_{+j},$$

where $\overline{R}_+$ and $\overline{E}_{+j}$ are, respectively, the mean of a random sample of r block effects and the mean of a random sample of r error effects. If we use this expression for $\overline{Y}_{+j}$ in the formula for $\hat{\theta}$, we get

$$\hat{\theta} = (-1)\left(\mu + \overline{R}_+ + \tau_1 + \overline{E}_{+1}\right) + (0.5)\left(\mu + \overline{R}_+ + \tau_2 + \overline{E}_{+2}\right)$$
$$+ (0.5)\left(\mu + \overline{R}_+ + \tau_3 + \overline{E}_{+3}\right),$$

which, on simplification, yields

$$= \left[(-1)\tau_1 + (0.5)\tau_2 + (0.5)\tau_3\right] + \left[(-1)\overline{E}_{+1} + (0.5)\overline{E}_{+2} + (0.5)\overline{E}_{+3}\right]$$
$$= \theta + \left[(-1)\overline{E}_{+1} + (0.5)\overline{E}_{+2} + (0.5)\overline{E}_{+3}\right].$$

Thus, the estimator $\hat{\theta}$ is the sum of θ and a linear combination of the means of error effects, so that $\hat{\theta}$ is not affected by block effects. We see that the block effects are eliminated because the coefficients -1, 0.5, 0.5 in $\hat{\theta}$ add to zero, so that the coefficient for $\overline{R}_+$ is zero. Indeed, straightforward calculations show that, in a RCB design, any linear combination of treatment means

$$\hat{\theta} = c_1\overline{Y}_{+1} + \cdots + c_t\overline{Y}_{+t},$$

where $\Sigma c_i = 0$, can be expressed as

$$\hat{\theta} = \begin{cases} \left(c_1\tau_1 + \cdots + c_t\tau_t\right) + \left(c_1\overline{E}_{+1} + \cdots + c_t\overline{E}_{+t}\right) & \text{if treatments are fixed} \\ \left(c_1 T_1 + \cdots + c_t T_t\right) + \left(c_1\overline{E}_{+1} + \cdots + c_t\overline{E}_{+t}\right) & \text{if treatments are random.} \end{cases} \tag{15.3}$$

As we know, $\hat{\theta}$ is the estimate of a contrast between the expected responses for treatments, and so Equation (15.3) implies that, in a RCB design, the estimate of a contrast between the expected responses for the treatments is not affected by the block effects. In other words, a RCB design can be used to eliminate (or adjust for) the effect of a confounding variable. Of course, such elimination is only possible if the model in Box 15.1 is a reasonable representation of reality.

If the treatment effects are fixed, we see from Equation (15.3) that the estimate of a contrast between expected responses for the treatments is a linear function of the means $(\overline{E}_{+1}, \ldots, \overline{E}_{+t})$ of independent samples from normally distributed populations. The coefficients of this linear function are $c_0 = \theta$ and $c_1, \ldots, c_t$. Consequently, as in the case of inferences about the overall mean in a one-way random-effects model (Section 14.3), inferences based on contrasts among treatment means in a RCB design can be made using the methods for inferences based on the means of independent random samples from normally distributed populations. In particular, the various inferential procedures described in Chapter 9 may be used to compare the treatment means in a RCB design.

When a linear combination of the expected responses for the treatments is not a contrast, its unbiased estimator will be affected by block effects. Consequently, if block effects are random, then the standard errors of the estimators will involve σ_R^2, the component of variance due to block effects. The exact procedures of Chapter 9 are not applicable, but the results in Boxes 14.8 and 14.9 can be used to make approximate inferences about such linear combinations, as the next example illustrates.

EXAMPLE **15.4**   Let's use the weight gain data in Example 15.1 to construct an approximate 95% confidence interval for the expected weight gain resulting from dietary supplement 2.

The model and ANOVA table for the data may be found in Example 15.3. We want to find a 95% confidence interval for $\theta = \mu + \tau_2$. An unbiased estimate of θ is the corresponding sample mean: $\hat{\theta} = \overline{Y}_{+2}$. The variance of $\hat{\theta}$ can be obtained from the results in Box 14.8 with $n = 1$, $a = 3$, $b = 4$, $\sigma_{\alpha B}^2 = 0$, $\sigma_B^2 = \sigma_R^2$, $c_1 = c_3 = 0$, and $c_2 = 1$. The variance of the unbiased estimate of $\hat{\theta}$ is

$$\sigma_{\hat{\theta}}^2 = \frac{1}{4}\sigma^2 + \frac{1}{4}\sigma_R^2.$$

From Table 15.4, we have the variance component estimates

$$\hat{\sigma}_R^2 = \frac{MS[R] - MS[E]}{3}, \quad \hat{\sigma}^2 = MS[E],$$

and so the variance of $\hat{\theta}$ can be estimated by the linear combination of mean squares

$$\hat{\sigma}_{\hat{\theta}}^2 = \frac{1}{12}MS[R] + \frac{1}{6}MS[E].$$

Thus, in contrast to Example 15.3, where $\hat{\theta}$ was a contrast between treatment means, the variance of $\hat{\theta}$ in this example involves both σ^2 and σ_R^2. Note that, in this case, $\hat{\theta}$ is not a contrast. An estimate of the standard error of $\hat{\theta}$ is

$$\hat{\sigma}_{\hat{\theta}} = \sqrt{\frac{1}{12}MS[R] + \frac{1}{6}MS[E]} = \sqrt{\frac{1}{12}(3.69) + \frac{1}{6}(1.24)} = 0.71.$$

Two mean squares are involved in the estimation of $\hat{\sigma}_{\hat{\theta}}$, and so an approximation to the associated degrees of freedom is the largest integer less than or equal to ν^*, where ν^* is defined in Box 14.9. With $\nu_1 = 3$, $\nu_2 = 6$, $p_1 = 1/12$, $p_2 = 1/6$,

$MS_1 = MS[R]$, and $MS_2 = MS[E]$, we get

$$v^* = \frac{\left(\dfrac{1}{12}MS[R] + \dfrac{1}{6}MS[E]\right)^2}{\left(\dfrac{1}{12}\right)^2 \dfrac{(MS[R])^2}{3} + \left(\dfrac{1}{6}\right)^2 \dfrac{(MS[E])^2}{6}}$$

$$= \frac{\left[\dfrac{1}{12}(3.69) + \dfrac{1}{12}(1.24)\right]^2}{\left(\dfrac{1}{12}\right)^2 \dfrac{(3.69)^2}{3} + \left(\dfrac{1}{6}\right)^2 \dfrac{(1.24)^2}{6}} = 6.84,$$

so that the required degrees of freedom may be approximated by the largest integer less than or equal to 6.84; that is, $v = 6$. An approximate 95% confidence interval for the expected response for dietary supplement 2 is

$$\overline{Y}_{+2} \pm t(6, 0.025)(0.26) = \frac{132.6}{4} \pm (2.447)(0.71) = 33.15 \pm 1.74$$

$$= (31.41, 34.89).$$

We can conclude with 95% confidence that dietary supplement 2 can be expected to produce a weight gain between 31.41 and 34.89 g. ∎

Note that there is an important difference between the procedures used to make inferences in Examples 15.3 and 15.4. In Example 15.3, the lower confidence bound is an exact bound, in the sense that, if the assumed model is correct, the confidence coefficient associated with the calculated bound is exactly 0.95. This is because, as already noted, the lower bound can be calculated using the exact procedures of Chapter 9 for independent samples from normal populations. In Example 15.4, the procedures of Chapter 9 are not applicable, because $\hat{\theta}$ is not a linear function of the means of independent samples from normal populations. The methods of inference described in Box 14.9 are based on the Satterthwaite approximation and are not exact even when the assumed model is correct. The appropriateness of the confidence coefficients and the p-values derived on the basis of the Satterthwaite approximation depends on the degrees of freedom associated with the various mean squares used in estimating the standard errors. In general, the larger the degrees of freedom, the better is the approximation.

Exercises

15.1   Refer to the RCB model in Box 15.1.

   a   Verify that contrasts among treatment means can be represented in the form in Equation (15.3).

   b   On the basis of (a), argue that the exact methods of inference in Chapter 9 can be used to compare sample means in data analysis using the RCB models in Box 5.1.

15.2   Refer to Example 15.3

a   Verify that the 95% lower confidence bound for the linear combination $\theta = (-1)\tau_1 + \frac{1}{2}(\tau_2 + \tau_3)$ constructed using the exact methods in Chapter 9 is the same as that calculated in Example 15.3.

b   Construct a 95% confidence interval for the differential expected weight gains for dietary supplements 2 and 3. Interpret the interval.

c   Explain why the Tukey multiple pairwise comparison method can be used to make inferences about the differences between the mean responses for the three dietary supplements. Use the Tikey method at the 0.05 significance level to compare the mean responses for the three means. Interpret the results.

15.3   An experiment to compare four methods of determining serum amylase values in patients with a particular disease was conducted as follows. Serum specimens were collected from six patients, and each specimen was divided into four parts. The four methods were randomly assigned to the four parts of each specimen, and serum amylase values were determined for each part using its assigned method. The serum amylase values obtained (enzyme units per ml of serum) were as follows:

Patient	Method 1	2	3	4
1	360	435	391	502
2	1035	1152	1002	1230
3	632	750	591	804
4	581	703	583	790
5	463	520	471	502
6	1131	1340	1144	1300

a   Write a model for the data. Explain all the terms in your model; give the reasons why you assume a particular term to be fixed or random.

b   Construct the appropriate ANOVA table for the data and perform all F-tests. Write the inferences that can be made from these tests.

c   Construct a 95% confidence interval for the difference between the true mean serum amylase values determined using methods 1 and 4.

d   On the basis of the interval constructed in (c), would it be reasonable to conclude that the two methods yield the same values on the average? Explain.

e   The population of subjects from which the study subjects were selected is known to have an average serum amylase value of 750 enzyme units per ml. Explain how an approximate test can be performed to see if methods 1 and 3 both produce values less than 750 on the average. [Hint: Construct simultaneous upper bounds for the population mean values for methods 1 and 3.]

f   Now suppose that, in this study, the methods were not randomized within patients. Instead, each method was assigned to six specimen parts randomly selected from the pool of 24 specimen parts. What would be wrong with this design?

15.4   Refer to the cell fluidity data in Example 4.10.

a   Explain why these data can be regarded as the result of a RCB design with two treatments.

b Write a RCB design model for the data. Explain all the terms in the model; give reasons why you assume a particular term to be fixed or random.

c Construct an appropriate ANOVA table and perform all F-tests.

d Construct a 95% confidence interval for the difference between the true mean cell fluidity of exposed and unexposed cells.

e Compare the interval obtained in (d) with that constructed in Example 4.11.

f Compare the independent sample t-test and the ANOVA F-test for treatments in a CR design (with $t = 2$) with the paired sample t-test and the ANOVA F-test for treatments in a RCB design (with $t = 2$). Comment on their similarity.

15.5 Professor Al Dudeck of the Department of Ornamental Horticulture, University of Florida, conducted a study to evaluate the performance of several cool-season grasses for the winter overseeding of golf greens in North Florida. In the study, a golf ball was rolled down an inclined plane, to induce a constant initial velocity, and then the distance traveled by the ball on the green was measured. This distance can be influenced by the slope of the green, and so the study was conducted in four randomized complete blocks; each block contained five plots with similar slopes. In each case, the ball was rolled down the slope of the plot. Five cultivars (A, Pennfine rye grass; B, Dasher rye grass; C, Regal rye grass; D, Marvel green supreme; E, Barry rye grass) were randomly assigned to the five plots in each block. For five balls rolled down the ramp on each plot, the average distances (m) from the base of the ramp to the stopping points are as follows:

Block	A	B	Variety C	D	E
1	2.764	2.568	2.506	2.612	2.238
2	3.043	2.977	2.533	2.675	2.616
3	2.600	2.183	2.334	2.164	2.127
4	3.047	3.028	2.895	2.724	2.697

a Write a model for the data; explain all the terms in your model.

b Set up the ANOVA table and draw conclusions based on appropriate F-tests.

c Perform pairwise multiple comparison of the treatment means using Duncan's procedure at the 0.05 level.

15.6 To evaluate the batch-to-batch variation of the total solids in brewer's yeast produced by a particular process, the total solids (1%) were independently determined by three analysts, with the following results:

Analyst	1	2	3	Batch 4	5	6	7	8
1	21.3	15.9	17.3	14.4	22.4	16.8	13.7	20.4
2	20.2	16.3	17.6	13.3	21.6	17.4	12.9	19.8
3	21.5	15.6	17.1	13.9	23.1	18.2	13.0	20.1

a Write a model for the data; explain all the terms.

b Estimate the proportion of the total variation in measured total solids that can be ascribed to batch-to-batch variation and the proportion that can be ascribed to analyst-to-analyst variation. Comment on the estimates.

15.3
RCB factorial experiments

Let's begin with an example of a RCB design with two crossed factors.

EXAMPLE 15.5 The response times (msec) of eight subjects after treatment with each of four therapies intended to increase mental awareness are as follows:

Therapy	Subject							
	1	2	3	4	5	6	7	8
No drugs	18.8	18.5	21.4	25.5	19.8	24.4	25.6	26.5
Drug A alone	13.5	9.8	12.1	22.9	8.3	14.9	23.0	15.5
Drug B alone	13.6	13.4	8.3	24.9	16.9	16.2	16.3	15.4
Drugs A and B in combination	10.6	12.6	11.7	16.8	4.0	13.4	14.9	12.6

In this study, the experimental design is a 2×2 factorial in randomized complete blocks. The treatments are the combinations of the levels of two factors: drug A and drug B. The treatment combinations are as follows:

Treatment description	Level of A	Level of B
No drugs (control therapy)	0	0
Drug A (single-drug therapy)	1	0
Drug B (single-drug therapy)	0	1
Drugs A and B (combination therapy)	1	1

■

Analysis of data from RCB factorial designs can be performed by methods analogous to those for single-factor RCB designs. The corresponding models assume that the effects of blocks and treatments are additive. Consequently, the models will not contain any terms representing interaction between the block and experimental factors. The following model corresponds to an $a \times b$ factorial experiment in a RCB design in which the experimental factor A is fixed and the blocks and experimental factor B are random

$$Y_{ijk} = \mu + R_i + \alpha_j + B_k + (\alpha B)_{jk} + E_{ijk},$$

$$i = 1, \ldots, r; \ j = 1, \ldots, a; \ k = 1, \ldots, b. \quad \textbf{(15.4)}$$

The notation here is as follows:

- Y_{ijk} is the measured response for treatment $a_j b_k$ in block i;
- μ is the overall mean;
- R_i is the random effect of block i;
- α_j is the fixed effect of the j-th level of A, and $\Sigma \alpha_j = 0$;
- B_k is the random effect of the k-th level of B;
- $(\alpha B)_{jk}$ is the joint effect of the j-th level of A and the k-th level of B;
- E_{ijk} is the experimental error.

The R_i are independent $N(0, \sigma_R^2)$; the B_k are independent $N(0, \sigma_B^2)$; the $(\alpha B)_{jk}$ are independent $N(0, \sigma_{\alpha B}^2)$; and the E_{ijk} are independent $N(0, \sigma^2)$.

Equation (15.4) is a three-way ANOVA model in which there is no interaction between blocks and the two experimental factors. The rules in Section 14.7 may be used to determine the expected mean squares for this model (Table 15.6).

TABLE 15.6

EMS for $a \times b$ RCB designs with random blocks, fixed effects of A, and random effects of B

Source	df	EMS
Blocks (R)	$r - 1$	$ab\sigma_R^2 + \sigma^2$
A	$a - 1$	$rb\psi_A^2 + r\sigma_{\alpha B}^2 + \sigma^2$
B	$b - 1$	$ra\sigma_B^2 + r\sigma_{\alpha B}^2 + \sigma^2$
AB	$(a-1)(b-1)$	$r\sigma_{\alpha B}^2 + \sigma^2$
Error	$(r-1)(ab-1)$	σ^2

The analysis of RCB factorial experiments can be based on statistical computing software or formulas similar to those for a one-factor RCB design. The next example illustrates the analysis of an $a \times b$ RCB design in which both experimental factors are fixed. In Exercise 15.10, you will be asked to analyze the results from a two-factor RCB factorial experiment in which both factors are random.

EXAMPLE 15.6   Consider the response time data in Example 15.5. Clearly, the levels of the two experimental factors—drug A and drug B—should be regarded as fixed. We want the comparison of treatments to be applicable to a population of subjects, and so the block effects should be regarded as random. A model for the observed response times is

$$Y_{ijk} = \mu + R_i + \alpha_j + \beta_k + (\alpha\beta)_{jk} + E_{ijk}, \qquad i = 0, 1; \ j = 0, 1; \ k = 1, \ldots, 8.$$

(15.5)

The notation here is as follows:

- Y_{ijk} is the measured response time for subject i treated with therapy $a_j b_k$;

- R_i is the random effect of subject i;
- α_j is the fixed effect of the j-th level of A, and $\Sigma\alpha_j = 0$;
- β_k is the fixed effect of the k-th level of B, and $\Sigma\beta_k = 0$;
- $(\alpha\beta)_{jk}$ is the joint effect of the j-th level of A and the k-th level of B; $\Sigma_j(\alpha\beta)_{jk} = 0$ and $\Sigma_k(\alpha\beta)_{jk} = 0$;
- E_{ijk} is the experimental error.

The R_i are independent $N(0, \sigma_R^2)$; and the E_{ijk} are independent $N(0, \sigma^2)$.

The EMS associated with this model can be obtained using the rules in Section 14.7 and are presented in Table 15.7 on the assumption of r blocks, a levels of A, and b levels of B. As we see from Table 15.7, all the effects in the model can be tested with F-tests in which MS[E] is the denominator of the test statistic.

TABLE   15.7

EMS for a RCB factorial experiment with random blocks and fixed effects of A and B

Source	df	EMS
Blocks (R)	$r - 1$	$ab\sigma_R^2 + \sigma^2$
A	$a - 1$	$rb\psi_A^2 + \sigma^2$
B	$b - 1$	$ra\psi_B^2 + \sigma^2$
AB	$(a - 1)(b - 1)$	$r\psi_{AB}^2 + \sigma^2$
Error	$(r - 1)(ab - 1)$	σ^2

The subclass numbers are equal, and so the appropriate ANOVA table can be constructed using Equation (13.18). A portion of the computer output from the ANOVA procedure of SAS is as follows.

```
-------------------------------------------------------------------
                    SAS Analysis of Response Times

                Class   Levels   Values
                SUBJ       8      1 2 3 4 5 6 7 8
                DRGA       2      0 1
                DRGB       2      0 1
            Number of observations in data set = 32


                    Analysis of Variance Procedure


Dependent Variable: RESPTIME
                          Sum of        Mean
Source            DF     Squares      Square   F Value   Pr > F
Model             10   839.63813    83.96381    10.31    0.0001
Error             21   171.04406     8.14496
Corrected Total   31  1010.68219
```

Source	DF	Anova SS	Mean Square	F Value	Pr > F
SUBJ	7	365.92469	52.27496	6.42	0.0004
DRGA	1	246.97531	246.97531	30.32	0.0001
DRGB	1	194.53781	194.53781	23.88	0.0001
DRGA*DRGB	1	32.20031	32.20031	3.95	0.0600

The following ANOVA table for the response time data can be constructed from this SAS output.

Source	df	SS	MS	F_c	p
Subjects (blocks)	7	365.92469	52.27496	6.42	0.0004
Drug A	1	246.97531	246.97531	30.32	0.0001
Drug B	1	194.53781	194.53781	23.88	0.0001
Drug $A \times$ Drug B	1	32.20031	32.20031	3.95	0.0600
Error	21	171.04406	8.14496		
Corrected total	31	1010.68219			

From the ANOVA table, we conclude that there is no demonstrable ($p = 0.06$) interaction between the two drugs. In other words, it appears that the effect of drug A does not change when it is administered in combination with drug B. Further analysis of the differences in observed response times can be based on the methods in Chapter 9.

The p-value of 0.0004 for block effects indicates that the subject-to-subject variation in response times is a significant ($\sigma_R^2 > 0$) contributor to the variability in the observed response times. ∎

Exercises

15.7 Using the rules in Section 14.7, calculate the EMS associated with a two-factor RCB design in each of the following cases. In each case, determine the form of the F ratios appropriate for testing the significance of the various effects in the model.

a The blocks are random; A and B are fixed.

b The blocks are random; A is fixed and B is random.

c The blocks are random; A and B are random.

15.8 Refer to the response time data in Example 15.5.

a Using Equation (13.18) or statistical computing software, verify the sums of squares in the ANOVA table in Example 15.5.

b Construct a 95% confidence interval for the difference between the expected response times for the control (no drugs) and treatment with drug A alone. Interpret the interval.

c   Use the Tukey 0.05-level multiple pairwise comparison method to compare the mean responses for the four therapies. Write your conclusions.

15.9   In a 2×2 RCB experiment, 48 yearling heifers were divided by initial weight into three blocks. Within blocks, the heifers were assigned to four pens at random. Pens within blocks were randomly assigned to four dietary treatments resulting from combining barley processed by two methods (whole and steam-rolled barley) with two sources of supplemental nitrogen (canola meal or urea). The heifers' mean dry matter intake (kg/day) in this experiment is as follows (Morgan, Gibson, Nelson, & Males, 1991):

		Barley
N-supplement	Whole	Steam-rolled
Canola	7.7	7.2
Urea	7.1	6.7

From their data, the researchers calculated that

$$MS[E] = 0.0588.$$

a   Perform a data analysis to test for the significance of the interaction and main effects of barley processing and N-supplement. Write appropriate conclusions.

b   On the basis of the results obtained in (a), perform further analyses to examine the effects of the levels of the two factors on dry matter intake. Write your conclusions.

15.10   An experiment was conducted to evaluate the effects of cooking time and cooking temperature on the tenderness of cakes whose recipe calls for baking at 250°C for 2 hours. Randomly choosing three temperatures in the range from 200°C to 300°C and three times in the range from 1 hr 45 min to 2 hr 15 min, the experimenters formulated nine temperature-time combinations for baking cakes. Because the effects of temperature and cooking may vary from oven to oven, they decided to perform the experiment with ovens as blocks. Five ovens were used, and nine cakes were baked in each oven (one at a time) using the nine temperature-time combinations. For each cake baked, a tenderness score was calculated by averaging the scores assigned by a panel of ten tasters. Each taster assigned scores on a seven-point scale, from 1 (least acceptable) to 7 (highly acceptable). The sums of squares needed to construct the ANOVA table from the tenderness scores are as follows:

Source	Sum of squares
Ovens	5.00
Temperature	11.50
Time	10.20
Temperature × time	5.45
Error	9.60

a   Write a model for the data. Explain all the terms in your model.

b   Construct the appropriate ANOVA table and the EMS values.

c   Perform all F-tests as appropriate and write a summary of your conclusions.

d   Estimate the components of variance in tenderness scores due to ovens, due to temperature, due to time, and due to interaction between time and temperature.

e   For a given oven, estimate the proportion of total variability due to temperature, due to time, and due to temperature-time interaction. Interpret the estimate.

15.11   A 2×3 experiment was conducted to study the effect of two factors—A, the height of the bed (low and high), and B, the row spacing (0.25, 0.5 and 1 m)—on the moisture content of soils in which a plant variety was grown under a specific type of irrigation practice. The experiment was conducted in 18 plots, divided into three blocks of six plots each. Six combinations of the levels of the two factors were assigned at random to the six plots in each block. The following table presents the measured percentage soil water contents at a depth of 90–100 cm:

Treatment	Levels of factors		Block			Total
	A	B	1	2	3	
1	L	0.25	30.7	24.7	16.9	72.3
2	L	0.50	22.7	15.0	13.6	51.3
3	L	1.00	23.0	17.1	15.2	55.3
4	H	0.25	28.4	19.6	16.2	64.2
5	H	0.50	20.1	16.8	15.1	52.0
6	H	1.00	18.4	13.9	14.3	46.6
Total			143.3	107.1	91.3	341.7

a   Write a model for the data. Explain all the terms in your model.

b   Construct an appropriate ANOVA table for analyzing the soil water content data.

c   Construct a 95% confidence interval for the difference between the expected moisture contents at low and high bed heights.

d   Use Duncan's 0.05-level multiple comparison method to compare the mean responses at the three row spacings. Write your conclusions.

15.4
The Latin square design

A RCB design is a simple experimental design that permits us to make comparisons between the expected responses for treatments after eliminating (adjusting for) the effect of a confounding variable. The design utilizes the levels of a confounding variable to form blocks of experimental units. By imposing the constraint that every treatment should be applied to exactly one experimental unit in every block, we can ensure that the design will eliminate the block effects when comparing treatment effects.

Simple designs for eliminating the effects of more than one confounding variable can be constructed analogously to the RCB design. In particular, the *Latin square design* (LS design) may be used to adjust for the effects of two confounding variables. The structure of a LS design for comparing t treatments can be displayed as t^2 experimental units arranged in a square array of t rows and t columns. The responses from experimental units in the same row are affected by a common confounding effect, called the *row effect*. Similarly, the responses from experimental units in the same column are affected by a common confounding effect called the *column effect*. The t treatments are applied to the t^2 experimental units in such a way that every treatment appears exactly once in each row and once in each column.

EXAMPLE **15.7**   We want to compare the effect of $t = 4$ fertilizers on the yield of a particular crop. The piece of land on which the experiment is to be conducted is such that the soil moisture content and the available amount of sunshine vary in perpendicular directions as shown in Figure 15.1

FIGURE   15.1
Soil moisture and sunshine gradients

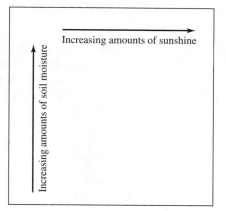

An LS design to eliminate the effects of the differences in soil moisture (confounding variable 1) and light exposure (confounding variable 2) can be constructed by dividing the piece of land into $t^2 = 16$ plots arranged in $t = 4$ rows and $t = 4$ columns. The division into plots should be such that the plots in a given row will have the same amount of light and the plots in a given column will have the same amount of soil moisture. The four fertilizer treatments are assigned to the 16 plots at random, subject to the restriction that every fertilizer should be applied to exactly one plot in each row and each column.

Treatments are randomly assigned to plots as follows. We start with an arrangement of four numbers in an array of four rows and four columns, such that each number appears exactly once in each row and exactly once in each column. Such an arrangement is called a 4 × 4 Latin square. Let's choose the following array, which we'll call Latin square 1.

1	2	3	4
2	3	4	1
3	4	1	2
4	1	2	3

In the second step, we arrange the first four integers—1, 2, 3, and 4—in a random sequence. Suppose this sequence is {2, 3, 1, 4}. Then we rearrange the columns of Latin square 1 by moving column 2 to the first position, column 3 to the second position, column 1 to the third position and column 4 to the fourth position; the result is Latin square 2:

2	3	1	4
3	4	2	1
4	1	3	2
1	2	4	3

Notice that the first column of Latin square 2 is the same as the second column of Latin square 1 and so on. In the third and final step, we select another random sequence of the first four integers and use it to rearrange the rows of Latin square 2 in the same way. If we select the sequence {2, 4, 1, 3} in the third step, we rearrange the rows of Latin square 2 to obtain Latin square 3:

3	4	2	1
1	2	4	3
2	3	1	4
4	1	3	2

The treatment assignments are determined by Latin square 3. For instance, treatment 4 is assigned to (row 1, column 2), (row 2, column 3), (row 3, column 4), and (row 4, column 1). On the basis of Latin square 3, each treatment will be assigned to exactly one plot in each row and each column.   ∎

As in the case of the RCB design, Latin square designs may be applied whenever we can assume that the confounding variables do not interact with the experimental factors—that is, when it is reasonable to assume that the effects of rows and columns do not interact with each other or with the treatments.

Let's look at the model for a LS design in which the row and column effects are random and the treatment effects are fixed. This model is based on the assumption that there is no interaction between rows, columns, and treatments. If k is the label of the treatment and Y_{ijk} denotes the response in the experimental unit common to row i and column j, then a model for Y_{ijk} is

$$Y_{ijk} = \mu + R_i + C_j + \tau_k + E_{ijk}, \qquad i = 1, \ldots, t;\ j = 1, \ldots, t. \qquad \textbf{(15.6)}$$

The notation here is as follows:

- μ is the overall mean;
- R_i is the random effect of the i-th row;
- C_j is the random effect of the j-th column;
- τ_k is the fixed effect of the k-th treatment, and $\Sigma \tau_k = 0$;
- E_{ijk} is the experimental error.

The R_i are independent $N(0, \sigma_R^2)$; the C_j are independent $N(0, \sigma_C^2)$, and the E_{ijk} are independent $N(0, \sigma^2)$. Also, the R_i, C_j, and E_{ijk} are mutually independent.

Equation (15.6) is a three-way ANOVA model, because the data represented by this model can be categorized into classes on the basis of three criteria: row, column, and treatment. In addition, even though four sources of variation (rows, columns, treatments, and error) affect the response, only three subscripts are used in the model, because at most one measurement is available for each combination of the levels of the three factors.

The rules for calculating EMS values in Section 14.7 are not applicable to LS designs because, when considered as three-way classification designs, they do not have equal subclass numbers; in each subclass, there will be either 0 or 1 measurement. For instance, in the Latin square design in Example 15.7, the responses can be classified using three criteria: the rows, representing the amounts of available sunshine; the columns, representing the soil moisture contents; and the treatments, representing the types of fertilizer. If n_{ijk} denotes the number of responses in the class defined by the i-th row, the j-th column, and the k-th treatment, then $n_{111} = 0$ because, as we see from Latin square 3, no response is measured in the class (row 1, column 1, treatment 1), and $n_{113} = 1$ because one response is available for treatment 3 in row 1 and column 1. Similarly, $n_{212} = 0$, whereas $n_{442} = 1$. The EMS for a Latin square design can be derived directly by the methods in Section B.4 of Appendix B. The structure of the EMS in a LS design is very similar to that in a RCB design. Table 15.8 shows the ANOVA table and the associated EMS for a Latin square design under the assumption that rows, columns, and treatments are fixed. (You will be asked to derive these EMS values in Exercise 15.6.)

If some of the factors are random, the corresponding EMS values can be obtained by minor modification of the expressions in Table 15.8—specifically, by replacing the ψ with σ in the EMS for each mean square representing a random effect. For instance, if the row effects are random, then the EMS associated with MS[R] is $t\sigma_R^2 + \sigma^2$.

Note that the degrees of freedom in Table 15.8 are as would be expected. There are t^2 responses, so that SS[TOT] has $t^2 - 1$ degrees of freedom. There are t rows, t columns, and t treatments, so that each of the three sums of squares SS[R], SS[C],

TABLE   15.8

ANOVA table and EMS for LS designs

Source	df	SS	MS	EMS values A and B fixed
Rows	$t-1$	SS[R]	MS[R]	$t\psi_R^2 + \sigma^2$
Columns	$t-1$	SS[C]	MS[C]	$t\psi_C^2 + \sigma^2$
Treatments	$t-1$	SS[T]	MS[T]	$t\psi_T^2 + \sigma^2$
Error	$(t-1)(t-2)$	SS[E]	MS[E]	σ^2
Total	t^2-1	SS[TOT]		

and SS[T] has $t-1$ degrees of freedom. Finally, the degrees of freedom for SS[E] may be calculated as the difference between the degrees of freedom for SS[TOT] and the sum of the degrees of freedom for rows, columns, and treatments

$$df_E = df_{TOT} - df_R - df_C - df_T$$
$$= (t^2 - 1) - (t - 1) - (t - 1) - (t - 1) = (t - 1)(t - 2).$$

The EMS values in Table 15.8 imply that MS[E] is the denominator for the test statistics for hypotheses about the row, column, and treatment effects. The calculations needed to construct the ANOVA table for a LS design are similar to those for a RCB design.

EXAMPLE 15.8   Researchers decide to compare four methods of determining impurities in a specimen sample. Because a measured impurity value can be affected both by the analyst who analyzes the specimen and by the batch of the raw material used to prepare the specimen, the experiment is based on a Latin square design with rows representing analysts and columns representing the batch of the raw material. Four analysts (who could be considered a random sample from a population of analysts) and a random sample of four batches of raw material are used in the study. The following table shows the experimental layout (the numbers 1–4 in parentheses denote the corresponding methods of determining impurity) and the measured percentage impurities:

| Analysts | Batches | | | | Total |
	1	2	3	4	
1	3.10 (2)	4.29 (3)	6.53 (4)	3.52 (1)	17.44
2	4.21 (3)	8.30 (4)	3.92 (1)	4.82 (2)	21.25
3	6.11 (1)	2.98 (2)	4.20 (3)	8.46 (4)	21.75
4	9.01 (4)	4.59 (1)	4.60 (2)	5.22 (3)	23.42
Total	22.43	20.16	19.25	22.02	83.86

The totals for the four methods are as follows:

Method	1	2	3	4
Total	18.14	15.50	17.92	32.30

A model for the impurity data is

$$Y_{ijk} = \mu + R_i + C_j + \tau_k + E_{ijk}, \qquad i = 1, \ldots, 4; \quad j = 1, \ldots, 4;$$

k is the identification number of the method used by the i-th analyst to analyze a specimen from the j-th batch. The other notation is as follows:

- Y_{ijk} is the impurity value obtained when the i-th analyst measures the impurity in a specimen from the j-th batch;
- μ is the overall mean;
- R_i is the random effect of the i-th analyst;
- C_j is the random effect of the j-th batch;
- τ_k is the fixed effect of the k-th method, and $\Sigma \tau_k = 0$;
- E_{ijk} is the experimental error.

The R_i are independent $N(0, \sigma_R^2)$; the C_j are independent $N(0, \sigma_C^2)$; and the E_{ijk} are independent $N(0, \sigma^2)$. The E_{ijk}, R_i, and C_j are mutually independent.

We have assumed the row and column effects to be random, because we want to treat the batches and the analysts as two random samples—each of size $n = 4$—from a population of batches and a population of analysts, respectively. Inferences resulting from such an assumption will apply to the populations of all batches and all analysts.

The calculations for Latin square ANOVA are similar to the calculations in CR and RCB designs. Each sum of squares is obtained by subtracting the correction factor,

$$CF = \frac{(\text{total of all responses})^2}{t^2}$$

from the sum of squares for the appropriate set of totals. For instance, the sum of squares for treatments is

$$SS[T] = \frac{(\text{treatment 1 total})^2}{4} + \frac{(\text{treatment 2 total})^2}{4}$$

$$+ \frac{(\text{treatment 3 total})^2}{4} + \frac{(\text{treatment 4 total})^2}{4} - CF$$

$$= \frac{(18.14)^2}{4} + \frac{(15.50)^2}{4} + \frac{(17.92)^2}{4} + \frac{(32.30)^2}{4} - \frac{(83.86)^2}{16}$$

$$= 43.9003.$$

The ANOVA table with the appropriate EMS values is as follows:

Source	df	SS	MS	EMS	F_c	p
Methods	3	43.9003	14.6334	$4\psi_M^2 + \sigma^2$	19.917	0.002
Analysts	3	4.7875	1.5958	$4\sigma_A^2 + \sigma^2$	2.172	0.192
Batches	3	1.7121	0.5707	$4\sigma_B^2 + \sigma^2$	< 1	
Error	6	4.4079	0.7347	σ^2		
Total	15	54.8078				

It appears that neither batch-to-batch variability nor analyst-to-analyst variability contributes significantly to the observed differences in the impurity measurements. However, there is a significant ($p = 0.002$) difference between the mean percentage impurities determined by the four methods. ■

Finally, note the following similarities between analyses for RCB and LS designs.

1   If the treatment effects are fixed, a contrast between expected responses for the treatments in a Latin square design can be estimated as a linear function of the means of independent samples from normally distributed populations. Therefore, as in a RCB design, inferences based on contrasts among treatment means in a Latin square design can be made by the methods in Chapter 9. In Exercise 15.13, you are asked to draw inferences about contrasts of expected responses in Example 15.8.

2   The analysis of a LS design in which the treatments are combinations of levels of several factors is similar to the analysis of a factorial experiment in randomized blocks. In Exercise 15.14, you are asked to analyze the results of a factorial experiment in a Latin square design.

Exercises

15.12   Use the impurity data in Example 15.8 to verify the entries in the ANOVA table.

15.13   In Example 15.8, suppose we want to compare new methods 2, 3, and 4 with the standard method 1, which is known to be accurate but time-consuming and expensive.

a   Let μ_i denote the expected values of the measured impurity for the i-th method and consider the three simple contrasts

$$\theta_{12} = \mu_1 - \mu_2; \qquad \theta_{13} = \mu_1 - \mu_3, \quad \text{and} \quad \theta_{14} = \mu_1 - \mu_4.$$

Perform a simultaneous test of the three null hypotheses

$$H_{012}: \theta_{12} = 0, \qquad H_{013}: \theta_{13} = 0, \quad \text{and} \quad H_{014}: \theta_{14} = 0$$

at $\alpha = 0.05$. Write your conclusions.

b Construct 95% simultaneous confidence intervals for θ_{12}, θ_{13}, and θ_{14}. Interpret the intervals.

c Perform a multiple pairwise comparison of the four treatment means using a 0.05-level Tukey test.

d One objective of the study is to see if any of the new methods produces results comparable to the results from the standard method. Of the three analyses in (a), (b), and (c), which is the most appropriate for this study objective? Explain.

e Using the procedure that you recommend in (d), identify which methods are most comparable to the standard method.

15.14 A study to compare four methods of treating physical discomfort in patients suffering from chronic asthma was designed as follows. Because the response to a drug (time to relief) varies from patient to patient, the subject effect on the response to a drug was regarded as a confounding factor. The order in which the medications are given may also affect a patient's response, and so the actual study was conducted in a Latin square design, in which the rows represented the order of presentation of treatments to the patient and the columns represented the patient receiving the drug. The four treatments were: AL, drug A at low dosage; AH, drug *A* at high dosage; BL, drug *B* at low dosage; BH, drug *B* at high dosage. The following data are the times to relief (min) observed in the study.

		Pati	ents		
Order	1	2	3	4	Total
1	14 (AH)	30 (AL)	15 (BH)	28 (BL)	87
2	25 (AL)	15 (BH)	22 (BL)	21 (AH)	83
3	17 (BL)	19 (AH)	26 (AL)	22 (BH)	84
4	7 (BH)	22 (BL)	19 (AH)	34 (AL)	82
Total	63	86	82	105	336

a Write a model for the data. Explain the meanings of all the terms in your model.

b Construct an appropriate ANOVA table.

c Do these data indicate that there is an interaction between drugs and dosages?

d What does your answer in (c) imply about the differential responses at high and low doses?

e Perform an appropriate analysis to identify the best treatments among the four. Write your conclusions.

f Use an appropriate method for simultaneous pairwise comparisons of the drug that had the shortest mean time to relief with the other three drugs.

15.15 A study to compare the tolerances (dose needed to produce a specified effect) of cats to four cardiac substances (A, B, C, D) was conducted using a LS design, in which the rows represented four combinations of two time periods (A.M., P.M.) and two technicians (I and II) and the columns represented the days on which the measurements were made. Each of the 16 cats was infused with a cardiac substance at a steady rate, and the dose (rate of infusion × time) at which the specified effect was observed was recorded. The following table shows the measured responses— specifically, $10 \log$ (dose in μg):

Observer, time	Day				Total
	1	2	3	4	
I, A.M.	3.26 (D)	4.15 (B)	3.02 (A)	3.67 (C)	14.10
I, P.M.	2.73 (B)	3.38 (D)	3.29 (C)	4.50 (A)	13.90
II, A.M.	3.45 (A)	4.09 (C)	2.66 (B)	3.51 (D)	13.71
II, P.M.	3.20 (C)	3.14 (A)	3.48 (D)	3.40 (B)	13.22
Total	12.64	14.76	12.45	15.08	54.93

a Write a model for the data. Explain the meanings of all the terms in your model.
b Construct an appropriate ANOVA table.
c Do these data indicate that there is a difference between the four cardiac substances?
d Perform a multiple pairwise comparison of the four cardiac substances and write your conclusions.

15.16 Derive the EMS values in Table 15.8.

15.5
Checking for violations of assumptions

The analyses of data from RCB and LS experiments described in this chapter are based on the following assumptions.

1 There is no interaction between blocks (or rows and columns) and treatments; that is, the model is additive.

2 The errors satisfy the usual ANOVA assumptions; that is, they have independent normal distributions with zero mean and variance σ^2.

Residual plots for checking additivity

To check for nonadditivity, which indicates the need for interaction terms in the model, we can examine a residual plot in which the residuals $R = Y - \hat{Y}$ are plotted against $\hat{Y}$, the predicted values. If an additive model is appropriate, then the residual plot should appear as a band that is symmetric about and parallel to the $\hat{Y}$ axis. Examples of residual plots for checking additivity are shown in Figure 15.2. For plotting purposes, it is best to calculate the predicted and residual values using appropriate statistical software. In the next example, a residual plot for checking additivity is constructed using the GLM and PLOT procedures of SAS; for more details, see Section 15.3 in the text by Younger (1997).

EXAMPLE 15.9 The following are the SAS codes needed to produce a residual plot for the weight gain data in Example 15.1.

FIGURE 15.2

Residual plots for checking additivity

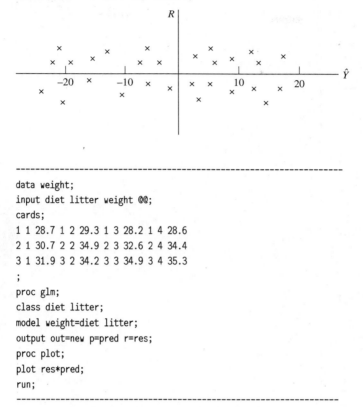

```
data weight;
input diet litter weight @@;
cards;
1 1 28.7 1 2 29.3 1 3 28.2 1 4 28.6
2 1 30.7 2 2 34.9 2 3 32.6 2 4 34.4
3 1 31.9 3 2 34.2 3 3 34.9 3 4 35.3
;
proc glm;
class diet litter;
model weight=diet litter;
output out=new p=pred r=res;
proc plot;
plot res*pred;
run;
```

The SAS output for the residual plot appears on the next page.

Considering that this output pertains to a rather small amount of data, we conclude that the plot does not provide any indication of nonadditivity. ∎

Residual plots can also be used to check for violations of the assumption that the errors are independently normally distributed and have a constant variance. The assumption of normally distributed errors is checked by the normal q-q plots described in Section 8.6. The equal variance assumption can be checked by plotting the residuals against the predicted values.

Tukey test for nonadditivity

Tukey (1949) suggested an F-test for testing the null hypothesis H_{0NA} that an additive model is appropriate against the alternative hypothesis H_{1NA} that an additive model is not appropriate. Though the Tukey test for additivity can be described in general terms (Box, Hunter, & Hunter, 1978), we'll focus on its application to RCB and LS designs (Box 15.2).

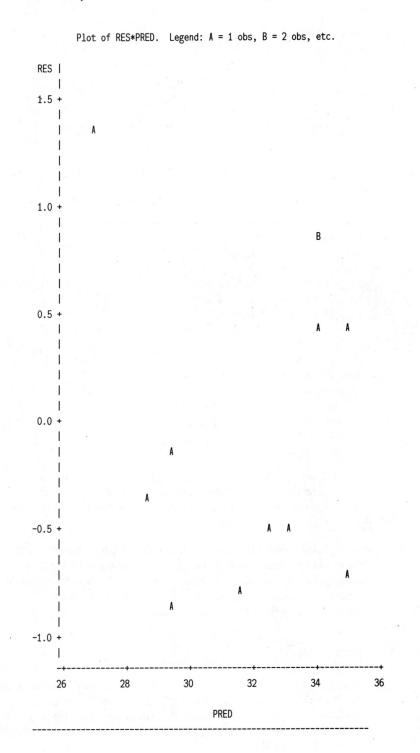

Plot of RES*PRED.   Legend: A = 1 obs, B = 2 obs, etc.

BOX **15.2**   ***The Tukey test for additivity in RCB and LS designs***

RCB design

Let Y_{ij} and SS[E] denote, respectively, the response to the j-th treatment in block i and the error sum of squares obtained from the RCB model in Box 15.1. Let

$$SS[NA] = \frac{\left(\sum\limits_{i=1}^{r} \sum\limits_{j=1}^{t} Y_{ij} Z_{ij} \right)^2}{\sum\limits_{i=1}^{r} \sum\limits_{j=1}^{t} Z_{ij}^2},$$

where

$$Z_{ij} = (\overline{Y}_{i+} - \overline{Y}_{++})(\overline{Y}_{+j} - \overline{Y}_{++}).$$

Then the null hypothesis H_{0NA} that the additive model is appropriate can be rejected at level α if

$$F_c = \frac{SS[NA]}{MSE[NA]} > F(1, \nu - 1, \alpha),$$

where ν denotes the degrees of freedom for SS[E] and

$$MSE[NA] = \frac{(SS[E] - SS[NA])}{(\nu - 1)}.$$

LS design

Let Y_{ijk} and SS[E] denote, respectively, the response to the k-th treatment in row i and column j and the error sum of squares based on Equation (15.6). Let

$$SS[NA] = \frac{\left(\sum\limits_{i=1}^{t} \sum\limits_{j=1}^{t} \sum\limits_{k=1}^{t} Y_{ijk} Z_{ijk} \right)^2}{\sum\limits_{i=1}^{t} \sum\limits_{j=1}^{t} \sum\limits_{k=1}^{t} Z_{ijk}^2},$$

where

$$Z_{ijk} = (\overline{Y}_{i++} - \overline{Y}_{+++})(\overline{Y}_{+j+} - \overline{Y}_{+++}) + (\overline{Y}_{i++} - \overline{Y}_{+++})(\overline{Y}_{++k} - \overline{Y}_{+++})$$
$$+ (\overline{Y}_{+j+} - \overline{Y}_{+++})(\overline{Y}_{++k} - \overline{Y}_{+++}).$$

The null hypothesis H_{0NA} that the additive model is appropriate can be rejected at level α if

$$F_c = \frac{SS[NA]}{MSE[NA]} > F(1, \nu - 1, \alpha),$$

where ν denotes the degrees of freedom for SS[E] and

$$MSE[NA] = \frac{(SS[E] - SS[NA])}{(\nu - 1)}.$$

For obvious reasons, the quantity SS[NA] in Box 15.2 is called the sum of squares for testing nonadditivity. The next example illustrates the Tukey test for nonadditivity in a RCB design.

EXAMPLE **15.10**   Let's apply the Tukey test for additivity to the weight gain data in Example 15.1, which arose from a RCB design with $r = 4$ blocks and $t = 3$ treatments. Straightforward calculations yield the following values for the block means

$$\overline{Y}_{1+} = 30.433, \quad \overline{Y}_{2+} = 32.800, \quad \overline{Y}_{3+} = 31.900, \quad \overline{Y}_{4+} = 32.767,$$

the treatment means

$$\overline{Y}_{+1} = 28.700, \quad \overline{Y}_{+2} = 33.150, \quad \overline{Y}_{+3} = 34.075,$$

and the overall mean

$$\overline{Y}_{++} = 31.975.$$

The ANOVA table for the data was constructed in Example 15.3. We have SS[E] = 7.41 with $df_E = 6$. The calculation of Z_{ij} and $Y_{ij}Z_{ij}$ in Box 15.2 for these data is summarized in Table 15.9. We have

$$\sum_{i=1}^{4}\sum_{j=1}^{3} Z_{ij}^2 = 60.96692, \qquad \sum_{i=1}^{4}\sum_{j=1}^{3} Y_{ij}Z_{ij} = 15.02988,$$

so that the sum of squares for testing H_{0NA} is

$$\text{SS[NA]} = \frac{(15.02988)^2}{60.96692} = 3.70.$$

The denominator of the test statistic is

$$\text{MSE[NA]} = \frac{\text{SS[E]} - \text{SS[NA]}}{df_E - 1} = \frac{7.41 - 3.70}{6 - 1} = \frac{3.71}{5} = 0.742,$$

TABLE 15.9
Values of Z_{ij} and $Y_{ij}Z_{ij}$ for weight gain data

i	j	Y_{ij}	$\overline{Y}_{i+} - \overline{Y}_{++}$	$\overline{Y}_{+j} - \overline{Y}_{++}$	Z_{ij}	$Y_{ij}Z_{ij}$
1	1	28.7	−1.542	−3.275	5.0501	144.93787
1	2	30.7	−1.542	1.175	−1.8119	−55.62533
1	3	31.9	−1.542	2.100	−3.2382	−103.29858
2	1	29.3	0.825	−3.275	−2.7019	−79.16567
2	2	34.9	0.825	1.175	0.9694	33.83206
2	3	34.2	0.825	2.100	1.7325	59.25150
3	1	28.2	−0.075	−3.275	0.2456	6.92592
3	2	32.6	−0.075	1.175	−0.0881	−2.87206
3	3	34.9	−0.075	2.100	−0.1575	−5.49675
4	1	28.6	0.792	−3.275	−2.5938	−74.18268
4	2	34.4	0.792	1.175	0.9306	32.01264
4	3	35.3	0.792	2.100	1.6632	58.71096

so that the calculated value of the test statistic is

$$F_c = \frac{3.70}{.742} = 4.987,$$

with one degree of freedom for the numerator and five degrees of freedom for the denominator. The p-value associated with the calculated F is .076, so that, on the basis of the Tukey test, there is no reason to conclude that an additive model is inappropriate for these data. ∎

Exercises

15.17 Refer to the serum amylase data in Exercise 15.3.

   a Use residual plot analyses to check for violation of additivity and ANOVA assumptions.

   b Use the Tukey test to check for additivity.

15.18 Refer to the data in Example 4.10.

   a Use residual plot analyses to check for violation of additivity and ANOVA assumptions.

   b Use the Tukey test to check for additivity.

15.19 Refer to the soil water content data in Exercise 15.11

   a Use residual plot analyses to check for violation of additivity and ANOVA assumptions.

   b Use the Tukey test to check for additivity.

15.20 Refer to the impurity data in Example 15.8.

   a Use residual plot analyses to check for violation of additivity and ANOVA assumptions.

   b Use the Tukey test to check for additivity.

15.21 Refer to time-to-relief data in Exercise 15.14

   a Use residual plot analyses to check for violation of additivity and ANOVA assumptions.

   b Use the Tukey test to check for additivity.

15.6
Randomized blocks with ordinal responses

In Section 8.8, we suggested the use of the Kruskal-Wallis test to check the equality of several populations based on ordinal data from CR designs. Similarly, to check the equality of treatments based on ordinal data from RCB designs, we can use the *Friedman rank sum test*, which utilizes the sums of ranks for each of the t treatments when the responses are ranked within blocks.

Just as the Kruskal-Wallis test is an extension of the Wilcoxon rank sum test to the comparison of more than two treatments, the Friedman rank sum test is an extension of the paired-sample sign test to the comparison of more than two treatments. Like the Wilcoxon and Kruskal-Wallis rank sum tests, the Friedman test is applicable under quite general conditions; the only requirement is that the rankings in any block be independent of the rankings in the other blocks. However, in many practical situations, working with an ANOVA-type model offers certain advantages, both in formulating the assumptions and in interpreting the conclusions. An ANOVA-type model under which the Friedman rank sum test can be used to verify the equality of treatment effects is summarized in Box 15.3.

BOX **15.3**   ***Assumptions underlying the Friedman test of equal treatment effects***

Let Y_{ij} denote the response for the j-th treatment in the i-th block ($i = 1, \ldots, r; j = 1, \ldots, t$). Then

$$Y_{ij} = \mu + \rho_i + \tau_j + E_{ij},$$

where μ is the overall mean; ρ_i is the fixed effect of the i-th block, and $\Sigma \rho_i = 0$; τ_j is the fixed effect of the j-th treatment, and $\Sigma \tau_j = 0$; and E_{ij} is the random error. The E_{ij} are independent and have a common continuous distribution.

The model in Box 15.3 is the same as the fixed-effects ANOVA model for a RCB design in Box 15.1, with one exception. We do not assume that the errors have a normal distribution with zero mean and a common variance. Thus, the model in Box 15.3 is less stringent than the fixed-effects ANOVA model for a RCB design.

The block effects in the model can be random. In that case, the model will be written with ρ_i replaced by R_i ($i = 1, \ldots, r$), where the R_i constitute a random sample from a population of block effects and the R_i and E_{ij} are independent. Thus, the mixed-effects model appropriate for the Friedman test does not require the assumption that R_i and E_{ij} have normal distributions.

The test statistic for the Friedman rank sum test is described in Box 15.4.

BOX **15.4**   ***The Friedman rank sum test***

Suppose we have data from a RCB design that satisfy the two-way additive model in Box 15.3. Let R_{ij} denote the rank of Y_{ij} when the responses in block i are ranked among themselves. Then, to test H_0: $\tau_1 = \cdots = \tau_t$ against the alternative hypothesis H_1 that $\tau_j \neq \tau_{j'}$ for at least one pair of integers $j \neq j'$, we can use the Friedman test statistic

$$W = \frac{W_1}{C},$$

where

$$W_1 = 12 \sum_{j=1}^{t} R_{+j}^2 - 3r^2 t(t+1)^2,$$

$$C = rt(t+1) - \frac{1}{t-1} \sum_{i=1}^{r} \sum_{k=1}^{g_i} t_{ik}(t_{ik}^2 - 1),$$

and $t_{i1}, \ldots, t_{ig_i}$ $(i = 1, \ldots, r)$ are the sizes of the g_i tied groups in block i. For an approximate test at level α, reject the null hypothesis if the calculated value of W exceeds $\chi^2(t-1, \alpha)$.

Like the Wilcoxon and Kruskal-Wallis tests, the Friedman rank sum test is based on the sums of treatment ranks. But unlike the former tests, the rankings in the Friedman test are formulated within blocks.

Most texts on nonparametric statistics (for example, Hollander & Wolfe, 1973; Sprent 1993) provide tables of critical values that can be used to perform exact tests based on the Friedman statistic. However, these tabulated critical values are computed under the assumption that tied values of rankings within blocks are not possible. If there are ties in the data, the *p*-values obtained from tabulated critical values tend to be conservative. Similar comments were made in Section 8.8 about the critical values for the Kruskal-Wallis test. StatXact can be used to determine the exact *p*-values or good approximations to the *p*-values of the Friedman tests.

EXAMPLE **15.11** In a wine-tasting trial, ten judges were asked to rank five brands of red wines on the basis of their overall palatability. The rankings are summarized in Table 15.10.

TABLE **15.10**
Rankings of five wines

Judge	A	B	C	D	E
1	3	2	1	4	5
2	2	3	1	4.5	4.5
3	2	2	2	5	4
4	2	3	1	5	4
5	3	1.5	1.5	4	5
6	2.5	2.5	1	4	5
7	4	1.5	1.5	3	5
8	1	2.5	2.5	5	4
9	3.5	2	1	3.5	5
10	3	4	1	2	5
Total	26	24	13.5	40	46.5

Since the responses are measured on an ordinal scale, the model in Box 15.3 cannot be used directly to describe the observed data. Very often, a model such as that in Box 15.3 is invoked for ordinal data by assuming that the observed ranks are based on the value of an underlying variable measured on an interval scale. For instance, we might assume that the judges' rankings result from a palatability score that each judge assigns (consciously or subconsciously) to each brand. In symbols, such an assumption implies that the rank assigned to the j-th brand by the i-th judge is determined by a palatability score Y_{ij}. Then the Friedman test for equality of treatment effects can be applied to the observed rankings if the Y_{ij} satisfy the model in Box 15.3—that is, if

$$Y_{ij} = \mu + R_i + \tau_j + E_{ij},$$

where μ is the overall mean; R_i is the random effect of the i-th judge, and the judges are a random sample from a population of judges; τ_j is the effect of the j-th treatment, and $\Sigma \tau_j = 0$; and E_{ij} is the random error. The E_{ij} are independent and have a common continuous distribution. The E_{ij} and R_i are independent.

We don't know whether it is reasonable to assume that palatability scores underlie the rankings of the wine. However, as already noted, the validity of the Friedman test does not depend on a model such as that in Box 15.3. All we need to assume is that the judges arrive at their rankings independently. Under such an assumption, the null hypothesis H_0 that there is no difference between the palatability of the five brands of wine can be tested using the Friedman test.

The calculations for the Friedman test are straightforward. The rank sums for the five treatments are

$$R_{+1} = 26, \ R_{+2} = 24, \ R_{+3} = 13.5, \ R_{+4} = 40, \quad \text{and} \quad R_{+5} = 46.5.$$

With $t = 5$ and $r = 10$, the value of W_1 in Box 15.4 is

$$W_1 = 12 \left(26^2 + 24^2 + 13.5^2 + 40^2 + 46.5^2 \right) - 3 \times 10^2 \times 5 \times (5+1)^2$$

$$= 62{,}358 - 54{,}000 = 8358$$

To compute C, we need to determine the tied groups and their sizes within each block (judge). Since judge 1 did not give any tied ranks, this judge's rankings resulted in $g_1 = 5$ tied groups, each of size 1. Therefore, within the first block ($i = 1$), the group sizes are $t_{11} = \cdots = t_{15} = 1$. These sizes contribute an amount

$$1(1^2 - 1) + \cdots + 1(1^2 - 1) = 0$$

to the sum $\sum\limits_{i=1}^{r} \sum\limits_{k=1}^{g_i} t_{ik}(t_{ik}^2 - 1)$ in the expression for C in Box 15.4. For judge 2 ($i = 2$), the ranks for brands D and E are tied, and so we have $g_2 = 4$ tied groups with group sizes $t_{21} = t_{22} = t_{23} = 1$, and $t_{24} = 2$. The resulting contribution to C is

$$1(1^2 - 1) + 1(1^2 - 1) + 1(1^2 - 1) + 2(2^2 - 1) = 6.$$

Note that, when calculating the contribution of group sizes, the t_{ij} that are equal to 1 can be ignored because, if $t_{ij} = 1$, then

$$t_{ij}(t_{ij}^2 - 1) = 1(1^2 - 1) = 0.$$

Similar calculations for judges 3–10 show that the contributions to C are: 24 for judge 3; 0 for judges 4 and 10; and 6 for judges 5–9. Therefore

$$C = 10 \times 5 \times (5 + 1) - \frac{1}{5 - 1} (0 + 6 + 24 + 0 + 6 + 6 + 6 + 6 + 6 + 0)$$

$$= 300 - 15 = 285.$$

Finally, the calculated value of the Friedman test statistic is

$$W_c = \frac{8358}{285} = 29.33.$$

We compare W_c to the critical values of χ^2 with $t - 1 = 5 - 1 = 4$ degrees of freedom. Clearly, the calculated value of the Friedman statistic is highly significant ($p < 0.0001$), and so there is good reason to conclude that the palatability of the five brands is not the same.   ∎

More details about the Friedman rank sum test and the associated inferential procedures can be found in texts on nonparametric methods such as Hollander and Wolfe (1973) and Sprent (1993).

Exercises

15.22   Refer to the weight gain data of Example 15.1

   a   Write a model under which the Friedman test can be used to test the null hypothesis that there is no difference between the effects of the three diets.

   b   Perform the Friedman test and write your conclusion.

   c   Compare your conclusion in (b) with that in Example 15.3. Comment on the validity of these two conclusions.

15.23   Refer to the serum amylase data in Exercise 15.3.

   a   Write a model under which the Friedman test can be used to compare the four methods.

   b   Perform the Friedman test and write your conclusion.

   c   Compare the conclusion in (b) and that based on a two-way ANOVA model. Comment on their validity.

15.24   Eggs produced by four diets were graded as AA, A, BB, and B by seven graders, with the following results.

		\multicolumn{7}{c}{Grader}					
		1	2	3	4	5	6	7
	1	AA	A	AA	AA	BB	A	AA
Diet	2	B	B	B	A	B	A	BB
	3	A	AA	A	A	A	AA	A
	4	B	A	BB	B	B	B	B

a Test the null hypothesis that there is no difference between the grades of the eggs produced by the four diets.

b What assumptions, if any, were needed to validate your test in (a)?

c Write an ANOVA-type model to describe the egg grade data. Clearly explain how your model relates to the actually observed data.

15.25 Refer to Example 5.3, where the sign test was used to compare the percentage membrane fractions of CPZ in red blood cells suspended in two solutions.

a Use the Friedman test to make the comparison previously performed by the sign test.

b Compare the result of your test in (a) with that obtained in Example 5.3. What conclusions can be based on your comparison?

c Compare and contrast the model appropriate for the test in Example 5.3 with that appropriate for the Friedman test.

15.7
Randomized blocks with dichotomous responses

In some studies involving RCB designs, the observed response will be categorical, and an analysis based on a two-way ANOVA or a Friedman test will not be appropriate. In Section 6.3, we described the McNemar test (Box 6.7) for comparing two treatments when the data are matched pairs (RCB design with two treatments) of categorical responses in two categories (dichotomous responses). In this section, we present the *Cochran test*, which can be used to compare t treatments when the study design is RCB and the responses are dichotomous. For more details on categorical data analysis, see Agresti (1990), for instance. The next example describes a RCB study in which the responses are dichotomous.

EXAMPLE **15.12** A study to compare the effect of a fungicide on six varieties of a particular species of plants was performed in a greenhouse. Ten greenhouse benches, each capable of holding six pots, were used. The six varieties were randomized to the pots within each bench, and the effect of the fungicide was observed after four weeks of application. The measured response was recorded as 1 if a plant remained free of fungal infection and 0 otherwise. The results are summarized in Table 15.11.

This is a RCB design, with benches as blocks and varieties as treatments. The responses are categorical, with two possible categories for each response. Thus, the null hypothesis H_0 that the fungicide is equally effective on the six varieties cannot be tested by either the RCB F-test or the Friedman test. ■

If there were only two varieties—say, variety 1 and variety 2—then the data in Example 15.12 could be treated as a random sample of $n = 10$ matched pairs of dichotomous responses. The data could be arranged in a table in the 2×2 format in Box 6.7, with 1 and 0 as response categories 1 and 2, respectively. If π_1 and π_2

TABLE 15.11

Effects of a fungicide on six varieties of plants

Bench	Variety						Total (Y_{i+})
	1	2	3	4	5	6	
1	1	0	0	1	0	0	2
2	0	0	0	0	0	0	0
3	0	1	0	0	1	0	2
4	1	1	1	0	0	0	3
5	1	0	1	0	1	0	3
6	1	1	0	0	0	1	3
7	1	0	1	0	0	0	2
8	1	0	1	1	0	0	3
9	1	1	0	1	1	0	4
10	1	1	1	0	0	0	3
Total (Y_{+j})	8	5	5	3	3	1	25

denote, respectively, the population proportion of plants of varieties 1 and 2 that remain clear of fungal infection—that is, that are in category 1—then the McNemar test in Box 6.7 could be used to test the null hypothesis H_0: $\pi_1 = \pi_2$.

In the general case, where we have t ($t \geq 2$) treatments per block, the null hypothesis of interest is H_0: $\pi_1 = \cdots = \pi_t$, where π_i is the population proportion of responses in category 1 for population i ($i = 1, \ldots, t$). The Cochran test in Box 15.5 can then be used to test H_0.

BOX 15.5

> ### The Cochran test for inferences about the equality of population proportions (matched samples)
>
> Let Y_{ij} denote a dichotomous response that takes the values 0 or 1, for the j-th treatment in the i-th block ($i = 1, \ldots, r; j = 1, \ldots, t$) in a RCB design with r blocks and t treatments. Let Y_{i+}, Y_{+j}, and Y_{++} denote the totals of the Y_{ij} in the i-th block, for the j-th treatment, and in the whole study, respectively.
>
> Let π_j denote the population proportion of responses that are equal to 1 in the j-th population. Then an approximate test of the null hypothesis H_0: $\pi_1 = \pi_2 = \cdots = \pi_t$ rejects H_0 at level α if the computed value
>
> $$Q_c = \frac{t(t-1)\sum_{j=1}^{t} Y_{+j}^2 - (t-1)Y_{++}^2}{tY_{++} - \sum_{i=1}^{r} Y_{i+}^2}$$
>
> exceeds $\chi^2(t-1, \alpha)$.

The Cochran test in Box 15.5 is a large-sample procedure. Let r^* denote the total number of blocks containing at least one 1 and one 0. Then the test in Box 15.5 can be used provided r^* and t satisfy the large-sample conditions (rules of thumb): $r^* \geq 4$ and $r^*t \geq 24$. If r^* and t do not satisfy these large-sample conditions, then critical values for the Cochran test can be obtained from tables such as those provided by Patil (1975). Also, StatXact can be used to obtain exact or approximate p-values for any sample size.

Blocks containing all 1s or all 0s will not affect the value of Q. Therefore, discarding any block that does not contain at least one 0 and at least one 1 will not change the value of Q.

As already noted, when there are only two treatments ($t = 2$), the Cochran test is equivalent to the McNemar test.

EXAMPLE 15.13   Consider the fungicide data in Example 15.12. Let π_i denote the population proportion of plants of variety i for which the fungicide is effective. If there is no difference in effectiveness of the fungicide on the varieties, then the null hypothesis H_0: $\pi_1 = \cdots = \pi_6$ will be true. This null hypothesis can be tested using the Cochran test.

We have $r = 10$, $t = 6$. The totals Y_{i+}, Y_{+j}, and Y_{++} are given in Table 15.11. For instance, $Y_{1+} = 2$, $Y_{+3} = 5$, and $Y_{++} = 25$.

Every block except block 2 contains at least one 0 and one 1. Therefore, $r^* = 10 - 1 = 9$. Since $r^*t = (9)(6) = 54$, the large-sample test in Box 15.5 can be used to test H_0.

We have

$$Q_c = \frac{6(6-1)(8^2 + 5^2 + \cdots + 1^2) - (6-1)(25)^2}{(6)(25) - (2^2 + 0^2 + \cdots + 3^2)}$$

$$= \frac{3990 - 3125}{150 - 73} = 11.24.$$

On comparing Q_c to the critical values of a χ^2-distribution with $t - 1 = 6 - 1 = 5$ degrees of freedom, we see that the calculated value of the test statistic is significant at the 0.05 level. Indeed, the exact p-value is 0.0468. Therefore, at the 5% level of significance, there is sufficient evidence to conclude that the fungicide has different effects on different varieties.   ∎

Exercises

15.26   Refer to the data from the case-control study in Exercise 6.21.

a   Use the Cochran test to determine if the proportion of estrogen users differs between the cases and controls.

b   Let Q_c denote the Cochran statistic calculated in (a). Calculate the McNemar's test statistic Z_c described in Box 6.7 and verify that

$$Q_c = Z_c^2.$$

c   On the basis of the result in (b) argue that, when $t = 2$, the McNemar and

Cochran tests will give identical results. [Hint: Verify that the α-level critical value of a χ^2-distribution is the same as the square of the $\alpha/2$-level critical value of a $N(0, 1)$-distribution.]

15.27 Each of 100 bacterial specimens was divided into three parts and grown in three types (A, B, C) of growth media. Of these specimens, 13 grew in all three media; six grew in media A and B; four grew in A and C; and two grew in B and C. The remaining 75 did not grow in any one of the media. Let π_A, π_B, and π_C denote, respectively, the probabilities that the bacteria will grow in growth media A, B, and C. Apply the Cochran test to the null hypothesis that $\pi_A = \pi_B = \pi_C$. State the assumptions you make and write the conclusions that can be drawn.

15.28 An experiment on the quality of watermelons grown under four conditions was conducted in a RCB design with six blocks. The results are summarized in the following table, where A denotes that the quality of the watermelon is acceptable and NA that the quality is not acceptable:

Block	Conditions			
	1	2	3	4
1	A	NA	NA	A
2	A	NA	A	A
3	A	A	NA	NA
4	A	A	NA	A
5	NA	A	A	A
6	A	A	NA	A

Use an appropriate statistical test to determine whether the data indicate a significant difference between the quality of watermelons grown under the four conditions.

15.8
Overview

Two block designs—the RCB design and the LS design—are discussed in this chapter. These designs are popular among researchers because they are easy to implement and are based on assumptions that are reasonable (at least approximately) in many experimental situations. The randomized complete block design is perhaps one of the most commonly used experimental designs.

The main disadvantage of these designs is the requirement that the effects of the experimental factors should not interact with the effects of the blocking variable. In a RCB design, the problem of block-treatment interaction can be overcome by replicating treatments more than once within blocks. However, this implies larger block sizes, and it may not be possible, in practice, to ensure that the experimental units are homogeneous within such larger blocks. To overcome this difficulty, we can use an incomplete block design, in which each block contains an appropriately selected subset of treatments. For more information about incomplete block designs, see Cochran and Cox (1957) and Steel and Torrie (1980).

16

Repeated-Measures Studies

16.1
Introduction

Thus far, most of our attention has focused on techniques suitable for data that contain one response per experimental unit. The only exceptions were in Section 4.3, where we considered the paired-difference t-test for pairs of responses observed on single experimental units, and in Section 14.6, where we discussed how to handle multiple independent responses obtained by subsampling of the experimental units. This chapter is devoted to describing *repeated-measures studies*. As the name suggests, the key distinguishing feature of a repeated measure study is the property that multiple responses are measured on each experimental unit. Let's begin with two examples.

EXAMPLE **16.1**   An investigator conducted a 3×4 factorial experiment to evaluate the effects of three methods of irrigation (factor A, with levels a_1, a_2, and a_3) and four methods

742

of fertilization (factor B, with levels b_1, b_2, b_3, and b_4) on the yield of a crop. Because it was convenient to apply irrigation methods to large areas and fertilization methods to smaller areas, the investigator used a *split-plot design*. The experimental field was first divided into six large areas, called *main plots* or main units. The three irrigation methods were randomly assigned to the six main plots in such a way that each method was assigned to exactly two main plots. Then four smaller areas, called *subplots* or subunits, were selected within each main plot. Within each main plot, the four fertilization methods were randomly assigned to the four subplots. Each main plot was irrigated by its assigned method of irrigation; within each main plot, each subplot was fertilized by its assigned fertilization method. Thus, each of the $6 \times 4 = 24$ subplots was exposed to a specific treatment combination $a_i b_j$. The responses (crop yields) were measured on the subplots.

At first sight, the crop yield study might look like a 3×4 factorial experiment in a completely randomized design with two replications per treatment combination. If this were the case, and the usual ANOVA assumptions were reasonable, the observed responses could be analyzed using the two-way ANOVA model

$$Y_{ijk} = \mu + \alpha_i + \beta_j + (\alpha\beta)_{ij} + E_{ijk}, \tag{16.1}$$

where Y_{ijk} is the k-th yield for treatment $a_i b_j$; the α_i, the β_j, and the $(\alpha\beta)_{ij}$ are the factorial effects of A and B; and the E_{ijk} are the random errors, which are assumed to be independent $N(0, \sigma^2)$. Unfortunately, in experiments such as these, the manner in which the treatment combinations are randomized to the subplots tends to undermine the assumption that the E_{ijk} are independent. Since the subplots within a main plot were treated with a common application of an irrigation method, there is often a correlation between responses from different subplots within the same main plot. A model that can accommodate correlations between the responses within main plots is likely to be a better model than the two-way ANOVA model in Equation (16.1). ∎

EXAMPLE 16.2 Potoff and Roy (1964) reported data on dental measurements (distances, mm, from the center of the pituitary to the maxillary fissure) of $n = 27$ children (11 girls and 16 boys) at the University of North Carolina Dental School. Each child was measured four times over a period of six years—at 8, 10, 12, and 14 years of age.

How can we analyze these data to evaluate the differences over time in the dental measurements of male and female children? One possibility is to use a general linear model in which gender (A) and age (B) are the explanatory factors. The model can be written in two ways. We can treat age as a qualitative factor with four levels and construct a model similar to Equation (16.1), with $A =$ gender (at $a = 2$ levels) and $B =$ age (at $b = 4$ levels); or we can treat age as a quantitative factor and use a general linear model based on an assumed mathematical relationship between age and dental measurements—for example

$$Y_i = \beta_0 + \beta_1 a_i + \beta b_i + E_i, \tag{16.2}$$

where Y_i, a_i and b_i denote, respectively, the dental measurement, the gender ($a_i = 0$ for female and $a_i = 1$ for male), and the age of the i-th child ($i = 1, 2, \ldots, 27$); the E_i are the errors and are assumed to be independent $N(0, \sigma^2)$.

Regardless of which model we choose, the assumption that the errors are independent is unlikely to be satisfied by the dental data, because a measurement of a child at a given age will most likely be correlated with repeated measurements of the same child at other ages. ■

There are many similarities between the crop yield study in Example 16.1 and the dental study in Example 16.2. Both have two experimental factors A and B: $A =$ irrigation method and $B =$ fertilization method in the crop yield study; $A =$ gender and $B =$ age in the dental study. The crop yield data are measured on subunits within main units, where the main units and subunits are plots of land in the experimental field. The dental measurements are also measured on subunits within main units; in this case, the main units are the observation periods (each of length six years) of the study subjects, and the subunits are the specific time points at which the responses are being measured. The subunits are separated by space in the crop yield study and by time in the dental study.

On the basis that different levels of factor A correspond to different main units, factor A is often called the *between-units factor* or the *between factor*. Similarly, different subunits within a main unit correspond to different levels of B, and so factor B is called the *within-units factor* or the *within factor*. The crop yield study and the dental study are both examples of a repeated–measures study, with one between-units factor and one within-units factor. Of course, not all repeated-measures studies will have only one between factor and one within factor. Furthermore, in some repeated-measures studies, the main units are arranged in blocks. The following is an example of a repeated-measures study in a randomized blocks design with two between and two within factors.

EXAMPLE **16.3**   Investigators wanted to evaluate vigilance performance as a function of four factors: A, prior training ($a_1 =$ trained; $a_2 =$ untrained); B, type of signal ($b_1 =$ auditory; $b_2 =$ visual); C, awareness ($c_1 =$ aware of the type of signal; $c_2 =$ not aware of the type of signal); and D, the signal environment ($d_1 =$ with distraction; $d_2 =$ without distraction). They designed a repeated-measures study with A and C as the between factors and B and D as the within factors. The study was conducted over a period of four weeks, with four new subjects participating each week (a total of 16 subjects). On the first day of each week, the four subjects were randomized to the four combinations of levels of A and C: a_1c_1, a_1c_2, a_2c_1, a_2c_2. Those who were assigned to the trained (a_1) groups were given four days of intensive training on how to watch for the specific visual and auditory signals used in the experiment. On the fifth day, each of the four subjects was exposed to the four conditions b_1d_1, b_1d_2, b_2d_1, and b_2d_2 in four sittings, each lasting 30 minutes. The time (msec) between signal exposure and signal detection (the response latency) was measured for each subject under each condition.

The measured response latencies were analyzed as data from a repeated-measures study with two between factors and two within factors. In this example, the subjects are the main units, the sittings are the subunits, and the weeks are the blocks. ■

The general objectives in the analysis of repeated-measures data are similar to those in the factorial analyses described in Chapter 14. The goal is to evaluate the factorial effects (main effects and interaction effects) of the between and within factors. The main difference between the usual factorial analyses and the repeated-measures analyses is in how the models are formulated and the necessary statistics are computed. In practice, two types of models—*univariate models* and *multivariate models*—are commonly used in repeated-measures analyses. Univariate models are models for individual responses in which the main unit effects are treated as random effects. As we'll see, the univariate models can be analyzed using the mixed model methods described in Chapter 14, under certain assumptions that are reasonable in some practical settings.

The multivariate models for repeated-measures data regard the multiple responses measured on a single main unit as the components of a single *multivariate response*. For instance, in the dental study, a multivariate model will regard the set of four measurements made on a particular child as a single 4-variate response for that child. Under such a setup, the dental data can be regarded as a collection of four-variate responses, one for each of the $n = 27$ children. A major advantage of multivariate models is that they are very flexible and can be used with less restrictive assumptions than their univariate counterparts. A disadvantage is that these models typically involve a large number of parameters, and so large quantities of data are necessary in order to use them effectively.

In this chapter, most of our attention is focused on univariate models. However, as we shall see, the interpretation of results from multivariate analyses is not much different from the interpretation of results from univariate analyses. Once you understand the implications of the assumptions in univariate models, you will have little difficulty in deciding when to use a multivariate model. Many computing software packages, including the GLM procedure in SAS, are currently available for multivariate analysis.

Exercises

16.1   In a study of a new drug, the total cholesterol levels of 23 subjects were measured at the end of six 4-week periods (Hirotsu, 1993). The results obtained are summarized in Table 16.1.

    a   Explain why the cholesterol measurements must be regarded as the results from a repeated-measures study.

    b   Identify the between and within factors and their levels.

    c   What are the main units and subunits in this study?

16.2   An experiment to evaluate the effects of two soil preparation methods and two weed control methods on the yield (bushels per acre) of three varieties of soybeans was conducted on three tracts of land. Each tract was divided into two plots, which were randomly assigned to the two soil preparation methods. Each plot was then subdivided into six smaller plots, and the six combinations of soybean varieties and

TABLE   16.1

Data for cholesterol study

Treatment	Subject	Period					
		1	2	3	4	5	6
Drug	1	317	280	275	270	274	266
	2	186	189	190	135	197	205
	3	377	395	368	334	338	334
	4	229	258	282	272	264	265
	5	276	310	306	309	300	264
	6	272	250	250	255	228	250
	7	219	210	236	239	242	221
	8	260	245	264	268	317	314
	9	284	256	241	242	243	241
	10	365	304	294	287	311	302
	11	298	321	341	342	357	335
	12	274	245	262	263	235	246
Placebo	13	232	205	244	197	218	233
	14	367	354	358	333	338	355
	15	253	256	247	228	237	235
	16	230	218	245	215	230	207
	17	190	188	212	201	169	179
	18	290	263	291	312	299	279
	19	337	337	383	318	361	341
	20	283	279	277	264	269	271
	21	325	257	288	326	293	275
	22	266	258	253	284	245	263
	23	338	343	307	274	262	309

weed control methods were randomly assigned to the smaller plots. The yield data obtained are summarized in Table 16.2.

a   Explain why the soybean yield data must be regarded as the results from a repeated-measures study.

b   Identify the between and within factors and their levels.

c   What are the main units and subunits in this study?

d   Are there any blocks in the study design? If so, identify them.

16.2
Univariate repeated-measures models

As we have seen, a repeated-measures model must allow for the possibility that responses in different subunits of the same main unit are correlated. There are many approaches to developing repeated-measures models with this desirable property; we will confine our attention to that used by Crowder and Hand (1990). This model

TABLE   16.2

Data on soybean yield

	Soil preparation method 1			Soil preparation method 2		
	Variety	Weed control	Yield	Variety	Weed control	Yield
Tract 1	1	1	25.7	1	1	19.1
	1	2	29.2	1	2	23.1
	2	1	21.0	2	1	19.7
	2	2	29.9	2	2	27.0
	3	1	19.2	3	1	26.7
	3	2	22.0	3	2	31.4
Tract 2	1	1	25.3	1	1	15.6
	1	2	30.2	1	2	19.6
	2	1	21.9	2	1	16.3
	2	2	29.6	2	2	23.8
	3	1	18.6	3	1	22.8
	3	2	23.2	3	2	26.7
Tract 3	1	1	25.5	1	1	14.9
	1	2	29.4	1	2	18.5
	2	1	21.9	2	1	15.2
	2	2	29.0	2	2	23.0
	3	1	17.9	3	1	21.6
	3	2	23.0	3	2	26.2

is presented in Box 16.1 for a repeated-measures study with two fixed factors A and B, where A is the within factor and B is the between factor. The model can easily be adapted for studies with any number of within and between factors, fixed or random.

Let $a_i b_j$ denote the treatment combination consisting of the i-th level of A and the j-th level of B, and let Y_{ij} denote the observed response to $a_i b_j$ in a particular main unit on which responses to the i-th level of A will be measured. Then the univariate repeated-measures model in Box 16.1 is based on the assumption that Y_{ij} can be expressed as a sum of three terms

$$Y_{ij} = \mu_{ij} + S_{ij} + E_{ij}.$$

The parameter μ_{ij} represents the expected response to $a_i b_j$. It can be interpreted as the mean obtained by averaging the responses over the conceptual population of all main units on which responses to $a_i b_j$ can be measured and also over the conceptual population of all possible measured values that can be observed within each such main unit.

The random variable S_{ij} accounts for the fact that expected responses to $a_i b_j$ may be different in different main units. In other words, $\mu_{ij} + S_{ij}$ is the expected response of the particular main unit to $a_i b_j$, and S_{ij} is the amount by which the expected response in this particular main unit differs from the overall mean response μ_{ij}. The model allows for the possibility that $S_{i1}, S_{i2}, \ldots, S_{ij}, \ldots, S_{ib}$ may be correlated.

The random quantity E_{ij} is the amount by which the observed response Y_{ij} differs from the expected response $\mu_{ij} + S_{ij}$ of the particular main unit. The presence of the term E_{ij} in Y_{ij} helps explain the variation in the observed response to $a_i b_j$, from observation to observation, within a main unit.

BOX **16.1**

A univariate repeated-measures model

Consider a repeated-measures study with one between factor A and one within factor B. Let Y_{ij} denote the response to the j-th level of B in a main unit corresponding to the i-th level of A ($i = 1, \ldots, a$; $j = 1, \ldots, b$). Then

$$Y_{ij} = \mu_{ij} + S_{ij} + E_{ij},$$

where μ_{ij} is the expected response to the combination of the i-th level of A and the j-th level of B; S_{ij} is a random variable that equals the difference between μ_{ij} and the expected response to the j-th level of B in this particular main unit; and E_{ij} is a random variable that equals the difference between Y_{ij} and the expected response for the j-th level of B in this particular main unit.

The assumptions underlying this model are as follows:

1 The $S_{i1}, S_{i2}, \ldots, S_{ij}, \ldots, S_{ib}$ have a multivariate normal distribution such that:

  a the mean of S_{ij} is zero;

  b the variance of S_{ij}, denoted by σ_{sj}^2, may depend on j;

  c the correlation coefficient between S_{ij} and $S_{ij'}$, denoted by $\rho_{jj'}$, may depend on j and j'.

2 The E_{ij} have a normal distribution with zero mean and variance σ_j^2.

3 The S_{ij} and E_{ij} are uncorrelated.

This model is adapted from Crowder and Hand (1990).

Most of the terminology relating repeated-measures studies was developed by behavioral scientists dealing with human or animal subjects as the main units in their investigations. Consequently, we often refer to a main unit as a *subject*, and to its effects S_{ij} as *subject effects*. The quantity E_{ij}, which measures the departure of a subject's observed response from the expected response for that particular subject, is called the *error effect*.

As already noted, the model in Box 16.1 applies to the observed responses in a single main unit. It can easily be extended to model all responses in the study, by combining the assumptions in Box 16.1 with the assumptions needed to describe how the responses from different subjects (main units) are related to each other. For example, if independent responses are observed from different subjects, then the assumptions in the overall model will include those in Box 16.1 plus the assumption that the S_{ij} and the E_{ij} corresponding to different subjects are uncorrelated. The model in Box 16.2 adapts Box 16.1 to the case where multiple independent responses may be available at some treatment combinations.

BOX **16.2**

> **Repeated-measures model with multiple responses for treatment combinations**
>
> Let Y_{ijk} denote the response to the j-th level of B in the k-th main unit assigned to the i-th level of A ($i = 1, \ldots, a$; $j = 1, \ldots, b$; $k = 1, \ldots, r_i$). Then
>
> $$Y_{ijk} = \mu_{ij} + S_{ijk} + E_{ijk},$$
>
> where μ_{ij} is the expected response for treatment combination $a_i b_j$; S_{ijk} is a random variable that equals the difference between μ_{ij} and the expected response at the j-th level of B in the k-th main unit assigned to the i-th level of A; and E_{ijk} is the amount by which the response at the j-th level of B for the k-th main unit assigned to a_i differs from its expected response.
>
> In addition to the assumptions in Box 16.1, the S_{ijk} and E_{ijk} associated with the measurements of different main units are assumed to be mutually independent.

The next example illustrates how the model in Box 16.2 can be used for the dental data in Example 16.2.

EXAMPLE **16.4** In the dental study in Example 16.2, a child's gender was the between factor with $a = 2$ levels—$a_1 = 0$ (female), $a_2 = 1$ (male)—and the time periods (of six years) over which the children were observed were the main units. Within each main unit, four measurements were made at $b = 4$ levels ($b_1 = 8$, $b_2 = 10$, $b_3 = 12$, and $b_4 = 14$) of the within factor, age. For the dental measurements of a single child, model in Box 16.1 can be expressed as

$$Y_{ij} = \mu_{ij} + S_{ij} + E_{ij}, \qquad i = 1, 2; \quad j = 1, \ldots, 4. \tag{16.3}$$

Here Y_{ij} is the dental measurement at the j-th age for a child of gender i ($i = 1$ for a female; $i = 2$ for a male); μ_{ij} is the expected dental measurement at the j-th age for a child of gender i; S_{ij} is a random variable that equals the difference between μ_{ij} and the expected dental measurement at the j-th age for this particular child; and E_{ij} is the amount by which the dental measurement at the j-th age for this particular child differs from the expected (average) dental measurement for all children of gender i and age j.

The S_{ij} and E_{ij} satisfy the assumptions in Box 16.1. The parameters corresponding to these assumptions for the dental measurements of two children over two six-year periods (main units) are shown in Display 16.1.

First, consider the assumptions about the subject effect S_{ij}. Among other things, these assumptions imply that, in the population of children from which the study

DISPLAY   16.1

(a) Parameters of S_{1j} and S_{2j}

	Age		Subject 1 ($i=1$)				Subject 2 ($i=2$)			
	Age		8	10	12	14	8	10	12	14
		j	1	2	3	4	1	2	3	4
Subject 1 ($i=1$)	8	1	1	ρ_{12}	ρ_{13}	ρ_{14}	0	0	0	0
	10	2	ρ_{21}	1	ρ_{23}	ρ_{24}	0	0	0	0
	12	3	ρ_{31}	ρ_{32}	1	ρ_{34}	0	0	0	0
	14	4	ρ_{41}	ρ_{42}	ρ_{43}	1	0	0	0	0
Subject 2 ($i=2$)	8	1	0	0	0	0	1	ρ_{12}	ρ_{13}	ρ_{14}
	10	2	0	0	0	0	ρ_{21}	1	ρ_{23}	ρ_{24}
	12	3	0	0	0	0	ρ_{31}	ρ_{32}	1	ρ_{34}
	14	4	0	0	0	0	ρ_{41}	ρ_{42}	ρ_{43}	1
Variance			σ_{s1}^2	σ_{s2}^2	σ_{s3}^2	σ_{s4}^2	σ_{s1}^2	σ_{s2}^2	σ_{s3}^2	σ_{s4}^2

(b) Parameters of E_{1j} and E_{2j}

	Age		Subject 1 ($i=1$)				Subject 2 ($i=2$)			
	Age		8	10	12	14	8	10	12	14
		j	1	2	3	4	1	2	3	4
Subject 1 ($i=1$)	8	1	1	0	0	0	0	0	0	0
	10	2	0	1	0	0	0	0	0	0
	12	3	0	0	1	0	0	0	0	0
	14	4	0	0	0	1	0	0	0	0
Subject 2 ($i=2$)	8	1	0	0	0	0	1	0	0	0
	10	2	0	0	0	0	0	1	0	0
	12	3	0	0	0	0	0	0	1	0
	14	4	0	0	0	0	0	0	0	1
Variance			σ_1^2	σ_2^2	σ_3^2	σ_4^2	σ_1^2	σ_2^2	σ_3^2	σ_4^2

children are selected, the correlation between dental measurements will depend on the ages at which the measurements are made. For instance, the correlation between the subject effects at ages of 8 and 10 ($j=1$ and $j=2$) is ρ_{12}, whereas that between the subject effects at ages of 8 and 14 ($j=1$ and $j=4$) is ρ_{14}. A question of practical interest is whether the model can be simplified by assuming that the correlations between subject effects do not depend on the ages at which measurements are made. In other words, is it reasonable to assume that the ρ_{ij} are all equal to a common value ρ? For the dental data, we might argue that the assumption of equal correlation is not reasonable; for example, we would expect ρ_{12} to be larger than ρ_{14}, because measurements made at ages 8 and 10 are closer to each other than the measurements made at ages 8 and 14.

The assumptions about S_{ij} also state that the variances of the subject effects can depend on age. For instance, the variance of these effects (in the population of children being studied) on the measurements at 8 years of age is σ_{s1}^2 whereas the corresponding variance at 14 is σ_{s4}^2. Again, whether this assumption can be simplified by assuming that $\sigma_{s1}^2 = \cdots = \sigma_{s4}^2 = \sigma_s^2$ depends on what is known about the nature of the dental measurements.

Like the errors in the ANOVA models considered in previous chapters, the E_{ij} in this model are assumed to have independent normal distributions. However, whereas ANOVA models assume a common error variance, the variances of the E_{ij} can depend on j in the present case.

Finally, the model in Box 16.1 implies that neither the subject effects nor the error effects from two different children (main units) are correlated; this means that the responses from two different main units are uncorrelated.

Equation (16.3) is a model for the dental measurements of a single child. Under the assumption that the measurements for different children are independent, an overall model can be expressed as

$$Y_{ijk} = \mu_{ij} + S_{ijk} + E_{ijk}, \quad i = 1, 2; \quad j = 1, \ldots, 4; \tag{16.4}$$

$$k = 1, \ldots, 11 \text{ if } i = 1 \text{ and } k = 1, \ldots, 16 \text{ if } i = 2.$$

Here Y_{ijk} is the dental measurement at the j-th age for the k-th child of gender i ($i = 1$ for a female; $i = 2$ for a male); μ_{ij} is the expected dental measurement at the j-th age for a child of gender i; S_{ijk} is a random variable that equals the difference between μ_{ij} and the expected dental measurement at the j-th age for the k-th child of gender i; E_{ijk} equals the amount by which the dental measurement at the j-th age for the k-th child of gender i differs from the child's expected dental measurement.

In addition to the assumptions in Box 16.1, we also assume that the S_{ijk} and E_{ijk} associated with the measurements of different children are mutually independent.   ■

Under the assumptions in Box 16.1, the variance of a response to the j-th level of the within factor is

$$\text{Var}\left(Y_{ij}\right) = \sigma_{sj}^2 + \sigma_j^2. \tag{16.5}$$

Therefore, unlike the fixed-effects ANOVA models discussed in previous chapters, the repeated-measures model in Box 16.1 allows for unequal (nonhomogeneous) variances of the observed responses. Furthermore, the assumptions in Box 16.1 imply that, in any main unit, the correlation coefficient between the responses to the j-th and j'-th levels of the within factor can be expressed as

$$\text{Corr}\left(Y_{ij}, Y_{ij'}\right) = \frac{\rho_{jj'}\sigma_{sj}\sigma_{sj'}}{\sqrt{(\sigma_{sj}^2 + \sigma_j^2)(\sigma_{sj'}^2 + \sigma_j^2)}}. \tag{16.6}$$

A key advantage of the model in Box 16.1 is that it can be used in a wide variety of repeated-measures settings. A disadvantage of the model is that it requires a large number of parameters to characterize the subject and error effects. In addition to the $2b$ variances $\sigma_{s1}^2, \ldots, \sigma_{sb}^2$ and $\sigma_1^2, \ldots, \sigma_b^2$, the model also needs $b(b-1)/2$ correlation coefficients $\rho_{jj'}$. For example, if the within factor has $b = 3$ levels, the model uses $2(3) + 3(3-1)/2 = 9$ parameters to specify the subject and error effects: three variances $\sigma_{s1}^2, \sigma_{s2}^2, \sigma_{s3}^2$; three variances $\sigma_1^2, \sigma_2^2, \sigma_3^2$; and three correlation coefficients $\rho_{12}, \rho_{13},$ and ρ_{23}. Note that only three of the six correlation coefficients need to be specified, because $\rho_{12} = \rho_{21}, \rho_{13} = \rho_{31},$ and $\rho_{23} = \rho_{32}$. The need to estimate a large number of parameters limits the practical use of the model in Box 16.1. Fortunately, a restricted version of this model that requires only two parameters to specify the subject and error effects proves reasonable in frequently used repeated-measures studies known as *split-plot experiments*. As the name suggests, the first split-plot experiments were agricultural field studies in which the levels of one or more factors were randomized to subplots within main plots; the crop yield study in Example 16.1 is a split-plot experiment. Currently, split-plot experiments are used in a wide variety of disciplines.

Exercises

16.3   Refer to the cholesterol data in Exercise 16.1.

a   Let μ_{ij} denote the expected cholesterol level at period j for a subject receiving treatment i ($i = 1$ for drug; $i = 2$ for placebo). Write an expression for μ_{ij} as a function of the effects of the treatment and period.

b   Let Y_{ijk} denote the measured total cholesterol at the j-th period for the k-th subject who received treatment i. Using the expression for μ_{ij} in (a), write a model for Y_{ijk} along the lines of the model in Box 16.2.

c   For these data, is it reasonable to assume that the correlations between the repeated measurements of a single subject are the same? Explain.

16.4   Refer to the soybean yield data in Exercise 16.2.

a   Let μ_{ij} denote the expected soybean yield when soil preparation method i and weed control method j are used. Write an expression for μ_{ij} as a function of the effects of soil preparation and weed control methods.

b   Let Y_{ijk} denote the measured soybean yield in the subplot in the k-th block that received the i-th soil preparation method and the j-th weed control method. Using the expression in (a), write a model for Y_{ijk} along the lines of the model in Box 16.2.

c   For these data, is it reasonable to assume that the correlations between the subplot effects within a main plot are the same? Explain.

16.5   Using the results in Box 9.1, verify that, for the model in Box 16.1, the variance of a response to the j-th level of the within factor is given by Equation (16.5)

16.3
Split-plot experiments

In a *split-plot design*, the treatments are combinations of the levels of within and between factors. Two of the most common examples are the *completely randomized split-plot design* and the *randomized complete block split-plot design*. The crop yield study in Example 16.1 is based on a completely randomized split-plot design, because the assignment of the levels of the between factor to the main units (main plots) corresponds to a completely randomized design. In a randomized complete block split-plot design, similarly, the assignment of the levels of the between factor corresponds to a randomized complete block design. Within each of the r blocks, the a levels of the between factor A are randomized to the main units (main plots). Within each main unit, the b levels of the between factor B are randomized to the b subunits (subplots). The responses for the combinations of levels of A and B are measured on each of the $a \times b \times r$ subnits.

Because they are randomized to the main plots, the levels of a between factor are called *main treatments*. Likewise, the levels of a within factor are called *subtreatments*. Here's an example of a randomized block split-plot design.

EXAMPLE **16.5**   In an experiment, three formulations of a diet (factor A) were compared on the basis of the amounts of a particular chemical in the diet that were absorbed by the kidneys of experimental rats. The investigators were also interested in comparing three techniques (factor B) for measuring the absorbed amounts. Four litters, each containing three rats, were used in the study. Within each litter, the animals were randomly assigned to the three formulations. After two weeks on the diets, the animals were sacrificed, and three sample specimens were selected from each animal's kidney. The three methods were randomly assigned to the three specimens, and the absorbed amounts were measured for the specimens using the assigned methods. The data (percentage of chemical absorbed) are given in Table 16.3.

This is a split-plot design in randomized complete blocks. Each of the $r = 4$ blocks (litters) contains $a = 3$ main plots (individual rats) and each main plot contains $b = 3$ subplots (specimens). The three formulations are the main treatments, and the three methods are the subtreatments.

TABLE **16.3**
Absorption data for diet study

Formulation	Method	Litter				Total
		1	2	3	4	
1	1	26.97	26.12	27.83	27.47	108.39
	2	22.60	22.91	19.83	21.63	86.97
	3	30.71	29.53	27.51	28.62	116.37
2	1	17.47	18.13	18.01	17.97	71.58
	2	16.90	16.31	16.52	15.93	65.66
	3	23.95	22.84	23.84	23.45	94.08
3	1	20.72	20.41	21.01	21.34	83.48
	2	24.32	25.06	25.92	25.33	100.63
	3	28.31	29.02	29.13	29.36	115.82

If the absorption amounts in the three specimens (subplots) from a single rat (main plot) are uncorrelated, then the data can be analyzed as data from a 3×3 factorial experiment in a RCB design. If, on the other hand, it is suspected that correlations might exist between measurements from the specimens of the same rat (main plot), then a model that allows for correlations between repeated measures should be used to analyze the absorption data. ■

There are numerous types of split-plot designs, including the Latin square split-plot design, in which the assignment of the main treatments to the main plots is based on a Latin square design.

A split-plot design can be conceptualized as consisting of two designs: a *main plot design* and a *subplot design*. The main plot design is the protocol used to assign the main treatments to the main units. In a completely randomized split-plot design, the main plot design is a completely randomized design; in a randomized complete block split-plot design, by contrast, the main plot design is a RCB design. The subplot design in a split-plot experiment is a collection of a RCB designs, where a is the number of main treatments. Each of these RCB designs has b treatments arranged in r blocks (main plots), where b is the number of subtreatments. Consider the following example.

EXAMPLE **16.6**   The design of the chemical absorption study in Example 16.5 consists of main plot and subplot components. In the RCB main plot design, the three formulations (levels of A) are the treatments, the litters are the blocks, and the rats (main units) are the experimental units. A possible layout of the main plot design in Example 16.5—that is, a random arrangement of the levels of A—is shown in Display 16.2.

DISPLAY **16.2**

Layout of the main plot design in Example 16.5

Block 1	Block 2	Block 3	Block 4
a_1	a_3	a_2	a_2
a_3	a_1	a_1	a_3
a_2	a_2	a_3	a_1

The subplot design is composed of $a = 3$ RCB designs, one for each formulation. The main units are the blocks and the subtreatments are the treatments in the subplot designs. A layout of the subplot designs in Example 16.5 is shown in Display 16.3. ■

The most frequently used model for split-plot experiments is obtained from Box 16.1 by including the additional assumption that the main plot effects (subject

DISPLAY   16.3

Layout of the subplot design in Example 16.5

Formulation 1

Block 1	Block 2	Block 3	Block 4
b_2	b_1	b_2	b_3
b_1	b_3	b_1	b_1
b_3	b_2	b_3	b_2

Formulation 2

Block 1	Block 2	Block 3	Block 4
b_2	b_3	b_1	b_2
b_1	b_1	b_2	b_1
b_3	b_2	b_3	b_3

Formulation 3

Block 1	Block 2	Block 3	Block 4
b_3	b_1	b_2	b_1
b_2	b_2	b_1	b_3
b_1	b_3	b_3	b_2

effects) are the same for all subplots within a main plot. If S_{ijk} is the subject effect on the j-th response of the k-th main plot receiving the i-th main treatment, the new assumption states that

$$S_{i1k} = S_{i2k} = \cdots = S_{ibk} = S_{ik}, \quad i = 1, 2, \ldots,$$

where S_{ik} is the common effect of the k-th subject (main plot) receiving the i-th main treatment. Under that assumption, the repeated-measures model in Box 16.1 reduces to the form

$$Y_{ijk} = \mu_{ij} + S_{ik} + E_{ijk},\tag{16.7}$$

where μ_{ij} and E_{ijk} are as in Box 16.1.

Box 16.3 shows the version of this model for a completely randomized split-plot experiment in which A and B are fixed factors.

BOX 16.3   *Completely randomized split-plot model with fixed treatment effects*

Consider a completely randomized split-plot experiment with a main treatments (levels of factor A) and b subtreatments (levels of factor B). Let r_i denote the number of replications of the i-th level of A, and let Y_{ijk} denote the k-th response for the treatment combination $a_i b_j$. Then

$$Y_{ijk} = \mu + \alpha_i + \beta_j + (\alpha\beta)_{ij} + S_{ik} + E_{ijk},$$

$$k = 1, \ldots, r_i; \quad i = 1, \ldots a; \quad j = 1, \ldots, b.$$

The notation here is as follows:

- μ is the overall mean;
- α_i is the fixed effect of the i-th level of A, and $\sum_{i=1}^{a} \alpha_i = 0$;
- β_j is the fixed effect of the j-th level of B, and $\sum_{j=1}^{b} \beta_j = 0$;
- $(\alpha\beta)_{ij}$ is the joint effect of the i-th level of A and the j-th level of B; $\sum_{i=1}^{a} (\alpha\beta)_{ij} = 0$ and $\sum_{j=1}^{b} (\alpha\beta)_{ij} = 0$;
- S_{ik} is the effect of the k-th main plot receiving the i-th level of A;
- E_{ijk} is the error effect.

The S_{ik} are independent $N(0, \sigma_s^2)$; the E_{ijk} are independent $N(0, \sigma^2)$; and the S_{ik} and E_{ijk} are mutually independent.

The completely randomized split-plot model can easily be modified to derive a model for a RCB split-plot experiment. To account for the fact that the main plots are arranged in blocks, we simply replace S_{ik} with $R_k + (SR)_{ik}$, where R_k is the effect of the k-th block (replication) and $(SR)_{ik}$ is the joint effect of the i-th main treatment and the k-th block. The resulting model is presented in Box 16.4.

BOX 16.4

Randomized complete block split-plot model with fixed treatment effects

Let Y_{ijk} denote the k-th response for the treatment combination $a_i b_j$ in a RCB split-plot experiment with a main treatments (levels of factor A), b subtreatments (levels of factor B), and r blocks. Then we assume the model

$$Y_{ijk} = \mu + \alpha_i + \beta_j + (\alpha\beta)_{ij} + R_k + (SR)_{ik} + E_{ijk},$$

$$k = 1, \ldots, r; \quad i = 1, \ldots, a; \quad j = 1, \ldots, b.$$

The notation here is as follows:

- μ is the overall mean;
- α_i is the fixed effect of the i-th level of A, and $\sum_{i=1}^{a} \alpha_i = 0$;
- β_j is the fixed effect of the j-th level of B, and $\sum_{j=1}^{b} \beta_j = 0$;
- $(\alpha\beta)_{ij}$ is the joint effect of the i-th level of A and the j-th level of B; $\sum_{i=1}^{a} (\alpha\beta)_{ij} = 0$, and $\sum_{j=1}^{b} (\alpha\beta)_{ij} = 0$;

- R_k is the random effect of the k-th block;
- $(SR)_{ik}$ is the joint effect of the i-th level of A and the k-th block;
- E_{ijk} is the error effect.

The R_k are independent $N(0, \sigma_r^2)$; the $(SR)_{ik}$ are independent $N(0, \sigma_{sr}^2)$; and the E_{ijk} are independent $N(0, \sigma^2)$. Also, the R_k, $(SR)_{ik}$, and E_{ijk} are mutually independent.

Comparison of the split-plot models in Box 16.4 and the corresponding two-factor models in Chapters 13 and 14 shows that the only difference is in the assumption about the errors. In the models in Chapters 13 and 14, the errors E_{ijk} are assumed to be uncorrelated, with a common variance σ^2. The errors $S_{ik} + E_{ijk}$ in the split-plot models also have a common variance but are not independent. To see this, note that, if the subject effects are the same for all subplots, then $\sigma_{s1}^2 = \cdots = \sigma_{sb}^2 = \sigma_s^2$ and $\rho_{jj'} = 1$. Therefore, the variance of a single response and the correlation between two responses within a main plot can be obtained by setting $\sigma_{sj}^2 = \sigma_s^2$, $\sigma_j^2 = \sigma^2$, and $\rho_{jj'} = 1$ in Equations (16.5) and (16.6). In the completely randomized split-plot model in Box 16.3, the variance of a single response is

$$\text{Var}(Y_{ijk}) = \sigma_s^2 + \sigma^2, \tag{16.8}$$

and the correlation coefficient between any two responses is

$$\rho = \begin{cases} \dfrac{\sigma_s^2}{\sigma_s^2 + \sigma^2} & \text{if the responses are in the same main plot} \\[2mm] 0 & \text{otherwise.} \end{cases} \tag{16.9}$$

The corresponding variance and correlation coefficient under the RCB split-plot model in Box 16.4 can be obtained by replacing the term σ_s^2 with $\sigma_r^2 + \sigma_{sr}^2$. The resulting variance and correlation coefficients are

$$\text{Var}(Y_{ijk}) = \sigma_r^2 + \sigma_{sr}^2 + \sigma^2 \tag{16.10}$$

and

$$\rho = \begin{cases} \dfrac{\sigma_r^2 + \sigma_{sr}^2}{\sigma_r^2 + \sigma_{sr}^2 + \sigma^2} & \text{if the responses are in the same main plot} \\[2mm] 0 & \text{otherwise.} \end{cases} \tag{16.11}$$

Thus, the split-plot models in Boxes 16.3 and 16.4 imply not only that the responses have a common variance but also that the structure of the correlations between the responses within a main plot can be represented by a matrix of b rows and *b* columns

$$\Sigma_w = \begin{bmatrix} 1 & \rho & \rho & \cdots & \rho & \rho \\ \rho & 1 & \rho & \cdots & \rho & \rho \\ \vdots & \vdots & \vdots & & \vdots & \vdots \\ \rho & \rho & \rho & \cdots & 1 & \rho \\ \rho & \rho & \rho & \cdots & \rho & 1 \end{bmatrix}, \qquad \textbf{(16.12)}$$

where the element common to the *i*-th row and the *j*-th column is the correlation coefficient between the responses to the *i*-th and *j*-th levels of the within factor. The assumption that the responses have a common variance is known as the *assumption of homogeneity*. The assumption that the correlations between responses within main units can be represented as a matrix Σ_w in which all off-diagonal elements have a common value ρ is known as the *assumption of compound symmetry*. Thus, the assumption of compound symmetry states that the correlation coefficient between any pair of repeated measures is the same. Furthermore, we know that a variance is never negative, and so Equations (16.9) and (16.11) imply that, in a split-plot model, any two repeated measures are positively correlated.

The assumption of compound symmetry

In any particular split-plot experiment, it is not simple to determine whether a correlation structure such as that in Equation (16.9) or (16.11) is reasonable. However, the assumption of positive correlation reflects the intuitively reasonable expectation that experimental units sharing a common property—such as subplots within a main plot—are likely to have similar influences on the responses. Also, the assumption of equal correlations appears reasonable in split-plot experiments, because the subtreatments are randomized within main plots and so the likelihood that a given subplot will influence the response to a particular subtreatment is the same for all subtreatments. Generally speaking, the assumption of equal nonnegative correlations between repeated measures is reasonable in split-plot experiments. On the other hand, in studies where no randomization of the subtreatments takes place, the assumption of equal correlation is often violated. For instance, in the dental study in Example 16.2, it is very likely that measurements of a child at short intervals will have a higher correlation than measurements at longer intervals.

The methods for random-effects and mixed-effects models in Chapter 14 are directly applicable to the split-plot models in Boxes 16.3 and 16.4. The EMS values needed to determine the form of the test statistics can be obtained by the rules in Section 14.7. When applying these rules, we should keep in mind that, in a completely randomized split-plot model, the main plots are nested in the levels of the main treatments. Thus, S_{ik} should be written as $S_{(i)k}$ in the rules for determining the EMS values.

The ANOVA tables associated with the two split-plot models are shown in Tables 16.4 and 16.5, where $r_+ = r_1 + \cdots + r_a$ is the total number of main plots; ψ_A^2,

TABLE 16.4

ANOVA table for a completely randomized split-plot experiment

Source	df	SS	MS	EMS
A	$a - 1$	SS[A]	MS[A]	$br\psi_A^2 + b\sigma_s^2 + \sigma^2$
Main plot error	$r_+ - a$	SS[$S(A)$]	MS[$S(A)$]	$b\sigma_s^2 + \sigma^2$
B	$b - 1$	SS[B]	MS[B]	$ar\psi_B^2 + \sigma^2$
AB	$(a - 1)(b - 1)$	SS[AB]	MS[AB]	$r\psi_{AB}^2 + \sigma^2$
Subplot error	$(r_+ - a)(b - 1)$	SS[E]	MS[E]	σ^2
Total	$br_+ - 1$	SS[TOT]		

TABLE 16.5

ANOVA table for RCB split-plot experiment

Source	df	SS	MS	EMS
Blocks	$r - 1$	SS[R]	$MS[R]$	$ab\sigma_r^2 + b\sigma_{sr}^2 + \sigma^2$
A	$a - 1$	SS[A]	MS[A]	$br\psi_A^2 + b\sigma_{sr}^2 + \sigma^2$
Main plot error	$(a - 1)(r - 1)$	SS[SR]	MS[SR]	$b\sigma_{sr}^2 + \sigma^2$
B	$b - 1$	SS[B]	MS[B]	$ar\psi_B^2 + \sigma^2$
AB	$(a - 1)(b - 1)$	SS[AB]	MS[AB]	$r\psi_{AB}^2 + \sigma^2$
Subplot error	$a(r - 1)(b - 1)$	SS[E]	MS[E]	σ^2
Total	$abr - 1$	SS[TOT]		

ψ_B^2, and ψ_{AB}^2, are parameters measuring the differences between the effects of the fixed factors and are defined as in Table 14.1.

As in other random-effects models, the statistics for testing the significance of the factorial effects of A and B can be determined by examining the expected mean squares (EMS) in the appropriate ANOVA table. In both tables, the denominator of the test statistic for testing the main effect of A is the mean square for the main plot error—MS[$S(A)$] or MS[SR]. The mean square for the subplot error—denoted by MS[E] in both tables—is the denominator for testing the main effect of B and interaction AB. Generally speaking, the rule for determining the denominators of the F-statistics in a split-plot ANOVA is to use the main plot error for testing any effect not involving the subtreatment factor. In all other cases, the subplot error is used as the denominator of the F-statistic.

The usual computing formulas for factorial designs with equal subclass numbers can be used to obtain the sums of squares in the ANOVA table for a split-plot experiment. Of course, whenever possible, it is better to implement the calculations using an appropriate computing package. See Section 16.2 in Younger (1997) for information on how to use SAS for this purpose.

The typical protocol for analyzing a split-plot experiment is exactly the same as that for a factorial experiment. The first step is to test the significance of the interaction. If interaction is found to be statistically significant, differences between the effects at the levels of a given factor are assessed by comparing the mean responses within the levels of the other factor. If interaction is not statistically significant, then the significances of the main effects are evaluated; that is, mean responses are compared across the levels of the other factors.

Because of the presence of two random terms in split-plot models, the standard errors of the differences between mean responses will depend on the type of difference being evaluated. A model-based approach—similar to that used to determine the standard error of the overall mean in Section 14.3—can be used for this purpose. Let's look at how we can determine the standard error of the difference between the mean responses for two treatment combinations containing a common level of A—that is, treatment combinations of the form $a_i b_j$ and $a_i b_m$—in the setting of a RCB split-plot design.

The observed mean response for the treatment combination $a_i b_j$ in a RCB split-plot experiment can be expressed as

$$\overline{Y}_{ij+} = \mu + \alpha_i + \beta_j + (\alpha\beta)_{ij} + \overline{R}_+ + \overline{S}_{i+} + \overline{E}_{ij+}, \tag{16.13}$$

where $\overline{R}_+, \overline{S}_{i+}$, and $\overline{E}_{ij+}$ are, respectively, the means of a random sample of r block effects, a random sample of r subject effects, and a random sample of r error effects. Therefore, the difference between the observed mean responses for treatments $a_i b_j$ and $a_i b_m$ is

$$\overline{Y}_{ij+} - \overline{Y}_{im+} = \beta_j - \beta_m + (\alpha\beta)_{ij} - (\alpha\beta)_{im} + \overline{E}_{ij+} - \overline{E}_{im+}.$$

Since $\overline{E}_{ij+}$ and $\overline{E}_{im+}$ are the means of independent samples of size r, the standard error of the difference $\hat{\theta} = \overline{E}_{ij+} - \overline{E}_{im+}$ can be obtained by taking $t = 2$, $n = r$, $c_0 = \beta_j - \beta_m + (\alpha\beta)_{ij} - (\alpha\beta)_{im}$, $c_1 = 1$ and $c_2 = -1$ in Box 9.1. Thus, the required standard error is

$$\text{SE}(\overline{Y}_{ij+} - \overline{Y}_{im+}) = \sqrt{(1)^2 \frac{\sigma^2}{r} + (-1)^2 \frac{\sigma^2}{r}} = \sqrt{\frac{2\sigma^2}{r}}. \tag{16.14}$$

From Table 16.5 the mean square error for subplots is seen to be an unbiased estimate of σ^2. Therefore, the standard error in Equation (16.14) can be estimated by

$$\widehat{\text{SE}}(\overline{Y}_{ij+} - \overline{Y}_{im+}) = \sqrt{\frac{2\text{MS}[E]}{r}}. \tag{16.15}$$

This estimated standard error can be used to make inferences about the differences between treatment combinations that contain a given common level of A. The hypothesis tests and confidence intervals will be based on t-tests with degrees of freedom equal to the degrees of freedom for MS[E]. It turns out that Equation (16.15) is also the estimate of the standard error of the corresponding difference between mean responses in a completely randomized split-plot experiment. Table 16.16 gives estimates of the standard errors of the difference between the mean responses for different types of treatment combinations. These standard errors can be used with both completely randomized and RCB split-plot designs. The standard errors are stated in terms of the mean squares in a completely randomized split-plot experiment. For a RCB split-plot design, replace MS[$S(A)$] by MS[SR].

The degrees of freedom associated with a standard error in the first three rows of Table 16.6 are the same as the degrees of freedom for the corresponding mean square error. The expressions for the standard errors in the last two rows include two mean squares, and so the associated degrees of freedom are approximated by a method such as that in Box 14.9.

TABLE 16.6

Estimated standard errors of differences between means in split-plot experiments

Description	Type of comparison	Symbol	Estimated SE
Different levels of A		$\overline{Y}_{i++} - \overline{Y}_{p++}$	$\sqrt{\dfrac{2\text{MS}[M(A)]}{rb}}$
Different levels of B		$\overline{Y}_{+j+} - \overline{Y}_{+m+}$	$\sqrt{\dfrac{2\text{MS}[E]}{ra}}$
Same level of A, different levels of B		$\overline{Y}_{ij+} - \overline{Y}_{im+}$	$\sqrt{\dfrac{2\text{MS}[E]}{r}}$
Different levels of A, same level of B		$\overline{Y}_{ij+} - \overline{Y}_{pj+}$	$\sqrt{\dfrac{2\{(b-1)\text{MS}[E] + \text{MS}[M(A)]\}}{rb}}$
Different levels of A, different levels of B		$\overline{Y}_{ij+} - \overline{Y}_{pm+}$	$\sqrt{\dfrac{2\{(b-1)\text{MS}[E] + \text{MS}[M(A)]\}}{rb}}$

EXAMPLE 16.7 In Example 16.5, a RCB split-plot experiment with three main treatments (formulations) and three subtreatments (methods) was described. The investigators were interested in comparing the formulations and methods on the basis of the absorption of a particular chemical in the kidneys of experimental rats. The ANOVA table for the absorption data is as follows:

Source	df	SS	MS	F	p
Litters (R)	3	0.3401	0.1134	0.11	0.9511
Formulations (A)	2	314.2317	157.1158	150.49	0.0001
Main plot error	6	6.2637	1.0440		
Method (B)	2	260.5736	130.2868	228.56	0.0001
Formulation $\times$ method (AB)	4	98.3084	24.5771	43.12	0.0001
Subplot error	18	10.2606	0.5700		
Total	35	689.9781			

The *F*-statistic for testing the interaction between formulation and method is 43.12 (the ratio of 24.5771, the mean square for *AB*, and 0.57, the mean square for subplot error), with 4 and 18 degrees of freedom for the numerator and denominator, respectively. The associated *p*-value of 0.0001 provides strong evidence that the absorption is influenced by the interaction effect between the methods and formulations. In other words, it appears that the differences between the expected absorption for the various formulations (methods) vary with the method (formulation) used. A comparison between formulations (methods) must be done separately for each method (formulation). These comparisons can be based on the standard errors in Table 16.6.

Let's compare the expected absorption for formulation 2 when methods 2 and 3 are used. This can be done by comparing the mean responses for the treatment combinations a_2b_2 and a_2b_3. The treatment combinations involve a common level of *A*, and so we use the standard error in the third row of Table 16.6. On the basis of the observed mean absorption of $\bar{y}_{22+} = 65.66/4 = 16.415$ and $\bar{y}_{23+} = 94.08/4 = 23.52$, a 95% confidence interval for the difference between the expected absorption for formulation 2 when methods 2 and 3 are used is

$$23.52 - 16.415 \pm t(18, 0.025)\sqrt{\frac{2 \times 0.57}{4}} = 7.11 \pm 1.12.$$

Thus, we can expect, with 95% confidence, that the absorption for formulation 2 measured by methods 2 and 3 can differ by as little as 5.99% and as much as 8.23%.

Now let's compare the mean responses to the treatment combinations a_2b_2 and a_3b_2—that is, the expected absorption for formulations 2 and 3 when method 2 is used. This comparison involves different levels of *B* and the same level of *A*. On the basis of Table 16.6, after adaptation for a RCB split-plot experiment, the appropriate standard error is

$$SE = \sqrt{\frac{2\{(b-1)MS[E] + MS[MR]\}}{rb}}$$

$$= \sqrt{\frac{2(2 \times 0.57 + 1.044)}{3 \times 4}} = 0.60.$$

The estimated standard error involves two mean squares, and so we approximate the associated degrees of freedom on the basis of Box 14.9. We know that the standard error can be expressed in the form $\sqrt{p_1 MS_1 + p_2 MS_2}$, where

$$p_1 = \frac{2}{rb} = \frac{2}{12},$$
$$MS_1 = MS[MR] = 1.044,$$
$$p_2 = \frac{2(b-1)}{rb} = \frac{4}{12},$$
$$MS_2 = MS[E] = 0.57,$$

and so the results in Box 14.9 are applicable with these value of p_1, p_2, MS_1, and MS_2 and with $v_1 = 6$ and $v_2 = 18$. The appropriate degrees of freedom are found to be $v = 18$. Thus, a 95% confidence interval for the difference between the expected absorption for formulations 2 and 3 when method 2 is used is

$$\frac{100.63}{4} - \frac{65.66}{4} \pm t(18, 0.025) \times 0.60 = 8.74 \pm 1.26. \quad \blacksquare$$

Note that the assumptions of homogeneity and compound symmetry are not necessary to validate the mixed model ANOVA F-tests for split-plot experiments data. Huynh and Feldt (1970) have shown that, in the general univariate repeated-measures model in Box 16.1, mixed-model F-tests are valid under the less restrictive assumption that the variance of the difference between a pair of responses within each main unit must be the same for all pairs and all main units. This condition—the *sphericity condition* or *sphericity assumption*—may be stated in more precise terms as follows

$$\text{Var}\left(Y_{ijk} - Y_{ij'k}\right) = \zeta, \tag{16.16}$$

where ζ is constant for all i, j, j', and k, and Y_{ijk} is the k-th response to the combination of the i-th level of the between factor and the j-th level of the within factor. The sphericity condition is less restrictive than the homogeneity and compound symmetry conditions because of the following circumstances:

1 Any model that satisfies the homogeneity and compound symmetry assumptions will also satisfy the sphericity condition.

2 There are models of the form in Box 16.1 that satisfy the sphericity condition but not the homogeneity and compound symmetry conditions.

In particular, the split-plot model in Box 16.3 satisfies the sphericity condition, because it satisfies the homogeneity and compound symmetry conditions. This follows because

$$Y_{ijk} - Y_{ij'k} = (\mu_{ij} + S_{ik} + E_{ijk}) - (\mu_{ij'} + S_{ik} + E_{ij'k})$$
$$= (\mu_{ij} - \mu_{ij'}) + (E_{ijk} - E_{ij'k}).$$

Because $(\mu_{ij} - \mu_{ij'})$ is a fixed quantity, the variance of the difference $Y_{ijk} - Y_{ij'k}$ is the same as the variance of $E_{ijk} - E_{ij'k}$. According to the assumptions in Box 16.3, E_{ijk} and $E_{ij'k}$ are independent with a common variance σ^2. The results in Box 9.1 imply that the variance of $E_{ijk} - E_{ij'k}$ is $2\sigma^2$.

Exercises

16.6 Refer to the RCB split-plot experiment in Example 16.5.

   a Complete the analysis in the example by making the necessary pairwise comparisons to evaluate the differences between the methods within formulations and between formulations within methods.

   b Using the GLM procedure in SAS or any other suitable statistical software, obtain the ANOVA table.

16.7   The crop yield data for the study in Example 16.1 are as follows:

Irrigation method	Fertilization method			
	1	2	3	4
1	8.41	10.92	12.02	13.24
	8.70	10.15	12.72	12.91
2	11.38	13.34	17.04	16.36
	10.90	13.78	18.93	19.35
3	4.24	5.55	8.25	10.42
	4.59	4.81	7.45	8.03

The associated ANOVA table takes the form:

Source	SS
Irrigation	287.08253
Main plot error	4.53176
Fertilizer	114.11835
Irrig × Fert	8.86624
Subplot error	6.08539
Total	420.68427

a   Write a model for the data and explain all the terms in your model.

b   State the conclusions that follow from the ANOVA table.

c   Is there a difference between the irrigation methods? If so, conduct a suitable multiple comparison to assess the differences.

d   Is there a difference between the four fertilization methods? If so, conduct a suitable multiple comparison to assess the differences.

e   The four fertilization methods are as follows:

  i   standard fertilizer;

  ii   standard fertilizer with added nitrogen;

  iii   standard fertilizer with added phosphate;

  iv   standard fertilizer with added nitrogen and phosphate.

  Form three orthogonal contrasts that can be used to measure the interaction and main effects of nitrogen and phosphate supplements to the standard fertilizer.

f   Construct a set of 95% simultaneous confidence intervals for the effects that can be estimated by the contrasts in (e). Interpret the intervals.

g   Using the GLM procedure of SAS or any other statistical software, do the calculations to construct the ANOVA table.

16.8   An experiment to compare the effect of storage time on the flavors of three brands of chocolate ice cream was conducted as follows. Twelve coolers, each with three spaces for storing half-gallon ice cream cartons, were randomly divided into three groups of four coolers each, and the three groups were randomly assigned to the three brands. Then, 12 half-gallon cartons of each brand were randomly divided into four groups, such that each group contained three cartons. Each group of three cartons was stored in a cooler assigned to the brand. One carton was randomly selected from each cooler at the end of one day, one week, and one month. Each selected carton was evaluated for flavor by five tasters, on a scale from 0 to 100, where higher ratings indicate higher acceptability of the flavor. The averages of the flavor ratings by 5 tasters are given in Table 16.7.

a Write a model for the data and explain all the terms in your model.

b Given the following statistics, perform a test to evaluate whether there is an interaction between brand and storage time: SS[brand] = 43.56; SS[storage time] = 518.72; SS[brand × storage time] = 19.61; SS[TOT] = 1667.64.

c Perform the analyses needed to determine the best storage time for each brand and state your conclusions.

d Using the GLM procedure of SAS or any other statistical software, generate the ANOVA table for the flavor data.

TABLE 16.7
Flavor ratings of ice cream after storage

Brand	Cooler	1 day	1 week	1 month
1	1	62	57	51
	2	66	46	58
	3	62	47	54
	4	47	45	40
2	1	59	52	56
	2	53	54	51
	3	54	52	43
	4	55	38	48
3	1	66	50	59
	2	63	54	51
	3	53	50	41
	4	56	47	57

Storage time column spans "1 day", "1 week", "1 month".

16.9 In a long-term site preparation study, one of the objectives was to evaluate the effects of four soil types (factor A) and 11 site preparation methods (factor B) on the five-year mean heights (feet) of pine plantations. Five locations (main plots) were selected within each soil type, and the 11 site preparation treatments were assigned randomly to 11 half-acre plots in each location. The data for five site preparation methods in two locations in each soil type are given in Table 16.8 (Parrish & Ware, 1989).

a Write a suitable model for the data. Explain the meaning of all the terms in your model.

TABLE 16.8

Five-year mean heights (ft.) of pine plantations

Soil type	Location	Site preparation method 1	2	3	4	5
1	1	12.375	9.478	11.032	6.876	7.600
1	2	9.237	10.052	8.590	9.253	9.039
2	1	14.542	14.038	14.217	13.051	13.361
2	2	11.329	11.644	9.354	8.839	8.624
3	1	8.854	7.264	6.245	8.737	5.857
3	2	12.086	13.701	10.144	8.952	10.118
4	1	7.670	9.554	5.500	4.068	6.210
4	2	11.812	16.134	10.348	10.479	12.100

b Given the following calculated values, construct the appropriate ANOVA table for testing the significance of interaction and main effects: SS[TOT] = 291.87671; SS[soil type] = 50.43173; SS[site preparation] = 46.10802; and SS[soil type $\times$ site preparation] = 20.94784.

c What conclusions can be drawn from the ANOVA table in (b)?

d The soil types are as follows:

i poorly drained; nonspodic;

ii somewhat poorly to moderately well drained; nonspodic;

iii poorly to moderately well drained; spodic with argillic horizon;

iv poorly to moderately well drained; spodic with no argillic horizon.

The five site preparation methods are as follows:

i fertilized, chopped;

ii fertilized, chopped, burned;

iii unfertilized, chopped;

iv unfertilized, chopped, burned;

v control (unprepared).

Use a multiple comparison method to compare the mean plant heights in the four soil types. State the conclusions that can be drawn from the comparisons.

e Construct a 95% confidence interval for a contrast that measures the difference between the average five-year heights of plants grown in treated sites that are fertilized and those that are not fertilized. Interpret the interval.

f Construct a 90% lower bound to the difference between the average five-year heights of plants grown in control (unprepared) sites and those grown in prepared sites. Interpret the bound.

16.10 Use the rules in Section 14.7 to derive the EMS values in Tables 16.4 and 16.5.

16.11 Derive the standard errors in Table 16.6.

16.12 A portion of the printout obtained when the GLM procedure of SAS is used to analyze the soybean data in Exercise 16.2 is as follows:

Source	DF	Sum of Squares
Model	15	750.61083
Error	20	3.83667
Corrected Total	35	754.44750
I (Blocks)	2	34.74500
J (Soil Prep. Meth.)	1	73.10250
I*J	2	37.00500
K (Variety)	2	0.44667
M (Weed Contrl)	1	255.46694
K*M	2	25.81556
J*K	2	323.84667
J*M	1	0.03361
J*K*M	2	0.14889

a   Write a model describing the effects of the blocks and factors on the measured soybean yield. Explain all the terms in your model.

b   Construct the appropriate ANOVA table and conduct the necessary F-tests to see if the data suggest significant interactions between the factors soybean variety, soil preparation method, and weed control method. Clearly state your conclusions.

c   Analyze the data further with the objective of recommending suitable methods of soil preparation and weed control for each of the three varieties.

d   Construct 95% confidence intervals for the expected yields of the three varieties when their recommended soil preparation and weed control methods are used.

16.13   Consider the paired-data analysis in Section 4.3.

a   Argue that the analysis can be viewed as a special case of a repeated-measures analysis, with only one level for the between factor and two levels for the within factor.

b   Use the notation in Box 16.1 to write the model for the paired data described in Section 4.3.

c   Use the model in (b) to show that the paired-data model satisfies the sphericity condition but not the homogeneity condition.

16.4
The sphericity assumption

As we have seen, the applicability of mixed-model ANOVA F-tests in repeated-measures analyses depends on whether the sphericity assumption is reasonable for the given data set. In cases where the sphericity assumption does not hold, two approaches are commonly used: Either the mixed-model univariate F-tests are modified, or we use a multivariate analysis, by treating the repeated measurements within main units as observations from multivariate normal distributions. The multivariate approach requires no assumptions about the correlation structure of the repeated measures and is always valid, provided the model in Box 16.1 is reasonable. However, some caution is necessary when using the multivariate approach because it tends to be much more conservative than the corresponding univariate procedures in situations where the sphericity assumption is satisfied (Rouanet & Lépine, 1970). In other words, when the sphericity assumption is satisfied, multivariate tests tend to detect far fewer differences than do their univariate counterparts. For more information on multivariate procedures and strategies for their implementation in the repeated-measures setting, see, for example, Crowder and Hand (1990), Kirk (1982), Maxwell and Delaney (1989), and Winer (1971). Here we confine our attention to the other option: modification of the mixed-model F-tests for cases where the sphericity assumption is doubtful.

First, note that no modification is needed for tests of effects that do not involve the within factor (for instance, a test of the main effect of A). To construct approximate tests of effects that do involve a within factor, we make use of an index ϵ, which Box (1954) introduced to measure the severity of violation of the sphericity assumption. If b is the number of levels of the within factor, then ϵ can take values between $1/(b - 1)$ and 1. When the sphericity assumption is satisfied, $\epsilon = 1$; as its value approaches $1/(b - 1)$, the violation of the sphericity assumption becomes more severe. Box's work implies that, regardless of whether the sphericity assumption is satisfied, approximate F-tests of the factorial effects can be performed by simple modification of the degrees of freedom for the F-tests associated with the split-plot model. All that is needed is to multiply by ϵ the degrees of freedom for each sum of squares for testing a within-factor effect. For example, we saw in Table 16.4 that, in the completely randomized split-plot model, the F-test for the main effect of the within factor B has $b - 1$ degrees of freedom for the numerator and $(r_+ - a)(b - 1)$ degrees of freedom for the denominator. The modified approximate F-test that does not require the sphericity assumption is the same as the corresponding F-test in the split-plot model but with the modified degrees of freedom $\epsilon(b - 1)$ and $\epsilon(r_+ - a)(b - 1)$ for the numerator and denominator, respectively. Clearly, when the sphericity condition holds ($\epsilon = 1$), the approximate F-tests are the same as the exact F-tests.

Unfortunately, practical implementation of the approximate F-tests is complicated, because the variances and correlation coefficients in the split-plot model must be known in order to calculate ϵ. In practice, ϵ is estimated from the observed data. Two of the most common estimates of ϵ are $\hat{\epsilon}$, suggested by Greenhouse and Geisser (1958), and $\tilde{\epsilon}$, suggested by Huynh and Feldt (1970). The actual computation of $\hat{\epsilon}$ and $\tilde{\epsilon}$ is laborious, but many statistical computing packages, including SAS, compute these values as default options.

Different researchers favor different strategies for analyzing repeated-measures data. Crowder and Hand (1990) offer a simple guideline: "If you have strong prior reasons for believing that sphericity holds then perform the standard unmodified F-test, and if not then perform a modified F-test (with $\hat{\epsilon}$ or $\tilde{\epsilon}$) or a multivariate test. There seems to be little to choose between the modified F-tests and the multivariate test." Crowder and Hand note one exception: If the number of subtreatments exceeds the number of replications per treatment combination then a multivariate analysis is the preferred option.

EXAMPLE **16.8** The data collected in the dental study in Example 16.2 are summarized in Table 16.9. These data were analyzed using the REPEATED option of the GLM procedure in SAS. In the analysis, the subjects (children) were treated as the main units, and gender and age as the between ($a = 2$ levels) and within ($b = 4$ levels) factors, respectively. Relevant portions of the SAS output obtained are as follows.

```
------------------------------------------------------------------
PART 1
Source: AGE
                                                  Adj  Pr > F
    DF   Type III SS   Mean Square   F Value   Pr > F    G - G     H - F
    3    209.43697391  69.81232464     35.35   0.0001   0.0001    0.0001


Source: AGE*GENDER
                                                  Adj  Pr > F
    DF   Type III SS   Mean Square   F Value   Pr > F    G - G     H - F
    3    13.99252946    4.66417649     2.36   0.0781   0.0878    0.0781


Source: Error(AGE)

    DF   Type III SS   Mean Square
    75   148.12784091   1.97503788


              Greenhouse-Geisser Epsilon = 0.8672
                 Huynh-Feldt Epsilon = 1.0156
------------------------------------------------------------------------
PART 2

            Tests of Hypotheses for Between Subjects Effects

Source          DF    Type III SS  Mean Square  F Value    Pr > F

GENDER           1       140.4649     140.4649     9.29    0.0054

Error           25       377.9148      15.1166
------------------------------------------------------------------------
PART 3
            Manova Test Criteria and Exact F Statistics for
                  the Hypothesis of no AGE Effect


Statistic                  Value        F     Num DF  Den DF  Pr > F

Wilks' Lambda          0.194794237  31.6911      3      23   0.0001
Pillai's Trace         0.805205763  31.6911      3      23   0.0001
Hotelling-Lawley Trace 4.133622111  31.6911      3      23   0.0001
Roy's Greatest Root    4.133622111  31.6911      3      23   0.0001
```

TABLE   16.9

Dental measurements of 27 children

Subject	Gender	Age			
k	i	$j = 1$	$j = 2$	$j = 3$	$j = 4$
1	1	21.0	20.0	21.5	23.0
2	1	21.0	21.5	24.0	25.5
3	1	20.5	24.0	24.5	26.0
4	1	23.5	24.5	25.0	26.5
5	1	21.5	23.0	22.5	23.5
6	1	20.0	21.0	21.0	22.5
7	1	21.5	22.5	23.0	25.0
8	1	23.0	23.0	23.5	24.0
9	1	20.0	21.0	22.0	21.5
10	1	16.5	19.0	19.0	19.5
11	1	24.5	25.0	28.0	28.0
1	2	26.0	25.0	29.0	31.0
2	2	21.5	22.5	23.0	26.5
3	2	23.0	22.5	24.0	27.5
4	2	25.5	27.5	26.5	27.0
5	2	20.0	23.5	22.5	26.0
6	2	24.5	25.5	27.0	28.5
7	2	22.0	22.0	24.5	26.5
8	2	24.0	21.5	24.5	25.5
9	2	23.0	20.5	31.0	26.0
10	2	27.5	28.0	31.0	31.5
11	2	23.0	23.0	23.5	25.0
12	2	21.5	23.5	24.0	28.0
13	2	17.0	24.5	26.0	29.5
14	2	22.5	25.5	25.5	26.0
15	2	23.0	24.5	26.0	30.0
16	2	22.0	21.5	23.5	25.0

```
          Manova Test Criteria and Exact F Statistics for
             the Hypothesis of no TIME*GENDER Effect
    H = Type III SS&CP Matrix for TIME*GENDER   E = Error SS&CP Matrix


                  S=1    M=0.5    N=10.5


Statistic                 Value        F     Num DF   Den DF  Pr > F


Wilks´ Lambda          0.739887394  2.69527     3       23   0.0696
Pillai´s Trace         0.260112606  2.69527     3       23   0.0696
Hotelling-Lawley Trace 0.351557018  2.69527     3       23   0.0696
Roy´s Greatest Root    0.351557018  2.69527     3       23   0.0696
=======================================================================
```

The severity of violation of the sphericity condition can be assessed by examining the estimated values of the sphericity index ϵ. From Part 1 of the output, we see that the Greenhouse-Geisser (G-G) $\hat{\epsilon}$ and the Huynh-Feldt (H-F) $\tilde{\epsilon}$ are both close to the upper bound of 1 ($\hat{\epsilon} = 0.8672$, $\tilde{\epsilon} = 1.0156$). Since small values of ϵ are associated with violations of the sphericity assumption, there is strong indication that the adjusted univariate approach or the multivariate approach may not be necessary for analysis of the dental data.

First, let's consider the adjusted univariate tests. Part 1 of the SAS output contains the relevant statistics and the associated p-values for unadjusted and adjusted tests of effects involving the within factor, age. The p-values for the adjusted tests may be summarized as follows:

Source	df	MS	F	p	p (G-G)	p (H-F)
Age (A)	3	69.81	35.35	0.0001	0.0001	0.0001
Gender × age (GA)	3	4.66	2.36	0.0781	0.0878	0.0781
Subplot error (E)	75	1.97				

The degrees of freedom for the unadjusted F-tests can be determined as indicated in Table 16.4. For example, there are 11 boys and 16 girls, and so the total number of main units is $r_{+} = 11 + 16 = 27$. Thus, the degrees of freedom for the subplot error sum of squares are $(r_{+} - a)(b - 1) = (27 - 2)(4 - 1) = 75$. Consequently, the denominator degrees of freedom are 75 for the unadjusted F-tests of the null hypothesis $H_{0[GA]}$ that there is no interaction between gender and age and the null hypothesis $H_{0[A]}$ that there is no main effect of age. The corresponding degrees of freedom for the adjusted F-test are $0.8672 \times 75 = 65$ when the G-G adjustment is used and $1 \times 75 = 75$, where we have replaced the estimated value of 1.0156 with the upper bound of 1 for its true value. The p-values associated with the adjusted tests of interaction are both larger than 0.05, and so we conclude that there is no demonstrable gender-age interaction and proceed to examine the main effects of these two factors. Both adjusted tests of the main effect of the within factor, age, yield small p-values; this suggests that the difference between the dental measurements varies with age. To test the main effect of the between factor, gender, we refer to Part 2 of the SAS output. Because the unadjusted test of the main effect is valid regardless of the validity of the sphericity assumption, Part 2 of the output shows the p-value for the unadjusted test only. The p-value of 0.0054 for the gender effect indicates that, on average, the dental measurements for males differ from those for females.

As already noted, an alternative to the adjusted univariate approach is to use a multivariate approach. Part 3 of the SAS output shows the results of four multivariate tests of the null hypotheses $H_{0[GA]}$ and $H_{0[A]}$, which are based on four different test statistics named after their original proponents: Wilks, Pillai, Hotelling and Lawley, and Roy. These tests will yield approximate p-values for the corresponding null hypotheses. However, the power functions of the four tests differ; that is, the tests differ in their ability to detect departures from the null hypotheses. Since the tests can lead to different conclusions from the same data, ideally, the test that one wants to use in a given application should be determined before starting data analyses. Unfortunately, there is no simple guideline for selecting the best test in a particular situation. My own preference is to use Wilks' lambda criterion, which possesses some good theoretical properties. On the other hand, many practitioners feel that it is best to look at the extent to which the conclusions from these four tests agree with each other. Fortunately, in certain circumstances—too complicated to describe here—the four tests will be equivalent and lead to the same conclusion; the selection of a particular test is not critical in those circumstances. As we see from the F-values for the four multivariate tests of $H_{0[GA]}$ and $H_{0[A]}$, these tests are equivalent in the present setting. They lead to the conclusion that interaction between gender and age is not significant ($p = 0.0696$) but there is a statistically significant ($p = 0.0001$) difference between the average dental measurements at the four ages. Thus, at least in this example, the conclusions of the adjusted univariate analyses and the multivariate analyses agree with each other. ■

In Example 16.8, a repeated-measures analysis of the dental measurements was used to conclude that there was no demonstrable interaction between age and gender, but the main effects of both of these factors were statistically significant. The next step is to conduct follow-up analyses of the association of children's age and gender with their dental measurements. In Chapters 13 and 14, we described how such follow-up analyses can be based on inferences about appropriately selected contrasts between the expected responses. Similar follow-up analyses are possible for repeated-measures data.

In the dental measurements example, only the main effects of age and gender were found to be significant. Accordingly, we might want to test contrasts of the expected responses for the two genders or the expected responses at the four ages. As before, let subscripts i and j identify the levels of gender and age, respectively, and let μ_{ij} denote the expected response for the combination of levels i and j. Consider the contrasts

$$\theta_1 = \frac{1}{4}(\mu_{11} + \mu_{12} + \mu_{13} + \mu_{14}) - \frac{1}{4}(\mu_{21} + \mu_{22} + \mu_{23} + \mu_{24}),$$

$$\theta_2 = \frac{1}{2}\left[(\mu_{24} + \mu_{14}) - (\mu_{21} + \mu_{11})\right], \tag{16.17}$$

$$\theta_3 = \frac{1}{2}\left[(\mu_{24} - \mu_{21}) - (\mu_{14} - \mu_{11})\right].$$

The contrast θ_1 is the difference between the average expected responses over the levels of the within factor for boys and girls. Thus, it is a component of the main effect of the between factor. We know that inferences about a between factor are not affected by the sphericity assumption, and so the significance of θ_1 can be tested by treating it as a contrast between the main treatments in a split-plot experiment. However, because the sum of squares for gender has only one degree of freedom, testing the significance of θ_1 is the same as testing the significance of the main effect of gender.

Contrasts θ_2 and θ_3 are components of the main effect of age and the effect of interaction between gender and age, respectively. This can be seen by observing that θ_2 is the difference between the averages of the expected responses (over the two levels of the between factor, gender) at ages 4 and 1. By contrast, θ_3 is the difference between boys and girls in the expected changes in dental measurements during the period starting at level $j = 1$ (8 years of age) and ending at level $j = 4$ (14 years of age). As already noted, both θ_2 and θ_3 are comparisons of levels of the within factor, and so inferences about these contrasts will be affected by violations of the sphericity assumption. Provided the model in Box 16.1 is reasonable, exact inferences about within contrasts can be based on suitably chosen linear combinations of the within responses, as the next example illustrates.

EXAMPLE 16.9   Let's consider the problem of making inferences about the contrast θ_2 defined in Equation (16.17). This contrast can be expressed as

$$\theta_2 = \frac{1}{2}\theta_2^* + \frac{1}{2}\theta_2^{**},$$

where $\theta_2^* = \mu_{14} - \mu_{11}$ is a contrast between the expected responses at $j = 4$ and $j = 1$ for female ($i = 1$) children, and $\theta_2^{**} = \mu_{24} - \mu_{21}$ is the corresponding contrast for male ($i = 2$) children. As before, let Y_{ijk} be the dental measurement at age j for the k-th child of gender i. Let $Z_{ik} = Y_{i4k} - Y_{i1k}$ denote the difference between the dental measurements at ages $j = 4$ and $j = 1$ of the k-th child of gender i. Then, the mean $\overline{Z}_1$ of the Z_{ik} for the $n_1 = 11$ female children is an unbiased estimate of θ_2^*. Similarly, θ_2^{**} can be estimated by $\overline{Z}_2$, the mean of an independent sample of the Z_{ik} for the male children. Thus, an unbiased estimate of θ_2 is $\hat{\theta}_2 = 0.5\overline{Z}_1 + 0.5\overline{Z}_2$. It can be shown (see Exercise 16.16) that, under the assumptions in Box 16.1, the Z_{ik} for female and male children can be treated as independent random samples of sizes $n_1 = 11$ and $n_2 = 16$ from two normally distributed populations such that:

a   the population means are θ_2^* and θ_2^{**};

b   the populations have a common variance, denoted by σ_z^2.

Therefore, we can make inferences about θ_2 by regarding it as a linear combination of $t = 2$ population means with coefficients $c_1 = c_2 = 0.5$ and using the procedure in Box 9.2. For example, the null hypothesis $H_0: \theta_2 = 2$ can be tested using the test statistic

$$t_{cal} = \frac{\hat{\theta}_2 - 2}{\hat{\sigma}_{\hat{\theta}_2}},$$

where

$$\hat{\sigma}_{\hat{\theta}_2} = \sqrt{\left(\frac{(0.5)^2}{11} + \frac{(0.5)^2}{16}\right)\hat{\sigma}_z^2}$$

and $\hat{\sigma}_z^2$ is the pooled estimate of the common variance σ_z^2.

The following table gives the values of the Z_{ik} for females and males, along with their means and variances:

Gender	Z_{ik}	Mean	Variance
Females	2.0 4.5 5.5 3.0 2.0 2.5 3.5 1.0 1.5 3.0 3.5	2.91	1.7409
Males	5.0 5.0 4.5 1.5 6.0 4.0 4.5 1.5 3.0 4.0 2.0 6.5 12.5 3.5 7.0 3.0	4.59	7.1406

From these Z_{ik}, we can calculate the estimate of θ_2, the pooled estimate of σ_z^2, and the estimated standard error $\hat{\sigma}_{\hat{\theta}}$ as

$$\hat{\theta}_2 = 0.5\bar{Z}_1 + 0.5\bar{Z}_2 = 0.5(2.91) + 0.5(4.59) = 3.755,$$

$$\hat{\sigma}_z^2 = \frac{(11-1)(1.7409) + (16-1)(7.1406)}{25} = 4.9807,$$

$$\hat{\sigma}_{\hat{\theta}} = \sqrt{\left(\frac{0.5^2}{11} + \frac{0.5^2}{16}\right)4.9807} = 0.4371,$$

so that the test statistic for H_0 is

$$t_{cal} = \frac{3.755 - 2}{0.4371} = 4.004,$$

which should be compared with the critical value of a t-distribution with 25 degrees of freedom. The p-value for a two-sided test of H_0 is less than 0.0001, and so there appears to be sufficient evidence that θ_2 is not equal to 2. A confidence interval for θ_2 can easily be constructed. ∎

In Example 16.9, the within comparison θ_2 was estimated by a linear combination of the means of two independent random samples from two normal populations with a common variance. The random samples were obtained by forming suitable linear combinations of the responses within each main unit. In general, it is always possible to determine linear combinations of the within responses such that a given contrast between the levels of the within factor can be estimated by a linear combination of the means of independent random samples from normal distributions with a common variance. The method described in Box 9.2 can be used to make inferences about a within contrast.

Exercises

16.14   Refer to Example 16.9.

a   Construct a 95% confidence interval for θ_2 and interpret the result.

b   Test to see if the dental measurements provide sufficient evidence to conclude that θ_2 is nonzero.

c   As noted already, θ_2 is a component of the main effect of age. Define and interpret two other comparisons that are components of the main effect of age and are orthogonal to θ_2.

d   Using the Bonferroni method or otherwise, construct a set of lower bounds for the three orthogonal comparisons in (c) with a simultaneous confidence level of at least 95%. Interpret the bounds.

16.15   In this exercise, the methods of factorial analysis (Chapters 13 and 14) will be combined with the methods for repeated-measures studies involving one within and one between factor, in order to analyze a repeated-measures study with multiple within and/or between factors.

The ability of children with learning disabilities to comprehend emotional facial expressions was assessed in a study. The study subjects consisted of 60 children between 8 and 12 years of age: 20 children with Type A learning disability; 20 children with Type B disability; and 20 children with no disability. Within each group, there were ten males and ten females. Each child was administered a battery of three tests; each test was designed to measure a child's ability to comprehend one aspect of facial expressions—for instance, the ability to comprehend facial identity. The experimenters assigned values 1 and 2 to the gender variable for females and males, respectively. The children's test scores are summarized in Table 16.10. The printout

T A B L E   16.10

Children's test scores

Subject	Group	Gender	Test 1	Test 2	Test 3
1	A	1	50	42	29
2	A	1	53	39	31
3	A	1	55	44	35
4	A	1	62	50	36
5	A	1	52	35	24
6	A	1	61	48	39
7	A	1	63	51	42
8	A	1	64	48	34
9	A	1	53	39	29
10	A	1	57	39	30
11	A	2	62	56	41
12	A	2	68	55	43
13	A	2	64	50	38
14	A	2	62	51	42
15	A	2	69	54	42
16	A	2	59	47	37
17	A	2	59	50	37
18	A	2	63	51	38
19	A	2	64	54	38
20	A	2	61	46	36

TABLE 16.10 (continued)

Children's test scores

Subject	Group	Gender	Test 1	Test 2	Test 3
21	B	1	37	44	31
22	B	1	36	46	28
23	B	1	49	51	38
24	B	1	33	47	27
25	B	1	45	55	38
26	B	1	37	46	28
27	B	1	47	54	37
28	B	1	42	53	36
29	B	1	35	37	23
30	B	1	44	53	36
31	B	2	41	48	33
32	B	2	44	49	31
33	B	2	45	54	42
34	B	2	43	54	36
35	B	2	40	49	33
36	B	2	48	54	36
37	B	2	37	46	29
38	B	2	52	57	42
39	B	2	34	43	25

Subject	Group	Gender	Test 1	Test 2	Test 3
40	B	2	43	44	32
41	N	1	43	21	60
42	N	1	50	29	53
43	N	1	38	21	50
44	N	1	40	19	47
45	N	1	46	27	57
46	N	1	49	23	52
47	N	1	39	17	48
48	N	1	43	24	48
49	N	1	49	26	55
50	N	1	48	28	58
51	N	2	44	24	52
52	N	2	45	24	52
53	N	2	43	29	54
54	N	2	51	30	58
55	N	2	47	29	57
56	N	2	46	23	53
57	N	2	46	29	55
58	N	2	48	26	54
59	N	2	36	19	49
60	N	2	41	22	47

obtained when the GLM procedure of SAS is applied to these data includes the following information:

PART 1

Univariate Tests of Hypotheses for Within Subject Effects
Source: TEST

					Adj Pr > F	
DF	Type III SS	Mean Square	F Value	Pr > F	G - G	H - F
2	2703.244444	1351.622222	323.00	0.0001	0.0001	0.0001

Source: TEST*GRP

					Adj Pr > F	
DF	Type III SS	Mean Square	F Value	Pr > F	G - G	H - F
4	14227.022222	3556.755556	849.97	0.0001	0.0001	0.0001

Source: TEST*GENDER

					Adj Pr > F	
DF	Type III SS	Mean Square	F Value	Pr > F	G - G	H - F
2	8.133333	4.066667	0.97	0.3817	0.3796	0.3817

Source: TEST*GRP*GENDER

					Adj Pr > F	
DF	Type III SS	Mean Square	F Value	Pr > F	G - G	H - F
4	14.333333	3.583333	0.86	0.4928	0.4901	0.4928

Source: Error(TEST)

| DF | Type III SS | Mean Square |
| 108 | 451.933333 | 4.184568 |

Greenhouse-Geisser Epsilon = 0.9705

Huynh-Feldt Epsilon = 1.0993

PART 2

Tests of Hypotheses for Between Subjects Effects

Source	DF	Type III SS	Mean Square	F Value	Pr > F
GRP	2	1896.844	948.422	17.06	0.0001
GENDER	1	432.450	432.450	7.78	0.0073
GRP*GENDER	2	308.133	154.067	2.77	0.0715
Error	54	3001.567	55.585		

Manova Test Criteria and Exact F Statistics for
the Hypothesis of no TEST Effect

Statistic	Value	F	Num DF	Den DF	Pr > F
Wilks' Lambda	0.089067914	271.026	2	53	0.0001
Pillai's Trace	0.910932086	271.026	2	53	0.0001
Hotelling-Lawley Trace	10.22738766	271.026	2	53	0.0001
Roy's Greatest Root	10.22738766	271.026	2	53	0.0001

```
              Manova Test Criteria and F Approximations for
                   the Hypothesis of no TEST*GRP Effect
Statistic                      Value       F  Num DF Den DF Pr > F


Wilks' Lambda           0.001653256  625.242   4     106  0.0001
Pillai's Trace          1.876816756  411.371   4     108  0.0001
Hotelling-Lawley Trace 72.50946347   942.623   4     104  0.0001
Roy's Greatest Root     64.23772953 1734.42    2      54  0.0001


              Manova Test Criteria and Exact F Statistics for
                   the Hypothesis of no TEST*GENDER Effect
Statistic                      Value       F  Num DF Den DF Pr > F


Wilks' Lambda           0.959974516  1.1049    2      53  0.3388
Pillai's Trace          0.040025484  1.1049    2      53  0.3388
Hotelling-Lawley Trace  0.041694319  1.1049    2      53  0.3388
Roy's Greatest Root     0.041694319  1.1049    2       3  0.3388


              Manova Test Criteria and F Approximations for
                 the Hypothesis of no TEST*GRP*GENDER Effect
Statistic                      Value       F  Num DF Den DF Pr > F


Wilks' Lambda           0.936513421  0.88349   4     106  0.4765
Pillai's Trace          0.063490698  0.88523   4     108  0.4754
Hotelling-Lawley Trace  0.067785958  0.88122   4     104  0.4779
Roy's Greatest Root     0.067721002  1.82847   2      54  0.1705
-----------------------------------------------------------------
```

a Write an appropriate model for the data. Explain all the terms in your model.

b Interpret the values of the Greenhouse-Geisser and Huynh-Feldt estimators of the sphericity index.

c Use the SAS output to construct an appropriate ANOVA table for a univariate analysis.

d State the conclusions that can be drawn from the ANOVA table in (c).

e Examine the results of multivariate analyses and comment on how they compare to the outcome of the univariate analyses.

f Let μ_{ijk} denote the expected response of a subject corresponding to group i ($i = 1$ for A, 2 for B, and 3 for N), gender j, and test k. Describe the effect that is being measured by the contrast

$$\theta_1 = \frac{1}{9}(\mu_{121} + \mu_{122} + \mu_{123} + \mu_{221} + \mu_{222} + \mu_{223} + \mu_{321} + \mu_{322} + \mu_{323})$$

$$- \frac{1}{9}(\mu_{111} + \mu_{112} + \mu_{113} + \mu_{211} + \mu_{212} + \mu_{213} + \mu_{311} + \mu_{312} + \mu_{313}).$$

g Construct a 95% confidence interval for θ_1 and interpret the interval.

h Describe the effects that are being measured by the contrasts

$$\theta_2 = \frac{1}{2}(\mu_{111} + \mu_{121}) - \frac{1}{2}(\mu_{311} + \mu_{321}),$$

$$\theta_3 = \frac{1}{2}(\mu_{211} + \mu_{221}) - \frac{1}{2}(\mu_{311} + \mu_{321}).$$

i Construct 95% confidence intervals for θ_2 and θ_3 and interpret them.

16.16 Refer to Example 16.9. Show that, if the assumptions in Box 16.1 are satisfied, then the Z_{ik} for males and females can be treated as independent random samples from normal distributions with a common variance. Determine the means of these populations.

16.5
Overview

In this chapter, we introduced analysis of repeated-measures designs as an application of general linear models with random and fixed effects. Repeated-measures experiments are studies in which multiple correlated measurements are made on each experimental unit. Two important examples of repeated-measures studies are split-plot experiments and studies in which each experimental subject is measured at several times over a time period.

The univariate mixed effects model approach and the multivariate mixed model approach are two common methods of analyzing repeated measures data. The univariate approach uses a general linear model for subunit responses under the assumption that subunit responses within a main unit may be correlated. The multivariate approach, on the other hand, uses a model in which the responses of subunits within a main unit are regarded as the components of a multivariate random variable.

Generally speaking, the univariate mixed model approach is better than the multivariate approach because of the ease with which it can be used in practice. However, the univariate model is more restrictive than the multivariate model because the appropriateness of the univariate model depends on an assumption known as the sphericity assumption. Some discussion of the sphericity assumption and its role in repeated-measures analyses also has been presented in this chapter.

APPENDIX A:
SELECTED REFERENCES

1 Agresti, A. (1990). *Categorical data analysis*. New York: Wiley.

2 Anderson, V. L., & McLean, R. A. (1974). *Design of experiments: A realistic approach*. New York: Dekker.

3 Antunes, C. M. F., Stolley, P. D., Rosenshein, N. B., Davies, J. L., Tonascia, J. A., Brown, C., Burnett, L., Rutledge, A., Pokempner, M., & Garcia, R. (1979). Endometrial cancer and estrogen use: Report of a large case-control study. *The New England Journal of Medicine, 300,* 9–13.

4 Augustin, M., & Clarke, P. T. (1991). Calcium ion activities of cooled and aged reconstituted and recombined milks. *Journal of Dairy Research, 58,* 219–229.

5 Barrett, J. P. (1974). The coefficient of determination: Some limitations. *The American Statistician, 28,* 1920.

6 Bartlett, M. S. (1947). The use of transformations. *Biometrics, 3,* 39–52.

7 Belsley, D. A., Kuh, E., & Welsch, R. E. (1980). *Regression diagnostics: Identifying influential data and the sources of collinearity*. New York: Wiley.

8 Box, G. E. P. (1954). Some theorems on quadratic forms applied in the study of analysis of variance problems: II. Effects of inequality of variance and of correlation between errors in the two-way classification. *Annals of Mathematical Statistics, 25,* 484–498.

9 Box, G. E. P., & Cox, D. R. (1964). An analysis of transformations. *Journal of Royal Statistical Society, B-26,* 211–243.

10 Box, G. E. P., Hunter, W. G., & Hunter, J. S. (1978). *Statistics for experimenters*. New York: Wiley.

11 Bradley, J. V. (1968). *Distribution-free statistical tests*. Upper Saddle River, NJ: Prentice-Hall.

12 Brandt, R. B., Waters, M. G., Muron, D. J., & Bloch, M. H. (1982). The glyoxalase system in rat blood. *Proceedings of the Society for Experimental Biology and Medicine, 169,* 463–469.

13 Casagrande, J. T., Pike, M. C., & Smith, P. G. (1978). An improved approximate formula for calculating sample sizes for comparing two binomial distributions. *Biometrics, 34,* 483–486.

14 Caceres, J. A. F. (1990). *Determinants of the difference in forage quality between Pensacola bahiagrass* (Pasapalum notatum Flugge) *and dwarf elephantgrass* (Pennisetum purpureum Schum). Unpublished dissertation, University of Florida, Gainesville.

15 Clark, D. A., Pincus, L. G., Oliphant, M., Hubbell, C., Oates, R. P., & Davey, F. R. (1982). HLA-A2 and chronic lung disease in neonates. *Journal of the American Medical Association, 248,* 1868–1869.

16 Cochran, W. G., & Cox, G. M. (1957). *Experimental designs* (2nd ed.). New York: Wiley.

17 Collings, B. J., & Hamilton, M. A. (1988). Estimating the power of the two-sample Wilcoxon test for location shift. *Biometrics, 44,* 847–860.

18 Cook, R. D., & Weisberg, S. (1982). *Residuals and influence in regression.* New York: Chapman & Hall.

19 Cornell, J. A. (1971). A review of multiple comparison procedures for comparing a set of *k* population means. *Soil and Crop Science Society of Florida Proceedings, 31,* 92–97.

20 Crocker, D. C. (1972). Some interpretations of the multiple correlation coefficient. *The American Statistician, 22,* 31–33.

21 Crow, E. L. (1956). Confidence intervals for a proportion. *Biometrika, 43,* 423–435.

22 Crowder, M. J., & Hand, D. J. (1990). *Analysis of repeated measures.* London: Chapman & Hall.

23 Cytel Software Corporation. (1995). *StatXact 3 for Windows.* Cambridge, MA: Cytel Software Corporation.

24 Daniel, W. W. (1990). *Applied Nonparametric Statistics.* Boston: PWS-Kent.

25 Davies, R. B., & Hutton, B. (1975). The effects of errors in independent variables in linear regression. *Biometrika, 62,* 383–391.

26 Deaton, M. L., Reynolds, M. R., Jr., & Myers, R. H. (1983). Estimation and hypothesis testing in regression in the presence of nonhomogenous error variances. *Communications in Statistics B, 12,* 45–66.

27 De Groot, M. H. (1970). *Optimal statistical decisions.* New York: McGraw-Hill.

28 Draper, N. R., & Smith, H. (1981). *Applied regression analysis* (2nd ed.). New York: Wiley.

29 Eysenck, M. W. (1974). Age differences in incidental learning. *Developmental Psychology, 10,* 936–941.

30 Fliess, J. L. (1981). *Statistical methods for rates and proportions.* New York: Wiley.

31 Freund, R. J., Littell, R. C., & Spector, P. C. (1986). *SAS system for linear models.* Cary, NC: SAS Institute.

32 Fuller, W. A. (1987). *Measurement error models.* New York: Wiley.

33 Gaylor, D. W. (1987). Linear-nonparametric upper limits for low dose extrapolation. *American Statistical Association: Proceedings of the Biopharmaceutical Section,* 63–66.

34 Gaylor, D. W., & Hopper, F. N. (1969). Estimating the degrees of freedom for linear combinations of mean squares by Satterthwaite's formula. *Technometrics, 11,* 691–706.

35 Geisser, S., & Greenhouse, S. (1958). An extension of Box's results on the use of *F*-distribution in multivariate analysis. *The Annals of Mathematical Statistics, 29,* 885–889.

36 Graybill, F. A. (1976). *Theory and applications of the linear model.* Pacific Grove, CA: Duxbury.

37 Hagi, M. A. (1978). *Radon flux measurements by charcoal canisters and a statistical analysis of sampling procedures.* Unpublished master's thesis, University of Florida, Gainesville.

38 Hirotsu, C. (1993). Beyond analysis of variance techniques: Some applications in clinical trials. *International Statistical Review, 61,* 183–201.

39 Hoaglin, D. C., Mosteller, F., & Tukey, J. W. (1983). *Understanding robust and exploratory data analysis.* New York: Wiley.

40 Hocking, R. R. (1976). The analysis and selection of variables in linear regression. *Biometrics, 32,* 1–40.

41 Holbrook, M. N., Denny, M. W., & Koehl, M. A. R. (1991). Intertidal trees: Consequences of aggregation on the mechanical and photosynthetic properties of

sea-palms (*Postelsia palmaeformis Ruprecht*). *Journal of Experimental Biology and Ecology, 146,* 39–67.

42 Hollander, M., & Wolfe, D. A. (1973). *Nonparametric statistical methods.* New York: Wiley.

43 Huynh, H., & Feldt, L. S. (1970). Conditions under which mean squares ratios in repeated measures designs have exact F-distributions. *Journal of the American Statistical Association, 65,* 1582–1589.

44 Idso, S. B., Kimball, B. A., & Hendrix, D. L. (1993). Air temperature modifies the size-enhancing effects of atmospheric CO_2 enrichment on sour orange tree leaves. *Environmental and Experimental Botany, 33,* 293–299.

45 Izquierdo, J. I. (1991). How does drosophila melanogaster overwinter? *Entomologia Experimentalis et Applicata, 59,* 51–58.

46 Johnson, N. L., & Kotz, S. (1970). *Distributions in statistics: Discrete distributions.* Boston: Houghton-Mifflin.

47 Johnson, N. L., Kotz, S., & Kemp, A. W. (1992). *Univariate discrete distributions* (2nd ed.). New York: Wiley.

48 Khuri, A. I. (1982). Direct products: A powerful tool for the analysis of balanced data. *Communications in Statistics, Theory, and Methods, 11,* 2903–2920.

49 Kirk, R. E. (1982). *Experimental design: Procedures for the behavioral sciences* (2nd ed.) Pacific Grove, CA: Brooks/Cole.

50 Kumar, M., Saleh, A., Rao, P. V., Morris, S., Myers, L., & Graham-Pole, J. (1997). High dose cytosine arabinoside and total body irradiation as conditioning for allogenic bone marrow transplantation: Toxicity associated with this regimen. To appear in *Bone Marrow Transplantation.*

51 Kyburg, H. E., & Smokler, H. E. (1964). *Studies in subjective probability.* New York: Wiley.

52 Lehmann, E. L. (1975). *Nonparametrics: Statistical methods based on ranks.* San Francisco: Holden-Day.

53 Lynch, J., Lauchili, A., & Epstein, E. (1991). Vegetative growth of common bean in response to phosphorus nutrition. *Crop Science, 31,* 380–387.

54 Mallows, C. P. (1973). Some comments on C_p. *Technometrics, 15,* 661–675.

55 Maxwell, S. E., & Delaney, H. D. (1989). *Designing experiments and analyzing data.* Belmont, CA: Wadsworth.

56 Mishel, M. H. (1981). The measurement of uncertainty in illness. *Nursing Research, 30,* 258–263.

57 Montalvo, J., Casado, M. A., Levassor, C., & Pineda, S. D. (1993). Species diversity patterns in Mediterranean grasslands. *Journal of Vegetation Science, 4,* 213–222.

58 Moore, J. (1975). *Total biochemical oxygen demand of dairy manures.* Unpublished doctoral dissertation, Department of Agricultural Engineering, University of Minnesota, St. Paul.

59 Morgan, E. K., Gibson, M. L., Nelson, M. L., & Males, J. R. (1991). Utilization of whole or steam rolled barley fed with forages to wethers and cattle. *Animal Feed Science and Technology, 33,* 59–91.

60 Myers, R. H. (1986). *Classical and modern regression with applications.* Pacific Grove, CA: Duxbury.

61 Nakabayashi, K., & Wada, H. (1991). Factors controlling Pg content in a dark brown forest soil. *Soil Science and Plant Nutrition, 37,* 93–99.

62 Noether, G. (1987). Sample size determination for some common nonparametric tests. *Journal of the American Statistical Association, 82,* 645–647.

63 O'Brien, R. G., & Muller, K. E. (1993). Unified power analysis for t-tests through multivariate hypotheses. In L. K. Edwards (Ed.), *Applied analysis of variance in the behavioural sciences.* New York: Dekker.

64 Oyarzun, P., Gerlagh, M., & Hoogland, A. E. (1993). Relation between cropping frequency of peas and other legumes and foot and root rot in peas. *Netherlands Journal of Plant Pathology, 99,* 35–44.

65 Parrish, R. S., & Ware, G. O. (1989). Analysis of a split-plot experiment using mixed model equations: A forest site preparation study. In *Application of mixed models in agriculture and related disciplines* (Southern Cooperative Series Bulletin 343, pp. 155–173). Baton Rouge, LA: Louisiana Agricultural Experiment Station.

66 Patil, K. D. (1975). Cochran's *Q*-test: Exact distribution. *Journal of the American Statistical Association, 70,* 186–189.

67 Pielou, E. C. (1969). *An introduction to mathematical ecology.* New York: Wiley.

68 Potoff, R. F., & Roy, S. N. (1964). A generalized multivariate analysis of variance model useful especially for growth curve problems. *Biometrika, 51,* 313–326.

69 Pott, E. B. (1992). *Relative bioavailability of copper and molybdenum sources and effect of molybdenum on molybdenum and copper excretion and tissue accumulation in lambs.* Unpublished doctoral dissertation, University of Florida, Gainesville.

70 Rao, P. V., Rao, P. S. C., Davidson, J. M., & Hammond, L. C. (1979). Use of goodness-of-fit tests for characterizing the spacial variability of soil properties. *Soil Science Society of America Journal, 43,* 274–278.

71 Rao, P. V., & Thornby, J. I. (1969). A robust point estimator in a generalized regression model. *The Annals of Mathematics Statistics, 40,* 1784–1790.

72 Rouanet, H., & Lepine, D. (1970). Comparison between treatments in a repeated measures design: ANOVA and multivariate methods. *British Journal of Mathematical and Statistical Psychology, 23,* 147–163.

73 SAS Institute, Inc. (1995). *SAS user's guide, version 6.2.* Cary, NC: SAS Institute.

74 Satterthwaite, F. E. (1946). An approximate distribution of estimates of variance components. *Biometrics Bulletin, 2,* 110–114.

75 Sawka, G. I., Collins, M. A., Brown, R. B., & Rao, P. V. (1988). Evaluation of selected Florida soils for onsite sewage disposal systems. *Proceedings of the Fifth National Symposium on Individual and Small Community Systems: American Society of Agricultural Engineers,* 384–393.

76 Scheff, H. (1959). *The analysis of variance.* New York: Wiley.

77 Schrier, S. L., & Junga, I. (1981). Entry and distribution of chlorpromazine and vinblastine into human erythrocytes during endocytosis. *Proceedings of the Society for Experimental Biology and Medicine, 168,* 159–167.

78 Searle, S. R. (1971). *Linear models.* New York: Wiley.

79 Searle, S. R. (1987). *Linear models for unbalanced data.* New York: Wiley.

80 Seber, G. A. F. (1977). *Linear regression analysis.* New York: Wiley.

81 Sen, P. K. (1968). Estimates of the regression coefficient based on Kendall's tau. *Journal of the American Statistical Association, 63,* 1379–1389.

82 Shapiro, X. X., & Wilk, X. X. (1980). *How to test for normality and other distributional assumptions.* Milwaukee, WI: American Society of Quality Control.

83 Smith, E. C., Valika, K. S., Woo, J. E., O'Donnell, D. L., Gordon, D. L., & Westerman, M. P. (1981). Serum phosphate abnormalities in sickle cell anemia. *Proceedings of the Society for Experimental Biology and Medicine, 168,* 254–258.

84 Smith, I. F., Latham, M. C., Azubuike, J. A., Butler, W. R., Phillips, L. S., Pond, W. G., & Enwonwu, C. O. (1981). Blood plasma levels of cortisol, insulin, growth hormone and somatomedin in children with marasmus, kwashiorkor, and intermediate forms of protein-energy malnutrition (41222). *Proceedings of the Society for Experimental Biology and Medicine, 167,* 607–611.

85 Sprent, P. (1993). *Applied nonparametric statistical methods* (2nd ed.). London: Chapman & Hall.

86 Steel, R. G. D., & Torrie, J. H. (1980). *Principles and procedures of statistics* (2nd ed.). New York: McGraw-Hill.

87 Tarla, F. N., Henry, P. R., Ammerman, C. B., Rao, P. V., & Miles, R. D. (1991). Effect of time and sex on tissue selenium concentrations in chicks fed practical diets supplemented with sodium selenite or calcium selenite. *Biological Trace Element Research, 31,* 11–20.

88 Tukey, J. W. (1949). One degree of freedom for nonadditivity. *Biometrics, 5,* 99–114.

89 Tukey, J. W. (1977). *Exploratory data analysis.* Boston: Addison-Wesley.

90 Underwood, A. J. (1981). Techniques of analysis of variance in experimental marine biology and ecology. *Oceanography and Marine Biology: An Annual Review, 19,* 513–605.

91 Volicer, B. J., & Bohannon, M. W. (1975). A hospital stress rating scale. *Nursing Research, 24,* 352–359.

92 Wackerly, D. D., Mendenhall, W., & Scheaffer, R. L. (1996). *Mathematical statistics with applications* (5th ed.). Boston: PWS-Kent.

93 Weisberg, S. (1981). A statistic for allocating C_p to individual cases. *Technometrics, 23,* 27–31.

94 Welch, B. L. (1951). On the comparison of several mean values: An alternative approach. *Biometrika, 38,* 330–336.

95 Winer, B. J., (1971). *Statistical principles in experimental design* (2nd ed.). New York: McGraw-Hill.

96 Younger, M. S. (1997). *SAS Companion.* Pacific Grove, CA: Duxbury.

APPENDIX B
NOTES ON
THEORETICAL STATISTICS

B.1
Introduction

In this appendix, we examine some key definitions and results in theoretical statistics, along with some examples of their use in deriving the statistical methods described in the text. The topics provide a glimpse into the content of graduate-level courses in theoretical statistics. Applied statisticians should have an appreciation for the theory that justifies the practical use of statistical techniques; good applications need to be supported by good theory. With that in mind, this appendix introduces some relevant theoretical material. It is not intended as a substitute for a rigorous exposition of statistical theory.

Section B.2 contains a brief introduction to the notions of random variables, random events, and probabilities of random events. In Section B.3, means, variances, and covariances are defined as the expected values of functions of real random variables. In Section B.4, some rules for computing the means, variances, and covariances of linear functions of random variables are given, while Section B.5 is devoted to the sampling distributions of some common statistics based on samples from normal populations. Section B.6 contains statements of some key results that are usually encountered in graduate-level statistics courses on the theory of linear models or the theory of least squares. Finally, Sections B.7–B.9 cover other miscellaneous topics.

Students who have had a senior-level course in theoretical statistics will be familiar with most of the results in Sections B.2–B.5; for proofs and illustrative examples of many of these results, see Wackerly, Mendenhall, and Scheaffer (1996). Graybill (1976) and Searle (1987) are good references for the more advanced topics in Section B.6, which involve matrix formulations of results on general linear models.

B.2
Random variables, random events, and probabilities

Random variables

A random variable can be defined in many ways, depending on the level of rigor required. For our purpose, we will be satisfied with the intuitive definition that a *random variable* is a variable whose value is determined by the observed characteristics of an item randomly selected from a statistical population. Capital letters—X, Y, and so on—are used to denote random variables, while the corresponding lower-case letters—x, y, and so on—denote their observed values. The *sample space* of a random variable Y is the set of all possible values that can be observed for Y. Sometimes the sample space of Y is referred to as the *support* of Y.

Often, it is useful to classify a random variable as univariate or multivariate. A *univariate random variable* is a variable that records a single value for each observation. This value can be measured on a nominal, an ordinal, or an interval scale. For instance, the random variable $Y =$ the height of a randomly selected tree is a univariate random variable that takes values measured on an interval scale, because the variable Y records a single numerical value—the height of the selected tree. Similarly, the random variable $Y =$ the gender of a randomly selected newborn baby is a univariate nominal random variable.

Discrete and *continuous random variables* are distinguished by their sample spaces. The sample space of a discrete univariate random variable is a finite or countably infinite set of values. The binomial and the Poisson random variables are discrete. The sample space of a continuous univariate random variable consist of intervals (finite or infinite) of a real line. Random variables such as normal variables, χ^2, t, and F are continuous.

A *multivariate random variable* is a column vector of $k \geq 2$ univariate random variables. For example

$$Y = \begin{bmatrix} Y_1 \\ Y_2 \\ Y_3 \end{bmatrix},$$

where Y_1, Y_2, and Y_3 are, respectively, the height, age, and maximum trunk circumference of a randomly selected tree, is a multivariate random variable with $k = 3$ components. A multivariate random variable with k univariate components is often referred to as a k-variate random variable.

Random events

An observable or a measurable characteristic of a random variable is called a *random event*, or simply, an *event*. An event is said to have occurred if the observed value of the corresponding random variable possesses the attribute defining the event. Consider, for example, the random variable $Y =$ the four-week height of a plant. Some events associated with Y are as follows:

1 The height is less than 2 m; that is, $Y < 2$.

2 The height is at least 2 m; that is, $Y \geq 2$.

3 The height is more than 6 m; that is, $Y > 6$.

4 The height is no more than 6 m; that is, $Y \leq 6$.

5 The height is no less than 2 m and no more than 6 m; that is, $2 \leq Y \leq 6$.

6 The height is either less than 2 m or more than 6 m; that is, $Y < 2$ or $Y > 6$.

Uppercase letters (with or without subscripts) will be used to denote events. The six events just described can be represented as $A_1, A_2, A_3, A_4, A_5,$ and A_6, where

$$A_1 = (Y < 2), \qquad A_2 = (Y \geq 2), \qquad A_3 = (Y > 6),$$

$$A_4 = (Y \leq 6), \qquad A_5 = (2 \leq Y \leq 6), \qquad A_6 = (Y < 2 \text{ or } Y > 6).$$

The events A_i, $i = 1, \ldots, 6$, are associated with a univariate random variable. An example of an event associated with a multivariate random variable is the event A that a randomly selected newborn is a male weighing 13 kg. If Y_1 is the gender and Y_2 is the weight of the newborn, then A is an event associated with the bivariate ($k = 2$) random variable

$$Y = \begin{bmatrix} Y_1 \\ Y_2 \end{bmatrix}.$$

The event A can be symbolically represented as $A = (Y_1 = \text{male and } Y_2 = 13)$.

Two events are said to be *complements* of each other if the occurrence of one implies, and is implied by, the nonoccurrence of the other. The complement of an event A is denoted by $\overline{A}$. As an example, consider the events A_1 and A_2. Since Y will be less than 2 if and only if Y is not at least 2, the event A_1 will occur if and only if the event A_2 does not occur. Consequently, A_1 and A_2 are complements; $A_1 = \overline{A}_2$ and $A_2 = \overline{A}_1$.

Two events that cannot occur together are said to be *mutually exclusive* or *disjoint*. Clearly, complementary events are mutually exclusive, because occurrence of an event implies that its complement has not occurred. However, two events can be mutually exclusive without being complementary. For instance, the events A_1 and A_3 are mutually exclusive, because it is not possible to observe $(Y < 2)$ and $(Y > 6)$ at the same time. These events are not complementary, because nonoccurrence of $(Y < 2)$ does not imply that $(Y > 6)$ has occurred. Finally, a pair of events can be neither complementary nor mutually exclusive. The events A_2 and A_3 are such events because they can occur simultaneously—for example, if $Y > 7$.

An event E is said to be the *intersection* of two events E_1 and E_2—in symbols, $E = E_1 \cap E_2$—if E occurs when and only when both E_1 and E_2 have occurred. It can be verified that A_5 is the intersection of A_2 and A_4; that is, $A_5 = A_2 \cap A_4$. The notion of intersection of two events can be extended to three or more events in an obvious manner. An event E is said to be the intersection of k events $E_1, \ldots, E_k$ if E occurs when and only when all of the k events $E_1, \ldots, E_k$ have occurred. We then write $E = E_1 \cap \cdots \cap E_k$.

An event that cannot occur is called a *null* event and is denoted by the symbol Φ. Because two mutually exclusive events cannot occur together, the intersection of an event and its complement is a null event; that is, $E \cap \overline{E} = \Phi$ for any event E.

The *union* of two events E_1 and E_2 is the event E that occurs if at least one of E_1 and E_2 occurs. In that case, we write $E = E_1 \cup E_2$. It can be verified that A_6 is the union of A_1 and A_3; that is, $A_6 = A_1 \cup A_3$. The union of k events $E_1, \ldots, E_k$ is the event E that occurs when and only when at least one of the k events $E_1, \ldots, E_k$ has occurred. We then write $E = E_1 \cup \cdots \cup E_k$.

An event that will definitely occur is called a *sure* event and denoted by S. The union of an event and its complement is a sure event; that is, $S = E \cup \overline{E}$ for every event E.

Probability

Let E be an event associated with a random variable Y. The *probability* of E, denoted by $P(E)$, is a numerical measure of the level of confidence with which we can predict that E will occur when the value of Y is observed once. The level of confidence can be defined in many ways. Two of the most frequently used definitions are as follows:

1   The *relative frequency* definition, according to which $P(E)$ is a number that approximates the proportion of times (relative frequency) that E would occur in a long sequence of repeated observations of Y.

2   The *subjective* definition, according to which $P(E)$ is a measure of an individual's willingness to bet in favor of the occurrence of E in a particular observation of Y.

For more details on the distinctions between the two definitions of probability, see Kyburg and Smokler (1964) and De Groot (1970). In this book, we use the relative frequency definition of probability. Any definition of probability should include three requirements, known as the *axioms of probability*.

Axioms of probability

Because relative frequency is a proportion, the probability of an event should be a number between 0 and 1. In symbols

Axiom 1. $0 \le P(E) \le 1$ for every event E.

We know that the null event never occurs, and the sure event always occurs. Consequently, the probabilities of Φ and S must equal 0 and 1, respectively. In symbols

Axiom 2. $P(\Phi) = 0$ and $P(S) = 1$.

If E_1 and E_2 are two mutually exclusive events—that is, if $E_1 \cap E_2 = \Phi$—then the total number of occurrences of their union—$E = E_1 \cup E_2$—will equal the sum of the number of occurrences of E_1 and the number of occurrences of E_2. Thus, the sum of the relative frequencies of two—in fact, any number of—mutually exclusive events must equal the relative frequency of their union. Therefore, if $E_1, \ldots, E_k, \ldots$ is a sequence of mutually exclusive events, then the probability of the union of the E_i must equal the sum of their probabilities. In symbols

Axiom 3. $P(E_1 \cup E_2 \cdots) = P(E_1) + P(E_2) + \cdots \quad (E_i \cap E_j = \Phi \, ; \, i \ne j).$

As noted in Chapter 3, probabilities play a fundamental role in the statistical design and analysis of research studies. A wide body of knowledge—the *calculus of probability*—concerns the rules for computing the probabilities of complex events from the probabilities of simpler events. The rules are derived under the assumption that probabilities satisfy the three axioms.

Conditional probability

Let E_2 be an event such that $P(E_2) > 0$. The *conditional probability* of E_1 given E_2 is defined as

$$P(E_1|E_2) = \frac{P(E_1 \cap E_2)}{P(E_2)}.$$

The quantity $P(E_1|E_2)$ should be interpreted as the probability that E_1 will occur given that E_2 has occurred. Under the relative frequency interpretation, $P(E_1|E_2)$ is an approximation to the long-run proportion of the times that the event E_1 will be observed among those independent observations in which E_2 has been observed.

Table B.1 shows a set of probabilities of observing various age-gender characteristics of a subject randomly selected from a hypothetical study population. The probabilities in Table B.1 can be interpreted as follows:

- In the long run, approximately 15% of the subjects randomly selected from the study population will be young males.

- In the long run, approximately 34% of the subjects randomly selected from the population will be young.

If B_1 and B_2 denote the events that the subject is a male and that the subject is young, respectively, then $P(B_1) = 0.42$, $P(B_2) = 0.34$, and $P(B_1 \cap B_2) = 0.15$. Therefore, the conditional probability that a randomly selected subject will be male, given that the subject is young, is

$$P(B_1|B_2) = \frac{0.15}{0.34} = 0.44.$$

The conditional probability of 0.44 can be interpreted as the long-run relative frequency of males among randomly selected subjects who are young.

TABLE   B.1

Probabilities of some age-gender combinations

Age	Gender Male	Female	Total
Young	0.15	0.19	0.34
Old	0.27	0.39	0.66
Total	0.42	0.58	1.00

Statistical independence

The concept of conditional probabilities leads naturally to the concept of statistical independence. If E_2 is an event of nonzero probability, then the two events E_1 and E_2 are said to be *statistically independent* if the probability of E_1 is the same as the conditional probability of E_1 given E_2. If the probability of E_2 is zero, then we say, by convention, that E_2 is independent of all other events. Thus, the two events E_1 and E_2 are independent if either E_1 or E_2 is the null event or

$$P(E_1|E_2) = P(E_1) \quad \text{and} \quad P(E_2|E_1) = P(E_2).$$

It can be verified that E_1 and E_2 are independent if and only if

$$P(E_1 \cap E_2) = P(E_1)P(E_2).$$

The events B_1 and B_2 defined earlier, are not independent, because the conditional probability of B_1 given B_2 (0.44) is not the same as the unconditional probability of B_1 (0.42). Because of the dependence of the two events, our confidence (as measured by the probability) that a randomly selected subject will be male will be higher (0.44 versus 0.42) if we know that the selected subject is young.

The notion of independence can be extended to k events, but we won't go into the details here. More important from our perspective is the notion of independence of a set of random variables. Again, we will be satisfied with an intuitive definition.

Two random variables Y_1 and Y_2 are said to be statistically independent if every event associated with Y_1 is independent of every event associated with Y_2. When two random variables are independent, the conditional probability of occurrence of an event associated with one variable does not depend on the observed value of the other random variable. As an example, let's return to Table B.1. If Y_1 and Y_2 are random variables defined as follows

$$Y_1 = \begin{cases} 1 & \text{if the subject is male} \\ 0 & \text{if the subject is female,} \end{cases}$$

$$Y_2 = \begin{cases} 1 & \text{if the subject is young} \\ 0 & \text{if the subject is male,} \end{cases}$$

the Y_1 and Y_2 are not independent, because the events B_1 and B_2 are not independent. Note that B_1 is an event associated with Y_1 and B_2 is an event associated with Y_2.

B.3
Expected values of real random variables

A random variable whose values are measured on an interval scale is called a *real random variable*. A real random variable can be univariate or multivariate. Some examples of univariate real random variables are $Y_1 =$ the number of germinating seeds and $Y_2 =$ the measured chemical influx in the brain of a treated rat. The random variable Y with $Y' = (Y_1, Y_2)$ (Y' is the transpose of Y), where $Y_1 =$ the age and $Y_2 =$ the height of a four-week-old plant is an example of a bivariate real random variable. On the other hand, consider the random variable $Y = 0$ if a newborn has low birth-weight and 1 otherwise; this is not a real random variable, because its values 0 and 1 are not measured on the interval scale. Rather, Y is a nominal univariate random variable, because its values are on a nominal scale.

Let $f(y)$ denote the probability density function of a continuous random variable Y or the probability function of a discrete random variable Y. In Section 2.3 we saw how a statistical population can be described using the probability or frequency function of a random variable.

Let $g(Y)$ denote a real-valued function of a univariate real random variable Y. Then the *expected value* or the *mean value* of $g(Y)$ is defined as

$$\mathcal{E}(g(Y)) = \begin{cases} \sum g(y)f(y) & \text{if } Y \text{ is discrete} \\ \int g(y)f(y)\,dy & \text{if } Y \text{ is continuous,} \end{cases} \tag{B.1}$$

where the summation and integration is carried over the entire range of values in the sample space of Y.

The mean and variance of a univariate real random variable Y are two important special cases of the expected value defined by Equation (B.1).

- The expected value of Y, denoted by μ_Y, is called the *mean* of Y. Thus, the mean of Y is

$$\mu_Y = \begin{cases} \sum yf(y) & \text{if } Y \text{ is discrete} \\ \int yf(y)\,dy & \text{if } Y \text{ is continuous.} \end{cases}$$

- The expected value of $g(Y) = (Y - \mu_Y)^2$, denoted by σ_Y^2, is called the *variance* of Y. Thus, the variance of Y is

$$\sigma_Y^2 = \begin{cases} \sum \left(y - \mu_Y\right)^2 f(y) & \text{if } Y \text{ is discrete} \\ \int \left(y - \mu_Y\right)^2 f(y)\,dy & \text{if } Y \text{ is continuous.} \end{cases}$$

A formula analogous to Equation (B.1) can be written for the expected value of a real-valued function of a multivariate real random variable. In this book, the most important expected values of functions of multivariate random variables are the means and variances of linear combinations of univariate random variables. As we'll see in the next section, these expected values can be expressed in terms of the means, variances, and covariances of univariate random variables.

Let $\mathbf{Y}' = (Y_1, Y_2)$, where Y_1 and Y_2 are univariate real random variables with means μ_{Y_1} and μ_{Y_2}. The covariance of Y_1 and Y_2 is defined as the expected value of $g(\mathbf{Y}) = (Y_1 - \mu_{Y_1})(Y_2 - \mu_{Y_2})$. In symbols, the covariance of Y_1 and Y_2 is

$$\sigma_{Y_1, Y_2} = \mathcal{E}((Y_1 - \mu_{Y_1})(Y_2 - \mu_{Y_2})).$$

The following useful result about the relationships between the means, variances, and covariances of two random variables can easily be established.

Theorem B.1   Let U_1 and U_2 be random variables with means μ_1 and μ_2, variances σ_1^2 and σ_2^2 and covariance σ_{12}.

a   Then

$$\sigma_1^2 = \mathcal{E}(U_1^2) - \mu_1^2.$$

b   Also
$$\sigma_{12} = \mathcal{E}(U_1 U_2) - \mu_1 \mu_2.$$

c   If U_1 and U_2 are independent, then
$$\mathcal{E}(U_1 U_2) = \mathcal{E}(U_1)\mathcal{E}(U_2).$$

Applications of Theorem B.1

Theorem B.1 has two important consequences.

1   The covariance of two independent random variables is zero. This follows from parts (b) and (c) of Theorem B.1.

2   Independent random variables are uncorrelated. This follows from the fact that the correlation coefficient equals the covariance divided by the product of the standard deviations

$$\rho = \frac{\sigma_{12}}{\sigma_1 \sigma_2}.$$

Matrix notation

Let U_i $(i = 1, \ldots, k)$ be univariate random variables with means μ_i, variances σ_i^2, and covariances σ_{ij} (σ_{ij} is the covariance of U_i and U_j, $i \neq j$), and let

$$U = \begin{bmatrix} U_1 \\ U_2 \\ \vdots \\ U_k \end{bmatrix}, \quad \mu = \begin{bmatrix} \mu_1 \\ \mu_2 \\ \vdots \\ \mu_k \end{bmatrix}, \quad \text{and} \quad \Sigma = \begin{bmatrix} \sigma_1^2 & \cdots & \sigma_{1j} & \cdots & \sigma_{1k} \\ \vdots & & \vdots & & \vdots \\ \sigma_{i1} & \cdots & \sigma_{ij} & \cdots & \sigma_{ik} \\ \vdots & & \vdots & & \vdots \\ \sigma_{k1} & \cdots & \sigma_{kj} & \cdots & \sigma_k^2 \end{bmatrix}.$$

Thus, U, μ, and Σ denote, respectively, the k-dimensional column vector representing the multivariate random variable with the U_i as its components; the k-dimensional column vector of the expected values of the components of U; and the $k \times k$ matrix of the variances and covariances of the components of U. We refer to μ as the expected value or the mean of U, and write

$$\mathcal{E}(U) = \mu.$$

Also, the matrix Σ is called the *covariance matrix* of U and is sometimes denoted by $\text{Cov}(U)$.

B.4

Expected values of linear combinations

Linear combinations of random variables play a key role in the development of statistical inferential procedures because many commonly used statistics can be either expressed as, or approximated by, linear combinations of random variables. Theorem B.2 lists some results that are useful for determining the means, variances, and covariances of linear combinations of random variables.

Theorem B.2 Let U_i ($i = 1, \ldots, k$) be random variables as defined in Section B.3. Given constants c_i and d_i, let L and M be linear functions defined as

$$L = c_0 + c_1 U_1 + c_2 U_2 + \cdots + c_k U_k = c_0 + \sum_{i=1}^{k} c_i U_i,$$

$$M = d_0 + d_1 U_1 + d_2 U_2 + \cdots + d_k U_k = d_0 + \sum_{i=1}^{k} d_i U_i.$$

1   Then we can state the following conclusions:

    a   The mean of L is

$$\mathcal{E}(L) = \mu_L = c_0 + c_1 \mu_1 + \cdots + c_k \mu_k = c_0 + \sum_{i=1}^{k} c_i \mu_i,$$

    b   The variance of L is

$$\mathrm{Var}(L) = \sigma_L^2 = \sum_{i=1}^{k} c_i^2 \sigma_i^2 + 2 \sum \sum_{1 \le i < j \le k} c_i c_j \sigma_{ij}.$$

    c   The covariance of L and M is

$$\mathrm{Cov}(L, M) = \sigma_{LM} = \sum_{i=1}^{k} c_i d_i \sigma_i^2 + \sum \sum_{1 \le i < j \le k} (c_i d_j + c_j d_i) \sigma_{ij}.$$

2   Let $Y_1, Y_2, \ldots, Y_k$ be uncorrelated random variables; that is, $\sigma_{ij} = 0$ for $i \ne j$.

    a   Then

$$\sigma_L^2 = c_1^2 \sigma_1^2 + \cdots + c_k^2 \sigma_k^2 = \sum_{i=1}^{k} c_i^2 \sigma_i^2.$$

    b   Also

$$\sigma_{LM} = c_1 d_1 \sigma_1^2 + \cdots + c_k d_k \sigma_k^2 = \sum_{i=1}^{k} c_i d_i \sigma_i^2.$$

Applications of Theorem B.2

Theorem B.2 contains some of the most frequently used results in theoretical statistics. Here are some examples of the uses of this theorem.

EXAMPLE B.1   In Example 3.15, we noted that the quantity

$$S'^2 = \frac{\sum_{i=1}^{n} (Y_i - \overline{Y})^2}{n},$$

where the Y_i are a random sample from a population with mean μ and variance σ^2, is a biased estimate of σ^2. Theorem B.2 can be used to verify this property.

First, we observe that $\overline{Y}$ is of the form

$$\overline{Y} = c_0 + c_1 U_1 + \cdots + c_k U_k,$$

where $k = n$, $c_0 = 0$, $c_i = 1/n$, $U_i = Y_i$ for $i = 1, \ldots, k$, and the Y_i are independent, with a common mean μ and variance σ^2. Accordingly, from parts 1(a) and 2(b) of Theorem B.2 we conclude that

$$\mathcal{E}(\overline{Y}) = n\left(\frac{1}{n}\right)\mu = \mu \text{ and } \mathrm{Var}(\overline{Y}) = n\left(\frac{1}{n}\right)^2 \sigma^2 = \frac{\sigma^2}{n}.$$

Next, we note that $\sum_{i=1}^{n}(Y_i - \overline{Y})^2 = \sum_{i=1}^{n} Y_i^2 - n\overline{Y}^2$, and so S'^2 can be expressed as the linear combination

$$S'^2 = c_0 + c_1 U_1 + \cdots + c_k U_k,$$

where $k = n + 1$, $c_0 = 0$, $c_i = 1/n$, $U_i = Y_i^2$ for $i = 1, \ldots, k - 1$; $c_k = -n$ and $U_k = \overline{Y}^2$. Now the mean and variance of Y_i and $\overline{Y}$ are, respectively, μ, σ^2 and μ, σ^2/n, and so, from part 1(a) of Theorem B.2, we conclude that

$$\mathcal{E}(U_i) = \sigma^2 + \mu^2, \quad i = 1, \ldots, k - 1; \quad \mathcal{E}(U_k) = \frac{\sigma^2}{n} + \mu^2.$$

Then an application of part 1(a) of Theorem B.2 yields

$$\mathcal{E}(S'^2) = n\left(\frac{1}{n}\right)(\sigma^2 + \mu^2) + (-1)\left(\frac{\sigma^2}{n} + \mu^2\right) = \frac{n-1}{n}\sigma^2,$$

and hence S'^2 is a biased estimate of σ^2.

Finally, we can also conclude from this example that the sample variance $S^2 = [n/(n-1)]S'^2$ is an unbiased estimator of σ^2. ∎

EXAMPLE **B.2**   When discussing the arcsin transformation in Section 8.7, we made use of the mean and variance of $\hat{\pi}$, the proportion of successes in a binomial experiment with n trials. The mean and variance of $\hat{\pi}$ can be derived using Theorem B.2.

Let U_i be the random variable that takes the value 1 if the result of the i-th binomial trial is a success and 0 if it is a failure. Since the binomial trials are independent, the random variables U_i are independent. Also, from Equation (2.8) the mean and variance of U_i are $\mu_i = \pi$ and $\sigma_i^2 = \pi(1 - \pi)$ respectively. The expressions,

$$\mathcal{E}(\hat{\pi}) = \pi$$

$$\mathrm{Var}(\hat{\pi}) = \frac{\pi(1 - \pi)}{n},$$

for the mean and variance of $\hat{\pi}$ can be obtained by writing the proportion of successes as

$$\hat{\pi} = c_0 + c_1 U_1 + \cdots + c_n U_n,$$

where $c_0 = 0$, $c_i = 1/n$, $i = 1, \ldots, n$, and letting $\mu_i = \pi$, $\sigma_i^2 = \pi(1 - \pi)$, $i = 1, \ldots, n$ in parts 1(a) and 2(b) of Theorem B.2.   ∎

EXAMPLE B.3   The mean and variance of the sampling distribution of the linear function $\hat{\theta}$ in Box 9.1 can be derived from Theorem B.2.   ∎

EXAMPLE B.4   In our discussion of the ANOVA F-test in Section 8.3, we noted that MS[E], the mean square for error, is an unbiased estimate of σ^2, whereas MS[T], the mean square for treatment, is an unbiased estimate of a quantity larger than σ^2. Both of these properties can be derived using Theorem B.2.

First let's derive the important relationship

$$SS[TOT] = SS[T] + SS[E].$$

From the computing formula in step 2 of Section 8.4, it is easy to verify that

$$SS[TOT] = \sum_{i=1}^{k} \sum_{j=1}^{n} (Y_{ij} - \overline{Y}_{++})^2.$$

Adding and subtracting the term $\overline{Y}_{i+}$ inside the parentheses on the right-hand side and expanding the square, we get

$$SS[TOT] = \sum_{i=1}^{k} \sum_{j=1}^{n} (Y_{ij} - \overline{Y}_{i+} + \overline{Y}_{i+} - \overline{Y}_{++})^2$$

$$= \sum_{i=1}^{k} \sum_{j=1}^{n} (Y_{ij} - \overline{Y}_{i+})^2 + \sum_{i=1}^{k} \sum_{j=1}^{n} (\overline{Y}_{i+} - \overline{Y}_{++})^2$$

$$+ 2 \sum_{i=1}^{k} \sum_{j=1}^{n} (Y_{ij} - \overline{Y}_{i+})(\overline{Y}_{i+} - \overline{Y}_{++}).$$

Let S_i^2 denote the variance of the i-th sample. Then

$$(n - 1)S_i^2 = \sum_{j=1}^{n} (Y_{ij} - \overline{Y}_{i+})^2,$$

so that the first term on the right-hand side in the expression for SS[TOT] is the same as SS[E]. The second term on the right-hand side equals $n \sum_{i=1}^{k} (\overline{Y}_{i+} - \overline{Y}_{++})^2$, which is the same as SS[T]. Finally, the third term on the right-hand side is zero, because

$$\sum_{i=1}^{k} \sum_{j=1}^{n} (Y_{ij} - \overline{Y}_{i+})(\overline{Y}_{i+} - \overline{Y}_{++}) = \sum_{i=1}^{k} (\overline{Y}_{i+} - \overline{Y}_{++}) \sum_{j=1}^{n} (Y_{ij} - \overline{Y}_{i+})$$

$$= \sum_{i=1}^{k} (\overline{Y}_{i+} - \overline{Y}_{++}) \left(\sum_{j=1}^{n} Y_{ij} - n\overline{Y}_{i+} \right) = 0,$$

where we have used the relationship $\sum_{j=1}^{n} Y_{ij} = n\overline{Y}_{i+}$.

Next, we find the expected values of SS[TOT] and SS[E]. Observe that SS[E] can be expressed as

$$\text{SS[E]} = \sum_{i=1}^{t}(n-1)S_i^2 = c_0 + \sum_{i=1}^{t} c_i U_i,$$

where $c_0 = 0$, $c_i = n - 1$, $U_i = S_i^2$, $i = 1, \ldots, t$. We know that S_i^2 is an unbiased estimate of σ^2; that is, $\mathcal{E}(S_i^2) = \sigma^2$. Hence, from part 1(a) of Theorem B.2

$$\mathcal{E}(\text{SS[E]}) = (n-1)t\sigma^2.$$

The expected value of SS[TOT] will depend on whether the treatment effects are fixed or random. Suppose the treatments are fixed. Then the model can be expressed as

$$Y_{ij} = \mu + \tau_i + E_{ij},$$

so that the Y_{ij} are independent random variables with means $\mu + \tau_i$ and variance σ^2. Then $\mathcal{E}(Y_{ij}^2) = (\mu + \tau_i)^2 + \sigma_i^2$. Furthermore, by writing the overall mean response as

$$\overline{Y}_{++} = \mu + \overline{E}_{++}$$

and observing that $\overline{E}_{++}$ is the mean of nt independent random variables with zero mean and a common variance σ^2, we see that $\overline{Y}_{++}$ has mean μ and variance σ^2/nt. Hence, using part (a) of Theorem B.1 we get

$$\mathcal{E}(\overline{Y}_{++}^2) = \mu^2 + \frac{\sigma^2}{nt}.$$

From Theorem B.2 and the relationship $\sum \tau_i = 0$, we now conclude that

$$\mathcal{E}(\text{SS[TOT]}) = \mathcal{E}\left(\sum_{i=1}^{t}\sum_{j=1}^{n} Y_{ij}^2 - nt\overline{Y}_{++}^2\right)$$

$$= \sum_{i=1}^{t}\sum_{j=1}^{n} \mathcal{E}\left(Y_{ij}^2\right) - nt\mathcal{E}\left(\overline{Y}_{++}^2\right)$$

$$= \sum_{i=1}^{t}\sum_{j=1}^{n} \left[(\mu + \tau_i)^2 + \sigma^2\right] - nt\left(\mu^2 + \frac{\sigma^2}{nt}\right)$$

$$= (nt-1)\sigma^2 + n\sum_{i=1}^{t}\tau_i^2.$$

Thus, the expected value of SS[T] is

$$\mathcal{E}(\text{SS[T]}) = \mathcal{E}(\text{SS[TOT]}) - \mathcal{E}(\text{SS[E]})$$

$$= (nt-1)\sigma^2 + n\sum_{i=1}^{t}\tau_i^2 - t(n-1)\sigma^2$$

$$= (t-1)\sigma^2 + n\sum_{i=1}^{t}\tau_i^2,$$

and the expected value of the mean square for treatments is

$$\mathcal{E}(\text{MS[T]}) = \sigma^2 + n\psi_T^2,$$

where

$$\psi_T^2 = \frac{1}{t-1} \sum_{i=1}^{t} \tau_i^2.$$

We know that ψ_T^2 is nonnegative and equals zero if and only if H_0: $\tau_1 = \cdots = \tau_t = 0$. Hence, MS[T] is an unbiased estimate of σ^2 if and only if H_0 is true. Otherwise, MS[T] estimates a quantity larger than σ^2.

This derivation of the expected mean squares in the one-way ANOVA setting is based on the assumptions that the treatments are equally replicated and the treatment effects are fixed. The modifications for the case of unequal replication and/or random treatment effects are straightforward. In the unequal-replication case, we simply replace n by n_i and carry out the algebraic simplifications needed to obtain the relevant expression in Box 14.2. In the random-effects case, the derivation of the expected value of SS[TOT] must take into account that the responses satisfy the model

$$Y_{ij} = \mu + T_i + E_{ij},$$

where the T_i and E_{ij} are independent random variables with zero mean and variances σ_T^2 and σ^2, respectively. Hence, the random variables Y_{ij} have mean μ and variance $\sigma_T^2 + \sigma^2$, so that $\mathcal{E}(Y_{ij}^2) = \sigma_T^2 + \sigma^2 + \mu^2$. ∎

EXAMPLE B.5 Theorem B.2 can be used to find the variances and covariances of the least squares estimators of the regression parameters in the simple linear regression model. The key step in this derivation is the recognition that both $\hat{\beta}_0$ and $\hat{\beta}_1$ in Box 10.3 can be expressed as linear combinations of observed responses. First, let's consider $\hat{\beta}_1$. We have

$$\hat{\beta}_1 = \frac{\sum(x_i - \bar{x})(Y_i - \bar{Y})}{S_{xx}}$$

$$= \frac{\sum(x_i - \bar{x})Y_i - \bar{Y}\sum(x_i - \bar{x})}{S_{xx}},$$

where the lowercase letters for the x_i indicate that these are constants, not random variables. We know that $\sum(x_i - \bar{x}) = 0$, and so

$$\hat{\beta}_1 = \sum \frac{(x_i - \bar{x})}{S_{xx}} Y_i.$$

Since the x_i are fixed constants, the least squares estimator of the slope takes the form

$$\hat{\beta}_1 = c_0 + \sum_{i=1}^{n} c_i U_i,$$

where $c_0 = 0$, $c_i = (x_i - \bar{x})/S_{xx}$ and $U_i = Y_i$. Straightforward substitution in the expressions for the mean and variance of L in parts 1(a) and 2(a) of Theorem B.2

shows that $\mathcal{E}(\hat{\beta}_1) = \beta_1$ and $\mathrm{Var}(\hat{\beta}_1) = \sigma^2/S_{xx}$. The mean and variance of $\hat{\beta}_0$ and the covariance of $\hat{\beta}_0$ and $\hat{\beta}_1$ can be derived from Theorem B.2 using the expression

$$\hat{\beta}_0 = d_0 + \sum_{i=1}^n d_i U_i,$$

where $d_0 = 0$ and $d_i = (1/n) - c_i \bar{x}$. These derivations are left as exercises.   ■

Matrix notation

Theorem B.3 Let U be a k-variate random variable with expected value μ and covariance matrix Σ. Let $L = c_0 + c'U$ and $M = d_0 + d'U$, where $c' = (c_1, \ldots, c_k)$ and $d' = (d_1, \ldots, d_k)$ are k-dimensional row vectors and the c_i and d_i are given constants.

a   Then

$$\mu_L = c_0 + c'\mu, \qquad \mu_M = d_0 + d'\mu.$$

b   Also

$$\sigma_L^2 = c'\Sigma c, \qquad \sigma_M^2 = d'\Sigma d.$$

c   Finally

$$\sigma_{LM} = c'\Sigma d.$$

Applications of Theorem B.3

Theorem B.3 is a matrix formulation of Theorem B.2. As an exercise, use Theorem B.3 to verify the details in Examples B.4 and B.5.

B.5
Sampling from normal distributions

Some of the common sampling distributions that occur when random samples are selected from normally distributed populations are listed in the following theorem.

Theorem B4   Let X, Y, U_1, U_2, $\ldots$, U_k be *independent* random variables such that X has a $\chi^2(p)$-distribution, Y has a $\chi^2(q)$-distribution, and U_i has a $N(\mu_i, \sigma_i^2)$ distribution.

a   The standardized variate

$$Z = \frac{(U_1 - \mu_1)}{\sigma_1}$$

has a standard normal distribution; that is, Z has a $N(0, 1)$ distribution.

b   Let

$$L = c_0 + c_1 U_1 + \cdots + c_k U_k,$$

μ_L, and σ_L^2 be as in parts 1(a) and 2(a) of Theorem B.2. Then L has a normal distribution with mean μ_L and variance σ_L^2.

c   The random variable

$$Z^2 = \frac{(U_1 - \mu_1)^2}{\sigma_1^2}$$

has a $\chi^2(1)$-distribution.

d   The random variable

$$T = \frac{Z}{\sqrt{(X/p)}} = \frac{(U_1 - \mu_1)\sqrt{p}}{\sigma_1 \sqrt{X}}$$

has a $t(p)$-distribution.

e   The random variable

$$F = \frac{(X/p)}{(Y/q)} = \frac{qX}{pY}$$

has an $F(p, q)$-distribution.

Applications of Theorem B.4

Let's consider three examples.

EXAMPLE **B.6**   Suppose that $Y_1, Y_2, \ldots, Y_n$ is a random sample from a normal distribution with mean μ and variance σ^2. We can use Theorem B.4 to derive the $100(1 - \alpha)\%$ confidence interval for μ given in Box 4.2.

Let $\overline{Y}$ and S^2 denote, respectively, the mean and the variance of the sample. In Example B.17, we'll show that $\overline{Y}$ has a normal distribution with mean μ and variance σ^2/n, that $(n - 1)S^2/\sigma^2$ has a χ^2-distribution with $n - 1$ degrees of freedom, and that $\overline{Y}$ and S^2 are independent. It follows from this conclusion and part (d) of Theorem B.4 that

$$T = \frac{(\overline{Y} - \mu)}{\sqrt{(\sigma^2/n)(S^2/\sigma^2)}} = \frac{(\overline{Y} - \mu)\sqrt{n}}{S}$$

has a t-distribution with $n - 1$ degrees of freedom. Now let $t(n - 1, \alpha/2)$ be the $\alpha/2$-level critical value of a t-distribution with $n - 1$ degrees of freedom. Then

$$\Pr\left\{-t(n - 1, \alpha/2) \leq \frac{(\overline{Y} - \mu)\sqrt{n}}{S} \leq t(n - 1, \alpha/2)\right\} = 1 - \alpha,$$

from which it follows that

$$\Pr\left\{\overline{Y} - t(n - 1, \alpha/2)\frac{S}{\sqrt{n}} \leq \mu \leq \overline{Y} + t(n - 1, \alpha/2)\frac{S}{\sqrt{n}}\right\} = 1 - \alpha.$$

Thus, the confidence level associated with the two-sided confidence interval in Box 4.2 is $1 - \alpha$. ∎

EXAMPLE B.7   Theorem B.4 can be used to derive the expressions for the $100(1 - \alpha)\%$ prediction bounds in Box 4.10.

Let $\overline{Y}$ and S^2 be as in Example B.6, and let $\overline{Y}_f$ denote the mean of an independent future random sample of size m from the same normal population. Then

$$L = \overline{Y}_f - \overline{Y}$$

is a linear function ($c_0 = 0$; $c_1 = \cdots = c_m = 1/m$; $c_{m+1} = \cdots = c_{m+n} = -1/n$) of $m + n$ independent normally distributed random variables with a common mean μ and a common variance σ^2. It follows from Theorem B.2 that the mean and variance of L are $\mu_L = 0$ and $\sigma_L^2 = (1/m + 1/n)\sigma^2$, respectively. Now, $\overline{Y}_f$ is the mean of a future independent sample, and S^2 depends on the current sample only, and so we can conclude that $\overline{Y}_f$ is independent of S^2. Furthermore, as noted in Example B.6, S^2 is independent of $\overline{Y}$. Finally, we know that L is a function of $\overline{Y}$ and $\overline{Y}_f$ only, and so L and S^2 are independent. From part (d) of Theorem B.4, it follows that

$$T = \frac{L}{\sqrt{\left(\dfrac{1}{m} + \dfrac{1}{n}\right) S^2}}$$

has a t-distribution with $n - 1$ degrees of freedom. Thus

$$\Pr\left\{-t(n - 1, \, \alpha/2) \leq \frac{\overline{Y}_f - \overline{Y}}{\sqrt{\left(\dfrac{1}{m} + \dfrac{1}{n}\right) S^2}} \leq t(n - 1, \, \alpha/2)\right\} = 1 - \alpha,$$

so that

$$\Pr\left\{\overline{Y} - t(n - 1, \, \alpha/2)\sqrt{\left(\frac{1}{m} + \frac{1}{n}\right) S^2} \leq \overline{Y}_f \leq \overline{Y} + t(n - 1, \, \alpha/2)\sqrt{\left(\frac{1}{m} + \frac{1}{n}\right) S^2}\right\}$$
$$= 1 - \alpha.$$

The last result provides justification for the two-sided prediction interval in Box 4.10.

Similarly, the fact that $(n - 1)S^2/\sigma^2$ has a χ^2 distribution with $n - 1$ degrees of freedom can be used to justify the χ^2-procedures described in Box 4.6. ∎

EXAMPLE B.8   A simple application of Theorem B.4 will yield the result in Box 7.3. Let Y_1 and Y_2 be independent and normally distributed, with respective means $\mu_0 + \Delta$ and μ_0 and a common variance σ^2. Then, by Theorem B.4, the linear combination $L = Y_1 - Y_2$ has a normal distribution with mean $\mu_L = \Delta$ and variance $\sigma_L^2 = 2\sigma^2$, and the standardized variate $(L - \Delta)/\sigma_L$ has a standard normal distribution. Therefore

$$\Pr\{Y_1 \geq Y_2\} = \Pr\{L \geq 0\} = \Pr\left\{\frac{L - \Delta}{\sigma_L} \geq \frac{0 - \Delta}{\sigma_L}\right\} = A\left(-\frac{\Delta}{\sigma\sqrt{2}}\right). ∎$$

B.6
Theory of linear models

After a brief description of a general linear model, we'll look at some examples of how it can be used to unify the statistical models that underly multiple linear regression analysis, analysis of variance, and analysis of covariance.

The general linear model

The n dimensional random variable (observation vector) Y is said to satisfy a general linear model (GLM) if

$$Y = X\beta + E$$

where

a X is a $n \times k$ matrix of known elements (the design matrix);

b β is a k-dimensional column vector of unknown parameters (the regression parameters);

c E is an n-dimensional random variable (the error vector), with mean and covariance matrix

$$\mathcal{E}(E) = O \text{ and } \text{Cov}(E) = I\sigma^2,$$

respectively; that is, the errors are uncorrelated, with a common variance σ^2.

Examples of general linear models

Important examples of general linear models include regression models, one-sample models, analysis of variance models, and analysis of covariance models.

EXAMPLE **B.9** *Regression models*

A matrix formulation of the simple linear linear regression model was described in Section 11.3. We have

$$Y = \begin{bmatrix} Y_1 \\ \vdots \\ Y_i \\ \vdots \\ Y_n \end{bmatrix}, \quad X = \begin{bmatrix} 1 & x_1 \\ \vdots & \vdots \\ 1 & x_i \\ \vdots & \vdots \\ 1 & x_n \end{bmatrix}, \quad \beta = \begin{bmatrix} \beta_0 \\ \beta_1 \end{bmatrix}, \quad E = \begin{bmatrix} E_1 \\ \vdots \\ E_i \\ \vdots \\ E_n \end{bmatrix}.$$

For a matrix formulation of a multiple linear regression model, see Box 11.3. ∎

EXAMPLE **B.10** *The one-sample model*

Let $Y_1, \ldots, Y_i, \ldots, Y_n$ be a random sample from a normal distribution with mean μ and variance σ^2. Then the Y_i satisfy the one-sample model in Box 4.8. The one-sample model is an example of general linear model with $k = 1$ regression parameter $\beta_0 = \mu$. The matrices in the general linear formulation are

$$Y = \begin{bmatrix} Y_1 \\ \vdots \\ Y_i \\ \vdots \\ Y_n \end{bmatrix}, \quad X = \begin{bmatrix} 1 \\ \vdots \\ 1 \\ \vdots \\ 1 \end{bmatrix}, \quad E = \begin{bmatrix} E_1 \\ \vdots \\ E_i \\ \vdots \\ E_n \end{bmatrix}, \quad \beta = \begin{bmatrix} \beta_0 \end{bmatrix}. \quad \blacksquare$$

EXAMPLE **B.11**   *Analysis of variance model*

All of the models in Chapters 12 and 13 are examples of general linear models as defined in Box 12.1. Here we confine our attention to one-way ANOVA models.

First, consider a completely randomized experiment with $t = 3$ treatments, each replicated $n = 2$ times. Let Y_{ij} denote the j-th response to be observed for the i-th treatment. Then, from our discussions in Section 12.2, it follows that the one-way ANOVA model in Equation (12.5)

$$Y_{ij} = \mu + \tau_i + E_{ij}$$

can be expressed as a general linear model with two dummy variables X_1 and X_2, which take values x_{1ij} and x_{2ij} where $x_{uij} = 1$ if $u = i$ and 0 otherwise. The resulting general linear model

$$Y_{ij} = \beta_0 + \beta_1 x_{1ij} + \beta_2 x_{2ij} + E_{ij} \qquad (i = 1, 2, 3; \ j = 1, 2)$$

will have $k = 3$ regression parameters. If μ_1, μ_2, μ_3 are the means of the three populations of responses, then

$$\beta_0 = \mu + \tau_3 = \mu_3; \quad \beta_1 = \tau_1 - \tau_3 = \mu_1 - \mu_3; \quad \beta_2 = \tau_2 - \tau_3 = \mu_2 - \mu_3.$$

The matrices $Y, X, \beta,$ and E can be expressed as

$$Y = \begin{bmatrix} Y_{11} \\ Y_{12} \\ Y_{21} \\ Y_{22} \\ Y_{31} \\ Y_{32} \end{bmatrix}; \quad X = \begin{bmatrix} 1 & 1 & 0 \\ 1 & 1 & 0 \\ 1 & 0 & 1 \\ 1 & 0 & 1 \\ 1 & 0 & 0 \\ 1 & 0 & 0 \end{bmatrix}; \quad \beta = \begin{bmatrix} \beta_0 \\ \beta_1 \\ \beta_2 \end{bmatrix}; \quad E = \begin{bmatrix} E_{11} \\ E_{12} \\ E_{21} \\ E_{22} \\ E_{31} \\ E_{32} \end{bmatrix}.$$

Note that there is more than one way in which an ANOVA model can be expressed as a general linear model. For instance, an alternative expression for the one-way model is

$$Y_{ij} = \beta_1 x_{1ij} + \beta_2 x_{2ij} + \beta_3 x_{3ij} + E_{ij},$$

where $x_{uij} = 1$ if $i = u$ and 0 otherwise. Again, this is a model with $k = 3$ regression parameters. The Y and E matrices are the same as in the previous expression, but the X matrix and the interpretation of the regression parameters are different. It can be verified that

$$
X = \begin{bmatrix} 1 & 0 & 0 \\ 1 & 0 & 0 \\ 0 & 1 & 0 \\ 0 & 1 & 0 \\ 0 & 0 & 1 \\ 0 & 0 & 1 \end{bmatrix}; \quad \beta = \begin{bmatrix} \beta_1 \\ \beta_2 \\ \beta_3 \end{bmatrix} = \begin{bmatrix} \mu_1 \\ \mu_2 \\ \mu_3 \end{bmatrix}. \quad \blacksquare
$$

EXAMPLE B.12 *Analysis of covariance model*

Equation (12.14) in Example 12.3 is a general linear model for expressing the relationship of a subject's posttest score to his or her pretest score and treatment group. The model has $k = 4$ regression parameters and is an example of a class of linear models known as analysis of covariance models. The construction of the associated Y, X, β, and E matrices is left as an exercise. $\blacksquare$

The method of least squares

Given β, let $\phi(\beta) = (Y - X\beta)'(Y - X\beta)$ denote the sum of squares of the difference between the observed and expected responses. Any $\hat{\beta}$ that minimizes $\phi(\beta)$ is called a *least squares estimator* of β. The minimum sum of squares $\phi(\hat{\beta})$ is called the *error sum of squares* and is denoted by SS[E]. The vector $\hat{Y} = X\hat{\beta}$ is called the *least squares predictor* of Y. The k equations

$$
X'Y = X'X\hat{\beta}
$$

are called the *normal equations*.

Theorem B.5 The following are some useful mathematical properties of least squares estimators.

a Any solution of the normal equations is a least squares estimator of β.

b Every solution of the normal equations gives the same least squares predictor of Y; that is, if $\hat{\beta}_1$ and $\hat{\beta}_2$ are two solutions of the normal equations, then $X\hat{\beta}_1 = X\hat{\beta}_2$.

c Let $\hat{\beta}$ be a solution of the normal equations. Then

$$
\text{SS[E]} = (Y - X\hat{\beta})'(Y - X\hat{\beta}) = Y'Y - \hat{\beta}'X'Y.
$$

Theorem B.5 states that any solution of the normal equations can be used to determine the least squares prediction equation. Thus, when the normal equations have more than one solution, the primary objective in least squares prediction is to find one solution of the normal equations. This is usually done by finding the so-called generalized inverse of the $X'X$ matrix. Alternatively, the general linear model can always be defined in such a way that the normal equations have a unique solution. Theorem B.5 also provides an expression for SS[E] in terms of least squares estimators of the regression parameters.

Applications of Theorem B.5

Once again, let's consider some examples.

EXAMPLE B.13 The matrix formulation of the simple linear regression model was described in Example B.9. Straightforward calculations show that

$$X'Y = \begin{bmatrix} \sum Y_i \\ \sum x_i Y_i \end{bmatrix}; \quad X'X = \begin{bmatrix} n & \sum x_i \\ \sum x_i & \sum x_i^2 \end{bmatrix}.$$

Solving the normal equations by straightforward algebra, it is easy to show that $\hat{\beta}_0$ and $\hat{\beta}_1$ in Box 10.3 are the least squares estimators of β_0 and β_1, respectively. Using Theorem B.5, we can easily verify that

$$\text{SS[E]} = Y'Y - \hat{\beta}' X'Y$$
$$= \sum Y_i^2 - \hat{\beta}_0 \left(\sum Y_i \right) - \hat{\beta}_1 \left(\sum x_i Y_i \right)$$
$$= \text{SS[TOT]} - \hat{\beta}^2 S_{xx}.$$

The reader should compare this expression for SS[E] with that in Box 10.3. ∎

EXAMPLE B.14 Let $Y_1, \ldots, Y_i, \ldots, Y_n$ be a random sample from a normal distribution with mean μ and variance σ^2. A matrix representation of the general linear model for these data was described in Example B.10. Thus, $X'Y = \sum Y_i$ and $X'X\hat{\beta} = n\hat{\beta}_0 = n\hat{\mu}$, so that $\sum Y_i = n\hat{\mu}$ is the normal equation for estimating the regression parameter $\beta_0 = \mu$. It follows that $\hat{\mu} = \overline{Y}$ is the unique solution of the normal equation, and the sample mean is the least squares estimator of the population mean.

From Theorem B.5

$$\text{SS[E]} = \phi(\hat{\mu}) = Y'Y - \hat{\mu} X'Y = \sum Y_i^2 - \frac{\left(\sum Y \right)^2}{n},$$

so that the error sum of squares due to fitting the general linear model is $(n-1)S^2$, where S^2 is the sample variance. ∎

EXAMPLE B.15 Let Y_{ij} denote the j-th response observed for the i-th treatment in a completely randomized design with $t = 2$ treatments, each replicated n times. Let X_{uij} be the dummy variable (see Section 12.2) that takes the value 1 if $i = u$ and 0 otherwise. Then the one-way ANOVA model for Y_{ij} can be expressed as

$$Y_{ij} = \beta_0 + \beta_1 x_{1ij} + \beta_2 x_{2ij} + E_{ij}, \quad i = 1, 2, 3; \ j = 1, \ldots, n.$$

Structurally, the matrices $Y, X, E,$ and β resemble the corresponding matrices in Example B.11. The number of replications is n instead of 2. We have

$$X'Y = \begin{bmatrix} Y_{++} \\ Y_{1+} \\ Y_{2+} \end{bmatrix}; \quad X'X = \begin{bmatrix} 3n & n & n \\ n & n & 0 \\ n & 0 & n \end{bmatrix}.$$

Thus, the normal equations for estimating the regression parameters are

$$Y_{++} = 3n\hat{\beta}_0 + n\hat{\beta}_1 + n\hat{\beta}_2,$$

$$Y_{1+} = n\hat{\beta}_0 + n\hat{\beta}_1,$$

$$Y_{2+} = n\hat{\beta}_0 + n\hat{\beta}_2.$$

Solving these normal equations is a straightforward exercise in algebra. It can be verified that the equations have the unique solution:

$$\hat{\beta}_0 = \overline{Y}_{3+}, \quad \hat{\beta}_1 = \overline{Y}_{1+} - \overline{Y}_{3+}, \quad \hat{\beta}_2 = \overline{Y}_{2+} - \overline{Y}_{3+}.$$

This solution, when used in Theorem B.5, shows that the error sum of squares can be expressed as

$$SS[E] = SS[TOT] - SS[T]$$

where SS[TOT] and SS[T] are the one-way ANOVA total and treatment sums of squares defined in steps 2 and 3 of Section 8.4.   ∎

Sampling distributions of $\hat{\beta}$

Theorem B.6 lists some key properties of the sampling distributions associated with the least squares estimators of β, on the assumption that the normal equations have a unique solution.

Theorem B.6   Suppose that the normal equations have a unique solution $\hat{\beta}$.

a   There exists a unique $k \times k$ matrix S such that

$$SX'X = I.$$

The matrix S is called the *inverse* of $X'X$ and is denoted by $(X'X)^{-1}$.

b   The least squares estimator can be expressed as

$$\hat{\beta} = SX'Y.$$

c   The estimator $\hat{\beta}$ has a k-variate normal distribution with mean β and covariance matrix $S\sigma^2$.

d   The random variable $SS[E]/\sigma^2$ has a $\chi^2(n-k)$-distribution.

e   The estimator $\hat{\beta}$ and $SS[E]/\sigma^2$ are independently distributed.

f   Let C be an $l \times k$ matrix such that $l \leq k$ and $Cb = O$ if and only if $b = O$. Then there exists a unique $l \times l$ matrix M such that

$$MCSC' = CSC'M = I.$$

M is called the inverse of CSC'.

g Let SS[E]$_r$ denote the minimum of $\phi(\boldsymbol{\beta})$ where the minimum is taken over all $\boldsymbol{\beta}$ such that $\boldsymbol{C\beta} = \boldsymbol{h}$. Then

$$F_c = \frac{\text{MS}[H_0]}{\text{MS}[E]},$$

where

$$\text{MS}[H_0] = \frac{\text{SS}[E]_r - \text{SS}[E]}{l} \quad \text{and} \quad \text{MS}[E] = \frac{\text{SS}[E]}{n - k},$$

has a noncentral F-distribution with l degrees of freedom for the numerator, $n - k$ degrees of freedom for the denominator, and noncentrality parameter

$$\lambda = \frac{(\boldsymbol{C\beta} - \boldsymbol{h})'\boldsymbol{M}(\boldsymbol{C\beta} - \boldsymbol{h})}{2\sigma^2}.$$

Applications of Theorem B.6

Two examples are of particular interest here.

EXAMPLE **B.16**   In Example B.15, we derived the normal equations for estimating the parameters in a one-way ANOVA model with $t = 3$ treatments and n replications per treatment. As noted there, the normal equations have a unique solution.

By direct matrix multiplication, it can be verified that

$$\boldsymbol{S} = \begin{bmatrix} \dfrac{1}{n} & -\dfrac{1}{n} & -\dfrac{1}{n} \\[2mm] -\dfrac{1}{n} & \dfrac{2}{n} & \dfrac{1}{n} \\[2mm] -\dfrac{1}{n} & \dfrac{1}{n} & \dfrac{2}{n} \end{bmatrix}.$$

is the inverse of $\boldsymbol{X}'\boldsymbol{X}$. This S-matrix, when used with part (b) of Theorem B.6, yields the same solution of the normal equations as in Example B.15.

Suppose we wish to test the null hypothesis H_0: $\mu_1 = \mu_2 = \mu_3$. If the general linear model is such that $\beta_0 = \mu_3$, $\beta_1 = \mu_1 - \mu_3$, $\beta_2 = \mu_2 - \mu_3$, then testing the null hypothesis of equal treatment means is equivalent to testing H_0: $\beta_1 = \beta_2 = 0$. In the notation of Theorem B.6, this null hypothesis can be expressed as H_0: $\boldsymbol{C\beta} = \boldsymbol{h}$, where

$$\boldsymbol{C} = \begin{bmatrix} 0 & 1 & 0 \\ 0 & 0 & 1 \end{bmatrix} \quad \text{and} \quad \boldsymbol{h} = \begin{bmatrix} 0 \\ 0 \end{bmatrix}.$$

To determine the test statistic, we first note from Example B.15 that SS[E] = SS[TOT] − SS[T], with $n - k = 3n - 3 = 3(n - 1)$ degrees of freedom. Next we note that, under H_0, the general linear model reduces to the one-sample model in Example B.10, so that, as shown in Example B.14, SS[E]$_r = (n - 1)S^2 = $ SS[TOT]. Also, knowing that $\boldsymbol{C}$ is a 2×3 matrix, we can take $l = 2$ in Theorem B.6. Thus

$$\text{SS}[H_0] = \text{SS}[TOT] - (\text{SS}[TOT]) - \text{SS}[T]) = \text{SS}[T],$$

with $l = 2$ degrees of freedom. The test statistic for H_0 takes the form

$$F_c = \frac{\text{MS[T]}}{\text{MS[E]}},$$

where

$$\text{MS[T]} = \frac{\text{SS[T]}}{2} \quad \text{and} \quad \text{MS[E]} = \frac{\text{SS[E]}}{3(n-1)}.$$

As we can easily verify, the test based on F_c is the same as the one-way ANOVA F-test in Section 8.3.

Finally, the noncentrality parameter, which is useful for determining the power of the ANOVA F-test, can be determined as follows. We have

$$CSC' = \begin{bmatrix} 0 & 1 & 0 \\ 0 & 0 & 1 \end{bmatrix} \begin{bmatrix} \dfrac{1}{n} & -\dfrac{1}{n} & -\dfrac{1}{n} \\[6pt] -\dfrac{1}{n} & \dfrac{2}{n} & \dfrac{1}{n} \\[6pt] -\dfrac{1}{n} & \dfrac{1}{n} & \dfrac{2}{n} \end{bmatrix} \begin{bmatrix} 0 & 0 \\ 1 & 0 \\ 0 & 1 \end{bmatrix} = \begin{bmatrix} \dfrac{2}{n} & \dfrac{1}{n} \\[6pt] \dfrac{1}{n} & \dfrac{2}{n} \end{bmatrix},$$

and the inverse of CSC' is

$$M = \begin{bmatrix} \dfrac{2n}{3} & -\dfrac{n}{3} \\[6pt] -\dfrac{n}{3} & \dfrac{2n}{3} \end{bmatrix}.$$

Therefore

$$(C\boldsymbol{\beta} - h)'M(C\boldsymbol{\beta} - h) = \begin{bmatrix} \beta_1 & \beta_2 \end{bmatrix} \begin{bmatrix} \dfrac{2n}{3} & -\dfrac{n}{3} \\[6pt] -\dfrac{n}{3} & \dfrac{2n}{3} \end{bmatrix} \begin{bmatrix} \beta_1 \\ \beta_2 \end{bmatrix}$$

$$= \frac{2n}{3}(\beta_1^2 + \beta_2^2 - \beta_1\beta_2).$$

Substituting $\beta_1 = \mu_1 - \mu_3$, $\beta_2 = \mu_2 - \mu_3$ in the last expression on the right-hand side and using Theorem B.6, we find that the noncentrality parameter is

$$\lambda = \frac{3n\sigma_\mu^2}{2\sigma^2},$$

where σ_μ^2 is defined in Equation (9.29). The reader should compare this result with Equation (9.32), which gives the noncentrality parameter associated with the one-way ANOVA F-test with t treatments and n replications per treatment. ∎

EXAMPLE **B.17**   For the one-sample problem in Example B.14, $k = 1$, $\boldsymbol{\beta} = \mu$, $\hat{\boldsymbol{\beta}} = \overline{Y}$, $\text{SS[E]} = (n-1)S^2$, and $X'X$ is a 1×1 matrix with element n, so that S is a 1×1 matrix with element $1/n$. Theorem B.6 leads to the following conclusions.

1 Part (c) implies that $\overline{Y}$ has a normal distribution with mean μ and variance σ^2/n.

2 Part (d) implies that $(n-1)S^2/\sigma^2$ has a $\chi^2(n-1)$-distribution.

3 From part (e), it follows that $\overline{Y}$ and S^2 are independently distributed. ∎

B.7

Confidence interval for the population median

The $100(1-\alpha)\%$ confidence interval in Box 5.1 for the population median η can be derived as follows. Let η denote the median of the population. As we know, a test of the null hypothesis H_0: $\eta = \eta_0$ against the research hypothesis H_1: $\eta \neq \eta_0$ can be performed by either the test-statistic approach or the confidence-interval approach. A hypothesized value η_0 will be rejected by a two-sided α-level sign test if and only if η_0 is outside the $100(1-\alpha)\%$ confidence interval for η. In other words, a $100(1-\alpha)\%$ confidence interval for η is made up of all those values η_0 such that an α-level sign test of the null hypothesis H_0: $\eta = \eta_0$ will not be rejected in favor of the research hypothesis H_1: $\eta \neq \eta_0$. Now, from the information in Box 5.1, we see that the values η_0 of η that will result in the acceptance of H_0: $\eta = \eta_0$ at level α are precisely those η_0 such that the calculated value of the test statistic B will be within the interval $(n - b_{n,\alpha/2},\ b_{n,\alpha/2})$—that is, those η_0 for which $n - b_{n,\alpha/2} < B_c < b_{n,\alpha/2}$. To see how a change in η_0 will cause a change in b_c, we can proceed as follows.

Let $y_{(1)} < y_{(2)} < \cdots < y_{(n)}$ denote the order statistics of the sample, and define the counting function $c(u)$ by

$$c(u) = \begin{cases} 1 & \text{if } u > 0 \\ 0 & \text{if } u \leq 0. \end{cases}$$

Clearly, for each i, $c(y_{(i)} - \eta_0) = 1$ if $y_{(i)} > \eta_0$ and 0 if $y_{(i)} \leq \eta_0$. The calculated value of the test statistic is equal to the number of $y_{(i)} - \eta_0$ that exceed zero, and so we can write

$$B_c = \sum_{i=1}^{n} c(y_{(i)} - \eta_0).$$

This expression represents B_c as a sum of n components, each of which can take one of two values, 0 and 1. Now, for any η_0 less than or equal to the smallest observation in the sample—that is, for $\eta_0 \leq y_{(1)}$—every one of the n components of B_c will be 1 so that $B_c = n$. If η_0 is in the interval $(y_{(1)}, y_{(2)}]$—that is, if $y_{(1)} < \eta_0 \leq y_{(2)}$—then all but the first component of B_c will equal 1, so that B_c will be equal to $n - 1$. In general, if $y_{(i)} < \eta_{(0)} \leq y_{(i+1)}$, then B_c will equal $n - i$. Thus, as η_0 increases, the corresponding B_c will decrease in jumps of 1 from the largest value n when $\eta_0 \leq y_{(1)}$ to the smallest value 0 when $\eta_0 > y_{(n)}$. If we adopt the convention that $y_{(0)} = -\infty$ and $y_{(n+1)} = +\infty$, B_c can be expressed in terms of η_0 as

$$B_c = n - i \quad \text{if} \quad y_{(i)} < \eta_0 \leq y_{(i+1)} \quad i = 0, 1, \ldots, n.$$

This expression can be used to show that the B_c will satisfy $n - b_{n,\alpha/2} < B_c < b_{n,\alpha/2}$ if and only if $y_{(L)} < \eta_0 < y_{(U)}$, where L and U are as defined in Box 5.1. In other words, the interval in Box 5.1 consists of those η_0 that will be accepted by the α-level sign test.

B.8
Variance-stabilizing transformations

The transformations in Table 8.4 are often referred to as *variance-stabilizing trans-formations*, because their primary purpose is to ensure that the transformed data sat-isfy the equal-variance component of the ANOVA assumptions. Variance-stabilizing transformations appropriate for data in some commonly occurring situations may be determined (at least approximately) by means of a general formula. This formula assumes a known relationship between the population variance and the population mean.

Let Y denote a typical response, and let μ and σ^2 denote, respectively, the mean and variance of Y. Assume that σ^2 and μ satisfy the relationship

$$\sigma^2 = h(\mu),$$

where $h(\mu)$ is a known function of μ. Then, the transformation

$$g(y) = C \int \frac{1}{\sqrt{h(y)}} dy,$$

where C is an arbitrary constant, is a variance-stabilizing transformation of Y.

For an example of the use of this transformation, recall from Example B.2 that the relationship between the variance and mean of the proportion of successes in n independent binomial trials is of the form $\sigma^2 = \mu(1 - \mu)/n$, where $\mu = \pi$. Taking $h(y) = y(1 - y)/n$ in the expression for $g(y)$, we get

$$g(y) = C \int \frac{\sqrt{n}}{\sqrt{y(1 - y)}} dy = C\sqrt{n} \arcsin \sqrt{y}.$$

Because C is arbitrary, we can let $C = 1/\sqrt{n}$ and apply the transformation $g(y) = \arcsin \sqrt{y}$ to the proportion of successes in independent binomial trials.

As an exercise, use the expression for $g(y)$ to derive the log and square-root transformations described in Section 8.7.

B.9
A useful lower bound for σ_μ^2

Let $\mu_1, \mu_2, \ldots, \mu_t$ be a set of t population means. In Section 9.8, we utilized the quantity

$$\sigma_\mu^2 = \frac{1}{t} \sum_{i=1}^{t} (\mu_i - \overline{\mu})^2$$

as a quantitative criterion for measuring departures from the null hypothesis of equal population means. A property of σ_μ^2 that was used in Section 9.8 was that, if the difference between at least one pair of means is as large as Δ, then $\sigma_\mu^2 \geq \Delta^2/2t$. The following is an outline of a proof of this result.

Without loss of generality, we can assume that $|\mu_1 - \mu_2| \geq \Delta$. Also, to emphasize that t values of μ_i are used in the definition of $\overline{\mu}$ and σ_μ^2, let us write $\overline{\mu}_t$ and σ_t^2 instead of $\overline{\mu}$ and σ_μ^2 respectively. Then using the easily verifiable relationships:

$$\overline{\mu}_t = \frac{(t-1)\overline{\mu}_{t-1} + \mu_t}{t},$$

$$\sum_{i=1}^{t}(\mu_i - \overline{\mu}_{t-1}) = \mu_t - \overline{\mu}_{t-1},$$

$$\sum_{i=1}^{t}(\mu_i - \overline{\mu}_{t-1})^2 = (t-1)\sigma_{t-1}^2 + (\mu_t - \overline{\mu}_{t-1})^2,$$

it can be shown that

$$\sigma_t^2 = \frac{t-1}{t}\sigma_{t-1}^2 + \left(\frac{1}{t} - \frac{1}{t^2}\right)(\mu_t - \overline{\mu}_{t-1})^2$$

$$\geq \frac{t-1}{t}\sigma_{t-1}^2$$

$$\geq \left(\frac{t-1}{t}\right)\left(\frac{t-2}{t-1}\right)\sigma_{t-2}^2 \geq \cdots \geq \left(\frac{t-1}{t}\right)\left(\frac{t-2}{t-1}\right)\cdots\left(\frac{2}{3}\right)\sigma_2^2$$

$$= \frac{2}{t}\sigma_2^2.$$

Finally, since

$$\sigma_2^2 = \frac{1}{4}(\mu_1 - \mu_2)^2 \quad \text{and} \quad |\mu_1 - \mu_2| \geq \Delta,$$

we have the inequality

$$\sigma_t^2 \geq \frac{2}{t}\sigma_2^2 \geq \frac{2}{t}\frac{\Delta^2}{4} = \frac{\Delta^2}{2t},$$

which proves the required result.

APPENDIX C
STATISTICAL TABLES

TABLE C.1

Areas under the standard normal distribution

z	0.00	0.01	0.02	0.03	0.04	0.05	0.06	0.07	0.08	0.09
0.0	0.5000	0.4960	0.4920	0.4880	0.4840	0.4801	0.4761	0.4721	0.4681	0.4641
0.1	0.4602	0.4562	0.4522	0.4483	0.4443	0.4404	0.4364	0.4325	0.4286	0.4247
0.2	0.4207	0.4168	0.4129	0.4090	0.4052	0.4013	0.3974	0.3936	0.3897	0.3859
0.3	0.3821	0.3783	0.3745	0.3707	0.3669	0.3632	0.3594	0.3557	0.3520	0.3483
0.4	0.3446	0.3409	0.3372	0.3336	0.3300	0.3264	0.3228	0.3192	0.3156	0.3121
0.5	0.3085	0.3050	0.3015	0.2981	0.2946	0.2912	0.2877	0.2843	0.2810	0.2776
0.6	0.2743	0.2709	0.2676	0.2643	0.2611	0.2578	0.2546	0.2514	0.2483	0.2451
0.7	0.2420	0.2389	0.2358	0.2327	0.2296	0.2266	0.2236	0.2206	0.2177	0.2148
0.8	0.2119	0.2090	0.2061	0.2033	0.2005	0.1977	0.1949	0.1922	0.1894	0.1867
0.9	0.1841	0.1814	0.1788	0.1762	0.1736	0.1711	0.1685	0.1660	0.1635	0.1611
1.0	0.1587	0.1562	0.1539	0.1515	0.1492	0.1469	0.1446	0.1423	0.1401	0.1379
1.1	0.1357	0.1335	0.1314	0.1292	0.1271	0.1251	0.1230	0.1210	0.1190	0.1170
1.2	0.1151	0.1131	0.1112	0.1093	0.1075	0.1056	0.1038	0.1020	0.1003	0.0985
1.3	0.0968	0.0951	0.0934	0.0918	0.0901	0.0885	0.0869	0.0853	0.0838	0.0823
1.4	0.0808	0.0793	0.0778	0.0764	0.0749	0.0735	0.0721	0.0708	0.0694	0.0681
1.5	0.0668	0.0655	0.0643	0.0630	0.0618	0.0606	0.0594	0.0582	0.0571	0.0559
1.6	0.0548	0.0537	0.0526	0.0516	0.0505	0.0495	0.0485	0.0475	0.0465	0.0455
1.7	0.0446	0.0436	0.0427	0.0418	0.0409	0.0401	0.0392	0.0384	0.0375	0.0367
1.8	0.0359	0.0351	0.0344	0.0336	0.0329	0.0322	0.0314	0.0307	0.0301	0.0294
1.9	0.0287	0.0281	0.0274	0.0268	0.0262	0.0256	0.0250	0.0244	0.0239	0.0233
2.0	0.0228	0.0222	0.0217	0.0212	0.0207	0.0202	0.0197	0.0192	0.0188	0.0183
2.1	0.0179	0.0174	0.0170	0.0166	0.0162	0.0158	0.0154	0.0150	0.0146	0.0143
2.2	0.0139	0.0136	0.0132	0.0129	0.0125	0.0122	0.0119	0.0116	0.0113	0.0110
2.3	0.0107	0.0104	0.0102	0.0099	0.0096	0.0094	0.0091	0.0089	0.0087	0.0084
2.4	0.0082	0.0080	0.0078	0.0075	0.0073	0.0071	0.0069	0.0068	0.0066	0.0064
2.5	0.0062	0.0060	0.0059	0.0057	0.0055	0.0054	0.0052	0.0051	0.0049	0.0048
2.6	0.0047	0.0045	0.0044	0.0043	0.0041	0.0040	0.0039	0.0038	0.0037	0.0036
2.7	0.0035	0.0034	0.0033	0.0032	0.0031	0.0030	0.0029	0.0028	0.0027	0.0026
2.8	0.0026	0.0025	0.0024	0.0023	0.0023	0.0022	0.0021	0.0021	0.0020	0.0019
2.9	0.0019	0.0018	0.0018	0.0017	0.0016	0.0016	0.0015	0.0015	0.0014	0.0014
3.0	0.0013	0.0013	0.0013	0.0012	0.0012	0.0011	0.0011	0.0011	0.0010	0.0010

TABLE   C.2
Critical values of *t*-distributions

df	α = 0.2	α = 0.15	α = 0.1	α = 0.05	α = 0.025	α = 0.01	α = 0.005	α = 0.001
1	1.3764	1.9626	3.0777	6.3138	12.7062	31.8205	63.6567	318.3088
2	1.0607	1.3862	1.8856	2.9200	4.3027	6.9646	9.9248	22.3271
3	0.9785	1.2498	1.6377	2.3534	3.1824	4.5407	5.8409	10.2145
4	0.9410	1.1896	1.5332	2.1318	2.7764	3.7469	4.6041	7.1732
5	0.9195	1.1558	1.4759	2.0150	2.5706	3.3649	4.0321	5.8934
6	0.9057	1.1342	1.4398	1.9432	2.4469	3.1427	3.7074	5.2076
7	0.8960	1.1192	1.4149	1.8946	2.3646	2.9980	3.4995	4.7853
8	0.8889	1.1081	1.3968	1.8595	2.3060	2.8965	3.3554	4.5008
9	0.8834	1.0997	1.3830	1.8331	2.2622	2.8214	3.2498	4.2968
10	0.8791	1.0931	1.3722	1.8125	2.2281	2.7638	3.1693	4.1437
11	0.8755	1.0877	1.3634	1.7959	2.2010	2.7181	3.1058	4.0247
12	0.8726	1.0832	1.3562	1.7823	2.1788	2.6810	3.0545	3.9296
13	0.8702	1.0795	1.3502	1.7709	2.1604	2.6503	3.0123	3.8520
14	0.8681	1.0763	1.3450	1.7613	2.1448	2.6245	2.9768	3.7874
15	0.8662	1.0735	1.3406	1.7531	2.1314	2.6025	2.9467	3.7328
16	0.8647	1.0711	1.3368	1.7459	2.1199	2.5835	2.9208	3.6862
17	0.8633	1.0690	1.3334	1.7396	2.1098	2.5669	2.8982	3.6458
18	0.8620	1.0672	1.3304	1.7341	2.1009	2.5524	2.8784	3.6105
19	0.8610	1.0655	1.3277	1.7291	2.0930	2.5395	2.8609	3.5794
20	0.8600	1.0640	1.3253	1.7247	2.0860	2.5280	2.8453	3.5518
21	0.8591	1.0627	1.3232	1.7207	2.0796	2.5176	2.8314	3.5272
22	0.8583	1.0614	1.3212	1.7171	2.0739	2.5083	2.8188	3.5050
23	0.8575	1.0603	1.3195	1.7139	2.0687	2.4999	2.8073	3.4850
24	0.8569	1.0593	1.3178	1.7109	2.0639	2.4922	2.7969	3.4668
25	0.8562	1.0584	1.3163	1.7081	2.0595	2.4851	2.7874	3.4502
26	0.8557	1.0575	1.3150	1.7056	2.0555	2.4786	2.7787	3.4350
27	0.8551	1.0567	1.3137	1.7033	2.0518	2.4727	2.7707	3.4210
28	0.8546	1.0560	1.3125	1.7011	2.0484	2.4671	2.7633	3.4082
29	0.8542	1.0553	1.3114	1.6991	2.0452	2.4620	2.7564	3.3962
30	0.8538	1.0547	1.3104	1.6973	2.0423	2.4573	2.7500	3.3852
40	0.8507	1.0500	1.3031	1.6839	2.0211	2.4233	2.7045	3.3069
80	0.8461	1.0432	1.2922	1.6641	1.9901	2.3739	2.6387	3.1953
100	0.8452	1.0418	1.2901	1.6602	1.9840	2.3642	2.6259	3.1737
200	0.8434	1.0391	1.2858	1.6525	1.9719	2.3451	2.6006	3.1315

TABLE C.3

Critical values of χ^2-distributions

df	0.999	0.995	0.99	0.975	$\alpha =$ 0.95	0.90	0.85	0.80
1	0.0000	0.0000	0.0002	0.0010	0.0039	0.0158	0.0358	0.0642
2	0.0020	0.0100	0.0201	0.0506	0.1026	0.2107	0.3250	0.4463
3	0.0243	0.0717	0.1148	0.2158	0.3518	0.5844	0.7978	1.0052
4	0.0908	0.2070	0.2971	0.4844	0.7107	1.0636	1.3665	1.6488
5	0.2102	0.4117	0.5543	0.8312	1.1455	1.6103	1.9938	2.3425
6	0.3811	0.6757	0.8721	1.2373	1.6354	2.2041	2.6613	3.0701
7	0.5985	0.9893	1.2390	1.6899	2.1673	2.8331	3.3583	3.8223
8	0.8571	1.3444	1.6465	2.1797	2.7326	3.4895	4.0782	4.5936
9	1.1519	1.7349	2.0879	2.7004	3.3251	4.1682	4.8165	5.3801
10	1.4787	2.1559	2.5582	3.2470	3.9403	4.8652	5.5701	6.1791
11	1.8339	2.6032	3.0535	3.8157	4.5748	5.5778	6.3364	6.9887
12	2.2142	3.0738	3.5706	4.4038	5.2260	6.3038	7.1138	7.8073
13	2.6172	3.5650	4.1069	5.0088	5.8919	7.0415	7.9008	8.6339
14	3.0407	4.0747	4.6604	5.6287	6.5706	7.7895	8.6963	9.4673
15	3.4827	4.6009	5.2293	6.2621	7.2609	8.5468	9.4993	10.3070
16	3.9416	5.1422	5.8122	6.9077	7.9616	9.3122	10.3090	11.1521
17	4.4161	5.6972	6.4078	7.5642	8.6718	10.0852	11.1249	12.0023
18	4.9048	6.2648	7.0149	8.2307	9.3905	10.8649	11.9463	12.8570
19	5.4068	6.8440	7.6327	8.9065	10.1170	11.6509	12.7727	13.7158
20	5.9210	7.4338	8.2604	9.5908	10.8508	12.4426	13.6039	14.5784
21	6.4467	8.0337	8.8972	10.2829	11.5913	13.2396	14.4393	15.4446
22	6.9830	8.6427	9.5425	10.9823	12.3380	14.0415	15.2788	16.3140
23	7.5292	9.2604	10.1957	11.6886	13.0905	14.8480	16.1219	17.1865
24	8.0849	9.8862	10.8564	12.4012	13.8484	15.6587	16.9686	18.0618
25	8.6493	10.5197	11.5240	13.1197	14.6114	16.4734	17.8184	18.9398
26	9.2221	11.1602	12.1981	13.8439	15.3792	17.2919	18.6714	19.8202
27	9.8028	11.8076	12.8785	14.5734	16.1514	18.1139	19.5272	20.7030
28	10.3909	12.4613	13.5647	15.3079	16.9279	18.9392	20.3857	21.5880
29	10.9861	13.1211	14.2565	16.0471	17.7084	19.7677	21.2468	22.4751
30	11.5880	13.7867	14.9535	16.7908	18.4927	20.5992	22.1103	23.3641
40	17.9164	20.7065	22.1643	24.4330	26.5093	29.0505	30.8563	32.3450
60	31.7383	35.5345	37.4849	40.4817	43.1880	46.4589	48.7587	50.6406
80	46.5199	51.1719	53.5401	57.1532	60.3915	64.2778	66.9938	69.2069
100	61.9179	67.3276	70.0649	74.2219	77.9295	82.3581	85.4406	87.9453

TABLE   C.3 (continued)
Critical values of χ^2-distributions

df	$\alpha =$ 0.20	0.15	0.10	0.05	0.025	0.01	0.005	0.001
1	1.6424	2.0723	2.7055	3.8415	5.0239	6.6349	7.8794	10.8276
2	3.2189	3.7942	4.6052	5.9915	7.3778	9.2103	10.5966	13.8155
3	4.6416	5.3170	6.2514	7.8147	9.3484	11.3449	12.8382	16.2662
4	5.9886	6.7449	7.7794	9.4877	11.1433	13.2767	14.8603	18.4668
5	7.2893	8.1152	9.2364	11.0705	12.8325	15.0863	16.7496	20.5150
6	8.5581	9.4461	10.6446	12.5916	14.4494	16.8119	18.5476	22.4577
7	9.8032	10.7479	12.0170	14.0671	16.0128	18.4753	20.2777	24.3219
8	11.0301	12.0271	13.3616	15.5073	17.5345	20.0902	21.9550	26.1245
9	12.2421	13.2880	14.6837	16.9190	19.0228	21.6660	23.5894	27.8772
10	13.4420	14.5339	15.9872	18.3070	20.4831	23.2092	25.1882	29.5883
11	14.6314	15.7671	17.2750	19.6751	21.9200	24.7249	26.7568	31.2641
12	15.8120	16.9893	18.5493	21.0260	23.3366	26.2169	28.2995	32.9095
13	16.9848	18.2020	19.8119	22.3620	24.7356	27.6882	29.8195	34.5282
14	18.1508	19.4062	21.0641	23.6848	26.1189	29.1412	31.3193	36.1233
15	19.3107	20.6030	22.3071	24.9958	27.4884	30.5779	32.8013	37.6973
16	20.4651	21.7930	23.5418	26.2962	28.8453	31.9999	34.2672	39.2524
17	21.6146	22.9770	24.7690	27.5871	30.1910	33.4087	35.7185	40.7902
18	22.7595	24.1555	25.9894	28.8693	31.5264	34.8053	37.1565	42.3124
19	23.9004	25.3288	27.2036	30.1435	32.8523	36.1909	38.5823	43.8202
20	25.0375	26.4976	28.4120	31.4104	34.1696	37.5662	39.9968	45.3147
21	26.1711	27.6620	29.6151	32.6706	35.4789	38.9322	41.4011	46.7970
22	27.3014	28.8224	30.8133	33.9244	36.7807	40.2894	42.7957	48.2679
23	28.4288	29.9792	32.0069	35.1725	38.0756	41.6384	44.1813	49.7282
24	29.5533	31.1325	33.1962	36.4150	39.3641	42.9798	45.5585	51.1786
25	30.6752	32.2825	34.3816	37.6525	40.6465	44.3141	46.9279	52.6197
26	31.7946	33.4295	35.5632	38.8851	41.9232	45.6417	48.2899	54.0520
27	32.9117	34.5736	36.7412	40.1133	43.1945	46.9629	49.6449	55.4760
28	34.0266	35.7150	37.9159	41.3371	44.4608	48.2782	50.9934	56.8923
29	35.1394	36.8538	39.0875	42.5570	45.7223	49.5879	52.3356	58.3012
30	36.2502	37.9902	40.2560	43.7730	46.9792	50.8922	53.6720	59.7031
40	47.2685	49.2438	51.8051	55.7585	59.3417	63.6907	66.7660	73.4020
60	68.9721	71.3411	74.3970	79.0819	83.2977	88.3794	91.9517	99.6072
80	90.4053	93.1058	96.5782	101.8795	106.6286	112.3288	116.3211	124.8392
100	111.6667	114.6588	118.4980	124.3421	129.5612	135.8067	140.1695	149.4493

TABLE C.4

Critical values of *F*-distributions ($df1$ = numerator df, $df2$ = denominator df)

$df2$	α	1	2	3	4	5	6	7	8	9	10
4	0.100	4.545	4.325	4.191	4.107	4.051	4.010	3.979	3.955	3.936	3.920
	0.050	7.709	6.944	6.591	6.388	6.256	6.163	6.094	6.041	5.999	5.964
	0.025	12.218	10.649	9.979	9.605	9.364	9.197	9.074	8.980	8.905	8.844
	0.010	21.198	18.000	16.694	15.977	15.522	15.207	14.976	14.799	14.659	14.546
5	0.100	4.060	3.780	3.619	3.520	3.453	3.405	3.368	3.339	3.316	3.297
	0.050	6.608	5.786	5.409	5.192	5.050	4.950	4.876	4.818	4.772	4.735
	0.025	10.007	8.434	7.764	7.388	7.146	6.978	6.853	6.757	6.681	6.619
	0.010	16.258	13.274	12.060	11.392	10.967	10.672	10.456	10.289	10.158	10.051
6	0.100	3.776	3.463	3.289	3.181	3.108	3.055	3.014	2.983	2.958	2.937
	0.050	5.987	5.143	4.757	4.534	4.387	4.284	4.207	4.147	4.099	4.060
	0.025	8.813	7.260	6.599	6.227	5.988	5.820	5.695	5.600	5.523	5.461
	0.010	13.745	10.925	9.780	9.148	8.746	8.466	8.260	8.102	7.976	7.874
7	0.100	3.589	3.257	3.074	2.961	2.883	2.827	2.785	2.752	2.725	2.703
	0.050	5.591	4.737	4.347	4.120	3.972	3.866	3.787	3.726	3.677	3.637
	0.025	8.073	6.542	5.890	5.523	5.285	5.119	4.995	4.899	4.823	4.761
	0.010	12.246	9.547	8.451	7.847	7.460	7.191	6.993	6.840	6.719	6.620
8	0.100	3.458	3.113	2.924	2.806	2.726	2.668	2.624	2.589	2.561	2.538
	0.050	5.318	4.459	4.066	3.838	3.687	3.581	3.500	3.438	3.388	3.347
	0.025	7.571	6.059	5.416	5.053	4.817	4.652	4.529	4.433	4.357	4.295
	0.010	11.259	8.649	7.591	7.006	6.632	6.371	6.178	6.029	5.911	5.814
9	0.100	3.360	3.006	2.813	2.693	2.611	2.551	2.505	2.469	2.440	2.416
	0.050	5.117	4.256	3.863	3.633	3.482	3.374	3.293	3.230	3.179	3.137
	0.025	7.209	5.715	5.078	4.718	4.484	4.320	4.197	4.102	4.026	3.964
	0.010	10.561	8.022	6.992	6.422	6.057	5.802	5.613	5.467	5.351	5.257
10	0.100	3.285	2.924	2.728	2.605	2.522	2.461	2.414	2.377	2.347	2.323
	0.050	4.965	4.103	3.708	3.478	3.326	3.217	3.135	3.072	3.020	2.978
	0.025	6.937	5.456	4.826	4.468	4.236	4.072	3.950	3.855	3.779	3.717
	0.010	10.044	7.559	6.552	5.994	5.636	5.386	5.200	5.057	4.942	4.849
11	0.100	3.225	2.860	2.660	2.536	2.451	2.389	2.342	2.304	2.274	2.248
	0.050	4.844	3.982	3.587	3.357	3.204	3.095	3.012	2.948	2.896	2.854
	0.025	6.724	5.256	4.630	4.275	4.044	3.881	3.759	3.664	3.588	3.526
	0.010	9.646	7.206	6.217	5.668	5.316	5.069	4.886	4.744	4.632	4.539
12	0.100	3.177	2.807	2.606	2.480	2.394	2.331	2.283	2.245	2.214	2.188
	0.050	4.747	3.885	3.490	3.259	3.106	2.996	2.913	2.849	2.796	2.753
	0.025	6.554	5.096	4.474	4.121	3.891	3.728	3.607	3.512	3.436	3.374
	0.010	9.330	6.927	5.953	5.412	5.064	4.821	4.640	4.499	4.388	4.296

TABLE   C.4 (continued)
Critical values of *F*-distributions

*df*2	α	11	12	13	14	15	16	17	18	19	20
4	0.100	3.907	3.896	3.886	3.878	3.870	3.864	3.858	3.853	3.849	3.844
	0.050	5.936	5.912	5.891	5.873	5.858	5.844	5.832	5.821	5.811	5.803
	0.025	8.794	8.751	8.715	8.684	8.657	8.633	8.611	8.592	8.575	8.560
	0.010	14.452	14.374	14.307	14.249	14.198	14.154	14.115	14.080	14.048	14.020
5	0.100	3.282	3.268	3.257	3.247	3.238	3.230	3.223	3.217	3.212	3.207
	0.050	4.704	4.678	4.655	4.636	4.619	4.604	4.590	4.579	4.568	4.558
	0.025	6.568	6.525	6.488	6.456	6.428	6.403	6.381	6.362	6.344	6.329
	0.010	9.963	9.888	9.825	9.770	9.722	9.680	9.643	9.610	9.580	9.553
6	0.100	2.920	2.905	2.892	2.881	2.871	2.863	2.855	2.848	2.842	2.836
	0.050	4.027	4.000	3.976	3.956	3.938	3.922	3.908	3.896	3.884	3.874
	0.025	5.410	5.366	5.329	5.297	5.269	5.244	5.222	5.202	5.184	5.168
	0.010	7.790	7.718	7.657	7.605	7.559	7.519	7.483	7.451	7.422	7.396
7	0.100	2.684	2.668	2.654	2.643	2.632	2.623	2.615	2.607	2.601	2.595
	0.050	3.603	3.575	3.550	3.529	3.511	3.494	3.480	3.467	3.455	3.445
	0.025	4.709	4.666	4.628	4.596	4.568	4.543	4.521	4.501	4.483	4.467
	0.010	6.538	6.469	6.410	6.359	6.314	6.275	6.240	6.209	6.181	6.155
8	0.100	2.519	2.502	2.488	2.475	2.464	2.455	2.446	2.438	2.431	2.425
	0.050	3.313	3.284	3.259	3.237	3.218	3.202	3.187	3.173	3.161	3.150
	0.025	4.243	4.200	4.162	4.130	4.101	4.076	4.054	4.034	4.016	3.999
	0.010	5.734	5.667	5.609	5.559	5.515	5.477	5.442	5.412	5.384	5.359
9	0.100	2.396	2.379	2.364	2.351	2.340	2.329	2.320	2.312	2.305	2.298
	0.050	3.102	3.073	3.048	3.025	3.006	2.989	2.974	2.960	2.948	2.936
	0.025	3.912	3.868	3.831	3.798	3.769	3.744	3.722	3.701	3.683	3.667
	0.010	5.178	5.111	5.055	5.005	4.962	4.924	4.890	4.860	4.833	4.808
10	0.100	2.302	2.284	2.269	2.255	2.244	2.233	2.224	2.215	2.208	2.201
	0.050	2.943	2.913	2.887	2.865	2.845	2.828	2.812	2.798	2.785	2.774
	0.025	3.665	3.621	3.583	3.550	3.522	3.496	3.474	3.453	3.435	3.419
	0.010	4.772	4.706	4.650	4.601	4.558	4.520	4.487	4.457	4.430	4.405
11	0.100	2.227	2.209	2.193	2.179	2.167	2.156	2.147	2.138	2.130	2.123
	0.050	2.818	2.788	2.761	2.739	2.719	2.701	2.685	2.671	2.658	2.646
	0.025	3.474	3.430	3.392	3.359	3.330	3.304	3.282	3.261	3.243	3.226
	0.010	4.462	4.397	4.342	4.293	4.251	4.213	4.180	4.150	4.123	4.099
12	0.100	2.166	2.147	2.131	2.117	2.105	2.094	2.084	2.075	2.067	2.060
	0.050	2.717	2.687	2.660	2.637	2.617	2.599	2.583	2.568	2.555	2.544
	0.025	3.321	3.277	3.239	3.206	3.177	3.152	3.129	3.108	3.090	3.073
	0.010	4.220	4.155	4.100	4.052	4.010	3.972	3.939	3.909	3.883	3.858

*df*1

TABLE C.4 (continued)
Critical values of F-distributions

$df2$	α	1	2	3	4	5	6	7	8	9	10
13	0.100	3.136	2.763	2.560	2.434	2.347	2.283	2.234	2.195	2.164	2.138
	0.050	4.667	3.806	3.411	3.179	3.025	2.915	2.832	2.767	2.714	2.671
	0.025	6.414	4.965	4.347	3.996	3.767	3.604	3.483	3.388	3.312	3.250
	0.010	9.074	6.701	5.739	5.205	4.862	4.620	4.441	4.302	4.191	4.100
14	0.100	3.102	2.726	2.522	2.395	2.307	2.243	2.193	2.154	2.122	2.095
	0.050	4.600	3.739	3.344	3.112	2.958	2.848	2.764	2.699	2.646	2.602
	0.025	6.298	4.857	4.242	3.892	3.663	3.501	3.380	3.285	3.209	3.147
	0.010	8.862	6.515	5.564	5.035	4.695	4.456	4.278	4.140	4.030	3.939
15	0.100	3.073	2.695	2.490	2.361	2.273	2.208	2.158	2.119	2.086	2.059
	0.050	4.543	3.682	3.287	3.056	2.901	2.790	2.707	2.641	2.588	2.544
	0.025	6.200	4.765	4.153	3.804	3.576	3.415	3.293	3.199	3.123	3.060
	0.010	8.683	6.359	5.417	4.893	4.556	4.318	4.142	4.004	3.895	3.805
16	0.100	3.048	2.668	2.462	2.333	2.244	2.178	2.128	2.088	2.055	2.028
	0.050	4.494	3.634	3.239	3.007	2.852	2.741	2.657	2.591	2.538	2.494
	0.025	6.115	4.687	4.077	3.729	3.502	3.341	3.219	3.125	3.049	2.986
	0.010	8.531	6.226	5.292	4.773	4.437	4.202	4.026	3.890	3.780	3.691
17	0.100	3.026	2.645	2.437	2.308	2.218	2.152	2.102	2.061	2.028	2.001
	0.050	4.451	3.592	3.197	2.965	2.810	2.699	2.614	2.548	2.494	2.450
	0.025	6.042	4.619	4.011	3.665	3.438	3.277	3.156	3.061	2.985	2.922
	0.010	8.400	6.112	5.185	4.669	4.336	4.102	3.927	3.791	3.682	3.593
18	0.100	3.007	2.624	2.416	2.286	2.196	2.130	2.079	2.038	2.005	1.977
	0.050	4.414	3.555	3.160	2.928	2.773	2.661	2.577	2.510	2.456	2.412
	0.025	5.978	4.560	3.954	3.608	3.382	3.221	3.100	3.005	2.929	2.866
	0.010	8.285	6.013	5.092	4.579	4.248	4.015	3.841	3.705	3.597	3.508
19	0.100	2.990	2.606	2.397	2.266	2.176	2.109	2.058	2.017	1.984	1.956
	0.050	4.381	3.522	3.127	2.895	2.740	2.628	2.544	2.477	2.423	2.378
	0.025	5.922	4.508	3.903	3.559	3.333	3.172	3.051	2.956	2.880	2.817
	0.010	8.185	5.926	5.010	4.500	4.171	3.939	3.765	3.631	3.523	3.434
20	0.100	2.975	2.589	2.380	2.249	2.158	2.091	2.040	1.999	1.965	1.937
	0.050	4.351	3.493	3.098	2.866	2.711	2.599	2.514	2.447	2.393	2.348
	0.025	5.871	4.461	3.859	3.515	3.289	3.128	3.007	2.913	2.837	2.774
	0.010	8.096	5.849	4.938	4.431	4.103	3.871	3.699	3.564	3.457	3.368
21	0.100	2.961	2.575	2.365	2.233	2.142	2.075	2.023	1.982	1.948	1.920
	0.050	4.325	3.467	3.072	2.840	2.685	2.573	2.488	2.420	2.366	2.321
	0.025	5.827	4.420	3.819	3.475	3.250	3.090	2.969	2.874	2.798	2.735
	0.010	8.017	5.780	4.874	4.369	4.042	3.812	3.640	3.506	3.398	3.310

TABLE C.4 (continued)
Critical values of F-distributions

$df2$	α	11	12	13	14	15	16	17	18	19	20
							$df1$				
13	0.100	2.116	2.097	2.080	2.066	2.053	2.042	2.032	2.023	2.014	2.007
	0.050	2.635	2.604	2.577	2.554	2.533	2.515	2.499	2.484	2.471	2.459
	0.025	3.197	3.153	3.115	3.082	3.053	3.027	3.004	2.983	2.965	2.948
	0.010	4.025	3.960	3.905	3.857	3.815	3.778	3.745	3.716	3.689	3.665
14	0.100	2.073	2.054	2.037	2.022	2.010	1.998	1.988	1.978	1.970	1.962
	0.050	2.565	2.534	2.507	2.484	2.463	2.445	2.428	2.413	2.400	2.388
	0.025	3.095	3.050	3.012	2.979	2.949	2.923	2.900	2.879	2.861	2.844
	0.010	3.864	3.800	3.745	3.698	3.656	3.619	3.586	3.556	3.529	3.505
15	0.100	2.037	2.017	2.000	1.985	1.972	1.961	1.950	1.941	1.932	1.924
	0.050	2.507	2.475	2.448	2.424	2.403	2.385	2.368	2.353	2.340	2.328
	0.025	3.008	2.963	2.925	2.891	2.862	2.836	2.813	2.792	2.773	2.756
	0.010	3.730	3.666	3.612	3.564	3.522	3.485	3.452	3.423	3.396	3.372
16	0.100	2.005	1.985	1.968	1.953	1.940	1.928	1.917	1.908	1.899	1.891
	0.050	2.456	2.425	2.397	2.373	2.352	2.333	2.317	2.302	2.288	2.276
	0.025	2.934	2.889	2.851	2.817	2.788	2.761	2.738	2.717	2.698	2.681
	0.010	3.616	3.553	3.498	3.451	3.409	3.372	3.339	3.310	3.283	3.259
17	0.100	1.978	1.958	1.940	1.925	1.912	1.900	1.889	1.879	1.870	1.862
	0.050	2.413	2.381	2.353	2.329	2.308	2.289	2.272	2.257	2.243	2.230
	0.025	2.870	2.825	2.786	2.753	2.723	2.697	2.673	2.652	2.633	2.616
	0.010	3.519	3.455	3.401	3.353	3.312	3.275	3.242	3.212	3.186	3.162
18	0.100	1.954	1.933	1.916	1.900	1.887	1.875	1.864	1.854	1.845	1.837
	0.050	2.374	2.342	2.314	2.290	2.269	2.250	2.233	2.217	2.203	2.191
	0.025	2.814	2.769	2.730	2.696	2.667	2.640	2.617	2.596	2.576	2.559
	0.010	3.434	3.371	3.316	3.269	3.227	3.190	3.158	3.128	3.101	3.077
19	0.100	1.932	1.912	1.894	1.878	1.865	1.852	1.841	1.831	1.822	1.814
	0.050	2.340	2.308	2.280	2.256	2.234	2.215	2.198	2.182	2.168	2.155
	0.025	2.765	2.720	2.681	2.647	2.617	2.591	2.567	2.546	2.526	2.509
	0.010	3.360	3.297	3.242	3.195	3.153	3.116	3.084	3.054	3.027	3.003
20	0.100	1.913	1.892	1.875	1.859	1.845	1.833	1.821	1.811	1.802	1.794
	0.050	2.310	2.278	2.250	2.225	2.203	2.184	2.167	2.151	2.137	2.124
	0.025	2.721	2.676	2.637	2.603	2.573	2.547	2.523	2.501	2.482	2.464
	0.010	3.294	3.231	3.177	3.130	3.088	3.051	3.018	2.989	2.962	2.938
21	0.100	1.896	1.875	1.857	1.841	1.827	1.815	1.803	1.793	1.784	1.776
	0.050	2.283	2.250	2.222	2.197	2.176	2.156	2.139	2.123	2.109	2.096
	0.025	2.682	2.637	2.598	2.564	2.534	2.507	2.483	2.462	2.442	2.425
	0.010	3.236	3.173	3.119	3.072	3.030	2.993	2.960	2.931	2.904	2.880

TABLE   C.4 (continued)
Critical values of *F*-distributions

df2	α	1	2	3	4	5	6	7	8	9	10
							df1				
22	0.100	2.949	2.561	2.351	2.219	2.128	2.060	2.008	1.967	1.933	1.904
	0.050	4.301	3.443	3.049	2.817	2.661	2.549	2.464	2.397	2.342	2.297
	0.025	5.786	4.383	3.783	3.440	3.215	3.055	2.934	2.839	2.763	2.700
	0.010	7.945	5.719	4.817	4.313	3.988	3.758	3.587	3.453	3.346	3.258
24	0.100	2.927	2.538	2.327	2.195	2.103	2.035	1.983	1.941	1.906	1.877
	0.050	4.260	3.403	3.009	2.776	2.621	2.508	2.423	2.355	2.300	2.255
	0.025	5.717	4.319	3.721	3.379	3.155	2.995	2.874	2.779	2.703	2.640
	0.010	7.823	5.614	4.718	4.218	3.895	3.667	3.496	3.363	3.256	3.168
26	0.100	2.909	2.519	2.307	2.174	2.082	2.014	1.961	1.919	1.884	1.855
	0.050	4.225	3.369	2.975	2.743	2.587	2.474	2.388	2.321	2.265	2.220
	0.025	5.659	4.265	3.670	3.329	3.105	2.945	2.824	2.729	2.653	2.590
	0.010	7.721	5.526	4.637	4.140	3.818	3.591	3.421	3.288	3.182	3.094
28	0.100	2.894	2.503	2.291	2.157	2.064	1.996	1.943	1.900	1.865	1.836
	0.050	4.196	3.340	2.947	2.714	2.558	2.445	2.359	2.291	2.236	2.190
	0.025	5.610	4.221	3.626	3.286	3.063	2.903	2.782	2.687	2.611	2.547
	0.010	7.636	5.453	4.568	4.074	3.754	3.528	3.358	3.226	3.120	3.032
30	0.100	2.881	2.489	2.276	2.142	2.049	1.980	1.927	1.884	1.849	1.819
	0.050	4.171	3.316	2.922	2.690	2.534	2.421	2.334	2.266	2.211	2.165
	0.025	5.568	4.182	3.589	3.250	3.026	2.867	2.746	2.651	2.575	2.511
	0.010	7.562	5.390	4.510	4.018	3.699	3.473	3.304	3.173	3.067	2.979
40	0.100	2.835	2.440	2.226	2.091	1.997	1.927	1.873	1.829	1.793	1.763
	0.050	4.085	3.232	2.839	2.606	2.449	2.336	2.249	2.180	2.124	2.077
	0.025	5.424	4.051	3.463	3.126	2.904	2.744	2.624	2.529	2.452	2.388
	0.010	7.314	5.179	4.313	3.828	3.514	3.291	3.124	2.993	2.888	2.801
50	0.100	2.809	2.412	2.197	2.061	1.966	1.895	1.840	1.796	1.760	1.729
	0.050	4.034	3.183	2.790	2.557	2.400	2.286	2.199	2.130	2.073	2.026
	0.025	5.340	3.975	3.390	3.054	2.833	2.674	2.553	2.458	2.381	2.317
	0.010	7.171	5.057	4.199	3.720	3.408	3.186	3.020	2.890	2.785	2.698
60	0.100	2.791	2.393	2.177	2.041	1.946	1.875	1.819	1.775	1.738	1.707
	0.050	4.001	3.150	2.758	2.525	2.368	2.254	2.167	2.097	2.040	1.993
	0.025	5.286	3.925	3.343	3.008	2.786	2.627	2.507	2.412	2.334	2.270
	0.010	7.077	4.977	4.126	3.649	3.339	3.119	2.953	2.823	2.718	2.632
120	0.100	2.748	2.347	2.130	1.992	1.896	1.824	1.767	1.722	1.684	1.652
	0.050	3.920	3.072	2.680	2.447	2.290	2.175	2.087	2.016	1.959	1.910
	0.025	5.152	3.805	3.227	2.894	2.674	2.515	2.395	2.299	2.222	2.157
	0.010	6.851	4.787	3.949	3.480	3.174	2.956	2.792	2.663	2.559	2.472

TABLE C.4 (continued)
Critical values of F-distributions

$df2$	α	11	12	13	14	15	16	17	18	19	20
						$df1$					
22	0.100	1.880	1.859	1.841	1.825	1.811	1.798	1.787	1.777	1.768	1.759
	0.050	2.259	2.226	2.198	2.173	2.151	2.131	2.114	2.098	2.084	2.071
	0.025	2.647	2.602	2.563	2.528	2.498	2.472	2.448	2.426	2.407	2.389
	0.010	3.184	3.121	3.067	3.019	2.978	2.941	2.908	2.879	2.852	2.827
24	0.100	1.853	1.832	1.814	1.797	1.783	1.770	1.759	1.748	1.739	1.730
	0.050	2.216	2.183	2.155	2.130	2.108	2.088	2.070	2.054	2.040	2.027
	0.025	2.586	2.541	2.502	2.468	2.437	2.411	2.386	2.365	2.345	2.327
	0.010	3.094	3.032	2.977	2.930	2.889	2.852	2.819	2.789	2.762	2.738
26	0.100	1.830	1.809	1.790	1.774	1.760	1.747	1.735	1.724	1.715	1.706
	0.050	2.181	2.148	2.119	2.094	2.072	2.052	2.034	2.018	2.003	1.990
	0.025	2.536	2.491	2.451	2.417	2.387	2.360	2.335	2.314	2.294	2.276
	0.010	3.021	2.958	2.904	2.857	2.815	2.778	2.745	2.715	2.688	2.664
28	0.100	1.811	1.790	1.771	1.754	1.740	1.726	1.715	1.704	1.694	1.685
	0.050	2.151	2.118	2.089	2.064	2.041	2.021	2.003	1.987	1.972	1.959
	0.025	2.494	2.448	2.409	2.374	2.344	2.317	2.292	2.270	2.251	2.232
	0.010	2.959	2.896	2.842	2.795	2.753	2.716	2.683	2.653	2.626	2.602
30	0.100	1.794	1.773	1.754	1.737	1.722	1.709	1.697	1.686	1.676	1.667
	0.050	2.126	2.092	2.063	2.037	2.015	1.995	1.976	1.960	1.945	1.932
	0.025	2.458	2.412	2.372	2.338	2.307	2.280	2.255	2.233	2.213	2.195
	0.010	2.906	2.843	2.789	2.742	2.700	2.663	2.630	2.600	2.573	2.549
40	0.100	1.737	1.715	1.695	1.678	1.662	1.649	1.636	1.625	1.615	1.605
	0.050	2.038	2.003	1.974	1.948	1.924	1.904	1.885	1.868	1.853	1.839
	0.025	2.334	2.288	2.248	2.213	2.182	2.154	2.129	2.107	2.086	2.068
	0.010	2.727	2.665	2.611	2.563	2.522	2.484	2.451	2.421	2.394	2.369
50	0.100	1.703	1.680	1.660	1.643	1.627	1.613	1.600	1.588	1.578	1.568
	0.050	1.986	1.952	1.921	1.895	1.871	1.850	1.831	1.814	1.798	1.784
	0.025	2.263	2.216	2.176	2.140	2.109	2.081	2.056	2.033	2.012	1.993
	0.010	2.625	2.562	2.508	2.461	2.419	2.382	2.348	2.318	2.290	2.265
60	0.100	1.680	1.657	1.637	1.619	1.603	1.589	1.576	1.564	1.553	1.543
	0.050	1.952	1.917	1.887	1.860	1.836	1.815	1.796	1.778	1.763	1.748
	0.025	2.216	2.169	2.129	2.093	2.061	2.033	2.008	1.985	1.964	1.944
	0.010	2.559	2.496	2.442	2.394	2.352	2.315	2.281	2.251	2.223	2.198
120	0.100	1.625	1.601	1.580	1.562	1.545	1.530	1.516	1.504	1.493	1.482
	0.050	1.869	1.834	1.803	1.775	1.750	1.728	1.709	1.690	1.674	1.659
	0.025	2.102	2.055	2.014	1.977	1.945	1.916	1.890	1.866	1.845	1.825
	0.010	2.399	2.336	2.282	2.234	2.192	2.154	2.119	2.089	2.060	2.035

TABLE C.5

Probabilities associated with $B(n, .5)$ distributions $\Pr\{B(n, .5) \geq a\}$

a	7	8	9	10	11	12	n 13	14	15	16	17	18
6	0.062	0.145	0.254	0.377	0.500	0.613	0.709	0.788	0.849	0.895	0.928	0.952
7	0.008	0.035	0.090	0.172	0.274	0.387	0.500	0.605	0.696	0.773	0.834	0.881
8	0.000	0.004	0.020	0.055	0.113	0.194	0.291	0.395	0.500	0.598	0.685	0.760
9	0.000	0.000	0.002	0.011	0.033	0.073	0.133	0.212	0.304	0.402	0.500	0.593
10	0.000	0.000	0.000	0.001	0.006	0.019	0.046	0.090	0.151	0.227	0.315	0.407
11	0.000	0.000	0.000	0.000	0.000	0.003	0.011	0.029	0.059	0.105	0.166	0.240
12	0.000	0.000	0.000	0.000	0.000	0.000	0.002	0.006	0.018	0.038	0.072	0.119
13	0.000	0.000	0.000	0.000	0.000	0.000	0.000	0.001	0.004	0.011	0.025	0.048
14	0.000	0.000	0.000	0.000	0.000	0.000	0.000	0.000	0.000	0.002	0.006	0.015
15	0.000	0.000	0.000	0.000	0.000	0.000	0.000	0.000	0.000	0.000	0.001	0.004
16	0.000	0.000	0.000	0.000	0.000	0.000	0.000	0.000	0.000	0.000	0.000	0.001
17	0.000	0.000	0.000	0.000	0.000	0.000	0.000	0.000	0.000	0.000	0.000	0.000

a	19	20	21	22	23	n 24	25	26	27	28	29	30
10	0.500	0.588	0.668	0.738	0.798	0.846	0.885	0.916	0.939	0.956	0.969	0.979
11	0.324	0.412	0.500	0.584	0.661	0.729	0.788	0.837	0.876	0.908	0.932	0.951
12	0.180	0.252	0.332	0.416	0.500	0.581	0.655	0.721	0.779	0.828	0.868	0.900
13	0.084	0.132	0.192	0.262	0.339	0.419	0.500	0.577	0.649	0.714	0.771	0.819
14	0.032	0.058	0.095	0.143	0.202	0.271	0.345	0.423	0.500	0.575	0.644	0.708
15	0.010	0.021	0.039	0.067	0.105	0.154	0.212	0.279	0.351	0.425	0.500	0.572
16	0.002	0.006	0.013	0.026	0.047	0.076	0.115	0.163	0.221	0.286	0.356	0.428
17	0.000	0.001	0.004	0.008	0.017	0.032	0.054	0.084	0.124	0.172	0.229	0.292
18	0.000	0.000	0.001	0.002	0.005	0.011	0.022	0.038	0.061	0.092	0.132	0.181
19	0.000	0.000	0.000	0.000	0.001	0.003	0.007	0.014	0.026	0.044	0.068	0.100
20	0.000	0.000	0.000	0.000	0.000	0.001	0.002	0.005	0.010	0.018	0.031	0.049
21	0.000	0.000	0.000	0.000	0.000	0.000	0.000	0.001	0.003	0.006	0.012	0.021
22	0.000	0.000	0.000	0.000	0.000	0.000	0.000	0.000	0.001	0.002	0.004	0.008
23	0.000	0.000	0.000	0.000	0.000	0.000	0.000	0.000	0.000	0.000	0.001	0.003
24	0.000	0.000	0.000	0.000	0.000	0.000	0.000	0.000	0.000	0.000	0.000	0.001

TABLE   C.6

Critical values for Wilcoxon signed rank test

This table gives values of n, w^+, and α such that $\Pr\left\{W^+ \geq w^+\right\} = \alpha$.

n	w^+	α	n	w^+	α	n	w^+	α	n	w^+	α	n	w^+	α
4	9	0.125	10	40	0.116	12	60	0.055	13	86	0.001	15	85	0.084
	10	0.062		41	0.097		61	0.046		87	0.001		86	0.076
5	12	0.156		42	0.080		62	0.039		88	0.001		87	0.068
	13	0.094		43	0.065		63	0.032		89	0.000		88	0.060
	14	0.062		44	0.053		64	0.026		90	0.000		89	0.053
	15	0.031		45	0.042		65	0.021		91	0.000		90	0.047
6	17	0.109		46	0.032		66	0.017	14	73	0.108		91	0.042
	18	0.078		47	0.024		67	0.013		74	0.097		92	0.036
	19	0.047		48	0.019		68	0.010		75	0.086		93	0.032
	20	0.031		49	0.014		69	0.008		76	0.077		94	0.028
	21	0.016		50	0.010		70	0.006		77	0.068		95	0.024
7	22	0.109		51	0.007		71	0.005		78	0.059		96	0.021
	23	0.078		52	0.005		72	0.003		79	0.052		97	0.018
	24	0.055		53	0.003		73	0.002		80	0.045		98	0.015
	25	0.039		54	0.002		74	0.002		81	0.039		99	0.013
	26	0.023		55	0.001		75	0.001		82	0.034		100	0.011
	27	0.016	11	47	0.120		76	0.001		83	0.029		101	0.009
	28	0.008		48	0.103		77	0.000		84	0.025		102	0.008
8	27	0.125		49	0.087		78	0.000		85	0.021		103	0.006
	28	0.098		50	0.074	13	64	0.108		86	0.018		104	0.005
	29	0.074		51	0.062		65	0.095		87	0.015		105	0.004
	30	0.055		52	0.051		66	0.084		88	0.012		106	0.003
	31	0.039		53	0.042		67	0.073		89	0.010		107	0.003
	32	0.027		54	0.034		68	0.064		90	0.008		108	0.002
	33	0.020		55	0.027		69	0.055		91	0.007		109	0.002
	34	0.012		56	0.021		70	0.047		92	0.005		110	0.001
	35	0.008		57	0.016		71	0.040		93	0.004		111	0.001
	36	0.004		58	0.012		72	0.034		94	0.003		112	0.001
9	33	0.125		59	0.009		73	0.029		95	0.003		113	0.001
	34	0.102		60	0.007		74	0.024		96	0.002		114	0.000
	35	0.082		61	0.005		75	0.020		97	0.002		115	0.000
	36	0.064		62	0.003		76	0.016		98	0.001		116	0.000
	37	0.049		63	0.002		77	0.013		99	0.001		117	0.000
	38	0.037		64	0.001		78	0.011		100	0.001		118	0.000
	39	0.027		65	0.001		79	0.009		101	0.000		119	0.000
	40	0.020		66	0.000		80	0.007		102	0.000		120	0.000
	41	0.014	12	55	0.117		81	0.005		103	0.000			
	42	0.010		56	0.102		82	0.004		104	0.000			
	43	0.006		57	0.088		83	0.003		105	0.000			
	44	0.004		58	0.076		84	0.002	15	83	0.104			
	45	0.005		59	0.065		85	0.002		84	0.094			

Adapted from Table A.4 in M. Hollander and D. Wolf, *Nonparametric statistical methods.* New York: Wiley (1973).

TABLE C.7

Critical values for Wilcoxon rank sum test

This table gives values of n_1, n_2, w, and α such that $\Pr\{W \geq w\} = \alpha$.

n_1	n_2	w	α	n_1	n_2	w	α	n_1	n_2	w	α	n_1	n_2	w	α
3	3	14	0.100	5	4	27	0.056		5	47	0.009			93	0.023
		15	0.050			28	0.032	8	5	47	0.047			96	0.010
4	3	16	0.114			30	0.008			49	0.023	10	8	95	0.051
		17	0.057	6	4	30	0.057			51	0.009			98	0.027
		18	0.029			31	0.033	9	5	50	0.056			102	0.010
5	3	19	0.071			33	0.010			53	0.021	9	9	105	0.047
		20	0.036	7	4	33	0.055			55	0.009			108	0.025
		21	0.018			35	0.021	10	5	54	0.050			112	0.009
6	3	22	0.048			36	0.012			56	0.028	10	9	111	0.047
		23	0.024	8	4	36	0.055			59	0.010			114	0.027
		24	0.012			38	0.024	6	6	50	0.047			118	0.011
7	3	24	0.058			40	0.008			52	0.021	10	10	127	0.053
		25	0.033	9	4	39	0.053			54	0.008			131	0.026
		27	0.007			41	0.025	7	6	54	0.051			136	0.009
8	3	27	0.042			43	0.010			56	0.026				
		28	0.024	10	4	42	0.053			58	0.011				
		29	0.012			44	0.027	8	6	58	0.054				
9	3	29	0.050			46	0.012			61	0.021				
		30	0.032	11	4	45	0.052			63	0.010				
		32	0.009			47	0.028	9	6	62	0.057				
10	3	31	0.056			49	0.013			65	0.025				
		33	0.024	12	4	48	0.052			68	0.009				
		34	0.014			51	0.021	10	6	67	0.047				
11	3	34	0.044			53	0.010			69	0.028				
		35	0.030	13	4	51	0.051			72	0.011				
		37	0.011			54	0.022	7	7	66	0.049				
12	3	36	0.051			56	0.011			68	0.027				
		38	0.024	14	4	54	0.051			71	0.009				
		40	0.009			57	0.023	8	7	71	0.047				
13	3	38	0.055			60	0.009			73	0.027				
		40	0.029	15	4	57	0.050			76	0.010				
		42	0.012			60	0.024	9	7	76	0.045				
14	3	41	0.046			63	0.010			78	0.027				
		43	0.024	5	5	36	0.048			81	0.011				
		45	0.010			37	0.028	10	7	80	0.054				
15	3	43	0.050			39	0.008			83	0.028				
		45	0.028	6	5	40	0.041			87	0.009				
		48	0.009			41	0.026	8	8	84	0.052				
4	4	24	0.057			43	0.009			87	0.025				
		25	0.029	7	5	43	0.053			90	0.010				
		26	0.014			45	0.024	9	8	89	0.057				

Adapted from Table A.6 in M. Hollander and D. Wolf, *Nonparametric statistical methods*. New York: Wiley (1973).

TABLE   C.8

Confidence intervals for proportions

For given values of n and f, the table gives $\hat{\pi}_L$, $\hat{\pi}_U$, the $100(1-\alpha)\%$ confidence limits for π.
If the table does not contain the required combination of n and f, then enter the table with n
and $n-f$; the corresponding confidence limits are the confidence limits for $1-\pi$.

n	f	$1-\alpha=0.95$ $\hat{\pi}_L$	$\hat{\pi}_U$	$1-\alpha=0.99$ $\hat{\pi}_L$	$\hat{\pi}_U$	n	f	$1-\alpha=0.95$ $\hat{\pi}_L$	$\hat{\pi}_U$	$1-\alpha=0.99$ $\hat{\pi}_L$	$\hat{\pi}_U$
5	0	0.0000	0.5218	0.0000	0.6534	12	0	0.0000	0.2209	0.0000	0.3187
	1	0.0051	0.7164	0.0010	0.8149		1	0.0221	0.3848	0.0004	0.4771
	2	0.0527	0.8534	0.9220	0.9172		2	0.0209	0.4841	0.0090	0.5730
6	0	0.0000	0.4593	0.0000	0.5865		3	0.0549	0.5718	0.0303	0.6553
	1	0.0042	0.6412	0.0008	0.7460		4	0.0992	0.6511	0.0624	0.7275
	2	0.0433	0.7772	0.0187	0.8564		5	0.1516	0.7233	0.1034	0.7915
	3	0.1181	0.8819	0.0663	0.9337		6	0.2109	0.7891	0.1522	0.8478
7	0	0.0000	0.4096	0.0000	0.5309	13	0	0.0000	0.2058	0.0000	0.2983
	1	0.0036	0.5787	0.0007	0.6849		1	0.0019	0.3603	0.0004	0.4490
	2	0.0367	0.7096	0.0158	0.7970		2	0.0192	0.4545	0.0083	0.5410
	3	0.0990	0.8159	0.0553	0.8823		3	0.0504	0.5381	0.0278	0.6206
8	0	0.0000	0.3694	0.0000	0.4843		4	0.0909	0.6143	0.0571	0.6913
	1	0.0032	0.5265	0.0006	0.6315		5	0.1386	0.6842	0.0942	0.7546
	2	0.0319	0.6509	0.0137	0.7422		6	0.1922	0.7487	0.1383	0.8113
	3	0.0852	0.7551	0.0475	0.8303	14	0	0.0000	0.1926	0.0000	0.2803
	4	0.1570	0.8430	0.0999	0.9001		1	0.0018	0.3387	0.0004	0.4241
9	0	0.0000	0.3363	0.0000	0.4450		2	0.0178	0.4281	0.0076	0.5123
	1	0.0028	0.4825	0.0006	0.5850		3	0.0466	0.5080	0.0257	0.5891
	2	0.0281	0.6001	0.0121	0.6926		4	0.0839	0.5811	0.0526	0.6580
	3	0.0749	0.7007	0.0416	0.7809		5	0.1276	0.6486	0.0866	0.7202
	4	0.1370	0.7880	0.0868	0.8539		6	0.1766	0.7144	0.1724	0.8276
10	0	0.0000	0.3085	0.0000	0.4113		7	0.2303	0.7697	0.2052	0.8413
	1	0.0025	0.4450	0.0005	0.5443	15	0	0.0000	0.1810	0.0000	0.2644
	2	0.0252	0.5561	0.0109	0.6482		1	0.0217	0.3195	0.0003	0.4016
	3	0.0667	0.6525	0.0370	0.7351		2	0.0166	0.4046	0.0071	0.4863
	4	0.1216	0.7376	0.0768	0.8091		3	0.0433	0.4809	0.0239	0.5605
	5	0.1871	0.8129	0.1283	0.8717		4	0.0779	0.5510	0.0488	0.6273
11	0	0.0000	0.2384	0.0000	0.3420		5	0.1182	0.6162	0.0801	0.6882
	1	0.0023	0.4128	0.0005	0.5085		6	0.1634	0.6771	0.1169	0.7439
	2	0.0228	0.5178	0.0098	0.6084		7	0.2127	0.7341	0.1587	0.7948
	3	0.0602	0.6097	0.0333	0.6934						
	4	0.1093	0.6921	0.0689	0.7668						
	5	0.1675	0.7662	0.1145	0.8307						

T A B L E   C.9
Critical values for Kruskal-Wallis test

Sample sizes			Critical value	α	Sample sizes			Critical value	α
n_1	n_2	n_3			n_1	n_2	n_3		
2	1	1	2.7000	0.500				4.7000	0.101
2	2	1	3.6000	0.200	4	4	1	6.6667	0.010
2	2	2	4.5714	0.067				6.1667	0.022
			3.7143	0.200				4.9667	0.048
3	1	1	3.2000	0.300				4.8667	0.054
3	2	1	4.2857	0.100				4.1667	0.082
			3.8571	0.133				4.0667	0.102
3	2	2	5.3572	0.029	4	4	2	7.0364	0.006
			4.7143	0.048				6.8727	0.011
			4.5000	0.067				5.4545	0.046
			4.4643	0.105				5.2364	0.052
3	3	1	5.1429	0.043				4.5545	0.098
			4.5714	0.100				4.4455	0.103
			4.0000	0.129	4	4	3	7.1439	0.010
3	3	2	6.2500	0.011				7.1364	0.011
			5.3611	0.032				5.5985	0.049
			5.1389	0.061				5.5758	0.051
			4.5556	0.100				4.5455	0.099
			4.2500	0.121				4.4773	0.102
3	3	3	7.2000	0.004	4	4	4	7.6538	0.008
			6.4889	0.011				7.5385	0.011
			5.6889	0.029				5.6923	0.049
			5.6000	0.050				5.6538	0.054
			5.0667	0.086				4.6539	0.097
			4.6222	0.100				4.5001	0.104
4	1	1	3.5714	0.200	5	1	1	3.8571	0.143
4	2	1	4.8214	0.057	5	2	1	5.2500	0.036
			4.5000	0.076				5.0000	0.048
			4.0179	0.114				4.4500	0.071
4	2	2	6.0000	0.014				4.2000	0.095
			5.3333	0.033				4.0500	0.119
			5.1250	0.052	5	2	2	6.5333	0.008
			4.4583	0.100				6.1333	0.013
			4.1667	0.105				5.1600	0.034
4	3	1	5.8333	0.021				5.0400	0.056
			5.2083	0.050				4.3733	0.090
			5.0000	0.057				4.2933	0.122
			4.0556	0.093	5	3	1	6.4000	0.012
			3.8889	0.129				4.9600	0.048
4	3	2	6.4444	0.008				4.8711	0.052
			6.3000	0.011				4.0178	0.095
			5.4444	0.046				3.8400	0.123
			5.4000	0.051	5	3	2	6.9091	0.009
			4.5111	0.098				6.8218	0.010
			4.4444	0.102				5.2509	0.049
4	3	3	6.7455	0.010				5.1055	0.052
			6.7091	0.013				4.6509	0.091
			5.7909	0.046				4.4945	0.101
			5.7273	0.050	5	3	3	7.0788	0.009
			4.7091	0.092				6.9818	0.011

TABLE **C.9** (continued)
Critical values for Kruskal-Wallis test

n_1	n_2	n_3	Critical value	α	n_1	n_2	n_3	Critical value	α
5	3	3	5.6485	0.049	5	5	1	6.8364	0.011
			5.5152	0.051				5.1273	0.046
			4.5333	0.097				4.9091	0.053
			4.4121	0.109				4.1091	0.086
5	4	1	6.9545	0.008				4.0364	0.105
			6.8400	0.011	5	5	2	7.3385	0.010
			4.9855	0.044				7.2692	0.010
			4.8600	0.056				5.3385	0.047
			3.9873	0.098				5.2462	0.051
			3.9600	0.102				4.6231	0.097
5	4	2	7.2045	0.009				4.5077	0.100
			7.1182	0.010	5	5	3	7.5780	0.010
			5.2727	0.049				7.5429	0.010
			5.2682	0.050				5.7055	0.046
			4.5409	0.098				5.6264	0.051
			4.5182	0.101				4.5451	0.100
5	4	3	7.4449	0.010				4.5363	0.102
			7.3949	0.011	5	5	4	7.8229	0.010
			5.6564	0.049				7.7914	0.010
			5.6308	0.050				5.6657	0.049
			4.5487	0.099				5.6429	0.050
			4.5231	0.103				4.5229	0.099
5	4	4	7.7604	0.009				4.5200	0.101
			7.7440	0.011	5	5	5	8.0000	0.009
			5.6571	0.049				7.9800	0.010
			5.6176	0.050				5.7800	0.049
			4.6187	0.100				5.6600	0.051
			4.5527	0.102				4.5600	0.100
5	5	1	7.3091	0.009				4.5000	0.102

From W. H. Kruskal and W. A. Wallis, Use of ranks in one-criterion analysis of variance. *Journal of the American Statistical Association, 47*, 583–621 (1952); Addendum, *Ibid.*, 48, 907–911 (1953) and W. W. Daniel, *Applied nonparametric statistics* (2nd ed.). Boston: PWS (1990).

TABLE C.10

Critical values for Bonferroni t-test ($\alpha = 0.05$)

df	2	3	4	5	6	7	8	9	10
					Number of contrasts (k)				
1	25.452	38.188	50.923	63.657	76.390	89.123	101.856	114.589	127.321
2	6.205	7.649	8.860	9.925	10.886	11.769	12.590	13.360	14.089
3	4.177	4.857	5.392	5.841	6.232	6.580	6.895	7.185	7.453
4	3.495	3.961	4.315	4.604	4.851	5.068	5.261	5.437	5.598
5	3.163	3.534	3.810	4.032	4.219	4.382	4.526	4.655	4.773
6	2.969	3.287	3.521	3.707	3.863	3.997	4.115	4.221	4.317
7	2.841	3.128	3.335	3.499	3.636	3.753	3.855	3.947	4.029
8	2.752	3.016	3.206	3.355	3.479	3.584	3.677	3.759	3.833
9	2.685	2.933	3.111	3.250	3.364	3.462	3.547	3.622	3.690
10	2.634	2.870	3.038	3.169	3.277	3.368	3.448	3.518	3.581
11	2.593	2.820	2.981	3.106	3.208	3.295	3.370	3.437	3.497
12	2.560	2.779	2.934	3.055	3.153	3.236	3.308	3.371	3.428
13	2.533	2.746	2.896	3.012	3.107	3.187	3.256	3.318	3.372
14	2.510	2.718	2.864	2.977	3.069	3.146	3.214	3.273	3.326
15	2.490	2.694	2.837	2.947	3.036	3.112	3.177	3.235	3.286
16	2.473	2.673	2.813	2.921	3.008	3.082	3.146	3.202	3.252
17	2.458	2.655	2.793	2.898	2.984	3.056	3.119	3.173	3.222
18	2.445	2.639	2.775	2.878	2.963	3.034	3.095	3.149	3.197
19	2.433	2.625	2.759	2.861	2.944	3.014	3.074	3.127	3.174
20	2.423	2.613	2.744	2.845	2.927	2.996	3.055	3.107	3.153
21	2.414	2.601	2.732	2.831	2.912	2.980	3.038	3.090	3.135
22	2.405	2.591	2.720	2.819	2.899	2.965	3.023	3.074	3.119
23	2.398	2.582	2.710	2.807	2.886	2.952	3.009	3.059	3.104
24	2.391	2.574	2.700	2.797	2.875	2.941	2.997	3.046	3.091
25	2.385	2.566	2.692	2.787	2.865	2.930	2.986	3.035	3.078
26	2.379	2.559	2.684	2.779	2.856	2.920	2.975	3.024	3.067
27	2.373	2.552	2.676	2.771	2.847	2.911	2.966	3.014	3.057
28	2.368	2.546	2.669	2.763	2.839	2.902	2.957	3.004	3.047
29	2.364	2.541	2.663	2.756	2.832	2.894	2.949	2.996	3.038
30	2.360	2.536	2.657	2.750	2.825	2.887	2.941	2.988	3.030
40	2.329	2.499	2.616	2.704	2.776	2.836	2.887	2.931	2.971
80	2.284	2.445	2.555	2.639	2.705	2.761	2.809	2.850	2.887
100	2.276	2.435	2.544	2.626	2.692	2.747	2.793	2.834	2.871
200	2.258	2.414	2.520	2.601	2.665	2.718	2.764	2.803	2.839

TABLE   C.10 (continued)
Critical values for Bonferroni *t*-test ($\alpha = 0.01$)

df	| Number of contrasts (k)									
	2	3	4	5	6	7	8	9	10
1	127.321	190.984	254.647	318.309	381.971	445.633	509.295	572.957	636.619
2	14.089	17.277	19.962	22.327	24.464	26.429	28.258	29.975	31.599
3	7.453	8.575	9.465	10.215	10.869	11.453	11.984	12.471	12.924
4	5.598	6.254	6.758	7.173	7.529	7.841	8.122	8.376	8.610
5	4.773	5.247	5.604	5.893	6.138	6.352	6.541	6.713	6.869
6	4.317	4.698	4.981	5.208	5.398	5.563	5.709	5.840	5.959
7	4.029	4.355	4.595	4.785	4.944	5.082	5.202	5.310	5.408
8	3.833	4.122	4.334	4.501	4.640	4.759	4.864	4.957	5.041
9	3.690	3.954	4.146	4.297	4.422	4.529	4.622	4.706	4.781
10	3.581	3.827	4.005	4.144	4.259	4.357	4.442	4.518	4.587
11	3.497	3.728	3.895	4.025	4.132	4.223	4.303	4.373	4.437
12	3.428	3.649	3.807	3.930	4.031	4.117	4.192	4.258	4.318
13	3.372	3.584	3.735	3.852	3.948	4.030	4.101	4.164	4.221
14	3.326	3.530	3.675	3.787	3.880	3.958	4.026	4.086	4.140
15	3.286	3.484	3.624	3.733	3.822	3.897	3.963	4.021	4.073
16	3.252	3.444	3.581	3.686	3.773	3.846	3.909	3.965	4.015
17	3.222	3.410	3.543	3.646	3.730	3.801	3.862	3.917	3.965
18	3.197	3.380	3.510	3.610	3.692	3.762	3.822	3.874	3.922
19	3.174	3.354	3.481	3.579	3.660	3.727	3.786	3.837	3.883
20	3.153	3.331	3.455	3.552	3.630	3.697	3.754	3.804	3.850
21	3.135	3.310	3.432	3.527	3.604	3.669	3.726	3.775	3.819
22	3.119	3.291	3.412	3.505	3.581	3.645	3.700	3.749	3.792
23	3.104	3.274	3.393	3.485	3.560	3.623	3.677	3.725	3.768
24	3.091	3.258	3.376	3.467	3.540	3.603	3.656	3.703	3.745
25	3.078	3.244	3.361	3.450	3.523	3.584	3.637	3.684	3.725
26	3.067	3.231	3.346	3.435	3.507	3.567	3.620	3.666	3.707
27	3.057	3.219	3.333	3.421	3.492	3.552	3.604	3.649	3.690
28	3.047	3.208	3.321	3.408	3.479	3.538	3.589	3.634	3.674
29	3.038	3.198	3.310	3.396	3.466	3.525	3.575	3.620	3.659
30	3.030	3.189	3.300	3.385	3.454	3.513	3.563	3.607	3.646
40	2.971	3.122	3.227	3.307	3.372	3.426	3.473	3.514	3.551
80	2.887	3.026	3.122	3.195	3.254	3.304	3.346	3.383	3.416
100	2.871	3.007	3.102	3.174	3.232	3.280	3.322	3.358	3.390
200	2.839	2.971	3.062	3.131	3.187	3.234	3.274	3.309	3.340

TABLE C.11

Critical values for studentized range

Error df	α	\multicolumn{10}{c}{t (number of treatment means)}
		2	3	4	5	6	7	8	9	10	11
5	0.05	3.64	4.60	5.22	5.67	6.03	6.33	6.58	6.80	6.99	7.17
	0.01	5.70	6.98	7.80	8.42	8.91	9.32	9.67	9.97	10.24	10.48
6	0.05	3.46	4.34	4.90	5.30	5.63	5.90	6.12	6.32	6.49	6.65
	0.01	5.24	6.33	7.03	7.56	7.97	8.32	8.61	8.87	9.10	9.30
7	0.05	3.34	4.16	4.68	5.06	5.36	5.61	5.82	6.00	6.16	6.30
	0.01	4.95	5.92	6.54	7.01	7.37	7.68	7.94	8.17	8.37	8.55
8	0.05	3.26	4.04	4.53	4.89	5.17	5.40	5.60	5.77	5.92	6.05
	0.01	4.75	5.64	6.20	6.62	6.96	7.24	7.47	7.68	7.86	8.03
9	0.05	3.20	3.95	4.41	4.76	5.02	5.24	5.43	5.59	5.74	5.87
	0.01	4.60	5.43	5.96	6.35	6.66	6.91	7.13	7.33	7.49	7.65
10	0.05	3.15	3.88	4.33	4.65	4.91	5.12	5.30	5.46	5.60	5.72
	0.01	4.48	5.27	5.77	6.14	6.43	6.67	6.87	7.05	7.21	7.36
11	0.05	3.11	3.82	4.26	4.57	4.82	5.03	5.30	5.35	5.49	5.61
	0.01	4.39	5.15	5.62	5.97	6.25	6.48	6.67	6.84	6.99	7.13
12	0.05	3.08	3.77	4.20	4.52	4.75	4.95	5.12	5.27	5.39	5.51
	0.01	4.32	5.05	5.50	5.84	6.10	6.32	6.51	6.67	6.81	6.94
13	0.05	3.06	3.73	4.15	4.45	4.69	4.88	5.05	5.19	5.32	5.43
	0.01	4.26	4.96	5.40	5.73	5.98	6.19	6.37	6.53	6.67	6.79
14	0.05	3.03	3.70	4.11	4.41	4.64	4.83	4.99	5.13	5.25	5.36
	0.01	4.21	4.89	5.32	5.63	5.88	6.08	6.26	6.41	6.54	6.66
15	0.05	3.01	3.67	4.08	4.37	4.59	4.78	4.94	5.08	5.20	5.31
	0.01	4.17	4.48	5.25	5.56	5.80	5.99	6.16	6.31	6.44	6.55
16	0.05	3.00	3.65	4.05	4.33	4.56	4.74	4.90	5.03	5.15	5.26
	0.01	4.13	4.79	5.19	5.49	5.72	5.92	6.08	6.22	6.35	6.46
17	0.05	2.98	3.63	4.02	4.30	4.52	4.70	4.86	4.99	5.11	5.21
	0.01	4.10	4.74	5.14	5.43	5.66	5.85	6.01	6.15	6.27	6.38
18	0.05	2.97	3.61	4.00	4.28	4.49	4.67	4.82	4.96	5.07	5.17
	0.01	4.07	4.70	5.09	5.38	5.60	5.79	5.94	6.08	6.20	6.31
19	0.05	2.96	3.59	3.98	4.25	4.47	4.65	4.79	4.92	5.04	5.14
	0.01	4.05	4.67	5.05	5.33	5.55	5.73	5.89	6.02	6.14	6.25
20	0.05	2.95	3.58	3.96	4.23	4.45	4.62	4.77	4.90	5.01	5.11
	0.01	4.02	4.64	5.02	5.29	5.51	5.69	5.84	5.97	6.09	6.19
24	0.05	2.92	3.53	3.90	4.17	4.37	4.54	4.68	4.81	3.92	5.01
	0.01	3.96	4.55	4.91	5.17	5.37	5.54	5.69	5.81	5.92	6.02
30	0.05	2.89	3.49	3.85	4.10	4.30	4.46	4.60	4.72	4.82	4.92
	0.01	3.89	4.45	4.80	5.05	5.24	5.40	5.54	5.65	5.76	5.85
40	0.05	2.86	3.44	3.79	4.04	4.23	4.39	4.52	4.63	4.73	4.82
	0.01	3.82	4.37	4.70	4.93	5.11	5.26	5.39	5.50	5.60	5.69
60	0.05	2.83	3.40	3.74	3.98	4.16	4.31	4.44	4.55	4.65	4.73
	0.01	3.76	4.28	4.59	4.82	4.99	5.13	5.25	5.36	5.45	5.53
120	0.05	2.80	3.36	3.68	3.92	4.10	4.24	4.36	4.47	4.56	4.64
	0.01	3.70	4.20	4.50	4.71	4.87	5.01	5.12	5.21	5.30	5.37
∞	0.05	2.77	3.31	3.63	3.86	4.03	4.17	4.29	4.39	4.47	4.55
	0.01	3.64	4.12	4.40	4.60	4.76	4.88	4.99	5.08	5.16	5.23

Abridged from Table 29 in E. S. Pearson and H. O. Hartley (eds), *Biometrika tables for statisticians* (2nd ed.). New York: Cambridge University Press (1958), Vol. 1. Reproduced with the permission of the editors and the trustees of *Biometrika*, presented in R. L. Ott, *An introduction to statistical methods and data analysis*. Belmont, CA: Wadsworth (1993).

TABLE C.11 (continued)
Critical values for studentized range

Error df	α	|	t (number of treatment means)							
		12	13	14	15	16	17	18	19	20
5	0.05	7.32	7.47	7.60	7.72	7.83	7.93	8.03	8.12	8.21
	0.01	10.70	10.89	11.08	11.24	11.40	11.55	11.68	11.81	11.93
6	0.05	6.79	6.92	7.03	7.14	7.24	7.34	7.43	7.51	7.59
	0.01	9.48	9.65	9.81	9.95	10.08	10.21	10.32	10.43	10.54
7	0.05	6.43	6.55	6.66	6.76	6.85	6.94	7.02	7.10	7.17
	0.01	8.71	8.86	9.00	9.12	9.24	9.35	9.46	9.55	9.65
8	0.05	6.18	6.29	6.39	6.48	6.57	6.65	6.73	6.80	6.87
	0.01	8.18	8.31	8.44	8.55	8.66	8.76	8.85	8.94	9.03
9	0.05	5.98	6.09	6.19	6.28	6.36	6.44	6.51	6.58	6.64
	0.01	7.78	7.91	8.03	8.13	8.23	8.33	8.41	8.49	8.57
10	0.05	5.83	5.93	6.03	6.11	6.19	6.27	6.34	6.40	6.47
	0.01	7.49	7.60	7.71	7.81	7.91	7.99	8.08	8.15	8.23
11	0.05	5.71	5.81	5.90	5.98	6.06	6.13	6.20	6.27	6.33
	0.01	7.25	7.36	7.46	7.56	7.65	7.73	7.81	7.88	7.95
12	0.05	5.61	5.71	5.80	5.88	5.95	6.02	6.09	6.15	6.21
	0.01	7.06	7.17	7.26	7.36	7.44	7.52	7.59	7.66	7.73
13	0.05	5.53	5.63	5.71	5.79	5.86	5.93	5.99	6.05	6.11
	0.01	6.90	7.01	7.10	7.19	7.27	7.35	7.42	7.48	7.55
14	0.05	5.46	5.55	5.64	5.71	5.79	5.85	5.91	5.97	6.03
	0.01	6.77	6.87	6.96	7.05	7.13	7.20	7.27	7.33	7.39
15	0.05	5.40	5.49	5.57	5.65	5.72	5.78	5.85	5.90	5.96
	0.01	6.66	6.76	6.84	6.93	7.00	7.07	7.14	7.20	7.26
16	0.05	5.35	5.44	5.52	5.59	5.66	5.73	5.79	5.84	5.90
	0.01	6.56	6.66	6.74	6.82	6.90	6.97	7.03	7.09	7.15
17	0.05	5.31	5.39	5.47	5.54	5.61	5.67	5.73	5.79	5.84
	0.01	6.48	6.57	6.66	6.73	6.81	6.87	6.94	7.00	7.05
18	0.05	5.27	5.35	5.43	5.50	5.57	5.63	5.69	5.74	5.79
	0.01	6.41	6.50	6.58	6.65	6.73	6.79	6.85	6.91	6.97
19	0.05	5.23	5.31	5.39	5.46	5.53	5.59	5.65	5.70	5.75
	0.01	6.34	6.43	6.51	6.58	6.65	6.72	6.78	6.84	6.89
20	0.05	5.20	5.28	5.36	5.43	5.49	5.55	5.61	5.66	5.71
	0.01	6.28	6.37	6.45	6.52	6.59	6.65	6.71	6.77	6.82
24	0.05	5.10	5.18	5.25	5.32	5.38	5.44	5.49	5.55	5.59
	0.01	6.11	6.19	6.26	6.33	6.39	6.45	6.51	6.56	6.61
30	0.05	5.00	5.08	5.15	5.21	5.27	5.33	5.38	5.43	5.47
	0.01	5.93	6.01	6.08	6.14	6.20	6.26	6.31	6.36	6.41
40	0.05	4.90	4.98	5.04	5.11	5.16	5.22	5.27	5.31	5.36
	0.01	5.76	5.83	5.90	5.96	6.02	6.07	6.12	6.16	6.21
60	0.05	4.81	4.88	4.94	5.00	5.06	5.11	5.15	5.20	5.24
	0.01	5.60	5.67	5.73	5.78	5.84	5.89	5.93	5.97	6.01
120	0.05	4.71	4.78	4.84	4.90	4.95	5.00	5.04	5.09	5.13
	0.01	5.44	5.50	5.56	5.61	5.66	5.71	5.75	5.79	5.83
∞	0.05	4.62	4.68	4.74	4.80	4.85	4.89	4.93	4.97	5.01
	0.01	5.29	5.35	5.40	5.45	5.49	5.54	5.57	5.61	5.65

TABLE C.12
Critical values for Duncan test

Error df	α	\multicolumn{14}{c}{r (number of ordered steps between means)}
		2	3	4	5	6	7	8	9	10	12	14	16	18	20
1	.05	18.0	18.0	18.0	18.0	18.0	18.0	18.0	18.0	18.0	18.0	18.0	18.0	18.0	18.0
	.01	90.0	90.0	90.0	90.0	90.0	90.0	90.0	90.0	90.0	90.0	90.0	90.0	90.0	90.0
2	.05	6.09	6.09	6.09	6.09	6.09	6.09	6.09	6.09	6.09	6.09	6.09	6.09	6.09	6.09
	.01	14.0	14.0	14.0	14.0	14.0	14.0	14.0	14.0	14.0	14.0	14.0	14.0	14.0	14.0
3	.05	4.50	4.50	4.50	4.50	4.50	4.50	4.50	4.50	4.50	4.50	4.50	4.50	4.50	4.50
	.01	8.26	8.5	8.6	8.7	8.8	8.9	8.9	9.0	9.0	9.0	9.1	9.2	9.3	9.3
4	.05	3.93	4.01	4.02	4.02	4.02	4.02	4.02	4.02	4.02	4.02	4.02	4.02	4.02	4.02
	.01	6.51	6.8	6.9	7.0	7.1	7.1	7.2	7.2	7.3	7.3	7.4	7.4	7.5	7.5
5	.05	3.64	3.74	3.79	3.83	3.83	3.83	3.83	3.83	3.83	3.83	3.83	3.83	3.83	3.83
	.01	5.70	5.96	6.11	6.18	6.26	6.33	6.40	6.44	6.5	6.6	6.6	6.7	6.7	6.8
6	.05	3.46	3.58	3.64	3.68	3.68	3.68	3.68	3.68	3.68	3.68	3.68	3.68	3.68	3.68
	.01	5.24	5.51	5.65	5.73	5.83	5.81	5.95	6.00	6.0	6.1	6.2	6.2	6.3	6.3
7	.05	3.35	3.47	3.54	3.58	3.60	3.61	3.61	3.61	3.61	3.61	3.61	3.61	3.61	3.61
	.01	4.95	5.22	5.37	5.45	5.53	5.61	5.69	5.73	5.8	5.8	5.9	5.9	6.0	6.0
8	.05	3.26	3.39	3.47	3.52	3.55	3.56	3.56	3.56	3.56	3.56	3.56	3.56	3.56	3.56
	.01	4.74	5.00	5.14	5.23	5.32	5.40	5.47	5.51	5.5	5.6	5.7	5.7	5.8	5.8
9	.05	3.20	3.34	3.41	3.47	3.50	3.52	3.52	3.52	3.52	3.52	3.52	3.52	3.52	3.52
	.01	4.60	4.86	4.99	5.08	5.17	5.25	5.32	5.36	5.4	5.5	5.5	5.6	5.7	5.7
10	.05	3.15	3.30	3.37	3.43	3.46	3.47	3.47	3.47	3.47	3.47	3.47	3.47	3.47	3.48
	.01	4.48	4.73	4.88	4.96	5.06	5.13	5.20	5.24	5.28	5.36	5.42	5.48	5.54	5.55
11	.05	3.11	3.27	3.35	3.39	3.43	3.44	3.45	3.46	3.46	3.46	3.46	3.46	3.47	3.48
	.01	4.39	4.63	4.77	4.86	4.94	5.01	5.06	5.12	5.15	5.24	5.28	5.34	5.38	5.39
12	.05	3.08	3.23	3.33	3.36	3.40	3.42	3.44	3.44	3.46	3.46	3.46	3.46	3.47	3.48
	.01	4.32	4.55	4.68	4.76	4.84	4.92	4.96	5.02	5.07	5.13	5.17	5.22	5.23	5.26
13	.05	3.06	3.21	3.30	3.35	3.38	3.41	3.42	3.44	3.45	3.45	3.46	3.46	3.47	3.47
	.01	4.26	4.48	4.62	4.69	4.74	4.84	4.88	4.94	4.98	5.04	5.08	5.13	5.14	5.15
14	.05	3.03	3.18	3.27	3.33	3.37	3.39	3.41	3.42	3.44	3.45	3.46	3.46	3.47	3.47
	.01	4.21	4.42	4.55	4.63	4.70	4.78	4.83	4.87	4.91	4.96	5.00	5.04	5.06	5.07
15	.05	3.01	3.16	3.25	3.31	3.36	3.38	3.40	3.42	3.43	3.44	3.45	3.46	3.47	3.47
	.01	4.17	4.37	4.50	4.58	4.64	4.72	4.77	4.81	4.84	4.90	4.94	4.97	4.99	5.00

TABLE   C.12 (continued)
Critical values for Duncan test

Error df	α	\multicolumn{14}{c}{r (number of ordered steps between means)}
		2	3	4	5	6	7	8	9	10	12	14	16	18	20
16	.05	3.00	3.15	3.23	3.30	3.34	3.37	3.39	3.41	3.43	3.44	3.45	3.46	3.47	3.47
	.01	4.13	4.34	4.45	4.54	4.60	4.67	4.72	4.76	4.79	4.84	4.88	4.91	4.93	4.94
17	.05	2.98	3.13	3.22	3.28	3.33	3.36	3.38	3.40	3.42	3.44	3.45	3.46	3.47	3.47
	.01	4.10	4.30	4.41	4.50	4.56	4.63	4.68	4.72	4.75	4.80	4.83	4.86	4.88	4.89
18	.05	2.97	3.12	3.21	3.27	3.32	3.35	3.37	3.39	3.41	3.43	3.45	3.46	3.47	3.47
	.01	4.07	4.27	4.38	4.46	4.53	4.59	4.64	4.68	4.71	4.76	4.79	4.82	4.84	4.85
19	.05	2.96	3.11	3.19	3.26	3.31	3.35	3.37	3.39	3.41	3.43	3.44	3.46	3.47	3.47
	.01	4.05	4.24	4.35	4.43	4.50	4.56	4.61	4.64	4.67	4.72	4.76	4.79	4.81	4.82
20	.05	2.95	3.10	3.18	3.25	3.30	3.34	3.36	3.38	3.40	3.43	3.44	3.46	3.46	3.47
	.01	4.02	4.22	4.33	4.40	4.47	4.53	4.58	4.61	4.65	4.69	4.73	4.76	4.78	4.79
22	.05	2.93	3.08	3.17	3.24	3.29	3.32	3.35	3.37	3.39	3.42	3.44	3.45	3.46	3.47
	.01	3.99	4.17	4.28	4.36	4.42	4.48	4.53	4.57	4.60	4.65	4.68	4.71	4.74	4.75
24	.05	2.92	3.07	3.15	3.22	3.28	3.31	3.34	3.37	3.38	3.41	3.44	3.45	3.46	3.47
	.01	3.96	4.14	4.24	4.33	4.39	4.44	4.49	4.53	4.57	4.62	4.64	4.67	4.70	4.72
26	.05	2.91	3.06	3.14	3.21	3.27	3.30	3.34	3.36	3.38	3.41	3.43	3.45	3.46	3.47
	.01	3.93	4.11	4.21	4.30	4.36	4.41	4.46	4.50	4.53	4.58	4.62	4.65	4.67	4.69
28	.05	2.90	3.04	3.13	3.20	3.26	3.30	3.33	3.35	3.37	3.40	3.43	3.45	3.46	3.47
	.01	3.91	4.08	4.18	4.28	4.34	4.39	4.43	4.47	4.51	4.56	4.60	4.62	4.65	4.67
30	.05	2.89	3.04	3.12	3.20	3.25	3.29	3.32	3.35	3.37	3.40	3.43	3.44	3.46	3.47
	.01	3.89	4.06	4.16	4.22	4.32	4.36	4.41	4.45	4.48	4.54	4.58	4.61	4.63	4.65
40	.05	2.86	3.01	3.10	3.17	3.22	3.27	3.30	3.33	3.35	3.39	3.42	3.44	3.46	3.47
	.01	3.82	3.99	4.10	4.17	4.24	4.30	4.34	4.37	4.41	4.46	4.51	4.54	4.57	4.59
60	.05	2.83	2.98	3.08	3.14	3.20	3.24	3.28	3.31	3.33	3.37	3.40	3.43	3.45	3.47
	.01	3.76	3.92	4.03	4.12	4.17	4.23	4.27	4.31	4.34	4.39	4.44	4.47	4.50	4.53
100	.05	2.80	2.95	3.05	3.12	3.18	3.22	3.26	3.29	3.32	3.36	3.40	3.42	3.45	3.47
	.01	3.71	3.86	3.93	4.06	4.11	4.17	4.21	4.25	4.29	4.35	4.38	4.42	4.45	4.48
∞	.05	2.77	2.92	3.02	3.09	3.15	3.19	3.23	3.26	3.29	3.34	3.38	3.41	3.44	3.47
	.01	3.64	3.80	3.90	3.98	4.04	4.09	4.14	4.17	4.20	4.26	4.31	4.34	4.38	4.41

From D. B. Duncan, Multiple range and multiple F tests. *Biometrics, 11:* 1–42 (1955). With permission from the Biometric Society and the author. Presented in R. L. Ott, *An introduction to statistical methods and data analysis.* Belmont, CA: Wadsworth (1993).

TABLE C.13

Critical values for S

This table gives, for each n, the smallest values for which $P(S \geq s) \leq \alpha$.

			α		
n	0.005	0.010	0.025	0.050	0.100
4	8	8	8	6	6
5	12	10	10	8	8
6	15	13	13	11	9
7	19	17	15	13	11
8	22	20	18	16	12
9	26	24	20	18	14
10	29	27	23	21	17
11	33	31	27	23	19
12	38	36	30	26	20
13	44	40	34	28	24
14	47	43	37	33	25
15	53	49	41	35	29
16	58	52	46	38	30
17	64	58	50	42	34
18	69	63	53	45	37
19	75	67	57	49	39
20	80	72	62	52	42
21	86	78	66	56	44
22	91	83	71	61	47
23	99	89	75	65	51
24	104	94	80	68	54
25	110	100	86	72	58
26	117	107	91	77	61
27	125	113	95	81	63
28	130	118	100	86	68
29	138	126	106	90	70
30	145	131	111	95	75
31	151	137	117	99	77
32	160	144	122	104	82
33	166	152	128	108	86
34	175	157	133	113	89
35	181	165	139	117	93
36	190	172	146	122	96
37	198	178	152	128	100
38	205	185	157	133	105
39	213	193	163	139	109
40	222	200	170	144	112

From L. Kaarsemaker and A. van Wijngaarden, Tables for use in rank correlation, *Statistica Neerlandica, 7*, 41–54 (1953). Presented in W. W. Daniel, *Applied nonparametric statistics* (2nd ed.). Boston: PWS (1990).

APPENDIX D
ANSWERS TO
SELECTED EXERCISES

Chapter 1

1.1 There are two populations—one consisting of growth rate measurements of treated rats and the other consisting of the measurements for the untreated rats.

1.3 There are four populations—one for each of the four methods of fertilizing the crop. Each of these populations is a collection of all yields that will be observed when the corresponding fertilization method is used.

1.5 There are two populations corresponding to the two nematocides. Each measurement in a population represents the observed number of nematodes in soil three months after application of the corresponding nematocide.

1.7 The population consist of the set of all measurements of mother's age and birth-weight of her baby in the population of all pregnant mothers.

1.9 The population can be modeled as a collection of 0's and 1's with 1 and 0 representing fourth grade children with and without ADD symptoms.

1.11 The sample is a collection of measurements representing the disease status of a sample of n trees. A measurement equals 1 if the corresponding tree has the disease and 0 otherwise.

1.13 **a** Select a representative sample of 10 homes of each type. Measure the indoor and outdoor radiation levels in each of the selected homes.

        b Select a representative sample of 30 homes. Measure the indoor and outdoor radiation levels in each of the selected homes.

Chapter 2

2.1 **a** Univariate **b** conceptual **c** infinite **d** quantitative **e** continuous.

2.3 **a** Univariate **b** conceptual **c** infinite **d** quantitative **e** continuous.

2.5   **a** Univariate   **b** conceptual   **c** infinite   **d** quantitative   **e** discrete.

2.7   **a** Bivariate   **b** conceptual   **c** infinite   **d** quantitative   **e** continuous.

2.9   **a** Univariate   **b** conceptual   **c** infinite   **d** qualitative   **e** discrete.

2.11   **a** 0.8   **b** 0.5.

2.13   **a** **i** 0.5   **ii** .90   **iii** .25.
  b **i** Approximately 50% of the randomly selected plants will have more than 18 leaves;   **ii** Approximately 90% of the randomly selected plants will have between 17 and 20 leaves;   **iii** Approximately 25% of the randomly selected plants will have less than 18 leaves.

2.15

(y_1, y_2)	(Y,L)	(Y,N)	(Y,H)	(M,L)	(M,N)	(M,H)
$f(y_1, y_2)$	0.01	0.16	0.03	0.06	0.32	0.08

(y_1, y_2)	(O,L)	(O,N)	(O,H)
$f(y_1, y_2)$	0 .04	0.16	0.14

2.17   **a**

y_1	Y	M	O
$f(y_1)$	0.20	0.46	0.34

  b 0.80.

2.19   **a** 0.42   **b** 0.19   **c** 0.40.

2.21   **a** Hard to tell   **b** Variety B.

2.23   **a** $\mu = 2.5$. The average size of the population of litters is 2.5.
  b In this population, no more than half the measurements are less than 2.5 and no more than half the measurements are more than 2.5.

2.31   **a** $\eta_{.25} = 26$, $\eta_{.50} = 42$, $\eta_{.75} = 60$.   **b** Mean is higher than the median.

2.33   **a** The scores 270 and 430 are two standard deviations above and below the mean.
  b The scores 230 and 470 are three standard deviations above and below the mean.

2.35   **a** At least 75% of the yields are between 100 and 140 bushels per acre.
  b At least 56% of the yields are between 105 and 135 bushels per acre.

2.37 $\sigma = 1.3$, $CV = 0.0703$.

2.39   **a** $\mu = 1.11, \sigma = 0.6617$;   **b** $\mu = 1.49, \sigma = 0.9539$;   **c** $\sigma_{12} = 0.0261$.

2.41   **a** Patient responses can be regarded as independent identical events.   **b** 0.6471.

2.43   **a** 0.0001   **b** 0.0150   **c** 0.8844.

2.45   **a** (1) The probability of a caterpillar occupying a small area in a shoot of a tree is proportional to the area. (2) The probability of finding more than one caterpillar in a small area of a shoot of a tree is negligible. (3) Caterpillars occupying small areas in the shoots of a tree are independent events.
  b In a large population of trees, there will be, on the average, two caterpillars per tree.

2.47   **a** Less than 0.001   **b** 0.3085   **c** 1.356.

2.51   **a** 0.155   **b** 6.

Chapter 3

3.7 **b** Due to small sample sizes, it is hard to judge the shape of the population distributions on the basis of histograms. However, the histograms indicate that the population distributions might be skewed to the right and that the mean milk production for cows on the supplemental diet might be higher than that for the control diet. The variances seem to be about the same for the two populations.

c The mean and standard deviation for control diet are: $\bar{y}_1 = 11, 180, s_1 = 1214.5$ and that for the supplemented diet is $\bar{y}_2 = 12, 130, s_2 = 1214.1$.

3.9 Children in district 3 seem to do better than the other children. Performances in districts 2 and 4 are very similar and below that in the other two districts. The distributions of scores in districts 1, 2, and 3 appear to be skewed to the right, implying that there are some very high scores in these districts.

3.11 **a** Sampling distribution of $\overline{Y}$ when the target population is $\{0, 1, 2, 6, 7, 9\}$:

$\bar{y}$	1.00	2.33	2.67	3.00	3.33	3.67	4.00	4.33
$f(\bar{y})$	0.05	0.05	0.10	0.10	0.10	0.05	0.05	0.05

$\bar{y}$	4.67	5.00	5.33	5.67	6.00	7.33
$f(\bar{y})$	0.05	0.10	0.10	0.10	0.05	0.05

Sampling distribution of $\overline{Y}$ when the target population is $\{4, 8, 12, 20, 24, 32\}$:

$\bar{y}$	8.00	10.67	12.00	13.33	14.67	16.00	17.33
$f(\bar{y})$	0.05	0.05	0.10	0.10	0.10	0.10	0.10

$\bar{y}$	18.67	20.00	21.33	22.67	25.33
$f(\bar{y})$	0.10	0.10	0.10	0.05	0.05

c The mean and standard deviation of the sampling distribution of $\overline{Y}$ are: $\mu_{\overline{Y}} = 4.17$ and $\sigma_{\overline{Y}} = 1.08$ for the population in Example 3.7; $\mu_{\overline{Y}} = 4.17$ and $\sigma_{\overline{Y}} = 1.50$ for the first population; and $\mu_{\overline{Y}} = 16.67$ and $\sigma_{\overline{Y}} = 4.31$ for the second population. The mean is the same for the target population and the sampling distribution of the sample mean whereas the standard deviation of the sampling distribution of the sample mean is smaller than that for the target population in all three cases.

3.13 **a**

$\tilde{y}$	2	3	5	6
$f(\tilde{y})$	0.20	0.30	0.30	0.20

b $\mu_{\tilde{Y}} = 4, \sigma_{\tilde{Y}} = 1.48$.

c The probability that $\tilde{Y}$ will differ from μ by an amount less than 0.25 is zero, whereas the same probability for $\overline{Y}$ is 0.20. Similarly, probabilities of other events can be compared to argue that, for this population, the sample mean is more likely to be closer to the target population mean than the sample median.

3.15 **a** The sampling distribution of $\overline{Y}$ is a normal distribution with mean $\mu_{\overline{Y}} = 1.5$ and standard deviation $\sigma_{\overline{Y}} = 0.1265$.

 b 0.9845.

3.17 **a** 0.001. **b** The calculated value is $t_c = 4.841$.

 c Let $\mu = 25$. Then the calculated value in (b) is the observed value of a $t(14)$ random variable. From (a) we know that the probability of observed value of $t(14)$ being larger than 3.79 is 0.001. Now 4.841 is larger than 3.79. Hence, if $\mu = 25$, the probability of the calculated value of t will be as large or larger than 4.841 is less than 0.001. If $\mu < 25$, this probability will be even smaller than 0.001. Thus are two possibilities: Either the assumption that $\mu \leq 25$ is true and a very rare (probability ≤ 0.001) event has occurred or our assumption is not true. If we rule out the possibility of observing an event of probability ≤ 0.001, we are led to conclude that $\mu \geq 25$.

3.19 **a** 43.8202 **b** 47.5 **c** similar to the answer to Exercise 3.17(c).

3.23 **a** Estimation **b** hypothesis testing **c** prediction.

3.25 **a** In Exercise 3.11, we verified that the mean of the sampling distribution of $\overline{Y}$ is the same as the mean of the corresponding target population. This verifies that the sample mean is an unbiased estimator of the population mean in all three cases.

3.29 **a** $\hat{\mu}_L = 135.4$.

 b We may conclude with 95% confidence that the average height of the population of all Douglas fir trees is at least 135.4 cm.

 c Yes.

3.31 **a** We can conclude with 99% confidence that at most 15% of the exposed population will suffer from toxic effects.

 b We can say that the fraction is less than 15%. This does not mean that it is more than 10%, but it does mean that it is less than 20%.

3.33 **a** When testing H_{01} the research hypothesis states that the mean is more than 190. Thus the research hypothesis is a statement about lower bound for the population mean. Thus H_{01} can be verified on the basis of a lower bound calculated from the data. When testing H_{02}, the research hypothesis concerns an upper bound for the population mean, so that we cannot verify on the basis of a lower confidence bound.

 b We cannot reject the null hypothesis (that is, we cannot conclude $\mu > 190$) because the calculated lower bound of 135.4 implies that $\mu > 135.4$ but does not imply that $\mu > 190$. For example, it is possible that $\mu = 150$.

 c Not enough evidence to reject H_{02} because the 99% upper bound for μ is 153.8, which is larger than 100.

3.37 The following table shows the powers of the test when $\alpha = 0.05, 0.01$ and $\mu = 12, 11, 10, 9$.

			μ	
Level	12	11	10	9
0.05	.6147	.9871	1.0000	1.0000
0.01	.3484	.9391	.9997	1.0000

3.41 **a** A 99% prediction interval is (1.03 , 4.65).

3.43 **a** $1.32 \leq \theta \leq 3.82$

b At the 0.10 level, we can reject $H_0 : \mu_1 - \mu_2 = 0$ and conclude that the true mean N-losses are not the same for the two fertilizers.

c $p \leq 0.002$.

Chapter 4

4.1 **a** Interval   **b** ordinal   **c** nominal   **d** nominal   **e** interval   **f** ordinal   **g** nominal.

4.3 **a** A graphical description will show the graph of a normal distribution with mean μ and standard deviation $\sigma = 4$.

b In $100\pi\%$ of a large number of repetitions of the experiment, the sample mean $\overline{Y}$ will have a value between $\mu - 2$ and $\mu + 2$.

c $\pi = 0.8858$.

d **i** $13.22 \leq \mu \leq 18.19$ **ii** At the 0.05 level, there is insufficient evidence to conclude that $\mu < 18$.

4.5 **a** The indices are a random sample from a normal distribution.

b $57.1 \leq \mu \leq 64.3$.

c The psychologist can conclude with 90% confidence, that μ has a value between 57.1 and 64.3.

d $p < 0.001$.

e If H_0 is true, the probability of observing a sample mean at least as favorable to H_1 as the one observed is less than 0.001. Thus, the observed sample provides strong evidence in favor of H_1.

f A confidence interval approach for testing H_0 at level α requires a $100(1 - \alpha)\%$ lower confidence bound for μ. The lower end-point of the 90% confidence interval in (b) is a 95% lower confidence bound. Since the lower bound in (b) is more than 50, we can reject H_0 at the 0.05 level.

4.7 **a** Let $\theta = \mu_1 - \mu_2$. Then $\hat{\theta}_L = 1.10$.

b We may conclude with 95% confidence that the true mean N-loss for fertilizer UN is at least 1.10 more than that for fertilizer U.

c $\hat{\theta}_L = 1.12$

4.9 **a** The lymphocyte count data are independent random samples from two normal distributions with a common variance.

b Let $\theta = \mu_H - \mu_G$. Then $\hat{\theta}_L = -397.7$.

c We may conclude with 95% confidence that the mean lymphocyte count for the population of Guernsey cows is at least 397.7 more than that for for the population of Holstein cows.

d $\hat{\theta}_L = 340$.

e The two sample variances are $s_H^2 = 2,296,497$ and $s_G^2 = 751,619$. We see that the variance for the Holstein cows is three times as large as the variance for Guernsey cows. Thus, the assumption of equal population variance is suspect in this case.

4.11 **a** Let μ_d denote the average PVR increase in the population of all treated dogs. Then we want to test null hypothesis $H_0 : \mu_d \leq 0$ against the research hypothesis $H_1 : \mu_d > 0$. We have $t_c = 1.884$ with 7 degrees of freedom which corresponds to a p-value of $p = 0.051$.

b $-0.01 \leq \mu_d \leq 0.08$.

4.13 **a** The standard deviation of the sample is an estimate of σ, the variance of the conceptual population of number of trials needed to learn to recognize the signal.

b $2.15 \leq \sigma \leq 5.71$.

4.15 **a** $0.456 \leq \sigma_B/\sigma_G \leq 1.497$.

4.21 Let Y_j denote the jth sample measurement $(j = 1, \ldots, 15)$. Then the model is $Y_j = \mu + E_j$ where μ is the average ear length in the population of all plants of this hybrid variety, E_j is the random error in Y_j, and the E_j are independent $N(0, \sigma^2)$.

4.23 Let Y_{ij} $(i = 1, 2; j = 1, \ldots, 11$ if $i = 1$, and $j = 1, \ldots, 16$ if $i = 2)$ denote the dental measurement for the jth child of ith gender $(i = 1$ for girl, $i = 2$ for boy). Then the model is $Y_{ij} = \mu_i + E_{ij}$, where μ_i is the mean dental measurement in the population of all children of gender i, E_{ij} is the random error in Y_{ij}, and the E_{ij} are independent $N(0, \sigma^2)$.

4.25 **a** $\overline{Y}_L = 1.67$.

4.27 **a** 95% normal limits for serum creatinine level in the population of subjects with no renal disease is $(0.41, 1.79)$.

b It is estimated that only 5% of the disease-free population will have creatinine levels outside the normal limits. Thus the probability of creatinine level above 1.79 for a randomly selected subject from a disease-free population is 0.025.

c On the basis of the upper normal limit, it is reasonable to suspect that this subject may have renal disease.

Chapter 5

5.1 **a** The groups are $\{L, L, L, L\}, \{M, M, M\}, \{H\}$.

b

Subject	1	2	3	4	5	6	7	8
Rank	2.5	2.5	8	6	2.5	6	2.5	6

5.5 **a** $H_0 : \eta \geq 13$, $H_1 : \eta < 13$, $B_c = 4$, and $n^* = 15$.

b There is not a $b(15, 0.05)$ entry in Table C.5 (Appendix C). However $b(15, 0.059) = 11$, and $n^* - B_c = 15 - 11 = 4$, so that H_0 can be rejected at the level 0.059.

c The p value from the corresponding t-test is less than 0.001, which is much smaller than 0.059. The t-test provides stronger evidence in favor of H_1, but its validity depends on the assumption of a normal distribution for the target population.

5.7 **a** Let Y_j denote the jth differential CPZ value $(j = 1, \ldots, 8)$. Then the model is $Y_j = \eta + E_j$, where η is the median of the distribution of the population of all differential CPZ values, E_j is the error in Y_j, and the E_j are a random sample from a symmetric continuous population with zero median.

   b $W_c^+ = 5$, and $N - w(8, 0.055) = 36 - 30 = 6$. Reject H_0 at level 0.055.

   c $0.027 \leq p \leq 0.055$.

5.9 **a** An approximate 90% confidence interval for η is $W_{(61)} \leq \eta \leq W_{(150)} \Rightarrow 2.5 \leq \eta \leq 5.4$.

5.11 **a** Let Y_{ij} be defined as in the answer to Exercise 4.13. Then the model is $Y_{ij} = \eta_i + E_{ij}$, where η_i is the mean dental measurement in the population of all children of gender i, E_{ij} is the random error in Y_{ij}, and the E_{ij} are a random sample from a continuous population with zero mean.

   b $W_c = 286.5$, $\mu_W = 224$, $\sigma_W = 20.21$ and $w(11, 16, 0.05) = 257.2$. Reject H_0 : $\mu_1 \geq \mu_2$ in favor of H_1 : $\mu_1 < \mu_2$ at the 0.05 level.

   c $1.5 \leq \mu_2 - \mu_1 \leq 5.0$.

   d The corresponding confidence interval based on the two-sample t-test procedure is $1.6 \leq \mu_2 - \mu_1 \leq 5.2$. This interval is very close to the one based on the rank sum procedure.

5.13 **a** Conclusion based on the Wilcoxon rank sum remains the same if 5.1 is changed to 25.1.

   b Conclusion based on the t-test changes if 5.1 is replaced by 25.1.

   c Wilcoxon rank sum test is less sensitive to presence of outliers in the data than the procedures based on independent sample t-test. This is because extreme observations have less influence on ranks than on the actual measurements.

Chapter 6

6.3 **a** Let π denote the proportion cured in the population of all treated patients. Then a 90% confidence interval for π is $0.80 \leq \pi \leq 0.92$

   b For testing H_0 : $\pi \geq 0.85$ the p-value is $p = 0.8897$. Hence there is not sufficient evidence to reject H_0.

6.5 **a** An estimate of the sensitivity of the test is 86.7%. Thus the test will be positive in 86.7% of the patients who have the disease.

   b On the basis of Table C.8 (Appendix C), a 95% confidence interval for sensitivity is seen to be (0.595 , 0.983).

   c Let π denote the sensitivity of the test. Then we want to test the null hypothesis H_0 : $\pi \leq 0.80$. The p-value for this test is $p = 0.398$. Thus there is not sufficient evidence in support of the claim that the sensitivity of the test is more than 80%.

6.7 **a** A 99% confidence interval for π_1 is (based on Table C.8) $0.31 \leq \pi_1 \leq 0.92$.

   b The p-value for testing H_0 : $\pi_1 \leq .6$ is $p = 0.217$.

   c A 95% confidence interval for π is (based on Table C.8) $0.27 \leq \pi \leq 0.79$.

6.9 **a** Let π_1, π_2, π_3 and π_4 denote the probabilities in the round and yellow, round and green, angular and yellow, and angular and green classes, respectively. A 95% confidence interval for π_1 is $0.48 \leq \pi_1 \leq 0.56$.

b Test the null hypothesis $H_0 : \pi_1 = \frac{9}{16}, \pi_2 = \frac{3}{16}, \pi_3 = \frac{3}{16}, \pi_4 = \frac{1}{16}$. The calculated value of the test statistic is $\chi_c^2 = 2.382$ with 3 degrees of freedom. The associated p-value is 0.497, so that there is not sufficient evidence to conclude that the given data contradict Mendelian theory.

6.11 **a** $\hat{\lambda} = 8.8$.

b The expected frequency in the ith category is $50\hat{\pi}_i$, where $\hat{\pi}_i$ is the estimated probability for category i under the assumption of a Poisson distribution with rate parameter 8.8.

Category	1	2	3	4	5	6	7	8	Total
Exp. frequency	6.4	4.9	6.1	6.7	6.6	5.8	4.6	8.9	50

c The calculated value of the test statistic is $\chi_c^2 = 5.061$ with 6 degrees of freedom (eight categories and one estimated parameter). The associated p-value is 0.536. The data do not contradict the assumption of Poisson distribution.

d Expected frequencies in each of the eight categories is close to or larger than 5.

6.13 **a** Let π_T and π_{UT} denote, respectively, the proportion of leaf- and stem-infested plants in the treated and untreated samples. Also let $\theta = \pi_{UT} - \pi_T$. Then a 95% lower bound for θ is $\hat{\theta}_L = 0.40$.

b We may conclude with 95% confidence, that the leaf- and stem- infestation rate in the untreated population is at least 40% more than that in the treated population.

c Let π_I and π_{UI} denote, respectively, the proportion of infected and uninfected plants in the untreated plants. We want to test the null hypothesis $H_0 : \pi_I = 0.80$. The calculated value of the test statistic is $z_c = .80$. The associated p-value is 0.4237.

6.15 **a** Let π_F and π_M denote, respectively, the population proportion of male and female children with low esteem. Then we want to test null hypothesis $H_0 : \pi_F \leq \pi_M$. The calculated value of the test statistic is $z_c = 2.03$ with the corresponding p-value 0.0212. It is reasonable to reject H_0 and conclude that the proportion of females with low esteem is higher than the corresponding proportion of males.

b The calculated value of the test statistic is $\chi_c^2 = 6.7031$ with 2 degrees of freedom. The associated p-value is between 0.05 and 0.025. We may conclude, at the 0.05 level, that the two distributions are different.

6.17 Let π_1 and π_2 denote the probability of germination for varieties 1 and 2 respectively. The null hypothesis $H_0 : \pi_1 \leq \pi_2$ can be tested using the Fisher exact test. The p value is $p = 0.1849$. There is not sufficient evidence to conclude that the germination rate for variety 1 is higher than that for variety 2.

6.19 Let π_A and π_B denote, respectively, the probability that a customer prefers brand A and brand B. Confidence intervals for $\theta = \pi_A - \pi_B$ can be constructed using the McNemar procedure. We have $\hat{\theta} = 0.395$, $\hat{\sigma}_{\hat{\theta}} = 0.0512$ and a 99% confidence interval for θ is $0.263 \leq \theta \leq 0.527$.

6.21   Let π_{CA} and π_{CO} denote the probability of exposure for a case and the probability of exposure for a control, respectively. Confidence intervals for $\theta = \pi_{CA} - \pi_{CO}$ can be constructed using the McNemar procedure. We have $\hat{\theta} = 0.2995$, $\hat{\sigma}_{\hat{\theta}} = 0.0460$ and a 99% confidence lower bound for θ is $\hat{\theta}_L = 0.18$. With 99% confidence, we may conclude that the probability that a case is a past estrogen user is 18% higher than the corresponding probability for a control.

Chapter 7

7.1   **a**   Designs 1 and 2 are for experimental studies because the investigators select the levels of at least one factor (OC use). Designs 3 is for an observational study because the level of the factor (OC use) is observed as a part of the study.

   b   Response variable is a variable that takes the values 0 or 1 to indicate presence or absence of bacteriuria. OC use and age are explanatory variables.

   c   Response variable is nominal. The explanatory variable indicating OC use is also nominal, but the explanatory variable that indicates the age group of the study subject is ordinal.

   d   OC use is the only factor (with two levels) and age group indicator is a potential confounding variable.

   e   Subjects are observational units.

7.3   **a**   This is an experimental study because investigators selected the levels of the factors at which responses will be measured.

   b   Response variable is $Y = (Y_1, Y_2, Y_3)$, where Y_1, Y_2, Y_3 are the subject's weight at 4, 8, and 12 months after he or she has been in the study.

   c   Experimental factors are gender (qualitative with two levels—male, female), calorie intake (qualitative with two levels—low, high) and amount of exercise (qualititative with two levels—low, high).

   d   Potential confounding variables are age (quantitative) and bone structure (qualitative with three levels—small, medium, large).

   e   There are eight treatments corresponding to the eight possible combinations of the levels of three factors. For instance, one of the treatments is the combination representing a male whose calorie intake and exercise levels are both low.

   f   Subjects are experimental units, the three time periods at which measurements are made on a subject are observational units.

7.5   **a**   Since there are six treatments, each block of a randomized complete design will consist of six subjects satisfying the block description. The six subjects are randomly assigned to the six treatments.

   b   There are four replications per treatment.

7.9   The required sample size can be determined using the result in Box 7.1 with $\Delta = 5$, $\sigma = 3.34$, $\alpha = 0.05$, and $\beta = .10$. A sample size of $n = 9$ per treatment will meet the given specifications.

7.11   Use the method described in Box 7.4 with $b = 2$, $\alpha = 0.05$, $\sigma = 3.34$, and $W = 4$. The required sample size is $n = 23$.

7.13 Let's calculate the sample size assuming a confidence level of 90%. Equation (7.6) with $\alpha = 0.10$, $\sigma = .5$, and $B = 0.05$ yields $n = 271$.

7.15 We use the method in Box 7.4 with $b = 1$, $\alpha = 0.01$, $\sigma = .3517$, and $W = 1$. The required sample size is $n = 7$.

7.17 Setting $A(-\Delta/(\sigma\sqrt{2})) = 0.65$ we get $\Delta/\sigma = 0.5424$. Using this value of Δ/σ, the sample size can be determined as described in Example 7.15.

Chapter 8

8.1 Design 1. There are two treatments and 10 replications per treatment.

8.3 $Y_{1+} = 8.93$, $Y_{2+} = 13.75$, $Y_{3+} = 18.00$, $\quad$ $Y_{++} = 40.68$, $\overline{Y}_{1+} = 2.2325$, $\overline{Y}_{2+} = 3.4375$, $\overline{Y}_{3+} = 4.50$, $\overline{Y}_{++} = 3.39$, $S_1^2 = 0.9051$, $S_2^2 = 0.2121$, $S_3^2 = 0.5426$.

8.5 **c** $\hat{\sigma}^2 = .5533$

     e   At least two of the three levels of glucose correspond to different expected amounts of insulin release.

     g $F_c = 9.30$. Reject H_0 ($p < .01$).

8.9

Source	df	SS	MS	F_c	p
Treatments	2	10.2967	5.1483	9.31	.006
Error	9	4.9793	0.5533		
Total	11	15.2760			

Since the p-value is quite small, it is reasonable to conclude that at least two of the three glucose concentrations correspond to different amounts of expected insulin release.

8.11 **a**   This is an observational study because the treatments are not assigned to the experimental units. There are three levels of the single factor—the group of a subject.

     b

Source	df	SS	MS	F_c	p
Treatments	2	5.4465	2.7232	12.22	.001
Error	13	2.8961	0.2228		
Total	15	8.3426			

     c   At least two of the patient groups have different TRCP values ($p = .001$).

8.13

Source	df	SS	MS	F_c	p
Treatments	11	84.3875	7.6716	13.4330	< .001
Error	24	13.7064	0.5711		
Total	35	98.0939			

8.15

Source	df	SS	MS	F_c	p
Treatments	2	0.2453	0.1228	6.5412	< .01
Error	15	0.2816	0.0188		
Total	17	0.5269			

8.17

Source	df	SS	MS	F_c	p
Treatments	6	60,775.34	10,129.22	5.937	< .001
Error	96	163,766.00	1706.00		
Total	102	224551.34			

8.19 Recall that Design 1 is CRD. Let Y_{ij} denote the measured height of plant j in group (time period) i. Then the model can be written as

$$Y_{ij} = \mu + \tau_i + E_{ij}, \qquad i = 1, 2, 3; \ j = 1, 2, \ldots 8,$$

where μ is the overall mean plant height, τ_i is the effect of time period i ($\tau_1 + \tau_2 + \tau_3 = 0$), and E_{ij} is the random error. The E_{ij} are independent $N(0, \sigma^2)$.

8.25
$$Y_{i1} + Y_{i2} + \cdots + Y_{in} = (\mu + \tau_i + E_{i1}) + (\mu + \tau_i + E_{i1}) + \cdots + (\mu + \tau_i + E_{i1})$$
$$= n\mu + n\tau_i + (E_{i1} + E_{i2} + \cdots + E_{in})$$
$$= n\mu + n\tau_i + E_{i+}$$

Hence,

$$\overline{Y}_{i+} = \frac{1}{n}(Y_{i1} + Y_{i2} + \cdots + Y_{in}) = \frac{1}{n}(n\mu + n\tau_i + E_{i+}) = \mu + \tau_i + \overline{E}_{i+}.$$

8.27 Residual plots suggest that ANOVA assumptions may not be appropriate for these data. The residual vs. predicted plot suggests unequal error variances while the shape of the normal q-q plot indicates a heavy-tailed error distribution. However, the sample size is small and conclusions from residual plots based on small amounts of data should be interpreted with caution.

8.29 The normal $q-q$ and $e\sqrt{p}$ plots indicate no serious violations of ANOVA assumptions. However, sample size is small and conclusions based on residual plots must be interpreted with caution.

8.31 The $e\sqrt{p}$ plot indicates variance increasing with mean. Log-transforming the data might be useful.

8.33 The mean and standard deviation show a straight line relationship. Log-transforming the data seems to get the data closer to samples form normal distribution while alleviating the unequal variance problem somewhat.

8.35 Log-transforming the data seems to alleviate the violation of ANOVA assumption problem.

8.37 $K_c = 7.0385$. From Table C.9 (Appendix C) we see that $0.011 < p < 0.049$. We may reject at the 0.05 level the null hypothesis H_0 that the population distributions are the same.

8.39 $K_c = 9.9832$. The sample sizes are outside those in Table C.9 (Appendix C); so compare with critical values of $\chi^2(2)$ distribution. We see that $p < .001$.

8.41 $K_c = 5.6$. From Table C.9 (Appendix C), $p = 0.05$.

8.43 $K_c = 7.5219$. From $\chi^2(3)$ table, $.05 < p < .10$.

8.45 $\chi_c^2 = 20.8125$, $df = 6$. Therefore, $p < .005$ and we conclude that the terrain type distributions are different in at least two counties.

8.47 $\chi_c^2 = 9.3752$, $df = 2$. Therefore, $p < .01$ and we conclude that the survival rates are different under at least two of the three treatments.

Chapter 9

9.1   **a** $dc_1 + dc_2 + \cdots + dc_t = d(c_1 + c_2 + \cdots + c_t) = 0$ implies that both $\hat{\theta}_1$ and $\hat{\theta}_2$ are contrasts.

     c   Since the test statistics have identical values, the p-values will also be identical.

     d   Testing $H_0 : c_1\mu_1 + \cdots + c_t\mu_t = 0$ is the same as testing $H_0 : d(c_1\mu_1 + \cdots + c_t\mu_t) = 0$.

9.3   **a**   The 95% confidence interval is $-0.2118 \le \mu_2 - \mu_1 \le 0.5518$. Since the interval contains 0, there is insufficient evidence (at the 0.05 level) to conclude that the mean soil bulk densities for the two grazing practices are different.

     b   **i** $\hat{\theta}_1$ estimates the difference between mean soil bulk density for continuous grazing and the average of the means resulting from the other two types of grazing practices.   **ii** $\hat{\theta}_2$ estimates the difference between mean soil bulk density for continuous grazing and that for two-week grazing with one week rest.

     c   There can be *at most* two mutually orthogonal contrasts between three means.

     d $\{\hat{\theta}_1, \hat{\theta}_3, \}$ and $\{\hat{\theta}_2, \hat{\theta}_3\}$ are two orthogonal sets. The first set provides a meaningful subdivision of the treatment sum of squares.

     e   For testing $\hat{\theta}_1$: $F_c = 24.749$, $p < .025$; For testing $\hat{\theta}_2$: $F_c = 10.140$, $p < .025$; Each test indicates that the corresponding contrast is significantly different from zero.

9.5   **a**   Number the combinations of level of P release (L, M, or H), type of inoculation (N or Y), and level of foliar P (N or Y) as follows:

Treat	(L,N,N)	(L,Y,N)	(M,N,N)	(M,Y,N)	(H,N,N)	(H,Y,N)
Symbol	$\overline{Y}_1$	$\overline{Y}_2$	$\overline{Y}_3$	$\overline{Y}_4$	$\overline{Y}_5$	$\overline{Y}_6$
Treat	(L,N,Y)	(L,Y,Y)	(M,N,Y)	(M,Y,Y)	(H,N,Y)	(H,Y,Y)
Symbol	$\overline{Y}_7$	$\overline{Y}_8$	$\overline{Y}_9$	$\overline{Y}_{10}$	$\overline{Y}_{11}$	$\overline{Y}_{12}$

i $\hat{\theta}_1 = \frac{1}{6}(\overline{Y}_1 + \cdots + \overline{Y}_6) - \frac{1}{6}(\overline{Y}_7 + \cdots + \overline{Y}_{12})$ **ii** $\hat{\theta}_2 = \frac{1}{6}(\overline{Y}_1 + \overline{Y}_7 +$
$\overline{Y}_3 + \overline{Y}_9 + \overline{Y}_5 + \overline{Y}_{11}) - \frac{1}{6}(\overline{Y}_2 + \overline{Y}_8 + \overline{Y}_4 + \overline{Y}_{10} + \overline{Y}_6 + \overline{Y}_{12})$ **iii** $\hat{\theta}_3 =$
$\frac{1}{4}(\overline{Y}_1 + \overline{Y}_2 + \overline{Y}_7 + \overline{Y}_8) - \frac{1}{6}(\overline{Y}_5 + \overline{Y}_6 + \overline{Y}_{11} + \overline{Y}_{12})$ **iv** $\hat{\theta}_4 = \frac{1}{2}(\overline{Y}_1 - \overline{Y}_7) -$
$\frac{1}{2}(\overline{Y}_2 + \overline{Y}_8)$

b $-0.4366 \le \theta_1 \le 0.6032, -2.3366 \le \theta_2 \le -1.2968, -3.5617 \le \theta_3 \le -2.2883,$
$-2.3505 \le \theta_4 \le 0 - 0.5495.$

9.7   **a** $\hat{\theta} = (\overline{Y}_7 - \overline{Y}_1) - (\overline{Y}_8 - \overline{Y}_2) = -1\overline{Y}_1 + \overline{Y}_2 + \overline{Y}_7 - \overline{Y}_8$

b   Use the Scheffe method because the contrast is suggested by the data. The 0.01 Scheffe CCV equals 3.947. Since $|\hat{\theta}| = 1.6$ is less than CCV, we conclude, at the 0.01 level, that the data do not support the conclusion that at the low P release, the difference between the effects of foliar application depends on whether the plants have been inoculated.

9.9   **a**

Contrast	Estimate	Bonferroni CCV	Scheffe CCV
$\hat{\theta}_1$	0.2300	0.1437	0.1348
$\hat{\theta}_2$	0.1450	0.1166	0.1558
$\hat{\theta}_3$	0.1700	0.1166	0.1558
$\hat{\theta}_4$	0.2425	0.1437	0.1348

b   The Fisher 2-sided t-test is equivalent to the corresponding F-test.
c   The Scheffe method does not depend on the number of contrasts tested.

9.11

Contrast	Estimate	Bonferroni	Scheffe	Tukey	SNK	Duncan
$\overline{Y}_H - \overline{Y}_L$	2.2675	S	S	S	S	S
$\overline{Y}_H - \overline{Y}_M$	1.0625	NS	NS	NS	NS	NS
$\overline{Y}_M - \overline{Y}_L$	2.2050	NS	NS	NS	S	S

(Significance columns: Bonferroni, Scheffe, Tukey, SNK, Duncan)

9.15   Least number of significances are found in (a) and so (a) must be the result of the most conservative Scheffe procedure. The next higher number of significances are found in (d); hence this is the result of Tukey method. Similarly, (b) and (c) are the results of the Duncan and Fisher methods, respectively.

9.17   **a** $0.1059 \le \theta_1 \le 0.3540, 0.0016 \le \theta_2 \le 0.2884, 0.0266 \le \theta_3 \le 0.3134, 0.1184 \le \theta_4 \le 0.3665.$

b $0.1255 \le \theta_1 \le 0.3345, 0.0492 \le \theta_3 \le 0.2908.$

c $-0.0042 \le \mu_1 - \mu_2 \le 0.2942, \ 0.0208 \le \mu_2 - \mu_3 \le 0.3191, \ 0.1659 \le \mu_1 - \mu_3 \le 0.4641.$

9.19   Bonferroni intervals will be narrower because $t(30, 0.05/20)$ is less than $q(10, 30, 0.05)$.

9.23   **a**   From the given values of the μ_i we get $\sigma_\mu^2 = 0.0156$. Hence, if we estimate σ^2 by $MS[E] = 0.0057$ we can express the noncentrality parameter as $\lambda = 4.11n$. The required sample size can be determined using the trial and error method.

b Given $\Delta = 0.1$ and $1 - \beta = .80$. Determine sample size with $\sigma_\mu^2 = \Delta^2/(2t) = 0.0167$.

c Take $A(\frac{-\Delta}{\sigma\sqrt{2}}) = 0.90$ so that $\Delta/\sigma = 1.813$. The appropriate noncentrality parameter can be determined using the value 1.813 for Δ/σ.

d Use the Bonferroni method in the confidence interval approach with $\theta_1 = \mu_1 - \mu_2$ and $\theta_2 = \mu_1 - 2\mu_2 + \mu_3$. If we estimate σ^2 by MS[E], we have $t = 3, k = 2$, $B = 1.0$, $\sigma^2 = 0.0057$, and $C = 6$.

e Use the Tukey method with the confidence interval approach. We have $C = 2$, $B = 0.5$, $\sigma^2 = 0.0057$, and $\alpha = .05$.

9.25 Since the power increases with σ_μ^2, the actual power will be greater than the power calculated with σ_μ^2 set equal to its smallest possible value. For a given value of Δ, the smallest possible value of σ_μ^2 is $\Delta^2/(2t)$.

Chapter 10

10.1 **a** $Y_{ij} = \mu + \tau_i + E_{ij}$ ($i = 1, 2, 3, 4; j = 1, 2, 3, 4$), where Y_{ij} is the measured chemical influx for the j-th rat receiving the i-th dose; μ is the overall mean chemical influx, τ_i is the effect of the i-th dose, E_{ij} is the random error in Y_{ij}, and the E_{ij} are independent $N(0, \sigma^2)$.

b

Source	df	SS	MS	F	p
Dose	3	2152.25	717.42	160.92	0.0001
Error	12	53.50	4.46		
Total	15	2205.75			

c Prediction interval is (40.85, 51.14).   **d** Lower bound is 9.00.

10.3 **a** It appears that a straight line is not adequate for representing the relationship between weight gain and mother's age.

b In the suggested analysis, there will be no degrees of freedom for error.

10.5 **b** Both models express an observed response as the sum of the expected response and random error. The random errors are assumed to be independent $N(0, \sigma^2)$ in both models. However, the assumptions about the expected responses are different in the two models. In the ANOVA model the only assumption about the expected responses is that they might be different at different depths. In the simple linear regression model, there is an additional assumption that the expected response changes linearly with depth.

c $H_0 : \beta_1 = 0$.   **e** Yes.

10.7 **c** It is very likely that the investigators used data for crop intervals that were available to them. Thus the crop intervals were not fixed by the investigators.

10.9 **a** $\hat{\beta}_0 = 24.85, \hat{\beta}_1 = 4.49$ **b** $\hat{y}_{100} = 45.56, \hat{y}_{150} = 47.36$

c   Residual sum of squares for the least squares line is the *smallest* sum of squares that one can get by fitting a straight line to the data. Thus the sum of squares corresponding to any other line, including the visually estimated line, will be larger than $SS[E]$. It can be verified that the residual sum of squares for the visually estimated line ($\hat{y} = 24 + 5x$) is 103.72 which is larger than the the residual sum of squares of 63.37 for the least squares line.

d   The residual sum of squares obtained by fitting the ANOVA model is the smallest residual sum of squares that can be obtained by fitting a set of models that includes simple linear models as a subset. Thus the residual sum of squares for ANOVA cannot be larger than the residual sum of squares obtained by fitting the simple linear regression model. For the given data the ANOVA residual sum of squares equals 53.50, which is smaller than 63.37.

10.11   **a**   Estimated expected change in mother's weight gain per one year increase in her age is $\hat{\beta}_1 = 1.08$ kg. The estimated expected change per one month increase in age is $1.08/12 = 0.09$ kg.

b   8.51 kg.

c $MSE = 3.0869$ estimates the variance of the population of measured weight gains of all mothers of a given age.

10.13   **a** $44.31 \le \mu(100) \le 46.81$. We may conclude with 95% confidence that the expected chemical influx when drug dose is 100 nM is a value in the interval (44.31 , 46.81).

b   The change in the expected influx that results from changing drug dose from d_2 to d_1 is $\theta = \beta_1(\log d_2 - \log d_1) = \beta_1 \log(d_2/d_1)$. When $d_2 = 4d_1$ the expected change equals $\beta_1(\log 4) = 1.37\beta_1$. The null hypothesis $H_0 : \theta \le 5$ can be tested on the basis of a lower confidence bound for θ. Such a bound can be constructed using the method described in Box 10.5(4).

c   (43.7 , 48.3). This interval is wider than that in (a). The additional assumption of linear regression required for constructing the interval in (a) results in an interval with a shorter length.

d   90% confidence intervals for the expected influx at drug doses 100 and 150 are, respectively, (44.54 , , 46.59) and (46.27 , 48.45). We have 90% confidence in the validity of each interval because the method used for constructing the interval insures that each interval has 90% chance of containing the corresponding expected chemical influx.

e $\hat{\mu}(10) = 34.48$.

f   Test $H_0 : \beta_1 = 0$. We have $t_c = 21.755$; $p < .001$.

g   Test $H_0 : \beta_0 \ge 25$ vs. $H_1 : \beta_0 < 25$.

10.17   **a**   **i**

Source	DF	Sum of Squares	Mean Square	F Value	Prob>F
Regression	1	1348.73020	1348.73020	22.032	0.0003
Error	14	857.01980	61.21570		
Total	15	2205.75000			

ii

Source	DF	Sum of Squares	Mean Square	F Value	Prob>F
Regression	1	2142.37651	2142.37651	473.278	0.0001
Error	14	63.37349	4.52668		
Total	15	2205.75000			

b $R^2 = .6115$ when $X = d$ and $R^2 = .9713$ when $X = \log d$. On the basis of the observed coefficients of determination, the model with $X = \log d$ fits the data better.

c For the model with $X = d$, $s^2_{obs} = 147.05$ and $s^2_{pred} = 89.9150$. Hence $s^2_{pred}/s^2_{obs} = 89.9150/147.05 = 0.6115$.

10.19

Source	DF	Sum of Squares	Mean Square	F Value	Prob>F
Regression	1	93.73135	93.73135	30.365	0.0015
Error	6	18.52113	3.08686		
Total	7	112.25249			

R-square	0.8350

10.21 Use the prediction interval formulas in Box 10.9 with $k = 2, x_1 = 100$, and $x_2 = 200$.

10.23 **a** In the altitude range of $0.64 \le x \le 1.72$, the Working-Hotelling band is the region contained between the lines

$$y = 23.35 - 8.17x \pm 2.864 \sqrt{\left(\frac{1}{12} + \frac{(x - 1.13)^2}{1.6584}\right) 7.0433}.$$

b The line with the maximum (negative) slope is the line that intersects the upper bound at $x = 0.64$ and the lower bound at $x = 1.72$. This line joins the points $(0.64, 21.76)$ and $(1.72, 5.19)$. The slope of this line is

$$\beta = \frac{5.19 - 21.76}{1.72 - 0.64} = -15.34.$$

Similarly, the line with the minimum (negative) slope is the line that intersects the lower bound at $x = 0.64$ and the upper bound at $x = 1.72$. That is, the line joins the points $(0.64, 14.50)$ and $(1.72, 13.42)$. The slope of this line is

$$\beta = \frac{13.42 - 14.50}{1.72 - 0.64} = -1.$$

c Every straight line with slope between -15.34 and -1.00 is a straight line that lies entirely within 90% confidence band provided $0.64 \le x \le 1.72$.

10.29 **a** **i** $\hat{y} = -10.9904 + 1.0833x$;   **ii** $\hat{y} = 11.7097 + 0.7708x$.

b $r_{XY} = \sqrt{(1.0833)(0.7708)} = 0.9138$, where the positive sign is used because the estimated slopes are positive.

c The correlation coefficient is unit-free, and so will be the same as in (b).

d Since $r_{XY}^2 = (0.9138)^2 = 0.8350$, 83% of the variability in the observed weight gains of mothers can be explained by the linear relationship between mother's expected weight gain and her age at pregnancy. $1 - r_{XY}^2 = 0.165$; the simple linear regression model reduced the variability in the observed weight gain by 16.5%.

e Test $H_0 : \rho_{XY} = 0$. We have $t_{r_{XY}} = 5.510$ and $p = 0.0015$. Thus data support the conclusion that there is a linear relationship between expected weight gain and age at pregnancy.

10.33 **a** $Z_{XY} = 0.4153$. A 95% lower bound for ψ_{XY} is $0.4153 - z(.05)(1/\sqrt{73}) = 0.2228$. Using the inverse Fisher transformation, a 95% lower confidence bound for ρ_{XY} is 0.2192. A 95% lower confidence bound for ρ_{XY}^2 is 0.0480.

b At the 0.05 level, there is insufficient evidence to support the claim.

c Construct a 95% confidence interval for ρ_{XY}^2.

10.35 **b** The magnitude of the observed correlation coefficient in itself is not an indicator of the magnitude of the population correlation coefficient. Sample size plays an important part in deciding whether an observed correlation coefficient is statistically significant.

10.37 **a** The required sample size can be determined using the confidence interval approach with $t = 5, a = 150, b = 200, c_0 = 1, c_1 = 175, \alpha = .10, \sigma = 0.5$ or 0.6 in Equation (10.19). An SAS program and the resulting output is shown below.

```
-----------------------------------------------------------------
INPUT
data a;
input a b c0 c1  sigma t n @@;
lines;
150 200 1 175  .5  5 8 150 200 1 175  .5  5 7
150 200 1 175  .5  5 6 150 200 1 175  .5  5 5
150 200 1 175  .5  5 4 150 200 1 175  .5  5 3
150 200 1 175  .5  5 2 150 200 1 175  .5  5 1
150 200 1 175  .6  5 8 150 200 1 175  .6  5 7
150 200 1 175  .6  5 6 150 200 1 175  .6  5 5
150 200 1 175  .6  5 4 150 200 1 175  .6  5 3
150 200 1 175  .6  5 2 150 200 1 175  .6  5 1
;
data b;
  set a;
   R  = b-a;
   nu = n*t -2;
   x  = tinv(.95, nu );
   B  = x*(((t+1)*(c0)**2*(R**2) + 12*(t-1)*(c1 -
           0.5*c0*(a+b))**2)/(n*t*(t+1)*(R**2)))**0.5*(sigma);
put x;
put B;
proc print;
 var n t nu sigma x B;
run;
```

```
-------------------------------------------------------------------
OUTPUT
          OBS   N   T   NU   SIGMA      X         B

           1    8   5   38   0.5     1.68595   0.13329
           2    7   5   33   0.5     1.69236   0.14303
           3    6   5   28   0.5     1.70113   0.15529
           4    5   5   23   0.5     1.71387   0.17139
           5    4   5   18   0.5     1.73406   0.19387
           6    3   5   13   0.5     1.77093   0.22863
           7    2   5    8   0.5     1.85955   0.29402
           8    1   5    3   0.5     2.35336   0.52623
           9    8   5   38   0.6     1.68595   0.15994
          10    7   5   33   0.6     1.69236   0.17164
          11    6   5   28   0.6     1.70113   0.18635
          12    5   5   23   0.6     1.71387   0.20566
          13    4   5   18   0.6     1.73406   0.23265
          14    3   5   13   0.6     1.77093   0.27435
          15    2   5    8   0.6     1.85955   0.35282
          16    1   5    3   0.6     2.35336   0.63147
-------------------------------------------------------------------
```

Thus, we will need between 2 (corresponding to $\sigma = .5$) and 3 replications (corresponding to $\sigma = .6$) at each design point.

b   Can use the Bonferroni approach with $k = 2$. Proceed as in (a) to determine the sample sizes corresponding to $c_1 = 160$, $\alpha = 0.05$ and $c_1 = 170$, $\alpha = 0.05$. The maximum of these two sample sizes is the required sample size.

c   Proceed as in (a) by replacing W in Equation (10.19) with

$$W^* = 2t(nt - 2, \alpha/2) \left[1 + \frac{(t + 1)c_0^2 R^2 + 12(t - 1)(c_1 - m_0 c_0)^2}{nt(t + 1)R^2} \right]^{.5} \sigma.$$

d   Proceed as in (b) by replacing W in Equation (10.19) with W^* defined in (c).

10.39   Plot the observed chemical influxes against dose. If the variances of responses within dosages appear substantially different, then use a weighted regression analysis with weights equal to the inverses of the within dosage variance. For example, the weights associated with the four responses corresponding to $d = 1$ are $w_i = 1/(2.16)^2$, $i = 1, 2, 3, 4$.

10.41   Since multiple responses are not available for several values of x, there is no obvious way of estimating the variances from the data at these points.

10.45   Let $\beta_0 = 1$ and $\beta_1 = -1.2$. Then $P(x)$ will be negative for any $x > 0.84$.

10.47   **a**   Let $P(x)$ denote the probability of pulmonary toxicity for a patient aged x years. Then we assume,

$$P(x) = \frac{e^{\beta_0 + \beta_1 x}}{1 + e^{\beta_0 + \beta_1 x}}.$$

b $\hat{\beta}_0 = -1.6033$ implies that the estimated probability of PT for a patient of "age 0" is

$$\hat{P}(0) = \frac{e^{-1.6033}}{1 + e^{-1.6033}} = 0.1675.$$

A positive estimate for β_1 indicates that the probability of pulmonary toxicity increases with age.

c
$$\hat{P}(x) = \frac{e^{-1.6033+0.0901x}}{1 + e^{-1.6033+0.0901x}}.$$

d The estimated probability of PT for a patient aged 15 years is $\hat{P}(15) = 0.4374$.

e The p-values of 0.0008 for β_0 and 0.0226 for β_1 imply that each of the two null hypotheses, $H_{01} : \beta_0 = 0$ and $H_{02} : \beta_1 = 0$, can be rejected at level $\alpha = 0.05$. Indeed, since both p-values are less than 0.025, the two null hypothesis can be rejected simultaneously at the $\alpha = 0.05$ level.

f With 95% confidence we can conclude that the probability of PT for a 13-year-old patient is somewhere between 19% and 41%.

Chapter 11

11.1 **a** Unlike the linear and quadratic regression functions, a cubic function has two turning points (bends). On the basis of Figure 11.1, it appears that the most appropriate function for modeling $\bar{}(Y|c)$ should have at most one turning point.

b The function can be expressed as $\mu(c) = \beta_0 + \beta_1 x_1 + \beta_2 x_2 + \beta_3 x_3$ where $x_i = c^i$, $i = 1, 2, 3$.

11.3 **a** (a) is additive; (b), (c), and (d) are interactive.

b β_1 is the change in expected value of log(oxygen demand) when A, the biological oxygen demand, is increased by one unit, holding all other explanatory variables fixed.

c $\beta_1 + \beta_2 + \beta_3$ is the change in expected value of log(oxygen demand) when A, B, and C are simultaneously increased by one unit, holding D and E fixed.

d Using the notation in Example 11.4, we can verify that

$$\mu(\Delta A, \Delta B, c, d, e) = \mu(\Delta A, b, c, d, e) + \mu(a, \Delta B, c, d, e) + \beta_4.$$

11.5 Let Y_i be the response that is observed when $X_j = x_{ij}$, $i = 1, \ldots, n, j = 1, \ldots, p$. Also, the E_i are independent $N(0, \sigma^2)$. Then the multiple linear regression model corresponding to the regression function in (a) can be expressed as

$$Y_i = \beta_0 + \beta_1 x_{i1} + \cdots + \beta_5 x_{i5} + E_i, \quad i = 1, \ldots, n,$$

where $x_{i1}, x_{i2}, x_{i3}, x_{i4}$, and x_{i5} denote the values of the variables A, B, C, D, and E respectively. Similarly, the regression function corresponding to 11.2(b) has the form

$$Y_i = \beta_0 + \beta_1 x_{i1} + \cdots + \beta_6 x_{i6} + E_i, \quad i = 1, \ldots, n,$$

where $x_{i1}, x_{i2}, x_{i3}, x_{i4}, x_{i5}$, and x_{i6} denote the values of the variables A, B, C, AB, AC, and BC respectively.

11.9

$$X\beta = \begin{bmatrix} \beta_0 + 4.2\beta_1 + 6.9\beta_2 \\ \beta_0 + 3.6\beta_1 + 2.3\beta_2 \\ \beta_0 + 4.7\beta_1 + 7.6\beta_2 \\ \beta_0 + 5.2\beta_1 + 5.8\beta_2 \\ \beta_0 + 4.6\beta_1 + 8.1\beta_2 \\ \beta_0 + 5.0\beta_1 + 8.6\beta_2 \\ \beta_0 + 5.4\beta_1 + 1.7\beta_2 \\ \beta_0 + 7.0\beta_1 + 6.9\beta_2 \end{bmatrix} \qquad X'y = \begin{bmatrix} 21.10 \\ 110.09 \\ 125.13 \end{bmatrix}.$$

11.11

a In both models, β_0 is the expected response when $X_1 = X_2 = 0$. In model 1, β_2 is the expected change in Y when X_1 is increased by one unit and X_2 is held fixed at *any value*. In model 2, β_2 is the expected change in Y when X_1 is increased by one unit and X_2 is held fixed at *zero*.

b $\hat{y} = -0.8611 + 0.8001x_1 - 0.0788x_2$.

c $SS[E]$ should equal the sum of squares of the residuals.

d From the ANOVA table we see that $p = .0012$. Hence H_0 can be rejected at a level as small as 0.001. It is reasonable to conclude that at least one of the two variables X_1 and X_2 is useful for predicting Y.

e $\hat{y} = -0.5978 + 0.7436x_1 - 0.1261x_2 + .0100x_1x_2$

f The required residuals are obtained by first determining the predicted values using the prediction equation in (e) and then subtracting these values from the corresponding observed responses. For example, the residual corresponding to the first observation is $y_1 - \hat{y}_1 = 1.9422 - 1.7 = -0.2442$.

g Residual plots are similar for the two models. Thus model 1 should be preferred because it contains fewer parameters.

h The error variance estimates are 0.06367 for model 1 and 0.07887 for model 2. Which model should be preferred should be based on considerations such as the fit of the model and the underlying assumptions. Estimated error variance does not give useful information about the adequacy of a statistical model.

11.15 $X'y$ for both models is the same as the one given in answer to Exercise (11.9). The $X'X$ for model 1 and model 2 are, respectively,

$$\begin{bmatrix} 8.00 & 39.70 & 47.90 \\ 39.70 & 204.05 & 240.88 \\ 47.90 & 240.88 & 334.37 \end{bmatrix}, \qquad \begin{bmatrix} 8.000 & 39.700 & 47.900 & 240.880 \\ 39.700 & 204.050 & 240.880 & 1250.308 \\ 47.900 & 240.880 & 334.370 & 1685.888 \\ 240.880 & 1250.308 & 1685.888 & 8742.413 \end{bmatrix}.$$

11.17

a The p-value of 0.0004 for X_1 in model 1, given in Exercise (11.11), implies that the null hypothesis $H_0 : \beta_1 = 0$ can be rejected at a level as small as $\alpha = 0.001$.

b A 95% confidence interval for β_2 is $(-0.1744, 0.0168)$. We may conclude with 95% confidence that for every increase of one unit in the value of X_1 and no change in X_2, we can expect a change in the range of -0.1744 to 0.0168 in the expected value of Y.

c A 99% confidence interval for $\theta = \beta_1 - \beta_2$ is (0.1369, 1.3055). We may conclude with 99% confidence that the difference between the expected change in Y resulting from increasing X_1 alone by one unit and X_2 alone by 1 unit is in the range of 0.1369 and 1.3055.

d The lower bound is 2.5847. At the 95% confidence level, the expected response at $X_1 = 5$ and $X_2 = 3$ is at least 2.5847.

e We need a 95% prediction interval for Y when $X_1 = 5$ and $X_2 = 3$. This interval is (2.1269, 3.6181).

f The interval in (e) will be wider because this interval is designed to cover a single observation, which equals the expected response plus a random error.

g We need to test for interaction between X_1 and X_2. That is, we need to test $H_0 : \beta_3 = 0$. The p-value, which is given in Exercise (11.11), equals 0.8582. The p-value is quite large, and so we conclude that the data indicates interaction.

11.19 a $\hat{\sigma}_{\hat{\beta}_0} = 0.9335$, $\hat{\sigma}_{\hat{\beta}_1} = 4.4829$, $\hat{\sigma}_{\hat{\beta}_2} = 4.2871$.

b For testing H_{01}, the calculated value of the test statistic is $t_{c1} = 11.434$ with 4 degrees of freedom. The corresponding calculated value for testing H_{02} is $t_{c2} = -6.67$. The two-sided p-values for testing H_{01} and H_{02} are, respectively, less than 0.001 and 0.0013. Thus both hypotheses can be rejected.

c The required confidence interval can be constructed using property 5 in Box 11.4. We have $c_0 = c_1 = 0$ and $c_1 = 2c$. The intersection of the confidence band and the perpendicular line at a given value c is the 95% confidence interval for the rate at which the expected adsorption changes at c.

11.23 If Equation (11.29) is true then $\beta_1 = \beta_3$ and $\beta_2 = \beta_3$ so that $\beta_1 = \beta_2 = \beta_3$. Conversely, if $\beta_1 = \beta_2 = \beta_3$ is true then Equation (11.29) is also true. Another set of restrictions that is equivalent to the null hypothesis in Equation (11.21) is $\beta_1 - \beta_2 = 0$ and $\beta_3 - \beta_2 = 0$.

11.25 a $SS[R] = 269,023.3238$.

b

Source	df	SS	MS	F_c
Regression	5	269,023.323	53,804.6648	3.007
Error	19	339,971.716	17,893.2428	
Total	24	608,995.039		

c $Fc = 3.007$ with 5 and 19 degrees of freedom for the numerator and denominator respectively. The p-value is between 0.025 and 0.05. We may conclude at the 0.05 level that at least one of the 5 independent variables in the model is useful for predicting stress in hospitalized patients.

d $R(\beta_3, \beta_4, \beta_5 | \beta_0, \beta_1, \beta_2) = 87,654.2499$. This is the additional variability in the responses that can be explained by adding the independent variables $X_1 X_2, X_1^2$, and X_2^2 to a model containing X_1 and X_2.

e $F_c = 1.633$ with 3 and 19 degrees of freedom. Since the p-value is larger than 0.10, we conclude that there is not sufficient evidence to indicate that addition of $X_1 X_2, X_1^2$, and X_2^2 to a model containing X_1 and X_2 is useful for predicting the stress of hospitalized patients.

f $R(\beta_4, \beta_5 | \beta_0, \beta_1, \beta_2, \beta_3) = 7859.3857$. Interpretation is similar to the interpretation in (d).

g $F_c = 0.22$ with 2 and 19 degrees of freedom. There is insufficient evidence to reject H_0. Conclusion follows the line of reasoning in (e).

11.27 **a** $s_{obs}^2 = 0.2307$, $s_{pred}^2 = 0.1898$. **b** $R_{Y.X_1,X_2,X_3} = 0.9701$.

c Let $\hat{Y}$ denote the predicted value of Y. Simple correlation coefficient between Y and $\hat{Y}$ equals

$$r_{Y\hat{Y}} = \frac{s_{y\hat{y}}}{\sqrt{s_y^2 s_{\hat{y}}^2}} = \frac{0.18982}{\sqrt{(0.23067)(18982)}} = 0.9071.$$

d The correlation coefficient will be less because the correlation coefficient in (c) is the correlation between the observed values and the predicted values based on the *best fitting* equation.

e $R_{YX_3.X_1X_2} = 0.8317$.

f There appears to be a strong linear relationship between the unexplained (by X_1 and X_2) portions of Y and X_3. Thus additional useful information about Y can be obtained by adding a linear term involving percent nitrogen content.

11.29 **b** The R^2 and C_p values for the three subsets are $R^2 = 0.5150$, $Cp = 45.1767$ for $\{X_1\}$, $R^2 = 0.6428$, $Cp = 28.4358$ for $\{X_1, X_2\}$, and $R^2 = 0.7238$, $Cp = 4$ for $\{X_1, X_2, X_3\}$. Thus $\{X_1, X_2, X_3\}$ is the best subset among the three.

Chapter 12

12.1 We will show that $H_{0\beta}$ is equivalent to $H_{0\mu}$. If we let $\beta_1 = \cdots = \beta_{t-1} = 0$ in Equation (12.11) we get $\mu_1 = \cdots = \mu_t$ so that $H_{0\beta}$ implies $H_{0\mu}$. Similarly, by substituting $\mu_1 = \cdots = \mu_t$ in Equation (12.11) we get $\beta_0 = \beta_1 + \beta_0 = \cdots = \beta_{t-1} + \beta_0$, from which it follows that $\beta_1 = \cdots = \beta_{t-1} = 0$. Thus $H_{0\mu}$ implies $H_{0\beta}$.

12.3 **a**

Source	df	SS	MS	F
Treatments	2	1752.27	876.13	6.706
Error	27	3527.60	130.65	
Total	31	5279.87		

b The two ANOVA tables are identical, showing that the one-way ANOVA model and the general linear model give identical results.

12.5 **a** Number the treatments as in Table 8.1 and let X_j be equal to 1 if response is for treatment j and 0 otherwise ($j = 1, 2, 3, 4$). Let Y_i be the ith response ($i = 1, \ldots, 20$). Then

$$Y_i = \beta_0 + \beta_1 x_{i1} + \beta_2 x_{i2} + \beta_3 x_{i3} + \beta_4 x_{i4} + E_i,$$

where x_{ij} is the value of X_j for the ith response and the E_i are independent $N(0, \sigma^2)$.

b Let μ_i denote the expected response for treatment i. Then $\beta_0 = \mu_5$, $\beta_1 = \mu_1 - \mu_5$, $\beta_2 = \mu_2 - \mu_5$, $\beta_3 = \mu_3 - \mu_5$, $\beta_4 = \mu_4 - \mu_5$.

c

Parameter	Estimate	Standard error	t for H_0 :Parameter $= 0$
β_0	34.29	0.5220	65.6897
β_1	−0.27	0.7382	−0.3657
β_2	−2.39	0.7382	−3.2376
β_3	−3.45	0.7382	−4.6735
β_4	0.02	0.7382	0.0271

d An $\alpha = 0.05$ level test of the null hypothesis $H_{0i} : \beta_i = 0$ can be performed by comparing the corresponding calculated value of t in (c) with $t(15, 0.025)$.

12.7 a The model is

$$Y_i = \beta_0 + \beta_1 x_i + E_i, \qquad i = 1, \ldots, 24$$

where Y_i is the frequency of painful crises for patient i and $x_i = 1$ or $x_i = 0$ according to whether the phosphate level for patient i is high or normal. The E_i are independent $N(0, \sigma^2)$.

b $\hat{\beta} = -0.1573$. A 90% lower confidence bound for β is $\hat{\beta}_L = -0.1956$.

c The parameter β equals the difference between the mean number of painful crises in the populations of patients with normal and high phosphate levels. That is, $\beta = \mu_{normal} - \mu_{high}$. We may conclude with 90% confidence that μ_{high} is at least 0.1956 more than μ_{normal}.

12.9 a Let Y_i and x_{ij} denote, respectively, the posttest score and the value of the dummy variable X_j associated with subject i ($i = 1, \ldots, 30; j = 1, 2$). Then the model can be expressed as

$$Y_i = \beta_0^* + \beta_1^* x_{i1} + \beta_2^* x_{i2} + E_i,$$

where the E_i are independent $N(0, \sigma^2)$.

b
$$\beta_0^* = \mu = \frac{3\beta_0 + \beta_1 + \beta_2}{3}$$

$$\beta_1^* = \tau_2 = \frac{2\beta_2 - \beta_1}{3}$$

$$\beta_2^* = \tau_1 = \frac{2\beta_1 - \beta_2}{3}.$$

c Entries in Part 2 can be obtained by using (b) to express the estimators of the parameters of the model in (a) as linear combinations of the estimators of the parameters of the model in Equation (12.2).

12.11 a Both are calculated values of the test statistics for testing $H_0 : \beta = 0$. One is for an F test whereas the other is for an equivalent t test. We know that the square of the statistic for t test is the same as the statistic for the equivalent F test.

b No. The two output are for two different models.

c This question can be answered in the same manner as Exercise 12.4. The main difference is that in Exercise 12.4 there is only one expected response corresponding to each treatment, whereas in this example the expected response for a particular treatment will depend on the covariate value. For instance, the expected response for treatment 1 when the covariate Z equals z is $\beta_0 + \beta_1 + \beta z$. The expected responses are linear combinations of four parameters rather than three. For instance, when computing the estimate and its standard error for the expected response for treatment 1, the coefficients in the linear combination are $c_0 = 0, c_1 = 1, c_2 = 0$ and $c_3 = z$).

d The required 95% confidence interval is (49.71 , 55.89).

e The 90% simultaneous lower prediction bounds for the three treatments are 45.13 for treatment 1, 27.14 for treatment 2, and 27.98 for treatment 3.

12.13 a Parallelism of the regression lines implies that the difference between chemical uptake is the same for all initial weights. This is precisely what is meant when we say that there is no interaction between chemical uptake and initial weight.

b The model that generated the output has the form $Y = \beta_0 + \beta_1 x_1 + \beta_2 z + \beta_3 x_1 z + E$, where Y is chemical uptake, x_1 is dummy variable for diet, z is initial weight and E is random error. For Type I sum of squares for diet, the full and reduced models are, respectively, $Y = \beta_0 + \beta_1 x_1 + E$ and $Y = \beta_0 + E$. For the corresponding Type III sum of squares, the full and reduced models are $Y = \beta_0 + \beta_1 x_1 + \beta_2 z + \beta_3 x_1 z + E$ and $Y = \beta_0 + \beta_2 z + \beta_3 x_1 z + E$, respectively. The p-value for Type I SS refers to the research hypothesis that diet by itself is a useful predictor of chemical uptake. The Type III SS p-value refers to the research hypothesis that the addition of x_1 to a model containing x_1 and $x_1 z$ improves the model's predictive power.

e $\hat{\theta}(x) = -10.3676 - 1.5980x$.

f $\hat{\theta}(10) = -26.3476, \hat{\theta}(13) = -31.1416, \hat{\theta}(16) = -35.9356$

g A set of 90% simultaneous lower bounds for $\theta(z), (z = 10, 13, 16)$ is $\{-30.5533, -33.0904, -39.6835\}$.

12.15 Let $\mu(\Delta X_1, x_2, x_3, x_4)$ denote the differential expected response resulting form changing X_1 from 1 to 0 while holding all other variables fixed. Then

$$\mu(\Delta X_1, x_2, x_3, x_4) = \beta_1 + \beta_{12} x_2 + \beta_{13} x_3.$$

Thus β_{13} measures the change in the differential expected response to the two diets resulting from a 1 unit increase in the value of X_3. Thus β_{13} is a measure of the interaction between X_1 and X_3 because $\beta_{13} \neq 0$ implies that $\mu(\Delta X_1, x_2, x_3, x_4)$ depends on x_3.

12.17 a Absence of interaction between gender and treatment can be demonstrated by showing that the difference between the expected responses for males and females at a given level x_2 of X_2 does not depend on x_2. That is, $\mu(1, x_2, x_3, x_4) - \mu(0, x_2, x_3, x_4)$ is independent of x_2. Similarly, presence of interaction between age and pretreatment implies that the difference $\mu(x_1, x_2, x + 1, x_4) - \mu(x_1, x_2, x, x_4)$ may depend on the value of x_4.

c The p-value of 0.4311 for AGE*PRESCR indicates that in the presence of other variables in the model, adding the term for interaction between age and pretreatment HbAlc does not significantly improve the predictive ability of the model. Also the p-value for Age is fairly large ($p = .3604$) suggesting that age is not a useful variable for predicting posttreatment HbAlc, provided the other variables are retained in the model. Thus it appears that at least one of the two variables—X_3 and X_3X_4—can be deleted from the model without loss of predictive ability. To decide whether both variables can be deleted from the model requires a test of the null hypothesis $H_0 : \beta_3 = \beta_{34} = 0$.

Chapter 13

13.1 **a** Each contrast should be tested using a one-sided t-test at the level $\alpha^* = \alpha/2$. For example, if we need a 0.05 level test, each contrast should be tested at the level $\alpha^* = 0.025$.

  b The calculated value of the test statistic is $t_{c1} = 2.79$ for θ_1 and $t_{c2} = -3.122$ for θ_2. Since $t(8, 0.025) = 2.306$, we can reject $H_{0\theta_1}$ and $H_{0\theta_2}$ at the level $\alpha = 0.05$.

  c Bonferroni 95% confidence intervals are $0.0058 \le \theta_1 \le 0.8342$ and $-0.8842 \le \theta_2 \le -0.0558$.

  d $\theta = \mu_{FW} - \mu_{SW} - \mu_{FC} + \mu_{SC}$.

  e Test $H_0 : \theta = 0$ vs. $H_1 : \theta \ne 0$. We have $t_c = 4.1804$ with 8 degrees of freedom. We may conclude at the 0.05 significance level that the differential weight gain between flowing and still waters is not the same at the two temperatures.

13.3 **a** There are two factors: Age (A) at two levels—$a_1 =$ young and $a_2 =$ old; Processing (B) at five levels $b_1 =$ counting, $b_2 =$ rhyming, $b_3 =$ adjective, $b_4 =$ imagery, and $b_5 =$ intentional.

  b This is a 2×5 complete factorial experiment. There are 10 treatment combinations denoted by a_ib_j, $i = 1, 2$; $j = 1, \ldots, 5$. For example, a_1b_3 corresponds to the treatment in which a young subject processes at the adjective level.

  c Two hypotheses that can be tested are:

  1 The difference between the recall abilities of the young and old subjects depends on the level of processing.

  2 On the average there is a difference between the recall abilities of young and old subjects.

13.7 The plots will lead to the same conclusion as that from Figure 13.1. This is because the definition of interaction does not depend on which factor is used to compare the simple effects.

13.9 **a** The ANOVA table is:

Source	df	SS	MS	F_c	p
Treatments	3	1086.00	362.00	37.358	$< .0001$
Error	33	319.77	9.69		
Total	36	1405.77			

The p-value is quite small and so we conclude that the four groups differ with respect to the mean reduction in their cholesterol levels.

b Contrast for interaction between age and gender: $\hat{\mu}[A \times G] = \frac{1}{2}(\bar{y}_{1+} - \bar{y}_{2+} - \bar{y}_{3+} + \bar{y}_{4+})$.

c The F_c value for testing $\mu[A \times G] = 0$ is 11.595 which is highly significant ($p = 0.0017$). Hence we conclude that the differences between cholesterol reduction for males and females are not the same for the two age groups.

d A 95% confidence interval is (1.41 , 5.59).

13.11 **a** Contrast for age: $\hat{\mu}[A] = \frac{1}{2}(\bar{y}_{3+} - \bar{y}_{1+} + \bar{y}_{4+} - \bar{y}_{2+})$; contrast for gender: $\hat{\mu}[G] = \frac{1}{2}(\bar{y}_{3+} - \bar{y}_{4+} + \bar{y}_{1+} - \bar{y}_{2+})$.

b The contrast for main effect of age is highly significant ($F_c = 104.38$; $p \leq .0001$), but the contrast for the main effect of gender is not significant ($F_c = 2.13$; $p = .1539$). However, no meaningful conclusions can be drawn from these significances because interaction between age and gender was found to be significant in Exercise 13.9(c).

c The Bonferroni simultaneous 95% confidence intervals for the three contrasts are $0.707 \leq \mu[A \times G] \leq 6.293$, $-13.293 \leq \mu[A] \leq -7.707$, and $-1.293 \leq \mu[G] \leq 4.293$.

13.13 **a** The p-value for testing interaction is 0.967. Hence, there is insufficient evidence of interaction between temperature and humidity.

b The p-values for testing the two main effects are 0.301 for temperature and 0.01 for humidity. For a Bonferroni simultaneous test at level 0.01, each of the comparisons should be tested at level $\alpha = 0.01/3 = 0.003$. Thus there is insufficient evidence that any one of the three factorial effects is different from zero.

13.17 **a** Using Equation (13.12) (with appropriate modifications of the denominators of the summands) we get $SS[TRT] = 1086$, $SS[A] = 955.9415$, $SS[G] = 10.8823$ and $SS[AG] = 119.1762$. These sums of squares are not the same as those calculated earlier. This is because Equation (13.12) is not appropriate for calculating the sum of squares when the sample sizes are unequal.

b The ANOVA table is

Source	df	SS	MS	F_c	p
Age	1	1011.47	1011.47	104.382	
Gender	1	20.64	20.64	2.130	
Age×Gender	1	112.36	112.36	11.595	0.002
Error	33	319.77	9.69		
Total	36	1405.77			

c The ANOVA table in (b) provides strong evidence ($p = 0.002$) of interaction between age and gender.

13.19 **a** Tables of components of simple effects of B:

Type of interaction	Components of the simple effect	Level of B (j)		
		$j = 1$	$j = 2$	$j = 3$
None	$\mu_1[AB_j]$	-10	-10	-10
	$\mu_2[AB_j]$	-20	-20	-20
Quantitative	$\mu_1[AB_j]$	-10	-10	-10
	$\mu_2[AB_j]$	-20	-30	-40
Qualitative	$\mu_1[AB_j]$	-10	-10	-10
	$\mu_2[AB_j]$	-20	-80	30

b As was in Table 13.7, the simple effects of B are equal across rows when there is no interaction; they are unequal with same signs when there is a quantitative interaction; and they are unequal with different signs in the case of qualitative interaction.

13.21 a $-0.78 \le \mu[A_1B] \le -0.16$, $-0.42 \le \mu[A_2B] \le 0.20$, $0.11 \le \mu[A_3B] \le -0.73$.

b The conclusions from the simultaneous intervals are the same as those resulting from one-at-a-time 0.10 level tests.

c Tukey multiple comparison at level $\alpha = 0.05$:

Still water: $a_1b_1 \; a_3b_1 \; a_2b_1$,

Flowing water: $a_1b_1 \; a_2b_1 \; a_3b_1$.

13.23 a

Source	df	SS	MS	F_c	p
Treatment	11	1872.56	170.23	8.524	0.0001
Error	24	479.33	19.97		
Total	35	2351.8889			

b On the basis of the plot of treatment means, it appears that there is interaction between pretreatment and variety. The differences between varieties appear to be similar for pretreatments 1 and 2 and different from those for pretreatment 3.

c

Source	df	SS	MS	F_c	p
PRETRT	2	924.89	462.19	23.144	0.0001
VARIETY	3	525.44	175.15	8.771	0.0004
PRETRT*VARIETY	6	422.72	70.45	3.528	0.0120
Error	24	479.34	19.97		
Total	35	2351.89			

d The ANOVA table supports the conclusion in (b) that there is interaction between pretreatment and variety. Further analysis of the data should look at (1) the differences between varieties within pretreatments and (2) the differences between pretreatments within varieties.

13.25 Let Y_{ijk} denote the percentage reduction in cholesterol level for the kth subject of ith gender ($i = 1$ female, $i = 2$ male) and jth age group ($j = 1$ young, $j = 2$ old). Then the model is

$$Y_{ijk} = \mu + \alpha_i + \beta_j + (\alpha\beta)_{ij} + E_{ijk}, \qquad i = 1, 2; j = 1, 2; k = 1, 2, \ldots, n_{ij};$$

where n_{ij} is the number of subjects in the ith gender and jth age groups, μ is the overall mean reduction in the cholesterol level, α_i is the effect of the ith gender group, β_j is the effect of the jth age group, $(\alpha\beta)_{ij}$ is the combined effect of the ith gender and jth age groups. The E_{ij} are errors with independent $N(0, \sigma^2)$ distributions. The parameters α_i, β_j, $(\alpha\beta)_{ij}$ satisfy the constraints

$$\alpha_1 + \alpha_2 = 0, \qquad \beta_1 + \beta_2 = 0$$

and

$$(\alpha\beta)_{1j} + (\alpha\beta)_{2j} = 0, j = 1, 2; (\alpha\beta)_{i1} + (\alpha\beta)_{i2} = 0, i = 1, 2.$$

13.29 **a** A way of plotting interaction is to have two plots, one for each level of B. In each plot the 12 means corresponding to the level of B are plotted against the levels of A and the adjacent means corresponding to the same level of C are joined by straight lines. Such a plot is useful for examining the presence of AC interaction at each level of B. For the data in Example 13.14, these plots suggest that AC is present at b_2 but not at b_1.

  b Let Y_{ijkl} denote the fungus growth at the ith time period, jth environment, kth growth medium and lth replicate. The the model is

$$Y_{ijkl} = \mu + \alpha_i + \beta_j + \gamma_k + (\alpha\beta)_{ij} + (\alpha\gamma)_{ik} + (\beta\gamma)_{jk} + (\alpha\beta\gamma)_{ijk} + E_{ijkl},$$
$$i = 1, 2; \ j = 1, 2, 3; \ k = 1, \ldots, 4,$$

where the quantities on the right hand side of the model have the usual interpretations.

  c 95% confidence intervals for $\mu[AB_2C_j] - \mu[AB_1C_j$ ($j = 1, 2, 3$), the difference between the two environmental effects for the three growth media, are: $(-1.27, 1.69)$ for medium 1, $(1.31, 4.27)$ for medium 1, and $(-3.10, -0.14)$ for medium 3. These confidence intervals can be used to perform hypotheses tests at the $\alpha = 0.05$ level. For instance, the interval for growth medium 1 suggests that there is not sufficient evidence to conclude (at the 0.05 level) that the two environments produce different amounts of fungus growth. In medium 2, there is evidence to conclude that environment 2 produced higher growth than environment 1 whereas the opposite is true in medium 3.

  d The following are the results for 0.05 level Tukey multiple comparison of the means at the two environments.(Note: Two means with a common letter in the parentheses are not significantly different.)
Environment 1: $\bar{y}_{+11}(A)$, $\bar{y}_{+12}(AB)$, $\bar{y}_{+13}(B)$;
Environment 2: $\bar{y}_{+11}$, $\bar{y}_{+13}(A)$, $\bar{y}_{+12}(A)$.

13.31 **b** The plots in (a) give some indication that there might be a second order interaction between treatment, gender, and age, and first order interaction between treatment and gender. There is strong indication of a first order interaction between age and gender.

c   The F tests provide little evidence ($p = 0.087$) of second order interaction. Also, among the first order interactions, only the interaction between treatment and age is significant ($p = .009$). Thus, gender does not interact with treatment or age. A test of the main effect of gender indicates that there is insufficient evidence ($p = .288$) of a difference between the mean test scores for males and females.

d   Analysis does not contradict what is suggested by the plots.

f   Multiple comparison suggested in (e) is appropriate because of the presence of interaction between treatment and age.

13.33   **a**   With the usual notation and assumptions, the model can be written as

$$Y_{ijk} = \mu + \alpha_i + \beta_j + (\alpha\beta)_{ij} + E_{ijk}$$

b   The reduced models for the three null hypotheses are:

$$H_{0A} : \quad Y_{ijk} = \mu + \beta_j + (\alpha\beta)_{ij} + E_{ijk}$$

$$H_{0B} : \quad Y_{ijk} = \mu + \alpha_i + (\alpha\beta)_{ij} + E_{ijk}$$

$$H_{0AB} : \quad Y_{ijk} = \mu + \alpha_i + \beta_j + E_{ijk}$$

c   **1**   The Type III SS for the variable FERT is the difference between the regression sums of squares for the full model and the reduced model under H_{0A}.

2   The Type III SS for the variable IRRIG is the difference between the regression sums of squares for the full model and the reduced model under H_{0B}.

3   The Type III SS for the variable FERT*IRRIG is the difference between the regression sums of squares for the full model and the reduced model under H_{0AB}.

13.35   **b**   Let Y_{ijk} denote the change in the blood sugar for the kth subject receiving the ith type of drug using the jth method of administration. The model is

$$Y_{ijk} = \mu + \alpha_i + \beta_{(i)j}, +E_{ijk}, \qquad i = 1, 2, 3; \ j = 1, 2; \ k = 1, 2, 3,$$

where μ is the overall mean change in blood sugar, α_i is the effect of the ith drug ($\alpha_1 + \alpha_2 + \alpha_3 = 0$), $\beta_{(i)j}$ is the effect of the jth method of administering ith drug, and E_{ijk} is the random error. The E_{ijk} are independent $N(0, \sigma^2)$.

c

Source	df	SS	MS	F	p
Brand	2	611.44	305.72	50.95	0.0001
Method(Brand)	3	72.17	24.06	4.01	0.0344
Error	12	72.00	6.00		
Total	17	755.61			

d   Because there are significant ($p = 0.0344$) differences between methods within brands, we should analyze the simple effects of methods within brands.

e   Bonferroni simultaneous 90% confidence intervals for differences between methods within brands are $(-0.81, 8.81)$ for brand 1, $(-7.47, 2.14)$ for brand 2 and $(0.19, 9.81)$ for brand 3. Based on the simultaneous confidence intervals, we may conclude at the 0.05 level of significance, that the two methods of insulin administration give different results and that there is insufficient evidence to infer that there is a difference between the results for the two methods within the other two brands.

f The result of a 0.05 level Tukey multiple comparison test is as follows. (Means connected with a common underline are not significantly different at the 0.05 level.)

$$\underline{\bar{y}_{1++} \quad \bar{y}_{2++}} \quad \bar{y}_{3++}$$

When averaged over the two methods, the responses for insulin produced a mean that is significantly different from the mean for the other two drugs. There is no difference between the means for the two drugs.

Chapter 14

14.1 **a** Let Y_{ij} denote the percentage protein content of seeds from the ith F_2 generation plant ($i = 1, \ldots, 10$) in the jth ($j = 1, 2, 3$) plot. Then the model is

$$Y_{ij} = \mu + T_i + E_{ij},$$

where μ is the overall mean protein content of seeds from all F_3 generation plants, T_i is the effect of the ith randomly selected F_2 generation plant, and E_{ij} is the random error. The T_i are independent $N(0, \sigma_T^2)$, the E_{ij} are independent $N(0, \sigma^2)$ and the T_i and the E_{ij} are independent.

    b $H_0 : \sigma_T^2 = 0$. **c** $H_0 : \mu = 40$. **d** σ^2.

14.3 **a** Let Y_{ij} denote the cation-exchange capacity of the jth sample ($j = 1, \ldots, 4$) from the ith ($i = 1, \ldots, 5$) core. Then the model is

$$Y_{ij} = \mu + T_i + E_{ij},$$

where μ is the overall mean cation-exchange capacity for the sampled region, T_i is the effect of the ith randomly selected core, and E_{ij} is the random error. The T_i are independent $N(0, \sigma_T^2)$, the E_{ij} are independent $N(0, \sigma^2)$ and the T_i and the E_{ij} are independent.

    b $H_0 : \mu > 18$.

14.5 **a**

Source	df	MS	F_c	p	EMS
F_2 Plants	9	8.22	4.25	0.0034	$\sigma^2 + 3\sigma_T^2$
Error	20	1.94			σ^2
Total	39	10.16			

    b Yes, because there is sufficient evidence to reject $H_0 : \sigma_T^2 = 0$ ($p = 0.0034$).

    c $\hat{\sigma}^2 = 1.94$, $\hat{\sigma}_T^2 = 2.10$. Thus the plant-to-plant variability is about the same as variability within plants.

d $\widehat{CV} = 0.05$. The standard deviation of the measured percentage protein contents is about 5% of the mean.

e $14.2 \leq \rho_I \leq 82.9$. We may conclude with 95% confidence that between 14.2% and 82.9% of the variability in protein content measurements is due to variability among F_2 plants.

g There is insufficient evidence ($t_c = -1.095$, $p = 0.849$) to indicate that the average protein content of seeds produced by F_3 generation plants is more than 40%.

14.7 **a**

Source	df	SS	MS	F_c	p
Core	4	9.441	2.360	2.06	0.1488
Error	12	13.718	1.143		
Total	16	23.159			

On the basis of the above ANOVA, there is insufficient evidence ($p = .1488$) to conclude that the cation-exchange capacity varies over the region.

b $\hat{\sigma}^2 = 1.143$, $\hat{\sigma}_T^2 = 0.387$. The core-to-core variability is about third of the variability among samples selected from the same core.

c $\hat{\mu} = 19.24$.

d Variance of $\hat{\mu}$ can be expressed as $\sigma_{\hat{\mu}}^2 = (61\sigma_T^2 + 17\sigma^2)/17^2$. An estimate of $\hat{\sigma}_{\hat{\mu}}$ is 0.1489.

e Upper bound can be calculated using the above estimate of μ and its standard error. The degrees of freedom for the appropriate t-distribution needs to be estimated using the Satterthwaite method described in Box 14.9.

14.9 **a** Sample sizes can be determined as in Example 14.8. Among the possible combinations of t (number of F_2 generation plants) and n (the number of plots per plant) are $(t, n) = (4, 5)$, $(5, 4)$. The decision of which combination to use depends on cost considerations.

b Use Equation (14.9). $(t, n) = (3, 2)$.

c Use Equation (14.11). $(t, n) = (8, 2)$.

d $\hat{\sigma}_{\overline{Y}_{++}} = .72$. In the new study, we can expect the width of a 95% confidence interval for μ to be approximately equal to $2t(7; .025)(0.72) = 3.4$. Thus, the mean percentage protein content can be estimated with an error less than 1.7%.

14.11 For a given t and n we can estimate the variance of $\overline{Y}_{+++}$ by using the estimates $\hat{\sigma}^2 = 1.14$ and $\hat{\sigma}_T^2 = 0.39$ (see answer to Exercise 14.7) in Equation (14.7). Determine t and n such that the estimated variance of $\overline{Y}_{+++}$ is less than 0.2. Some feasible combinations of t and n are $(t, n) = (4, 3)$, $(5, 2)$, and $(8, 1)$.

14.15 **a** Use the results in Box 14.8. A 95% confidence interval for the given contrast is $(-0.45, 3.29)$. We can conclude at the 95% confidence level that the average of the expected responses for fertilization methods 1 and 2 can be as much as 0.45 below and 3.29 above the expected response for fertilization method 3.

b Use Bonferroni method. Since there are three differences ($\alpha_1 - \alpha_2, \alpha_1 - \alpha_3, \alpha_2 - \alpha_3$), each confidence interval should have confidence level $1 - 0.05/3 = 0.983$.

14.17 **a** Let Y_{ijk} denote the radon flux measurement on the kth ($k = 1, 2$) plot by the jth ($j = 1, \ldots, 6$) technician using the ith ($i = 1, 2$) method. The model is

$$Y_{ijk} = \mu + \alpha_i + B_{(i)j} + E_{ijk}$$

where μ is the mean radon flux over the region, α_i is the fixed effect of method i ($\alpha_1 + \alpha_2 = 0$), $B_{(i)j}$ is the random effect of the jth technician using the ith method, and E_{ijk} is the random error. The $B_{(i)j}$ are independent $N(0, \sigma_B^2)$, the E_{ijk} are independent $N(0, \sigma^2)$ and the $B_{(i)j}$ and E_{ijk} are mutually independent.

b

Source	df	SS	MS	F_c	p
Method (A)	1	7.594	7.594	1.59	0.2362
Tech(Meth) ($B(A)$)	10	47.811	4.781	7.49	0.0009
Error	12	7.655	0.638		
Total	23	63.060			

The ANOVA indicates no significant difference between the two methods ($p = .2362$) and a significant variability ($p = .0009$) between analysts.

c $\hat{\sigma}^2 = 0.638$, $\hat{\sigma}_{B(A)}^2 = 2.072$.

d $\overline{Y}_{1++}$ estimates $\mu + \alpha_1$, the expected radon flux measurement from method 1.

f (39.08, 43.08).

g A 95% confidence upper bound for $\alpha_1 - \alpha_2$ equals 2.74.

14.19 **a** Let Y_{ijk} denote the measure amylase value for the k-th specimen analyzed by the jth technician in the ith lab. The model is

$$Y_{ijk} = \mu + \alpha_i + B_{(i)j} + E_{ijk}, \qquad i = 1, \ldots, 4; j = 1, \ldots, 4; k = 1, \ldots, 10,$$

where the terms in the model have the usual meanings.

b Any of the multiple pairwise comparison methods described in Chapter 9 can be used with the following modification. Since the variance of $\overline{Y}_{i++}$ is estimated by $MS[B(A)]/nb$, we need to replace n, $MS[E]$ and v in Section 9.5 with $nb = 40$ and $MS[B(A)] = 1.63$ and $df_{MS[B(A)]} = 12$, respectively.

c $\hat{\rho} = 0.056$. Approximately 5.6% of the variability in the measured amylase values is due to between-technician variability.

d A set of 90% Bonferroni confidence intervals is

$$\{-0.486 \le \theta_1 \le 0.886; \; -1.410 \le \theta_2 \le -0.440; \; -0.486 \le \theta_3 \le 0.886\}.$$

14.21 **b** Since $EMS[A] = \sigma_S^2 + s_1 \sigma_R^2 + s_1 s_2 \sigma_T^2$, an estimator for the standard error of $\overline{Y}_{+++}$ is $\hat{\sigma}_{\overline{Y}_{+++}} = \sqrt{MS[A]/(ts_1 s_2)}$.

Chapter 15

15.1 **a** Assume treatment effects are fixed. The case of random treatment effects is similar. The contrast can be expressed as

$$\hat{\theta} = \sum_{j=1}^{t} c_j \overline{Y}_{+j}$$

$$= \sum_{j=1}^{t} c_j \left(\mu + \overline{R}_+ + \tau_j + \overline{E}_{+j} \right)$$

$$= (\mu + \overline{R}_+) \left(\sum_{j=1}^{t} c_j \right) + \sum_{j=1}^{t} \left(c_j \tau_j + c_j \overline{E}_{+j} \right),$$

which reduces Equation (5.3) because $\sum c_j = 0$.

b In (a) $\hat{\theta}$ is a linear function (with $c_0 = \sum c_j \tau_j$) of the means of independent random samples from normal distributions.

15.3 a Let Y_{ij} denote the serum amylase value of the ith patient measured using the jth method. The model is

$$Y_{ij} = \mu + R_i + \tau_j + E_{ij}, \qquad i = 1, \ldots, 6; j = 1, 2, 3, 4,$$

where μ is the overall mean, R_i is the random effect of patient i, τ_j is the effect of method j and E_{ij} is the random error. The R_i are independent normal $N(0, \sigma_R^2)$, the E_{ij} are independent normal $N(0, \sigma^2)$, and R_i and E_{ij} are independent.

b

Source	df	SS	MS	F_c	p
Patient	5	2,157,788.3	431,557.7	263.2	0.0001
Method	3	116,979.3	38,993.1	23.78	0.0001
Error	15	24,593.7	1,639.6		
Total	23	2,299,361.3			

There is a significant ($p = 0.0001$) patient-to-patient variability in the measured serum values. Also, the data indicate ($p = 0.0001$) that the expected serum amylase values determined by the four methods are not the same.

c A 95% confidence interval for $\tau_4 - \tau_1$ is (104.5, 204.1).

d No. The interval does not contain zero.

e The required upper confidence bounds can be calculated as in Example 15.4.

f If serum amylase values varied with patients, not randomizing methods within patients will cause the differences due to methods to be confounded with the differences due to patients.

15.5 b

Source	df	SS	MS	F_c	p
Block	3	1.0503	0.3501	25.743	0.0001
Variety	4	0.4513	0.1128	8.294	0.0019
Error	12	0.1629	0.0136		
Total	19	1.6645			

   c The LSD-values for the Duncan method have the form

$$LSD_{ij}(D) = R(p, \nu, \alpha)\sqrt{MS[E]/n},$$

where $p = 2, 3, 4, 5$; $\nu = 12$, $\alpha = 0.05$, $MS[E] = 0.013574$ and $n = 4$.

15.7 The EMS values can be determined by expressing the model as

$$Y_{ijkl} = \mu + R_i + \alpha_j + \beta_k + (\alpha\beta)_{jk} + E_{ijkl},$$

$$i = 1, \ldots, r; j = 1, \ldots, a; k = 1, \ldots, b; l = 1.$$

The following table shows the EMS values in the three cases.

Source	EMS (a)	EMS (b)	EMS (c)
Block	$ab\sigma_R^2 + \sigma^2$	$ab\sigma_R^2 + \sigma^2$	$ab\sigma_R^2 + \sigma^2$
A	$rb\psi_A^2 + \sigma^2$	$rb\psi_A^2 + r\sigma_{\alpha B}^2 + \sigma^2$	$rb\sigma_A^2 + r\sigma_{AB}^2 + \sigma^2$
B	$ra\psi_B^2 + \sigma^2$	$ra\psi_B^2 + r\sigma_{\alpha B}^2 + \sigma^2$	$ra\sigma_B^2 + r\sigma_{AB}^2 + \sigma^2$
AB	$r\psi_{AB}^2 + \sigma^2$	$r\sigma_{\alpha B}^2 + \sigma^2$	$r\sigma_{AB}^2 + \sigma^2$
Error	σ^2	σ^2	σ^2

15.9 The study design is a randomized blocks design with 4 treatments and three blocks of 4 plots (pens) each. Each mean dry matter intake is the average of the responses from three pens receiving a particular treatment combination. The ANOVA table for the data follows (NA means there is insufficient information to calculate the value).

Source	df	SS	MS	
Blocks	2	NA	NA	$4\sigma_R^2 + \sigma^2$
Treatment	3	1.5225	0.5075	$6\psi_A^2 + \sigma^2$
Barley (A)	1	0.9075	0.9075	$6\psi_A^2 + \sigma^2$
N-supp (B)	1	0.6075	0.6075	$6\psi_B^2 + \sigma^2$
AB	1	0.0075	0.0075	$3\psi_{AB}^2 + \sigma^2$
Error	6	0.3528	0.0588	σ^2
Total	11	NA		

15.11

Source	df	SS	MS	
Blocks	2	236.89	118.45	$4\sigma_R^2 + \sigma^2$
Treatment	5	151.48		$6\psi_A^2 + \sigma^2$
Height (A)	1	14.40	14.40	$6\psi_A^2 + \sigma^2$
Spacing (B)	2	127.85	63.93	$6\psi_B^2 + \sigma^2$
AB	2	9.23	4.62	$3\psi_{AB}^2 + \sigma^2$
Error	10	44.60	4.45	σ^2
Total	17	432.97		

15.13 **a** If we use the Bonferroni method with $k = 3$ in Box 9.7, only H_{014} will be rejected. Thus, at the 0.05 level the data provide sufficient evidence in favor of the conclusion that methods 1 and 4 have different expected responses. there is insufficient evidence to infer that μ_1 is different from μ_2 or μ_3.

b $-1.33 \le \theta_{12} \le 2.65$, $-1.94 \le \theta_{13} \le 2.05$, $-5.53 \le \theta_{14} \le -1.55$.

c A 0.05-level Tukey test will declare the expected response for method 4 as different from the other three.

d The analysis in (b) is most appropriate among the three because in this analysis, we estimate ranges of plausible values for the differences of interest.

e Based on the intervals in (b), it appears that method 4 is not comparable to the standard method. There is very little we can say about the comparability of methods 2 and 3 to 1 except that there is insufficient evidence to rule out comparability.

15.15

Source	df	SS	MS
OBS	1	0.0716	0.0716
TIME	1	0.0298	0.0298
OBS*TIME	1	0.0053	0.0053
DAY	3	1.4275	0.4758
SUBST	3	0.2622	0.0874
ERROR	6	1.8093	1.8093
Total	15	3.6055	

15.17 **b** The sums of squares needed for Tukey test for additivity are $SS[NA] = 7799.805$, $MSE[NA] = 1700.993$. The calculated F is $F_c = 4.585$, $df_1 = 1$, $df_2 = 14$, which is not significant at the 0.05 level.

15.19 **b** The sums of squares needed for Tukey test for additivity are $SS[NA] = 25.8077$, $MSE[NA] = 2.3478$. The calculated F for the Tukey test for additivity is $F_c = 10.992$, $df_1 = 1$, $df_2 = 9$, which is significant at the 0.01 level. Thus it appears that there is a block treatment interaction.

15.21 **b** The sums of squares needed for Tukey test for additivity are $SS[NA] = 2.6356$, $MSE[NA] = 4.4479$. The calculated F is $F_c = 0.593$, $df_1 = 1$, $df_2 = 5$, which is not significant at the 0.05 level.

15.23 **b.c** The calculated value of the Friedman statistic is $W_c = 14.8$. When compared to the critical values of a $\chi^2(3)$-distribution, the test yields a p-value less than 0.001. Hence the null hypothesis of equal treatment effects can be rejected. This conclusion is the same as that based on an ANOVA F-test, but the test is valid under less stringent assumptions.

15.25 The calculated value of the Friedman statistic is $W_c = 0.1429$. When compared to the critical values of a $\chi^2(1)$-distribution, the resulting p-value is greater than .5. Hence there is not enough evidence to conclude that the two solutions lead to different results. The sign test and the Friedman test lead to same conclusions. The model for the Friedman test implies the model for the sign test, but the converse is not true. However, when there are only two groups, the tests are equivalent.

15.27   The calculated value of the Cochran statistic is $Q_c = 2$. When compared to the critical values of a $\chi^2(2)$-distribution, the resulting p-value is 0.3679. Hence there is not enough evidence to conclude that $\pi_A = \pi_B = \pi_C$.

Chapter 16

16.1   **a**   This is a repeated measures study because multiple measurements made on the same subject are likely to be correlated.

   b   Treatment (with two levels: drug, placebo) is the between factor, and period (with six levels) is the within factor.

   c   Each 24 week period in which measurements are made on a study subject is a main unit. The six four-week periods within a main unit are subunits.

16.3   **a** $\mu_{ij} = \mu + \alpha_i + \beta_j + (\alpha\beta)_{ij}$ where μ is the overall mean cholesterol level, α_i is the effect of the ith treatment, β_j is the effect of the jth period, and $(\alpha\beta)_{ij}$ is the joint effect of ith treatment and the jth period.

   b $Y_{ijk} = \mu + \alpha_i + \beta_j + (\alpha\beta)_{ij} + S_{ijk} + E_{ijk}$, where S_{ijk} is the random variable that equals the difference between μ_{ij} and the expected cholesterol level during the jth period for the kth subject receiving treatment i, and E_{ijk} is a random variable that equals the difference between the expected cholesterol level at the jth period for the kth subject receiving treatment i and his or her measured cholesterol level.

   c   No. We would expect the correlation between measurements taken close together in time to be higher than that between measurements taken further apart.

16.5 Y_{ij} is in the form of $\hat{\theta}$ in Box 9.1 with $t = 2$, $c_0 = \mu_{ij}, c_1 = c_2 = 1; n_1 = n_2 = 1;$ $\overline{Y}_1 = S_{ij}, \overline{Y}_2 = E_{ij}; \sigma_1^2 = \sigma_{sj}^2, \sigma_2^2 = \sigma_j^2.$

16.7   **b**

Source	df	SS	MS	F_c	p
IRRIG	2	287.0825	143.5413	90.023	0.0019
MAIN PLOT ERROR	3	4.5318	1.5106		
FERT	3	114.1184	38.0395	56.259	0.0001
IRRIG*FERT	6	8.8662	1.4777	2.186	0.1404
SUBPLOT ERROR	9	6.0854	0.6762		
TOTAL	23	420.6843			

There is no evidence ($p = .1404$) of interaction between irrigation method and fertilization method. However, the main effects of fertilization method and irrigation method are both significant ($p = .0001$ and $p = .0019$, respectively) indicating differences in the expected responses due to different fertilization and irrigation methods.

   c   The Tukey method at the 0.05 level indicates that the three irrigation methods produce different expected responses.

   d   The Tukey method at the 0.05 level yield the following result for the four fertilization methods: $\underline{1}$, $\underline{2}$, $\underline{3}$, $\underline{4}$.

e The three contrasts can be specified as

$$\hat{N} = (-1)\overline{Y}_{+1+} + (1)\overline{Y}_{+2+} + (-1)\overline{Y}_{+3+} + (1)\overline{Y}_{+4+}$$

$$\hat{P} = (-1)\overline{Y}_{+1+} + (-1)\overline{Y}_{+2+} + (1)\overline{Y}_{+3+} + (1)\overline{Y}_{+4+};$$

$$\hat{NP} = (1)\overline{Y}_{+1+} + (-1)\overline{Y}_{+2+} + (-1)\overline{Y}_{+3+} + (1)\overline{Y}_{+4+}.$$

g A set of 95% simultaneous Bonferroni confidence intervals for N, P and NP is

$$\{0.41 \le N \le 4.34, \quad 6.35 \le P \le 10.29, \quad -0.90 \le NP \le 3.04\}.$$

16.9 b

Source	df	SS	MS	F_c	p
Soil type	3	50.4317	16.8106	.4541	.7287
Main plot error	4	148.0893	37.0223		
Site prep	4	46.1080	11.5270	7.013	.0018
Soil type × Site prep	12	20.9478	1.7457	1.062	.4460
Subplot error	16	26.2999	1.6437		
Total	35	291.8767			

c There is no evidence ($p = .446$) of interaction between soil type and site preparation method, no evidence ($p = .7287$) of differences due to soil types, but a statistically significant ($p = .002$) difference between the expected responses due to site preparation methods.

d The Tukey method at the $\alpha = 0.05$ level provides no evidence of difference between any pair of soil type means. Note that comparison of soil type means is comparison of main plot means. Hence the appropriate mean square error is the *MS* for the main plot error.

e An unbiased estimate of the contrast is

$$\hat{\theta} = \overline{Y}_{+1+} + \overline{Y}_{+2+} - \overline{Y}_{+3+} - \overline{Y}_{+4+}.$$

This is a comparison of subplot means. Hence the appropriate mean square error is the *MS* for the subplot error. A 95% confidence interval is $2.24 \le \theta \le 6.18$.

f An unbiased estimate of the contrast is

$$\hat{\theta} = \overline{Y}_{+5+} - (1/4)\left(\overline{Y}_{+1+} + \overline{Y}_{+2+} + \overline{Y}_{+3+} + \overline{Y}_{+4+}\right).$$

This is a comparison of subplot means. The lower bound is $\hat{\theta}_L = -1.73$.

16.15 a Let Y_{ijkl} denote the score in the kth ($k = 1, 2, 3$) test for the lth ($l = 1, \ldots, 10$) child of gender j ($j = 1, 2$) in the ith ($i = 1, 2, 3$) group. Then the model is

$$Y_{ijkl} = \mu + \alpha_i + \beta_j + (\alpha\beta)_{ij} + S_{ijl} + \gamma_k + (\alpha\gamma)_{ik} + (\beta\gamma)_{jk} + (\alpha\beta\gamma)_{ijk} + E_{ijkl},$$

where the terms in the model have the usual interpretations.

b Both Greenhouse-Geisser and Huynh-Feldt estimators are close to 1, and so we conclude that the sphericity assumption is reasonable.

c

Source	df	MS	F_c	p
GRP	2	948.422	17.06	.0001
GENDER	1	432.450	7.78	.0073
GRP×GENDER	2	154.067	2.77	.0716
Main plot error	54	55.585		.4460
TEST	2	1351.622	323.00	.0001
TEST×GRP	4	3556.756	849.97	.0001
TEST×GENDER	2,	4.067	0.97	.3817
TEST×GRP×GENDER	4	3.583	0.86	.4832
Sub plot error	108	4.185		

e   Conclusions from multivariate analyses are similar to those from univariate analyses.

f θ_1 is the difference between the expected responses (scores) averaged over groups and tests.

g θ_1 is a comparison of the main plot means. Hence the appropriate mean square error is the the main plot error. A 95% confidence interval is $0.92 \le \theta_1 \le 5.44$.

h θ_2 is the difference between the expected scores in test 1 for groups A and N; θ_3 is the difference between the expected scores in test 1 for groups B and N.

i   Both θ_2 and θ_3 are difference between treatment means for a given main treatment. 95% confidence interval for θ_1 and θ_2 are $12.54 \le \theta_2 \le 18.36$; and $-5.91 \le \theta_3 \le -0.09$.

INDEX